STRIKE-SLIP DEFORMATION, BASIN FORMATION, AND SEDIMENTATION

*Based on a Symposium
Sponsored by the Society of
Economic Paleontologists and Mineralogists*

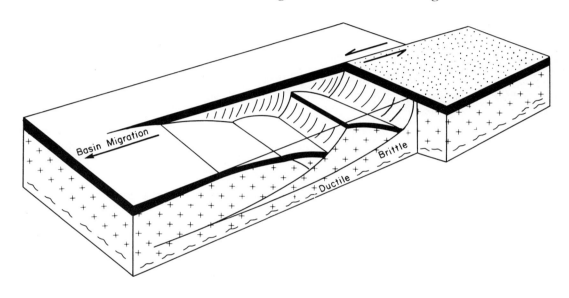

*Edited by
Kevin T. Biddle
Exxon Production Research Company, P.O. Box 2189, Houston, Texas
and
Nicholas Christie-Blick
Lamont-Doherty Geological Observatory of Columbia University, Palisades, New York*

Copyright ©1985 by

SOCIETY OF ECONOMIC PALEONTOLOGISTS AND MINERALOGISTS
*R. L. Ethington, Editor of Special Publications
Special Publication No. 37
In Honor of John C. Crowell*

Tulsa, Oklahoma, U.S.A.   ISBN: 0-918985-58-7   *December, 1985*

A Publication of

The Society of Economic Paleontologists and Mineralogists

a division of

The American Association of Petroleum Geologists

PREFACE

This Special Publication is an outgrowth of a Research Symposium entitled "Strike-Slip Deformation, Basin Formation, and Sedimentation," held at the 1984 joint meeting of the Society of Economic Paleontologists and Mineralogists and American Association of Petroleum Geologists in San Antonio, Texas. Fifteen of the seventeen papers presented at San Antonio are included in this volume, along with five additional contributions.

The idea for a symposium on strike-slip basins came about in the summer of 1982. In the decade or so since the publication of John Crowell's seminal papers (1974a, b) on basins in southern California, a significant body of new geological and geophysical data on strike-slip basins had been acquired, and there had been significant progress in understanding the mechanisms by which basins form and deform in strike-slip settings. Some of this work, especially new studies in the Cenozoic of New Zealand, and in Paleozoic basins of western Europe, was reported in an important volume on "Sedimentation in Oblique-Slip Mobile Zones," edited by Peter Ballance and Harold Reading (1980). That volume also includes a brief introduction by the editors to the history of investigations in such settings, and a summary article by Reading (1980) on the characteristics and recognition of strike-slip fault systems. The purpose of this Special Publication is to augment Ballance and Reading (1980) with emphasis on the relations between deformation patterns along strike-slip faults, the mechanisms by which basins form, and the configuration of sedimentary facies within such basins. Three papers also discuss aspects of the thermal history of strike-slip basins, a topic of particular interest to petroleum geologists. As much as synthesizing our current understanding of strike-slip basins, we hope that the volume will set the stage for future research, particularly the integration in a variety of plate-tectonic settings of geophysical and geological studies with modern techniques of basin modelling. Examples are drawn from North America, Spitsbergen, the Caribbean, South America, Europe, the Middle East, and southeast Asia. Most are from the Cenozoic and late Mesozoic, because it is generally in young basins that the geologic record is best preserved.

The volume is organized into three sections, entitled Overview, Extensional Settings, and Contractional Settings, together with a Glossary of terms having to do with strike-slip deformation, basin formation, and sedimentation. This organization is somewhat artificial, because many regions characterized by overall shortening locally contain extensional basins, for example, and a given basin may at different times or in different places experience both extensional and contractional deformation. However, it seemed more useful to group the contributions according to tectonic setting than by age or geographic distribution.

OVERVIEW

The Overview consists of five papers. Christie-Blick and Biddle summarize the characteristics and controls on structural patterns along strike-slip faults, and the processes and settings of basin formation, in the context of progress achieved over the past decade. Aydin and Nur discuss the origin and significance of stepovers, which are fundamental features of many strike-slip systems, and are responsible for some pull-apart basins and push-up ranges. Pitman and Andrews investigate the subsidence and thermal history of small pull-apart basins, and reaffirm the importance in such basins of lateral heat loss in accelerating lithospheric cooling and subsidence during strike slip. Using both outcrop and subsurface data, Harding et al. summarize the structural styles, plate-tectonic settings, and hydrocarbon traps of divergent wrench faults, a variety of fault little-discussed in the literature, and along which strike slip is accompanied by a component of extension transverse to the fault. The final paper of the Overview is a comparison of the tectonic framework and depositional patterns in three strike-slip basins, the Hornelen Basin (Devonian) of Norway, and the Ridge and Little Sulphur Creek Basins (Neogene) of California. In this paper, Nilsen and McLaughlin demonstrate numerous similarities between the basins, the first two of which are classic examples, and they provide a list of characteristics that may be useful for recognizing strike-slip basins.

EXTENSIONAL SETTINGS

Six papers are included in the section on basins in Extensional Settings. The first two, by Link et al., and by Cemen et al. have to do with Quaternary and Neogene sedimentation along strike-slip faults in the western part of the Basin and Range Province of Nevada and California. In that area, strike-slip deformation is accompanied by pronounced regional extension on both high-angle and low-angle faults. Many of the strike-slip faults act as tears in extensional allochthons, and they separate blocks that experienced different amounts of extension. Manspeizer reviews Pleistocene and Holocene sedimentation along the classic Dead Sea Rift, emphasizing the interaction of rift tectonics and climate in controlling patterns of sedimentation. Guiraud and Seguret discuss the structural, sedimentary, and thermal history of the Late Jurassic to Early Cretaceous Soria Basin of northern Spain in the light of mathematical and microtectonic models. The Soria Basin is thought to have developed in a solitary overstep during an interval of continental extension that preceded the separation of the Iberian Peninsula from North America. The paper by Zalan et al. deals with an offshore example of rifting and strike slip associated with the Cretaceous opening of the South Atlantic Ocean—the Piaui Basin of the northern Brazilian continental margin. The case for strike-slip tectonics in an Early Proterozoic basin is made by Aspler and Donaldson for the Nonacho Basin, a remarkable succession thought to be more than 40 km thick in the Northwest Territories of Canada.

CONTRACTIONAL SETTINGS

The final section on Contractional Settings includes eight papers. Mann et al. discuss the history and controls on Cenozoic deformation associated with a complex restraining bend along the boundary between the North American and Caribbean plates in Jamaica. Şengör et al. summarize the phenomenon of tectonic escape in zones of continental convergence, using the late Cenozoic history of Turkey as an example. Patterns of deformation, basin formation, and sedimentation in such settings tend to be exceedingly complex, and Şengör and his colleagues provide valuable insights into the controls on basin development. Large-scale strike-slip deformation also accompanied the Mesozoic and Cenozoic convergence and accretion of oceanic and island-arc terranes to western North America. Eisbacher reviews the deformational and sedimentary history of the Cordilleran orogen in western Canada, and Johnson presents new sedimentological data from forearc (or intra-arc) strike-slip basins of Eocene age in adjacent Washington. Anadón et al. document the structural and stratigraphic evidence for convergent strike-slip faulting during Paleogene time along the southeastern margin of the Ebro Basin in northern Spain. The Ebro Basin is a complex foreland basin associated with southward thrusting in the Pyrenees. An example of a pull-apart basin in a contractional setting is provided by Royden's discussion of the Vienna Basin (Miocene) of Austria and Czechoslovakia. The basin is superimposed on the Carpathian nappes, and associated with tear faults that in Miocene time separated areas of active northward thrusting from those where thrusting had already been completed. Little or no heating of the lithosphere occurred during basin formation, because extension was restricted to shallow crustal levels, and potential hydrocarbon source rocks are relatively immature. The final two papers, by Steel et al. and by Miall, concern the Cretaceous and Cenozoic evolution of the orogenic belt and associated sedimentary basins of western Spitsbergen, and the Eurekan orogen of the north-eastern Arctic islands of Canada. Successive intervals of transtension and transpression suggested by plate-tectonic models for these orogens are related by the authors to observed patterns of sedimentation and deformation.

ACKNOWLEDGMENTS

It takes the help of many people to bring an SEPM Special Publication into print. We particularly wish to thank all of the authors for their hard work and co-operation, and the many reviewers, whose names may be found in the acknowledgments of the individual papers. We also thank John E. Warme, 1984 SEPM President, and the SEPM staff for their support and encouragement. Logistical support was provided by Exxon Production Research Company, and by an ARCO Foundation Fellowship to Christie-Blick.

REFERENCES

BALLANCE, P. F., AND READING, H. G., eds., 1980, Sedimentation in Oblique-Slip Mobile Zones: International Association of Sedimentologists Special Publication No. 4, 265 p.

CROWELL, J. C., 1974 a, Sedimentation along the San Andreas fault, California, in Dott, R. H., Jr., and Shaver, R. H., eds., Modern and Ancient Geosynclinal Sedimentation: Society of Economic Paleontologists and Mineralogists Special Publication No. 19, p. 292–303.

————. 1974 b, Origin of late Cenozoic basins in southern California, in Dickinson, W. R., ed., Tectonics and Sedimentation: Society of Economic Paleontologists and Mineralogists Special Publication No. 22, p. 190–204.

READING, H. G., 1980, Characteristics and recognition of strike-slip fault systems, in Ballance, P. F., and Reading, H. G., eds., Sedimentation in Oblique-Slip Mobile Zones: International Association of Sedimentologists Special Publication No. 4, p. 7–26.

KEVIN T. BIDDLE
NICHOLAS CHRISTIE-BLICK
Co-Editors

DEDICATION TO JOHN C. CROWELL

    John Crowell began studies of sedimentation along strike-slip faults as part of his dissertation research at the University of California, Los Angeles (1947). Structural and stratigraphic relations in the Miocene Ridge Basin, near the intersection of the San Gabriel and San Andreas faults, proved fertile in developing concepts of strike-slip deformation, basin formation, and sedimentation, especially the use of geological piercing points to establish the large magnitude of strike slip in southern California (1952, 1962). A major synthesis of the geology of the Ridge Basin, edited by John Crowell and Martin Link (1982), and including contributions from numerous authors, provides what is probably the best-described example of a strike-slip basin in the world. Two of John's most influential papers appeared in 1974 in Special Publications 19 and 22 of the Society of Economic Paleontologists and Mineralogists, one on "Sedimentation along the San Andreas fault, California," and the other on the "Origin of late Cenozoic basins in southern California." For his work in structural geology, tectonics, sedimentation, and ancient climates, John has received numerous honors, including membership in the U.S. National Academy of Sciences and the American Academy of Arts and Sciences. In view of his outstanding contributions over a period of 40 years to the tectonics of southern California and the geology of strike-slip basins, it is with great pleasure that we dedicate this volume to John C. Crowell.

NICHOLAS CHRISTIE-BLICK

KEVIN T. BIDDLE

CONTENTS

PREFACE iii
DEDICATION TO JOHN C. CROWELL v

OVERVIEW

DEFORMATION AND BASIN FORMATION ALONG STRIKE-SLIP FAULTS . . . . *Nicholas Christie-Blick*
and *Kevin T. Biddle*   1
THE TYPES AND ROLE OF STEPOVERS IN STRIKE-SLIP TECTONICS . . . . *Atilla Aydin and Amos Nur*   35
SUBSIDENCE AND THERMAL HISTORY OF SMALL PULL-APART BASINS . . . *W. C. Pitman, III and J. A. Andrews*   45
STRUCTURAL STYLES, PLATE-TECTONIC SETTINGS, AND HYDROCARBON TRAPS OF DIVERGENT (TRANSTENSIONAL)
WRENCH FAULTS . . . . *T. P. Harding, R. C. Vierbuchen and Nicholas Christie-Blick*   51
COMPARISON OF TECTONIC FRAMEWORK AND DEPOSITIONAL PATTERNS OF THE HORNELEN STRIKE-SLIP BASIN OF
NORWAY AND THE RIDGE AND LITTLE SULPHUR CREEK STRIKE-SLIP BASINS OF CALIFORNIA . . . *Tor H. Nilsen*
and *Robert J. McLaughlin*   79

EXTENSIONAL SETTINGS

WALKER LAKE BASIN, NEVADA: AN EXAMPLE OF LATE TERTIARY (?) TO RECENT SEDIMENTATION IN A BASIN
ADJACENT TO AN ACTIVE STRIKE-SLIP FAULT . . . . *Martin H. Link, Michael T. Roberts and Mark S. Newton*   105
CENOZOIC SEDIMENTATION AND SEQUENCE OF DEFORMATIONAL EVENTS AT THE SOUTHEASTERN END OF FURNACE
CREEK STRIKE-SLIP FAULT ZONE, DEATH VALLEY REGION, CALIFORNIA . . . . *Ibrahim Cemen, L. A. Wright,*
R. E. Drake and F. C. Johnson   127
THE DEAD SEA RIFT: IMPACT OF CLIMATE AND TECTONISM ON PLEISTOCENE AND HOLOCENE SEDIMENTATION
Warren Manspeizer   143
A RELEASING SOLITARY OVERSTEP MODEL FOR THE LATE JURASSIC-EARLY CRETACEOUS (WEALDIAN) SORIA
STRIKE-SLIP BASIN (NORTHERN SPAIN) . . . . *Michel Guiraud and Michel Seguret*   159
THE PIAUI BASIN: RIFTING AND WRENCHING IN AN EQUATORIAL ATLANTIC TRANSFORM BASIN
Pedro V. Zalan, Eric P. Nelson, John E. Warme and Thomas L. Davis   177
THE NONACHO BASIN (EARLY PROTEROZOIC), NORTHWEST TERRITORIES, CANADA; SEDIMENTATION AND
DEFORMATION IN A STRIKE-SLIP SETTING . . . . *Lawrence B. Aspler and J. A. Donaldson*   193

CONTRACTIONAL SETTINGS

NEOTECTONICS OF STRIKE-SLIP RESTRAINING BEND SYSTEM, JAMAICA . . . . *Paul Mann, Grenville Draper*
and *Kevin Burke*   211
STRIKE-SLIP FAULTING AND RELATED BASIN FORMATION IN ZONES OF TECTONIC ESCAPE: TURKEY AS A CASE
STUDY . . . . *A. M. C. Şengör, Naci Görür and Fuat Şaroğlu*   227
PERICOLLISIONAL STRIKE-SLIP FAULTS AND SYNOROGENIC BASINS, CANADIAN CORDILLERA . . . *G. H. Eisbacher*   265
EOCENE STRIKE-SLIP FAULTING AND NON-MARINE BASIN FORMATION IN WASHINGTON . . . . *Samuel Y. Johnson*   283
PALEOGENE STRIKE-SLIP DEFORMATION AND SEDIMENTATION ALONG THE SOUTHEASTERN MARGIN OF THE EBRO
BASIN . . . . *Pere Anadón, Lluís Cabrera, Joan Guimerà and Pere Santanach*   303
THE VIENNA BASIN: A THIN-SKINNED PULL-APART BASIN . . . . *Leigh H. Royden*   319
THE TERTIARY STRIKE-SLIP BASINS AND OROGENIC BELT OF SPITSBERGEN . . . . *Ron Steel, John Gjelberg,*
William Helland-Hansen, Karen Kleinspehn, Arvid Nøttvedt and Morten Rye-Larsen   339
STRATIGRAPHIC AND STRUCTURAL PREDICTIONS FROM A PLATE-TECTONIC MODEL OF AN OBLIQUE-SLIP OROGEN:
THE EUREKA SOUND FORMATION (CAMPANIAN-OLIGOCENE), NORTHEAST CANADIAN ARCTIC ISLANDS
Andrew D. Miall   361

GLOSSARY

GLOSSARY—STRIKE-SLIP DEFORMATION, BASIN FORMATION, AND SEDIMENTATION . . . . *Kevin T. Biddle*
and *Nicholas Christie-Blick*   375

Cover illustration—Modified from Figure 13 of Manspeizer, 1985 this volume. The original figure is based on a 1984 personal communication from K. Arbenz to W. Manspeizer.

DEFORMATION AND BASIN FORMATION ALONG STRIKE-SLIP FAULTS[1]

NICHOLAS CHRISTIE-BLICK,
Department of Geological Sciences and
Lamont-Doherty Geological Observatory of Columbia University
Palisades, New York 10964;

AND

KEVIN T. BIDDLE
Exxon Production Research Company
P.O. Box 2189
Houston, Texas 77252-2189

ABSTRACT: Significant advances during the decade 1975 to 1985 in understanding the geology of basins along strike-slip faults include the following: (1) paleomagnetic and other evidence for very large magnitude strike slip in some orogenic belts; (2) abundant paleomagnetic evidence for the pervasive rotation of blocks about vertical axes within broad intracontinental transform boundaries; (3) greater appreciation for the wide range of structural styles along strike-slip faults; (4) new models for the evolution of strike-slip basins; and (5) a body of new geophysical and geological data for specific basins. In the light of this work, and as an introduction to the remainder of the volume, the purpose of this paper is to summarize the major characteristics of and controls on structural patterns along strike-slip faults, the processes and tectonic settings of basin formation, and distinctive stratigraphic characteristics of strike-slip basins.

Strike-slip faults are characterized by a linear or curvilinear principal displacement zone in map view, and in profile, by a subvertical fault zone that ranges from braided to upward-diverging within the sedimentary cover. Many strike-slip faults, even those involving crystalline basement rocks, may be detached within the middle to upper crust. Two prominent characteristics are the occurrence of en echelon faults and folds, within or adjacent to the principal displacement zone, and the co-existence of faults with normal and reverse separation. The main controls on the development of structural patterns along strike-slip faults are (1) the degree to which adjacent blocks either converge or diverge during strike slip; (2) the magnitude of displacement; (3) the material properties of the sediments and rocks being deformed; and (4) the configuration of pre-existing structures. Each of these tends to vary spatially, and, except for the last, to change through time. It is therefore not surprising that structural patterns along strike-slip faults differ in detail from simple predictions based on the instantaneous deformation of homogeneous materials. In the analysis of structural style, it is important to attempt to separate structures of different ages, and especially to distinguish structures due to strike-slip deformation from those predating or post-dating that deformation. Distinctive aspects of structural style for strike-slip deformation on a regional scale include evidence for simultaneous shortening and extension, and for random directions of vergence in associated thrusts and nappes.

Sedimentary basins form along strike-slip faults as a result of localized crustal extension, and, especially in zones of continental convergence, of localized crustal shortening and flexural loading. A given basin may alternately experience both extension and shortening through variations in the motion of adjacent crustal blocks, or extension in one direction (or in one part of the basin) may be accompanied by shortening in another direction (or in another part of the basin). The directions of extension and shortening also tend to vary within a given basin, and to change through time; and the magnitude of extension may be depth-dependent. Theoretical studies and observations from basins where strike-slip deformation has ceased suggest that many strike-slip basins experience very little thermally driven post-rift subsidence. Strike-slip basins are typically narrow (less than about 50 km wide), and they rapidly lose anomalous heat by accentuated lateral as well as vertical conduction. Detached or thin-skinned basins also tend to be cooler after rifting has ended than those resulting from the same amount of extension of the entire lithosphere. In some cases, subsidence may be arrested or its record destroyed as a result of subsequent deformation. Subsidence due to extension, thermal contraction, or crustal loads is amplified by sediment loading.

The location of depositional sites is determined by (1) crustal type and the configuration of pre-existing crustal structures; (2) variations in the motion of lithospheric plates; and (3) the kinematic behavior of crustal blocks. The manner in which overall plate motion is accommodated by discrete slip on major faults, and by the rotation and internal deformation of blocks between those faults is especially important. Subsidence history cannot be determined with confidence from present fault geometry, which therefore provides a poor basis for basin classification. Every basin is unique, and palinspastic reconstructions are useful even if difficult to undertake.

Distinctive aspects of the stratigraphic record along strike-slip faults include (1) geological mismatches within and at the boundaries of basins; (2) a tendency for longitudinal as well as lateral basin asymmetry, owing to the migration of depocenters with time; (3) evidence for episodic rapid subsidence, recorded by thick stratigraphic sections, and in some marine basins by rapid deepening; (4) the occurrence of abrupt lateral facies changes and local unconformities; and (5) marked differences in stratigraphic thickness, facies geometry, and occurrences of unconformities from one basin to another in the same region.

INTRODUCTION

Strike-slip deformation occurs where one crustal or lithospheric block moves laterally with respect to an adjacent block. In reality, most "strike-slip" faults accommodate oblique displacements along some segments or during part of the time they are active; and most are associated with an assemblage of related structures including both normal and reverse faults. A component of oblique slip is also required for the formation of sedimentary basins. However,

the title of this paper and that of the volume were chosen to emphasize tectonics and sedimentation in regions where strike-slip deformation is prominent.

In this paper, we follow Mann et al. (1983) in using the term "strike-slip basin" for any basin in which sedimentation is accompanied by significant strike slip. We acknowledge that at any given time some strike-slip basins are hybrids associated with regional crustal extension or shortening, and most are composite, influenced by varying tectonic controls during their evolution (for various recent perspectives about sedimentary basins in a plate-tectonic framework see Green, 1977; Dickinson, 1978; Bally and Snelson, 1980; Klemme, 1980; Bois et al., 1982; Dewey,

[1]Lamont-Doherty Geological Observatory Contribution No. 3910.

1982; Reading, 1982; Kingston et al., 1983a; Perrodon and Masse, 1984). Strike-slip basins also occur in a wide range of plate-tectonic settings including (1) intracontinental and intraoceanic transform zones; (2) divergent plate boundaries and extensional continental settings; and (3) convergent plate boundaries and contractional continental settings.

Among numerous articles on aspects of strike-slip deformation, basin formation, and sedimentation published in the last decade or so, several have been influential (Wilcox et al., 1973; Crowell, 1974a, b; Freund, 1974; Sylvester and Smith, 1976; Segall and Pollard, 1980; a collection of articles edited by Ballance and Reading, 1980a; Reading, 1980; Aydin and Nur, 1982a; Mann et al., 1983). Sylvester (1984) provides a compilation of classic papers on the mechanics, structural style, and displacement history of strike-slip faults, with emphasis on examples from California. In view of these existing summaries, we draw attention here to significant recent advances in understanding the large-scale characteristics of strike-slip faults and strike-slip basins, including several topics not discussed or only briefly mentioned in Ballance and Reading (1980a) and Sylvester (1984).

The remainder of this summary paper introduces two broad themes, which are elaborated in the articles that follow: (1) the characteristics of and controls on structural patterns along strike-slip faults; and (2) the processes and tectonic settings of basin formation, and distinctive aspects of the stratigraphic record of strike-slip basins.

PROGRESS DURING THE DECADE 1975–1985

Examples of advances in understanding the geology of strike-slip basins during the decade 1975 to 1985 include the following: (1) paleomagnetic and other evidence suggesting the very large magnitude of strike slip in some orogenic belts; (2) abundant paleomagnetic evidence where continents are intersected by diffuse transform plate boundaries for pervasive rotation of blocks about vertical axes, with implications for processes of deformation, basin formation and palinspastic reconstruction of strike-slip basins; (3) greater appreciation for the range of structural styles along strike-slip faults, both on the continents and in the ocean basins, and for the processes by which those styles arise; (4) new theoretical and empirical models for the evolution of strike-slip basins; and (5) new geophysical and geological data for many strike-slip basins, some of which are reported in this volume.

Large-Magnitude Strike Slip in Orogenic Belts

The role of strike slip in the evolution of orogenic belts and associated sedimentary basins has been recognized for several decades (references in Ballance and Reading, 1980b; and Sylvester, 1984), but the possible magnitude of such deformation may have been underestimated. A combination of paleomagnetic, faunal, and other geological data now indicate that in some complex orogens such as the North American Cordillera, individual elements of the tectonic collage (Helwig, 1974), termed terranes, have moved thousands not merely hundreds of kilometers with respect to each other along the trend of the orogen (Jones et al., 1977;

Irving, 1979; Beck, 1980; Coney et al., 1980; Irving et al., 1980; Champion et al., 1984; Eisbacher, 1985 this volume). Although part of this longitudinal displacement in the Cordillera pre-dates accretion, a significant component occurred after the terranes were sutured to North America. The Stikine Terrane, for example, appears to have been displaced northward by 13° to 20° since early Cretaceous time, or on the order of 1,500 km with respect to the North American craton (Jones et al., 1977; Irving et al., 1980; Chamberlain and Lambert, 1985; Eisbacher, 1985 this volume).

Large-scale strike-slip deformation also permits lateral tectonic escape in zones of continental convergence, such as between India and Eurasia (Molnar and Tapponnier, 1975), or on a smaller scale, in Turkey (Şengör et al., 1985 this volume). Molnar and Tapponnier (1975) estimated that about one third to half of the relative plate motion between India and Eurasia since the onset of continental collision in Eocene-Oligocene time (at least 1,500 km) could be accounted for by a comparable amount of strike-slip faulting in China and Mongolia.

We do not imply that such huge displacements characterize all orogenic belts, or that available paleomagnetic results are in every case without ambiguity, but only that the possibility of large-scale strike slip should be seriously entertained unless precluded by firm data. In the Appalachian-Caledonide orogen, for example, paleomagnetic studies initially suggested cumulative sinistral offset of as much as 2,000 km in late Paleozoic time (Kent and Opdyke, 1978, 1979; van der Voo et al., 1979; van der Voo and Scotese, 1981; Kent, 1982; van der Voo, 1982, 1983; Perroud et al., 1984). The timing, magnitude, and sense of displacement have recently been questioned on the basis of (1) new determinations of the Early Carboniferous pole for cratonic North America (Irving and Strong, 1984; Kent and Opdyke, 1985); (2) doubt about the age of the magnetization directions measured in some of the samples from eastern North America and Scotland (Donovan and Meyerhoff, 1982; Roy and Morris, 1983; Cisowski, 1984); and (3) the difficulty of finding appropriate faults on which to distribute the displacement (Ludman, 1981; Bradley, 1982; Donovan and Meyerhoff, 1982; Parnell, 1982; Winchester, 1982; Smith and Watson, 1983; Briden et al., 1984; Haszeldine, 1984). Where large displacements have occurred, however, even relatively young basins may have been dismembered and strung out over huge distances, and some geological mismatches may be resolved only by considering the history of an entire orogenic belt.

Rotations About Vertical Axes

The rotation of blocks about vertical axes and the bending of segments of orogenic belts have long been postulated on structural grounds (Carey, 1955, 1958; Albers, 1967; Freund, 1970; Garfunkel, 1974; Dibblee, 1977). Paleomagnetic data now confirm that such rotations tend to be pervasive in strike-slip regimes over a wide range of scales, especially where continents are intersected by diffuse transform plate boundaries (Figs. 1, 2), and the data suggest additional constraints on the timing of rotation and on the kinematics of deformation (Beck, 1980; Cox, 1980; Lu-

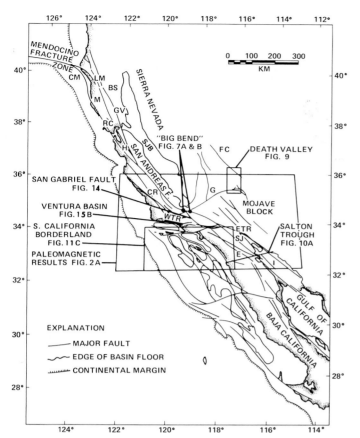

FIG. 1.—Major faults in the diffuse transform plate boundary of California and adjacent parts of Baja California, together with selected basins and elevated blocks mentioned in the text: BS, Bartlett Springs fault zone; CM, Cape Mendocino; CR, Coast Ranges; E, Elsinore fault; ETR, eastern Transverse Ranges; FC, Northern Death Valley-Furnace Creek fault zone; G, Garlock fault; GV, Green Valley fault zone; H, Hayward fault zone; I, Imperial fault; LM, Lake Mountain fault zone; M, Maacama fault zone; RC, Healdsburg-Rodgers Creek fault zone; SJ, San Jacinto fault; SJB, San Joaquin Basin; WTR, western Transverse Ranges (from King, 1969; Crowell, 1974a; Herd, 1978; Nilsen and McLaughlin, 1985 this volume). Map shows locations of areas and cross sections illustrated in Figures 2A, 7, 9, 10A, 11C, 14, 15B.

yendyk et al., 1980, 1985; Ron et al., 1984; Ron and Eyal, 1985).

Paleomagnetic evidence from Neogene rocks of southern California indicates that blocks such as the western Transverse Ranges (WTR in Fig. 1), bounded by east-striking left-slip faults, have experienced net clockwise rotations of between 35° and 90°, with sites near one major right-slip fault being rotated by more than 200° (near DB in Fig. 2A; Luyendyk et al., 1985). In contrast, the Mojave block, located between the San Andreas and Garlock faults, and characterized by northwest-striking right-slip faults, seems to have been rotated counterclockwise by about 15° ± 11° since 6 Ma (c in Fig. 2A; Morton and Hillhouse, 1985), but with large variations in declination over distances of 30 to 120 km from one sub-block to another. In the Cajon Pass region (f in Fig. 2A), there has been no significant rotation since 9.5 Ma (Weldon et al., 1984). These results generally support the tectonic models of Freund (1970), Garfunkel (1974), and Dibblee (1977), in which major crustal blocks

deform internally like a set of dominoes. Garfunkel (1974) suggested as much as 30° of rotation for the Mojave block, approximately twice that determined paleomagnetically, but he probably overestimated by a factor of two to three the magnitude of displacement on the strike-slip faults (Dokka, 1983).

The dimensions of the rotating blocks are uncertain for three reasons: (1) As noted above, blocks tend to deform internally, producing dispersion in declination anomalies; (2) In some cases, rotation seems to have occurred during deposition or eruption of the sedimentary and volcanic rocks studied, so that there is a systematic variation in the magnitude of rotation with age (Luyendyk et al., 1985); (3) Similar rotations may characterize blocks that were actually behaving independently. It is also likely that the boundaries of crustal fragments have changed through time, as displacement occurred sequentially along different strike-slip faults. Some blocks may have undergone, at different times, both clockwise and counterclockwise rotation. A possible example is the San Gabriel block between the San Gabriel and San Andreas faults (SB and e in Fig. 2A), although available data are inconclusive (Ensley and Verosub, 1982; Luyendyk et al., 1985). Small blocks can rotate at an alarming speed. Plio-Pleistocene sediments of the Vallecito-Fish Creek Basin in the western Imperial Valley (d in Fig. 2A) have been rotated 35° since 0.9 Ma (Opdyke et al., 1977; Johnson et al., 1983), and there is thus no assurance that rotation accumulates at a uniform rate, any more than does displacement.

Paleomagnetic and structural data from northern Israel indicate that Neogene intraplate deformation was accommodated by block rotation and strike-slip deformation similar to but on a smaller scale than that documented in California (Fig. 2B; Ron et al., 1984; Ron and Eyal, 1985). Domains of left-slip faults (Galilee and Carmel regions) have rotated clockwise by 23° to 35° since the Cretaceous, and domains of right-slip faults (Galilee and Tiberias regions), counterclockwise by 23° to 53° since the Cretaceous and Miocene, respectively. Sites associated with east-striking normal faults in Galilee yield the expected Cretaceous direction. Strike slip on individual faults west of the Dead Sea fault zone (Sea of Galilee, Fig. 2B) is measurable in hundreds of meters, beginning in late Miocene to early Pliocene time (Ron and Eyal, 1985). The normal faults are predominantly of post-middle Pliocene age.

The significance of these paleomagnetic results is that strike-slip basins can no longer be viewed solely in terms of bends, oversteps or junctions along strike-slip faults, to list some popular models; rotations may play an important role in basin formation. In addition, we should expect facies and paleogeographic elements not only to be offset along strike-slip faults, but also to be systematically misaligned. A further implication of block rotations, discussed below, is that the blocks and bounding faults are detached at some level in the crust or upper mantle (Terres and Sylvester, 1981; Dewey and Pindell, 1985).

Structural Style of Strike-Slip Faults

There is growing appreciation for the wide range of structural styles along strike-slip faults, both on the conti-

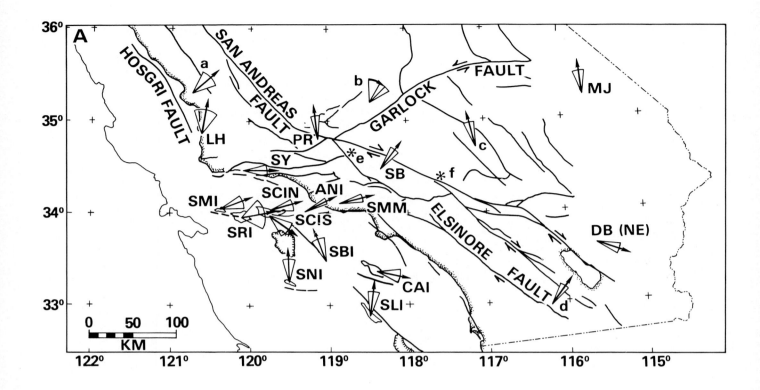

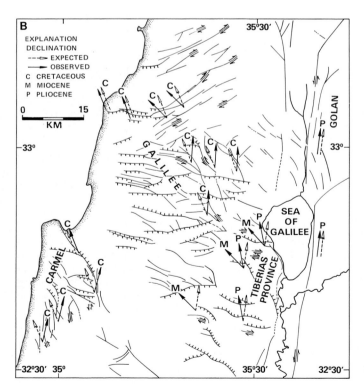

FIG. 2.—A) Paleomagnetic declinations measured in rocks of Neogene age (older than 13 Ma, except where indicated) and Quaternary sediments (site d) in southern California (see Fig. 1 for location). For each site, except e and f, the mean declination is shown along with the 95% confidence limit on the mean. Most of the data and the figure are from Luyendyk et al. (1985). Additional published data; a, Greenhaus and Cox (1979); b, apparent rotation between about 80 and 20 Ma (in comparison with the late Cretaceous pole of Irving, 1979; Kanter and McWilliams, 1981); c, mean counterclockwise rotation of 15° ± 11° since 6 Ma, but with large variations in declination from one sub-block to another (Morton and Hillhouse, 1985); d, 35° clockwise rotation since 0.9 Ma (Opdyke et al., 1977; Johnson et al., 1983); e, no major rotation since 8.5 Ma; the data can be interpreted in terms of clockwise rotation prior to 7.5 Ma, and counterclockwise rotation after that time (Ensley and Verosub, 1982); f, no significant rotation since 9.5 Ma (Weldon et al., 1984). Abbreviations for sites: ANI, Anacapa Island; CAI, Catalina Island; DB (NE) northeastern Diligencia Basin; LH, Lions Head; MJ, Mojave; PR, Plush Ranch Formation, Lockwood Valley; SB, Soledad Basin; SBI, Santa Barbara Island; SCIN, north Santa Cruz Island; SCIS, south Santa Cruz Island; SLI, San Clemente Island; SMI, San Miguel Island; SMM, Santa Monica Mountains; SNI, San Nicolas Island; SRI, Santa Rosa Island; SY, Santa Ynez Range; a, Morro Rock-Islay Hill Complex; b, southern Sierra Nevada; c, Mojave block; d, Vallecito-Fish Creek Basin; e, Ridge Basin; f, Cajon Pass. Faults are from Jennings (1975).

B) Paleomagnetic declinations measured in rocks of Cretaceous, Miocene, and Pliocene age in northern Israel; strike-slip faults, with sense of slip; and normal faults, indicated by hachures on downthrown side (simplified from Ron et al. 1984; Ron and Eyal, 1985). Domains associated with left-slip faults have undergone clockwise rotation, whereas those associated with right-slip faults have experienced counterclockwise rotation; domains associated predominantly with normal faults yield the expected declination.

nents and in the ocean basins, and for the processes by which those styles arise. Wilcox et al. (1973) recognized that one of the main controls on structural style along continental strike-slip faults cutting appreciable thicknesses of sediment is the degree to which adjacent blocks either converge or diverge during deformation. Little attention has been paid subsequently to the structures associated with divergence (see Harding et al., 1985 this volume). In addition, although experimental models, such as those of Wilcox et al. (1973), constitute a useful point of departure for analyzing the development of structures along strike-slip faults, they cannot account adequately for the considerably larger length scale, lower strain rate, greater strength, complex variations in block motion, syn-deformational sedimentation, or material heterogeneity inherent in natural examples (for further discussion, see Hubbert, 1937).

Since transform faults were first recognized in the ocean basins (Wilson, 1965; Sykes, 1967), a wealth of information has accumulated about their physiographic and structural characteristics, particularly from investigations with deep-towed echo-sounding and side-scan sonar instruments such as Deep Tow, GLORIA, Sea Beam, and Sea MARC 1 (Lonsdale, 1978; Macdonald et al., 1979; Searle, 1979, 1983; Bonatti and Crane, 1984; Fox and Gallo, 1984), the deep-towed ANGUS camera (Karson and Dick, 1983), and the deep-sea submersible ALVIN (Choukroune et al., 1978; Karson and Dick, 1983). The gross crustal structure across fracture zones has been determined by a variety of geophysical means, including seismic refraction experiments (e.g., Detrick and Purdy, 1980; Sinha and Louden, 1983), and multichannel seismic reflection profiles (e.g., Mutter and Detrick, 1984). Fossil oceanic fracture zones have also been studied in ophiolites such as the Coastal Complex of Newfoundland (Karson, 1984). Sediment accumulations in oceanic transform zones are generally thinner than about 1 km, except in areas such as the northern Gulf of California, where there is an abundant supply of terrigenous detritus (Crowell, 1981a; Kelts, 1981). The main structural features of the transforms are therefore expressed largely in igneous and metaigneous rocks, and they differ in detail from the styles of many continental strike-slip faults.

The morphology of oceanic transforms is in part a function of the offset and spreading rate of associated ridges (Fox and Gallo, 1984), and of changes in spreading direction (Menard and Atwater, 1969; Macdonald et al., 1979; Bonatti and Crane, 1984). For example, slow-slipping transforms (full rate of 1.5 to 5 cm/yr), such as those of the North Atlantic, are characterized by prominent linear topography and aligned closed basins oriented transverse to offset ridges. The relief increases from about 1,500 m for small-offset (<30 km) transforms to several thousand meters for those with large offset (>100 km), and gradients of valley walls are typically 20° to 30°, with scarps locally near vertical (Fox and Gallo, 1984). Ridge-flank topography tends to deviate in the direction of strike-slip within a few kilometers of the transform. In contrast, many fast-slipping transforms (full rate of 12 to 18 cm/yr), such as those of the eastern Pacific, are characterized by broad zones (7 to 150 km wide), composed of numerous small-offset transform segments, linked by oblique extensional do-mains. Relief ranges from several hundred to a few thousand meters (Fox and Gallo, 1984). The pronounced topography of some transforms may be an expression of changes in plate motion, and of convergent strike-slip deformation. For example, near its eastern intersection with the mid-Atlantic ridge, the Romanche fracture zone shallows to little more than 1,000 m below sea level, and the presence of reefal limestone shows that it was locally emergent at about 5 Ma (Bonatti and Crane, 1984). About 400 km farther east, there is evidence from a seismic reflection profile for reverse faulting and folding on the north flank of the fracture zone (Lehner and Bakker, 1983).

The investigation of oceanic transform faults is currently a major frontier in the geological and geophysical sciences, but in view of the emphasis of this volume on strike-slip deformation and basin formation in continental settings, in the remainder of this paper we focus primarily on continental examples.

Models for Strike-Slip Basins

Models for strike-slip basins are of two kinds: (1) theoretical models, derived from relatively simple assumptions about the thermal and mechanical properties of the lithosphere, but which can be compared with natural basins (Rodgers, 1980; Segall and Pollard, 1980; and in this volume, Aydin and Nur, 1985; Guiraud and Seguret, 1985; Pitman and Andrews, 1985; Royden, 1985); and (2) empirical models, which represent a distillation of geological and geophysical data from several different basins (Crowell, 1974a, b; Aydin and Nur, 1982a; Mann et al., 1983; and in this volume, Nilsen and McLaughlin, 1985; Şengör et al., 1985).

One theoretical approach for investigating the development of a pull-apart basin (Fig. 11A, B below; Burchfiel and Stewart, 1966; Crowell, 1974a, b), the simplest type of basin along a strike-slip fault, is to model the state of stress, secondary fracturing, and vertical displacements produced by infinitesimal lateral displacements on over-stepping discontinuities. It is assumed for the sake of simplicity that the discontinuities are planar, vertical, and parallel, that the crust is composed of an isotropic, homogeneous, linear elastic material, and that the far-field stress is spatially uniform (Rodgers, 1980; Segall and Pollard, 1980; based on earlier studies by Chinnery, 1961, 1963; Weertman, 1965). Although such models reproduce the first-order characteristics of pull-apart basins located between overstepping strike-slip faults, they do not account satisfactorily for protracted deformation, material heterogeneity, inelastic behavior, variations in the state of stress with depth, or changes in fault geometry with depth; and currently available models are not applicable to more complex geometry and kinematic history.

Another approach (e.g., Pitman and Andrews, 1985 this volume; Royden, 1985 this volume) is to consider the subsidence and thermal history of pull-apart basins using models similar to the stretching model of McKenzie (1978), but incorporating such effects as finite rifting times, accentuated lateral heat flow from narrow basins, and depth-dependent extension (Jarvis and McKenzie, 1980; Royden and

Keen, 1980; Steckler, 1981; Steckler and Watts, 1982; Cochran, 1983). In the McKenzie model, subsidence results from instantaneous, uniform extension of the lithosphere (rifting), and from subsequent cooling by vertical conduction of heat. Theoretical studies and observations from basins where strike-slip deformation has ceased suggest that many strike-slip basins experience very little thermally driven post-rift subsidence. Basins along strike-slip faults are typically narrow (less than about 50 km wide; Aydin and Nur, 1982a; and in this volume, Aspler and Donaldson, 1985; Cemen et al., 1985; Guiraud and Seguret, 1985; Johnson, 1985; Link et al., 1985; Manspeizer, 1985; Nilsen and McLaughlin, 1985; Royden, 1985; Şengör et al., 1985), and much of the thermal anomaly decays during the rifting stage (Pitman and Andrews, 1985 this volume). In some cases, the absence of evidence for post-rift thermal subsidence may also be due to subsequent deformation of the basin (Reading, 1980; Mann et al., 1983; Nilsen and McLaughlin, 1985 this volume). An example of a large pull-apart basin that may have experienced some thermal subsidence after extension ceased is the Magdalen Basin (Carboniferous), approximately 100 km by 200 km, and situated in the Gulf of St. Lawrence between Newfoundland and New Brunswick (Bradley, 1982; Mann et al., 1983). From an analysis of subsidence and heat-flow data for the Vienna Basin (Miocene), Royden (1985 this volume) has shown that extension was confined to shallow crustal levels, and that the basin formed as a pull-apart between tear faults of the Carpathian nappes (Fig. 10B below). Detached or thin-skinned basins such as the Vienna Basin, also tend to be cooler during post-rift subsidence than those produced by the same amount of extension of the entire lithosphere.

Theoretical models are useful for analyzing processes of basin formation along strike-slip faults, although they are difficult to evaluate owing to the complexities of the real world. Examples of uncertainties are as follows: (1) lithospheric and crustal thickness prior to strike-slip deformation; (2) for predominantly extensional basins, the magnitude of extension, and the variation of extension with depth and from one part of a basin to another; (3) the relative contributions of lithospheric stretching and igneous intrusion in accommodating extension; (4) the manner in which complex fault geometry evolved, and the possible role in basin subsidence of flexural loading due to crustal shortening; (5) for starved basins, poor paleobathymetry; and for non-marine basins, the lack of a suitable datum from which to measure subsidence; (6) age control, particularly in non-marine basins; (7) the degree to which lithification occurred by physical compaction or by cementation of externally derived minerals; (8) the magnitudes of background heat flow and of heat production within the basin sediments; and (9) the relative contributions of vertical and lateral heat conduction, and of fluid motion in thermal history. In spite of these uncertainties, however, modelling studies have great potential when applied to areas for which there is abundant geological and geophysical data.

Recent empirical models for strike-slip basins (e.g., Aydin and Nur, 1982a; Mann et al., 1983) emphasize the mechanisms by which basins evolve along a single strike-slip fault or at fault oversteps, but no attempt has been made to improve on the qualitative models of Crowell (1974a, b) for more complicated basins involving intersecting strike-slip faults and significant block rotation. In spite of notable improvements in palinspastic reconstructions for orogenic belts, such restorations are commonly difficult to undertake in individual basins, which may have been deformed and dislocated by subsequent strike slip.

Geophysical, Stratigraphic, and Sedimentological Data From Strike-Slip Basins

Seismic-reflection profiles provide an important tool for basin analysis, augmenting seismic refraction, gravity, and magnetic data, which suggest only the broadest features of basin geometry. Where outcrops and wells or boreholes are few, reflection data, calibrated by available information from wells, are indispensible in developing time-stratigraphy and determining the internal geometry of the basin fill (Vail et al., 1977), and they can provide important insights into the large-scale relations between structural evolution and sedimentation. During the past decade, there has been tremendous progress in techniques for the acquisition and processing of multichannel seismic data, considerable amounts of which have been obtained by industry and academic institutions in areas affected by strike-slip deformation. Unfortunately, much of this information remains relatively inaccessible to a large part of the scientific community. One exception is a series of three volumes edited by A. W. Bally (1983) that includes several seismic profiles across strike-slip zones (D'Onfro and Glagola, 1983; Harding, 1983; Harding et al., 1983; Lehner and Bakker, 1983; Roberts, 1983). In the future, data such as these, when coupled with standard field observations, should significantly improve our understanding of the development of strike-slip basins.

Deep seismic-reflection studies, such as those undertaken in a variety of tectonic settings in the United States by COCORP and CALCRUST, in Canada by LITHOPROBE, and in Europe by BIRPS, should also provide important information about the large-scale characteristics of crustal structure associated with strike slip. By running the seismic recorders for longer than in conventional seismic experiments, it is possible to obtain images of geological structure through the crust to the Moho, and into the upper mantle (White, 1985). Direct sampling of the deep crust may also be possible through programs such as Continental Scientific Drilling.

Continued refinement of the correlation between biostratigraphy, magnetostratigraphy and the numerical timescale (e.g., Armstrong, 1978; Harland et al., 1982; Palmer, 1983; Salvador, 1985) permits improved resolution of times, durations, and rates of sedimentation. Of special significance for the analysis of strike-slip basins in complex continental settings is the more precise correlation obtainable between non-marine, marginal marine and marine successions (e.g., Royden, 1985; Şengör et al., 1985, both in this volume).

Modern facies analysis has drawn attention to the numerous factors influencing basin fill, such as basin geometry and subsidence history, climate, sediment source, drainage patterns, transport and depositional mechanisms, depositional setting, and geological age. Examples of facies analysis in strike-slip basins are numerous papers in the volumes edited by Ballance and Reading (1980a) and by Crowell and Link (1982); and articles in this volume by Aspler and Donaldson (1985), Johnson (1985), Link et al. (1985), and Nilsen and McLaughlin (1985).

STRUCTURES ALONG STRIKE-SLIP FAULTS

Major Characteristics

Strike-slip faults are characterized by a linear or curvilinear principal displacement zone in map view (Fig. 3), because significant lateral displacement cannot be accommodated where there are discontinuities or abrupt changes in fault orientation without pervasive deformation within one or both of the juxtaposed blocks. For example, angular deviations of as little as 3° between 12 to 13 km-long fault segments along the southern San Andreas fault zone may be responsible for young topographic features, as well as for the spatial distribution of aseismic triggered slip (Bilham and Williams, 1985). As viewed in profile, and in places in outcrop, most prominent strike-slip faults involve igneous and metamorphic basement rocks as well as supracrustal sediments and sedimentary rocks. Such faults are commonly termed "wrench faults," particularly in the literature of petroleum geology (e.g., Kennedy, 1946; Anderson, 1951; Moody and Hill, 1956; Wilcox et al., 1973;

Harding et al., 1985 this volume). Typically, they consist of a relatively narrow, sub-vertical principal displacement zone at depth, and within the sedimentary cover, of braided splays that diverge and rejoin both upwards and laterally (Fig. 4). Arrays of upward-diverging fault splays are known as "flower structures" (attributed to R. F. Gregory by Harding and Lowell, 1979), or less commonly, "palm tree structures" (terminology of A. G. Sylvester and R. R. Smith; Sylvester, 1984). Some strike-slip faults terminate at depth (or upward) against low-angle detachments that may be located entirely within the sedimentary section or involve basement rocks as well. Examples are high-angle to low-angle tear faults and lateral ramps of foreland thrust and fold belts (Dahlstrom, 1970; Butler, 1982; Royden et al., 1982; Royden, 1985 this volume), and tear faults associated with pronounced regional extension, as in the Basin and Range Province of the western United States (Wright and Troxel, 1970; Davis and Burchfiel, 1973; Guth, 1981; Wernicke et al., 1982; Stewart, 1983; Cheadle et al., 1985).

The distinction between wrench faults and tear faults according to whether they are "thick-skinned" or "thin-skinned" (e.g., Sylvester, 1984) is somewhat arbitrary. Several authors have suggested that crustal blocks and associated wrench faults in central and southern California are decoupled near the base of the seismogenic crust (10–15 km) from a deeper aseismic shear zone that accommodates mo-

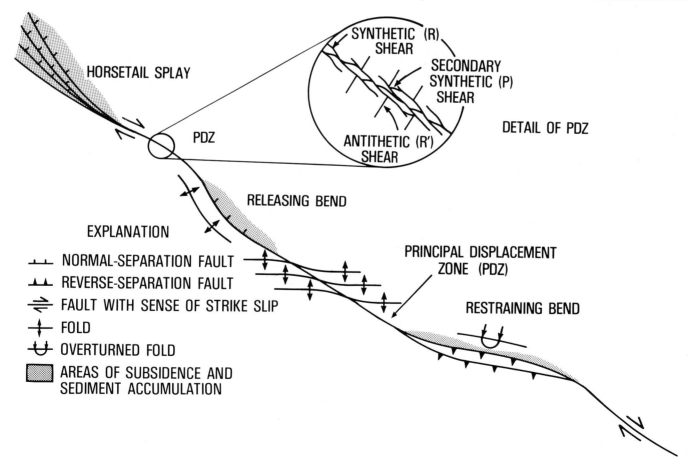

FIG. 3.—The spatial arrangement, in map view, of structures associated with an idealized right-slip fault.

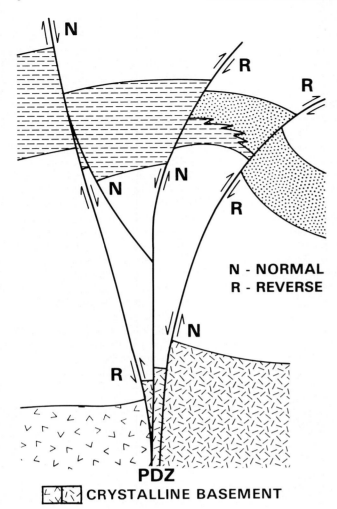

N - NORMAL
R - REVERSE

PDZ

 CRYSTALLINE BASEMENT

MAJOR CHARACTERISTICS

- **BASEMENT - INVOLVED**
- **PDZ SUB-VERTICAL AT DEPTH**
- **UPWARD DIVERGING & REJOINING SPLAYS**

JUXTAPOSED ROCKS

- **CONTRASTING BASEMENT TYPE**
- **ABRUPT VARIATIONS IN THICKNESS & FACIES IN A SINGLE STRATI-GRAPHIC UNIT**

SEPARATION IN ONE PROFILE

- **NORMAL- & REVERSE-SEPARATION FAULTS IN SAME PROFILE**
- **VARIABLE MAGNITUDE & SENSE OF SEPARATION FOR DIFFERENT HORIZONS OFFSET BY THE SAME FAULT**

SUCCESSIVE PROFILES

- **INCONSISTENT DIP DIRECTION ON A SINGLE FAULT**
- **VARIABLE MAGNITUDE & SENSE OF SEPARATION FOR A GIVEN HORIZON ON A SINGLE FAULT**
- **VARIABLE PROPORTIONS OF NORMAL- & REVERSE-SEPARATION FAULTS**

TIME-STRATIGRAPHIC UNIT WITH VARIABLE SEDIMENTARY FACIES

FIG. 4.—The major characteristics, in transverse profile, of an idealized strike-slip fault.

tion between the Pacific and North American plates (e.g., Hadley and Kanamori, 1977; Lachenbruch and Sass, 1980; Yeats, 1981; Hill, 1982; Crouch et al., 1984; Turcotte et al., 1984; Nicholson et al., 1985a, b; Webb and Kanamori, 1985). Recent COCORP seismic reflection profiling in California suggests that the Garlock fault, long recognized as both a major wrench zone (Moody and Hill, 1956) and a tear fault bounding an extensional allochthon (Davis and Burchfiel, 1973; Burchfiel et al., 1983), may terminate downwards against a mid-crustal (9–21 km), low-angle reflecting horizon (Cheadle et al., 1985).

A prominent feature of many strike-slip faults is the occurrence of "en echelon" faults and folds within and adjacent to the principal displacement zone (Figs. 3, 5). The term en echelon refers to a stepped arrangement of relatively short, consistently overlapping or underlapping structural elements that are approximately parallel to each other, but oblique to the linear zone in which they occur (Biddle and Christie-Blick, 1985a this volume; modified from Goguel, 1948, p. 435; Cloos, 1955; Campbell, 1958; Harding and Lowell, 1979). En echelon has also been used for oblique

elements extending tens or even hundreds of kilometers from a principal displacement zone (e.g., Wilcox et al., 1973), where it is doubtful that they have much to do with strike-slip deformation (see Harding et al., 1985 this volume), and in several classic papers on thrust and fold belts for inconsistently overlapping elements arranged parallel to a zone of deformation (e.g., Rodgers, 1963; Gwinn, 1964; Armstrong, 1968; Fitzgerald, 1968; Dahlstrom, 1970). We prefer to describe the latter as a "relay" arrangement (Harding and Lowell, 1979), because it is geometrically different and because it characterizes distinctly different tectonic regimes, those associated with regional extension or with regional shortening rather than with strike slip (Harding and Lowell, 1979; Harding, 1984).

In strike-slip regimes, we also distinguish between en echelon arrangements of structures along a given principal displacement zone, and oversteps between different segments of the principal displacement zone ("en relais" of Harris and Cobbold, 1984; Biddle and Christie-Blick, 1985a this volume). Solitary oversteps (Guiraud and Seguret, 1985 this volume) and many multiple oversteps (Biddle and

Christie-Blick, 1985a this volume) do not constitute a linear zone and by our definition are not en echelon. This is not simply a matter of scale or semantics. Oversteps between different segments and en echelon structures (our usage) appear to have different origins (Aydin and Nur, 1985 this volume; see discussion below). We note, however, that there is no general agreement about such distinctions.

Idealized en echelon arrangements, such as those shown in Figure 3, are reproduced most closely in model studies involving clay, unconsolidated sand, sheets of paraffin wax, Plasticine, or wet tissue paper (Riedel, 1929; Cloos, 1955; Pavoni, 1961; Emmons, 1969; Morgenstern and Tchalenko, 1967; Tchalenko, 1970; Wilson, 1970; Lowell, 1972; Wilcox et al., 1973; Courtillot et al., 1974; Freund, 1974; Mandl et al., 1977; Graham, 1978; Groshong and Rodgers, 1978; Rixon, 1978; Gamond, 1983; Odonne and Vialon, 1983; Harris and Cobbold, 1984), in the experimental deformation of homogeneous rock samples under confining pressure (Logan et al., 1979; Bartlett et al., 1981), and in the deformation of alluvium during large earthquakes (Tchalenko, 1970; Tchalenko and Ambraseys, 1970; Clark, 1972, 1973; Sharp, 1976, 1977; Philip and Megard, 1977). Five sets of fractures are commonly observed (Fig. 5): (1) synthetic strike-slip faults or Riedel (R) shears; (2) antithetic strike-slip faults or conjugate Riedel (R') shears; (3) secondary synthetic faults or P shears; (4) extension or tension fractures (see Biddle and Christie-Blick, 1985a this volume); and (5) faults parallel to the principal displacement zone, or Y shears of Bartlett et al. (1981). In experimental deformation of Indiana limestone, Bartlett et al. (1981) have also described what they call X shears (not shown in Fig. 5). These are symmetrical with R' in relation to the principal displacement zone. The sense of offset for R and P shears is the same as that of the principal displacement zone, whereas R' and X shears have the opposite sense

of offset. The approximate orientation at which faults and associated folds develop under simple conditions is indicated in Figure 5. For the right-slip system illustrated, folds, P shears and X shears are right-handed (Campbell, 1958; Wilcox et al., 1973; Biddle and Christie-Blick, 1985a this volume), whereas R shears, R' shears and extension (tension) fractures are left-handed. As discussed below, however, geological examples tend to be more complicated, and even in the case of Holocene deformation (Fig. 7 below), observed arrangements of structures do not necessarily conform to those predicted by models or experiments. This is because rocks are heterogeneous, because structures develop sequentially rather than instantaneously, and because early-formed structures tend to be rotated during protracted deformation. The structural style is also affected by even a small component of extension or shortening across the principal displacement zone (Wilcox et al., 1973; Harding et al., 1985 this volume), a circumstance that favors fault-parallel folds rather than en echelon ones.

In addition to the occurrence of flower structures, strike-slip faults exhibit several characteristics in profile that result in part from the segmentation of wedge-shaped sediment or rock bodies with laterally variable facies, and in part from a component of convergence or divergence across different fault strands (Fig. 4). Examples are (1) the presence in a given profile across the principal displacement zone of both normal- and reverse-separation faults, and of variable proportions of normal and reverse faults in different profiles; (2) the tendency in a single profile for the magnitude and sense of separation of a given fault splay to vary from one horizon to another; and (3) the tendency in successive profiles for a given fault to dip alternately in one direction and then in the opposite direction, and to display variable separation (both magnitude and sense) for a single horizon.

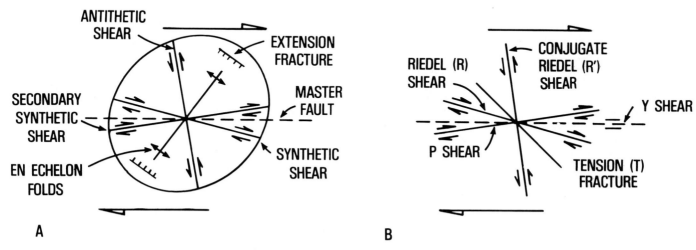

A

B

Fig. 5.—The angular relations between structures that tend to form in right-lateral simple shear under ideal conditions, compiled from clay-cake models and from geological examples. Arrangements of structures along left-slip faults may be determined by viewing the figure in reverse image. A) Terminology largely from Wilcox et al. (1973), superimposed on a strain ellipse for the overall deformation. B) Riedel shear terminology, modified from Tchalenko and Ambraseys (1970) and Bartlett et al. (1981). Extension fractures form when effective stresses are tensile (i.e., when pore-fluid pressure exceeds lithostatic pressure); tension fractures form when lithostatic loads become negative (J. T. Engelder, personal commun., 1985). In geological examples, faults with normal separation tend to develop parallel to the orientation of the extension and tension fractures in A and B; faults with reverse separation tend to develop parallel to the orientation of the fold in A. See text for further discussion of structures along strike-slip faults.

On a regional scale, distinctive aspects of strike-slip structural style are evidence for simultaneous shortening and extension, and for random directions of vergence in associated thrusts and nappes (e.g., Heward and Reading, 1980; Miall, 1985 this volume).

Controls on Structural Patterns

The main controls on the development of structural patterns along individual continental strike-slip faults are (1) the degree to which adjacent blocks either converge or diverge during strike-slip; (2) the magnitude of displacement; (3) the material properties of the sediments and rocks being deformed; and (4) the configuration of pre-existing structures. Each of these factors tends to vary along any given fault, and, except for the last, to change through time. Individual faults are commonly elements of broader regions of deformation, particularly along major intracontinental transform zones, and different faults may be active episodically at different times as a result of block rotation or the reorganization of block boundaries. Structural style may also be influenced directly by the rotation of small blocks that can produce segments of relative convergence and divergence even along straight strike-slip faults (Fig. 6), where adjacent blocks are for the most part neither converging nor diverging. The material properties of the sediments and rocks being deformed tend to change with time where, for example, strike slip is accompanied by sedimentation, so that sediments formerly near the surface become progressively more deeply buried; or where strike slip is accompanied by uplift, so that once buried strata are brought closer to the surface (as in the familiar mechanism of Karig, 1980, for raising "knockers" from great depths in accretionary prisms). Pre-existing structures can markedly influence the location and orientation of strike-slip faults, or simply complicate the overall structural pattern without being reactivated during strike-slip deformation.

Convergent, Divergent and Simple Strike-Slip Faults.—The terms "convergent wrench fault" and "divergent wrench fault" were introduced by Wilcox et al. (1973) to describe basement-involved strike-slip fault zones that, judging from the proportions of reverse-separation and normal-separation faults along and adjacent to the principal displacement zone, are thought to involve a significant component of shortening or extension, respectively. Convergent and divergent thus imply much the same as transpressional and transtensional of Harland (1971), although they refer to kinematics rather than to stress, and are for this reason preferred. We note that the use of these terms in the context of deformational process, rather than for description alone, presupposes that strike slip and transverse shortening or extension occurred simultaneously, not sequentially. In some cases, it may be difficult to make such distinctions. Faults along which there is no evidence for preferential convergence or divergence were described by Wilcox et al. (1973) as "simple parallel wrench faults," by Harding and Lowell (1979) as "side-by-side wrench faults," and by Mann et al. (1983) as "slip-parallel faults." Here, we use the term "simple strike-slip fault" in order to avoid confusion with a geometrical pattern in which a number of faults are actually parallel to

each other or arranged side-by-side, and because for any fault, slip is by definition parallel with the fault surface (see Biddle and Christie-Blick, 1985a this volume).

In comparison with simple strike slip, convergent strike slip leads to the development not only of abundant reverse faults, including low-angle thrust faults, but also of folds that are arranged both en echelon and parallel to the principal displacement zone (Fig. 3; Lowell, 1972; Wilcox et al., 1973; Sylvester and Smith, 1976; Lewis, 1980; and in this volume, Anadón et al., 1985; Mann, et al., 1985; Şengör et al., 1985; Steel et al., 1985). In cases of pronounced convergence, as in the western Transverse and Coast Ranges of California (WTR and CR in Fig. 1), the structural style becomes similar to that of a thrust and fold belt (Nardin and Henyey, 1978; Suppe, 1978; Yeats, 1981, 1983; Crouch et al., 1984; Davis and Lagoe, 1984; Wentworth et al., 1984; Namson et al., 1985). In contrast, along divergent strike-slip faults, folds are less well developed, and commonly consist of flexures arranged parallel rather than oblique to the principal displacement zone, although both types are known (Fig. 3; Wilcox et al., 1973; Nelson and Krausse, 1981; Harding et al., 1985 this volume).

Flower structures develop along both convergent strike-slip faults (Allen, 1957, 1965; Wilcox et al., 1973; Sylvester and Smith, 1976; Harding and Lowell, 1979; Harding et al., 1983; Harding, 1985), where they are associated with prominent antiforms and known as positive flower structures, and along divergent faults, where they are associated with synforms and termed negative (D'Onfro and Glagola, 1983; Harding, 1983, 1985; Harding et al., 1985 this volume). The occurrence of upward-diverging fault splays is thus not due essentially to convergence but to the propagation of faults upward through the sedimentary cover toward a free surface (Allen, 1965; Sylvester and Smith, 1976), an interpretation confirmed in experimental studies with unconsolidated sand (Emmons, 1969) and with homogeneous rock samples under confining pressure (Bartlett et al., 1981).

Strike-slip faults may be characterized by convergence or divergence for considerable distances (Namson et al., 1985; Harding et al., 1985 this volume), or only locally at restraining and releasing bends, fault junctions, and oversteps (Crowell, 1974b; Aydin and Nur, 1982b; Mann et al., 1985 this volume). Shortening and extension can occur simultaneously at oversteps associated with block rotation. A small-scale example is provided by a set of left-stepping en echelon R shears formed in a newly seeded carrot field during the Imperial Valley, California, earthquake of 15 October, 1979 (Fig. 6A; Terres and Sylvester, 1981). Elongate blocks of soil separated along furrows, became detached from the less rigid subsoil (a "Riedel flake" of Dewey, 1982), and rotated as much as 70°. The complex pattern of fractures, faults, folds and buckles, and gaps is related to motion on individual blocks. Şengör et al. (1985 this volume) describe an analogous rotated flake more than 70 km long, along the North Anatolian fault. Block rotation may also produce local convergence and divergence along relatively straight fault segments (Fig. 6B; Dibblee, 1977; Nicholson et al., 1985a, b). From an analysis of earthquake hypocentral locations and first-motion studies near the intersection of the San Andreas and San Jacinto faults, California, Ni-

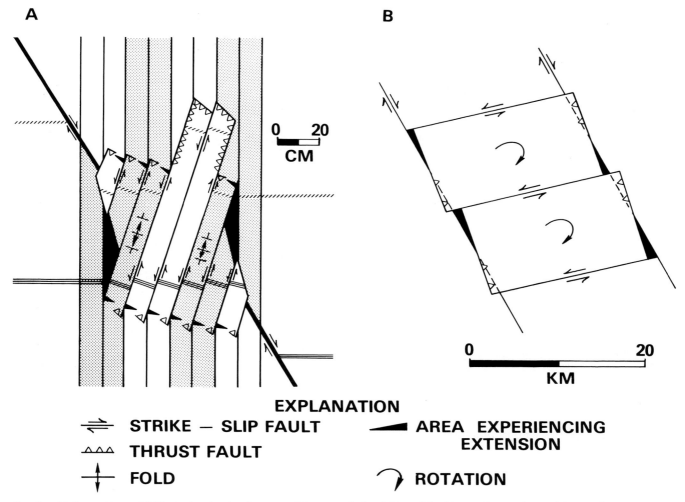

A

B

0 20
CM

0 20
KM

EXPLANATION

⇌ **STRIKE – SLIP FAULT**

▲▲▲ **THRUST FAULT**

╪ **FOLD**

◀ **AREA EXPERIENCING EXTENSION**

↷ **ROTATION**

FIG. 6.—A) Geometric model illustrating the development of fractures, faults, folds and buckles, and gaps observed in a surface rupture in the Imperial Valley, California (I in Fig. 1), following the earthquake of 15 October, 1979 (from Terres and Sylvester, 1981). The spaced parallel ruling represents the furrows in a newly seeded carrot field.

B) A model for the rotation of blocks near the intersection of the San Andreas and San Jacinto faults, California, from earthquake hypocentral locations and first-motion studies (from Nicholson et al., 1985a, b; and similar to a figure of Dibblee, 1977). During a large earthquake, there is right slip on one of the major bounding faults; during the inter-seismic interval, the major faults are locked, and right-lateral motion is taken up by clockwise rotation of small blocks, and by minor left-slip faults. Incompatible motion at the corners of blocks leads to the development of normal and reverse fault segments, which alternate along the major right-slip faults. If such a non-elastic process were to continue on a geological timescale, it would produce complex wedge-shaped basins in areas subject to extension, and crustal flakes detached above thrust faults merging with the right-slip faults.

cholson et al. (1985a, b) have shown that incompatible motion at the corners of blocks leads to the development of normal- and reverse-fault segments, which alternate along the major right-slip faults.

Displacement Magnitude.—Experimental work, sub-surface studies and field observations indicate that structural style along strike-slip faults is influenced qualitatively by the magnitude of displacement (Wilcox et al., 1973; Harding, 1974; Harding and Lowell, 1979; Bartlett et al., 1981). In experiments, folds and shear fractures tend to form sequentially, although not in the same abundance or in the same order for all materials under all conditions (Wilcox et al., 1973; Bartlett et al., 1981). In clay models, for example, folds and Riedel shears develop first, followed by P shears, and finally by the development of a relatively narrow through-going principal displacement zone (Morgenstern and Tchalenko, 1967; Tchalenko, 1970; Wilcox et al., 1973). In experiments with rock samples under confining pressure, R and P shears develop concurrently, and R shears tend to propagate toward the orientation of the principal displacement zone with increasing displacement (Bartlett et al., 1981). R′ and X shears form at this latter stage. The simultaneous development of both R and P shears cannot be explained by the Coulomb-Mohr failure criterion. The width of the zone of deformation increases rapidly during the initial development of folds and fractures, but tends to stabilize quickly because weakening of the sheared material leads to the concentration of subsequent deformation (Odonne and Vialon, 1983). Early-formed structures tend to rotate during progressive deformation (Tchalenko, 1970;

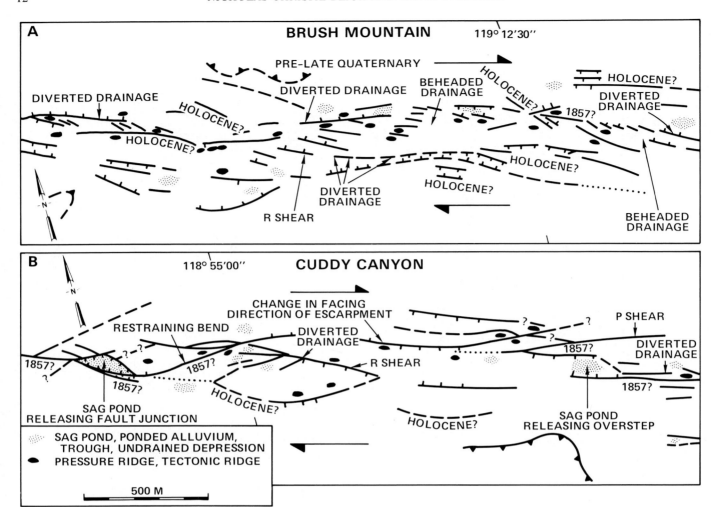

FIG. 7.—Holocene faults and tectonic landforms along the "big bend" of the San Andreas fault, California, A) south of Brush Mountain, and B) in Cuddy Canyon, east of Frazier Park (simplified from T. Davis and E. Duebendorfer, unpublished mapping, 1981, with permission from the authors; see Fig. 1 for location). Hachures on topographically lower side of fault traces; sawteeth on upper plate of thrust faults. The maps indicate the structural complexity that can develop in a few thousand years along a major strike-slip fault. The sediments at the surface are largely of Neogene and Quaternary age (A) and Quaternary age (B). The stipple pattern indicates the sites of basins (sag ponds, ponded alluvium, troughs, undrained depressions); elliptical black dots indicate elevated blocks (pressure ridges, tectonic ridges). Except for a few obvious examples of basins at releasing fault junctions or releasing oversteps, at the scale of these examples, there is no obvious relation between sites of sedimentation and fault geometry.

Wilcox et al., 1973; Rixon, 1978; Odonne and Vialon, 1983), particularly R′ shears, initially oriented at a high angle to the principal displacement zone. Thus the final orientations of folds and faults depend on the magnitude of displacement and on the time at which these structures formed during the deformation history.

Many of these features documented in experiments, can also be observed in geological examples. For instance, strike-slip faults with small lateral displacement in the basement are generally expressed within the sedimentary cover by discontinuous faults and folds (e.g., the Newport-Inglewood fault zone of the Los Angeles Basin with right slip of about 200 to 750 m; Harding, 1974). At the same scale, faults with large displacement (such as the San Andreas) consist of a through-going principal displacement zone (see Fig. 5 of Harding and Lowell, 1979). One difference, however, between experiments and natural examples is that in

nature, deformation is commonly accompanied by sedimentation, so that younger sediments record less offset than older ones. Faults within Holocene alluvium along the San Andreas fault zone tend to be discontinuous (Fig. 7; from T. Davis and E. Duebendorfer, unpublished mapping, 1981). Figure 7 also illustrates the rate at which structural complexities can arise. Given an immensely longer geological timescale, and the fact that rocks and sediments are heterogeneous even before deformation, we should not be surprised if the orientation and geometry of observed structures along a given strike-slip fault depart significantly from those predicted by idealized strain ellipse summaries (Fig. 5). This is of considerable importance to the petroleum geologist constructing structure maps for a prospective exploration target from limited data. Idealized models are useful, but they should be applied with caution. That such models commonly "match" geological observations is in part due to the large number of different fault orientations available (Fig. 5), and interpretations based on such comparisons are not necessarily correct. In addition, the absence of a simple strike-slip structural style does not eliminate the possibility that strike slip played a major role in the deformation.

Material Properties.—The character of a given strike-slip fault zone is also a function of the lithology of the juxtaposed rocks, the confining pressure, fluid pressure, and temperature conditions at which deformation occurred, and the rates of strain and recovery (Donath, 1970; Mandl et al., 1977; Sibson, 1977; Logan, 1979; Logan et al., 1979; Ramsay, 1980; Bartlett et al., 1981; Wise et al., 1984). These factors are likely to change during continuing strike-slip deformation as a result of lateral variations in depositional facies, or through sedimentation and burial, uplift and erosion, changes in provenance, or changes in heat flow.

Though difficult to predict, fault-zone characteristics are of significance to petroleum geologists because they influence the tendency of faults to behave as either conduits or barriers to fluid migration. Depending on the conditions of deformation, the accumulation of displacement along a fault may be accompanied by the development of gouge, which can act as a barrier to fluid migration (e.g., Pittman, 1981). However, displacement may also promote fracturing and additional avenues for leakage. As with other faults, strike-slip faults may juxtapose rocks with significantly different permeability, thus forming either seals or conduits for the migration of petroleum across the fault surface (see Downey, 1984, for a general treatment of hydrocarbon seals).

Pre-existing Structures.—In regions where strike-slip faults are present, pre-existing structures are of two kinds, "essential" and "incidental." As applied to the analysis of strike-slip deformation, essential structures are defined here as those which significantly influence the location and orientation of faults and folds during strike slip. Incidental structures are inherited and contribute little to that deformation. Both types of structure are elements of the overall structural pattern of the region, but incidental ones should be excluded from the analysis strike-slip structural style.

Essential structures commonly influence patterns of strike-slip deformation on the continents. The process can be observed today in the Upper Rhine graben (Fig. 8), formed by extension in middle Eocene to early Miocene time in response to Alpine deformation (Illies, 1975; Illies and Greiner, 1978; Şengör et al., 1978). In Pliocene to Holocene time, normal faults parallel to the graben were reactivated as strike-slip faults during uplift of the Alps. Earthquake first-motion studies suggest that much of the current deformation is due to left slip (Fig. 8; Ahorner, 1975), and this sense of displacement is consistent with measurements of in-situ stress (Illies and Greiner, 1978). Other examples of essential structures are discussed in this volume by Cemen et al. (1985), Harding et al. (1985), Mann et al. (1985), Royden (1985), Şengör et al. (1985), and Zalan et al. (1985). In the oceans, the location of some major transform faults is controlled by weaknesses within the continents prior to continental separation (e.g., the Atlantic equatorial megashear zone; Bonatti and Crane, 1984; Zalan et al., 1985 this volume).

Some basement-involved folds in the Death Valley area, southeastern California, illustrate the concept of incidental pre-existing structures (Figs. 1, 9). Central Death Valley developed in late Miocene through Holocene time as an east-tilted half graben bounded to the north and south by northwest-striking right-slip faults (the northern Death Valley–Furnace Creek fault zone and the southern Death Valley fault zone; Stewart, 1983). Although the type example of a pull-apart basin (Fig. 9B; Burchfiel and Stewart, 1966; Mann et al., 1983), Death Valley is not a particularly good example, because it is only one of many grabens in the Basin and Range Province that formed in response to late Cenozoic regional extension, not as a result of a bend or an overstep in a strike-slip fault (see reviews of the Basin and Range Province by Stewart, 1978; Eaton, 1982; and Fig. 1 of Cemen et al., 1985 this volume). The strike-slip faults appear to be tear faults within an extensional allochthon separating areas that experienced different amounts of extension (Wright and Troxel, 1970; Wernicke et al., 1982; Stewart, 1983). There is no evidence for strike slip within and parallel to the *central* segment of Death Valley (Wright and Troxel, 1970), although the basin undoubtedly opened obliquely, as shown in Figure 9B, and is at present extending in an approximately northwest-southeast direction (Sbar, 1982). Such a strike-slip fault was shown on an earlier map by Hill and Troxel (1966; trend S_1 in the strain ellipse insert of Fig. 9A), and reproduced by Mann et al. (1983). Evidence cited to support that interpretation, in addition to the presence of oblique striae on fault surfaces in the Black Mountains, consists of spectacular northwest-plunging folds within the basement (trend A in Fig. 9A). Hill and Troxel (1966) described the folds as en echelon, but regional arguments, including evidence for the pressure and temperature conditions at the time of folding, suggest that they are probably incidental oblique structures predating strike-slip deformation. Folds with similar orientation are present in metamorphic rocks of both the Panamint Range, west of central Death Valley, and in the Funeral Mountains, north of (and approximately parallel to) the northern Death Valley-Furnace Creek fault zone. Metamorphism in the Funeral Mountains occurred, probably in late Mesozoic time, at a temperature of 600° to 700° C, and at a pressure of 7.2 to 9.6 kb (Labotka, 1980). In the Panamint Range, folding occurred at 70 to 80 Ma during retrograde metamorphism, and at a temperature of about 450° C, following prograde metamorphism at 400° to 700° C (Labotka, 1981; Labotka and Warasila, 1983). These metamorphic conditions and the age of folding inferred in the Panamint Range are incompatible with the Miocene and younger tectonic denudation that accompanied strike-slip deformation in the Death Valley area (see Cemen et al., 1985 this volume). Although low-angle normal faults follow the top of the basement, the folds themselves are incidental structures.

BASINS ALONG STRIKE-SLIP FAULTS

Some basins along strike-slip faults developed as a direct response to strike-slip deformation. Others owe their origin to a different tectonic regime, in which strike slip played only a subsidiary role in basin development. Examples of

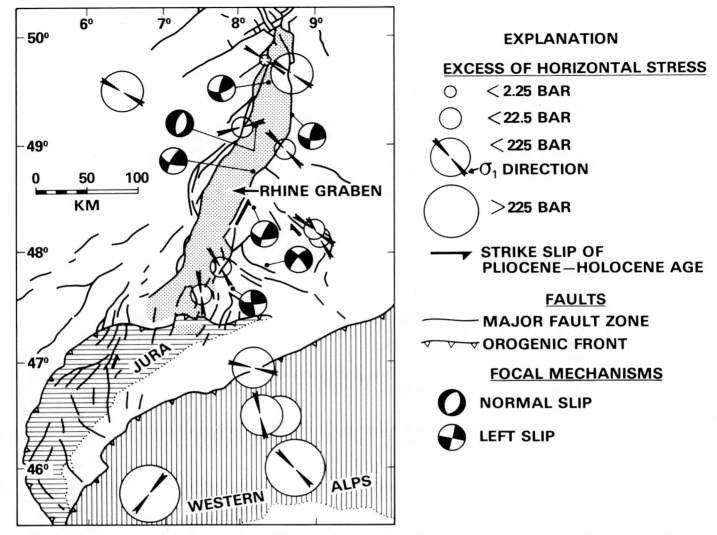

FIG. 8.—The Rhine graben and adjacent segments of the Jura and western Alps; with selected examples of the excess horizontal stress from in situ determinations; and earthquake focal mechanisms (modified from Ahorner, 1975; Illies and Greiner, 1978). Normal faults of Eocene to Miocene age were reactivated as left-slip faults in Pliocene to Holocene time, during uplift of the Alps.

the latter are (1) some intracontinental grabens, foreland basins, and forearc basins associated with strike-slip faults, which though active during sedimentation, had little influence on sedimentation patterns; and (2) various basins that were subsequently reactivated by strike-slip deformation. Basins of these types are considered here only to the extent that the geology is related to strike-slip deformation.

One of the most obvious characteristics of a strike-slip basin is its present geometry in map view, particularly the geometry of any bounding faults, and this has been a useful point of departure for various classification schemes and for most discussions about processes of basin formation (Carey, 1958; Kingma, 1958; Lensen, 1958; Quennell, 1958; Burchfiel and Stewart, 1966; Clayton, 1966; Belt, 1968; Freund, 1971; Crowell, 1974a, b, 1976; Ballance, 1980; Aydin and Nur, 1982a, b; Burke et al., 1982; Crowell and Link, 1982; Fralick, 1982; Mann and Burke, 1982; Mann et al., 1983; Mann and Bradley, 1984; see glossary in Bid-

dle and Christie-Blick, 1985a this volume). However, with the exception of very young sedimentary accumulations, the geometry, location, and perhaps orientation of a given basin have undoubtedly changed with time, and some prominent faults may be younger than much of the sedimentary fill. In many cases, present geometry may thus provide only limited information about either the kinematic history or the ultimate controls on basin evolution, and the utility of some of the classification schemes is questionable.

In this section, we focus on processes of basin formation, on the tectonic setting of depositional sites, and on certain distinctive characteristics of the stratigraphic record along strike-slip faults. We emphasize as others have before (Bally and Snelson, 1980) that every basin has a unique history, and that simple models may provide only a superficial summary of basin development. No strike-slip basin can be considered thoroughly understood unless its history can be reconstructed by a series of well-constrained palinspastic

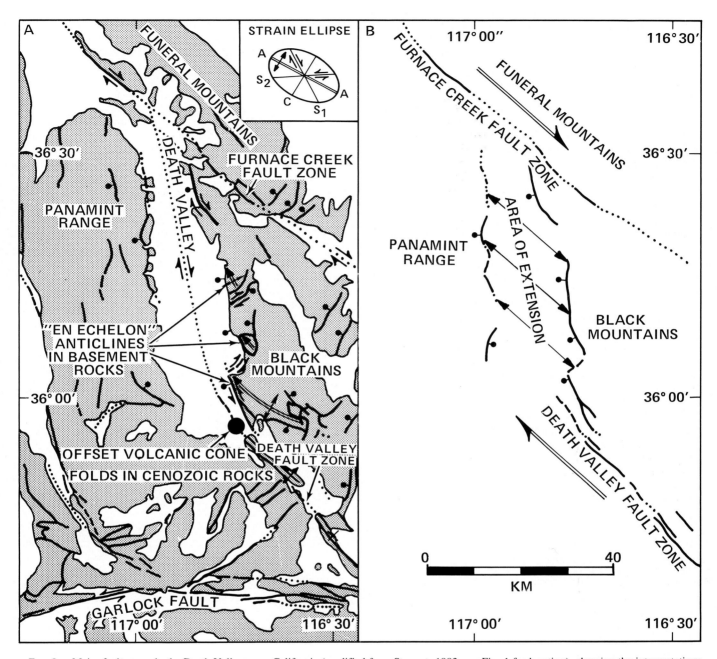

Fig. 9.—Major fault zones in the Death Valley area, California (modified from Stewart, 1983; see Fig. 1 for location), showing the interpretations of A) Hill and Troxel (1966), and B) Burchfiel and Stewart (1966). Shading (A) indicates outcrops of Proterozoic to Tertiary sedimentary and volcanic rocks; unshaded area represents Quaternary alluvial deposits. Evidence for strike slip along the northern Death Valley-Furnace Creek and southern Death Valley fault zones includes en echelon folds in Cenozoic rocks and an offset volcanic cone, together with regional stratigraphic arguments.

A) A buried strike-slip fault is inferred in the central north-trending segment of Death Valley on the basis of oblique striae on fault surfaces in the Black Mountains, and of "en echelon" anticlines in basement rocks (Hill and Troxel, 1966). The insert compares the orientations of observed structures with an idealized strain ellipse for the overall deformation; right slip inferred parallel to direction C is incompatible with orientations summarized in Figure 5.

B) Death Valley interpreted as a pull-apart along an oblique segment of a strike-slip fault system (Burchfiel and Stewart, 1966). Indicators of crustal stress and regional seismicity indicate continued extension in an approximately northwest-southeast direction parallel with the Furnace Creek and southern Death Valley fault zones (Sbar, 1982). See text for further explanation.

maps. Many of the examples discussed are taken from California, a classic region with which we are most familiar, but where appropriate, we draw attention to articles in this volume dealing with strike-slip basins in other parts of the world.

Processes of Basin Formation

Sedimentary basins form along strike-slip faults as a result of localized crustal extension, and especially in zones of continental convergence, of localized crustal shortening. Individual basins vary greatly in size (e.g., Aydin and Nur, 1982a; Mann et al., 1983), but they tend to be smaller than those produced by regional extension (many intracontinental grabens) or regional shortening (foreland and forearc basins). In addition, a given basin may alternately experience both extension and shortening on a timescale of thousands to millions of years, through variations in the motion of adjacent crustal blocks (e.g., the Ventura Basin, California, Fig. 15 below; and in this volume, Miall, 1985; Nilsen and McLaughlin, 1985; Steel et al., 1985); or extension in one direction (or in one part of the basin) may be accompanied by shortening in another direction (or in another part of the basin). A possible example of the latter is the occurrence of positive flower structures in the Mecca Hills on the northeastern side of the Salton Trough, southern California, a basin that can be related to extension in a northwestward direction in the overstep between the San Andreas and Imperial faults (Fig. 1; Sylvester and Smith, 1976; Crowell, 1981b; Fig. 6 of Harding et al., 1985 this volume). The directions of extension and shortening also tend to vary within a given basin and to change through time, especially where crustal blocks rotate (as seen on a small scale in Fig. 6 A), where there are significant differences in the rate of internal strain of adjacent crustal blocks (see Figs. 10, 11, and 12 of Şengör et al., 1985 this volume), or when there are changes in lithospheric plate motion.

In the simplest strike-slip basins (e.g., the "pull-apart hole" and "sharp pull-apart basin" of Crowell, 1974a, b), the bounding blocks are torsionally rigid and deform only at their edges, and subsidence is due to extension only in a direction parallel to the regional strike of the fault(s). Examples include many small pull-apart basins (Aydin and Nur, 1982a; Mann et al., 1983), and some detached or thin-skinned basins such as the Vienna Basin (Fig. 10B; Royden, 1985 this volume). In the vicinity of the Salton Trough, however, the crustal structure inferred from seismic refraction and gravity data (Fuis et al., 1984) suggests that significant crustal thinning has occurred outside the overstep between the San Andreas and Imperial faults (Fig. 10A). The interpretation of the gravity data is not unique, but the relatively flat gravity profile across the Salton Trough requires that the upper surface of the subbasement (lined area) largely mirror the contact between the sedimentary rocks and basement (Fuis et al., 1984). The subbasement was modelled with a density of 3.1 gm/cm^3, and refraction data suggest a P-wave velocity of greater than or equal to 7.2 km/s, consistent with significant amounts of mafic igneous rocks intruded into the lower part of the crust. The depth to the top of the subbasement decreases abruptly from 16 km at the Salton Sea to approximately 10 km at the U.S.-Mexico border, about 30 km to the south, indicating that the gross crustal structure is for the most part related to the opening of the Gulf of California, and is not simply inherited from an earlier phase of regional extension.

The profile shown in Figure 10A is probably characteristic of junctures between continental transform systems and divergent plate boundaries, but some strike-slip basins are detached. Examples are known from (1) areas of pronounced regional shortening, such as the Vienna Basin, which formed adjacent to tear faults of the Carpathian nappes (Fig. 10B; Royden, 1985 this volume), and the St. George Basin, located in the forearc of the Bering Sea, Alaska (Marlow and Cooper, 1980); and (2) areas subject to marked regional extension, such as the West Anatolian extensional province of Turkey (Fig. 18 of Şengör et al., 1985 this volume), and the Basin and Range Province of the western United States (Cemen et al., 1985; Link et al., 1985, both in this volume). Detached strike-slip basins may also prove to be relatively common in intracontinental transform zones, particularly those located along former convergent plate boundaries. In central and southern California, for example, the presence of mid-crustal detachments is suggested by (1) earthquakes with low-angle nodal planes, and the alignment of earthquake hypocenters parallel to the gently dipping base of the seismogenic crust (Nicholson et al., 1985a, b; Webb and Kanamori, 1985); (2) a pronounced change, near the junction of the San Andreas and San Jacinto faults (SJ in Fig. 1), in the patterns of seismicity with depth, with shallow seismicity suggesting the rotation of small crustal blocks (Fig. 6B; Nicholson et al., 1985a, b); (3) regional patterns of earthquake travel-time residuals (Hadley and Kanamori, 1977); (4) the distribution of upper crustal seismic velocities (Hearn and Clayton, 1984; Nicholson et al., 1985a); (5) geodetic constraints on the effective elastic thickness of the upper crust (Turcotte et al., 1984); (6) paleomagnetically determined patterns of large-scale block rotation (Fig. 2A; Luyendyk et al., 1985); (7) deep seismic-reflection studies that reveal the presence of low-angle reflecting horizons (Cheadle et al., 1985); and (8) evidence from surface geology and shallow seismic reflection profiles for numerous Neogene and Quaternary low-angle faults (Yeats, 1981, 1983; Crouch et al., 1984; Wentworth et al., 1984; Namson et al., 1985). We note that it is not yet clear how much of the California margin is detached, although the scale probably exceeds that of individual basins. Manspeizer (1985 this volume) interprets the Dead Sea Basin as detached on a smaller scale above listric normal faults bounded by the overstepping strands of the Dead Sea fault zone (his Fig. 13, after a concept of K. Arbenz, and reproduced on the cover of this book).

As discussed above, lithospheric or crustal extension produces a thermal anomaly, and subsequent cooling leads to additional subsidence (McKenzie, 1978; Royden and Keen, 1980; Steckler and Watts, 1982). In narrow basins, and in those for which extension is not "instantaneous," but occurs over an interval of more than about 10 m.y., a significant fraction of the thermal anomaly decays during the rifting stage, increasing the amount of syn-rift subsidence at the expense of the post-rift (Jarvis and McKenzie, 1980;

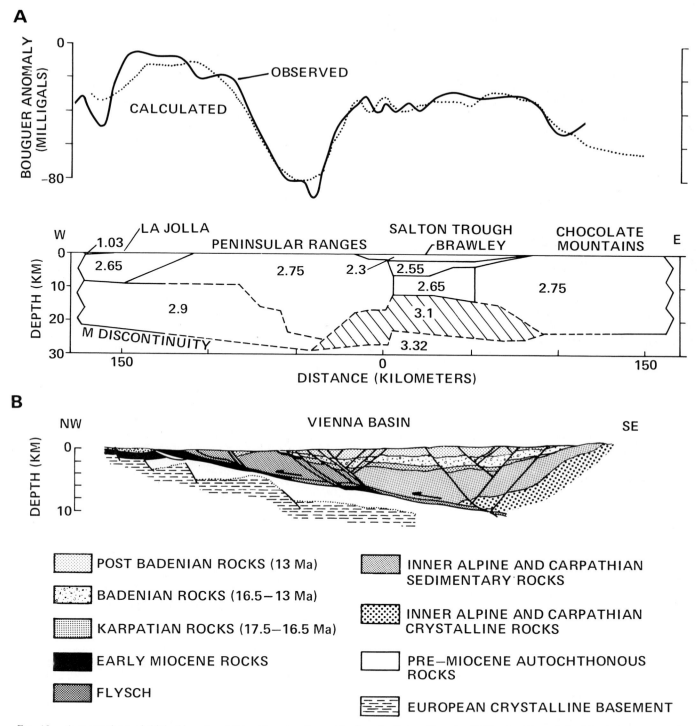

Fig. 10.—A comparison of thick-skinned and thin-skinned pull-apart basins: the Salton Trough, California (A), and the Vienna Basin, Austria and Czechoslovakia (B).

A) Cross section of southern California from La Jolla to the Chocolate Mountains (from Fuis et al., 1984, see Fig. 1 for location). The observed gravity anomaly is compared with the anomaly calculated from the model (densities in gm/cm³). Solid boundaries are those controlled by seismic refraction data; dashed lines indicate boundaries adjusted to fit the gravity data. Sub-basement (lined area, density 3.1 gm/cm³) beneath the Salton Trough provides most of the gravitational compensation for sedimentary rocks (densities 2.3 and 2.55 gm/cm³) and inferred metasedimentary rocks (density 2.65 gm/cm³). The San Andreas and Imperial faults are located near the east and west edges of the block with density of 2.65 gm/cm³.

B) Cross section of the Miocene Vienna Basin, a detached pull-apart basin superimposed partly on nappes of the outer Carpathian flysch belt, and partly on nappes of the inner Carpathians (section 3 from Fig. 6 of Royden, 1985 this volume). Tertiary thrusts are indicated by arrows; Miocene normal faults displace Miocene rocks at the surface; normal faults confined to the autochthon are mainly Jurassic syn-sedimentary faults associated with Mesozoic rifting.

Steckler, 1981; Cochran, 1983; Pitman and Andrews, 1985 this volume). The post-rift thermal anomaly is also less for detached basins than for those produced by an equivalent amount of extension of the entire lithosphere, and detached basins thus experience reduced post-rift subsidence. Extensional strike-slip basins tend to be short-lived, but the rifting stage is preferentially represented in the stratigraphic record because most such basins are small, in some cases detached, and in many cases subject to subsequent uplift and erosion.

Another potentially important mechanism for basin subsidence along strike-slip faults, in addition to crustal extension, is loading due to the local convergence of crustal blocks (e.g., the Ventura Basin of southern California; Fig. 15 below; Burke et al., 1982; Yeats, 1983). Similar basins bounded by major thrust faults are discussed by Şengör et al. (1985 this volume) and by Steel et al. (1985 this volume). As in foreland basins (Beaumont, 1981), patterns of subsidence are likely to be influenced by lithospheric flexure, a mechanism of regional isostatic compensation in which loads are supported by broad deflection of the lithosphere as a result of lithospheric rigidity (Forsyth, 1979; Watts, 1983), but to our knowledge, no attempt has yet been made to model flexure in strike-slip basins. Some strike-slip basins experience both extension and shortening (e.g., a pull-apart basin adjacent to a convergent strike-slip fault). Flexural effects are reduced when the lithosphere is warm and relatively weak, and we presume that in such circumstances subsidence induced by loading would be localized.

Subsidence due to extension, thermal contraction or crustal loads is amplified by sediment loading (Steckler and Watts, 1982). Simple isostatic considerations indicate that for typical crustal and sediment densities, sediment loading accounts for about half the total subsidence. Loading by water is important only to the extent that sea level or lake level varies.

Although it is relatively simple to construct models of the processes outlined above for hypothetical basins, it is clearly difficult to unravel the processes from the geological record of a given strike-slip basin. A fruitful avenue of research will be to use a combination of geological and geophysical data from well studied basins to test these theoretical concepts.

Tectonic Setting of Depositional Sites

The location of depositional sites along strike-slip faults is controlled by several factors operating at a variety of length scales and time scales. Along intracontinental transform zones, such as in California, and the Dead Sea fault zone, these include (1) crustal type and the configuration of pre-existing crustal structures, especially the distribution, orientation and dimensions of zones of weakness, such as faults; (2) variations in the overall motion of adjacent lithospheric plates; and (3) the kinematic behavior of crustal blocks within the transform zone.

Pre-existing Crustal Structure.—Pre-existing crustal structure influences the location of major strike-slip faults, and the way in which crustal blocks between those faults move and deform to produce basins. The San Andreas fault, for example, originated in southern California by about 8 Ma, with most of the 320 km of right slip accumulating since about 5.5 Ma during oblique opening of the Gulf of California (Fig. 1; Karig and Jensky, 1972; Crowell, 1981b; Curray and Moore, 1984). Strike-slip faulting is only the latest event in a long geologic history, extending back into Proterozoic time, and including the assembly, amalgamation and accretion of various suspect terranes to North America during Mesozoic time; Mesozoic to Miocene arc magmatism, associated forearc and perhaps back-arc sedimentation; and Oligocene to Miocene Basin and Range extension (Hamilton, 1978; Crowell, 1981a, b; Dokka and Merriam, 1982; Champion et al., 1984). The present Gulf is the result of seafloor spreading beginning in latest Miocene time, but possibly along the line of a somewhat older Miocene tectonic depression that has been termed the proto-Gulf of California (Karig and Jensky, 1972; Terres and Crowell, 1979; Crowell, 1981a, b; but see Curray and Moore, 1984, for a modified interpretation). With local exceptions, such as in the vicinity of the Transverse Ranges (ETR and WTR in Fig. 1), the San Andreas fault and related transform faults to the south in the Gulf of California approximately parallel the boundaries of older tectonic elements, and occupy a position along the eastern edge of the Cretaceous Peninsular Ranges and Baja California Batholith (Hamilton, 1978).

Vink et al. (1984) suggested that the Gulf of California and the associated transform system formed just within the continent because continental lithosphere is weaker than oceanic lithosphere, and that rifting close to a continent-ocean boundary invariably follows a continental pathway. A similar model has been proposed by Steckler and ten Brink (1985) to explain the development in mid-Miocene time of the Dead Sea fault, at the expense of extension in the Gulf of Suez. According to their model, the hinge zone along the Mediterranean margin between thinned and unthinned continental crust acted as a barrier to continued northward propagation of the Red Sea rift.

The distribution of Neogene basins in California is probably influenced to a certain extent by crustal composition and thickness, and by the location of older basins (Blake et al., 1978). For example, the San Joaquin Basin (SJB in Fig. 1), located along the eastern side of the San Andreas Fault in central California, is superimposed on a forearc basin of late Jurassic to mid-Miocene age, which in turn overlies the boundary between Sierran-arc and ophiolitic basement (Blake et al., 1978; Hamilton, 1978; Bartow, 1984; Bartow and McDougall, 1984; Namson et al., 1985).

Relative Plate Motion.—Another factor that influences the development of sedimentary basins along a continental transform zone is the overall relative motion of the adjacent lithospheric plates (Crowell, 1974b; Mann et al., 1983). Such motion may be strictly transform, or it may involve a component of either convergence or divergence (Harland, 1971). In practice, however, it is generally difficult to predict the behavior of individual blocks solely on this basis, owing to the complex jostling that takes place in a broad transform plate boundary, even when there is little variation in relative plate movement. The character of the plate boundary is also sensitive to the migration of any unstable triple junctions, and to changes in the position of the instantaneous relative-motion pole. Either of these eventualities may be associated with the reorganization of plate-boundary geometry, and with the transfer of slices of one plate to another (Crowell, 1979).

Mann et al. (1983) argued that the gross distribution of regions of extension and shortening along the San Andreas and Dead Sea transform systems can be explained by comparing the orientation of the major faults with the interplate slip lines suggested by Minster et al. (1974). According to Mann et al., active strike-slip basins occur preferentially where the principal displacement zone is divergent with respect to overall plate motion (e.g., the Wagner and Delfin Basins of the northern Gulf of California; and the Dead Sea Basin); push-up blocks (the Transverse Ranges, California, and the Lebanon Ranges) occur where the principal displacement zone is convergent. Such statements are probably valid where most of the plate motion is taken up along a single fault (e.g., Dead Sea fault zone), but less certain where the motion is accommodated by a number of major faults, and especially where crustal blocks rotate or deform internally.

Estimates of relative motion between the Pacific and North American plates, and of Quaternary slip rates for strike-slip faults in California suggest that there should be pronounced shortening across the Transverse Ranges of California (ETR and WTR in Fig. 1), an observation qualitatively supported by abundant geological observations (Nardin and Henyey, 1978; Jackson and Yeats, 1982; Yeats, 1983; Crouch et al., 1984). About half to two thirds of the Pacific-North American relative motion (56 mm/yr; Minster et al., 1974; Minster and Jordan, 1978, 1984) occurs on the San Andreas fault (Sieh and Jahns, 1984; Weldon and Humphreys, 1985). Recent estimates of the slip rate are 25 mm/yr during the Quaternary in southern California (Weldon and Sieh, 1985), and 34 mm/yr during the Holocene in central California (Sieh and Jahns, 1984). The remaining displacement is thought to be taken up by the San Jacinto and Elsinore faults (about 10 and 1 mm/yr, respectively, for Quaternary time; Sharp, 1981; Ziony and Yerkes, 1984), and by offshore faults such as the San Gregorio-Hosgri (6 to 13 mm/yr for the late Pleistocene and Holocene; Weber and Lajoie, 1977; fault not shown in Fig. 1), assuming that Quaternary plate motion has been much the same as the average motion for the past 5 to 10 m.y. In assessing these data, Weldon and Humphreys (1985) have concluded that the magnitude of Quaternary shortening across the Transverse Ranges is considerably smaller than expected. They suggest that plate motion may have been taken up not merely by translation on the major faults but in part by counterclockwise rotation about a pole 650 km southwest of the "big bend" of the San Andreas fault (Fig. 1), an interpretation that contrasts with paleomagnetic evidence for predominantly clockwise rotations west of the San Andreas on a longer time scale (Fig. 2A; Luyendyk et al., 1985). The rates of both strike-slip faulting and regional shortening are uncertain (see Bird and Rosenstock, 1984, for a different model), but available data indicate only a qualitative relation between the orientation of interplate slip lines and patterns of uplift and subsidence, even for tectonic features as prominent as the Transverse Ranges. In zones of continental convergence, such as Turkey (Şengör et al., 1985 this volume), patterns of deformation are even more complicated.

The Neogene history of California also provides examples of plate boundary reorganization resulting from the migration of unstable triple junctions and changes in the position of the instantaneous relative-motion pole. The transform plate margin originated in California at about 30 Ma, following the impingement of the Pacific plate against the North American plate (Atwater, 1970; Atwater and Molnar, 1973). The transform system lengthened, probably intermittently, by northward motion of the northern trench-transform-transform triple junction and by generally southward motion of the southern transform-trench-ridge triple junction. At the same time, perhaps because of irregularities along the plate boundary (Crowell, 1979), slices of the North American plate were incorporated within the evolving transform system and offset differentially along the plate margin (Crouch, 1979). By about 5.5 Ma, the southern triple junction reached the mouth of the proto-Gulf of California, and much of the transform motion was taken up by the San Andreas fault, effectively transferring the Peninsular Ranges and Baja California to the Pacific plate (Curray and Moore, 1984). This reorganization of the plate boundary may be related to a small change in the relative plate motion determined by Page and Engebretson (1984). Another consequence of the inferred change in plate motion may be the onset in Pliocene and Pleistocene time of the shortening, described above, across the Transverse Ranges and Coast Ranges of California, and an acceleration in the subsidence rates for several basins (Yeats, 1978). In the Ventura Basin, for example, where considerable geochronological precision is possible, subsidence rates are estimated to have increased from approximately 250 m/m.y. at about 4 Ma to between 2,000 and 4,000 m/m.y. in the past million years. Note that these figures of Yeats (1978) incorporate inferred changes in water depth, but no corrections for compaction, loading or eustatic changes in sea level.

A modern analogue for the geological complexities at a trench-transform-transform triple junction, and a mechanism by which tectonic slices are transferred from one plate to another, has been described by Herd (1978). In the vicinity of the Cape Mendocino triple junction, the San Andreas fault appears to split into two subparallel fault zones, approximately 70 to 100 km apart. The western zone, the San Andreas proper, terminates at Cape Mendocino (CM in Fig. 1). The eastern zone, consisting of the right-stepping Hayward, Healdsburg-Rodgers Creek, Maacama and Lake Mountain fault zones (H, RC, M, and LM in Fig. 1), appears to be very youthful, and extends northward onto the continental shelf about 150 km north of Cape Mendocino.

Kinematic Behavior of Crustal Blocks.—The kinematic behavior of fault-bounded crustal blocks within a transform zone has long been considered the principal control on the development of strike-slip basins (Lensen, 1958; Kingma, 1958; Quennell, 1958; Crowell, 1974a, b, 1976; Reading, 1980; Aydin and Nur, 1982a; Mann et al., 1983). The general idea is that subsidence tends to occur where strike slip is accompanied by a component of divergence, as a result, for example, of a bend or an overstep in the fault trace ("pull-apart basin" of Burchfiel and Stewart, 1966) or through extension near a fault junction ("fault-wedge basin" of Crowell, 1974b). Uplift occurs where there is a component of convergence, although an overridden block may be depressed by the overriding one. Examples of strike-slip basins with different geometry and of different size are illustrated in Figure 11. Crowell (1974b) described bends associated predominantly with stretching and subsidence or with shortening and uplift as "releasing" and "restraining" bends, respectively, and here, we extend the use of the terms

"releasing" and "restraining" to kinematically equivalent oversteps and fault junctions. As recognized by Crowell (1974a, b), the kinematic behavior of individual crustal blocks is superimposed on broader patterns of plate interaction, and in complexly braided fault systems, many basins experience episodic subsidence as a result of changes in fault geometry and/or block motion. In some cases, therefore, the stratigraphic record of a strike-slip basin may not be related in a simple way to the present configuration of faults, and it may even be difficult to predict contemporary patterns of subsidence and uplift from fault geometry.

Consider, for example, basins and horsts associated with fault junctions. Some of the numerous possible patterns of subsidence and uplift at a simple fault junction are illustrated in Figure 12, modified from Figure 11 of Crowell (1974b). As originally discussed by Crowell, and reproduced in subsequent summary articles (e.g., Bally and Snelson, 1980; Reading, 1980, 1982; Freund, 1982), the wedge between the faults shown in Figure 12A is said to be "compressed and elevated" where the "faults converge," and "extended" where the "faults diverge." It is clear, however, that faults that converge in one direction diverge in the opposite direction. Assuming deformation only along the edges of blocks, and rotation only to the extent required by fault curvature, uplift and subsidence are related to the orientation (dip and strike) of the faults with respect to the overall slip vectors of the blocks (horizontal in Fig. 12B), and to the amount of extension or shortening associated with each fault. Although distinct in cross section, note that similar map-view configurations of basins and horsts can arise at either releasing or restraining junctions. More complicated arrangements can be envisaged at junctions with both releasing and restraining characteristics, and where blocks are internally deformed or rotated (see Fig. 6). Segmentation of the block between the branching faults may produce grabens within a horst or horsts within a graben, as in the Dead Sea fault zone (Garfunkel, 1981). In general, we expect basins and horsts to evolve continuously by a combination of fault slip and rotations about both vertical and horizontal axes, but there are also discontinuous changes due to episodic slip and rotation, and to the propagation of new faults. These are needed from time to time to eliminate tectonic knots, or complexities that inhibit further strike-slip deformation. Thus basins with thick sedimentary accumulations may at times experience uplift, and horsts with little preserved sediment may subside. Many strike-slip basins are actually slices of basins offset along one or more younger strike-slip faults.

An important aspect of strike-slip deformation and basin formation, but one for which documentation is acquired with difficulty, is the manner in which complexities such as fault bends, oversteps and junctions arise or are subsequently removed. Some curvature in strike-slip faults is attributable to a near pole of plate rotation (e.g., the Dead Sea fault zone; Garfunkel, 1981), and some bends may result from crustal heterogeneity or from local variations in the distribution of stress influencing the path of fault propagation. In eastern Jamaica, for example, a prominent right-hand restraining bend between the left-lateral east-striking Plan-

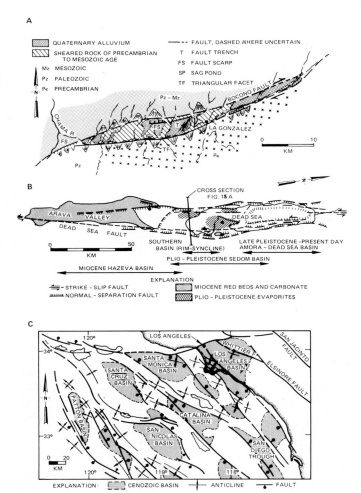

FIG. 11.—A comparison in map view of strike-slip basins of different ages, geometry, and scale.

A) The La González Basin, a lazy Z-shaped pull-apart basin of Pliocene (?) to Quaternary age along the Boconó fault zone, Venezuela (from Schubert, 1980).

B) The Dead Sea Rift, a rhomboidal pull-apart basin of Miocene to Holocene age (from Zak and Freund, 1981). Since Miocene time, the depocenter has migrated northward from the Arava Valley to the site of the present Dead Sea. For cross section, see Figure 15 A.

C) Selected faults, anticlines, and Cenozoic strike-slip basins in the southern California borderland (modified from Moore, 1969; Junger, 1976; Howell et al., 1980; see Figs. 1 and 2A for location). Many of these basins differ from the La González and Dead Sea Basins in being bounded by strike-slip faults of different orientations in a broad transform zone.

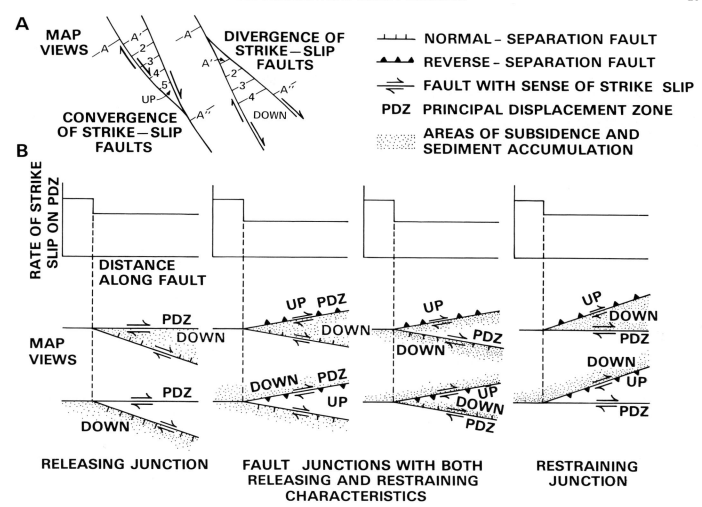

FIG. 12.—A) Sketch maps showing uplift of the tip of a fault wedge with convergence of right-slip faults, and subsidence of the tip with divergence (from Crowell, 1974b).

B) An alternative version of A, showing a range of possible results of slip along a bifurcating right-slip fault. See text for futher explanation.

tain Garden and Duanvale fault zones appears to have nucleated in Miocene time on northwest-striking normal faults that bounded a Paleogene graben (Fig. 2 of Mann et al., 1985 this volume). Other bends in strike-slip faults are due to the deformation of initially straight faults, as a result of (1) incompatible slip at a fault junction; (2) rotations within one or more adjacent blocks; or (3) intersection of a strike-slip fault with a zone of greater extensional or convergent strain. The development of the "big bend" of the San Andreas fault (Fig. 1) was attributed by Bohannon and Howell (1982) to incompatible displacement in late Cenozoic time on the San Andreas and Garlock faults (320 km of right slip, and 65 km of left slip, respectively; Smith, 1962; Smith and Ketner, 1970; Crowell, 1981b). As discussed above, this deformation was accompanied by counterclockwise rotation of the Mojave block between the faults (Fig. 2A; Morton and Hillhouse, 1985). Bends may also develop as a result of small-scale block rotation of the sort documented by Nicholson et al. (1985a, b; Fig. 6B). Şengör et al. (1985 this volume) describe fault bends due to the intersection of

strike-slip faults with zones of accentuated extensional strain at the western end of the North Anatolian fault, and of accentuated convergent strain near the junction of the North Anatolian and East Anatolian faults in eastern Turkey (see their Figs. 10, 11, and 12).

Oversteps, and branching and braiding are fundamental features of many strike-slip fault zones and fault systems, and they develop by a number of mechanisms: (1) bending of initially straight faults; (2) direct and indirect interaction between faults; (3) segmentation of curved faults; (4) faulting within a weak zone oblique to possible failure planes; and (5) reactivation of pre-existing extension fractures (Crowell, 1974b; Freund, 1974; Segall and Pollard, 1980; 1983; Mann et al., 1983; Aydin and Nur, 1982a, 1985 this volume). Mann et al. (1983) proposed that pull-aparts evolve from incipient to mature ("extremely developed") basins through a sequence of closely related states. According to them, basins tend to form at releasing bends, and develop by way of spindle-shaped and "lazy S" (or "lazy Z") basins such as the La González Basin, Venezuela (Fig. 11A;

Schubert, 1980), to rhomboidal basins such as the Dead Sea Rift (Fig. 11B; Zak and Freund, 1981; Manspeizer, 1985 this volume), and eventually, in some cases, to long narrow troughs floored by oceanic crust (e.g., the Cayman Trough of the northern Caribbean). In this model, oversteps arise by the propagation of secondary strike-slip faults in the vicinity of the releasing bend. A possible example of this branching process is the junction between the San Gabriel and San Andreas faults, the site of the Ridge Basin (Fig. 1). Between about 12 and 5 Ma, 13 km of sediment was deposited at a right bend along the San Gabriel fault, which at the same time experienced as much as 60 km of right slip (Crowell, 1982; Nilsen and McLaughlin, 1985 this volume). Beginning between 5 and 6 Ma, the San Gabriel fault ceased to be active, and strike-slip deformation was taken over by the San Andreas fault.

The interaction of parallel faults propagating from opposite directions is inherent in the models of Rodgers (1980) and Segall and Pollard (1980) for the evolution of pull-apart basins, and a possible example, the Soria Basin of northern Spain, is discussed in this volume by Guiraud and Seguret (1985). Theoretical studies indicate that significant interaction should occur between strike-slip faults if they are separated by less than twice the depth of faulting (Segall and Pollard, 1980). For strike-slip faults in California, where seismicity is observed to depths of 10 to 15 km, interaction is expected if faults are closer than 20 to 30 km. The other mechanisms listed above for generating oversteps are discussed elsewhere in this volume by Aydin and Nur (1985), and only briefly mentioned here. A consistent sense of overstepping along some curved fault zones suggests a relation between the sense of step and curvature (Aydin and Nur, 1985 this volume). An example of overstepping faults within an inappropriately oriented weak zone is the series of transform faults in the Gulf of California (Fig. 2 of Mann et al., 1983). The reactivation of pre-existing extension fractures appears to be largely a small-scale phenomenon (Segall and Pollard, 1983).

In the light of the foregoing discussion of the kinematics of strike-slip basins, we here consider the interpretation of one of the key geological elements needed to undertake a palinspastic reconstruction: piercing points of known age with which to derive the displacement history of the major strike-slip faults (e.g., Crowell, 1962, 1982, for offsets across the San Andreas and San Gabriel faults, California; and Freund et al., 1970, for offsets across the Dead Sea fault). Owing to the paucity of suitable geologic "lines," such reconstructions are commonly difficult. In the simplest case, the time of earliest movement on a strike-slip fault is given approximately by the age of the youngest rocks offset by the maximum amount (T_1 in Fig. 13A). For the San Andreas fault in southern California, this is late Miocene (Crowell, 1981a). The timing for the Dead Sea fault is less well constrained as approximately mid-Miocene, but definitely younger than basaltic dikes dated as about 20 Ma (Garfunkel, 1981), and the nature of this sort of uncertainty is shown diagrammatically in Figure 13A. Another difficulty arises if displacement varies along the fault as well as increasing with time. Strike-slip faults commonly branch and intersect, and in places accommodate differential ex-

tension or shortening. Figure 13 illustrates alternative interpretations of the displacement history of a hypothetical fault, given piercing points of known age (T_0, T_1, T_2, T_3), known displacement magnitude (D_0, D_1, D_2, D_3), and known location along one side of the fault (L_0, L_1, L_2, L_3). In one interpretation (Fig. 13A), the fault moves episodically over a long interval, beginning between times T_1 and T_2, and ending at T_4. For rocks of a given age, the magnitude of displacement is the same at all points along the fault. In the second interpretation (Fig. 13B), faulting is confined to a short interval between times T_3 and T_4. Piercing points of different ages indicate different magnitudes of displacement, because displacement varies along the fault. In reality, a given fault may exhibit displacement that is both variable and episodic, with the rate of movement changing from one interval of geologic time to another (not shown in Fig. 13). A given data set may also be incompatible with the extreme interpretations presented. The point of the illustration, however, is that in areas of complex deformation it is important to interpret the offset history of a given strike-slip fault in a regional plate-tectonic context, incorporating other stratigraphic clues to the timing of faulting.

Distinctive Aspects of the Stratigraphic Record

Strike-slip basins are present in many different plate tectonic settings, and they are filled with sediments deposited in a variety of marine and non-marine environments, subject to a range of climatic conditions. In spite of these obvious differences, however, certain aspects of the stratigraphic record appear to be distinctive. These are (1) geological mismatches within and at the boundaries of basins, that is, features which document the occurrence of strike slip; (2) a tendency for longitudinal as well as lateral basin asymmetry, owing to the migration of depocenters with time; (3) evidence for episodic rapid subsidence, recorded by thick stratigraphic sections, and in some marine basins by rapid deepening; (4) the development of pronounced topographic relief, which is associated with abrupt lateral facies changes and local unconformities at basin margins; and (5) marked differences in stratigraphic thickness, facies geometry, and the occurrence of unconformities from one basin to another in the same region.

Geological Mismatches.—A geological mismatch occurs where rocks juxtaposed by a fault require a considerable amount of displacement to have taken place on the fault or on another structure cut by the fault. Such mismatches occur at sutures, in thrust and fold belts, and across some low-angle normal faults in extensional allochthons. They are also common in regions deformed by major strike-slip faults. The segment of the San Gabriel fault between the southern part of the Ridge Basin and the eastern Ventura Basin is a good example (Fig. 14; Crowell, 1982). Lateral facies relations, clast-size trends, and paleocurrent data suggest that conglomerate of the upper Miocene Modelo Formation of the Ventura Basin was derived from a nearby source to the northeast. The conglomerate consists of distinctive clasts of gabbro, norite, anorthosite, and gneiss as large as 1.5 m in diameter for which no nearby source is known across the

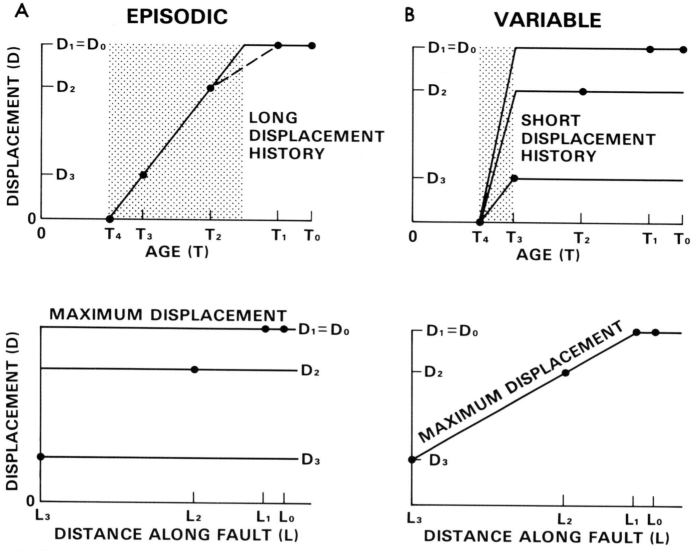

FIG. 13.—Alternative interpretations of the displacement history of a hypothetical fault, given piercing points of known age (T₀, T₁, T₂, T₃), known displacement magnitude (D₀, D₁, D₂, D₃), and known location along one side of the fault (L₀, L₁, L₂, L₃).

A) The fault moves episodically over a long interval, beginning between time T₁ and time T₂, and ending at T₄. For rocks of a given age, the magnitude of displacement is the same at all points along the fault. Note that by connecting displacement-age pairs, one might infer (perhaps incorrectly) that faulting began as early as time T₁, and that the slip rate increased after T₂ (dashed line).

B) Movement along the fault is confined to a short interval between times T₃ and T₄. Piercing points of different ages indicate different magnitudes of displacement, because displacement varies along the fault.

San Gabriel fault. Instead, that region is underlain by thick Miocene and older sedimentary rocks overlying a basement terrane quite different from that represented in the Modelo clasts. Immediately across the fault from the Modelo conglomerate, the Violin Breccia, also upper Miocene and also marine, was derived from a predominantly gneissic source to the southwest, where the basement is still largely covered by Miocene and older sedimentary rocks. The mismatch between sediments and suitable source rocks on both sides of the San Gabriel fault can be resolved, however, by removing between 35 and 60 km of right slip (Crowell, 1982). We emphasize that although evidence of this sort is important for documenting the magnitude or even occur-

rence of lateral offsets, the presence of similar geology on opposite sides of a fault at a particular locality does not necessarily preclude strike-slip deformation, a phenomenon that Crowell (1962) termed regional trace slip.

Basin Asymmetry.—Many sedimentary basins are asymmetrical, especially if faults occur preferentially along one side, and as in the case of grabens formed by regional extension (Harding, 1984), the sense of asymmetry in strike-slip basins may change from one profile to another (e.g., the Gulf of Elat; Ben-Avraham et al., 1979). The faults bounding strike-slip basins may be characterized by either normal separation (e.g., the Dead Sea Rift; Fig. 15A) or reverse separation (e.g., the Ventura Basin, California; Fig.

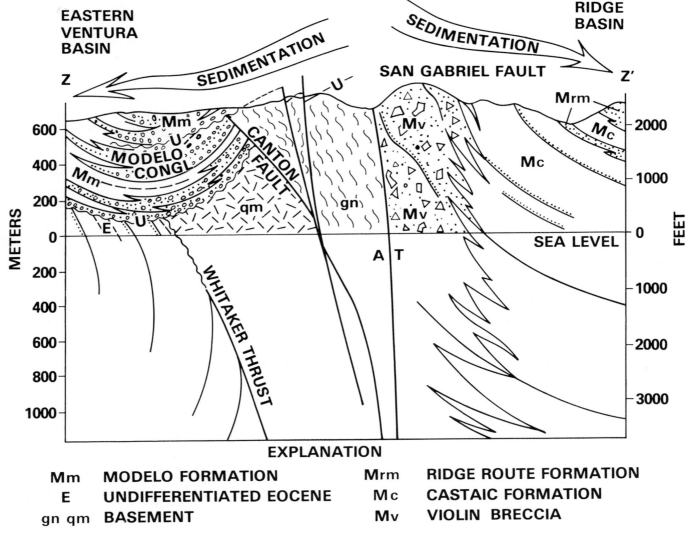

Fig. 14.—An example of a geological mismatch across a strike-slip fault: the San Gabriel fault, California (from Crowell, 1982; see Fig. 1 for location). The Modelo Conglomerate, derived from the northeast, is faulted against the Violin Breccia, derived from the southwest. T, displacement toward the observer; A, displacement away from the observer.

15B), or as discussed in the section on structural style, both normal and reverse faults may be present in the same basin (see Nilsen and McLaughlin, 1985 this volume). A particularly distinctive feature of strike-slip basins is the tendency for longitudinal as well as lateral asymmetry. The depocenter of the Dead Sea Basin, for example, has migrated northward more than 100 km from the site of the Arava Valley in the Miocene to the present Dead Sea (Fig. 11B; Zak and Freund, 1981; Manspeizer, 1985 this volume). Other basins with longitudinal asymmetry described in this volume are the Ridge Basin, California, and Hornelen Basin, Norway (Nilsen and McLaughlin, 1985), the Soria Basin of northern Spain (Guiraud and Seguret, 1985), the Nonacho Basin of Canada (Aspler and Donaldson, 1985), and possibly the Central Basin of Spitsbergen (Steel et al., 1985).
Episodic Rapid Subsidence.—Strike-slip basins are characterized by extremely rapid rates of subsidence (Fig. 16),

even more rapid than many grabens and foreland basins, and where there is an abundant sediment supply, by very thick stratigraphic sections in comparison with lateral basin dimensions (S. Y. Johnson, 1985; Nilsen and McLaughlin, 1985, both in this volume). For example, about 13 km of sediment accumulated in the Ridge Basin in only 7 m.y. (Crowell and Link, 1982), and 5 km of sediment was deposited in the Vallecito-Fish Creek Basin in about 3.4 m.y. (N. M. Johnson et al., 1983). The Ventura Basin subsided nearly 4 km in the past 1 m.y. (Fig. 15B; Yeats, 1978). Marine basins and some deep lakes tend to become temporarily starved of sediment, a situation that promotes the accumulation of fine-grained organic-rich sediments suitable for the generation of petroleum (Graham et al., 1985; Link et al., 1985 this volume). Depending on local patterns of deformation, however, the subsidence in strike-slip basins is also episodic, and may end abruptly. The Vallecito-

Fish Creek Basin has been uplifted more than 5 km in the past 0.9 m.y. (N. M. Johnson et al., 1983).

Local Facies Changes and Unconformities.—Although not diagnostic of a strike-slip setting, many strike-slip basins form adjacent to uplifted blocks with pronounced topographic relief. As described in many of the papers that follow, this leads to very coarse sedimentary facies along some basin margins and to abrupt lateral facies changes. Local vertical movements of blocks result in localized unconformities.

Contrasts Between Basins.—Again, because of local tectonic controls, patterns of sedimentation vary markedly from one basin to another within the same region (see the description of Eocene sedimentation in Washington by Johnson, 1985 this volume). In the case of basins for which original geometry has been obscured by subsequent defor-

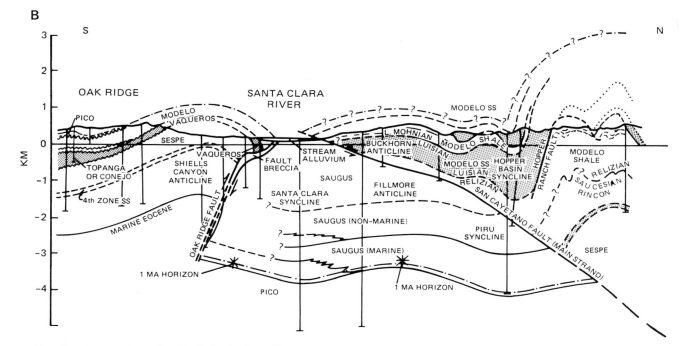

FIG. 15.—A comparison of strike-slip basins in profile.
A) The Dead Sea Rift, bounded by faults with normal separation (from Zak and Freund, 1981; see Fig. 11B for location).
B) The Ventura Basin, California, bounded by faults with reverse separation (from Yeats, 1983; see Fig. 1 for location).

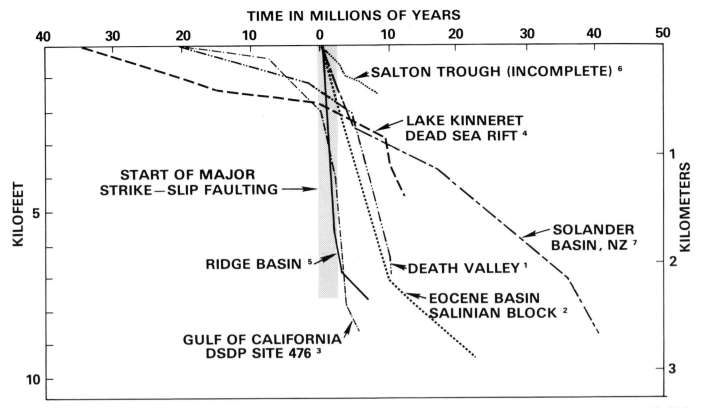

FIG. 16.—Tectonic subsidence curves for selected strike-slip basins. Tectonic subsidence is calculated by correcting cumulative stratigraphic thicknesses for the effects of sediment loading, sediment compaction, and water-depth changes. For each basin, the patterned area in the vicinity of the origin represents the approximate time of onset of major strike-slip deformation. Sources of data: 1, Hunt and Mabey (1966); 2, Graham (1976); 3, Curray et al. (1982); 4, Ben-Avraham et al. (1981); 5, Crowell and Link (1982), and papers therein; 6, Kerr et al. (1979); 7, Norris and Carter (1980).

mation (e.g., many Paleozoic and Precambrian examples), regional stratigraphic comparisons may provide some of the more important clues that sedimentation was accompanied by strike-slip deformation (Heward and Reading, 1980; Aspler and Donaldson, 1985 this volume).

ECONOMIC IMPORTANCE

Oil and gas are the most important resource exploited in strike-slip basins. Huff (1980) estimated that as much as 133 billion barrels has been discovered, but we regard this figure as rather high. The Dead Sea, known as Lake Asphaltitis to the ancient Greeks, was mentioned by the Latin authors Josephus and Tacitus as a source of asphalt used for calking ships and preparing medicines (Nissenbaum, 1978). In modern times, some of the most productive basins have been in California. About 10 billion barrels of oil have been produced from the Los Angeles, Ventura, Santa Maria, Salinas, and Cuyama Basins alone (Taylor, 1976). Current production from these basins is about 600,000 barrels of oil per day, and the ultimate recovery is estimated to be as much as 20 billion barrels of oil equivalent. The Point Arguello oil field, discovered in the offshore Santa Maria Basin in 1981, is the largest U.S. discovery since Prudhoe Bay, Alaska, with between 2.2 and 2.5 billion barrels of oil in place, and total recoverable reserves of be-

tween 300 and 500 million barrels (Crain et al., 1985). Other northern California offshore basins are promising (Crouch and Bachman, 1985). Additional reserves may be discovered in provinces affected by strike-slip deformation elsewhere in the world, such as the southern Caribbean region (Leonard, 1983), the Bering Sea (Marlow, 1979; Fisher, 1982), the Dead Sea (Wilson et al., 1983), and parts of China (Li Desheng, 1984). Individual strike-slip basins range from very rich to non-productive, depending on such factors as the presence of source rocks, thermal history and maturation, migration potential, reservoir quality and distribution, occurrence of traps and seal, and preservation of trapped hydrocarbons (Kingston et al., 1983b; Biddle and Christie-Blick, 1985b). Most important is the timing of maturation, migration and trap formation, because strike-slip basins tend to be short-lived. Examples of commodities, other than petroleum, found in strike-slip basins are ground water, coal and lignite, geothermal energy, and subsurface brines.

CONCLUSIONS

(1) Structural patterns along strike-slip faults differ in detail from simple predictions. The geometry of faults and folds along a given fault zone is generally a result of protracted, episodic deformation of heterogeneous sediments

and sedimentary rocks, involving the rotation of crustal blocks at a variety of scales, as well as strike slip, together with varying degrees of convergence and divergence, all superimposed on a pre-existing structural grain. On a regional scale, distinctive aspects of the structural style are the occurrence of en echelon structures, evidence for simultaneous shortening and extension, and random directions of vergence in associated thrusts and nappes.

(2) Basins form along strike-slip faults as a result of localized crustal extension and/or localized crustal shortening. The main processes leading to subsidence are mechanical thinning, thermal contraction, and loading due to the convergence of crustal blocks, amplified by the effects of sediment loading. The rifting stage is preferentially represented in the stratigraphic record of many strike-slip basins, because they are typically narrow, and in some cases detached (and therefore subject to less post-rift thermal subsidence), and many basins experience subsequent uplift and erosion. The importance of lithospheric flexure depends on the amount of extension involved in basin formation. Flexural effects are probably reduced during rifting.

(3) The location of depositional sites along strike-slip faults is controlled by crustal type and the configuration of pre-existing crustal structures, variations in the motion of adjacent lithospheric plates, and the kinematic behavior of crustal blocks. A key factor influencing the development of sedimentary basins is the manner in which overall plate motion is accommodated by discrete slip on major faults, and by the rotation and internal deformation of blocks. Subsidence history cannot be determined with confidence from present fault geometry, which therefore provides a poor basis for basin classification.

(4) Distinctive aspects of the stratigraphic record along strike-slip faults are geological mismatches; a tendency for basins to be asymmetrical both longitudinally and laterally; thick stratigraphic sections representing short intervals of time; the occurrence of abrupt lateral facies changes and local unconformities; and marked differences in stratigraphic thickness, facies geometry, and occurrence of unconformities from one basin to another in the same region.

(5) A major frontier for research in strike-slip basins is that of integrated geophysical, geological and modelling studies in a variety of plate-tectonic settings. The geophysical work should include standard seismic reflection, seismic refraction, gravity, magnetic and heat flow measurements, together with paleomagnetic studies to establish the magnitude and timing of rotations, and the boundaries of the blocks experiencing rotation. The geological work should include structural, stratigraphic and sedimentological studies using outcrop, borehole and seismic reflection data, together with investigations of diagenesis and paleothermometry. Modelling should be directed at using all available data in quantitative tests of our notions of the processes that control the development of strike-slip basins.

ACKNOWLEDGMENTS

This paper is an outgrowth of work conducted for Exxon Production Research Company, and we are grateful to that company for permission to publish and for support in the preparation of the manuscript. Christie-Blick acknowledges additional logistical support from an ARCO Foundation Fellowship. For stimulating discussions at various times about deformation and basins along strike-slip faults and for preprints of articles, we thank numerous geoscientists, but especially Tod Harding and Richard Vierbuchen (Exxon); Roger Bilham, Gerard Bond, Kathy Crane, Dennis Kent, Michelle Kominz, Craig Nicholson, Leonardo Seeber, Michael Steckler, Anthony Watts, and Patrick Williams (Lamont-Doherty Geological Observatory); John Crowell, Bruce Luyendyk and Arthur Sylvester (University of California, Santa Barbara); James Helwig and Jay Namson (ARCO); Ray Weldon (Caltech); M. J. Cheadle (University of Cambridge); William Dickinson (University of Arizona); and all the participants of the 1984 Research Symposium of the Society of Economic Paleontologists and Mineralogists on Strike-Slip Deformation, Basin Formation, and Sedimentation. Thom Davis and Ernie Duebendorfer kindly allowed us to use their unpublished mapping of neotectonic features along the San Andreas Fault to construct Figure 7. The manuscript benefitted from constructive reviews by Gerard Bond, Kevin Burke, Kathy Crane, James Helwig, Dennis Kent, Craig Nicholson, Harold Reading, and Arthur Sylvester.

REFERENCES

AHORNER, L., 1975, Present-day stress field and seismotectonic block movements along major fault zones in central Europe: Tectonophysics, v. 19, p. 233–249.

ALBERS, J. P., 1967, Belt of sigmoidal bending and right-lateral faulting in the western Great Basin: Geological Society of America Bulletin, v. 78, p. 143–156.

ALLEN, C. R., 1957, San Andreas fault zone in San Gorgonio Pass, southern California: Geological Society of America Bulletin, v. 68, p. 315–350.

———, 1965, Transcurrent faults in continental areas, in A Symposium on Continental Drift: Philosophical Transactions of Royal Society of London, Series A, v. 258, p. 82–89.

ANADÓN, P., CABRERA, L., GUIMERÀ, J., AND SANTANACH, P., 1985, Paleogene strike-slip deformation and sedimentation along the southeastern margin of the Ebro Basin, in Biddle K. T., and Christie-Blick, N., eds., Strike-Slip Deformation, Basin Formation, and Sedimentation: Society of Economic Paleontologists and Mineralogists Special Publication No. 37, p. 303–318.

ANDERSON, E. M., 1951, The dynamics of faulting and dyke formation with applications to Britain: Edinburgh, Oliver and Boyd, 2nd edition, 206 p.

ARMSTRONG, R. L., 1968, Sevier orogenic belt in Nevada and Utah: Geological Society of America Bulletin, v. 79, p. 429–458.

———, 1978, Pre-Cenozoic Phanerozoic time scale—Computer file of critical dates and consequences of new and in-progress decay-constant revisions, in Cohee, G. V., Glaessner, M. F., and Hedberg, H. D., eds., Contributions to the Geologic Time Scale: American Association of Petroleum Geologists Studies in Geology No. 6, p. 73–91.

ASPLER, L. B., AND DONALDSON, J. A., 1985, The Nonacho Basin (Early Proterozoic), Northwest Territories, Canada: Sedimentation and deformation in a strike-slip setting, in Biddle K. T., and Christie-Blick, N., eds., Strike-Slip Deformation, Basin Formation, and Sedimentation: Society of Economic Paleontologists and Mineralogists Special Publication No. 37, p. 193–209.

ATWATER, T., 1970, Implications of plate tectonics for the Cenozoic tectonic evolution of western North America: Geological Society of America Bulletin, v. 81, p. 3513–3536.

ATWATER, T., AND MOLNAR, P., 1973, Relative motion of the Pacific and North American plates deduced from sea-floor spreading in the Atlantic, Indian, and South Pacific Oceans, in Kovach, R. L., et al., eds., Proceedings, Tectonic Problems of the San Andreas Fault System:

Stanford University Publications in Geological Sciences, v. 13, p. 136–148.

AYDIN, A., AND NUR, A., 1982a, Evolution of pull-apart basins and their scale independence: Tectonics, v. 1, p. 91–105.

———, 1982b, Evolution of pull-apart basins and push-up ranges: Pacific Petroleum Geologists Newsletter: American Association of Petroleum Geologists, Pacific Section, Nov. 1982, p. 2–4.

———, 1985, The types and role of stepovers in strike-slip tectonics, *in* Biddle K. T., and Christie-Blick, N., eds., Strike-Slip Deformation, Basin Formation, and Sedimentation: Society of Economic Paleontologists and Mineralogists Special Publication No. 37, p. 35–44.

BALLANCE, P. F., 1980, Models of sediment distribution in non-marine and shallow marine environments in oblique-slip fault zones, *in* Ballance, P. F., and Reading, H. G., eds., Sedimentation in Oblique-Slip Mobile Zones: International Association of Sedimentologists Special Publication No. 4, p. 229–236.

BALLANCE, P. F., AND READING, H. G., eds., 1980a, Sedimentation in Oblique-Slip Mobile Zones: International Association of Sedimentologists Special Publication No. 4, 265 p.

———, 1980b, Sedimentation in oblique-slip mobile zones: an introduction, *in* Ballance, P. F., and Reading, H. G., eds., Sedimentation in Oblique-Slip Mobile Zones: International Association of Sedimentologists Special Publication No. 4, p. 1–5.

BALLY, A. W., ed., 1983, Seismic Expression of Structural Styles: American Association of Petroleum Geologists Studies in Geology Series, No. 15, 3 volumes.

BALLY, A. W., AND SNELSON, S., 1980, Realms of subsidence, *in* Miall, A. D., ed., Facts and Principles of World Petroleum Occurrence: Canadian Society of Petroleum Geologists Memoir 6, p. 9–94.

BARTLETT, W. L., FRIEDMAN, M., AND LOGAN, J. M., 1981, Experimental folding and faulting of rocks under confining pressure. Part IX. Wrench faults in limestone layers: Tectonophysics, v. 79, p. 255–277.

BARTOW, J. A., 1984, Geologic map and cross sections of the southeastern margin of the San Joaquin Valley, California: United States Geological Survey Miscellaneous Investigations Series Map I–1496, Scale 1:125,000.

BARTOW, J. A., AND McDOUGALL, K., 1984, Tertiary stratigraphy of the southeastern San Joaquin Valley, California: United States Geological Survey Bulletin 1529-J, 41 p.

BEAUMONT, C., 1981, Foreland basins: Geophysical Journal of Royal Astronomical Society, v. 65, p. 291–329.

BECK, M. E., JR., 1980, Paleomagnetic record of plate-margin tectonic processes along the western margin of North America: Journal of Geophysical Research, v. 85, p. 7115–7131.

BELT, E. S., 1968, Post-Acadian rifts and related facies, eastern Canada, *in* Zen, E-An, White, W. S., Hadley, J. B., and Thompson, J. B., Jr., eds., Studies of Appalachian Geology, Northern and Maritime: New York, Wiley Interscience, p. 95–113.

BEN-AVRAHAM, Z., ALMAGOR, G., AND GARFUNKEL, Z., 1979, Sediments and structure of the Gulf of Elat (Aqaba)—northern Red Sea: Sedimentary Geology, v. 23, p. 239–267.

BEN-AVRAHAM, Z., GINZBURG, A., AND YUVAL, Z., 1981, Seismic reflection and refraction investigations of Lake Kinneret—central Jordan Valley, Israel: Tectonophysics, v. 80, p. 165–181.

BIDDLE, K. T., AND CHRISTIE-BLICK, N., 1985a, Glossary—Strike-slip deformation, basin formation, and sedimentation, *in* Biddle, K. T., and Christie-Blick, N., eds., Strike-Slip Deformation, Basin Formation, and Sedimentation: Society of Economic Paleontologists and Mineralogists Special Publication No. 37, p. 375–386.

BIDDLE, K. T., AND CHRISTIE-BLICK, N. H., 1985b, Basin formation, structural traps, and controls on hydrocarbon occurrence along wrench-fault zones: Offshore Technology Conference Paper OTC 4872, p. 291–295.

BILHAM, R., AND WILLIAMS, P., 1985, Sawtooth segmentation and deformation processes on the southern San Andreas fault, California: Geophysical Research Letters, v. 12, p. 557–560.

BIRD, P., AND ROSENSTOCK, R. W., 1984, Kinematics of present crust and mantle flow in southern California: Geological Society of America Bulletin, v. 95, p. 946–957.

BLAKE, M. C., JR., CAMPBELL, R. H., DIBBLEE, T. W., JR., HOWELL, D. G., NILSEN, T. H., NORMARK, W. R., VEDDER, J. C., AND SILVER, E. A., 1978, Neogene basin formation in relation to plate-tectonic evolution of San Andreas fault system, California: American Association

of Petroleum Geologists Bulletin, v. 62, p. 344–372.

BOHANNON, R. G., AND HOWELL, D. G., 1982, Kinematic evolution of the junction of the San Andreas, Garlock, and Big Pine faults, California: Geology, v. 10, p. 358–363.

BOIS, C., BOUCHE, P., AND PELET, R., 1982, Global geologic history and distribution of hydrocarbon reserves: American Association of Petroleum Geologists Bulletin, v. 66, p. 1248–1270.

BONATTI, E., AND CRANE, K., 1984, Oceanic fracture zones: Scientific American, v. 250, p. 40–51.

BRADLEY, D. C., 1982, Subsidence in Late Paleozoic basins in the northern Appalachians: Tectonics, v. 1, p. 107–123.

BRIDEN, J. C., TURNELL, H. B., AND WATTS, D. R., 1984, British paleomagnetism, Iapetus Ocean, and the Great Glen Fault: Geology, v. 12, p. 428–431.

BURCHFIEL, B. C., AND STEWART, J. H., 1966, "Pull-apart" origin of the central segment of Death Valley, California: Geological Society of America Bulletin, v. 77, p. 439–442.

BURCHFIEL, B. C., WALKER, D., DAVIS, G. A., AND WERNICKE, B., 1983, Kingston Range and related detachment faults—a major "breakaway" zone in the southern Great Basin: Geological Society of America Abstracts with Programs, v. 15, p. 536.

BURKE, K., MANN, P., AND KIDD, W., 1982, What is a ramp valley?: 11th International Congress on Sedimentology, Hamilton, Ontario, International Association of Sedimentologists, Abstracts of Papers, p. 40.

BUTLER, R. W. H., 1982, The terminology of structures in thrust belts: Journal of Structural Geology, v. 4, p. 239–245.

CAMPBELL, J. D., 1958, En echelon folding: Economic Geology, v. 53, p. 448–472.

CAREY, S. W., 1955, The orocline concept in geotectonics: Royal Society of Tasmania Proceedings, v. 89, p. 255–288.

———, 1958, A tectonic approach to continental drift, *in* Carey, S. W., convenor, Continental Drift: A Symposium: Hobart, University of Tasmania, p. 177–355.

CEMEN, I., WRIGHT, L. A., DRAKE, R. E., AND JOHNSON, F. C., 1985, Cenozoic sedimentation and sequence of deformational events at the southeastern end of the Furnace Creek strike-slip fault zone, Death Valley region, California, *in* Biddle, K. T., and Christie-Blick, N., eds., Strike-Slip Deformation, Basin Formation, and Sedimentation: Society of Economic Paleontologists and Mineralogists Special Publication No. 37, p. 127–141.

CHAMBERLAIN, V. E., AND LAMBERT, R. St J., 1985, Cordilleria, a newly defined Canadian microcontinent: Nature, v. 314, p. 707–713.

CHAMPION, D. E., HOWELL, D. G., AND GROMME, C. S., 1984, Paleomagnetic and geologic data indicating 2500 km of northward displacement for the Salinian and related terranes, California: Journal of Geophysical Research, v. 89, p. 7736–7752.

CHEADLE, M. J., CZUCHRA, B. L., BYRNE, T., ANDO, C. J., OLIVER, J. E., BROWN, L. D., KAUFMAN, S., MALIN, P. E., AND PHINNEY, R. A., 1985, The deep crustal structure of the Mojave Desert, California, from COCORP seismic reflection data: Tectonics, in press.

CHINNERY, M. A., 1961, The deformation of the ground around surface faults: Seismological Society of America Bulletin, v. 51, p. 355–372.

———, 1963, The stress changes that accompany strike-slip faulting: Seismological Society of America Bulletin, v. 53, p. 921–932.

CHOUKROUNE, P., FRANCHETEAU, J., AND LE PICHON, X., 1978, In situ structural observations along Transform Fault A in the FAMOUS area Mid-Atlantic Ridge: Geological Society of America Bulletin, v. 89, p. 1013–1029.

CISOWSKI, S. M., 1984, Evidence for early Tertiary remagnetization of Devonian rocks from the Orcadian basin, northern Scotland, and associated transcurrent fault motion: Geology, v. 12, p. 369–372.

CLARK, M. M., 1972, Surface rupture along the Coyote Creek Fault, *in* The Borrego Mountain Earthquake of April 9, 1968: United States Geological Survey Professional Paper 787, p. 55–86.

———, 1973, Map showing recently active breaks along the Garlock and associated faults, California: United States Geological Survey Miscellaneous Investigations Series Map I–741.

CLAYTON, L., 1966, Tectonic depressions along the Hope Fault, a transcurrent fault in North Canterbury, New Zealand: New Zealand Journal of Geology and Geophysics, v. 9, p. 95–104.

CLOOS, E., 1955, Experimental analysis of fracture patterns: Geological Society of America Bulletin, v. 66, p. 241–256.

COCHRAN, J. R., 1983, Effects of finite rifting times on the development of sedimentary basins: Earth and Planetary Science Letters, v. 66, p. 289–302.

CONEY, P. J., JONES, D. L., AND MONGER, J. W. H., 1980, Cordilleran suspect terranes: Nature, v. 288, p. 329–333.

COURTILLOT, V., TAPPONIER, P., AND VARET, J., 1974, Surface features associated with transform faults—A comparison between observed examples and an experimental model: Tectonophysics, v. 24, p. 317–329.

COX, A., 1980, Rotation of microplates in western North America, *in* Strangway, D. W., ed., The Continental Crust and its Mineral Deposits: Geological Association of Canada Special Paper 20, p. 305–321.

CRAIN, W. E., MERO, W. E., AND PATTERSON, D., 1985, Geology of the Point Arguello discovery: American Association of Petroleum Geologists Bulletin, v. 69, p. 537–545.

CROUCH, J. K., 1979, Neogene tectonic evolution of the California continental borderland and western Transverse Ranges: Geological Society of America Bulletin, Part I, v. 90, p. 338–345.

CROUCH, J. K., BACHMAN, S. B., AND SHAY, J. T., 1984, Post-Miocene compressional tectonics along the central California margin, *in* Crouch, J. K., and Bachman, S. B., eds., Tectonics and Sedimentation Along the California Margin: Society of Economic Paleontologists and Mineralogists, Pacific Section, v. 38, p. 37–54.

CROUCH, J., AND BACHMAN, S., 1985, California basins gain attention: American Association of Petroleum Geologists Explorer, April, p. 1, 8–9.

CROWELL, J. C., 1962, Displacement along the San Andreas fault, California: Geological Society of America Special Paper No. 71, 61 p.

———, 1974a, Sedimentation along the San Andreas fault, California, *in* Dott, R. H., Jr., and Shaver, R. H., eds., Modern and Ancient Geosynclinal Sedimentation: Society of Economic Paleontologists and Mineralogists Special Publication No. 19, p. 292–303.

———, 1974b, Origin of late Cenozoic basins in southern California, *in* Dickinson, W. R., ed., Tectonics and Sedimentation: Society of Economic Paleontologists and Mineralogists Special Publication No. 22, p. 190–204.

———, 1976, Implications of crustal stretching and shortening of coastal Ventura Basin, California, *in* Howell, D. G., ed., Aspects of the Geologic History of the California Continental Borderland: American Association of Petroleum Geologists, Pacific Section, Miscellaneous Publication 24, p. 365–382.

———, 1979, The San Andreas fault system through time: Journal of Geological Society of London, v. 136, p. 293–302.

———, 1981a, An outline of the tectonic history of southeastern California, *in* Ernst, W. G., ed., The Geotectonic Development of California (Rubey Volume 1): Englewood Cliffs, New Jersey, Prentice-Hall, p. 584–600.

———, 1981b, Juncture of San Andreas transform system and Gulf of California rift: Oceanologica Acta, Proceedings, 26th International Geological Congress, Paris, Geology of Continental Margins Symposium, p. 137–141.

———, 1982, The tectonics of Ridge Basin, southern California, *in* Crowell, J. C., and Link, M. H., eds., Geologic History of Ridge Basin, Southern California: Society of Economic Paleontologists and Mineralogists, Pacific Section, p. 25–42.

CROWELL, J. C., AND LINK, M. H., eds., 1982, Geologic History of Ridge Basin, Southern California: Society of Economic Paleontologists and Mineralogists, Pacific Section, 304 p.

CURRAY, J. R., AND MOORE, D. G., 1984, Geologic history of the mouth of the Gulf of California, *in* Crouch, J. K., and Bachman, S. B., eds., Tectonics and Sedimentation Along the California Continental Margin: Society of Economic Paleontologists and Mineralogists, Pacific Section, v. 38, p. 17–36.

CURRAY, J. R., AND MOORE, D. G., et al., 1982, Baja California passive margin transect: Sites 474, 475 and 476, *in* Curray, J. R., Moore, D. G., et al., Initial Reports of Deep Sea Drilling Project, v. 64, Part 1, p. 35–210.

DAHLSTROM, C. D. A., 1970, Structural geology in the eastern margin of the Canadian Rocky Mountains: Bulletin of Canadian Petroleum Geology, v. 18, p. 332–406.

DAVIS, G. A., AND BURCHFIEL, B. C., 1973, Garlock fault: an intracontinental transform structure, southern California: Geological Society of America Bulletin, v. 84, p. 1407–1422.

DAVIS, T., AND LAGOE, M., 1984, Cenozoic structural development of the north-central Transverse Ranges and southern margin of the San Joaquin Valley: Geological Society of America Abstracts with Programs, v. 16, p. 484.

DETRICK, R. S., JR., AND PURDY, G. M., 1980, The crustal structure of the Kane fracture zone from seismic refraction studies: Journal of Geophysical Research, v. 85, p. 3759–3777.

DEWEY, J. F., 1982, Plate tectonics and the evolution of the British Isles: Journal of Geological Society of London, v. 139, p. 371–412.

DEWEY, J. F., AND PINDELL, J. L., 1985, Neogene block tectonics of eastern Turkey and northern South America: Continental applications of the finite difference method: Tectonics, v. 4, p. 71–83.

DIBBLEE, T. W., JR., 1977, Strike-slip tectonics of the San Andreas fault and its role in Cenozoic basin evolvement, *in* Late Mesozoic and Cenozoic Sedimentation and Tectonics in California: San Joaquin Geological Society Short Course, p. 26–38.

DICKINSON, W. R., 1978, Plate tectonic evolution of sedimentary basins, *in* Dickinson, W. R., and Yarborough, H., eds., Plate tectonics and hydrocarbon accumulation: American Association of Petroleum Geologists Continuing Education Course Note Series No. 1, p. 1–62.

D'ONFRO, P., AND GLAGOLA, P., 1983, Wrench fault, southeast Asia, *in* Bally, A. W., ed., Seismic Expression of Structural Styles, American Association of Petroleum Geologists Studies in Geology Series 15, v. 3, p. 4.2–9 to 4.2–12.

DOKKA, R. K., 1983, Displacements on late Cenozoic strike-slip faults of the central Mojave Desert, California: Geology, v. 11, p. 305–308.

DOKKA, R. K., AND MERRIAM, R. H., 1982, Late Cenozoic extension of northeastern Baja California, Mexico: Geological Society of America Bulletin, v. 93, p. 371–378.

DONATH, F. A., 1970, Some information squeezed out of rock: American Scientist, v. 58, p. 54–72.

DONOVAN, R. N., AND MEYERHOFF, A. A., 1982, Comment on 'Paleomagnetic evidence for a large (~2,000 km) sinistral offset along the Great Glen Fault during Carboniferous time:' Geology, v. 10, p. 604–605.

DOWNEY, M. W., 1984, Evaluating seals for hydrocarbon accumulations: American Association of Petroleum Geologists Bulletin, v. 68, p. 1752–1763.

EATON, G. P., 1982, The Basin and Range Province: Origin and tectonic significance: Annual Review of Earth and Planetary Sciences, v. 10, p. 409–440.

EISBACHER, G. H., 1985, Pericollisional strike-slip faults and syn-orogenic basins, Canadian Cordillera, *in* Biddle, K. T., and Christie-Blick, N., eds., 1985, Strike-Slip Deformation, Basin Formation, and Sedimentation: Society of Economic Paleontologists and Mineralogists Special Publication No. 37, p. 265–282.

EMMONS, R. C., 1969, Strike-slip rupture patterns in sand models: Tectonophysics, v. 7, p. 71–87.

ENSLEY, R. A., AND VEROSUB, K. L., 1982, Biostratigraphy and magnetostratigraphy of southern Ridge Basin, central Transverse Ranges, California, *in* Crowell, J. C., and Link, M. H., eds., Geologic History of Ridge Basin, Southern California: Society of Economic Paleontologists and Mineralogists, Pacific Section, p. 13–24.

FISHER, M. A., 1982, Petroleum geology of Norton Basin, Alaska: American Association of Petroleum Geologists Bulletin, v. 66, p. 286–301.

FITZGERALD, E. L., 1968, Structure of British Columbia foothills: American Association of Petroleum Geologists Bulletin, v. 52, p. 641–664.

FORSYTH, D. W., 1979, Lithospheric flexure: Reviews of Geophysics and Space Physics, v. 17, p. 1109–1114.

FOX, P. J., AND GALLO, D. G., 1984, A tectonic model for ridge-transform-ridge plate boundaries: Implications for the structure of oceanic lithosphere: Tectonophysics, v. 104, p. 205–242.

FRALICK, P. W., 1982, Wrench basin type inferred from sedimentation style: Examples from the Upper Paleozoic Cumberland Basin, Maritime, Canada: 11th International Congress on Sedimentology, Hamilton, Ontario, International Association of Sedimentologists, Abstracts of Papers, p. 38.

FREUND, R., 1970, Rotation of strike slip faults in Sistan, southeast Iran: Journal of Geology, v. 78, p. 188–200.

———, 1971, The Hope fault, a strike-slip fault in New Zealand: New Zealand Geological Survey Bulletin, v. 86, p. 1–49.

———, 1974, Kinematics of transform and transcurrent faults: Tectonophysics, v. 21, p. 93–134.

————, 1982, The role of shear in rifting, *in* Pálmason, G., ed., Continental and Oceanic Rifts: American Geophysical Union Geodynamics Series, v. 8, p. 33–39.

FREUND, R., GARFUNKEL, Z., ZAK, I., GOLDBERG, M., WEISSBROD, T., AND DERIN, B., 1970, The shear along the Dead Sea rift: Philosophical Transactions of Royal Society of London, Series A, v. 267, p. 107–130.

FUIS, G. S., MOONEY, W. D., HEALY, J. H., McMECHAN, G. A., AND LUTTER, W. J., 1984, A seismic refraction survey of the Imperial Valley region, California: Journal of Geophysical Research, v. 89, p. 1165–1189.

GAMOND, J. F., 1983, Displacement features associated with fault zones: a comparison between observed examples and experimental models: Journal of Structural Geology, v. 5, p. 33–45.

GARFUNKEL, Z., 1974, Model for the late Cenozoic tectonic history of the Mojave Desert, California, and for its relation to adjacent regions: Geological Society of America Bulletin, v. 85, p. 1931–1944.

————, 1981, Internal structure of the Dead Sea leaky transform (rift) in relation to plate kinematics: Tectonophysics, v. 80, p. 81–108.

GOGUEL, J., 1948, Introduction à l'étude mécanique des déformations de l'écorce terrestre: Mémoires, Carte Géologique Détaillée de la France, 2nd edition, 530 p.

GRAHAM, R. H., 1978, Wrench faults, arcuate fold patterns and deformation in the southern French Alps: Proceedings of Geological Association, v. 89, p. 125–142.

GRAHAM, S. A., 1976, Tertiary sedimentary tectonics of the central Salinian block of California [unpubl. Ph.D. dissertation]: Stanford, Stanford University, 510 p.

GRAHAM, S. A., AND WILLIAMS, L. A., 1985, Tectonic, depositional, and diagenetic history of Monterey Formation (Miocene), central San Joaquin Basin, California: American Association of Petroleum Geologists Bulletin, v. 69, p. 385–411.

GREEN, A. R., 1977, The evolution of the Earth's crust and sedimentary basin development: Offshore Technology Conference Paper OTC 2885, p. 67–72.

GREENHAUS, M. R., AND COX, A., 1979, Paleomagnetism of the Morro Rock-Islay Hill Complex as evidence for crustal block rotations in central coastal California: Journal of Geophysical Research, v. 84, p. 2393–2400.

GROSHONG, R. H., AND RODGERS, D. A., 1978, Left-lateral strike-slip fault model, *in* Wickham, J., and Denison, R., eds., Structural Style of Arbuckle Region: Geological Society of America Field Trip No. 3.

GUIRAUD, M., AND SEGURET, M., 1985, A releasing solitary overstep model for the late Jurassic—early Cretaceous (Wealdian) Soria strike-slip basin (northern Spain), *in* Biddle, K. T., and Christie-Blick, N., eds., Strike-Slip Deformation, Basin Formation, and Sedimentation: Society of Economic Paleontologists and Mineralogists Special Publication No. 37, p. 159–175.

GUTH, P. L., 1981, Tertiary extension north of the Las Vegas Valley shear zone, Sheep and Desert Ranges, Clark County, Nevada: Geological Society of America Bulletin, Part I, v. 92, p. 763–771.

GWINN, V. E., 1964, Thin-skinned tectonics in the Plateau and northwestern Valley and Ridge Provinces of the central Appalachians: Geological Society of America Bulletin, v. 75, p. 863–900.

HADLEY, D., AND KANAMORI, H., 1977, Seismic structure of the Transverse Ranges, California: Geological Society of America Bulletin, v. 88, p. 1469–1478.

HAMILTON, W., 1978, Mesozoic tectonics of the western United States, *in* Howell, D. G., and McDougall, K. A., eds., Mesozoic Paleogeography of the Western United States: Pacific Coast Paleogeography Symposium 2, Society of Economic Paleontologists and Mineralogists, Pacific Section, p. 33–70.

HARDING, T. P., 1974, Petroleum traps associated with wrench faults: American Association of Petroleum Geologists Bulletin, v. 58, p. 1290–1304.

————, 1983, Divergent wrench fault and negative flower structure, Andaman Sea, *in* Bally, A. W., ed., Seismic Expression of Structural Styles: American Association of Petroleum Geologists Studies in Geology, Series 15, v. 3, p. 4.2–1 to 4.2–8.

————, 1984, Graben hydrocarbon occurrences and structural style: American Association of Petroleum Geologists Bulletin, v. 68, p. 333–362.

————, 1985, Seismic characteristics and identification of negative flower structures, positive flower structures, and positive structural inversion: American Association of Petroleum Geologists Bulletin, v. 69, p. 582–600.

HARDING, T. P., AND LOWELL, J. D., 1979, Structural styles, their plate-tectonic habitats, and hydrocarbon traps in petroleum provinces: American Association of Petroleum Geologists Bulletin, v. 63, p. 1016–1058.

HARDING, T. P., GREGORY, R. F., AND STEPHENS, L. H., 1983, Convergent wrench fault and positive flower structure, Ardmore Basin, Oklahoma, *in* Bally, A. W., ed., Seismic Expression of Structural Styles: American Association of Petroleum Geologists Studies in Geology, Series 15, v. 3, p. 4.2–13 to 4.2–17.

HARDING, T. P., VIERBUCHEN, R. C., AND CHRISTIE-BLICK, N., 1985, Structural styles, plate-tectonic settings, and hydrocarbon traps of divergent (transtensional) wrench faults, *in* Biddle, K. T., and Christie-Blick, N., eds., Strike-Slip Deformation, Basin Formation, and Sedimentation: Society of Economic Paleontologists and Mineralogists Special Publication no. 37, p. 51–77.

HARLAND, W. B., 1971, Tectonic transpression in Caledonian Spitsbergen: Geological Magazine, v. 108, p. 27–42.

HARLAND, W. B., COX, A. V., LLEWELLYN, P. G., PICTON, C. A. G., SMITH, A. G., AND WALTERS, R., 1982, A Geologic Time Scale: Cambridge, Cambridge University Press, 131 p.

HARRIS, L. B., AND COBBOLD, P. R., 1984, Development of conjugate shear bands during simple shearing: Journal of Structural Geology, v. 7, p. 37–44.

HASZELDINE, R. S., 1984, Carboniferous North Atlantic paleogeography: stratigraphic evidence for rifting, not megashear or subduction: Geological Magazine, v. 121, p. 443–463.

HEARN, T. M., AND CLAYTON, R. W., 1984, Crustal structure and tectonics in Southern California (abs.): EOS, v. 65, p. 992.

HERD, D. G., 1978, Intracontinental plate boundary east of Cape Mendocino, California: Geology, v. 6, p. 721–725.

HELWIG, J., 1974, Eugeosynclinal basement and a collage concept of orogenic belts, *in* Dott, R. H., Jr., and Shaver, R. H., eds., Modern and Ancient Geosynclinal Sedimentation: Society of Economic Paleontologists and Mineralogists Special Publication No. 19, p. 359–376.

HEWARD, A. P., AND READING, H. G., 1980, Deposits associated with a Hercynian to late Hercynian continental strike-slip system, Cantabrian Mountains, northern Spain, *in* Ballance, P. F., and Reading, H. G., eds., Sedimentation in Oblique-Slip Mobile Zones: International Association of Sedimentologists Special Publication No. 4, p. 105–125.

HILL, D. P., 1982, Contemporary block tectonics: California and Nevada: Journal of Geophysical Research, v. 87, p. 5433–5450.

HILL, M. L., AND TROXEL, B. W., 1966, Tectonics of Death Valley region, California: Geological Society of America Bulletin, v. 77, p. 435–438.

HOWELL, D. G., CROUCH, J. K., GREENE, H. G., McCULLOCH, D. S., AND VEDDER, J. G., 1980, Basin development along the late Mesozoic and Cainozoic California margin: a plate tectonic margin of subduction, oblique subduction and transform tectonics, *in* Ballance, P. F., and Reading, H. G., eds., Sedimentation in Oblique-Slip Mobile Zones: International Association of Sedimentologists Special Publication No. 4, p. 43–62.

HUBBERT, M. K., 1937, Theory of scale models as applied to the study of geologic structures: Geological Society of America Bulletin, v. 48, p. 1459–1520.

HUFF, K. F., 1980, Frontiers of world exploration, *in* Miall, A. D., ed., Facts and Principles of World Petroleum Occurrence: Canadian Society of Petroleum Geologists Memoir 6, p. 343–362.

HUNT, C. B., AND MABEY, D. R., 1966, Stratigraphy and structure, Death Valley, California: United States Geological Survey Professional Paper 494-A, 162 p.

ILLIES, J. H., 1975, Intraplate tectonics in stable Europe as related to plate tectonics in the Alpine System: Geologische Rundschau, v. 64, p. 677–699.

ILLIES, J. H., AND GREINER, G., 1978, Rhinegraben and the Alpine System: Geological Society of America Bulletin, v. 89, p. 770–782.

IRVING, E., 1979, Paleopoles and paleolatitudes of North America and speculations about displaced terrains: Canadian Journal of Earth Sciences, v. 16, p. 669–694.

IRVING, E., AND STRONG, D. F., 1984, Palaeomagnetism of the Early Carboniferous Deer Lake Group, western Newfoundland: no evidence for mid-Carboniferous displacement of "Acadia:" Earth and Planetary Science Letters, v. 69, p. 379–390.

IRVING, E., MONGER, J. W. H., AND YOLE, R. W., 1980, New paleomagnetic evidence for displaced terranes in British Columbia, *in* Strangway, D. W., ed., The Continental Crust and its Mineral Deposits: Geological Association of Canada Special Paper 20, p. 441–456.

JACKSON, P. A., AND YEATS, R. S., 1982, Structural evolution of Carpinteria Basin, western Transverse Ranges, California: American Association of Petroleum Geologists Bulletin, v. 66, p. 805–829.

JARVIS, G. T., AND MCKENZIE, D. P., 1980, Sedimentary basin formation with finite extension rates: Earth and Planetary Science Letters, v. 48, p. 42–52.

JENNINGS, C. W., compiler, 1975, Preliminary fault and geologic map of southern California, in Crowell, J. C., ed, San Andreas Fault in Southern California: California Division of Mines and Geology Special Report 118, Scale 1:750,000.

JOHNSON, N. M., OFFICER, C. B., OPDYKE, N. D., WOODWARD, G. D., ZEITLER, P. K., AND LINDSAY, E. H., 1983, rates of late Cenozoic tectonism in the Vallecito-Fish Creek basin, western Imperial Valley, California: Geology, v. 11, p. 664–667.

JOHNSON, S. Y., 1985, Eocene strike-slip faulting and nonmarine basin formation in Washington, *in* Biddle, K. T., and Christie-Blick, N., eds., Strike-Slip Deformation, Basin Formation, and Sedimentation: Society of Economic Paleontologists and Mineralogists Special Publication No. 37, p. 283–302.

JONES, D. L., SILBERLING, N. J., AND HILLHOUSE, J., 1977, Wrangellia—A displaced terrane in northwestern North America: Canadian Journal of Earth Sciences, v. 14, p. 2565–2577.

JUNGER, A., 1976, Tectonics of the southern California borderland, *in* Howell, D. G., ed., Aspects of the Geologic History of the California Continental Borderland: American Association of Petroleum Geologists, Pacific Section, Miscellaneous Publication 24, p. 486–498.

KANTER, L. R., AND MCWILLIAMS, M. O., 1982, Rotation of the southernmost Sierra Nevada, California: Journal of Geophysical Research, v. 87, p. 3819–3830.

KARIG, D. E., 1980, Material transport within accretionary prisms and the "knocker" problem: Journal of Geology, v. 88, p. 27–39.

KARIG, D. E., AND JENSKY, W., 1972, The proto-Gulf of California: Earth and Planetary Science Letters, v. 17, p. 169–174.

KARSON, J. A., 1984, Variations in structure and petrology in the Coastal Complex, Newfoundland: anatomy of an oceanic fracture zone, *in* Gass, I. G., Lippard, S. J., and Shelton, A. W., eds., Ophiolites and Oceanic Lithosphere: Geological Society of London, Special Publication No. 13, p. 131–144.

KARSON, J. A., AND DICK, H. J. B., 1983, Tectonics of ridge-transform intersections at the Kane fracture zone: Marine Geophysical Researches, v. 6, p. 51–98.

KELTS, K., 1981, A comparison of some aspects of sedimentation and translational tectonics from the Gulf of California and the Mesozoic Tethys, northern Penninic margin: Eclogae Geologicae Helvetiae, v. 74, p. 317–338.

KENNEDY, W. Q., 1946, The Great Glen Fault: Quarterly Journal of Geological Society of London, v. 102, p. 41–76.

KENT, D. V., 1982, Paleomagnetic evidence for post-Devonian displacement of the Avalon platform (Newfoundland): Journal of Geophysical Research, v. 87, p. 8709–8716.

KENT, D. V., AND OPDYKE, N. D., 1978, Paleomagnetism of the Devonian Catskill red beds: Evidence for motion of the coastal New England–Canadian Maritime region relative to cratonic North America: Journal of Geophysical Research, v. 83, p. 4441–4450.

———, 1979, The Early Carboniferous paleomagnetic field of North America and its bearing on tectonics of the northern Appalachians: Earth and Planetary Science Letters, v. 44, p. 365–372.

———, 1985, Multicomponent magnetizations from the Mississippian Mauch Chunk Formation of the central Appalachians and their tectonic implications: Journal of Geophysical Research, v. 90, p. 5371–5383.

KERR, D. R., PAPPAJOHN, S., AND PETERSON, G. L., 1979, Neogene stratigraphic section at Split Mountain, eastern San Diego County, California, *in* Crowell, J. C., and Sylvester, A. G., eds., Tectonics of the Juncture Between the San Andreas Fault System and the Salton Trough, Southeastern California: Santa Barbara, California, Department of Geological Sciences, University of California, p. 111–123.

KING, P. B., 1969, Tectonic map of North America: United States Geological Survey, Scale 1:5,000,000.

KINGMA, J. T., 1958, Possible origin of piercement structures, local unconformities, and secondary basins in the Eastern Geosyncline, New Zealand: New Zealand Journal of Geology and Geophysics, v. 1, p. 269–274.

KINGSTON, D. R., DISHROON, C. P., AND WILLIAMS, P. A., 1983a, Global basin classification system: American Association of Petroleum Geologists Bulletin, v. 67, p. 2175–2193.

———, 1983b, Hydrocarbon plays and global basin classification: American Association of Petroleum Geologists Bulletin, v. 67, p. 2194–2198.

KLEMME, H. D., 1980, Petroleum basins—classifications and characteristics: Journal of Petroleum Geology, v. 3, p. 187–207.

LABOTKA, T. C., 1980, Petrology of a medium-pressure regional metamorphic terrane, Funeral Mountains, California: American Mineralogist, v. 65, p. 670–689.

———, 1981, Petrology of an andalusite-type regional metamorphic terrane, Panamint Mountains, California: Journal of Petrology, v. 22, p. 261–296.

LABOTKA, T. C., AND WARASILA, R., 1983, Ages of metamorphism in the central Panamint Mountains, California: Geological Society of America Abstracts with Programs, v. 15, p. 437.

LACHENBRUCH, A. H., AND SASS, J. H., 1980, Heat flow and energetics of the San Andreas fault zone: Journal of Geophysical Research, v. 85, p. 6185–6222.

LEHNER, P., AND BAKKER, G., 1983, Equatorial fracture zone (Romanche fracture), *in* Bally, A. W., ed., Seismic Expression of Structural Styles: American Association of Petroleum Geologists Studies in Geology, Series 15, v. 3, p. 4.2–25 to 4.2–29.

LENSEN, G. J., 1958, A method of graben and horst formation: Journal of Geology, v. 66, p. 579–587.

LEONARD, R., 1983, Geology and hydrocarbon accumulations, Columbus Basin, offshore Trinidad: American Association of Petroleum Geologists, v. 67, p. 1081–1093.

LEWIS, K. B., 1980, Quaternary sedimentation on the Hikurangi oblique-subduction and transform margin, New Zealand, *in* Ballance, P. F., and Reading, H. G., eds., Sedimentation in Oblique-Slip Mobile Zones: International Association of Sedimentologists Special Publication No. 4, p. 171–189.

LI DESHENG, 1984, Geologic evolution of petroliferous basins on continental shelf of China: American Association of Petroleum Geologists Bulletin, v. 68, p. 993–1003.

LINK, M. H., ROBERTS, M. T., AND NEWTON, M. S., 1985, Walker Lake Basin, Nevada: An example of late Tertiary (?) to Recent sedimentation in a basin adjacent to an active strike-slip fault, *in* Biddle, K. T., and Christie-Blick, N., eds., Strike-Slip Deformation, Basin Formation, and Sedimentation: Society of Economic Paleontologists and Mineralogists Special Publication No. 37, p. 105–125.

LOGAN, J. M., 1979, Brittle phenomena: Reviews of Geophysics and Space Physics, v. 17, p. 1121–1132.

LOGAN, J. M., FRIEDMAN, M., HIGGS, N. G., DENGO, C., AND SHIMAMOTO, T., 1979, Experimental studies of simulated gouge and their application to studies of natural fault zones: United States Geological Survey Open-File Report 79-1239, p. 305–343.

LONSDALE, P., 1978, Near-bottom reconnaissance of a fast-slipping transform zone at the Pacific-Nazca plate boundary: Journal of Geology, v. 86, p. 451–472.

LOWELL, J. D., 1972, Spitsbergen Tertiary orogenic belt and the Spitsbergen fracture zone: Geological Society of America Bulletin, v. 83, p. 3091–3102.

LUDMAN, A., 1981, Significance of transcurrent faulting in eastern Maine and location of the suture between Avalonia and North America: American Journal of Science, v. 281, p. 463–483.

LUYENDYK, B. P., KAMERLING, M. J., AND TERRES, R., 1980, Geometric model for Neogene crustal rotations in southern California: Geological Society of America Bulletin, Part I, v. 91, p. 211–217.

LUYENDYK, B. P., KAMERLING, M. J., TERRES, R. R., AND HORNAFIUS, J. S., 1985, Simple shear of southern California during Neogene time suggested by paleomagnetic declinations: Journal of Geophysical Research, in press.

MACDONALD, K. C., KASTENS, K., SPIESS, F. N., AND MILLER, S. P., 1979, Deep tow studies of the Tamayo transform fault: Marine Geophysical Researches, v. 4, p. 37–70.

MANDL, G., DE JONG, L. N. G., AND MALTHA, A., 1977, Shear zones in granular material: Rock Mechanics, v. 9, p. 95–144.

MANN, P., AND BRADLEY, D., 1984, Comparison of basin types in active and ancient strike-slip zones (abs.): American Association of Petroleum Geologists Bulletin, v. 68, p. 503.

MANN, P., AND BURKE, K., 1982, Basin formation at intersections of conjugate strike-slip faults: examples from southern Haiti: Geological Society of America Abstracts with Programs, v. 14, p. 555.

MANN, P., HEMPTON, M. R., BRADLEY, D. C., AND BURKE, K., 1983, Development of pull-apart basins: Journal of Geology, v. 91, p. 529–554.

MANN, P., DRAPER, G., AND BURKE, K., 1985, Neotectonics of a strike-slip restraining bend system, Jamaica, in Biddle, K. T., and Christie-Blick, N., eds., Strike-Slip Deformation, Basin Formation, and Sedimentation: Society of Economic Paleontologists and Mineralogists Special Publication No. 37, p. 211–226.

MANSPEIZER, W., 1985, The Dead Sea Rift: Impact of climate and tectonism on Pleistocene and Holocene sedimentation, in Biddle, K. T., and Christie-Blick, N., eds., Strike-slip Deformation, Basin Formation, and Sedimentation: Society of Economic Paleontologists and Mineralogists Special Publication No. 37, p. 143–158.

MARLOW, M. S., 1979, Hydrocarbon prospects in Navarin basin province, northwest Bering Sea shelf: Oil and Gas Journal, October 29, v. 77, p. 190–196.

MARLOW, M. S., AND COOPER, A. K., 1980, Mesozoic and Cenozoic structural trends under southern Bering Sea shelf: American Association of Petroleum Geologists Bulletin, v. 64, p. 2139–2155.

MCKENZIE, D., 1978, Some remarks on the development of sedimentary basins: Earth and Planetary Science Letters, v. 40, p. 25–32.

MENARD, H. W., AND ATWATER, T., 1969, Origin of fracture zone topography: Nature, v. 222, 1037–1040.

MIALL, A. D., 1985, Stratigraphic and structural predictions from a plate-tectonic model of an oblique-slip orogen: The Eureka Sound Formation (Campanian-Oligocene), northeast Canadian Arctic islands, in Biddle, K. T., and Christie-Blick, N., eds., Strike-Slip Deformation, Basin Formation, and Sedimentation: Society of Economic Paleontologists and Mineralogists Special Publication No. 37, p. 000–000.

MINSTER, J. B. AND JORDAN, T. H., 1978, Present-day plate motions: Journal of Geophysical Research, v. 83, p. 5331–5354.

―――, 1984, Vector constraints on Quaternary deformation of the western United States east and west of the San Andreas fault, in Crouch, J. K., and Bachman, S. B., eds., Tectonics and Sedimentation Along the California margin: Society of Economic Paleontologists and Mineralogists, Pacific Section, v. 38, p. 1–16.

MINSTER, J. B., JORDAN, T. H., MOLNAR, P., AND HAINES, E., 1974, Numerical modelling of instantaneous plate tectonics: Geophysical Journal of Royal Astronomical Society, v. 36, p. 541–576.

MOLNAR, P., AND TAPPONNIER, P., 1975, Cenozoic tectonics of Asia: Effects of a continental collision: Science, v. 189, p. 419–426.

MOODY, J. D., AND HILL, M. J., 1956, Wrench-fault tectonics: Geological Society of America Bulletin, v. 67, p. 1207–1246.

MOORE, D. G., 1969, Reflection profiling studies of the California continental Borderland: Structure and Quarternary turbidite basins: Geological Society of America Special Paper 107, 138 p.

MORGENSTERN, N. R., AND TCHALENKO, J. S., 1967, Microscopic structures in kaolin subjected to direct shear: Géotechnique, v. 17, p. 309–328.

MORTON, J. L., AND HILLHOUSE, J. W., 1985, Paleomagnetism and K-Ar ages of Miocene basaltic rocks in the western Mojave Desert, California: manuscript.

MUTTER, J. C., AND DETRICK, R. S., 1984, Multichannel seismic evidence for anomalously thin crust at Blake Spur fracture zone: Geology, v. 12, p. 534–537.

NAMSON, J. S., DAVIS, T. L., AND LAGOE, M. B., 1985, Tectonic history and thrust-fold deformation style of seismically active structures near Coalinga, California: United States Geological Survey Professional Paper, submitted.

NARDIN, T. R., AND HENYEY, T. L., 1978, Pliocene-Pleistocene diastrophism of Santa Monica and San Pedro shelves, California continental borderland: American Association of Petroleum Geologists Bulletin, v. 62, p. 247–272.

NELSON, W. J., AND KRAUSSE, H.-F., 1981, The Cottage Grove fault system in southern Illinois: Illinois Institute of Natural Resources, State Geological Survey Division, Circular 522, 65 p.

NICHOLSON, C., SEEBER, L., WILLIAMS, P., AND SYKES, L. R., 1985a, Seismicity and fault kinematics through the eastern Transverse Ranges, California: Block rotation, strike-slip faulting and shallow-angle thrusts: Journal of Geophysical Research, in press.

NICHOLSON, C., SEEBER, L., WILLIAMS, P. L., AND SYKES, L. R., 1985b, Seismic deformation along the southern San Andreas fault, California: Implications for conjugate slip rotational block tectonics: Tectonics, submitted.

NILSEN, T. H., AND MCLAUGHLIN, R. J., 1985, Comparison of tectonic framework and depositional patterns of the Hornelen strike-slip basin of Norway and the Ridge and Little Sulphur Creek strike-slip basins of California, in Biddle, K. T., and Christie-Blick, N. eds., Strike-Slip Deformation, Basin Formation, and Sedimentation: Society of Economic Paleontologists and Mineralogists Special Publication No. 37, p. 80–103.

NISSENBAUM, A., 1978, Dead Sea Asphalts—Historical aspects: American Association of Petroleum Geologists Bulletin, v. 62, p. 837–844.

NORRIS, R. J., AND CARTER, R. M., 1980, Offshore sedimentary basins at the southern end of the Alpine fault, New Zealand, in Ballance, P. F., and Reading, H. G., eds., Sedimentation in Oblique-Slip Mobile Zones: International Association of Sedimentologists Special Publication No. 4, p. 237–265.

ODONNE, F., AND VIALON, P., 1983, Analogue models of folds above a wrench fault: Tectonophysics, v. 99, p. 31–46.

OPDYKE, N. D., LINDSAY, E. H., JOHNSON, N. M., AND DOWNS, T., 1977, The paleomagnetism and magnetic polarity stratigraphy of the mammal-bearing section of Anza Borrego State Park, California: Quaternary Research, v. 7, p. 316–329.

PAGE, B. M., AND ENGEBRETSON, D. C., 1984, Correlation between the geologic record and computed plate motions for central California: Tectonics, v. 3, p. 133–155.

PALMER, A. R., 1983, The Decade of North American Geology 1983 time scale: Geology, v. 11, p. 503–504.

PARNELL, J. T., 1982, Comment on 'Paleomagnetic evidence for a large (2,000 km) sinistral offset along the Great Glen fault during Carboniferous time:' Geology, v. 10, p. 605.

PAVONI, N., 1961, Die Nordanatolische Horizontalverschiebung: Geologische Rundschau, v. 51, p. 122–139.

PERRODON, A., AND MASSE, P., 1984, Subsidence, sedimentation and petroleum systems: Journal of Petroleum Geology, v. 7, p. 5–26.

PERROUD, H., VAN DER VOO, R., AND BONHOMMET, N., 1984, Paleozoic evolution of the Armorica plate on the basis of paleomagnetic data: Geology, v. 12, p. 579–582.

PHILIP, H., AND MEGARD, F., 1977, Structural analysis of the superficial deformation of the 1969 Pariahuanca earthquakes (central Peru): Tectonophysics, v. 38, p. 259–278.

PITMAN, W. C., III, AND ANDREWS, J. A., 1985, Subsidence and thermal history of small pull-apart basins, in Biddle, K. T., and Christie-Blick, N., eds., Strike-Slip Deformation, Basin Formation, and Sedimentation: Society of Economic Paleontologists and Mineralogists Special Publication No. 37, p. 45–49.

PITTMAN, E. D., 1981, Effect of fault-related granulation on porosity and permeability of quartz sandstones, Simpson Group (Ordovician), Oklahoma: American Association of Petroleum Geologists Bulletin, v. 65, p. 2381–2387.

QUENNELL, A. M., 1958, The structural and geomorphic evolution of the Dead Sea Rift: Quarterly Journal of Geological Society of London, v. 114, p. 1–24.

RAMSAY, J. G., 1980, Shear zone geometry: a review: Journal of Structural Geology, v. 2, p. 83–99.

READING, H. G., 1980, Characteristics and recognition of strike-slip fault systems, in Ballance, P. F., and Reading, H. G., eds., Sedimentation in Oblique-Slip Mobiles Zones: International Association of Sedimentologists Special Publication No. 4, p. 7–26.

―――, 1982, Sedimentary basins and global tectonics: Proceedings of

Geological Association, v. 93, p. 321–350.

RIEDEL, W., 1929, Zur Mechanik geologischer Brucherscheinungen: Zentrablatt für Mineralogie, Geologie und Pälaeontologie, v. 1929 B, p. 354–368.

RIXON, L. K., 1978, Clay modelling of the Fitzroy Graben: Bureau of Mineral Resources Journal of Australian Geology and Geophysics, v. 3, p. 71–76.

ROBERTS, M. T., 1983, Seismic example of complex faulting from northwest shelf of Palawan, Phillipines, *in* Bally, A. W., ed., Seismic Expression of Structural Styles: American Association of Petroleum Geologists Studies in Geology, Series 15, v. 3, p. 4.2–18 to 4.2–24.

RODGERS, D. A., 1980, Analysis of pull-apart basin development produced by en echelon strike-slip faults, *in* Ballance, P. F., and Reading, H. G., eds., Sedimentation in Oblique-Slip Mobile Zones: International Association of Sedimentologists Special Publication No. 4, p. 27–41.

RODGERS, J., 1963, Mechanics of Appalachian foreland folding in Pennsylvania and West Virginia: American Association of Petroleum Geologists Bulletin, v. 47, p. 1527–1536.

RON, H., FREUND, R., GARFUNKEL, Z., AND NUR, A., 1984, Block rotation by strike-slip faulting: structural and paleomagnetic evidence: Journal of Geophysical Research, v. 89, p. 6256–6270.

RON, H., AND EYAL, Y., 1985, Intraplate deformation by block rotation and mesostructures along the Dead Sea transform, northern Israel: Tectonics, v. 4, p. 85–105.

ROY, J. L., AND MORRIS, W. A., 1983, A review of paleomagnetic results from the Carboniferous of North America; the concept of Carboniferous geomagnetic field horizon markers: Earth and Planetary Science Letters, v. 65, p. 167–181.

ROYDEN, L. H., 1985, The Vienna Basin: A thin-skinned pull-apart basin, *in* Biddle, K. T., and Christie-Blick, N., eds., Strike-Slip Deformation, Basin Formation, and Sedimentation: Society of Economic Paleontologists and Mineralogists Special Publication No. 37, p. 319–338.

ROYDEN, L., AND KEEN, C. E., 1980, Rifting processes and thermal evolution of the continental margin of eastern Canada determined from subsidence curves: Earth and Planetary Science Letters, v. 51, p. 343–361.

ROYDEN, L. H., HORVÁTH, F., AND BURCHFIEL, B. C., 1982, Transform faulting, extension, and subduction in the Carpathian Pannonian region: Geological Society of America Bulletin, v. 93, p. 717–725.

SALVADOR, A., 1985, Chronostratigraphic and geochronometric scales in COSUNA stratsigraphic correlation charts of the United States: American Association of Petroleum Geologists Bulletin, v. 69, p. 181–189.

SBAR, M. L., 1982, Delineation and interpretation of seismotectonic domains in western North America: Journal of Geophysical Research, v. 87, p. 3919–3928.

SCHUBERT, C., 1980, Late-Cenozoic pull-apart basins, Boconó fault zone, Venezuelan Andes: Journal of Structural Geology, v. 2, p. 463–468.

SEARLE, R. C., 1979, Side-scan sonar studies of North Atlantic fracture zones: Journal of Geological Society of London, v. 136, p. 283–292.

———, 1983, Multiple, closely spaced transform faults in fast-slipping fracture zones: Geology, v. 11, p. 607–610.

SEGALL, P., AND POLLARD, D. D., 1980, Mechanics of discontinuous faults: Journal of Geophysical Research, v. 85, p. 4337–4350.

———, 1983, Nucleation and growth of strike slip faults in granite: Journal of Geophysical Research, v. 88, p. 555–568.

ŞENGÖR, A. M. C., BURKE, K., AND DEWEY, J. F., 1978, Rifts at high angles to orogenic belts: Tests for their origin and the Upper Rhine Graben as an example: American Journal of Science, v. 278, p. 24–40.

ŞENGÖR, A. M. C., GÖRÜR, N., AND ŞAROĞLU, F., 1985, Strike-slip faulting and related basin formation in zones of tectonic escape: Turkey as a case study, *in* Biddle, K. T., and Christie-Blick, N., eds., Strike-Slip Deformation, Basin Formation, and Sedimentation: Society of Economic Paleontologists and Mineralogists Special Publication No. 37, p. 227–264.

SHARP, R. V., 1976, Surface faulting in Imperial Valley during the earthquake swarm of January–February, 1975, Seismological Society of America Bulletin, v. 66, p.1145–1154.

———, 1977, Map showing the Holocene surface expression of the Brawley Fault, Imperial County, California: United States Geological

Survey Miscellaneous Field Studies Map MF-838.

———, 1981, Variable rates of late Quaternary strike slip on the San Jacinto fault zone, southern California: Journal of Geophysical Research, v. 86, p. 1754–1762.

SIBSON, R. H., 1977, Fault rocks and fault mechanisms: Journal of Geological Society of London, v. 133, p. 191–213.

SIEH, K. E., AND JAHNS, R. H., 1984, Holocene activity of the San Andreas fault at Wallace Creek, California: Geological Society of America Bulletin, v. 95, p. 883–896.

SINHA, M. C., AND LOUDEN, K. E., 1983, The Oceanographer fracture zone–I. Crustal structure from seismic refraction studies: Geophysical Journal of Royal Astronomical Society, v. 75, p. 713–736.

SMITH, D. I., AND WATSON, J., 1983, Scale and timing of movements on the Great Glen fault, Scotland: Geology, v. 11, 523–526.

SMITH, G. I., 1962, Large lateral displacement on Garlock fault, California, as measured from offset dike swarm: American Association of Petroleum Geologists Bulletin, v. 46, p.85–104.

SMITH, G. I., AND KETNER, K. B., 1970, Lateral displacement on the Garlock fault, southeastern California, suggested by offset sections of similar metasedimentary rocks: United States Geological Survey Professional Paper 700-D, p. 1–9.

STECKLER, M. S., 1981, Thermal and mechanical evolution of Atlantic-type margins [unpubl. Ph.D. thesis]: New York, Columbia University, 261 p.

STECKLER, M. S., AND TEN BRINK, U. S., 1985, Replacement of the Gulf of Suez rift by the Dead Sea transform: The role of the hinge zones in rifting (abs.): EOS, v. 66, p. 364.

STECKLER, M. S., AND WATTS, A. B., 1982, Subsidence history and tectonic evolution of Atlantic-type continental margins, *in* Scrutton, R. A., ed., Dynamics of Passive Margins: American Geophysical Union Geodynamics Series, v. 6, p. 184–196.

STEEL, R., GJELBERG, J., HELLAND-HANSEN, W., KLEINSPEHN, K., NØTTVEDT, A., AND RYE-LARSEN, M., 1985, The Tertiary strike-slip basins and orogenic belt of Spitsbergen, *in* Biddle, K. T., and Christie-Blick, N., eds., Strike-Slip Deformation, Basin Formation, and Sedimentation: Society of Economic Paleontologists and Mineralogists Special Publication No. 37, p. 227–264.

STEWART, J. H., 1978, Basin-range structure in western North America: A review, *in* Smith, R. B., and Eaton, G. P., eds., Cenozoic tectonics and regional geophysics of the western Cordillera: Geological Society of America Memoir 152, p. 1–31.

———, 1983, Extensional tectonics in the Death Valley area, California: Transport of the Panamint Range structural block 80 km northwestward: Geology, v. 11, p. 153–157.

SUPPE, 1978, Cross section of southern part of northern Coast Ranges and Sacramento Valley, California: Geological Society of America Map and Chart Series, MC-28B.

SYKES, L. R., 1967, Mechanism of earthquakes and nature of faulting on the mid-oceanic ridges: Journal of Geophysical Research, v. 72, p. 2131–2153.

SYLVESTER, A. G., compiler, 1984, Wrench fault tectonics: American Association of Petroleum Geologists Reprint Series, No. 28, 374 p.

SYLVESTER, A. G., AND SMITH, R. R., 1976, Tectonic transpression and basement-controlled deformation in San Andreas fault zone, Salton Trough, California: American Association of Petroleum Geologists Bulletin, v. 60, p. 2081–2102.

TAYLOR, J. C., 1976, Geologic appraisal of the petroleum potential of offshore southern California: the borderland compared to onshore coastal basins: United States Geological Survey Circular 730, 43 p.

TCHALENKO, J. S., 1970, Similarities between shear zones of different magnitudes: Geological Society of America Bulletin, v. 81, p. 1625–1640.

TCHALENKO, J. S., AND AMBRASEYS, N. N., 1970, Structural analyses of the Dasht-e Bayaz (Iran) earthquake fractures: Geological Society of America Bulletin, v. 81, p. 41–60.

TERRES, R., AND CROWELL, J. C., 1979, Plate tectonic framework of the San Andreas-Salton Trough juncture, *in* Crowell, J. C., and Sylvester, A. G., eds., Tectonics of the Juncture Between the San Andreas Fault System and the Salton Trough, Southeastern California: Santa Barbara, California, Department of Geological Sciences, University of California, p. 15–25.

TERRES, R. R., AND SYLVESTER, A. G., 1981, Kinematic analysis of rotated fractures and blocks in simple shear: Seismological Society of American Bulletin, v. 71, p. 1593–1605.

TURCOTTE, D. L., LIU, J. Y., AND KULHAWY, F. H., 1984, The role of an intracrustal asthenosphere on the behavior of major strike-slip faults: Journal of Geophysical Research, v. 89, p. 5801–5816.

VAIL, P. R., MITCHUM, R. M., JR., TODD, R. G., WIDMIER, J. M., THOMPSON, S., JR., SANGREE, J. B., BUBB, J. N., AND HATELID, W. G., 1977, Seismic stratigraphy and global changes of sea level, in Payton, C. E., ed., Seismic Stratigraphy—Applications to Hydrocarbon Exploration: American Association of Petroleum Geologists Memoir 26, p. 49–212.

VAN DER VOO, R., 1982, Pre-Mesozoic paleomagnetism and plate tectonics: Annual Review of Earth and Planetary Sciences, v. 10, p. 191–220.

————, 1983, Paleomagnetic constraints on the assembly of the Old Red continent: Tectonophysics, v. 91, p. 271–283.

VAN DER VOO, R., AND SCOTESE, C. R., 1981, Paleomagnetic evidence for a large (~2,000 km) sinistral offset along the Great Glen Fault during Carboniferous time: Geology, v. 9, p. 583–589.

VAN DER VOO, R., FRENCH, A. N., AND FRENCH, R. B., 1979, A paleomagnetic pole position from the folded Upper Devonian Catskill red beds, and its tectonic implications: Geology, v. 7, p. 345–348.

VINK, G. E., MORGAN, W. J., AND ZHAO, W.-L., 1984, Preferential rifting of continents: A source of displaced terranes: Journal of Geophysical Research, v. 89, p. 10,072–10,076.

WATTS, A. B., 1983, The strength of the Earth's crust: Marine Technology Society Journal, v. 17, p. 5–17.

WEBB, T. H., AND KANAMORI, H., 1985, Earthquake focal mechanisms in the eastern Transverse Ranges and San Emigdio Mountains, southern California, and evidence for a regional decollement: Seismological Society of America Bulletin, v. 75, p. 737–757.

WEBER, G. E., AND LAJOIE, K. R., 1977, Late Pleistocene and Holocene tectonics of the San Gregorio fault zone between Moss Beach and Point Ano Nuevo, San Mateo County, California: Geological Society of America Abstracts with Programs, v. 9, p. 524.

WEERTMAN, J., 1965, Relationship between displacements on a free surface and the stress on a fault: Seismological Society of America Bulletin, v. 55, p. 945–953.

WELDON, R., AND HUMPHREYS, G., 1985, A kinematic model of southern California: Tectonics, in press.

WELDON, R. J., II, AND SIEH, K. E., 1985, Holocene rate of slip and tentative recurrence interval for large earthquakes on the San Andreas fault, Cajon Pass, southern California: Geological Society of American Bulletin, v. 96, p. 793–812.

WELDON, R. J., WINSTON, D. S., KIRSCHVINK, J. L., AND BURBANK, D. W., 1984, Magnetic stratigraphy of the Crowder Formation, Cajon Pass, southern California: Geological Society of America Abstracts with Programs, v. 16, p. 689.

WENTWORTH, C. M., BLAKE, M. C., JR., JONES, D. L., WALTER, A. W., AND ZOBACK, M. D., 1984, Tectonic wedging associated with emplacement of the Franciscan assemblage, California Coast Ranges, in Blake, M. C., Jr., ed., Franciscan Geology of Northern California: Society of Economic Paleontologists and Mineralogists, Pacific Section, v. 43, p. 163–173.

WERNICKE, B., SPENCER, J. E., BURCHFIEL, B. C., AND GUTH, P. L., 1982, Magnitude of crustal extension in the southern Great Basin: Geology, v. 10, p. 499–502.

WHITE, R. S., 1985, Seismic reflection profiling comes of age: Geological Magazine, v. 122, p. 199–201.

WILCOX, R. E., HARDING, T. P., AND SEELY, D. R., 1973, Basic wrench tectonics: American Association of Petroleum Geologists Bulletin, v. 57, p. 74–96.

WILSON, G., 1970, Wrench movements in the Aristarchus region of the Moon: Proceedings of Geological Association, v. 81, p. 595–608.

WILSON, J. E., KASHAI, E. L., AND CROKER, P., 1983, Hydrocarbon potential of Dead Sea Rift Valley: Oil and Gas Journal, v. 81, No. 25, June 20, p. 147–154.

WILSON, J. T., 1965, A new class of faults and their bearing on continental drift: Nature, v. 207, p. 343–347.

WINCHESTER, J. A., 1982, Comment on 'Paleomagnetic evidence for a large (~2,000 km) sinistral offset along the Great Glen fault during Carboniferous time:' Geology, v. 10, p. 487–488.

WISE, D. U., DUNN, D. E., ENGELDER, J. T., GEISER, P. A., HATCHER, R. D., KISH, S. A., ODOM, A. L., AND SCHAMEL, S., 1984, Fault-related rocks: Suggestions for terminology: Geology, v. 12, p. 391–394.

WRIGHT, L. A., AND TROXEL, B. W., 1970, Summary of regional evidence for right-lateral displacement in the western Great Basin: Discussion: Geological Society of America Bulletin, v. 81, p. 2167–2173.

YEATS, R. S., 1978, Neogene acceleration of subsidence rates in southern California: Geology, v. 6, p. 456–460.

————, 1981, Quaternary flake tectonics of the California Transverse Ranges: Geology, v. 9, p. 16–20.

————, 1983, Large-scale Quaternary detachments in Ventura Basin, southern California: Journal of Geophysical Research, v. 88, p. 569–583.

ZALAN, P. V., NELSON, E. P., WARME, J. E., AND DAVIS, T. L., 1985, The Piaui Basin: Rifting and wrenching in an equatorial Atlantic transform basin, in Biddle, K. T., and Christie-Blick, N., eds., Strike-Slip Deformation, Basin Formation, and Sedimentation: Society of Economic Paleontologists and Mineralogists Special Publication No. 37, p. 000–000.

ZAK, I., AND FREUND, R., 1981, Asymmetry and basin migration in the Dead Sea rift, in Freund, R., and Garfunkel, Z., eds., The Dead Sea Rift: Tectonophysics, v. 80, p.27–38.

ZIONY, J. I., AND YERKES, R. F., 1984, Fault slip-rate estimation for the Los Angeles region: Challenges and opportunities (abs.): Seismological Society of America, Eastern Section, Earthquake Notes, v. 55, No. 1, p. 8.

THE TYPES AND ROLE OF STEPOVERS IN STRIKE-SLIP TECTONICS

ATILLA AYDIN
Department of Geosciences
Purdue University
West Lafayette, Indiana 47907;

AND

AMOS NUR
Department of Geophysics
Stanford University
Stanford, California 94305

ABSTRACT: Stepovers are fundamental features along strike-slip faults of various lengths. Two types of stepover between strike-slip faults are considered in this paper: (1) along-strike stepovers that are due to en echelon arrangement of faults in map view, and (2) down-dip stepovers that are due to en echelon arrangement of faults in cross section. Along-strike stepovers produce pull-apart basins and push-up ranges depending on the sense of stepover. Down-dip stepovers of both senses may produce strike-slip faults in orientations different from the initial major strike-slip faults that are arranged en echelon. Some possible mechanisms that produce stepovers and control the sense of stepover are (1) bending of initially straight faults, (2) faulting within a weak zone oriented slightly off a local failure plane, (3) segmentation of faults to accommodate curved fault traces, (4) horizontal slip across pre-existing extensional fractures or dip-slip faults that have steps, (5) a change of physical parameters such as elastic moduli and pore pressure, and (6) stress field resulting from fault interaction.

INTRODUCTION

There has been an increasing interest in stepovers associated with strike-slip faults in recent years. It is now believed that stepovers on, or en echelon segmentation of, strike-slip faults is the rule rather than the exception. Numerous examples of stepovers of a wide range of sizes along strike-slip faults from all over the world are described in the literature. A few recent publications (e.g., Aydin and Nur, 1982; Mann et al., 1983; Bahat, 1984) provide surveys of previously recognized stepovers as well as a number of new ones.

The objectives of this paper are (1) to identify basic types of stepover, (2) to describe the geometry of the secondary structures associated with pull-apart basins, (3) to emphasize the role of stepovers in strike-slip tectonics, and (4) to explore possible origins and mechanisms of stepovers on strike-slip faults. A rather different classification of en echelon strike-slip faults, based on the map traces of the faults, has been given by Sharp (1979). In spite of numerous field and theoretical studies of pull-apart basins and push-up ranges (for references see the papers cited above) the nature and the orientation of the secondary structures and the origins of stepovers remain obscure.

TYPES OF STEPOVERS

Stepovers associated with strike-slip faults may be classified into two major groups: (1) stepovers along the strike of faults, and (2) stepovers along the dip of faults. Along-strike stepovers are observed in map view. The traces of strike-slip faults jump right or left but the faults are continuous in the dip direction (Fig. 1A). Down-dip stepovers occur when steps are along the dip direction of otherwise continuous strike-slip faults (Fig. 1B). These two markedly different varieties are likely to be the idealized end members of a continuous range of stepover configurations. For simplicity, in this paper we will be concerned with the two end members.

Both types of stepover may occur on the same fault or fault zone. Figure 1C elucidates the kinematics of a strike-slip fault system that includes along-strike (top view) and down-dip (side view) stepovers. As suggested in the figure, the nature of the deformation at the two stepovers is such that the left-lateral motion between the two blocks is accommodated in the most efficient way. Before discussing the nature of the deformation at these two types of stepover, it is necessary to deal with the mechanism(s) by which strike-slip faults initiate and propagate. As will be shown in this paper, the nature and geometry of such deformations are dependent largely upon the geometry of strike-slip faults at both the initial and successive stages of the faulting.

Micromechanisms for the initiation and propagation of faults are complicated and the matter is far from being resolved. We are evaluating here only the implications of two basic mechanisms of initiation and growth of strike-slip faults for the formation of, and deformation at, stepovers without analyzing the details of the processes involved in faulting. These two mechanisms are (1) shear faulting and (2) horizontal slip on pre-existing fracture planes such as joints and high-angle dip-slip faults.

Consider the lateral propagation of two non-colinear strike-slip faults (Fig. 2A). Two possibilities are likely to occur in this situation. If the separation between the faults is large enough, the neighboring fault tips will pass each other and propagate farther, resulting in parallel faults. On the other hand, if the two faults are close enough so that they interact, the two faults hinder each other's growth and form en echelon geometry (Fig. 2B, C). The critical value of separation for interaction or non-interaction is controlled primarily by the lengths of the faults and the prevailing stress system. An upper bound value of the critical separation can be estimated by examining the spacing of parallel faults of comparable lengths with about one hundred percent overlap in the same deformation domain. Exceptions to this rule may occur as a result of strain hardening or softening behavior within shear zones and perhaps different frictional behaviors of the faults involved.

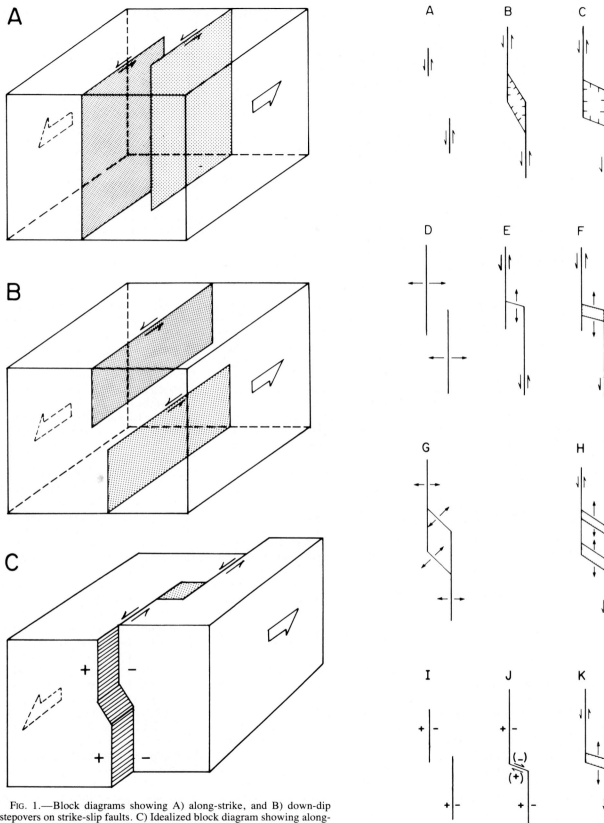

FIG. 1.—Block diagrams showing A) along-strike, and B) down-dip stepovers on strike-slip faults. C) Idealized block diagram showing along-strike (top view) and down-dip (side view) stepovers and the interaction products for both types of stepover.

The second mechanism, that is horizontal slip across pre-existing fracture surfaces, is quite common in nature (Dyer, 1983; Segall and Pollard, 1983). This mechanism requires that the geometry of pre-existing structures (joints, dip-slip faults, etc.) provide the initial stepover configuration. Indeed, the en echelon nature of many types of fractures such as joints (Segall and Pollard, 1983), veins and dikes (Delaney and Pollard, 1981), normal faults (Aydin and Pollard, 1982) and reverse or thrust faults (Dahlstrom, 1969) has been documented in the literature. Furthermore, it is known that opening-mode structures (joints, veins, dikes, and ocean ridges) possess a fairly narrow range of overlap-separation ratio as shown in Figures 2D and G (Pollard et al., 1982; Macdonald et al., 1984; Pollard and Aydin, 1984). In some cases the overlapped extensional fractures are connected by diagonal extensional fractures at either the tip of one of the en echelon fractures (similar to the diagonal connection in Figure 2E, except that it occurs before the strike-slip commences) or the closer tips of both fractures as shown in Figure 2G.

Figures 2I and J depict the en echelon arrangements of dip-slip faults as observed by Dahlstrom (1969) and by Aydin and Pollard (1982). The en echelon dip-slip faults are commonly connected either by other shorter dip-slip faults or by strike-slip (tear) faults. If dip-slip faults reactivate as strike-slip faults, the stepovers along the former will be inherited by the latter.

Although many of the examples described above represent along-strike stepovers in map views, joints and dip-slip faults are known to have discontinuities in the down-dip direction as well. The so-called step faults and ramps are examples of en echelon arrangements in the down-dip directions of dip-slip faults of normal and reverse types, respectively.

We will now return to the discussion of the deformation at along-strike and down-dip stepovers in the light of the possible mechanisms for strike-slip faulting and the implications of these mechanisms for the initial stepover configurations.

Along-Strike Stepovers

If the sense of an along-strike stepover is the same as the sense of fault slip, a pull-apart basin is formed. If, in contrast, the sense of the stepover is opposite to that of fault slip, a push-up range is formed. Pull-apart basins and push-up ranges exhibit a variety of extensional and compressional structures, respectively. The extensional deformation of pull-aparts of a wide range of sizes can be considered under two categories: opening and shearing modes.

The opening mode includes structures such as joints, fissures, and dikes. These structures occur either by themselves or with shearing mode structures. The best examples of nearly pure opening-mode structures are observed in highly brittle rocks such as the granites of Sierra Nevada, California (Moore, 1963; Segall and Pollard, 1983). There, the opening-mode deformation is induced by horizontal motion along pre-existing joints as shown in Figure 2D–H. Therefore, the initial geometry at the inception of the faulting is fixed by the en echelon geometry of the joints, which was briefly described above. Two slightly different arrangements either formed at the beginning of the faulting (Fig. 2E) or inherited from the previous joints (Fig. 2G), produce one or two pull-apart openings (Figs. 2F, H).

The orientation of cracks or fissures inferred from rhomb-shaped pull-apart openings varies considerably (Gamond, 1983; Bahat, 1984). Theoretical models based on elastic fracture mechanics suggest that the angle between strike-slip faults and the opening-mode structures (cracks and fissures) are influenced by the amount of separation between, and the overlap and lengths of, the en echelon strike-slip faults (Rodgers, 1980; Segall and Pollard, 1980), in addition to the relative magnitudes of the principal stresses (Cotterel and Rice, 1980; Nemat-Nasser and Horii, 1982). The angle is smaller for underlap and larger for overlap fault configuration. If the blocks bounding the fault system move rigidly, then the displacement on the strike-slip fault portions around the pull-apart should be horizontal or slightly oblique (Fig. 3A). An important question that arises then is to what depth does deformation associated with pull-apart formation extend? The depth of the pull-apart should be either equal to or shallower than the depth of the bounding strike-slip faults. Note that the former possibility requires some kind of detachment at the base of the pull-apart (see Fig. 3A), whereas the latter may require a wedge-shaped cavity because of decreasing opening at depth.

The shearing mode of deformation at pull-apart basins induces primarily normal faulting (see Fig. 3B), which in turn produces a deformation geometry similar to necking observed in uniaxial test specimens. A series of normal faults oblique to the trend of strike-slip faults accommodates extension and subsidence. The strike of these normal faults is expected to be the same as that of cracks and fissures in the opening mode discussed earlier. Figure 4 shows a series of horsts and grabens near Koehn Lake in the Cantil Valley, a composite pull-apart basin (Aydin and Nur, 1982) along the Garlock fault in southern California. Notice also a small pull-apart on a strand of the Garlock fault next to the horsts and grabens. It is interesting that the strikes of the faults involved in the horsts and grabens are about 35–45 degrees to the strike of the strike-slip fault, and that the orientation of the diagonal normal faults bounding the small pull-apart basin is comparable to the orientation of the normal faults bounding the horsts and grabens within the composite ba-

FIG. 2.—Mechanisms of strike-slip faulting and possible implications for stepover configurations and the geometry of pull-apart basins. A) Lateral propagation of non-colinear shear faults. B, C) Formation of stepovers and pull-apart basins. The angle between the diagonal normal faults and the en echelon strike-slip faults increases as the amount of overlap increases. D) En echelon joints. E) Initiation of horizontal slip across the en echelon joints and the formation of a diagonal joint connecting the two faults. F) Pull-apart opening as horizontal slip increases. G) En echelon joints connected by two diagonal joints at the closer tips of the en echelon joints. H) Occurrence of horizontal slip and formation of two openings. I) En echelon dip-slip faults. J) En echelon dip-slip faults connected by a shorter dip-slip fault or a strike-slip (tear) fault. K) Formation of pull-apart basin as a result of reactivation of dip-slip faults as strike-slip faults. Arrows indicate direction of motion. Barbs on down-thrown side for normal faults. Plus (+) and minus (−) signs on upthrown and downthrown sides, respectively, of dip-slip faults. Figures D, E, and F from Segall and Pollard (1983). Figure H proposed by Garfunkel (1981) and observed in the Sierra Nevada, California. Figure K is hypothetical.

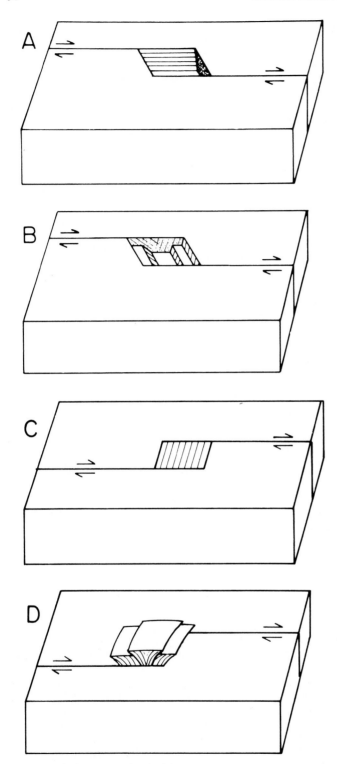

FIG. 3.—Major modes of deformation at extensional (A and B) and compressional (C and D) along-strike stepovers. A) Opening mode (line pattern shows approximate direction of motion on strike-slip fault and dotted pattern indicates extensional fracture surface). B) Shearing mode (lines show approximate direction of slip on oblique and normal faults). C) Cleavage or pressure solution formation (thin lines represent orientation). D) Reverse/thrust faulting (thin lines indicate oblique-slip direction on the bounding fault).

sin. However, higher values for intersection angles between the oblique-slip faults and the strike-slip faults also are common. For example, the northwestern boundary of the Lake Elsinore pull-apart basin in southern California (Crowell and Sylvester, 1979) appears to make a considerably larger angle with the strike of the Elsinore fault. It is also interesting to note that in some cases the normal faults at extensional stepovers dip in one direction only (Qidong and Peizhan, 1984). The reasons for the wide range of geometric relationships between the strike-slip faults and extensional structures need to be investigated further.

Rodgers (1980) and Segall and Pollard (1980) have considered the formation of secondary strike-slip faults in pull-apart basins. Secondary strike-slip faults, if their formation is favored, may occur in two sets, the orientations of which are different from that of the major strike-slip faults.

The accommodation of the dip-slip component of displacement on the portions of strike-slip faults bounding pull-apart basins appears to be complicated. Figure 5, adapted from Ben-Avraham et al. (1979), illustrates that the approximate dip angle of the faults bounding the Elat pull-apart basin (Fig. 5A) in the Gulf of Aqaba (Elat) varies considerably depending upon the position of the faults and the pull-apart (lines I and II in Figs. 5B and C, respectively). Usually the dip angles of the bounding faults are not symmetric (Ben-Avraham et al., 1979; and Schubert, 1982). However, depictions of approximately symmetric bounding faults can be found in the literature (Burchfiel and Royden, 1982).

The depth of extension at pull-apart basins deformed by the shearing mode also may be controlled by the depth of strike-slip faults, and some sort of detachment surface or zone may form within the basement underneath basins.

Deformation at push-up ranges is accommodated by either cleavage (Moore, 1963; Segall and Pollard, 1980) or folds and reverse thrust faults with or without cleavage (Sharp and Clark, 1972; Aydin and Page, 1984) as shown schematically in Figures 3C and 3D. The orientation of these compressional structures is nearly 90 degrees different from that of the extensional structures at a pull-apart basin along the same fault.

Down-Dip Stepovers

Down-dip stepovers are not well known simply because of the lack of field observations. Detailed mapping of strike-slip faults in shallow-level trenches (Sieh, 1978, 1984) and intermediate-level underground mines (Wallace and Morris, 1979) suggests that discontinuities, steps, and curvatures are as common in the down-dip direction of strike-slip faults as they are along the strike direction.

Detailed studies of seismicity associated with active strike-slip faults offer clues about the existence of deeper level irregularities on strike-slip faults. One of these studies was carried out by Reasenberg and Ellsworth (1982) on the aftershocks of the 1979 Coyote Lake earthquake, which occurred on the Calaveras fault about 12 km northeast of Gilroy, California. Based on the hypocentral trends and space-time evolution of the aftershocks, Reasenberg and Ellsworth (1982) were able to delineate two slip zones, which

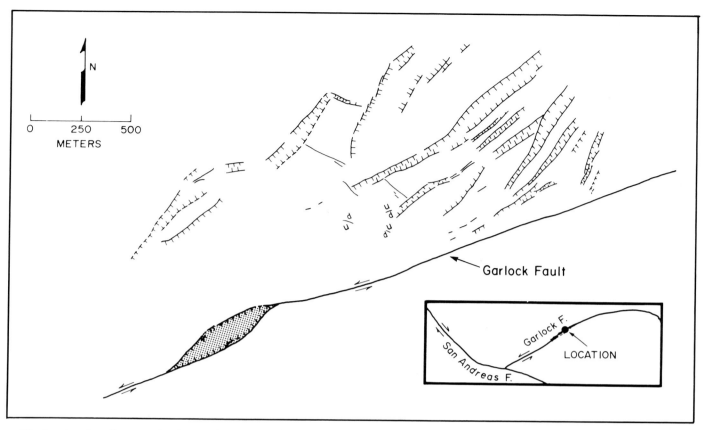

FIG. 4.—A series of horsts and grabens and a small pull-apart (shaded) within the Cantil Valley composite pull-apart basin along the Garlock fault (see inset for location). The orientation of the normal faults within the composite pull-apart basin is about parallel to the bounding diagonal faults of the small pull-apart. The angle between the trend of strike-slip faults and the bounding normal faults of the pull-apart basin is highly oblique (35–45°).

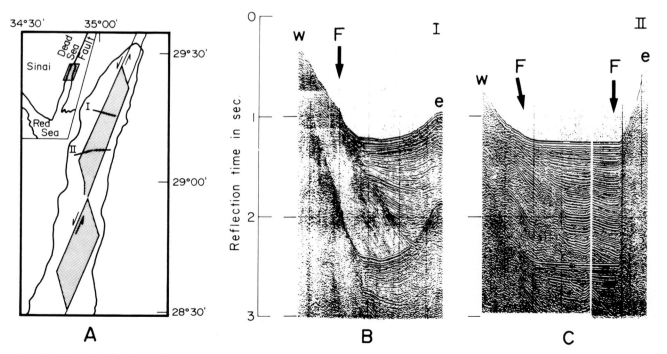

FIG. 5.—Seismic reflection profiles (B and C) across the Elat pull-apart basin (A) in the Gulf of Aqaba (see the location map in A) showing the dip-slip component on the overlapped portions of the strike-slip faults and their approximate dip angles (from Ben-Avraham et al., 1979).

they identified as right-stepping en echelon fault segments (I and II in Figure 6). One of the segments (II) lies 2 to 7 km deep and the other segment (I), which is about 2 km away from the first one, is restricted to an interval between 4 and 10 km deep (Fig. 6B). Thus, in addition to a right-hand along-strike step, a left-handed down-dip step can also be recognized in a section viewed from the south. There are some indications that similar down-dip steps may exist on other faults. For example, the data on the hypocentral distribution of the aftershocks of the 1966 Parkfield-Cholame earthquake, California (Eaton et al., 1970a; 1970b), may be interpreted as showing a down-dip step near Gold Hill. The nature of the deformation at down-dip stepovers is not known. The example from the Calaveras fault by Reasenberg and Ellsworth (1982) shows significant seismic activity in the stepover region (III in Fig. 6). The focal-mechanism solutions for these events indicate strike-slip faulting on planes oriented about 14 degrees from the major strike-slip fault segments.

A two-dimensional model for down-dip en echelon strike-slip faults would involve the traces of strike-slip faults in transverse section as antiplane cracks. In this sense, the problem is similar to that of interaction among normal faults that are en echelon in map view. Field observation and theoretical models for en echelon normal faults indicate that

the en echelon segments are commonly connected by normal faults of different orientation (Aydin and Pollard, 1982). Similarly, it is likely that two segments of down-dip en echelon strike-slip faults may be connected by one or more strike-slip faults of slightly different orientation. Furthermore, both right and left down-dip stepovers would result in the same interaction products for underlap cases because deformations for both senses are symmetric.

It should be pointed out that as the amount of overlap increases, the fault configuration may sweep out a range from down-dip steps to either along-strike steps or parallel faults with 100 percent overlap. It is difficult to predict interaction effects for fault configurations in between down-dip and along-strike stepovers. Actually, the aftershocks of the Coyote Lake earthquake suggest that the Calaveras fault has both along-strike and down-dip types of stepover there.

ORIGINS OF STEPOVERS

Why are strike-slip faults segmented? What influences the amount of overlap at stepovers and what controls the sense of stepover? It appears that in some cases, both senses of along-strike stepover may occur on the same fault as reported by Wallace (1973). In other cases, one sense of stepover may persist along some faults. Does one sense of

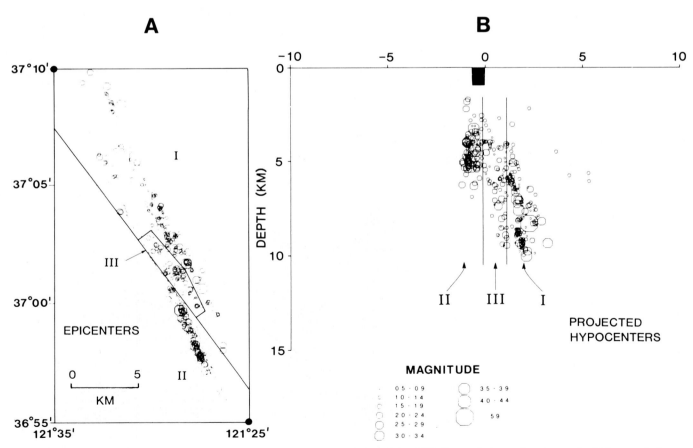

FIG. 6.—A) Epicenters, and B) hypocenters, projected on a transverse plane, of the 1979 Coyote Lake earthquake on the Calaveras fault near Gilroy, California (from Reasenberg and Ellsworth, 1982). Solid lines divide regions I, II, and III. Thin lines in A and solid rectangle in B mark the surface trace of the Calaveras fault.

stepover inducing pull-apart basins occur more commonly? A survey of stepovers around the world (Aydin and Nur, 1982) suggests that pull-aparts are more common than push-ups. However, push-ups may be more difficult to recognize, and may in fact be just as abundant as pull-aparts. The questions raised above have received very little attention in the literature. The answers hinge upon the mechanisms by which stepovers form. Below we briefly review some suggestions as to the origins of stepovers.

Bending of Initially Straight Faults.—It has been suggested by several geologists (Freund, 1971, 1974; Rogers, 1973; Crowell, 1974; Wallace, 1975; Allen, 1981) that originally straight strike-slip faults bend locally, thereby initiating steps as illustrated in Figure 7 (from Crowell, 1974). This hypothesis has not been tested as yet. A kink-like geometry, frequency, sense, and a wide range of scale of bending raise serious problems for the widespread occurrence of this mechanism.

Weak Zone Oriented Slightly off Local Plane.—A weak zone oblique to the direction of the principal stresses with an angle that is different from that of possible failure planes may produce short fault segments arranged en echelon within the weak zone (Fig. 8). The so-called R-faults in clay and shear box experiments (Wilcox et al., 1973) appear to be analogous to the strike-slip faults in fore- and back-arc environments (Dewey, 1980). However, as pointed out by Wallace (1973) this mechanism produces left steps for dextral faults and therefore apparently cannot produce pull-apart basins.

Curvature on Fault Traces.—Some major strike-slip faults that have measurably curved traces show a consistent sense of stepping along their entire length, suggesting a possible relationship between the curvature of the fault trace and the sense of step. For example, the senses of stepover are consistent along the left-lateral Garlock fault (left steps), California (see Clark, 1973), and along the right-lateral North Anatolian fault (right steps), Turkey (Fig. 9; see Ketin, 1969). The notion is that the individual segments tend to remain

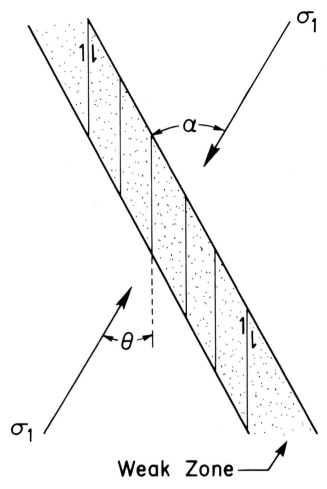

FIG. 8.—En echelon strike-slip faults formed in a weak zone (dotted pattern) off failure plane. σ_1, maximum compressive principal stress; α, angle between the weak zone and the direction of the maximum compressive principal stress; and θ, fault angle, or the angle between the failure plane and the direction of maximum principal stress.

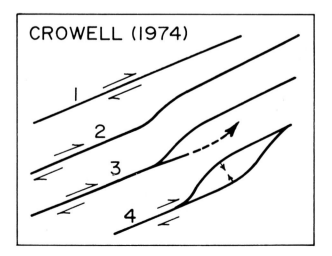

FIG. 7.—Bending of an initially straight fault leading to the formation of a step-like geometry along a strike-slip fault (from Crowell, 1974). 1–4 show evolution of fault geometry. Arrows indicate sense of motion.

straight, which is energetically more favorable, while following an overall curved path that is presumably required by either pre-existing boundaries or a changing state of stress.

Horizontal Slip Across Pre-existing Extensional Fractures or Dip-slip Faults.—If strike-slip faults are produced by horizontal slip across pre-existing extensional joints or high-angle dip-slip faults, then steps can be inherited from previous structures. Thus the problem of the origin of the steps on strike-slip faults that occur on the pre-existing extensional joints or dip-slip faults is related to the formation of steps on joints (Fig. 2D–H) and dip-slip faults (Fig. 2I–K).

Steps on extensional joints are understood better than those on shear fractures. It is believed that the formation of steps on joints is related to breakdown or segmentation of a continuous fracture plane. This process is controlled by the stress field associated with the crack tip, namely, the magnitude and sense (sign) of shear stress intensity (Lawn and Wilshaw, 1975; Pollard et al., 1982). Thus the formation of

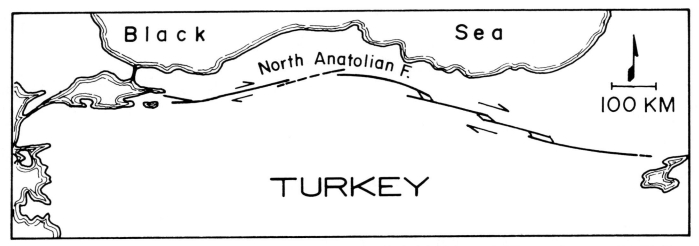

FIG. 9.—The curved trace of the North Anatolian fault, a right-lateral strike-slip fault in Turkey, and right sense of stepovers (simplified from Ketin, 1969).

steps on extensional fractures is controlled by a change in the state of stress and the sense of steps is the same as the sense of shear stress induced by such a change at the crack front. Figure 10 illustrates this concept for two cases: laterally and vertically propagating cracks. Figures 2D, E and G show a few possible varieties of stepover geometry associated with extensional fractures.

The extensional origin of some small-scale faults a few hundred meters long has been demonstrated by Segall and Pollard (1983). However, the sizes of continent-scale strike-slip faults are quite different from the sizes of regional extensional joints. If some of the large strike-slip faults originated from extensional fractures, the processes involved in this development and the associated stepover formation are certainly complicated and are not known.

ROLE OF STEPOVERS

Stepovers play an important role in strike-slip tectonics by (1) producing pull-apart basins and push-up ranges, which

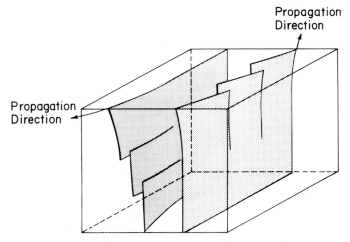

FIG. 10.—Breakdown of laterally and vertically propagating extensional fractures producing down-dip and along-strike stepovers (based on Delaney and Pollard, 1981).

are potential oil, gas, and geothermal exploration targets; (2) enhancing permeability of fault zones by pull-apart openings or tail cracks along the faults, thus facilitating the formation of hydrothermal ore deposits (McKinstry, 1948); (3) generating heterogeneities along faults and thus controlling seismicity in active fault environments (Johnson and Hadley, 1976; Bakun, 1980; Lindh and Boore, 1981; Reasenberg and Ellsworth, 1982); (4) localizing secondary structures, such as, joints, dikes, volcanoes, folds, cleavages, and dip-slip faults associated with strike-slip faulting (Segall and Pollard, 1980, 1983; Rispoli, 1981; Aydin and Nur, 1982; Aydin and Page, 1984); (5) enhancing rock rotations within the stepover area (Cox, 1980); and (6) controlling the location of back-arc extension (Dewey, 1980) and subduction-related volcanism (Craig Weaver, personal commun., 1984).

For a given fault, one sense of stepover produces basins, whereas the other sense results in mountain ranges. What controls the sense of stepover is totally unknown. It appears that several mechanisms produce stepovers between strike-slip faults or fault segments. Each mechanism may have somewhat unique initial and subsequent geometric features. A sound understanding of the mechanisms by which stepovers along strike-slip faults form is needed in order to improve upon our prediction and interpretation capability regarding the issues listed above.

CONCLUDING REMARKS

Stepovers are fundamental features of strike-slip faults and fault systems. Because of considerable stress concentration and consequent localization of the associated structures at stepover areas, they could play an important role as potential hosts of ore deposits and oil and gas accumulations.

Two types of stepover, along-strike and down-dip, are identified (Fig. 1A, B, C). A continuous range of stepover configurations between these two idealized types probably occurs in nature. It is inferred that down-dip stepovers are deformed primarily by strike-slip faults as shown in Figure

1C, and that both senses of down-dip stepover yield the same type of deformation, as opposed to extensional and compressional deformations at right and left along-strike stepovers. As the amount of overlap increases down-dip stepovers may become along-strike stepovers.

The genesis of stepovers on strike-slip faults is an unresolved problem. The validity and generality of the mechanisms listed above should be tested by theoretical and observational means before further progress in this area of tectonics and structural geology can be made.

ACKNOWLEDGMENTS

Comments by K. T. Biddle, D. A. Rodgers, and A. C. Tuminas have improved this manuscript.

REFERENCES

ALLEN, C. R., 1981, The modern San Andreas fault, *in* Ernst, W. G., ed., The Geotectonic Development of California: Engelwood Cliffs, New Jersey, Prentice-Hall, p. 511–534.

AYDIN, A., AND NUR, A., 1982, Evolution of pull-apart basins and their scale independence: Tectonics, v. 1, p. 11–21.

AYDIN, A., AND PAGE, B. M., 1984, Diverse Pliocene-Quaternary tectonics in a transform environment, San Francisco Bay region, California: Geological Society of America Bulletin, v. 95, p. 1303–1317.

AYDIN, A., AND POLLARD, D. D., 1982, Origin of zig-zag pattern of normal faults: Geological Society of America Abstracts with Programs, v. 14, p. 436.

BAHAT, D., 1984, New aspects of rhomb structures: Journal of Structural Geology, v. 5, p. 591–601.

BAKUN, W. J., 1980, Seismic activity on the southern Calaveras fault in central California: Seismological Society of America Bulletin, v. 70, p. 1181–1197.

BEN-AVRAHAM, Z., ALMAGOR, G., AND GARFUNKEL, Z., 1979, Sediments and structure of the Gulf of Elat (Aqaba)—North Red Sea: Sedimentary Geology, v. 23, p. 239–267.

BURCHFIEL, B. C., AND ROYDEN, L., 1982, Carpathian foreland fold and thrust belt and its relation to the Pannonian and other basins: American Association of Petroleum Geologist Bulletin, v. 66, p. 1179–1195.

CLARK, M. M., 1973, Map showing recently active breaks along the Garlock and associated faults, California: United States Geological Survey Miscellaneous Geologic Investigations Map I-741.

COTTERELL, B., AND RICE, J. R., 1980, Slightly curved or kinked cracks: International Journal of Fracture, v. 16, p. 155–169.

COX, A., 1980, Rotation of microplates in western North America, *in* Strangway, D. W., ed., The Continental Crust and its Mineral Deposits: Geological Association of Canada Special Paper 20, p. 305–321.

CROWELL, J. C., 1974, Origin of late Cenozoic basins in southern California, *in* Dickinson, W. R., ed., Tectonics and Sedimentation: Society of Economic Paleontologists and Mineralogists Special Publication 22, p. 190–204.

CROWELL, J. C., AND SYLVESTER, A. G., 1979, Introduction to the San Andreas–Salton Trough juncture, *in* Crowell, J. C., and Sylvester, A. G., eds., Tectonics of the Juncture Between the San Andreas Fault System and the Salton Trough, Southeastern California—A Guidebook: Santa Barbara, California, Department of Geological Sciences, University of California, p. 1–13.

DAHLSTROM, C. D. A., 1969, Balanced cross sections: Canadian Journal of Earth Sciences, v. 6, p. 743–757.

DELANEY, P. T., AND POLLARD, D. D., 1981, Deformation of host rocks and flow of magma during growth of minette dikes and breccia-bearing intrusions near Ship Rock, New Mexico: United States Geological Survey Professional Paper 1201, 61 p.

DEWEY, J. F., 1980, Episodicity, sequence, and style at convergent plate boundaries, *in* Strangway, D. W., ed., The Continental Crust and its Mineral Deposits: Geological Association of Canada Special Paper 20, p. 553–573.

DYER, J. R., 1983, Jointing in sandstones, Arches National Park, Utah

[unpub. Ph.D. thesis]: Stanford, California, Stanford University, 202 p.

EATON, J., O'NEILL, M. E., AND MURDOCK, J. N., 1970a, Aftershocks of the 1966 Parkfield-Cholame, California, earthquake: a detailed study: Seismological Society of America Bulletin, v. 60, p. 1151–1197.

EATON, J., LEE, W. H. K., AND PAKISER, L. C., 1970b, Use of microearthquakes in the study of the mechanics of earthquake generation along the San Andreas fault in central California: Tectonophysics, v. 9, p. 259–282.

FREUND, R., 1971, The Hope Fault, a strike-slip fault in New Zealand: New Zealand Geological Survey Bulletin, v. 86, p. 1–48.

———, 1974, Kinematics of transform and transcurrent faults: Tectonophysics, v. 21, p. 93–134.

GAMOND, J. F., 1983, Displacement features associated with fault zones; a comparison between observed examples and experimental models: Journal of Structural Geology, v. 5, p. 33–45.

GARFUNKEL, Z., 1981, Internal structure of the Dead Sea leaky transform (rift) in relation to plate kinematics: Tectonophysics, v. 80, p. 81–108.

JOHNSON, C. E., AND HADLEY, D. M., 1976, Tectonic implication of the Brawley earthquake swarm, Imperial Valley, California, January 1975: Seismological Society of America Bulletin, v. 66, p. 1133–1144.

KETIN, I., 1969, Kuzey Anadolu fayi hakkinda (in Turkish and German): Maden Tetkik Arama Enstitusu, Dergisi, No. 72, p. 1–27.

LAWN, B. R., AND WILSHAW, T. R., 1975, Fracture of Brittle Solids: London, England, Cambridge University Press, 204 p.

LINDH, A., AND BOORE, D. M., 1981, Control of rupture by fault geometry during the 1966 Parkfield earthquake: Seismological Society of America Bulletin, v. 71, p. 95–116.

MACDONALD, K., SEMPERE, J. C., AND FOX, P. J., 1984, East Pacific Rise from Siqueiros to Orozco Fracture Zones: Along-strike continuity of axial neovolcanic zone and structure and evolution of overlapping spreading centers: Journal of Geophysical Research, v. 89, p. 6049–6069.

MANN, P., HEMPTON, M. R., BRADLEY, D. C., AND BURKE, K., 1983, Development of pull-apart basins: Journal of Geology, v. 91, p. 529–554.

McKINSTRY, H. E., 1948, Mining Geology: New York, Prentice Hall, 680 p.

MOORE, J. G., 1963, Geology of the Mount Pinchot quadrangle, southern Sierra Nevada, California: United States Geological Survey Bulletin 1130, 152 p.

NEMAT-NASSER, S., AND HORII, H., 1982, Compression-induced nonplanar crack extension with application to splitting, exfoliation, and rock burst: Journal of Geophysical Research, v. 87, p. 6805–6821.

POLLARD, D. D., AND AYDIN, A., 1984, Propagation and linkage of oceanic ridge segments: Journal of Geophysical Research, v. 89, p. 10,017–10,028.

POLLARD, D. D., SEGALL, P., AND DELANEY, P. T., 1982, Formation and interpretation of dilatant echelon fractures: Geological Society of America Bulletin, v. 93, p. 1291–1303.

QIDONG, D., AND PEIZHEN, Z., 1984, Research on the geometry of shear fracture zones: Journal of Geophysical Research, v. 89, p. 5699–5710.

REASENBERG, P., AND ELLSWORTH, W. L., 1982, Aftershocks of the Coyote Lake, California, earthquake of August 6, 1979. A detailed study: Journal of Geophysical Research, v. 87, p. 10,637–10,655.

RISPOLI, R., 1981, Stress fields about strike-slip faults inferred from stylolites and tension gashes: Tectonophysics, v. 75, p. 29–36.

RODGERS, D. A., 1980, Analysis of pull-apart basin development produced by en echelon strike-slip faults, *in* Balance, P. F., and Reading, H. G., eds., Sedimentation in Oblique-Slip Mobile Zones: International Association of Sedimentology Special Publication, No. 4, p. 27–41.

ROGERS, T. H., 1973, Fault trace geometry within the San Andreas and Calaveras fault zones—A clue to the evolution of some transcurrent fault zones, *in* Kovach, R. L., and Nur, A., eds., Conference on Tectonic Problems of the San Andreas Fault System, Proceedings: Stanford University Publications in Geological Sciences, v. 13, p. 251–258.

SCHUBERT, C., 1982, Origin of Cariaco basin, southern Caribbean Sea: Marine Geology, v. 47, p. 345–360.

SEGALL, P., AND POLLARD, D. D., 1980, Mechanics of discontinuous faults: Journal of Geophysical Research, v. 85, p. 4337–4350.

SEGALL, P., AND POLLARD, D. D., 1983, Nucleation and growth of strike-slip faults in granite: Journal of Geophysical Research, v. 88, p. 555–568.

SHARP, R. V., 1979, The implication of surficial strike-slip fault patterns for simplification and widening with depth, *in* Conference VIII on Actual Fault Zones in Bedrock: United States Geological Survey Open-File Report 79–1239.

SIEH, K. E., 1978, Prehistoric large earthquakes produced by slip on the San Andreas Fault at Pallett Creek, California: Journal of Geophysical Research, v. 83, p. 3907–3939.

————, 1984, Lateral offsets and revised dates of large prehistoric earthquakes at Pallett Creek, southern California: Journal of Geophysical Research, v. 89, p. 7641–7670.

WALLACE, R. E., Surface fracture patterns along the San Andreas fault, *in* Kovach, R. L., and Nur, A., eds., Conference on Tectonic Problems of the San Andreas Fault System, Proceedings: Stanford University Publications in Geological Sciences, v. 13, p. 248–250.

————, 1975, The San Andreas fault in the Carrizo Plain-Temblor Range region, California: California Division of Mines and Geology Special Report 118, p. 241–250.

WALLACE, R. E., AND MORRIS, T., 1979, Characteristics of faults and shear zones as seen in mines at depths as much as 2.5 km below the surface, *in* Proceedings: Conference VIII, Analysis of actual fault zones in bedrock: United States Geological Survey Open-File Report 79-1239, p. 79–101.

WILCOX, R. E., HARDING, T. P., AND SEELY, D. R., 1973, Basic wrench tectonics: American Association of Petroleum Geologists Bulletin, v. 57, p. 74–996.

SUBSIDENCE AND THERMAL HISTORY OF SMALL PULL-APART BASINS[1]

W. C. PITMAN III AND J. A. ANDREWS

Department of Geological Sciences and
Lamont-Doherty Geological Observatory of Columbia University
Palisades, New York 10964

ABSTRACT: The very rapid subsidence, sediment accumulation, and hydrocarbon maturation observed in many small extensional or "pull-apart" basins can be explained using a McKenzie-type model. It has been shown that in basins of 100 km width or less, lateral heat loss is quite important and accelerates lithospheric cooling and subsidence. We show here that cooling that is simultaneous with stretching is very important for basins formed by stretching of lithospheric blocks that are 10 km to several tens of kilometers wide. In fact, for most of these very narrow basins, most of the anomalous heat introduced by stretching is also dissipated during the stretching event. We have calculated the effect of alternate short periods of stretching and cooling to approximate simultaneous stretching and cooling. The results show, for example, that for a block, initially 10 km wide and stretched uniformly at 3 cm/yr, sufficient subsidence will take place in 200,000 years to accumulate 4–5 km of sediment. A consequence of this rapid subsidence is initial sediment starvation. These results may be applicable to many of the small extensional basins associated with the San Andreas transform system.

INTRODUCTION

Recently, simple models have been developed that may be used to explain the subsidence and thermal history of basins created along pull-apart zones. McKenzie (1978) showed that the major features of basin development could be accounted for by a model in which the lithosphere is uniformly attenuated by stretching during a short time interval. The model gives a reasonable and quantitative explanation for the observations that either subsidence or uplift may take place during rifting and that rifting is always followed by a long protracted period of subsidence. The McKenzie model (and a similar model, Royden et al., 1980) was developed for slabs of infinite horizontal extent; there was no horizontal heat loss. Subsequently, Steckler (1981) showed that where basin width is less than about 100 km, the lateral heat loss through the sides of the basin is significant; cooling and hence subsidence in this case is more rapid than predicted by the McKenzie model. The narrower the basin the more rapid the cooling. These models, however, are two-stage models in which there is an abbreviated period of rifting without significant cooling, followed by a long period of cooling without rifting. For a slab of infinite horizontal extent, it was shown that if the rifting stage were less than 20 m.y., syn-rift heat loss could be ignored (Jarvis and McKenzie, 1980). However, Cochran (1983) has shown that for basins of finite width (<150 km), the loss of heat during rifting may also be important. We show that for basins less than 50 km wide this syn-rift heat loss is very important.

In this paper we make calculations that approximate the thermal and subsidence history of narrow basins, incorporating syn-rift cooling. We employ a version of the McKenzie model modified by Steckler (1981) to include lateral heat conduction through the sides of the basin. Our results approximate the effect of heat loss during the rifting stage. It will be seen that subsidence in these basins is quite rapid and hence early sediment starvation is to be expected.

Narrow pull-apart basins are common. They appear as failed rifts, at the edges of rotated terranes, and at offsets or at extensional segments of transform faults. The San Andreas fault system in southern and central California contains a number of these narrow pull-apart basins (Crowell, 1974). The Ventura, Santa Maria, and Cuyama Basins are three examples. We suggest that the model presented here is applicable to basins such as these.

THE MODEL

The purpose of this paper is to estimate the subsidence and thermal history of typical small basins using a model that takes into account the cooling that occurs during rifting. We have used as initial conditions those believed to have existed in the San Andreas transform zone in the mid-Tertiary. The model of Atwater (1970) for the Tertiary evolution of the San Andreas zone suggests that the mid- to late Tertiary lithosphere was anomalously thin (see also Dickinson and Snyder, 1978). This conclusion is supported by heat-flow data (Lachenbruch and Sass, 1980) and seismic data (Zandt and Furlong, 1982). Thus we assume an initial lithospheric thickness of 62.5 km, and an initial crustal thickness of 30 km. Other parameters are those of Parsons and Sclater (1977). The procedure employed is repeatedly to stretch and then to cool a lithospheric block using small time increments (i.e., 50,000 yr). At the end of each cooling period the new lithospheric thickness is calculated. This is the input for the next stretching phase. By making such calculations using small time intervals, an approximation of simultaneous stretching and cooling can be obtained. We use a two-dimensional model, assuming that the basin length (i.e., the dimension perpendicular to the stretching direction) is significantly greater than its width (i.e., the dimension in the stretching direction). The results are shown in Figures 1–3.

DISCUSSION

Subsidence

We have used the above technique to compare the subsidence in three hypothetical basins. A rate of 3 cm/yr is used as the stretching rate for each of the blocks. This is

---
[1]Lamont-Doherty Geological Observatory Contribution No. 3911

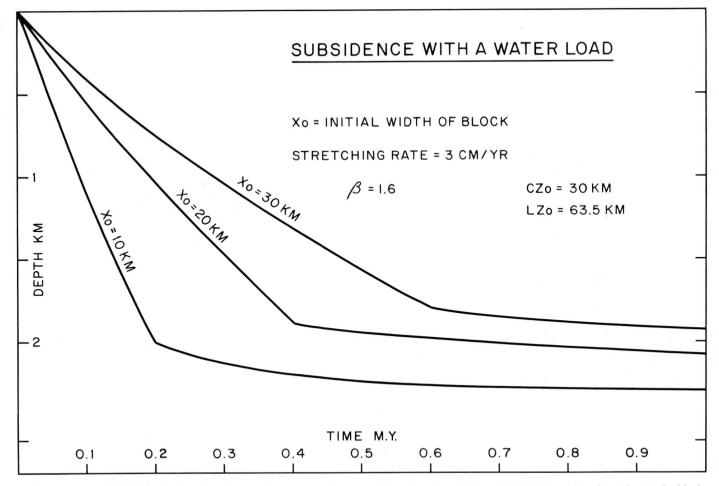

SUBSIDENCE WITH A WATER LOAD

Xo = INITIAL WIDTH OF BLOCK

STRETCHING RATE = 3 CM/YR

$\beta = 1.6$

CZo = 30 KM
LZo = 63.5 KM

Xo = 30 KM
Xo = 20 KM
Xo = 10 KM

DEPTH KM

TIME M.Y.

0.1   0.2   0.3   0.4   0.5   0.6   0.7   0.8   0.9

Fig. 1.—The subsidence history for three basins with a water load only is shown for three different initial block widths. In each case the block is stretched at 3 cm/yr. The break in each curve marks the time at which each of the blocks has been stretched by a factor (β) of 1.6. To the left of the break, the subsidence is caused by simultaneous cooling and stretching, calculated as explained in the text. To the right of the break, the subsidence is caused by cooling only. CZ_0 and LZ_0 are the initial crustal and lithospheric thicknesses, respectively.

somewhat less than the average rate of transform motion between North America and the Pacific plate, which has been on the order of 4–6 cm/yr for the past 35 m.y. (Atwater, 1970; Atwater and Molnar, 1973). The blocks are stretched to a maximum of 1.6 times their original width ($\beta = 1.6$); Andrews and M. S. Steckler (personal commun., 1985) have examined stratigraphic data from several basins of this type and concluded that the maximum β is about 1.6. Initial widths (X_o) of 10 km, 20 km, and 30 km were chosen. As a consequence, the duration of the stretching time increases in proportion to the increasing initial width of the block (Fig. 1). The subsidence history has been calculated for a point at the center of each basin. The subsidence that takes place during the stretching phase is caused by the isostatic response to the thinning of the crust and lithosphere and by the dissipation of the anomalous heat introduced by the thinning. The break in each of the three subsidence curves marks the time at which each of the blocks has been stretched to 1.6 times the initial width. Prior to this time, subsidence is very rapid, on the order of 500 cm/1,000 yr. After that time subsidence takes place by cooling

only, with the rate of subsidence much reduced. The basin formed by stretching the block with an initial width of 10 km has the most rapid subsidence during the stretching phase; the lateral heat flux is proportionally greater. The basin formed by stretching the 30 km block shows the slowest subsidence; the lateral heat loss is proportionally less. The widths of these basins are much less than the lithospheric thickness. Therefore, it is the basin width that controls the magnitude of the cooling time constant. That is, horizontal rather than vertical heat flux dominates cooling.

Sedimentation

Subsidence (with a water load; Fig. 1) varies, with increasing basin width, from 1 km at the end of 100,000 yr to 1 km at the end of 300,000 yr. In all cases the subsidence rate is greatest when stretching begins, and decreases thereafter. In Figure 1, the subsidence with a water load during the first 100,000 yr is 376 m, 548 m, and 1,075 m, respectively, for each of the three basins (decreasing initial width). Using a local Airy model for compensation, the amount of sediment that may be accumulated is:

$$T_s = T_w \frac{(\rho_m - \rho_w)}{(\rho_m - \rho_s)}$$

T_s = thickness of the sediments
T_w = water depth
ρ_m = 3.33 mantle density
ρ_s = 2.165 sediment density
ρ_w = 1.00 density of water

If sedimentation is able to keep pace with subsidence, then the amount of sediment that would be acquired during the first 100,000 yr would be 752 m, 1,096 m, and 2,150 m, respectively, requiring rates of 752 cm/1,000 yr, 1,096 cm/1,000 yr, and 2,150 cm/1,000 yr.

The difficulty in accumulating sediments at these rates is discussed below. On the basis of data and arguments presented by Ahnert (1970), Pitman and Golovchenko (in prep.) have shown that, to a first approximation, the rate of denudation ($d\bar{Y}_D/dt$) of a region is proportional to the average regional elevation ($\bar{Y}$):

$$\frac{d\bar{Y}_D}{dt} = K \times \bar{Y}$$

K = the denudation constant

For most regions the denudation constant K appears to be on the order of $10^{-4}/1,000$ yr. For example, the Appalachians, with an average regional elevation of $\bar{Y} = 500$ m, yield an average regional denudation of:

$$\frac{d\bar{Y}_D}{dt} = 5 \text{ cm}/1,000 \text{ yr}$$

The Himalayas, where $\bar{Y} = 5,000$ m, yield a rate of

$$\frac{d\bar{Y}_D}{dt} = 50 \text{ cm}/1,000 \text{ yr}$$

Both values are in reasonable agreement with available data (Judson and Ritter, 1964; Ahnert, 1970; R. W. Fairbridge, personal commun., 1985).

With this formulation we are able to estimate the required size and denudation rate of drainage basins that would provide sufficient sediment to fill the model basins. There is no geologic evidence to suggest that the California basins were formed within a truly mountainous region ($\bar{Y} > 1,000$ m). Therefore, we assume a maximum average regional elevation of 500 m for the drainage region feeding the model basins. This is probably a bit high. Assuming an erosional constant of $K = 10^{-4}/1,000$ yr, the regional denudation rate is $dY_D/dt = K \times \bar{Y} = 5$ cm/1,000 yr. Thus in the case of the basin with initial width $X_o = 30$ km (requiring a sedimentation rate of 752 cm/1,000 yr for sediment to fill the basin at all times), the area of the drainage basin must be 150 times the size of the depositional basin. For the basin with $X_o = 20$ km (requiring a sedimentation rate of 1,096 cm/1,000 yr), the area of the drainage basin must be 219 times the size of the depocenter. For the basin in which $X_o = 10$ km (requiring a sedimentation rate of 2,150 cm/1,000

yr), the area of the drainage basin must be 430 times the size of the depocenter.

If, for example, the basin with an initial width of 10 km were 50 km long, then the required drainage area would be 215,000 km^2, which is half the area of California. A potential drainage area one-tenth that size is more likely, in which case the basin would be initially sediment starved, unless the erosional constant K were high, say $10^{-3}/1,000$ yr.

To illustrate this effect we have calculated the total subsidence and sediment accumulation as a function of time for the basin in which $X_o = 10$ km (Fig. 2). The sedimentation rate is assumed to be constant at 100 cm/1,000 yr. In this case, the drainage area must be 20 times the size of the basin, which is more reasonable. Even with this rather high sedimentation rate, the basin deepens rapidly to a water depth of about 2 km. The basin gradually fills as the subsidence rate decreases, eventually filling at about 5 m.y. after the initiation of the rifting.

Thermal History

We next calculate the thermal history of the sediments. For ease of computation we assume that the sediment influx

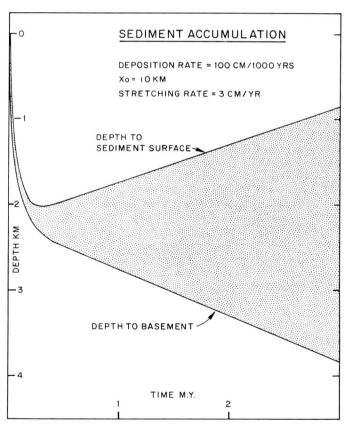

FIG. 2.—The depth to the basement and the depth to the sediment surface in a subsiding basin with limited sediment input. The subsidence curve is for a block with an initial width of $X_o = 10$ km being stretched at 3 cm/yr. The deposition rate is limited to 100 cm/1,000 yr. A local Airy model is assumed for the adjustment of basement to the sediment load. Note that deep-water conditions prevail and that it takes approximately 5 m.y. for the sediment surface to approach sea level.

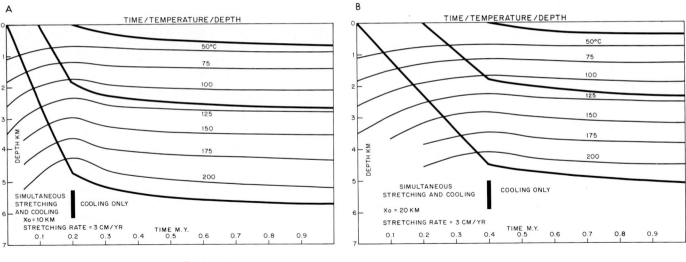

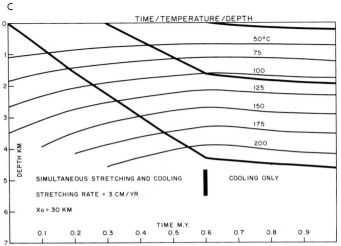

FIG. 3.—The time/temperature/depth history of the sediments for each of the basins in Figure 1: A) initial width is 10 km; B) initial width is 20 km. C) initial width is 30 km. In each case, sediment accumulation keeps pace with subsidence. The fine lines show the depth of the isotherms as a function of time. It is assumed that the sediment type does not vary with depth. The heavy lines show the burial and temperature history for sediments that enter the system at the beginning, middle, and end of the stretching phase. In each case, the isotherms do not rise very far during the stretching phase. This is because of the rapid simultaneous cooling. The effect of the simultaneous stretching and cooling is to cause rapid subsidence. Thus the sediments deposited during the early stages of the rifting, as well as the sediment already in place at the beginning of rifting, are quickly heated to high temperatures and have a high maturation potential.

was sufficient to keep pace with subsidence.

The heat flow through the basement in the center of each basin is calculated at the end of each cooling stage using the basic McKenzie model as modified by Steckler (1981), and these values are used to calculate the time-temperature-depth curves shown in Figure 3. In calculating the temperature in the sediments as a function of depth, it is assumed there are no heat sources in the sediments. The following equation gives the temperature $T(z,t)$ of the sediments at depth z and time t:

$$T(z,t) = T_o + \int \frac{Q(t)}{K(z)} \, dz$$

T_o = temperature at the surface
$Q(t)$ = heat flow through basement at time (t)

$K(z)$ = thermal conductivity of the sediments as function of depth z

The thermal conductivity, $K(z)$, of the sediments usually increases as the sediments are compacted. We have derived a general formula for the thermal conductivity of the sediments using data of Sclater and Christie (1980) for chalks and shales. The thermal conductivity is about 2.2×10^{-3} cal cm^{-1} sec^{-1} °C^{-1} at the surface and increases with depth asymptotically approaching a value of 5.5×10^{-3} cal cm^{-1} sec^{-1} °C^{-1}. The equation used is:

$$K(z) = (5.5 - 3.3 \, e^{-z/1.059}) \text{ cal cm}^{-1}\text{sec}^{-1} \text{ °C}^{-1}$$

Note that background heat flow is anomalously high because the lithosphere being stretched is initially about half normal thickness. Thus the initial heat flow is about twice its normal value, or about 2 H.F.U.

In each case the isotherms rise during the stretching and cooling phase and then deepen again during the "cooling only" stage. The effect of lateral flow of heat during the stretching is to dampen the rise of the isotherms. At all stages the lateral heat flow increases the rate of basin cooling. The burial and thermal history of the sediments that enter the basin at the beginning of the stretching phase, at the middle of the stretching phase, and at the termination of the stretching phase are shown by the heavy lines in Figure 3. Sediments that were in place at the time of the initiation of stretching or were deposited during the earliest stage of the stretching are rapidly heated to high temperatures (>150° C) and hence should mature quickly. Because of lateral heat loss, the subsidence curves and the isotherms soon approach equilibrium. Thus sediments deposited after the initial stage of rifting appear unlikely to reach temperatures sufficient for maturation.

CONCLUSIONS

(1) Basins formed by rapidly stretching narrow lithospheric blocks lose most of their anomalous heat during the stretching phase with very rapid syn-rift subsidence (i.e., on the order of 500 cm/1,000 yr).

(2) The rate of syn-rift subsidence is so high that these basins may be sediment starved for several million years.

(3) Because of the rapid syn-rift cooling, the thermal gradient remains remarkably stable throughout stretching and cooling.

(4) However, because of the rapid syn-rift subsidence, pre-rift and syn-rift sediments sink quickly through the field of isotherms. As a consequence, sediments deposited prior to and during the early stages of rifting are heated quickly to temperatures in excess of 150° C and therefore tend to be mature. It is worth noting that small extensional basins formed in strike-slip zones may undergo a subsequent phase of compressional deformation that may enhance entrapment (e.g., the Ventura Basin).

ACKNOWLEDGMENTS

The manuscript was reviewed by K. T. Biddle, N. Christie-Blick, M. A. Kominz, L. H. Royden, and C. R. Tapscott.

REFERENCES

AHNERT, F., 1970, Functional relationships between denudation, relief and uplift in large mid-latitude drainage basins: American Journal of Science, v. 270 p. 243–263.

ATWATER, T., 1970, Implications of plate tectonics for the Cenozoic tectonic evolution of western North America: Geological Society of America Bulletin, v. 81, p. 3513–3536.

ATWATER, T., AND MOLNAR, P., 1973, Relative motion of the Pacific and North American plates deduced from sea-floor spreading in the Atlantic, Indian and South Pacific Oceans in Kovach, R. L., and Nur, A., eds., Conference on Tectonic Problems of the San Andreas Fault System, Proceedings: Stanford University Publications in Geological Sciences, v. 13, p. 136–148.

COCHRAN, J. R., 1983, Effects of finite rifting times on the development of sedimentary basins: Earth and Planetary Science Letters, v. 66, p. 289–302.

CROWELL, J. C., 1974, Origin of late Cenozoic basins in southern California, in Dickinson, W. R., ed., Tectonics and Sedimentation: Society Economic Paleontologists and Mineralogists Special Publication No. 22, p. 190–204.

DICKINSON, W. R., AND SNYDER, W. S., 1978, Plate tectonics of the Laramide orogeny: Geological Society of America Memoir 151, p. 355–366.

JARVIS, G. T., AND MCKENZIE, D. P., 1980, Sedimentary basin formation with finite extension rates: Earth and Planetary Science Letters, v. 48, p. 42–52.

JUDSON, S., AND RITTER, D., 1964, Rates of regional denudations in the U.S.: Journal of Geophysical Research, v. 69, p. 3395–3401.

LACHENBRUCH, A. H., AND SASS, J. H., 1980, Heat flow and energetics of the San Andreas fault zone: Journal of Geophysical Research, v. 85, p. 6185–6223.

MCKENZIE, D. P., 1978, Some remarks on the development of sedimentary basins: Earth and Planetary Science Letters, v. 49, p. 25–32.

PARSONS, B., AND SCLATER, J. G., 1977, An analysis of the variation of ocean floor bathymetry and heat flow with age: Journal of Geophysical Research, v. 82, p. 803–827.

PITMAN, W. C., AND GOLOVCHENKO, X., The effects of sea level change on geomorphic processes: in prep.

ROYDEN, L., SCLATER, J. G., AND VON HERZEN, R. P., 1980, Continental margin subsidence and heat flow—important parameters in formation of petroleum hydrocarbons: American Association of Petroleum Geologists Bulletin, v. 64, p. 173–187.

SCLATER, J. G., AND CHRISTIE, P. A. F., 1980, Continental stretching—an explanation of the post mid-Cretaceous subsidence of the central North Sea basin: Journal of Geophysical Research, v. 85, p. 683–702.

STECKLER, M. S., 1981, The thermal and mechanical evolution of Atlantic type continental margins [unpubl. Ph.D. thesis]: New York, New York, Columbia University, 261 p.

ZANDT, G., AND FURLONG, K. P., 1982, Evolution and thickness of the lithosphere beneath coastal California: Geology, v. 10, p. 376–381.

STRUCTURAL STYLES, PLATE-TECTONIC SETTINGS, AND HYDROCARBON TRAPS OF DIVERGENT (TRANSTENSIONAL) WRENCH FAULTS

T. P. HARDING AND R. C. VIERBUCHEN
Exxon Production Research Company
P. O. Box 2189
Houston, Texas 77252-2189;

AND

NICHOLAS CHRISTIE-BLICK
Department of Geological Sciences and
Lamont-Doherty Geological Observatory
of Columbia University
Palisades, New York 10964

ABSTRACT: A divergent (transtensional) wrench fault is one along which strike-slip deformation is accompanied by a component of extension. Faulting dominates the structural style and can initiate significant subsidence and sedimentation. The divergent wrench fault differs from other types of wrench faults by having mostly normal separation on successive profiles, negative flower structures, and a different suite of associated structures. En echelon faults, most with normal separation, commonly flank the zone, and some exhibit evidence of external rotation about vertical axes and have evidence of superimposed strike slip. Flexures associated with the wrench fault are formed predominantly by vertical components of displacement, and most are drag and forced folds parallel to and adjacent to the wrench. Hydrocarbon traps can occur in fault slices within the principal strike-slip zone, at culminations of forced folds, in the flanking tilted fault blocks, and within less common en echelon folds oblique to the zone.

Divergent wrench faults occur at active plate boundaries, in extensional and contractional continental settings, and within plates far from areas of pronounced regional deformation. Along transform margins and within wrench systems, divergent wrench styles tend to develop where major strands or segments of strands bend or splay toward the orientation of associated normal faults (e.g., elements of the San Andreas system in the Mecca Hills, California), and where major strands are regionally oblique to interplate slip lines (e.g., Dead Sea transform, Middle East). The style also develops at releasing fault oversteps and fault junctions (e.g., Ridge Basin, California), and locally where crustal blocks rotate between bounding wrench faults. In extensional settings, divergent wrench faults may form within graben doglegs and oversteps (e.g., between the Rhine and Bresse grabens, northern Europe), and they may separate regions that experienced different magnitudes of extension (e.g., Andaman Sea area). Many oceanic fracture zones have divergent wrench characteristics. The style has also been recognized in magmatic arcs (e.g., the Great Sumatran fault) and in both backarc and peripheral foreland settings (e.g., Lake Basin fault zone, Montana) near convergent plate boundaries, and in intra-plate settings (e.g., Cottage Grove fault, Illinois; Scipio-Albion trend, Michigan).

INTRODUCTION AND PREVIOUS WORK

Divergent wrench zones are defined as those strike-slip systems within which the principal displacement zone and, in many cases, the adjacent associated structures are dominated by extensional characteristics. This article illustrates the structural style of divergent wrench zones and summarizes their known tectonic habitats and associated hydrocarbon traps.

Past discussions of wrench-fault styles have emphasized the contractional structures that accompany convergent strike slip (e.g., Wilcox et al., 1973; Sylvester and Smith, 1976). Several studies have shown individual structures attributable to extensional strike slip, but none has dealt with the full range of structural features that can be present. Nelson and Krausse (1981) presented the most thorough documentation, using an example from the Illinois basin where coal mining has provided unusually dense subsurface control. The results of that study are incorporated in our present work. The mechanics of oblique plate separation were discussed briefly by Harland (1971), and he first applied the term "transtension" to deformation in these zones. Wilcox et al. (1973) introduced the term "divergent wrench fault" and suggested that the structure along such faults is characterized by an increase in extensional block faulting and a decrease in en echelon folding. These authors described graben structures that were formed by divergent strike-slip

deformation of a clay-cake model. Courtillot et al. (1974) also reproduced extensional strike-slip structures with a clay-cake model of a transform fault between simulated spreading centers. Earlier, Bishop (1968) documented fault patterns in New Zealand, portions of which are now recognized as essential elements of the structural style. Harding (1983) has recently illustrated a divergent wrench fault on seismic reflection data from the Andaman Sea and has established criteria for differentiating these zones from other styles on seismic profiles (Harding, 1985). D'Onfro and Glagola (1983) have published seismic reflection profiles of another divergent wrench zone in southeast Asia.

Our overview of divergent wrench systems presented here is drawn from these previous studies, from new field work in the Mecca Hills of southern California, and from additional seismic reflection profiles.

EXTENSIONAL STRIKE-SLIP STYLES

Two important changes occur in the structures associated with wrench faults when strike slip is accompanied by significant divergence. First, lateral compression is decreased to the extent that en echelon folds and thrusts are present only diffusely or concentrated in localized settings. Second, regional extension enhances brittle deformation to the degree that the style is commonly dominated by faults.

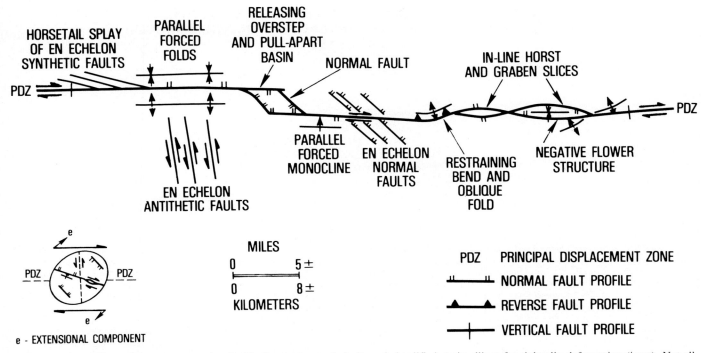

FIG. 1.—Assemblage of structures associated with divergent wrench faults and simplified strain ellipse for right-slip deformation (inset). Not all divergent wrench zones contain every feature and those structures present may be combined in different ways. The assemblage shown here is a composite of examples presented in this paper and others cited in the references.

Principal Strike-Slip Displacement Zone

A linear, throughgoing principal displacement zone is the fundamental element of the structural assemblage (PDZ in Fig. 1). This zone accounts for most of the lateral displacement between opposing blocks. Other faults in the assemblage tend to be shorter and to have smaller displacements. In some cases, the deformation consists solely of the principal displacement zone; in other instances, the zone is flanked on one or both sides by associated external structures.

Tchalenko and Ambraseys (1970) demonstrated that the principal displacement zone is composed mostly of synthetic strike-slip faults (i.e., Riedel shears; Riedel, 1929) and interconnecting P shears. Strike slip along P shears is in the same sense as along the synthetic faults, but P shears and synthetic faults step in opposite senses (Fig. 2). The prediction of structures and their orientations shown in Figure 2 is simplistic, and is most applicable to the early stages of a single tectonic event and to one wrench zone. Protracted or episodic deformation involving two or more major wrench faults can be considerably more complex. Orientations of structures in left-slip systems may be determined by viewing Figures 1 and 2 in reverse image.

The overall dip of the principal displacement zones is thought to be near-vertical at depth in many instances, and where there is definitive control, faults with normal separation dominate in transverse profiles (i.e., hanging-wall block apparently downthrown; Fig. 3). As with other types of wrench faults, the direction of fault dip may change along

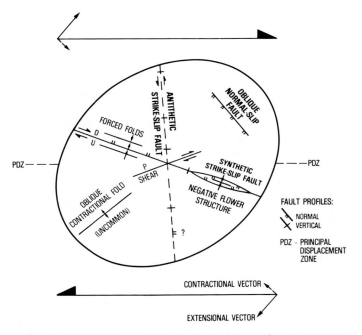

FIG. 2.—Strain ellipse with idealized orientations of main structures initiated along a right-lateral divergent wrench fault, compiled from clay-cake models and from geologic examples. Most of the structures and their orientations are present in other styles of wrench faults, but the comparative abundance of the various features differs significantly. Segments of the principal displacement zone (PDZ) that splay or bend to the right tend to have normal separation, but reverse faults can also be present depending on the way in which displacements are distributed along the PDZ.

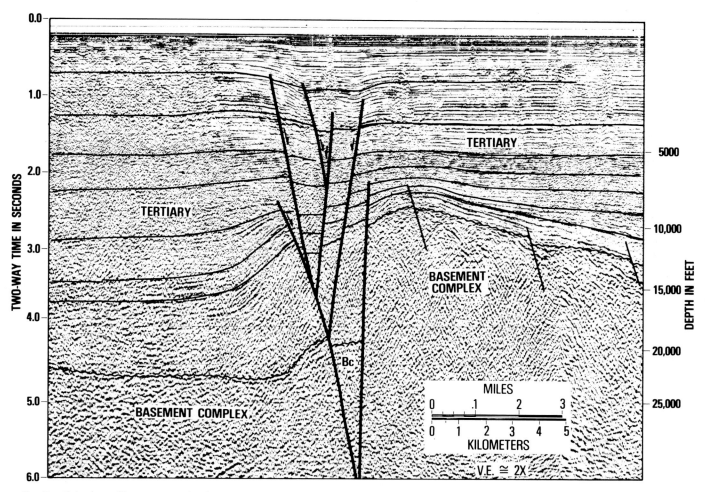

FIG. 3.—Seismic profile across a major fault zone in the Bering Sea. This area has undergone several deformations with different structural styles. Divergent wrench fault characteristics are most apparent in the younger Tertiary sediments above 2.0 seconds. Interpretation courtesy of W. A. Spindler (personal commun., 1984).

strike, and the apparently upthrown side can also change along strike or from one structural level to another. In some examples, the principal displacement zone appears to terminate abruptly at depth against a subhorizontal crustal detachment (Royden, 1985 this volume; Cheadle et al., in press).

Along some divergent wrench systems the principal displacement zone splays upward and outward at shallower levels (Emmons, 1969). The fault slices may form the core of a negative flower structure, which is defined as a shallow synform bounded by the upward and outward spreading strands of strike-slip faults that have mostly normal separation (above 2.0 seconds in Fig. 3; Harding and Lowell, 1979; D'Onfro and Glagola, 1983; Harding, 1983, 1985). The fold is termed a synform because at depth the flanks may have tilted independently and may never have been parts of an integrated flexure. Divergent wrench faults composed of a single strand may be more common, however.

Faults with reverse separation in transverse profile (i.e., hanging-wall block apparently upthrown) are present within some principal displacement zones. Contractional structures are thought to be most common where irregularities in the zone's trace cause opposing blocks to converge (e.g., restraining bend in Fig. 1; Crowell, 1974; Aydin and Nur, 1982). The contractional features commonly distinguish divergent wrench faults from normal-slip faults, which otherwise may have similar profile characteristics (Harding, 1985). However, successive intervals of normal slip and block rotation (in cross section) in areas of very great extension can over-steepen the initial normal faults so that they acquire reverse separation (Proffett, 1977). The presence of both normal and reverse separation then ceases to distinguish the two styles. Releasing bends (Crowell, 1974) and releasing oversteps (Christie-Blick and Biddle, 1985 this volume; see Fig. 1) may cause blocks to diverge and enhance the development of extensional strike-slip features. Several examples of these relationships are discussed below.

Most folds associated with the divergent wrench faults we have studied occur adjacent to and parallel with the principal displacement zone. In the absence of regional horizontal shortening, the more subtle vertical components of movement appear to be the major cause of fold formation.

The flexures resemble the forced folds that develop above normal fault blocks: anticlines or monoclinal knees adjacent to and parallel with the relatively higher side of the principal displacement zone; synclines or monoclinal ankle flexures adjacent to and parallel with the edge of the apparently down-dropped block. These flexures commonly bound negative flower structures at shallow levels (e.g., above 2.0 seconds in Fig. 3). Where the principal displacement zone consists of a single strand, a single monoclinal knee or ankle may face toward the relatively down-dropped block, or both high- and low-side flexures may be absent.

External Associated Structures

Deformation of the terrane flanking the principal displacement zone may be negligible. The regional dip in several examples continues unchanged across the zone, and external structures are diffuse or absent (see shallower reflections on Fig. 3). In other examples, faults oblique to the trace of the principal displacement zone are the most common type of associated structure.

External Faults.—The external faults, where present, occur in belts along one or both sides of the throughgoing wrench and are en echelon or oblique to the trend of the zone: left-stepping for right-lateral systems (Fig. 1) and right-stepping for left-lateral zones (Fig. 4). The strike of the external faults ranges from slightly oblique to the principal displacement zone (e.g., synthetic or Riedel strike-slip faults; Riedel, 1929) to nearly transverse (e.g., antithetic or conjugate Riedel faults; Fig. 2). The external fault sets commonly contain several fault orientations, but most external faults tend to be subparallel to the direction predicted for normal faults by our simple strain ellipse. In addition, where control is sufficient for determination, most of the external faults are characterized by normal separation and steep dips.

The fault set of the Lake Basin fault zone is an example of a distinct right-stepping en echelon pattern (Fig. 4). There is no principal displacement zone at the surface, but a left-lateral divergent wrench fault is believed to be present at depth (Chamberlain, 1919; Alpha and Fanshawe, 1954; Smith, 1965). A segment of the Dasht-e Bayaz earthquake system studied by Tchalenko and Ambraseys (1970) in central Iran similarly consists solely of en echelon fractures. The relationships there are unusually clear and these authors interpreted the pattern as representative of an initial stage in the development of the throughgoing principal displacement zone.

Individual dislocations along the Lake Basin fault zone strike at angles of 30° to 70° (most are between 45° and 60°) to the overall trend and have been interpreted as normal faults (Smith, 1965). Seismic control near the west end of the zone demonstrates steep fault dips with normal and, possibly, several reverse separations (Fig. 5). The variety of trends and separations probably results from the interaction of two factors: (1) pre-deformational anisotropies, and (2) external rotation and reactivation of the faults.

External rotation may significantly alter the structural pattern, and the style is, therefore, best expressed in the early stages of deformation. Attempts to identify it in later stages on the basis of observed or reconstructed fault patterns may be misleading. The effects of rotation have been demonstrated on a small scale by means of earthquake fractures in southern California (Terres and Sylvester, 1981; Fig. 6A in Christie-Blick and Biddle, 1985 this volume). There, clockwise external rotation of a planar anisotropy oriented oblique to the right-slip Imperial fault occurred during the earthquake of 15 October 1979 (see inset map in Fig. 6 for location of Imperial fault). The earthquake fractures approximate en echelon normal faults, and antithetic strike slip was superimposed on them by the rotation. Both reverse and normal separations are present. The sense of external rotation matched the sense of strike slip on the Imperial fault, and imposed the antithetic displacements in a mechanism roughly analogous to the slip generated by inclining a row of dominoes. Such a mechanism has been discussed by Freund (1970), and also by Luyendyk et al. (1980) to explain how paleomagnetically determined block rotations of 70° to 80° may be accommodated in the Miocene tectonics of southern California.

External Associated Folds.—Folds outside the principal displacement zone are generally sparse. Where developed, they are oblique or en echelon to the master fault: left-step-

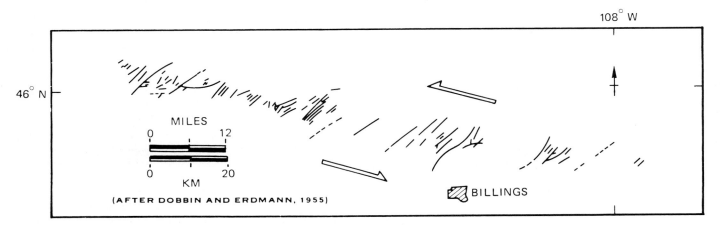

FIG. 4.—Surface map of Lake Basin fault zone, Montana. Orientations of most en echelon faults correspond with the idealized normal-fault direction in Figure 2 for left-lateral wrench faulting, but some faults are more nearly transverse and several are less oblique than predicted.

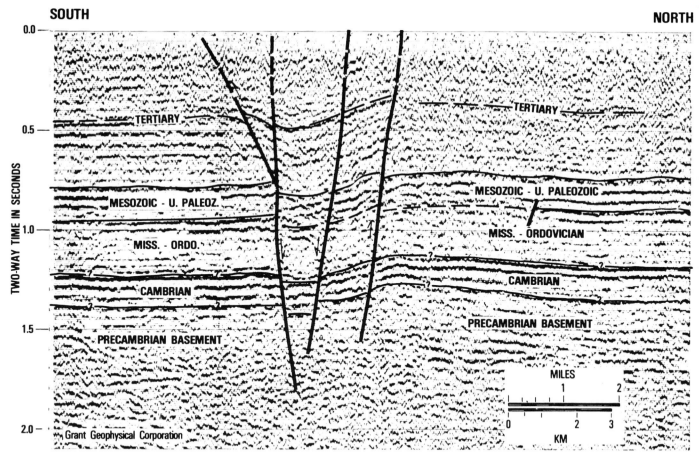

FIG. 5.—Seismic profile near west end of the Lake Basin fault zone. Zone includes both faults with normal separation and faults with reverse separation. Most faults dip steeply and they may converge at greater depth into a single, throughgoing principal displacement zone. The deformation has been dated as post-Paleocene and possibly post-Eocene by Alpha and Fanshawe (1954). Profile courtesy of Grant Geophysical Corporation, 1983; interpretation courtesy of G. H. Weisser (personal commun., 1983).

ping for left-lateral faults and right-stepping (Fig. 1) for right-lateral faults. Where we have observed external folds, they result from localized convergent deformation, for example, at a restraining bend (Fig. 1). They may also form during an earlier, convergent phase of strike slip that had the same sense of offset as the subsequent divergent phase.

Other folds and faults may be inherited from earlier deformation unrelated to strike slip or may be superimposed by later tectonic events. Some of these structures could appear by coincidence to be congruent with the structural relationships of a wrench zone, and recognition of this disparity is a prerequisite for the correct analysis and identification of style.

DIVERGENT WRENCH FAULTS OF THE MECCA HILLS

Plate-Tectonic Setting of the Mecca Hills Faults

The Mecca Hills faults are three subparallel, right-slip faults that splay northward from the San Andreas wrench system along the northeast margin of the Salton Trough (Fig. 6). The divergent wrench style consists here almost solely of the principal displacement zone. The present tectonic framework originated around 4.5 Ma and is considered to be a northwestward continuation of the Gulf of California ridge-and-transform system (see inset map in Fig. 6; Crowell and Sylvester, 1979; Crowell, 1981). The subsurface axis of the trough lies within a releasing overstep (Fig. 1) between the San Andreas and Imperial faults. Their side-stepped pattern is indicated by earthquake epicenters and the surface traces of active faults (Rockwell and Sylvester, 1979). Away from the San Andreas and Imperial faults, the Salton Trough is flanked by high basement terranes that form the limbs of a broad arch (J. C. Crowell, personal commun., 1979).

On the northeast flank of the basin, convergent wrench faults dominate the structural style of the San Andreas and other major northwest-striking faults (Sylvester and Smith, 1976). The strike of the Mecca Hills splay faults is midway between that of the convergent strands and the north-south orientation predicted for normal faults associated with the San Andreas system (compare orientations with the strain ellipse of Fig. 2). Structural features typical of divergent wrench faults have been observed along three of the splays: the Painted Canyon, Eagle Canyon, and Hidden Springs

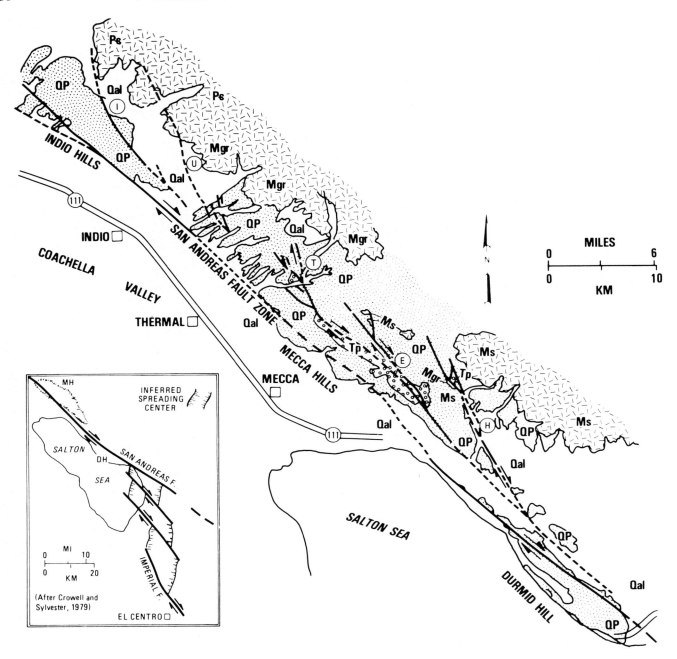

RECOGNIZED AND POTENTIAL DIVERGENT WRENCH FAULTS

(H) HIDDEN SPRINGS FAULT (FIG. 8)

(E) NORTHERN SEGMENT OF
EAGLE CANYON FAULT (FIG. 7)

(T) THERMAL CANYON SPLAY OF
PAINTED CANYON FAULT

(U) UNNAMED FAULT

(I) INDIO HILLS FAULT

Qal   RECENT ALLUVIUM

QP   PLIOCENE-PLEISTOCENE
NONMARINE SEDIMENTS

Tp   PLIOCENE NONMARINE SEDIMENTS

Mgr   MESOZOIC (?) GRANITIC ROCKS

Ms   MESOZOIC OROCOPIA SCHIST

Pϵ PRECAMBRIAN COMPLEX

(GENERALIZED AFTER ROGERS, 1965, AND JENNINGS, 1967)

faults (Fig. 6). The investigation of two faults to the north with similar orientations is incomplete.

Evidence of Right Slip on the Mecca Hills Faults

Fine-grained Pliocene rocks adjacent to the Painted Canyon fault are offset approximately 3 km from their source (A. G. Sylvester, personal commun., 1979). Along another portion of this fault, Hays (1957) reported distributions of rock colors that suggest about 0.8 km of right slip. In addition, the en echelon pattern of folds associated with the southeastern segments of the Painted Canyon and Eagle Canyon faults suggests right slip.

The distribution of Pliocene sedimentary rocks along the Eagle Canyon fault indicates right slip of 1.6 km (Hays, 1957). Along the southern segment of this fault the north-western boundary of an en echelon fold set lies approximately 1.3 km farther southeast than its position across the fault (Fig. 7). The folds are thought to have formed mainly during accumulation of the right slip (Sylvester and Smith, 1976), but they could have developed independently, which would negate their validity as offset evidence.

Along the Hidden Springs fault, the northeast limit of a distinctive basal breccia is apparently offset 1.2 km right-laterally, but according to Hays (1957), irregularities in the original depositional pattern could account for much of this displacement. Near Shaver's Well, displacement on the Hidden Springs fault dies out rapidly into horsetail splays (Fig. 8). Here, tongues of a schist breccia are offset as much as 45 m right-laterally on individual faults (Hays, 1957).

Structural Styles of the Mecca Hills Faults

Both the Painted Canyon and the Eagle Canyon faults have convergent strike-slip styles along their southern and central segments where they are closest to the San Andreas fault zone (Fig. 6). Along the central segment of the Eagle Canyon fault, beds are steeply dipping to overturned, the fault has a vertical or upthrust profile (segment with saw-tooth symbols in Fig. 7), and drag folds are tight where present. Farther south, the limbs of tight, en echelon folds are offset across the fault, and it is difficult to differentiate these external structures from features of the principal displacement zone.

Deformation along the northern, divergent trace of the Eagle Canyon fault (segment with double tick symbol in Fig. 7) is much less intense and less complex than that to the south. A major normal fault, oriented north-south within the extensional sector predicted for the regional right-slip deformation, is the main external structure in this area (left of mid-point of Fig. 7). Beds dip 5° to 15° basinward across the area, with only minimal disruption. The principal displacement zone consists of a single strand with normal separation (Fig. 9). At the locality of Figure 9 (shown in Fig.

7), the zone's dip appears to flatten with depth, but this may be only a local irregularity. A gentle upturn of beds within the hanging-wall block locally reverses the basinward dip and forms a low-side syncline (Fig. 10). The syncline is thought to reflect the vertical component of the deformation. An accompanying high-side, forced anticline lies opposite the syncline only 0.5 km to the southeast (Fig. 7), and drag folds that demonstrate shortening are noticeably absent within the principal displacement zone. To the northwest, the Eagle Canyon fault terminates into a large horsetail structure composed of closely spaced, subvertical to vertical faults. There is little or no folding or other distortion of the regional dip of beds within the horsetail.

The Thermal Canyon fault (T in Fig. 6) is the north-northwest-striking splay of the Painted Canyon fault and is the segment of that system that has a divergent style. The contractional structures that dominate the southern portions of the Painted Canyon fault appear to be absent, and the principal displacement zone has normal separation. Horizontal slickenside striae are the only structural indication of strike slip.

The Hidden Springs fault (H in Fig. 6) is consistently oriented within the extensional sector of the regional right-slip deformation. The principal displacement zone of this fault generally dips steeply to vertically, and along much of its length has an extensional strike-slip style (Fig. 8). The zone terminates northward into a complex horsetail composed of several major fault strands. In a tributary canyon just south of Shaver's Well, the main faults bound narrow, steeply dipping slices. One strand splays upward and outward, forming a fault architecture similar to a negative flower structure (Fig. 11). Other subvertical faults within the canyon abut the throughgoing north-northwest-striking faults at high angles. Both the main strands and the external faults have abundant horizontal slickenside striae, but there is little or no evidence of horizontal shortening within the offset beds.

South-southeast of the horsetail, the strand of the Hidden Springs fault illustrated in Figure 11 turns eastward into the contractional sector of the deformation (i.e., turns subparallel to the orientation predicted for contractional structures associated with right slip; compare with Fig. 2), and forms a restraining bend (segment with sawtooth symbols in Fig. 8). The profile of the principal displacement zone opposite this bend is that of a downward-steepening reverse fault, and beds immediately adjacent to the fault are shortened by drag folding (Fig. 12).

COTTAGE GROVE FAULT SYSTEM

Plate-Tectonic Setting of the Cottage Grove Fault System

The structural style of the Cottage Grove fault system is dominated by external en echelon faults and has been elab-

---

FIG. 6.—Late Cenozoic tectonic framework of Salton Trough, California, in the vicinity of the Mecca Hills (inset), and location of divergent wrench faults identified along the northeast side of the San Andreas fault. Lack of continuous outcrops adjacent to the San Andreas fault makes it difficult to demonstrate a direct connection with the divergent wrench faults, but a splay configuration is inferred from the tectonic relationships. Abbreviations in inset map: MH, Mecca Hills; DH, Durmid Hill.

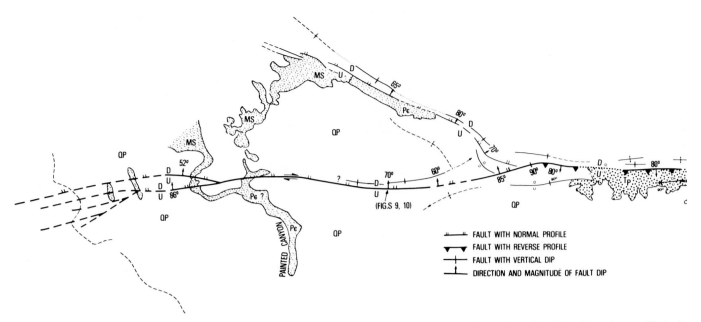

FIG. 7.—Generalized surface geologic map of the Eagle Canyon fault, Mecca Hills. Dating and correlation of stratigraphic units modified after

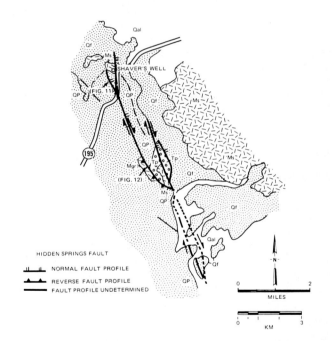

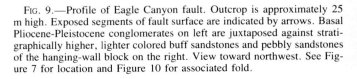

FIG. 8.—Generalized geologic map of northern segment of the Hidden Springs fault, Mecca Hills (after Jennings, 1967). Qf, Pleistocene fanglomerate; see Figure 6 for explanation of other symbols and for setting of fault.

FIG. 9.—Profile of Eagle Canyon fault. Outcrop is approximately 25 m high. Exposed segments of fault surface are indicated by arrows. Basal Pliocene-Pleistocene conglomerates on left are juxtaposed against stratigraphically higher, lighter colored buff sandstones and pebbly sandstones of the hanging-wall block on the right. View toward northwest. See Figure 7 for location and Figure 10 for associated fold.

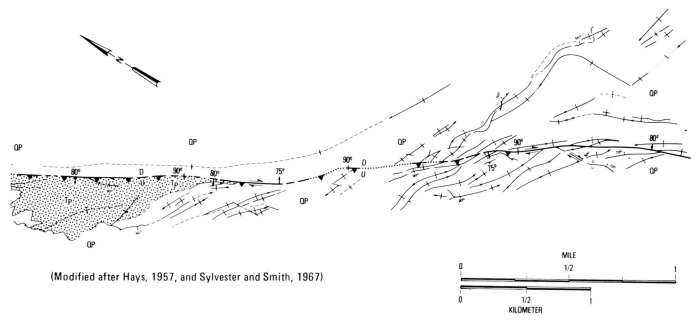

(Modified after Hays, 1957, and Sylvester and Smith, 1967)

Jennings (1967). See Figure 6 for explanation of symbols for rock units and setting of fault.

FIG. 10.—Drag syncline on eastern, relatively down-dropped side (foreground) of northern segment of the Eagle Canyon fault. Position of fault is indicated by arrow. Visible portion of cliff face is approximately 20 m high. Fault dips toward viewer and follows face of darker colored outcrop from left margin of photograph. View toward west-northwest. See Figure 7 for location and Figure 9 for profile of fault zone.

FIG. 11.—Portion of a divergent horsetail at the northwest termination of the Hidden Springs fault. Exposure of faults is 12 to 15 m high. Splays are closely spaced and steeply dipping; most have horizontal slickenside striae here or in adjacent outcrops. Displaced Pliocene-Pleistocene gravels exhibit little or no folding. Faults merge downward into single, subvertical zone at right base of photo. View toward south-southeast. See Figure 8 for location. T, displacement toward viewer; A, away from viewer.

orately documented by Nelson and Krausse (1981; see their illustrations for further details of the style). The system trends east-west for possibly 110 km across the southwest flank of the Illinois Basin (Fig. 13). The basin is a gentle, intracratonic downwarp. The broad arches and domes at the periphery of the downwarp were formed by mid-Ordovician time, except the Pascola arch, which is thought to have developed in the late Paleozoic (Buschbach and Atherton, 1979; Sloss, 1979). The basement warps and adjacent forelands (Appalachian and Arkoma Basins in Fig. 13) separated the Illinois Basin from orogenic belts that were active on the southeast (Appalachian thrust front) and southwest during the late Paleozoic.

The origin of the intracratonic sag is uncertain, but it may be related to thermal contraction of lithosphere, extended during the formation of an earlier graben system (Ervin and McGinnis, 1975; Sloss, 1979). Aeromagnetic and gravity surveys have delineated a deeply buried northeast-trending trough, the Reelfoot rift of Ervin and McGinnis (1975) or the Mississippi Valley graben of Kane et al. (1981). The

northeasternmost segment of this graben corresponds closely with the southward plunge of the Cambrian-Ordovician precursor of the Illinois Basin (Fig. 1 in Sloss, 1979). The Cottage Grove fault system may have developed along a segment of the older graben system (Heyl and Brock, 1961; Buschbach and Atherton, 1979).

Precise dating of the age of faulting is not possible because of the absence of stratigraphic section between the deformed Pennsylvanian beds and the surficial Pleistocene cover. Igneous dikes in the nearby fluorspar district (4 in Fig. 13) have been dated as 267 ± 20 Ma, or Early Permian (Zartman et al., 1967), and are similar compositionally to dikes intruded into extensional faults of the Cottage Grove system. Nelson and Krausse (1981) accept this as evidence of a late Paleozoic age for the wrench faulting.

Evidence of Right Slip on the Cottage Grove Fault System

Clark and Royds (1948) were the first of a number of authors (e.g., Heyl and Brock, 1961; Wilcox et al., 1973) to suggest that the Cottage Grove system contains structures that resulted from strike-slip deformation. Nelson and Krausse (1981) have recently summarized this structural evidence: (1) The orientations of structural elements generally repeat those predicted for right slip by the strain-ellipse and clay-cake models for right slip (compare Figs. 2 and 14; Wilcox et al., 1973). (2) The fault sets resemble the fracture systems developed along seismically active faults that have explicit evidence of lateral displacements (e.g., Fig. 8 in Tchalenko and Ambraseys, 1970). (3) The fault pattern and elements of the fold pattern are also similar to zones that have geologic evidence of historic strike slip (e.g., Bishop, 1967).

The boundaries of a Pennsylvanian stream deposit suggest as much as 1.6 km of right slip, but other channels limit the offset to a smaller amount. Nelson and Krausse (1981) have concluded that the maximum lateral displacement is on the order of several hundred meters.

Characteristics of the Principal Strike-Slip Displacement Zone

The principal displacement zone includes the main faults of the Cottage Grove fault system and has structural char-

acteristics typical of wrench zones. These are (1) subvertical faults, (2) inconsistent apparent upthrown side, (3) interchanging normal and reverse separations across the zone as a whole, and (4) narrow, steeply dipping, in-line horst and graben slices that form a braided swath and have normal and reverse separations on individual faults (Nelson and Krausse, 1981). Nelson and Krausse describe the master zone as typically tens to hundreds of meters wide, with vertical displacements as great as 60 m.

Discontinuities occur in the major strands of the principal displacement zone and are occupied by considerably shorter, northwest-striking en echelon faults (Fig. 15). These en echelon faults are steeply dipping, and most have normal separations. Some of the faults have horizontal slickenside striae and reverse separations that are interpreted by Nelson and Krausse (1981) as consequences of oblique slip. These authors, in an interpretation similar to ours and others (e.g., Tchalenko and Ambraseys, 1970), have interpreted the structure of the discontinuities as characteristic of an early stage in the development of the Cottage Grove system.

Folds associated with the known extent of the system lie immediately adjacent to the upthrown side of the principal displacement zone, and most trend subparallel with the zone (Fig. 16). The zone-parallel folds verge toward the appar-

FIG. 12.—Steep upthrust profile at restraining bend in trace of the Hidden Springs fault. Fault bounds a large block of Mesozoic(?) granite on the left. Exposed segment is approximately 6 m high. Pliocene-Pleistocene sandstones and gravels in footwall on right are dragged steeply upward in apparent response to a vertical, convergent component of displacement. View toward the northwest. See Figure 8 for location. T, displacement toward viewer; A, away from viewer.

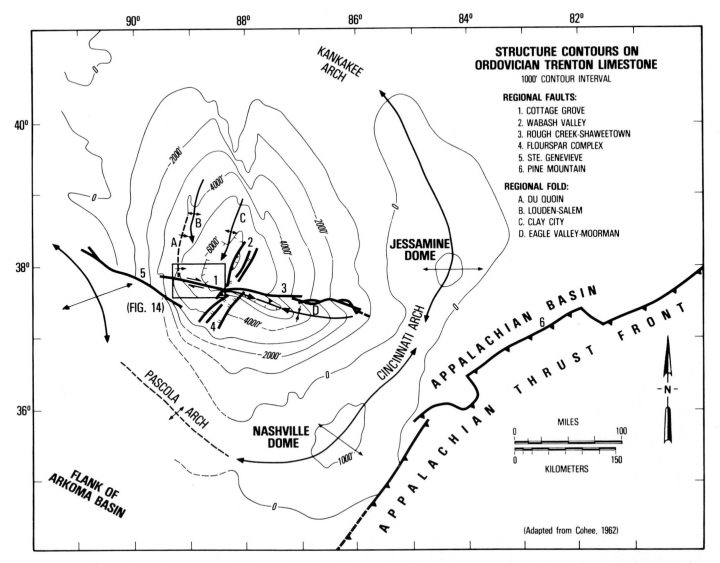

FIG. 13.—Intraplate tectonic setting of the Illinois Basin and Cottage Grove fault system (fault 1). Peripheral arches and flanks of foreland basins (Arkoma Basin, Appalachian Basin, and Black Warrior Basin south of Nashville Dome) separate the Illinois Basin from the convergent margin structures to the east (e.g., Appalachian Thrust Front) and south (south of Arkoma Basin).

ent down-dropped block (Figs. 15, 16). Nelson and Krausse (1981, p. 56) interpreted the structures as drape folds caused by vertical components of the deformation. Two small anticlines adjacent to the eastern portion of the principal displacement zone have a fundamentally different pattern. They lie oblique to the trend in a right-stepping en echelon pattern that is compatible with the right-slip deformation.

Characteristics of the External Structures

Left-stepping en echelon faults are present within a belt 5 to 16 km wide along both sides of the entire mapped trace of the principal displacement zone (Fig. 14). Individual faults are as long as 11 km and have a maximum vertical separation of 20 m (Nelson and Krausse, 1981). The orientation of these faults relative to the principal displacement zone varies from that predicted for en echelon normal faults to the more nearly transverse trend anticipated for antithetic

strike-slip faults (compare Figs. 2 and 14). Most dislocations are thought to be steeply dipping (60° to 90°) normal faults by Nelson and Krausse, and they consider faults on opposite sides of the principal displacement zone to have developed independently.

Nelson and Krausse (1981) also report that some of the external faults show evidence of oblique slip, such as oblique and horizontal slickenside striae, and changes in dip direction, separation sense (normal and reverse), and apparent upthrown side on individual dislocations. Strike slip may be dominant on a number of faults, and at several localities, reverse and normal separations occur on adjacent, parallel strands (Fig. 17). We suggest that, similar to the Lake Basin fault zone, external clockwise rotation of the blocks bounded by the en echelon faults occurred during wrench deformation and could have selectively rejuvenated some of the faults in a reverse- and strike-slip sense. In addition, other breaks may have originated as antithetic strike-slip faults.

Several anticlines are west of the principal displacement zone and trend oblique to its projection (Bremen anticline, Wine Hill dome, and Campbell Hill anticline in Fig. 16).

Their location and right-stepping en echelon map pattern suggest an extension of the zone of right-slip deformation. The folding may have been accentuated by shortening at a restraining bend formed where the system's inferred continuation trends more nearly due west.

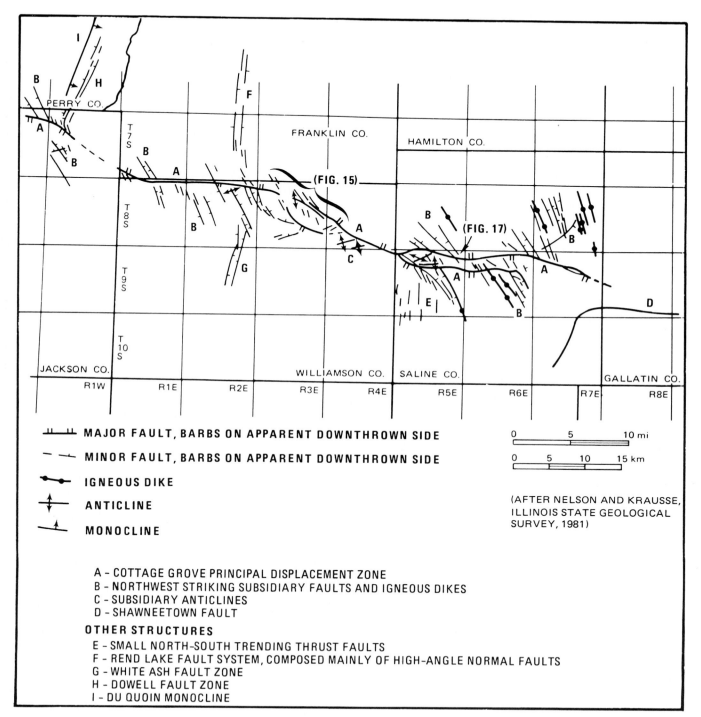

FIG. 14.—Main structural elements of the Cottage Grove fault system (subsidiary anticlines are incompletely shown; see Fig. 16). Structures D through I are considered not part of the Cottage Grove fault system by Nelson and Krausse (1981), and several are known to be of a different age. See Figure 13 for regional setting.

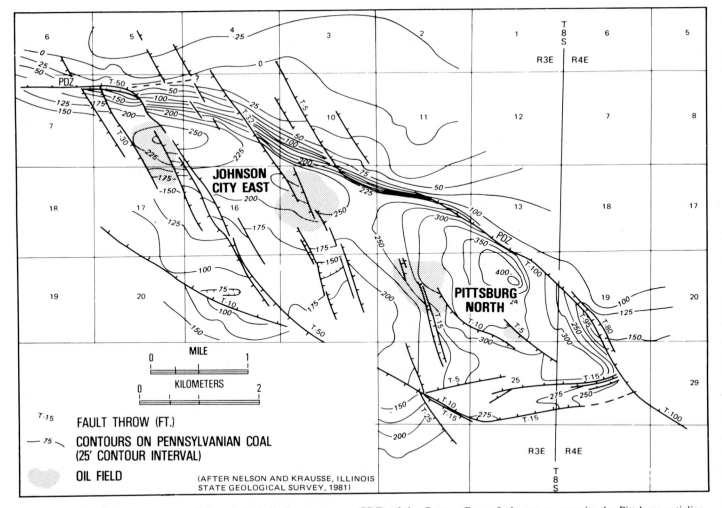

FIG. 15.—Detailed structure map of the principal displacement zone (PDZ) of the Cottage Grove fault system opposite the Pittsburg anticline (located in Figs. 14, 16). A gap in the PDZ occurs in Sections 8, 9, 10, 14, and 15 and is replaced by elements of an en echelon fault set. Other north-northwest-striking faults are external to the zone. The fold lies along the apparent high side of a projection of the PDZ and verges toward the relatively down-dropped, northeastern block.

ANDAMAN SEA FAULT

Plate-Tectonic Setting of the Andaman Sea Fault

The Andaman Sea fault illustrates the profile characteristics of divergent wrench faults over a much greater range of depth than the Mecca Hills and Cottage Grove examples and further demonstrates the manifestation of the style in seismic reflection data. The wrench fault lies within the Andaman Sea region, a marginal basin bounded on the west by the Andaman-Nicobar Ridge and subduction zone, and on the east by the magmatic arc terranes of the Malay Peninsula (Fig. 18; C.C.O.P., 1981). On the eastern margin of the sea, north-striking normal faults were active from the Oligocene to the early Miocene (Curray et al., 1979). In the central portion of the sea, magnetic anomalies in oceanic crust demonstrate that seafloor spreading has been occurring there since at least 13.5 Ma (Lawver and Curray, 1981). Geophysical data and the trend of the central rift valley indicate that segments of the spreading ridge trend east-northeast to north-northeast and are linked by right-slip transform faults that strike north-northwest (Fig. 18).

Evidence for Right Slip on the Andaman Sea Fault

The Andaman Sea fault trends north-northwest subparallel to right-slip faults of similar age elsewhere in the region (Fig. 19; Eguchi et al., 1979). The zone has a number of features characteristic of wrench faults: (1) a relatively straight, throughgoing trace; (2) changes in the apparent upthrown block that occur along strike (Fig. 19) and with depth (Fig. 20); (3) flanking folds that have a right-stepping en echelon pattern near the northern end of the fault; and (4) a possible restraining bend suggested by reverse-separation faults at the zone's southern end. Several of the en echelon folds appear to be offset 2.5 to 3 km right-laterally (folds A-A′ and B-B′ in Fig. 19).

Structural Style of the Andaman Sea Region

The structural assemblage of the Andaman Sea region repeats elements present in the above examples (compare Figs. 14, 16, and 19) and corroborates the observation that divergent strike slip has a unique and definable structural style. Deformation is characterized by a single major fault flanked by a set of discontinuous, oblique faults with normal separation (Fig. 19, and right side of Fig. 20).

The external fault set is asymmetric; dislocations are more numerous east of the principal displacement zone. Their profile characteristics are relatively simple and are repeated from one external fault to the next. The features suggestive of external rotation, cited for the Lake Basin and Cottage Grove zones, are absent. En echelon folds occur only near an apparent restraining bend formed where the wrench fault turns toward a more northwesterly course (north of profile 1 in Fig. 19). Elsewhere there are few if any structures

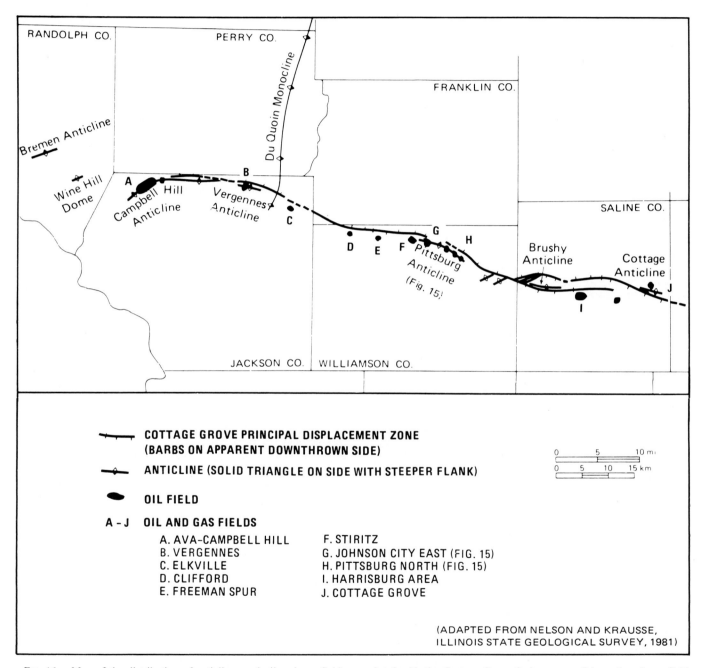

FIG. 16.—Map of the distribution of anticlines and oil and gas fields associated with the Cottage Grove fault system. Other oil and gas fields external to the system (most numerous to the north) are not shown. The inception of the Du Quoin monocline predates strike-slip deformation, and this structure is not considered to be part of the Cottage Grove system (Nelson and Krausse, 1981).

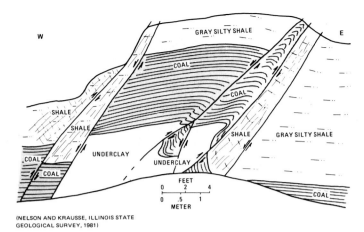

(NELSON AND KRAUSSE, ILLINOIS STATE
GEOLOGICAL SURVEY, 1981)

Fig. 17.—Detailed cross section of northwest-striking en echelon faults of the Cottage Grove fault system. Individual strands with reverse separation are tentatively interpreted as inverted normal faults that have undergone external rotation about a vertical axis. Other elements have retained at least part of their original normal separation. Beds within the fault zone are tightly folded. See Figure 14 for location.

indicative of regional shortening, and reflections from Neogene strata have consistent, low dips across the region (Figs. 20, 21).

The principal displacement zone is more complex and variable than the external normal faults (compare faults in Figs. 20 and 21). Its profile characteristics resemble those illustrated in Figures 3 and 5: (1) Faults splay upward and outward and most have normal separations; and (2) monoclinal flexures at the shoulders of the zone verge inward toward a central, down-dropped slice and bound a shallow synform (Fig. 21). Similar features characterize the seismic reflection profiles of another wrench fault in southeast Asia described by D'Onfro and Glagola (1983). A negative flower structure is present on profile 2 and includes a subsidiary fault with a reverse separation within Miocene strata on the zone's east side (Fig. 21). The reverse fault developed at a restraining bend within the zone's southern segment (Fig. 19). The shallow synform is not developed at profile 1, and a change in the dip separation at depth is the most distinctive expression of wrench faulting here (Fig. 20). The apparent upthrown block changes from the west side above 3.3 seconds to the east side below 3.5 seconds. The top of the basement in the western block is only approximated from regional control, but its displacement to deeper levels is corroborated by the presence of stratal reflections considerably below 3.5 seconds.

PLATE–TECTONIC SETTINGS FOR EXTENSIONAL STRIKE SLIP

Divergent wrench faults occur in a wide variety of plate-tectonic settings, including transform, divergent, and convergent plate boundaries; extensional and contractional continental settings; and within plates far from areas of pronounced regional deformation (Fig. 22). Evidence for extensional strike slip is least common, however, in regions dominated by crustal shortening, or where there is minimal basement-involved deformation (e.g., passive continental margins). The zones and settings evolve through time (not shown in the instantaneous sketches of Fig. 22), and the faults may exhibit extensional strike-slip during only part of their history. Divergent segments may occur within systems that elsewhere have convergent styles (e.g., Fig. 22 a, b) or may dominate the entire length of the wrench zone (e.g., Fig. 22 d, e).

Transform Plate Boundary and Intra-Wrench Settings

Divergent wrench styles tend to develop within transform or wrench systems (1) where the principal displacement zone bends or splays into the extensional sector of regional deformation (Fig. 22a); (2) at releasing fault oversteps and fault junctions (Fig. 22b); and (3) locally where crustal blocks rotate between bounding wrench faults (Fig. 22c).

The geometrically simplest habitat of divergent wrench style is a releasing splay or bend in a single wrench fault. A releasing splay develops where splay faults, when viewed down the trace of a right-slip zone, diverge away from the viewer on the right side of the master fault or converge with the zone in a direction away from the viewer on the left side (Fig. 22a). The converse is true for left-slip zones. The Mecca Hills faults are examples of right-slip releasing splays (Fig. 6).

A similar pattern of orientation change defines a releasing bend. The western segment on the Bocono fault adjacent to the La González Basin, northwestern Venezuela, is an example (Fig. 23a). Quaternary normal separation is on the order of several kilometers, and right slip is indicated by displaced drainages and stratigraphy, offset moraines, and seismicity (see references in Schubert, 1980). Other examples of releasing bends have been summarized recently by Mann et al. (1983).

The entire wrench system may have a divergent wrench style where a straight fault is consistently oblique to the regional interplate slip lines. The principal displacement zones of the Dead Sea transform south of Lebanon, for example, are predominantly of divergent wrench style judging from descriptions of surface features by Freund et al. (1970), Schulman and Bartov (1978), Garfunkel (1981), Eyal and Reches (1983), and Manspeizer (1985 this volume). Restoration of the 105 km of left slip documented along this zone includes a clockwise rotation of the eastern block (Figs. 4b to 9b in Freund et al., 1970), suggesting an oblique divergent plate motion of several degrees (see also Garfunkel, 1981). Plate-tectonic analyses of the Red Sea – Gulf of Aden area also imply extensional strike-slip along the Dead Sea transform (e.g., Le Pichon and Francheteau, 1978; Cochran, 1983; Mann et al., 1983). The inception of the folds in Israel that are oriented oblique to the Dead Sea transform, and interpreted by Freund (1965), Vroman (1967), and Wilcox et al. (1973) to be the result of wrench-related deformation, significantly predates the initiation of strike slip (Eyal and Reches, 1983). In addition, Eyal and Reches (1983) have determined that the stress field related to the Dead Sea transform is different from the stress field involved in the development of the folds. The folds are therefore not considered an essential element of the divergent wrench style south of Lebanon.

Releasing oversteps and releasing fault junctions (Fig. 22b) also have patterns that result in the oblique separation of blocks, and they are similar to those defined for releasing splays and bends. Indeed, Mann et al. (1983) argued that oversteps generally evolve from bends in an earlier throughgoing fault, although other origins are possible (Aydin and Nur, 1985 this volume). A possible example of a releasing overstep characterized by divergent wrench style is the Hula graben of the northern Dead Sea rift (Fig. 23c;

Freund et al., 1968). Numerous other examples of pull-apart basins are cited by Aydin and Nur (1982) and by Mann et al. (1983), but not all such basins are associated with divergent wrench faults. The southern San Andreas fault opposite the Salton trough, for example, is dominated by a convergent wrench fault style (Sylvester and Smith, 1976).

Extension occurs at releasing fault junctions in the manner discussed by Kingma (1958), Lensen (1959), and Crowell (1974). At the junction that bounds the Ridge Basin (Fig.

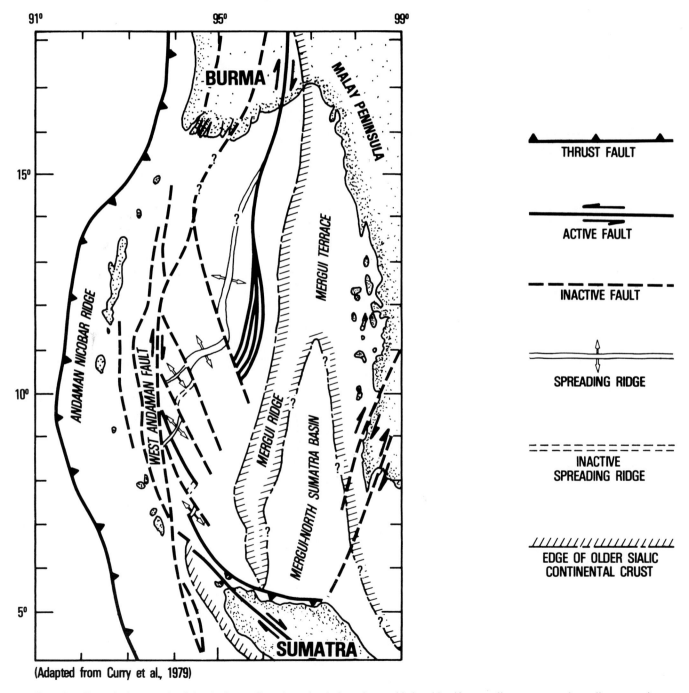

(Adapted from Curry et al., 1979)

FIG. 18.—Tectonic framework of the Andaman Sea. Arrowheads have been added to identify spreading centers and are diagrammatic.

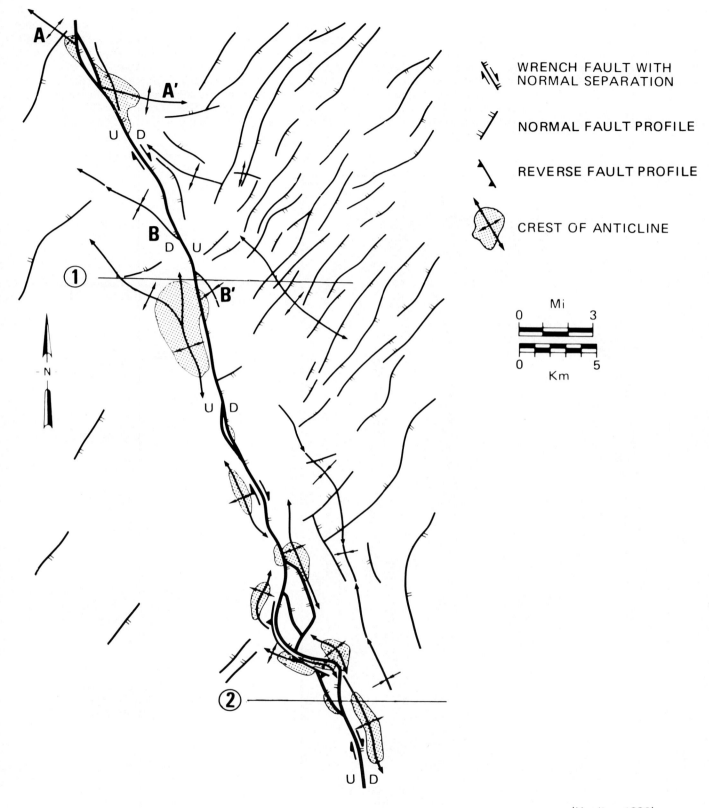

FIG. 19.—Tectonic map of structures mapped on a lower Miocene reflection and position of the Andaman Sea seismic profiles. A–A' and B–B' identify possibly correlative folds that appear to be offset in a right-lateral sense.

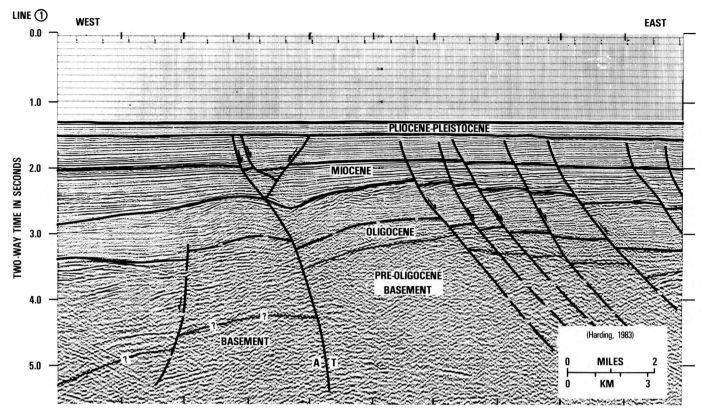

FIG. 20.—Interpreted seismic profile 1 across the Andaman Sea wrench fault. Deeper stratal reflections on west side of wrench fault (left center) indicate a change in the apparently down-dropped block at depth. Obliquely striking normal faults dip toward the right on eastern half of profile. See Figure 19 for structural setting of line. T, displacement toward viewer; A, displacement away from viewer.

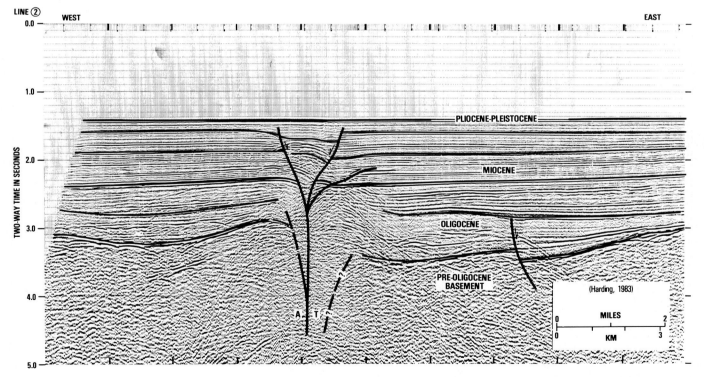

FIG. 21.—Interpreted seismic profile 2 across the Andaman Sea wrench fault. This profile illustrates the characteristics of a negative flower structure and changes in the development of the fault zone that occur along strike from profile 1 (compare with Fig. 20). See Figure 19 for structural setting of line. T, displacement toward viewer; A, displacement away from viewer.

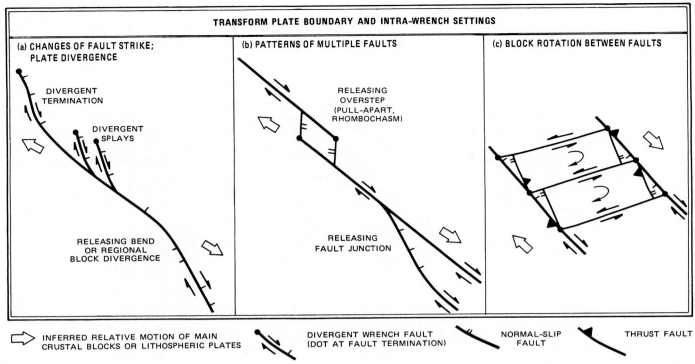

TRANSFORM PLATE BOUNDARY AND INTRA-WRENCH SETTINGS

(a) CHANGES OF FAULT STRIKE; PLATE DIVERGENCE

DIVERGENT TERMINATION

DIVERGENT SPLAYS

RELEASING BEND OR REGIONAL BLOCK DIVERGENCE

(b) PATTERNS OF MULTIPLE FAULTS

RELEASING OVERSTEP (PULL-APART, RHOMBOCHASM)

RELEASING FAULT JUNCTION

(c) BLOCK ROTATION BETWEEN FAULTS

INFERRED RELATIVE MOTION OF MAIN CRUSTAL BLOCKS OR LITHOSPHERIC PLATES

DIVERGENT WRENCH FAULT (DOT AT FAULT TERMINATION)

NORMAL-SLIP FAULT

THRUST FAULT

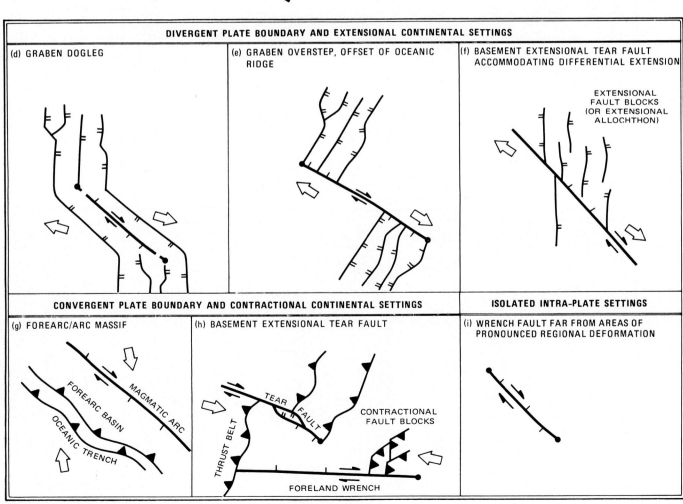

DIVERGENT PLATE BOUNDARY AND EXTENSIONAL CONTINENTAL SETTINGS

(d) GRABEN DOGLEG

(e) GRABEN OVERSTEP, OFFSET OF OCEANIC RIDGE

(f) BASEMENT EXTENSIONAL TEAR FAULT ACCOMMODATING DIFFERENTIAL EXTENSION

EXTENSIONAL FAULT BLOCKS (OR EXTENSIONAL ALLOCHTHON)

CONVERGENT PLATE BOUNDARY AND CONTRACTIONAL CONTINENTAL SETTINGS

ISOLATED INTRA-PLATE SETTINGS

(g) FOREARC/ARC MASSIF

FOREARC BASIN

MAGMATIC ARC

OCEANIC TRENCH

(h) BASEMENT EXTENSIONAL TEAR FAULT

THRUST BELT

TEAR FAULT

CONTRACTIONAL FAULT BLOCKS

FORELAND WRENCH

(i) WRENCH FAULT FAR FROM AREAS OF PRONOUNCED REGIONAL DEFORMATION

23b), the San Gabriel fault was active from about 12 Ma to 5 Ma and has predominantly normal separation (Crowell, 1982). However, the adjacent Liebre fault, active from 8 Ma to between 6 and 5 Ma and approximately parallel to the San Gabriel, is characterized by reverse separation. This difference in style illustrates the fallibility of using patterns predicted from Figures 1 and 2 in areas where two or more deformational events are superimposed or where more than one major strike-slip fault is involved. In addition, the fault patterns shown in Figure 23 do not always result in basin subsidence.

A possible scheme for the rotation of crustal blocks between wrench faults is shown in Figure 22c (scale is uncertain). Abundant paleomagnetic evidence demonstrates widespread, predominantly clockwise rotation of crustal blocks in western North America during Cenozoic time (Beck, 1980; Luyendyk et al., 1980). Any significant block rotations similar to those shown in Figure 22c should lead to the initiation of sedimentary basins and to alternations of convergent and divergent wrench style where block corners alternately rotate toward or away from the through-going fault (indicated with normal- and thrust-fault symbols). These structural complexities have been corroborated by earthquake hypocentral location and first-motion studies in the eastern Transverse Ranges, California (Nicholson et al., 1985). The structural patterns that result from such rotational deformation are not included in Figures 1 and 2.

Divergent Plate Boundary and Extensional Continental Settings

Well-known examples of divergent wrench faults within divergent plate boundaries and extensional continental settings are oriented oblique or transverse to the trend of normal fault systems. The most common are apparently localized in the vicinity of graben doglegs and oversteps (Fig. 22d; Harding, 1984). Doglegs are short, oblique graben segments connecting subparallel straightaways. Extensional strike slip can occur within the middle segment of the dogleg, particularly if the regional plate separation is oblique to that segment. Several transform faults in the northern Red Sea (see Fig. 1 in Coleman, 1974) may be interpreted in this manner. Freund (1982) has also observed that obliquely oriented strike-slip faults, not necessarily confined to doglegs, are an occasional part of graben structures. He has attributed the formation of these faults to rare instances of crustal attenuation parallel to the graben axis.

Graben oversteps occur where the ends of graben segments overlap and are not connected directly by normal faults (Fig. 22e). Continued extension ultimately causes strike slip along a line connecting the grabens (Courtillot et al., 1974). Zones of en echelon faults connect the overstepped ends of the Rhine and Bresse grabens and have been interpreted by Illies (see Fig. 10 in Illies, 1974) as left-lateral wrench zones or oblique-slip faults with left-slip components. The zones resemble the belts of external faults associated with diver-

gent wrench faults, and he has compared the deformation to an incipient ridge-ridge transform fault system.

A somewhat different class of divergent wrench fault occurs in broad areas of pervasive regional extension, such as the Basin and Range Province of southeastern California and adjacent Nevada (Figs. 22f, 23d). There, wrench faults act as major basement tear faults, separating terranes that in late Cenozoic time experienced different amounts of extension (Wernicke et al., 1982; Burchfiel et al., 1983). These faults, however, do not necessarily have a divergent style throughout their lengths. Specific examples are the Garlock fault (Davis and Burchfiel, 1973), the Las Vegas Valley shear zone (Guth, 1981), and the Furnace Creek fault zone (Wright and Troxel, 1970; Stewart, 1983; Cemen et al., 1985 this volume). The Andaman Sea fault (Fig. 19) may also be an extensional tear fault. In both the Basin and Range and Andaman Sea examples, the associated normal faults are oblique rather than en echelon because they are of regional extent, and not arranged solely along wrench zones.

Divergent wrench faults are also known from oceanic regimes. The western rift zone of southwestern Iceland, for example, is connected to the Reykjanes segment of the North Atlantic mid-oceanic spreading center by an obliquely oriented zone of right-stepping, en echelon fissure swarms (Tryggvason, 1982). The swarms are geometrically similar to the en echelon normal faults between the Rhine and Bresse grabens. Elsewhere, many oceanic transform zones consist of valleys and ridges within and parallel to the principal displacement zone and of external, oblique slip faults (Fox and Gallo, 1984). The structural style of such fracture zones is similar to that of some continental divergent wrench faults, but the style does not develop in the same way. The transform valley is thought to be a manifestation of restricted partial melting at the relatively cold ridge-transform intersection. This results in reduced segregation of basaltic melt and thinner oceanic crust. Oblique faulted topography adjacent to the transform arises from reorientation of the principal stress axes as a result of welding of the upper mantle across the transform (Fox and Gallo, 1984).

Convergent Plate Boundary and Contractional Continental Settings

We have recognized divergent wrench faults in three convergent margin and contractional continental settings thus far: arc massif, and both backarc and peripheral forelands (cf. Dickinson, 1974); additional settings are probable.

Regions of high heat flow and thin, weakened lithosphere, such as arc massifs, appear to have localized longitudinal wrench faults within convergent margin settings (Fig. 22g; Fitch, 1972). Along the axis of the central Barisan Mountains in Sumatra, the right-lateral Great Sumatran fault consists of a braided zone that includes in-line grabens, pull-apart basins, and linear topographic depressions. The structural depressions are bounded by major right-lateral faults and subordinate, but still important, normal faults

FIG. 22.—Cartoon of plate-tectonic settings and patterns of divergent strike-slip faulting. Examples are based on known occurrences (except style of fault in d, which is assumed), and are consistently drawn for a right-slip system to facilitate comparison. There is undoubtedly a greater variety of possible fault patterns and complexities than is shown. Scale is variable, and the patterns in one sketch may constitute local details of another.

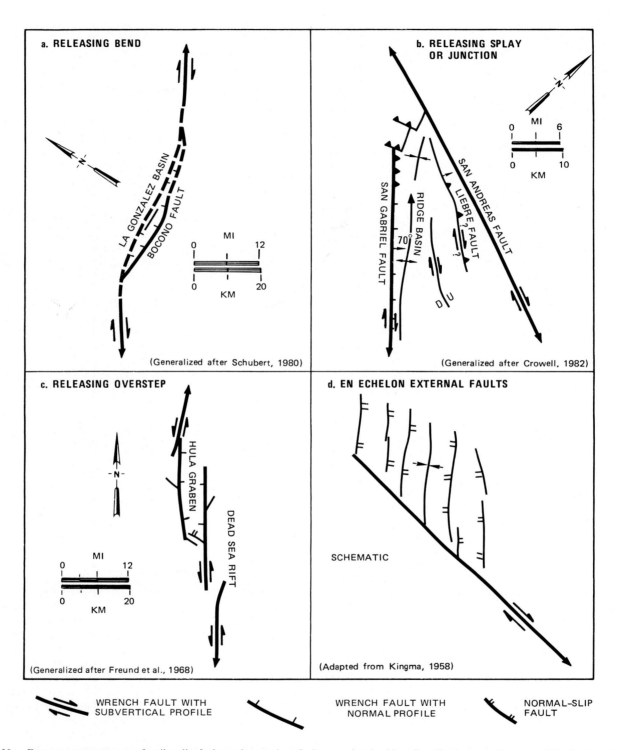

FIG. 23.—Four common patterns of strike-slip faults and secondary faults associated with strike slip that sometimes result in basin subsidence. Fault profile styles in (a) are from Giegengack (1977) and Schubert (1980). Fault profiles in (b) are generalized from descriptions in Crowell (1982); the reverse segment at northwest end of San Gabriel fault was superimposed on an earlier extensional profile. Fault profiles in (c) are from Horowitz (1973) and Schulman and Bartov (1978).

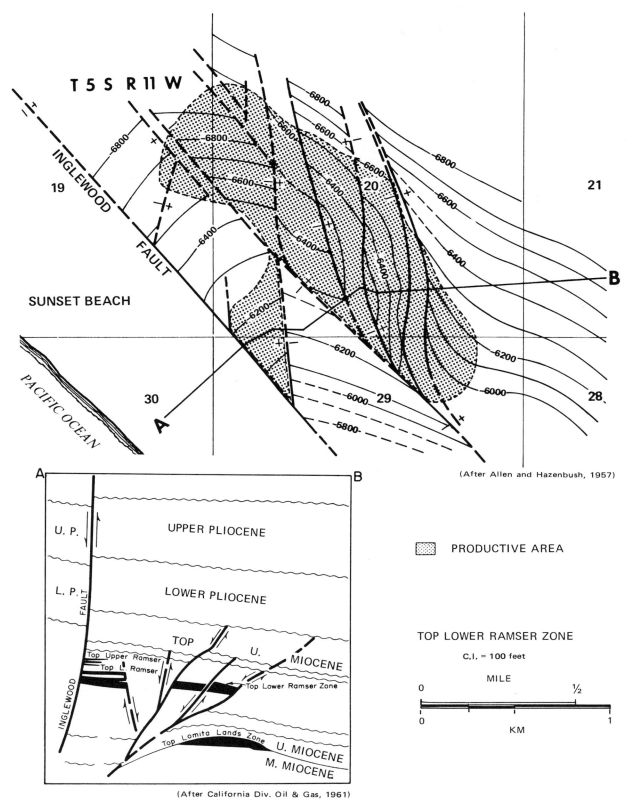

(After Allen and Hazenbush, 1957)

(After California Div. Oil & Gas, 1961)

Fig. 24.—Structure of the Sunset Beach oil field, Los Angeles Basin, California. Shallow, north-south, en echelon fault blocks are closed southward by throughgoing, northwest-striking faults, and westward by normal faults. The Inglewood fault is part of the regional Newport-Inglewood wrench zone, which has a maximum right slip of 760 m on structural markers near the top of the Miocene (Harding, 1973).

(see Fig. 9 in Posavec et al., 1973). Strike slip probably resulted from oblique encroachment between the Indian Ocean plate and the Sumatran plate in a setting roughly analogous to that shown in Figure 22g (Fitch, 1972). The extensional component of the strike slip is more difficult to explain, especially because of the dominance of contractional structures in the adjacent backarc (Hamilton, 1979, Plate I). The extension has been attributed to oceanward migration of the frontal parts of the Sumatran arc system by D. E. Karig and W. Jensky (in Hamilton, 1979). Dickinson (1974) related extension in arc massifs to crustal uplift and arching, and to volcano-tectonic subsidence. Recent unpublished work by T. R. Bultman and T. P. Harding indicates that contractional folds within the Sumatran backarc basins may owe their oblique orientation to the Great Sumatran fault more to pre-existing basement grain than to right-slip tectonics as previously reported by Wilcox et al. (1973).

Wrench faults oriented oblique to the trend of major tectonic zones within the region of convergence may also have a divergent style (foreland wrench in Fig. 22h). The Lake Basin fault zone (Figs. 4, 5) has this orientation within an orogenic foreland. Factors that control the strike-slip style of the oblique faults are poorly understood.

Several oblique faults in the peripheral foreland region of the outer West Carpathians appear to have acted as tears, separating eastern areas with continuing thrusting from areas in the west where thrusting was already complete (Royden, 1985 this volume). The Vienna Basin is located at a complex releasing overstep between at least three such faults (see pull-apart within extensional tear fault in Fig. 22h), each consisting of several branches with normal separation. Reflection seismic data, subsidence history, and heat-flow data suggest that the Vienna Basin is allochthonous, and that the bounding tear faults merge at depth with a gently dipping crustal detachment (Royden, 1985 this volume). Oblique tear faults, such as those of the Vienna Basin, are commonly late structures, and are invariably associated with belt-parallel extension, which is promoted by pronounced salients and re-entrants in the orogen (Dahlstrom, 1970).

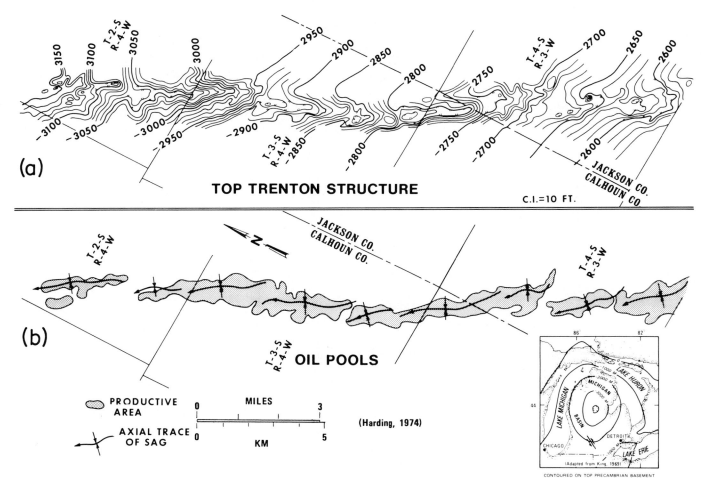

FIG. 25.—Structure map and setting (inset) of the Scipio-Albion trend, Michigan Basin, Michigan. (a) Map of central and northern parts of the field contoured on the top of the Ordovician Trenton Formation, which directly overlies the productive interval. (b) Pattern of sag axes determined from the structure contours.

Intraplate Occurrences

The Cottage Grove fault of the Illinois Basin illustrates the intracratonic setting of divergent wrench faults (Figs. 13, 22i), and the Scipio-Albion trend of the Michigan Basin provides another example (see the following discussion). Pre-existing zones of crustal weakness are recognized as a potentially important determinant of location and orientation. Convergent wrench faults are also known from mid-plate localities, but the cause of the mid-plate strike slip and controls on its style are unknown.

HYDROCARBON TRAPS ASSOCIATED WITH DIVERGENT WRENCH FAULTS

Three main types of structural traps associated with divergent wrench faults are known to be productive (see also D'Onfro and Glagola, 1983): (1) fault slices within the principal displacement zone, (2) forced folds parallel to the high side of the principal displacement zone, and (3) en echelon fault blocks flanking either or both sides of the zone. The forced folds have the greatest potential for providing both effective and abundant trap opportunities (e.g., Figs. 16, 19). Most hydrocarbon accumulations along the Cottage Grove fault system are located near anticlinal culminations along the apparent high side of the principal displacement zone, but they are productive in very low volumes in this example. Subsidiary traps commonly associated with anticlines, such as up-plunge reservoir terminations (pinchout, truncation, etc.) and cross faults, increase the potential for hydrocarbon prospects at the forced folds (e.g., Fig. 15). External, oblique anticlines, such as the Campbell Hill anticline in Figure 16, may also provide additional fold closures, but are less common.

En echelon normal-fault blocks are productive in the Newport-Inglewood trend, California, and in several other wrench zones (Fig. 24; Harding, 1973, 1974; Barrows, 1974). The fault blocks are developed where the principal displacement zone has the characteristics of a divergent wrench fault. Effective closure is dependent on block tilt away from fault intersections and on fault seals, usually at both the oblique normal faults and the principal displacement zone (Weber et al., 1978; Smith, 1980).

A vertical linear zone of fractured, dolomitic limestone forms the reservoir for the hydrocarbon accumulation along the Scipio-Albion trend, Michigan. A sag overlies the zone and is thought to reflect the distribution of fractures (Fig. 25). Both the sag and the productive limits of the fractured reservoir have a segmented right-stepping, en echelon pattern that resembles the orientation of synthetic faults at the principal displacement zone of a left-slip system. The series of sags trends obliquely down the regional dip (inset map in Fig. 25), and there is little other discernible deformation. Divergent wrench faulting is interpreted to have formed the unique sag and its en echelon pattern while at the same time enhancing fracturing and limiting deformation to a narrow swath (Harding, 1974). Graben slices within the principal displacement zone of the Cottage Grove fault system are also intensely fractured (Nelson and Krausse, 1981).

Although potential traps occur along divergent wrench faults, it is thought that the limited extent of external struc-

tures, the paucity of folds, and the dependence on sealing faults for closure are all significant limiting factors. Trends developed where blocks move in a convergent sense (Harding, 1974) or with neither convergence nor divergence (Harding, 1973, 1976) should have larger numbers of effective structural traps.

CONCLUSIONS

Our investigations indicate that some wrench zones contain a structural assemblage that is dominated by extensional features and that distinctive elements of this assemblage are repeated in a wide variety of tectonic settings. In the past, zones such as these may not have been properly recognized as wrench faults. The structural assemblage differs in important ways from the other wrench styles and constitutes a discrete structural style: (1) faulting dominates deformation adjacent to the principal displacement zone; (2) folds are for the most part oriented parallel to the principal displacement zone; and (3) the overall zone of deformation is typically relatively narrow. The structures may be combined in several ways. Some divergent wrench systems consist solely of a principal displacement zone with normal separation; other divergent wrench zones are composed of sets of en echelon normal faults; and still other systems have both elements. One result of their structural characteristics is that the potential for effective structural traps is not as great along divergent wrench faults as it is in regions of convergent or simple strike slip. The extensional characteristics of the style also make it difficult to differentiate divergent wrench faults from normal fault blocks, especially when interpreting seismic reflection data. In these instances it is necessary to determine both profile and map characteristics.

ACKNOWLEDGMENTS

This study was originally conducted for Exxon Production Research Company and we are grateful to that company for its permission to publish. Exxon Company, U.S.A., provided the Bering Sea and Lake Basin seismic reflection profiles and the maps of the Scipio-Albion trend. Esso Exploration, Inc., supplied the Andaman Sea example. K. T. Biddle, C. A. Dengo, R. P. George, D. W. Phelps, and A. C. Tuminas, of Exxon Production Research Company, reviewed the manuscript, and R. W. Wiener provided regional data for the Illinois Basin. G. C. Bond, Kristian Meisling, K. R. Schmitt, and D. M. Worral also made helpful suggestions. Logistical support at Lamont-Doherty Geological Observatory was provided by an ARCO Foundation Fellowship to Christie-Blick. Lamont-Doherty Geological Observatory Contribution No. 3912 (Christie-Blick).

REFERENCES

ALLEN, D. R., AND HAZENBUSH, G. C., 1957, Sunset oil field: California Division of Oil and Gas, California Oil Fields—Summary of Operations, v. 43, no. 2, p. 47–50.

ALPHA, A. G., AND FANSHAWE, J. R., 1954, Tectonics of northern Bighorn basin area and adjacent south-central Montana: Billings Geological Society Guidebook, 5th Annual Field Conference, Pryor Mountains-Northern Bighorn Basin, Montana, p. 72–79.

AYDIN, A., AND NUR, A., 1982, Evolution of pull-apart basins and their scale independence: Tectonics, v. 1, p. 91–105.

AYDIN, A., AND NUR, A., 1985, The types and role of stepovers in strike-slip tectonics, *in* Biddle, K. T., and Christie-Blick, N., eds., Strike-slip Deformation, Basin Formation and Sedimentation: Society of Economic Paleontologists and Mineralogists Special Publication No. 37, p. 35–44.

BARROWS, A. G., 1974, A review of the geology and earthquake history of the Newport-Inglewood structural zone, southern California: California Division Mines and Geology Special Report 114, 115 p.

BECK, M. E., JR., 1980, Paleomagnetic record of plate-margin tectonic processes along the western edge of North America: Journal of Geophysical Research, v. 85, p. 7115–7131.

BISHOP, D. G., 1968, The geometric relationships of structural features associated with major strike-slip faults in New Zealand: New Zealand Journal of Geology and Geophysics, v. 11, p. 405–417.

BURCHFIEL, B. C., WALKER, D., DAVIS, G. A., AND WERNICKE, B., 1983, Kingston Range and related detachment faults—a major "breakaway" zone in the southern Great Basin: Geological Society of America Abstracts with Programs, v. 15, p. 536.

BUSCHBACH, T. C., AND ATHERTON, E., 1979, History of the structural uplift of the southern margin of the Illinois basin, *in* Palmer, J. E., and Dutcher, R. R., eds., Depositional and Structural History of the Pennsylvanian System of the Illinois Basin, Part 2: Illinois State Geological Survey Guidebook Series 15a, Field trip 9, Ninth International Congress, Carboniferous Stratigraphy and Geology, p. 112–115.

CALIFORNIA DIVISION OF OIL AND GAS, 1961, California oil and gas fields maps and data sheets, part 2, Los Angeles–Ventura basins and central coastal regions: California Division of Oil and Gas, p. 496–913.

CEMEN, I., WRIGHT, L. A., DRAKE, R. E., AND JOHNSON, F. C., 1985, Cenozoic sedimentation and sequence of deformational events at the southeastern end of the Furnace Creek strike-slip fault zone, Death Valley region, California, *in* Biddle, K. T., and Christie-Blick, N., eds., Strike-Slip Deformation, Basin Formation, and Sedimentation: Society of Economic Paleontologists and Mineralogists Special Publication No. 37, p. 127–141.

CHAMBERLAIN, R. T., 1919, A peculiar belt of oblique faulting: Journal of Geology, v. 27, p. 602–613.

CHEADLE, M. J., CZUCHRA, B. L., BYRNE, T., ANDO, C. J., OLIVER, J. E., BROWN, L. O., KAUFMAN, S., MALM, P. E., AND PHINNEY, R. A., in press, The deep crustal structure of the Mojave Desert, California, from COCORP seismic reflection data: Tectonics.

CHRISTIE-BLICK, N., AND BIDDLE, K. T., 1985, Deformation and basin formation along strike-slip faults, *in* Biddle, K. T., and Christie-Blick, N., eds., Strike-Slip Deformation, Basin Formation, and Sedimentation: Society of Economic Paleontologists and Mineralogists Special Publication No. 37, p. 1–34.

CLARK, S. K., AND ROYDS, J. S., 1948, Structural trends and fault systems in Eastern Interior basin: American Association of Petroleum Geologists Bulletin, v. 32, p. 1728–1749.

COCHRAN, J. R., 1983, A model for development of Red Sea: American Association of Petroleum Geologists Bulletin, v. 67, p. 41–69.

COHEE, G. V., 1962, Tectonic map of the United States: United States Geological Survey and American Association of Petroleum Geologists, 1:2,500,000.

COLEMAN, R. G., 1974, Geologic background of the Red Sea, *in* Burk, C. A., and Drake, C. L., eds., The Geology of Continental Margins: New York, Springer-Verlag, p. 743–751.

COMMITTEE FOR COORDINATION OF JOINT PROSPECTING FOR MINERAL RESOURCES IN ASIAN OFFSHORE AREAS (C.C.O.P.), 1981, Studies in east Asian tectonics and resources (SEATAR): United Nations ESCAP, C.C.O.P. Technical Publication 7a, p. 37–50.

COURTILLOT, V., TAPPONNIER, P., AND VARET, J., 1974, Surface features associated with transform faults: a comparison between observed examples and an experimental model: Tectonophysics, v. 24, p. 317–329.

CROWELL, J. C., 1974, Origin of Late Cenozoic basins in southern California, *in* Dickinson, W. R., ed., Tectonics and Sedimentation: Society of Economic Paleontologists and Mineralogists Special Publication No. 22, p. 190–204.

———, 1981, Juncture of San Andreas transform system and Gulf of California rift: Oceanologica Acta, Proceedings, 26th International Geological Congress, Geology of Continental Margins Symposium, Paris, p. 137–141.

———, 1982, The tectonics of Ridge Basin, southern California, *in* Crowell, J. C., and Link, M. H., eds., Geologic History of Ridge Basin, Southern California: Society of Economic Paleontologists and Mineralogists, Pacific Section, p. 25–42.

CROWELL, J. C., AND SYLVESTER, A. G., eds., 1979, Tectonics of the Juncture Between the San Andreas Fault System and the Salton Trough, Southeastern California—A Guidebook: Santa Barbara, California, Department of Geological Sciences, University of California, 193 p.

CURRAY, J. R., MOORE, D. G., LAWVER, L. A., EMMEL, F. J., RAITT, R. W., HENRY, M., AND KIECKHEFEIZ, R., 1979, Tectonics of the Andaman Sea and Burma, *in* Watkins, J. S., Montadert, L., and Dickenson, P. W., eds., Geological and Geophysical Investigations of Continental Margins: American Association of Petroleum Geologists Memoir 29, p. 189–198.

DAHLSTROM, C. D. A., 1970, Structural geology in the eastern margin of the Canadian Rocky Mountains: Bulletin of Canadian Petroleum Geology, v. 18, p. 332–406.

DAVIS, G. A., AND BURCHFIEL, B. C., 1973, Garlock fault: an intracontinental transform structure, southern California: Geological Society of America Bulletin, v. 84, p. 1407–1422.

DICKINSON, W. R., 1974, Plate tectonics and sedimentation, *in* Dickinson, W. R., ed., Tectonics and Sedimentation: Society of Economic Paleontologists and Mineralogists Special Publication No. 22, p. 1–27.

DOBBIN, C. E., AND ERDMAN, C. E., 1955, Structure contour map of the Montana plains: United States Geological Survey Oil and Gas Investigation Map OM 178A, Scale 1:500,000.

D'ONFRO, P., AND GLAGOLA, P., 1983, Wrench fault, southeast Asia, *in* Bally, A. W., ed., Seismic Expression of Structural Styles, v. 3: American Association of Petroleum Geologists Studies in Geology, Series 15, p. 4.2–9 to 4.2–12.

EGUCHI, T., UYEDA, S., AND MAKI, T., 1979, Seismotectonics and tectonic history of the Andaman sea: Tectonophysics, v. 57, p. 35–51.

EMMONS, R. C., 1969, Strike-slip rupture patterns in sand models: Tectonophysics, v. 7, No. 1, p. 71–87.

ERVIN, C. P., AND McGINNIS, L. D., 1975, Reelfoot rift: reactivated precursor to the Mississippi embayment: Geological Society America Bulletin, v. 86, p. 1287–1295.

EYAL, Y., AND RECHES, ZE'EV, 1983, Tectonic analysis of the Dead Sea rift region since the Late-Cretaceous based on mesostructures: Tectonics, v. 2, p. 167–185.

FITCH, T. J., 1972, Plate convergence, transcurrent faults, and internal deformation adjacent to southeast Asia and the western Pacific: Journal of Geophysical Research, v. 77, p. 4432–4460.

FOX, P. J., AND GALLO, D. G., 1984, A tectonic model for ridge-transform-ridge plate boundaries: Implications for the structure of oceanic lithosphere: Tectonophysics, v. 104, p. 205–242.

FREUND, R., 1965, A model of the structural development of Israel and adjacent areas since Upper Cretaceous times: Geological Magazine, v. 102, p. 189–205.

———, 1970, Rotation of strike slip faults in Sistan, southeast Iran: Journal of Geology, v. 78, p. 188–200.

———, 1982, The role of shear in rifting, *in* Pálmason, G., ed., Continental and Oceanic Rifts: American Geophysical Union Geodynamics Series, v. 8, p. 33–39.

FREUND, R., GARFUNKEL, Z., ZAK, I., GOLDBERG, M., WEISSBROD, T., AND DERIN, B., 1970, The shear along the Dead Sea rift: Philosophical Transactions of Royal Society of London, Series A, v. 267, p. 107–130.

GARFUNKEL, Z., 1981, Internal structure of the Dead Sea leaky transform (rift) in relation to plate kinematics: Tectonophysics, v. 80, p. 81–108.

GIEGENGACK, R., 1977, Late Cenozoic tectonics of the Tabay-Estangues graben, Venezuelan Andes, *in* Espejo, A., ed., Memorias del V Congreso Geologico Venezolano: Ministerio de Energia y Minos, Caracas, Tomo II, p. 721–737.

GUTH, P. L., 1981, Tertiary extension north of the Las Vegas Valley shear zone, Sheep and Desert Ranges, Clark County, Nevada: Geological Society of America Bulletin, Part I, v. 92, p. 763–771.

HAMILTON, W., 1979, Tectonics of the Indonesian region: United States Geological Survey Professional Paper 1078, 345 p.

HARDING, T. P., 1973, Newport-Inglewood trend, California—an example of wrenching style of deformation: American Association of Petroleum Geologists Bulletin, v. 57, p. 97–116.

———, 1974, Petroleum traps associated with wrench faults: American

Association of Petroleum Geologists Bulletin, v. 58, p. 1290–1304.

———, 1976, Tectonic significance and hydrocarbon trapping consequences of sequential folding synchronous with San Andreas faulting, San Joaquin Valley, California: American Association of Petroleum Geologists Bulletin, v. 60, p. 356–378.

———, 1983, Divergent wrench fault and negative flower structure, Andaman Sea, *in* Bally, A. W., ed., Seismic Expression of Structural Styles, v. 3: American Association of Petroleum Geologists Studies in Geology, Series 15, p. 4.2–1 to 4.2–8.

———, 1984, Graben hydrocarbon occurrences and structural style: American Association of Petroleum Geologists Bulletin, v. 68, p. 333 to 362.

———, 1985, Seismic characteristics and identification of negative flower structures, positive flower structures and positive structural inversion: American Association of Petroleum Geologists Bulletin, v. 69, p. 582–600.

HARDING, T. P., AND LOWELL, J. D., 1979, Structural styles, their plate-tectonic habitats and hydrocarbon traps in petroleum provinces: American Association of Petroleum Geologists Bulletin, v. 63, p. 1016–1058.

HARLAND, W. B., 1971, Tectonic transpression in Caledonian Spitsbergen: Geological Magazine, v. 108, p. 27–42.

HAYS, W. H., 1957, Geology of the central Mecca Hills, Riverside County, California [unpubl. Ph.D. Thesis]: New Haven, Yale University, 324 p.

HEYL, A. V., JR., AND BROCK, M. R., 1961, Structural framework of the Illinois-Kentucky mining district and its relation to mineral deposits: United States Geological Survey Professional Paper 424-D, p. D3–D6.

HOROWITZ, A., 1973, Development of the Hula basin, Israel: Israel Journal Earth Sciences, v. 22, p. 107–139.

ILLIES, J. H., 1974, Taphrogenesis and plate tectonics, *in* Illies, J. H., and Fuchs, K., eds., Approaches to Taphrogenesis: Stuttgart, E. Schweizerbart'sche Verlagsbuchhandlung, p. 433–460.

JENNINGS, C., 1967, Salton Sea sheet, Geologic map of California, Olaf P. Jenkins edition: California Division of Mines and Geology, Scale 1:250,000.

KANE, M. F., HILDENBRAND, T. G., AND HENDRICKS, J. D., 1981, Model for the tectonic evolution of the Mississippi embayment and its contemporary seismicity: Geology, v. 9, p. 563–568.

KING, P. B., 1969, Tectonic map of North America: United States Geological Survey, Scale 1:5,000,000.

KINGMA, J. T., 1958, Possible origin of piercement structures, local unconformities, and secondary basins in the Eastern Geosyncline, New Zealand: New Zealand Journal of Geology and Geophysics, v. 1, p. 269–274.

LAWVER, L. A., AND CURRAY, J. R., 1981, Evolution of the Andaman Sea (abs.): EOS, Transactions of American Geophysical Union, v. 62, No. 45, p. 1044.

LENSEN, G. J., 1959, Secondary faulting and transcurrent splay-faulting at transcurrent fault intersections: New Zealand Journal of Geology and Geophysics, v. 2, p. 729–734.

LE PICHON, X., AND FRANCHETEAU, J., 1978, A plate-tectonic analysis of the Red Sea–Gulf of Aden area: Tectonophysics, v. 46, p. 369–406.

LUYENDYK, B. P., KAMERLING, M. J., AND TERRES, R., 1980, Geometric model for Neogene crustal rotations in southern California: Geological Society of America Bulletin, Part I, v. 91, p. 211–217.

MANN, P., HEMPTON, M. R., BRADLEY, D. C., AND BURKE, K., 1983, Development of pull-apart basins: Journal of Geology, v. 91, p. 529–554.

MANSPEIZER, W., 1985, The Dead Sea Rift: Impact of climate and tectonism on Pleistocene and Holocene sedimentation, *in* Biddle, K. T., and Christie-Blick, N., eds., Strike-Slip Deformation, Basin Formation, and Sedimentation: Society of Economic Paleontologists and Mineralogists Special Publication No. 37, p. 143–158.

NELSON, W. J., AND KRAUSSE, H.-F., 1981, The Cottage Grove fault system in southern Illinois: Illinois Institute of Natural Resources, State Geological Survey Division, Circular 522, 65 p.

NICHOLSON, C., SEEBER, L., WILLIAMS, P., AND SYKES, L. R., 1985, Seismicity and fault kinematics through the eastern Transverse Ranges, California: Block rotation, strike-slip faulting and shallow-angle thrusts: Journal of Geophysical Research, in press.

POSAVEC, M., TAYLOR, D., VAN LEEVEN, TH., AND SPECTOR, A., 1973, Tectonic controls of vulcanism and complex movements along the Su-

matran fault system: Geological Society of Malaysia Bulletin, v. 6, p. 43–60.

PROFFETT, J. M., JR., 1977, Cenozoic geology of the Yerington district, Nevada, and implications for the nature and origin of Basin and Range faulting: Geological Society of America Bulletin, v. 88, p. 247–266.

RIEDEL, W., 1929, Zur mechanik geologischer Brucherscheinungen: Zentralblatt für Mineralogie and Paleontologie, v. 1929B, p. 354–368.

ROCKWELL, T., AND SYLVESTER, A. G., 1979, Neotectonics of the Salton Trough, *in* Crowell, J. C., and Sylvester, A. G., eds. Tectonics of the Juncture Between the San Andreas and the Salton Trough, Southeastern California—A Guidebook: Santa Barbara, California, Department of Geological Sciences, University of California, p. 41–52.

ROGERS, T. H., 1965, Santa Ana sheet, Geologic Map of California, Olaf P. Jenkins edition: California Division of Mines and Geology, Scale 1:250,000.

ROYDEN, L. H., 1985, The Vienna Basin: A thin-skinned pull-apart basin, *in* Biddle, K. T., and Christie-Blick, N., eds., Strike-Slip Deformation, Basin Formation, and Sedimentation: Society of Economic Paleontologists and Mineralogists Special Publication No. 37, p. 319–338.

SCHUBERT, C., 1980, Late-Cenozoic pull-apart basins, Boconó fault zone, Venezuelan Andes: Journal of Structural Geology, v. 2, p. 463–468.

SCHULMAN, N., AND BARTOV, Y., 1978, Tectonics and sedimentation along the rift valley, excursion Y2: 10th International Congress on Sedimentology, Jerusalem, International Association of Sedimentologists, Guidebook, p. 37–94.

SLOSS, L. L., 1979, Plate-tectonic implications of the Pennsylvanian System in the Illinois basin, *in* Palmer, J. E., and Dutcher, R. R., eds., Depositional and Structural History of the Pennsylvanian System of the Illinois Basin, Part 2: Illinois State Geological Survey Guidebook, Series 15a, Field trip 9, Ninth International Congress, Carboniferous Stratigraphy and Geology, p. 107–112.

SMITH, D. A., 1980, Sealing and nonsealing faults in Louisiana Gulf Coast salt basin: American Association of Petroleum Geologists Bulletin, v. 64, p. 145–172.

SMITH, J. G., 1965, Fundamental transcurrent faulting in northern Rocky Mountains: American Association of Petroleum Geologists Bulletin, v. 49, p. 1398–1409.

STEWART, J. H., 1983, Extensional tectonics in the Death Valley area, California: Transport of the Panamint Range structural block 80 km northwestward: Geology, v. 11, p. 153–157.

SYLVESTER, A. G., AND SMITH, R. R., 1976, Tectonic transpression and basement-controlled deformation in San Andreas fault zone, Salton trough, California: American Association of Petroleum Geologists Bulletin, v. 60, p. 2081–2102.

TCHALENKO, J. S., AND AMBRASEYS, N. N., 1970, Structural analysis of the Dasht-e Bayāz (Iran) earthquake fractures: Geological Society of America Bulletin, v. 81, p. 41–60.

TERRES, R. R., AND SYLVESTER, A. G., 1981, Kinematic analysis of rotated fractures and blocks in simple shear: Seismological Society of America Bulletin, v. 71, p. 1593–1605.

TRYGGVASON, E., 1982, Recent ground deformation in continental and oceanic rift zones, *in* Pálmason, G., ed., Continental and Oceanic Rifts: American Geophysical Union Geodynamics Series, v. 8, p. 17–29.

VROMAN, A. J., 1967, On the fold pattern of Israel and the Levant: Geological Survey of Israel Bulletin, v. 43, p. 23–32.

WEBER, K. J., MANDL, G., PILAAR, W. F., LEHNER, F., AND PRECIOUS, R. G., 1978, The role of faults in hydrocarbon migration and trapping in Nigerian growth fault structures: Offshore Technology Conference Paper OTC 3356, p. 2643–2652.

WERNICKE, B., SPENCER, J. E., BURCHFIEL, B. C., AND GUTH, P. L., 1982, Magnitude of crustal extension in the southern Great Basin: Geology, v. 10, p. 499–502.

WILCOX, R. E., HARDING, T. P., AND SEELY, D. R., 1973, Basic wrench tectonics: American Association of Petroleum Geologists Bulletin, v. 57, p. 74–96.

WRIGHT, L. A., AND TROXEL, B. W., 1970, Summary of regional evidence for right-lateral displacement in the western Great Basin: Discussion: Geological Society of America Bulletin, v. 81, p. 2167–2173.

ZARTMAN, R. E., BROCK, M. R., HEYL, A. V., AND THOMAS, H. H., 1967, K-Ar and Rb-Sr ages of some alkaline intrusive rocks from central and eastern United States: American Journal Science, v. 265, no. 10, p. 848–870.

COMPARISON OF TECTONIC FRAMEWORK AND DEPOSITIONAL PATTERNS OF THE HORNELEN STRIKE–SLIP BASIN OF NORWAY AND THE RIDGE AND LITTLE SULPHUR CREEK STRIKE–SLIP BASINS OF CALIFORNIA

TOR H. NILSEN[1] AND ROBERT J. MCLAUGHLIN
U.S. Geological Survey
345 Middlefield Road
Menlo Park, California 94025

ABSTRACT: Thick nonmarine sequences with similar facies and geometry may accumulate in basins that develop adjacent to strike-slip faults. Herein we compare three basins of different age and size whose tectonic and depositional characteristics suggest a similar origin and history.

The Hornelen Basin developed during the Middle and possibly Early Devonian in western Norway. The basin is bounded on the north and south by east-striking faults, and the northern fault is considered to have been a zone of major right-slip movement. The basin is 60–70 km long, 15–25 km wide, and about 1,250 km^2 in areal extent; its cumulative fill of 25,000 m was deposited at an estimated rate of 2.5 m/1,000 yr. The Ridge Basin developed during the Miocene and Pliocene between the right-lateral San Gabriel and San Andreas faults in southern California. The basin is 30–40 km long, 6–15 km wide, and about 400 km^2 in areal extent; its cumulative fill of 7,000–11,000 m was deposited at an estimated rate of about 3 m/1,000 yr. The three Little Sulphur Creek Basins probably developed between 4 and 2 Ma along the east side of the right-lateral Maacama fault zone in northern California. These basins cumulatively are about 12 km long, 1.5 to 2 km wide, and about 15 km^2 in areal extent; their cumulative fill of 5,000 m was deposited at an estimated rate of about 2.5 m/1,000 yr.

Coarse sedimentary breccia, which constitutes a relatively small volume of the fill, was deposited in each of these basins along the active right-slip fault margin as talus, landslide, and small but steep debris-flow-dominated alluvial fans. Along other margins of the basins, a much larger volume of the fill accumulated as larger streamflow-dominated alluvial fans, braided-stream, meandering-stream, fan-delta, and deltaic deposits. Lacustrine deposits that include turbidites and, locally in Ridge Basin, chemical precipitates, accumulated in the centers of the basins. The basin floors are generally tilted toward the margins with active right-slip faults so that the basin axes and the depocenters are subparallel to, and shifted toward, this margin. Sediment was transported toward the basin center from surrounding highlands and then longitudinally down the basin axis. The basin fills were syn-depositionally faulted and post-depositionally folded into large plunging synclines. The basins lengthened over time and contain thicknesses of sedimentary rocks that are comparable to or greater than their widths.

INTRODUCTION

Sedimentary basins that develop along strike-slip faults have various shapes and sizes, and differ in the details of their tectonic and depositional history. These differences are related in part to the geometry and orientation of the principal bounding faults. Because the principal faults may be curved, braided, or en echelon, areas of compression (or transpression) may develop adjacent to or even within areas of extension (or transtension). Basins that are of pull-apart origin are common along strike-slip faults and may be filled with very thick sedimentary sequences.

The purpose of this paper is to compare the tectonic evolution, geometry, depositional framework, and geologic history of three well-studied basins that developed adjacent to major strike-slip faults. The three basins, the Hornelen Basin of western Norway, the Ridge Basin of southern California, and the Little Sulphur Creek Basins of northern California (Fig. 1), developed in continental settings and are filled chiefly, if not wholly, by nonmarine sedimentary rocks that include prominent lacustrine facies. Two of these three basins, the Ridge and Hornelen Basins, have long been used as models for sedimentation along strike-slip faults.

The three basins vary in size (Fig. 2), permitting interesting comparisons of their tectonic and depositional features. Although we have worked chiefly in the Little Sulphur Creek Basins, we feel confident in making the comparative analysis because of the excellent descriptions of the Hornelen Basin by R. J. Steel and his colleagues and of the Ridge Basin by J. C. Crowell, M. H. Link and their

colleagues, whose work we heavily rely upon. The depositional aspects of the basins are emphasized herein, particularly the distribution of various facies and the style of basin sedimentation. The largest and oldest basin, the Devonian Hornelen Basin, is discussed first, then the intermediate-size Ridge Basin, and finally the small Little Sulphur Creek Basins.

HORNELEN BASIN

Setting

The Hornelen Basin is the largest and northernmost of four east-trending Devonian sedimentary basins that crop out along the west coast of Norway between Nordfjord and Sognefjord (Fig. 1A; Kolderup, 1904; Nilsen, 1973; Bryhni, 1975; Steel, 1976). The basin was mapped and its stratigraphy and sedimentology studied by Kolderup (1927), Bryhni (1964a,b,c; 1978), Steel et al. (1977, 1979), Larsen and Steel (1978), Steel and Aasheim (1978), Steel and Gloppen (1980), Gloppen and Steel (1981), and Pollard et al. (1982). Fossils of plants (Kolderup, 1904, 1915a,b; 1916; Nathorst, 1915; Hoeg, 1936) and freshwater crossopterygian fishes (Kiaer, 1918; Jarvik, 1949) indicate a Middle Devonian age for the uppermost strata in the basin.

The basin developed after the Late Silurian Caledonian orogeny, when the northwestern European (Baltica) and northeastern North American cratons were sutured by the closure of Iapetus, the proto-Atlantic Ocean (Gee, 1975). The suturing in western Norway caused the detachment and eastward thrusting of nappes onto the edge of the craton (Gee, 1978). The Devonian basins have been primarily regarded as pull-apart basins that formed grabens and halfgrabens within a framework of post-Caledonian strike-slip

---

[1]Present address: RPI Pacific, Inc., 507 Seaport Court, Suite 101, Redwood City, California 94063

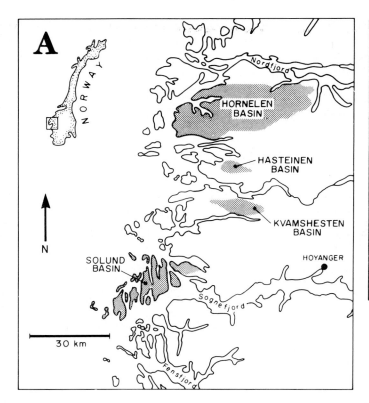

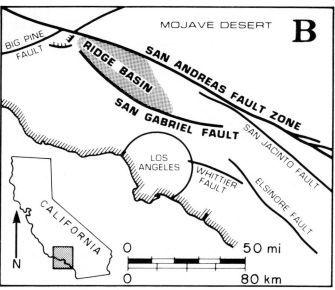

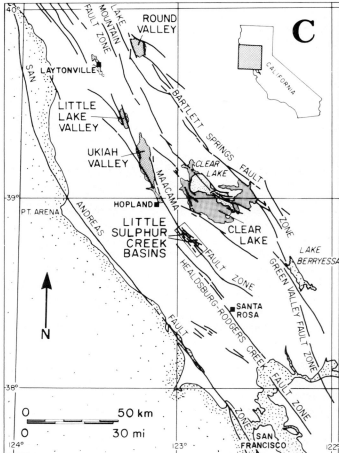

FIG. 1.—Index maps showing the locations of the Hornelen, Ridge, and Little Sulphur Creek Basins. A) Hornelen Basin and other Devonian basins of western Norway (modified from Nilsen, 1967, Fig. 1). B) Ridge Basin of southern California adjacent to the San Gabriel and San Andreas faults (modified from Link and Osborne, 1978, Fig. 1). C) Little Sulphur Creek Basins of northern California (modified from McLaughlin and Nilsen, 1982, Fig. 1).

HORNELEN BASIN

RIDGE BASIN

LITTLE SULPHUR CREEK BASINS

0 5 10 15 KM

FIG. 2.—Outline maps of the Hornelen, Ridge, and Little Sulphur Creek Basins showing their relative sizes.

faulting (Steel and Gloppen, 1980).

The Hornelen Basin is 60–70 km long, 15–25 km wide, and occupies an area of 1,250 km^2 (Fig. 3). The strata in the axial part of the basin dip consistently eastward and attain a total stratigraphic thickness of 25,000 m. The northern and southern margins of the basin are formed by high-angle faults, the southern fault striking slightly north of east and the northern fault almost due east (Figs. 3 and 4A, B). The Devonian sedimentary rocks rest unconformably on pre-Devonian basement rocks along the western margin of the basin and rest with fault contact on basement rocks along the eastern margin of the basin (Fig. 4B).

Conglomerate is common around most of the basin margins and generally thins and pinches out toward the center of the basin (Fig. 3). The axial deposits are generally finer grained and consist of fine-grained conglomerate, sandstone, siltstone, and shale. No evidence of marine sedimentation has been observed.

The basement rocks that underlie and surround Hornelen Basin consist of two distinct types: (1) to the north, east, and south, Precambrian metamorphic rocks that include chiefly schist, gneiss and quartzite; and (2) to the west and probably beneath the basin, Cambrian, Ordovician, and Silurian shale, graywacke, basalt, gabbro, and granitic rocks.

Structure

The Hornelen Basin forms a broad east-plunging syncline in which the synclinal axis is located close to the northern basin margin (Fig. 3). The basin fill is characterized by a step-like topography defined by basin-wide coarsening- and thickening-upward sequences (Fig. 4C). Beds along the

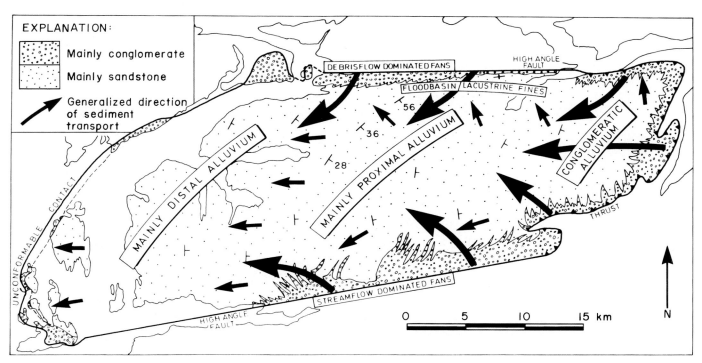

FIG. 3.—Simplified geologic and paleogeographic map of Hornelen Basin showing overall structural framework, major rock types, and generalized directions of sediment transport (from Steel and Aasheim, 1978, Fig. 1). For location, see figure 1A.

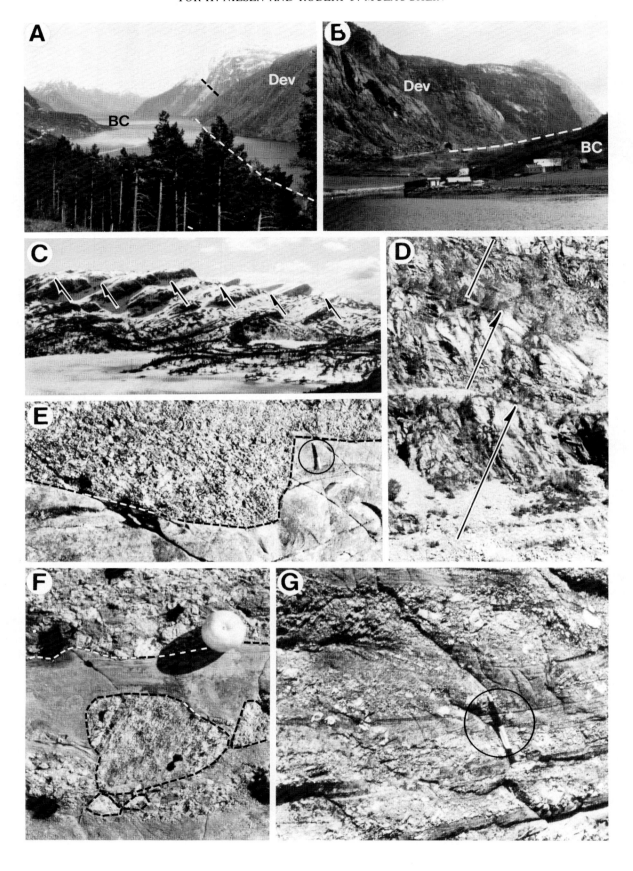

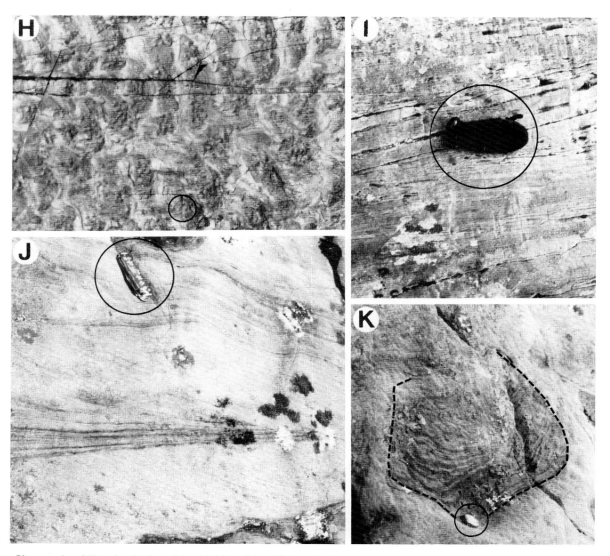

Fig. 4.—Photographs of Hornelen Basin and its chief depositional facies. A) view east along northern boundary fault (white dashed line) showing eastern faulted contact (dashed black line) of Devonian sedimentary rocks (Dev) on basement rocks (BC). B) view east along southern boundary fault (white dashed line) showing juxtaposition of Devonian sedimentary rocks (Dev) and basement rocks (BC). C) stacked 100-to-200-m-thick thickening- and coarsening-upward sequences (arrows) in the central part of the basin. D) stacked thickening- and coarsening-upward sequences (arrows) in streamflow-dominated alluvial-fan deposits along southern margin of basin. E) channeled base (dashed line) of subaerial conglomeratic debris-flow deposit of alluvial fan; pen circled for scale. F) poorly sorted and reverse-graded subaerial conglomeratic debris-flow of alluvial-fan deposit; apple for scale. G) well-stratified, interbedded conglomerate and sandstone of streamflow-dominated alluvial-fan deposit; pen circled for scale. H) wave-rippled, bioturbated, lacustrine siltstone deposited in axial part of basin; circle with diameter of 7 cm for scale. I) planar- and cross-stratified basin-axis fluvial deposits; mitten circled for scale. J) cross-stratified basin-axis fluvial deposits; knife circled for scale. K) synsedimentary deformation associated with upward expulsion of fluids and development of sandstone diapirs in basin-axis fluvial deposits; bottle cap circled for scale.

northern and southern margins of the basin, which are also organized into sequences of this type (Fig. 4D), are locally steeply dipping and have clearly been affected by post-depositional faulting along these margins. The faulted eastern margin of the basin has generally been interpreted as a thrust. The northern boundary fault can be traced both eastward and westward into the basement and the southern boundary fault can be traced westward into the basement (R. J. Steel, written commun., July, 1984).

Depositional Facies

The deposits of Hornelen Basin have been studied in considerable detail and the general patterns of sedimentation are well understood (Fig. 5A, B). Small, debris-flow-dominated alluvial fans are present along the entire northern margin of the basin adjacent to that basin-margin fault, and larger streamflow-dominated alluvial fans are present along the entire southern margin of the basin adjacent to that ba-

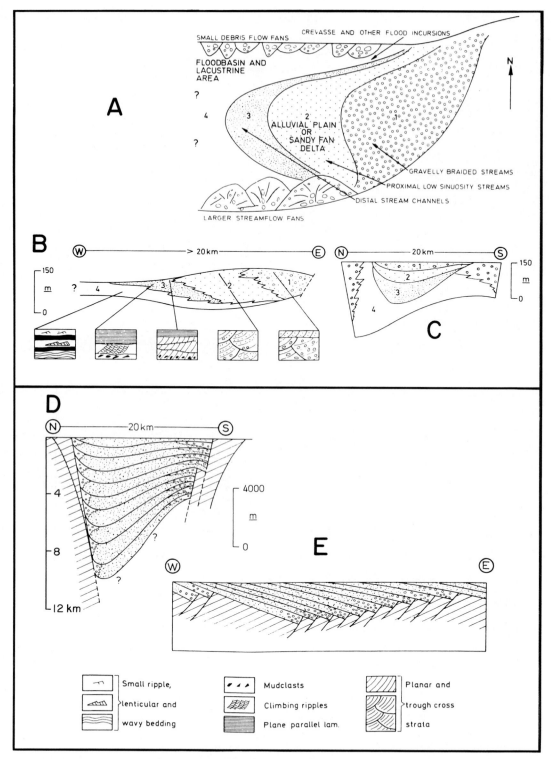

FIG. 5.—Structural and depositional framework of Hornelen Basin (modified from Steel and Gloppen, 1980, Fig. 9). A) general paleogeography and style of basin filling, showing three principal sources of detritus and distribution of depositional facies along the basin margins and in the basin axis. B) longitudinal east-west cross-section of the basin showing proximal-to-distal organization of depositional facies during progradation of a single 100-to-200-m-thick sequence. C) transverse north-south cross-section showing lateral relations between basin-margin alluvial-fan deposits and basin-axis fluvial and lacustrine deposits during progradation of a single 100-to-200-m-thick sequence. D) simplified transverse north-south cross-section showing vertical and horizontal stacking arrangement of multiple 100-to-200-m-thick sequences and asymmetry of the basin. E) inferred longitudinal east-west cross-section showing repetitive stacking of 100-to-200-m-thick sequences, eastward overstep of sedimentary sequence onto faulted basement surface, and eastward migration of basin depocenter.

sin-margin fault (Steel, 1976; Larsen and Steel, 1978; Steel and Aasheim, 1978; Gloppen and Steel, 1981). The axial part of the basin is dominated by fluvial and lacustrine facies; the fluvial sediments accumulated on a broad alluvial plain or sandy lacustrine fan-delta, fed by ephemeral low-sinuosity streams that prograded into lakes situated to the northwest (Steel and Aasheim, 1978).

The northern debris-flow-dominated alluvial fans generally have radii of 0.5–1.3 km, thicknesses of 100–200 m, and are characterized by relatively steep paleoslopes (Gloppen and Steel, 1981). The fans, which prograded into adjacent lacustrine environments, have a wedge-shaped geometry in which they thicken toward the basin-margin faults (Fig. 5C). These fans consist largely of subaerial and subaqueous conglomeratic debris-flow deposits characterized by poor sorting, poor rounding, general lack of stratification and clast orientation, packing that is either clast- or matrix-supported, reverse grading (Fig. 4F), and a positive correlation between bed thickness and maximum clast size (Gloppen and Steel, 1981). Angular clasts as large as 75 cm are locally present in these fan deposits. Bed thicknesses and the proportion of debris-flow deposits within the fans decrease southward. The fans are typically organized into overall thickening- and coarsening-upward sequences that are 100–200 m thick and similar thinner, smaller sequences that are 5–25 m thick (Gloppen and Steel, 1981).

The streamflow-dominated alluvial fans along the southern margin of the basin generally have radii of 2–4 km, thicknesses of 75–120 m, and are characterized by relatively gentle paleoslopes (Gloppen and Steel, 1981). Although these fan deposits are similar to the debris-flow-dominated fans in overall geometry and cyclical organization, they consist mostly of beds of conglomerate and conglomeratic sandstone characterized by better sorting, better rounding, parallel stratification and cross-stratification, well-developed clast orientation, packing that is generally clast-supported, normal grading, and a positive correlation between bed thickness and maximum clast size (Fig. 4G; Gloppen and Steel, 1981). The distal parts of these fans consist mostly of sheetflood deposits formed of laterally continuous alternating thin beds of nonchannelized fine-grained conglomerate and sandstone.

The axial deposits of Hornelen Basin are organized into about 200 basin-wide coarsening-upward sequences that average about 150 m in thickness and form the prominent step-like topography in the basin (Figs. 4C and 5D; Steel and Aasheim, 1978). These sequences can be divided in ascending order into four distinct facies: (1) floodplain and lacustrine deposits, dominantly mudstone, siltstone, and very fine grained sandstone that contain current, wave, and climbing ripple marks (Fig. 4H), wavy and lenticular bedding, mudcracks, shale rip-up clasts, syn-sedimentary slumps, trace fossils, and some thin-bedded turbidites; (2) alluvial deposits of several types, consisting for the most part of fine- to medium-grained, cross-stratified and parallel-stratified sandstone deposited by distal braided or meandering streams (Fig. 4I, J); (3) alluvial plain or sandy fan-delta deposits, dominated by pebbly or very coarse grained, cross-stratified or parallel-stratified sandstone; and (4) conglomeratic braided-stream deposits, dominated by

massive, planar-, and cross-stratified conglomerate and sandstone that fills channels and contains well-sorted and well-rounded clasts of varied composition (Steel and Aasheim, 1978; Pollard et al., 1982). These axial depositional bodies thin and fine westward, extend for east-west distances of at least 20 km, and prograded westward and northwestward into the basin. The bodies generally thicken toward the northern basin margin and the floodplain and lacustrine deposits are more widely distributed adjacent to the northern basin margin, suggesting that the basin floor was generally tilted northward.

Paleocurrents and Maximum Clast Size

Paleocurrent directions from the Hornelen Basin yield a pattern of lateral infilling of the basin from its northern and southern margins and westward transport down the axis of the basin (Figs. 3 and 5A; Brhyni, 1964a,b; Steel, 1976; Steel and Gloppen, 1980). Directions of sediment transport along the western margin of the basin indicate westward flow above the basal unconformity. Detailed paleocurrent studies of the small debris-flow-dominated alluvial fan deposits along the northern margin of the basin show patterns of sediment transport that are southward in proximal facies and westward in distal facies, concomitant with gradual decrease in the same directions of the average and maximum clast size of the conglomerate (Larsen and Steel, 1978; Steel and Aasheim, 1978; Gloppen and Steel, 1981). Analyses of the larger streamflow-dominated alluvial-fan deposits along the southern margin yield similar but reversed results, with the directions being toward the north and northwest (Fig. 5A; Steel and Gloppen, 1980; Gloppen and Steel, 1981). In the axial fluvial and lacustrine deposits, sediment transport was generally westward and the maximum clast sizes, where observed along correlative horizons, generally fine westward (Steel and Aasheim, 1978; Pollard et al., 1982).

Tectonic and Depositional History

Steel and Gloppen (1981) concluded that Hornelen Basin formed as a result of Devonian strike-slip faulting. They thought that the northern basin-margin fault was the dominant fault that controlled sedimentation within the basin (Fig. 3); on the basis of the composition of clasts within the northern debris-flow-dominated alluvial fans, and projections of the sizes of the drainage areas for these fans, they concluded that the northern basin-margin fault was a right-lateral strike-slip fault. The southern boundary fault is thought to be a normal fault, however, chiefly because the conglomerate clasts can be readily matched with adjacent source areas.

The systematic westward displacement and overlapping of the depositional loci of alluvial fans along the northern and southern basin margins and of the axial facies suggests that the basin depocenter systematically migrated eastward through time (Fig. 5). As a result of this migration, the maximum thickness of sedimentary strata at any one place probably never exceeded 8,000 m, even though the cumulative stratigraphic thickness for the entire basin is about 25,000 m. The basin floor generally sloped westward, as indicated by the general directions of sediment transport in the axial parts of the basin, but also had a northerly component of slope, as indicated by the concentration of lacustrine deposits along the northern margin of the basin (Fig. 5A,B,C). Although much coarse-grained sediment was transported into the basin as alluvial fans from its northern and southern margins, most of the sediment was transported axially into the basin from source areas to the east.

The 100-to-200-m-thick basin-wide coarsening-upward sequences in the fill of Hornelen Basin are thought by Steel and Gloppen (1980) to correspond to major episodes of horizontal displacement along basin-margin faults (Fig. 5D). The lowering of the basin floor along now buried west-dipping normal faults resulted in eastward migration of the basin depocenter (Fig. 5E); thinner coarsening-upward sequences were thought to result chiefly from vertical movement of the basin floor, which caused migration of the source area and progradation of depositional bodies into the basin. Abundant syn-sedimentary deformation, particularly in the basin-axis lacustrine and marginal lacustrine deposits, may record penecontemporaneous seismic activity (Fig. 4K).

RIDGE BASIN

Setting

The Ridge Basin is located between the San Andreas and San Gabriel faults in southern California (Fig. 1B) and was mapped by Clements (1937), Eaton (1939), and Crowell (1950, 1954a). Its geologic history has recently been summarized in detail in a volume edited by Crowell and Link (1982a) that has an accompanying geologic map by Crowell et al. (1982). The basin is 30 to 40 km long, 6 to 15 km wide, and covers an area of about 400 km^2 (Fig. 6). The basin developed in the Mohnian (late Miocene) and continued to fill with sediment through the early Pleistocene, after which it was deformed and uplifted (Crowell, 1974a,b, 1975, 1982a; Crowell and Link, 1982b; Link, 1983).

During the late Miocene, the San Gabriel fault was a major active strand of the San Andreas fault system (Crowell, 1952, 1954b, 1975), and the Ridge Basin formed east of a major bend in the fault. More than 60 km of right slip occurred along the San Gabriel fault from late Miocene into

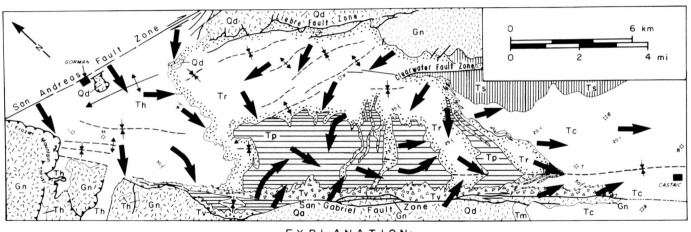

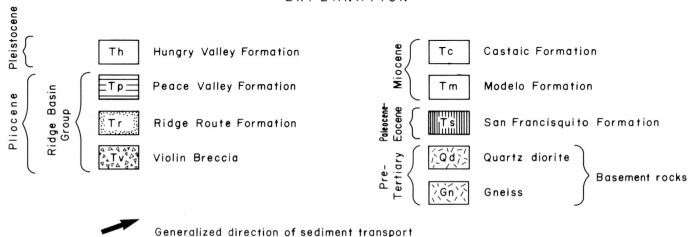

EXPLANATION:

FIG. 6.—Generalized geologic map of Ridge Basin (modified from Link and Osborne, 1978, Fig. 3, after Crowell, 1954a, 1975) showing generalized directions of sediment transport (from Link, 1982c, Fig. 2). For location see Figure 1B.

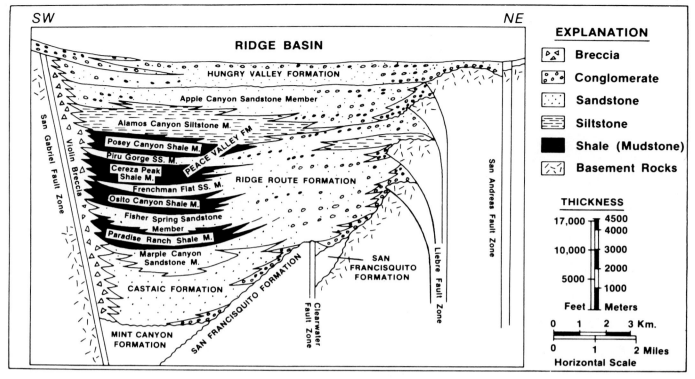

Fig. 7.—Generalized cross-section of Ridge Basin showing principal faults (from Crowell and Link, 1982b, Fig. 2).

Pliocene time. During the Pleistocene, the modern trace of the San Andreas fault developed along the northeastern flank of the Ridge Basin, and active right slip shifted to this fault. As much as 240 km of displacement has taken place on the San Andreas fault in this region, offsetting the Ridge Basin from its chief source area, which was originally located to the northeast.

The Ridge Basin is filled with more than 13,500 m of sedimentary rocks (Fig. 7). The nonmarine Miocene Mint Canyon Formation forms the lowest part of the basin fill and it is overlain by the marine upper Miocene Castaic Formation, which is about 2,200 m thick. The remainder of the basin fill has been assigned to the nonmarine Ridge Basin Group (Crowell, 1975), which conformably overlies and laterally interfingers with the Castaic Formation. The Ridge Basin Group, which is 9,000–11,000 m thick, contains marine deposits in its lower 600 m. It consists of four formations: (1) the Violin Breccia, which is confined to the western margin of the basin, is about 11,000 m thick, and interfingers northeastward with the Castaic Formation and other units of the Ridge Basin Group; (2) the Peace Valley Formation, about 8,000 m thick, dominantly shale and siltstone, but also including an analcimic and ferroan dolomitic mudstone, deposited chiefly in the central part of the basin; (3) the Ridge Route Formation, about 9,000 m thick, dominantly conglomerate and sandstone, deposited chiefly along the eastern margin of the basin; and (4) the Hungry Valley Formation, about 1,100 m thick, dominantly conglomerate and sandstone, deposited basin-wide as the final stage of basin filling (Link, 1982a).

Ridge Basin is bounded both on the southwest and northeast and is also underlain by granite, quartz diorite, quartz monzonite, and gneiss of Precambrian to Mesozoic age (Figs. 6 and 7). These types of rocks, which also extend southeastward into the Mojave Desert province east of the San Andreas fault, as well as sandstone of the underlying San Francisquito Formation and Mint Canyon Formation, formed the chief source of detritus deposited in the Ridge Basin. Thus, the sandstone deposits of Ridge Basin are chiefly arkosic in composition (Link, 1982b), reflecting the dominantly granitic provenance (Link, 1982c).

Structure

In northeast-southwest cross-section, the Ridge Basin has the appearance of a complex half-graben bounded by the San Gabriel fault on the southwest and the San Andreas fault on the northeast (Figs. 6 and 7). The Clearwater and Liebre fault zones are primarily high-angle reverse faults that were active during basin sedimentation. Paleomagnetic studies of the stratigraphic sequence of Ridge Basin by Ensley and Verosub (1982a,b) suggest that the Clearwater fault was tectonically active from 8.1 to 7.8 Ma, the southern branches of the Liebre fault were active from 7.3 to 6.1 Ma, and the northern branches of the Liebre fault were active from 6.0 to 5.0 Ma. They also concluded that the San Gabriel fault was active continuously from prior to 8.1 to 5.0 Ma and determined that the sedimentary rocks of Ridge Basin had not been rotated. The unconformable overlap of the Clearwater and Liebre faults by the base of the Hungry

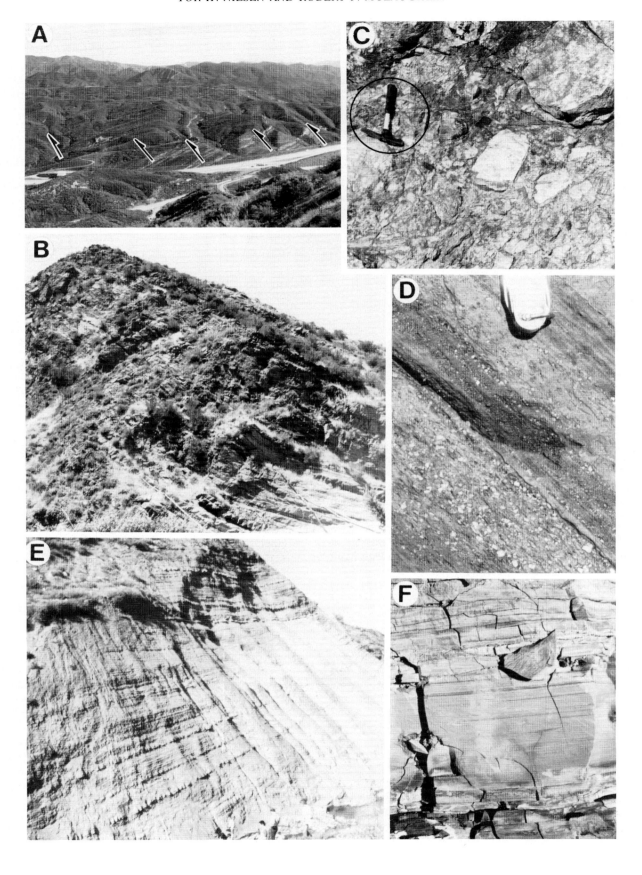

Fɪɢ. 8.—Photographs of Ridge Basin and its chief depositional facies. A) view eastward from ridge underlain by the Violin Breccia toward the axis of Ridge Basin showing stacked sequences (arrows) in the central part of the basin; interstate highway in middle of photograph for scale. B) view northwest of northeast-dipping beds of the Violin Breccia interfingering laterally from left to right with fine-grained and thin-bedded lacustrine deposits. C) typical appearance of proximal facies of the Violin Breccia; hammer circled for scale. D) reverse-graded distal subaqueous debris-flow deposits of the Violin Breccia; end of pocket knife for scale. E) thin-bedded lacustrine deposits of the Peace Valley Formation; geologists in lower right corner for scale. F) very thinly laminated analcimic lacustrine mudstone of the Peace Valley Formation; thickness of section shown is about 25 cm. G) channelized conglomeratic fluvial deposits of the Ridge Route Formation; note underlying bed with abundant climbing ripple lamination. H) deltaic sandstone member of the Ridge Route Formation that extends across most of Ridge Basin. I) large-scale basin-axis syn-sedimentary deformation. J) small-scale basin-axis syn-sedimentary deformation, J. C. Crowell for scale.

Valley Formation (Figs. 6 and 7) shows that the San Gabriel fault continued to move after the Liebre fault had become inactive, probably until the beginning of the Pleistocene (Crowell, 1975). The uppermost part of the basin fill, the Hungry Valley Formation, was deposited between 5.0 and 4.0 Ma, as the San Andreas fault zone became active. The Frazier Mountain thrust developed in the middle Pleistocene and subsequently became inactive and was folded (Crowell, 1982a).

The sedimentary fill of Ridge Basin forms a series of basin-wide northwest-dipping ridges (Fig. 8A). The fill has been folded into a broad asymmetric northwest-plunging syncline (Figs. 6 and 7) whose axis lies close to the southwestern flank of the basin, adjacent to outcrops of the Violin Breccia. The northwest-plunging strata of the Ridge Basin Group lap successively northwestward onto the basement so that the true thickness of the stratigraphic sequence at any point in the basin is less than 4,500 m, although the total measured stratigraphic sequence along the synclinal axis is more than 13,500 m (Crowell, 1975; Crowell and Link, 1982b).

Depositional Facies

The general character and distribution of depositional facies in the Ridge Basin have been summarized by Link and Osborne (1978) and by Link (1983). The facies consist chiefly of (1) marine mudstone and turbidites of the Castaic Formation and the lower part of the Ridge Basin Group in the southeastern axial part of the basin, (2) alluvial fans of the Violin Breccia along the southwestern margin of the basin, (3) alluvial-fan, fan-delta, sublacustrine deltas, and fluvial deposits of the Ridge Route Formation along the northeastern margin of the basin, (4) shallow- and deep-lacustrine deposits of the Peace Valley Formation in the axial part of the basin, and (5) alluvial-fan and fluvial deposits of the Hungry Valley Formation (Figs. 6 and 7).

The Castaic Formation (Crowell, 1954b) and the lower part of the Ridge Route Formation consist dominantly of marine slope deposits and basin-axis turbidites (Link, 1982b). The Castaic Formation is more than 2,500 m thick in the southeastern end of Ridge Basin (Stitt, 1982). The unit rests unconformably on the Mint Canyon Formation and records a late Miocene marine transgression into the basin; foraminifers indicate initial shallow-marine conditions followed by moderately deep-marine conditions, and finally, shoaling of the environments accompanied by partly restricted circulation and the development of a silled basin (McDougall, 1982; Stanton, 1982). Water depths increased southeastward from less than 100 m in the northwest to as much as 1,500 m in the southeast (McDougall, 1982). The slope deposits consist chiefly of poorly bedded mudstone that contains large slide blocks, are slump-folded, and are cut by channels or small canyons filled with sandstone, conglomerate, and coquina; the basin-axis turbidites consist of channel, interchannel, and depositional-lobe deposits (Link, 1982d; Stitt, 1982).

The Violin Breccia accumulated as talus and steep alluvial cones adjacent to the San Gabriel fault (Crowell, 1982b). The breccia crops out for a distance of 34 km adjacent to the fault but extends northeastward toward the center of the basin for a distance of only 1.5 km, where it pinches out into lacustrine shale and siltstone of the axial Peace Valley Formation or sandstone members of the Ridge Route Formation (Fig. 8B). The Violin Breccia is as thick as 11,000 m and its lower part is marine. The proximal part of the Violin Breccia consists of unbedded massive rubble that contains blocks as large as 2 m long (Fig. 8C). To the northeast, it gradually fines, becomes better stratified, contains more interbedded mudstone, and its clasts are better rounded. The reverse grading at the base of many of the coarse-grained beds suggests deposition by subaerial and subaqueous debris flows (Fig. 8D). A landslide block about a kilometer in length is interbedded in the breccia at one location. The Violin Breccia is organized into very distinctive coarsening- and thickening-upward sequences. The conglomerate contains mostly clasts of gneiss and granite derived from the Alamo and Frazier Mountain areas, which

are currently located along the northwestern margin of the Ridge Basin.

The basin-axis Peace Valley Formation consists chiefly of fine-grained lacustrine deposits that vary considerably in character because of fluctuations in lake level and transport of coarse-grained sediments into the lake along its margins (Smith, 1982a). The marginal lacustrine facies include shoreline, nearshore sandflat and mudflat, offshore bar, and fluvio-deltaic deposits, and the offshore lacustrine facies include turbidite and mudrock-carbonate deposits (Fig. 8E; Link and Osborne, 1978). Fossils from the Peace Valley Formation include molluscs (Young, 1982), ostracodes (Forester and Brouwers, 1982), algal stromatolites (Link et al., 1978), insects (Squires, 1979), and various invertebrate fossils that include freshwater fishes and bones of horses, turtles, camels, mastodons, ground sloths, and rhinoceroses (see summary in Welton and Link, 1982). Abundant plant fossils (Axelrod, 1982) and trace fossils (Smith et al., 1982) are present in both the Peace Valley Formation and intertongued coarser deposits of adjacent members of the Ridge Basin Group. Analcime, dolomite, and locally gypsum that are present within the Posey Canyon Shale Member and Alamos Canyon Siltstone Member of the Peace Valley Formation probably indicate brackish and alkaline conditions (Fig. 8F; Irvine, 1977; Smith, 1982b).

The Ridge Route Formation consists of more than 9,000 m of alluvial-fan, fluvial, and lacustrine shoreline deposits shed into the basin from source areas located on the northeast flank of Ridge Basin (Link, 1982e). The alluvial-fan deposits accumulated as conglomerate, breccia, and sandstone chiefly adjacent to basement that was uplifted along the Clearwater and Liebre fault zones (Fig. 7). The fan deposits grade basinward into fluvial deposits of meandering and braided types, which consist chiefly of cross-stratified conglomeratic sandstone and mudstone (Fig. 8G; Link, 1984). The shoreline deposits include nearshore, beach, delta, and fan-delta sequences that interfinger with the lacustrine beds of the Peace Valley Formation. A number of the prominent sandstone members of the Ridge Route Formation extend entirely across Ridge Basin to interfinger with the Violin Breccia (Figs. 6 and 8H); these units grade laterally into sublacustrine deltaic and turbidite complexes that were at times exposed as a result of falling lake levels (Harper and Osborne, 1982; Hollywood and Osborne, 1982; Link, 1982f; Wood and Osborne, 1982).

The youngest part of the Ridge Basin Group, the Hungry Valley Formation, is about 1,400 m thick and consists chiefly of a lower member deposited by braided streams and an upper member deposited as alluvial-fan deposits (Crowell, 1982c; Ramirez, 1983). The alluvial-fan deposits overlap the Liebre and San Gabriel fault zones, but are truncated by the San Andreas fault (Crowell, 1982c). The sedimentary rocks of the Hungry Valley Formation record final filling of Ridge Basin, and derivation of detritus from the Frazier Mountain area and various sources northeast of the San Andreas fault. Ramirez (1983) suggests that part of the San Bernardino Mountains, now located 220 km to the southeast of and across the San Andreas fault from Ridge Basin, formed the source area of much of the Hungry Valley Formation.

Paleocurrents and Maximum Clast Size

Paleocurrent measurements from the various stratigraphic units of the Ridge Basin Group indicate that (1) the Violin Breccia was derived from the southwestern margin of the basin across the San Gabriel fault, (2) the Ridge Route Formation was derived from the northeastern flank of the basin, with some units clearly shed off uplifts along the Clearwater and Liebre fault zones, (3) the axial deposits of the basin, which include parts of the Violin Breccia, Ridge Route Formation, and Peace Valley Formation, were generally transported southeastward down the axis of the basin, and (4) the younger Hungry Valley Formation was derived from the north, northwest, and west (Fig. 6; Link, 1982d). Paleowind directions were generally along the basin axis but toward the north-northwest (Link, 1982d).

The largest clasts in the Ridge Basin are found in the Violin Breccia adjacent to the San Gabriel fault margin. Angular blocks of basement rocks as large as several meters are found locally in the Violin Breccia, and the maximum sizes of the blocks generally increases toward the fault. Conglomerate along the northeastern margin of the basin contains generally more rounded fragments, although it is locally very coarse, including sedimentary breccia adjacent to the Clearwater and Liebre fault zones, and it generally increases in coarseness toward the northeastern basin margin. In the northwestern part of the basin and upper part of the Ridge Basin Group, conglomerate of the Hungry Valley Formation generally coarsens toward the northern basin margin.

Tectonic and Depositional History

The Ridge Basin developed as a stretched and sagged crustal wedge northeast of the San Gabriel fault in the area where the fault had a curvilinear trace (Crowell, 1982a). The San Gabriel fault was the principal zone of right slip during most of the history of the basin, and its curved trace led to uplift of a source area along the southwestern margin of the basin that migrated northwestward through time, shedding detritus that formed the Violin Breccia into the basin (Figs. 9, 10A; Crowell, 1974a, b; 1975; 1982b). The Clearwater and Liebre fault zones became active as the basin widened and grew, and most of the basin fill was transported into the basin by rivers draining source areas located to the northeast. The basin depocenter migrated northwestward through time as the source of the Violin Breccia moved northward along the southwest side of the San Gabriel fault (Fig. 10B); as a result, the successive alluvial fans that form the Violin Breccia become younger northwestward and form a shingled pattern that coincides with the northwestward migration of the depocenter. Within the axial part of the wedge-shaped basin, sediments were transported southeastward down the axis of the basin concurrently with northwestward migration of the depocenter. Abundant evidence of syn-sedimentary deformation in the basin fill suggests possible penecontemporaneous seismicity (Fig. 8I, J). When the principal locus of right slip transferred to the San Andreas fault, the other basin-margin faults became inactive, the basin was folded, uplifted, and transported northward about 220–240 km.

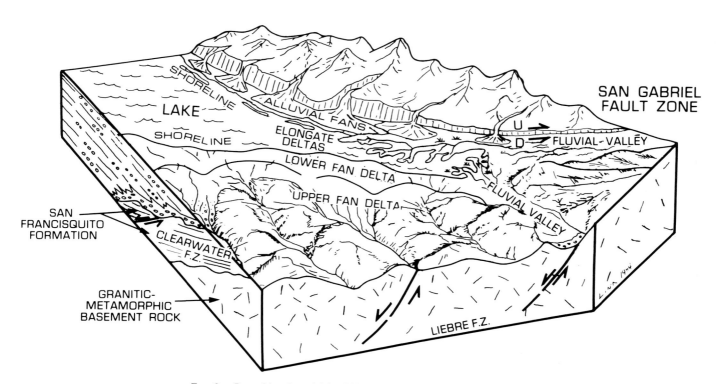

FIG. 9.—Depositional model for Ridge Basin (from Link, 1984, Fig. 13).

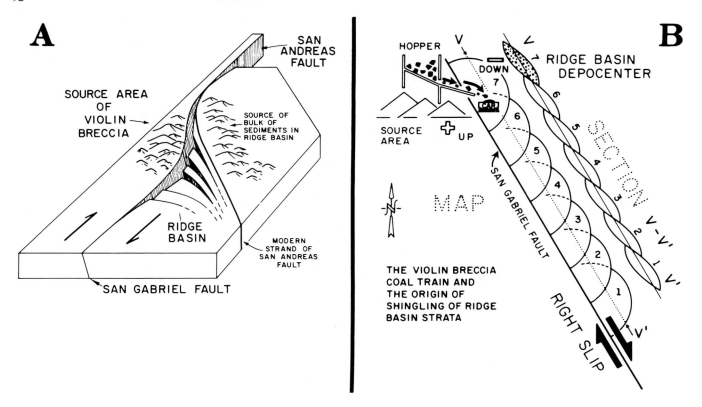

FIG. 10.—Tectonic and depositional models for origin and filling of Ridge Basin. A) block diagram showing origin of Ridge Basin east of the San Gabriel fault (from Crowell, 1974b, Fig. 5). B) model of Ridge Basin infilling and depocenter migration (from Crowell, 1982b, Fig. 9).

LITTLE SULPHUR CREEK BASINS

Setting

The Little Sulphur Creek Basins, which developed along the Maacama fault zone, form part of a large number of Neogene nonmarine sedimentary basins in northern California (Fig. 1C). The California Coast Ranges north of San Francisco are characterized by a wide zone of right-lateral shear associated with the San Andreas fault system (Mc-Laughlin, 1978, 1981). In addition to the San Andreas fault itself, the Healdsburg–Rogers Creek, Maacama, Green Valley, and Bartlett Springs fault zones accommodate major amounts of right slip. As a result of oblique pull-apart extension between branching splays of these faults, sedimentary basins or depressions developed in the Pliocene and Pleistocene that include the Ukiah Valley, Little Lake Valley, Round Valley, Clear Lake Basin, and the three Little Sulphur Creek Basins.

The Little Sulphur Creek Basins extend along the Maacama fault zone for about 13 km (Fig. 11; McLaughlin and Nilsen, 1980, 1982; Nilsen et al., 1980). They are bounded on the southwest by branching splays of the active right-lateral Maacama fault zone that strike N30°–40°W (Fig. 12A). Conjugate thrusts that strike west and northwest, and left-lateral fault splays that also strike northwest, intersect the main northwest-striking fault zone, separating the basin fill into three distinct 1.5-to-2.0-km-wide pull-apart basins (Fig. 11). The basins have a cumulative fill of about 5,000 m.

Recent seismicity (Bufe et al., 1981) and youthful physiographic fault features associated with the Maacama fault zone indicate that it is still active (Donnelly et al., 1976; Herd and Helley, 1977; Herd, 1978; Pampeyan et al., 1981). First-motion solutions for several earthquakes indicate right slip with maximum extension parallel to vectors oriented approximately N60°W–S60°E (Bufe et al., 1981). On the basis of offset of the Sonoma Volcanics, and of an offset of similar magnitude of late Mesozoic ophiolitic rocks (McLaughlin, 1981), maximum displacement along the Maacama fault zone is considered to be approximately 20 km or less.

The Little Sulphur Creek Basins are surrounded by complexly deformed late Mesozoic rocks that consist of (1) melange on the southeast and west, with abundant blocks of chert and blueschist, belonging to the Franciscan assemblage; (2) an intact terrane of weakly metamorphosed graywacke on the northeast and northwest, also assigned to the Franciscan assemblage; and (3) an ophiolite terrane on the southeast composed of ultramafic rocks, gabbro, diabase, mafic breccia, and basalt (Fig. 11). Almost all the sedimentary fill in the Little Sulphur Creek Basins appears to be derived from these three late Mesozoic terranes.

The age of the fill of the Little Sulphur Creek Basins is poorly known. However, on the basis of a mollusc-bearing clast of sandstone, regional stratigraphic and structural relations, and a probable genetic relation between the age of basin formation and the northward propagation of the San Andreas transform system during the Neogene (Blake et al.,

1978; Dickinson and Snyder, 1979), the basins probably began to develop between 3 and 4 Ma and have been filled and deformed since then.

Structure

The fill of the Little Sulphur Creek Basins has been deformed by tilting, folding, and probably by rotation between the basin-margin faults (Fig. 11). Tilting, indicated by very steeply dipping beds, is most prominent along the southwest side of the southeasternmost basin. In that area, conglomerate and breccia have vertical dips and are truncated against a major active strand of the Maacama fault

zone. The basin fill forms a series of stacked basin-wide sequences that form northwest-dipping ridges that have been folded into a broad syncline (Fig. 12B). Synclinal axes of the southeastern and central basins plunge to the northwest. The synclinal axis of the northwestern basin plunges both northwestward and southeastward. Prominent right-lateral drag folds are locally present along the fault zone; in addition to the drag folds, major synclinal folds in the basin fill have axes that trend N40°–69°W, oblique to the trend of the Maacama fault zone. Along the northeast margins of the basins, beds are only slightly tilted, without noticeable oversteepening or drag folding, suggesting lesser uplift and deformation along that margin.

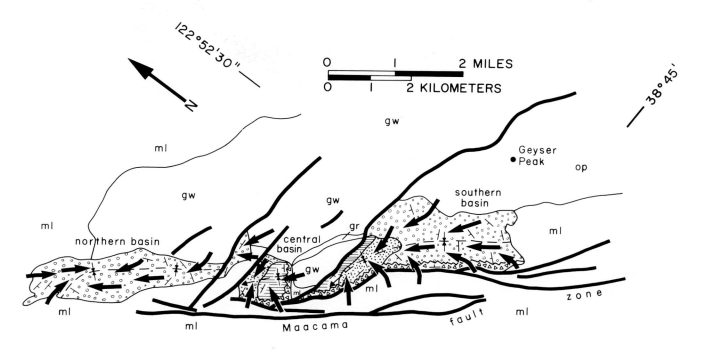

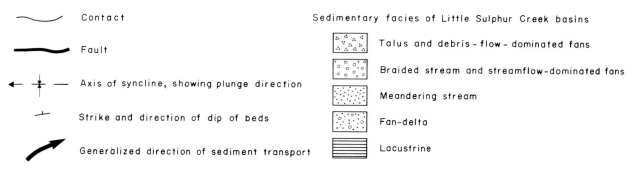

EXPLANATION:

— Contact

— Fault

←—+— Axis of syncline, showing plunge direction

⊥ Strike and direction of dip of beds

➤ Generalized direction of sediment transport

Sedimentary facies of Little Sulphur Creek basins

Talus and debris-flow-dominated fans

Braided stream and streamflow-dominated fans

Meandering stream

Fan-delta

Lacustrine

Franciscan rocks

gr, Greenstone
gw, Graywacke
ml, Melange
op, Ophiolite

FIG. 11.—Generalized geologic map of the Little Sulphur Creek area, showing the three Little Sulphur Creek Basins, principal faults, distribution of surrounding rock types, depositional facies, and generalized directions of sediment transport. For location, see Figure 1C.

F<small>IG</small>. 12.—Photographs of Little Sulphur Basins and their chief depositial facies. A) view northwest along trace of Maacama fault zone (dashed line) showing outcrops of basin-margin conglomerate on ridge crest to right. B) view northeastward toward the axis of the basin showing stacked sequences (arrows) in the central part of the basins. C) large angular block of chert in talus and debris-flow-dominated alluvial-fan deposits along southwestern margin of the basin. D) poorly sorted debris-flow-dominated alluvial-fan deposits along southwestern margin of basin. E) well-stratified conglomerate and sandstone deposited by streamflow-dominated alluvial fans and braided streams. F) cross-stratified sandstone and interbedded conglomerate deposited by braided streams. G) thin-bedded basin-axis lacustrine deposits. H) basin-axis lacustrine turbidites; hammer circled for scale. I) fining- and thinning-upward sequence probably deposited by meandering stream in axial part of basin. J) conglomeratic fan-delta deposits that prograded southwestward (to left) over mollusc-bearing lacustrine mudstone and siltstone. K) syn-sedimentary slump in lacustrine-margin deposits. L) mass of conglomerate injected upward about 35 cm into pebbly siltstone.

Depositional Facies

The distribution of nonmarine facies in the three Little Sulphur Creek Basins is complex and probably reflects an irregular paleogeography (Fig. 11). Conglomeratic braided-stream deposits dominate the sedimentary fill of the southern and northern basins. Other important deposits include fault-scarp talus and debris-flow breccia, meandering-stream, conglomeratic fan-delta, and lacustrine deposits (Fig. 13). Representative measured sections for each depositional facies were shown by McLaughlin and Nilsen (1982, Fig. 5).

The talus deposits include some conglomeratic debris-flows

and landslides. Talus deposits are most prominent along the southwestern margin of the southern basin adjacent to the Maacama fault, where angular blocks of chert and blueschist as long as several meters are enclosed in coarse, unsorted and angular breccia (Fig. 12C). Abundant debris-flow conglomerate (Fig. 12D) intertongues with the extremely coarse talus deposits, and both grade laterally northeastward into braided-stream deposits. Debris-flow conglomerate is also common along most basin margins, where it interfingers toward the basin axes with braided-stream deposits.

The braided-stream deposits consist chiefly of fining-up-

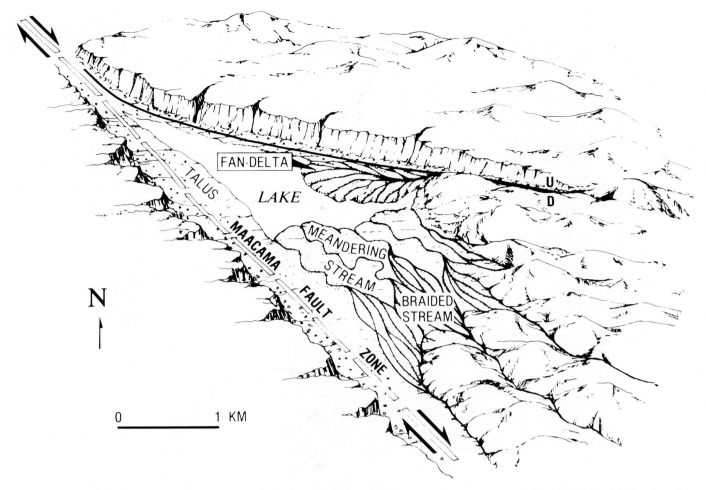

FIG. 13.—Depositional and tectonic model for the Little Sulphur Creek Basins showing the distribution of chief depositional facies (from Mc-Laughlin and Nilsen, 1982, Fig. 9).

ward couplets of conglomerate and sandstone (Fig. 12E, F). Shale and mudstone are generally absent. Each of the conglomerate and sandstone couplets, which range from about 0.5 m to as much as 5 m in thickness, is thought to represent a gravel bar or flood unit developed in a braided-stream channel. The base of each couplet is channeled into underlying conglomerate, conglomeratic sandstone, or sandstone. Typically, the conglomerate is massive or crudely parallel-stratified, and the overlying sandstone is massive, parallel-stratified, or low-angle cross-stratified. The conglomerate has well-developed imbrication and long-axis orientation, which is characteristic of streamflow deposits.

Lacustrine deposits of both deep and shallow origin crop out at the northwest end of the southern basin and in the central basin (Fig. 12G, H). Shoreline deposits contain mud cracks and other features indicative of periodic desiccation. The deeper-water lacustrine deposits consist of thin-bedded turbidite sandstone and siltstone that alternate with varie-gated mudstone and claystone.

The meandering-stream deposits are abundant in the northwestern part of the southern basin (Fig. 12I). They consist of fining-upward sequences that are about 5 to 10 m thick. The fining-upward sequences consist, in as-cending order, of a basal channeled conglomerate or sand-stone that is massive or crudely parallel-stratified, trough cross-stratified sandstone, ripple-laminated sandstone and siltstone, and mudstone. The locally thick mudstone units that are floodplain deposits contain abundant disseminated tubular structures and nodules of calcium carbonate inferred to be of paleosol origin, thin seams of coal, and abundant plant debris. Some of these sequences may also be repre-sentative of braided streams, with extensive floodplain shale intervals.

The conglomeratic fan-delta deposits of the central basin contain conglomeratic beds that grade laterally through a zone of foresets with amplitudes locally greater than 10 m into thinly interbedded sandstone and mudstone (Fig. 12J). The fan-delta deposits prograde basinward into fossiliferous lacustrine strata. The well-defined topset, foreset, and bot-tomset beds of the fan delta resemble those of classical Gil-bert-type deltas.

Paleocurrents and Maximum Clast Size

Based on a total of 121 paleocurrent measurements of clast imbrication and long-axis orientation, cross-stratifi-cation, and flute and groove casts, McLaughlin and Nilsen

(1982, Fig. 7) demonstrated a pattern of basin infilling from the basin margins and longitudinal transport along the basin axes, toward the northwest and southeast (Fig. 11). The paleocurrent pattern suggests the presence of source areas on most margins of the basin. The predominant transport in the southern and central basins was toward the northwest along the basin axes in the direction of plunge of the basin synclines; in the northern basin it was toward the northwest and southeast, parallel to the basin axis.

Based on many measurements of maximum clast size in the basins, McLaughlin and Nilsen (1982, Fig. 6) determined that the largest clasts are located along the southwestern flank of the southernmost basin adjacent to the active Maacama fault zone. Here large angular clasts several meters long are embedded in generally finer-grained, angular conglomerate. These large clasts appear to represent blocks that slid or slumped into the basin along either a syndepositionally active fault zone or an uplifted block of basement rocks along the basin margin.

Cobble-sized debris that is generally well-rounded is common in many other parts of the basins, and forms a chief component of braided- and meandering-stream, fan-delta, and alluvial-fan deposits. Based on the overall pattern of paleocurrents, maximum clast sizes, depositional facies, and clast composition, it is clear that the chief source area of the basin fill was located to the southeast (Fig. 11).

Tectonic and Depositional History

The depositional facies of the Little Sulphur Creek Basins and their relation to basin-margin faults strongly imply that deposition took place concurrently with uplift and with oblique slip along the Maacama fault zone (Fig. 14). The buttressing of the megabreccias against a major strand of the Maacama fault zone on the southwest side of the southern basin, and the sheared surfaces of numerous clasts in the megabreccias, suggests deposition as fault-scarp talus. Abundant syn-sedimentary slumps, dewatering structures, liquefaction features, and minor angular unconformities in finer-grained sediments, together with the megabreccias and structural setting, strongly suggest that basin-margin faults were seismically active during sedimentation, when much of the sedimentary fill was water-saturated (Fig. 12K, L).

Although very coarse sedimentary breccia was transported into the southern basin from the southwestern margin as talus cones, landslides, and debris flows, most detritus was transported into Little Sulphur Creek basins from the southeast, where braided-stream and debris-flow deposits form a proximal facies. The braided-stream deposits grade distally into meandering-stream deposits. Here, more extensive floodplains developed, and sand and silt were transported into a deep lake extending from the northwest end of the southern basin into the central basin. Turbidity currents were generated where the meandering streams entered into the lake when lake levels and/or stream discharge were high.

From the northeast margin of the central basin, a coarse-grained fan-delta prograded out into the lake, triggering turbidity currents along its front. The source of detritus for the fan-delta may have lain to the southeast, from which sediment was transported along the trend of a secondary west-

trending fault that probably connected to the Maacama fault. Abundant macrofossils in bottomset beds indicate the presence of good faunal habitats within the lake. The lake probably emptied toward the northwest, and varied greatly in size and depth, responding to changing tectonic, climatic, and outlet conditions.

The basins appear to have formed initially by east-west oblique pull-apart between en echelon and branching splays of the Maacama fault zone that strike N30°–40°W (Fig. 14). Subsequently, the basin fill was deformed by tilting against basin-margin faults, with the greatest uplift along the southwest side of the basins. Continued uplift and right-lateral shear formed synclinal folds whose axes trend N40°–60°W (Fig. 14), approximately parallel with the direction of maximum extension.

TECTONIC AND SEDIMENTOLOGIC COMPARISON

It is clear from our review of the tectonic and depositional framework of the Hornelen, Ridge, and Little Sulphur Creek Basins that, despite important differences in age, size, thickness of basin fill, and regional setting, the three basins are remarkably similar. In Figure 15, the basins are shown at roughly the same size and orientation to emphasize many of these similarities; the major controlling strike-slip fault is positioned along the left margin of each basin so that the style of infilling, distribution of depositional facies, and tectonic framework can be understood and compared.

Each basin is elongate parallel to the orientation of the major controlling strike-slip fault, and talus, landslide, and debris-flow-dominated alluvial fans were deposited along that margin of the basin. Each of the bounding strike-slip faults is dextral; although the amount of offset along the Hornelen margin is unknown, the San Gabriel and Maacama faults have known offsets of more than 60 km and about 20 km, respectively. In each basin, the major controlling strike-slip fault is throughgoing and forms a continuous bounding margin to the basin, although in the later history of Ridge Basin, the San Gabriel fault was unconformably overlapped by the Hungry Valley Formation as active strike-slip faulting shifted to the San Andreas fault. Because of the active strike slip, source areas along this margin migrated through time, resulting in a similar migration of the depocenter in the same direction.

The basins probably became longer and wider through time. Progressive basin-widening faulting along the margins of the basins opposite to the major strike-slip fault, either along high-angle reverse faults, oblique-slip faults, or normal faults, eventually yielded length-to-width ratios of between 3:1 and 5:1. This widening and lengthening process can be most clearly demonstrated in Ridge Basin, where the growth, cessation of movement, and eventual overlap of the Clearwater and Liebre faults can be documented in the field. In Ridge and Hornelen Basins, the migration of the depocenters in a direction parallel to the basin is also clearly demonstrable.

Each basin was subsequently folded into a broad, gently plunging syncline, although numerous subsidiary folds have been mapped, especially in Ridge Basin. The synclines generally plunge in the same direction that the depocenters

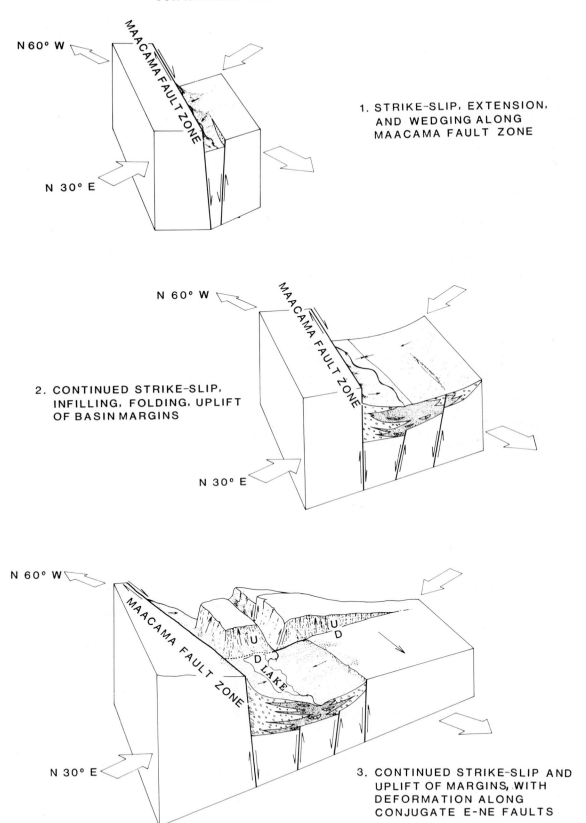

FIG. 14.—Tectonic and sedimentary evolution of Little Sulphur Creek Basins within the framework of active strike-slip movement along the Maacama fault (from McLaughlin and Nilsen, 1982, Fig. 8).

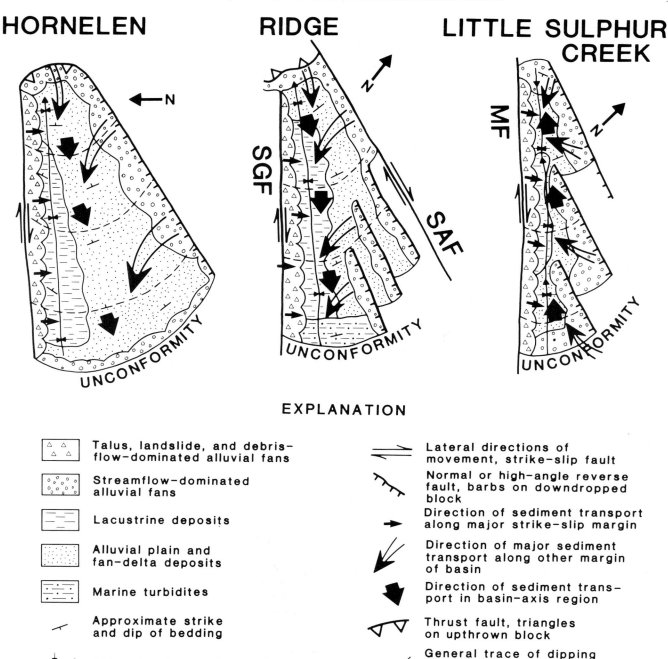

Fɪɢ. 15.—Tectonic and sedimentary comparison of Hornelen, Ridge, and Little Sulphur Creek Basins. Note that the scales are variable and that the three basins, of varying size, are drawn at roughly the same size for purposes of comparison. The basins have also been reoriented (see respective north arrows) to facilitate the comparison. Abbreviations: MF, Maacama fault; SAF, San Andreas fault; SGF, San Gabriel fault.

migrated, although the structural termination of each basin in the direction of plunge is different. Differently oriented thrust faults terminate the eastern and northwestern margins of Hornelen and Ridge Basins, respectively, and at the northwestern margin of the Little Sulphur Creek Basins, the plunge of the synclinal axis apparently reverses. The opposite end of the basins in each case is marked by a major unconformity.

The debris-flow-dominated alluvial-fan deposits adjacent to the major strike-slip-faulted margin form a narrow strip of coarser-grained deposits that extend generally less than 2 km into the basin-axis region. These alluvial fans were apparently small and steep, and in many areas shed debris directly into lakes of various depths. Subaerial debris-flow deposits can be traced laterally into sublacustrine debris-flow deposits.

The distribution of the lacustrine facies in each basin shifted toward the active strike-slip basin margin, suggesting that the basin floors were tilted toward that margin. This tilting or slope of the basin floor explains the formation of larger streamflow-dominated alluvial fans along the opposite basin margin. These streamflow-dominated fans transported better sorted, better rounded, and generally finer-grained detritus that was deposited on larger and more gently sloping fans. The fans graded laterally into alluvial-plain and fan-delta deposits that were transported into the lakes. Most sediment was introduced into each basin from this opposite margin, and some coarse-grained units extend all the way across the basins to interfinger with the debris-flow-dominated alluvial-fan deposits.

Basin-axis sediment transport is markedly opposite to the direction of depocenter migration in Hornelen and Ridge Basins but generally in the same direction as the depocenter migration in the Little Sulphur Creek Basins, except at their northwestern end, where the paleocurrent directions appear to reverse. It is possible that major uplifts to the south related to volcanic activity yielded a northwesterly plunge of the basin floor of the Little Sulphur Creek Basins.

Sedimentation in each of the basins was rapid, as is typical for many basins generated along strike-slip faults. We calculate average sedimentation rates of 2.5 m/1,000 yr for Hornelen Basin, 3 m/1,000 yr for Ridge Basin, and 2.5 m/1,000 yr for the Little Sulphur Creek Basins, making conservative assumptions about the duration of basin filling from published time scales (Palmer, 1983). These sedimentation rates are very high and reflect active syndepositional tectonism, basin development, uplift of surrounding source areas, subsidence of the basin floor, and lateral migration of basin depocenters. The history of each basin appears to be similar, including (1) rapid basin development adjacent to a major strike-slip fault during a phase of transtension, (2) basin lengthening and widening during continued active strike-slip faulting, and (3) rapid closure, folding and uplift of the basin fill during a phase of transpression. This last phase can be associated with shifting of strike-slip activity to another fault, as is the case with Ridge Basin, or to other complex changes in the style of strike-slip activity.

SUMMARY AND CONCLUSIONS

Marine sedimentary basins that have developed along strike-slip faults have been of great interest because of the large amounts of petroleum resources found in them (Moody, 1973; Harding, 1974, 1976). For example, many of the Tertiary basins in California associated with the San Andreas fault system, such as the Los Angeles and Ventura Basins, have been extremely productive (Blake et al., 1978; Howell et al., 1980). The Ridge Basin in southern California, although petroleum has not been produced from it, has been extensively explored, because the fine-grained lacustrine facies has a very high content of total organic matter (Link and Smith, 1982) and the basal part of its fill consists of marine deposits.

In this paper we have compared two other well-studied basins with Ridge Basin in an attempt to gain a better understanding of the nature and organization of the sedimentary infilling of basins that develop adjacent to strike-slip faults. Our comparison has yielded a surprisingly similar sedimentary and tectonic history in each case. We feel that the following list of similar characteristics of the basins is collectively significant for recognizing these basins as being of strike-slip origin and may have relevance for the general recognition of basins developed adjacent to strike-slip faults. Not all of the criteria, of course, are applicable only to strike-slip basins; many are also applicable to simple extensional rift basins as well.

(1) The basins were formed adjacent to important continental-margin or intracontinental strike-slip faults.

(2) Although the basins are fault-bounded along more than one margin, a major strike-slip fault forms the most prominent basin margin.

(3) The basins are elongate parallel to the major controlling strike-slip fault.

(4) The basins are synclinal in overall structure.

(5) The basins were initially formed by extension and subsequently compressed and folded, probably as a result of transtension followed by transpression.

(6) The basins are asymmetrical, with their structurally deepest parts close to and subparallel to the major syn-depositionally active strike-slip margin.

(7) The basins do not contain any record of syn-depositional volcanism or plutonism. Volcanism may, however, be common in some other similar basins.

(8) The basins, although small, are characterized by a wide variety of depositional facies, including talus, landslide, alluvial-fan, braided and meandering fluvial, deltaic, fan-delta, shoreline, shallow and deep lacustrine, turbidite, chemical precipitate, and algal limestone deposits.

(9) The basin fill is characterized dominantly by axial infilling, subparallel to the major strike-slip basin margin.

(10) The basin-margin deposits are distinctive. Along the syn-depositionally active strike-slip margin, small debris-flow-dominated alluvial fans that contain coarse sedimentary breccia and conglomerate were deposited. Along the inactive or less active margins, larger streamflow-dominated alluvial-fan and fluvial deposits that contain finer-grained conglomerate and little breccia were deposited.

(11) The basin fill is characterized by abrupt facies changes.

(12) The basin fill was derived from multiple basin-margin sources that changed through time as a result of continued lateral movement along the basin-margin faults.

(13) The basin fill has a varied and complex petrography as a result of the multiple sources that change through time and space.

(14) The basins contain very thick sedimentary sections compared to their area.

(15) The basin fill is characterized by high rates of sedimentation, roughly 2.5–3.0 m/1,000 yrs.

(16) The basins are characterized by depocenters that migrated in the same direction as that of the source terrane opposite the major strike-slip margin of the basin; this migration direction is generally opposite to that of the general flow of axial paleocurrents.

(17) The basin fill is characterized by abundant syn-sedimentary slumping and deformation, possibly in response to seismicity along the basin margins.

These criteria may define some of the chief characteristics of certain types of basins that develop adjacent to strike-slip faults. The criteria are clearly developed for nonmarine basins and most of the aspects of the sedimentary facies will not be applicable to shallow- and deep-marine basins such as those of offshore New Zealand described in Ballance and Reading (1980), the Southern California Borderland (Moore, 1969), and the Caribbean region (Schubert, 1982).

Basins along strike-slip faults can originate in a number of different ways, and the depositional and tectonic styles of these basins vary accordingly. For example, basins can develop as a result of extension adjacent to regions of fault curvature, as suggested by Crowell (1974) for the Ridge Basin. Basins can also develop in areas where strike-slip faults bifurcate and diverge, producing alternating uplifted and downdropped blocks, and in areas where strike-slip faults are oriented in en echelon fashion (Quennell, 1958; Rodgers, 1980). Aydin and Nur (1982) concluded from an analysis of many pull-apart basins in the world that the basin width is controlled by the initial fault geometry and that the basin length is a function of increasing fault displacements.

Nevertheless, there appear to be a number of additional modern and ancient pull-apart basins associated with strike-slip faults, which have characteristics similar to the three basins described herein. Late Cenozoic basins in Venezuela along the Bocono fault zone appear to be almost identical (Schubert, 1980). A number of basins in the San Francisco Bay region of California have been filled in an analogous way (Aydin and Page, 1984). The Lake Hazar and other basins in eastern Turkey, along the East Anatolian fault, are currently being filled in a very similar manner (Hempton et al., 1983; Hempton and Dunne, 1984). Thus, the general style of basin tectonism and sedimentation in the three basins that we have compared is probably applicable to many other basins. In particular, the general cycle for strike-slip basins postulated by Reading (1980) of (1) transtension to form the basin, (2) basin filling, and (3) transpression to fold and uplift the basin fill, is directly applicable to the basins we have considered herein.

ACKNOWLEDGMENTS

Nilsen acknowledges the following individuals for helpful field excursions: Inge Bryhni through the Hornelen Basin in 1964; Ron Steel through the Hornelen Basin in 1982; and John Crowell and Martin Link through the Ridge Basin on several occasions between 1973 and 1983. For our work in the Little Sulphur Creek Basins, we thank Henry Ohlin for assistance in field work at various times, E. A. Griffin for compilation of data, and G. G. Zuffa for help in measurement of sections and studies of sandstone petrography. We thank the many ranchers and property owners along Little Sulphur Creek for permitting us to work on their land. We also thank R. J. Steel, J. A. Bartow, R. H. Brady, M. H. Link, E. J. Helley, Harvey Kelsey and several other colleagues at the U.S. Geological Survey for helpful discussions of the Little Sulphur Creek Basins in the field. Finally, we thank J. A. Bartow, P. F. Ballance, Hugh McLean, R. J. Steel, and the editors of this volume for constructive reviews of this paper and Mary Milan for her patient and skillful typing.

REFERENCES

AXELROD, D. I., 1982, Vegetation and climate during middle Ridge Basin deposition, *in* Crowell, J. C., and Link, M. H., eds., Geologic History of Ridge Basin, Southern California: Society of Economic Paleontologists and Mineralogists, Pacific Section, p. 247–252.

AYDIN, ATILLA, AND NUR, AMOS, 1982, Evolution of pull-apart basins and their scale independence: Tectonics, v. 1, p. 91–105.

AYDIN, ATILLA, AND PAGE, B. M., 1984, Diverse Pliocene-Quaternary tectonics in a transform environment, San Francisco Bay Region, California: Geological Society of America Bulletin, v. 95, p. 1303–1317.

BALLANCE, P. F., AND READING, H. G., eds., 1980, Sedimentation in oblique-slip mobile zones: International Association of Sedimentologists Special Publication No. 4, 265 p.

BLAKE, M. C., JR., CAMPBELL, R. H., DIBBLEE, T. W., JR., HOWELL, D. G., NILSEN, T. H., NORMARK, W. R., VEDDER, J. C., AND SILVER, E. A., 1978, Neogene basin formation in relation to plate-tectonic evolution of San Andreas fault system, California: American Association of Petroleum Geologists Bulletin, v. 62, p. 344–372.

BRYHNI, INGE, 1964a, Relasjonen mellom senkaledonsk tektonikk og sedimentasjon ved Hornelens og Hasteinens devon: Norges Geologiske Undersøkelse, v. 233, p. 10–25.

———, 1964b, Migrating basins on the Old Red Continent: Nature, v. 202, p. 384–385.

———, 1964c, Sediment structures in the Hornelen series: Norsk Geologisk Tiddskrift, v. 44, p. 486–488.

———, 1975, The West Norwegian basins of Old Red Sandstone: Ninth International Congress of Sedimentology, Nice, France, Theme 5, p. 111–117.

———, 1978, Flood deposits in the Hornelen Basin, west Norway (Old Red Sandstone): Norsk Geologisk Tidsskrift, v. 58, p. 273–300.

BUFE, C. G., MARKS, S. M., LESTER, F. W., LUDWIN, R. S., AND STICKNEY, M. C., 1981, Seismicity of the Geysers-Clear Lake region, *in* McLaughlin, R. J., and Donnelly-Nolan, J. M., eds., Research in the Geysers-Clear Lake Geothermal Area, Northern California: United States Geological Survey Professional Paper 1141, p. 129–137.

CLEMENTS, THOMAS, 1937, Structure of southeastern part of Tejon quadrangle, California: American Association of Petroleum Geologists Bulletin, v. 21, p. 212–232.

CROWELL, J. C., 1950, Geology of Hungry Valley area, southern California: American Association of Petroleum Geologists Bulletin, v. 34, p. 1623–1646.

———, 1952, Probable large lateral displacement on the San Gabriel fault, southern California: American Association of Petroleum Geologists Bulletin, v. 36, p. 2026–2035.

———, 1954a, Geology of the Ridge Basin area, California: California Division Mines Bulletin 170, Map sheet 7.

———, 1954b, Strike-slip displacement of the San Gabriel fault, southern California: California Division of Mines Bulletin 170, Ch. 4, p. 49–52.

———, 1974a, Origin of late Cenozoic basins in southern California, *in* Dickinson, W. R., ed., Tectonics and Sedimentation: Society of Economic Paleontologists and Mineralogists Special Publication No. 22, p. 190–204.

———, 1974b, Sedimentation along the San Andreas fault, California, *in* Dott, R. H., Jr., and Shaver, R. H., eds., Modern and Ancient Geosynclinal Sedimentation: Society of Economic Paleontologists and Mineralogists Special Publication No. 19, p. 292–303.

———, 1975, The San Gabriel fault and Ridge Basin, *in* Crowell, J. C., ed., San Andreas Fault in Southern California: California Division of Mines and Geology Special Report 118, p. 208–233.

———, 1982a, The tectonics of Ridge Basin, Southern California, *in* Crowell, J. C., and Link, M. H., eds., Geologic History of Ridge Basin, Southern California: Society of Economic Paleontologists and Mineralogists, Pacific Section, p. 25–42.

———, 1982b, The Violin Breccia, Ridge Basin, *in* Crowell, J. C., and

Link, M. H., eds. Geologic History of Ridge Basin, Southern California: Society of Economic Paleontologists and Mineralogists, Pacific Section, p. 89–98.

———, 1982c, Pliocene Hungry Valley Formation, Ridge Basin, southern California, *in* Crowell, J. C., and Link, M. H., eds., Geologic History of Ridge Basin, Southern California: Society of Economic Paleontologists and Mineralogists, Pacific Section, p. 143–150.

CROWELL, J. C., AND LINK, M. H., eds. 1982a, Geologic history of Ridge Basin, southern California: Society of Economic Paleontologists and Mineralogists, Pacific Section, 304 p.

CROWELL, J. C., AND LINK, M. H., 1982b, Ridge Basin, southern California: Introduction, *in* Crowell, J. C., and Link, M. H., eds., Geologic History of Ridge Basin, Southern California: Society of Economic Paleontologists and Mineralogists, Pacific Section, p. 1–4.

CROWELL, J. C., ET AL., 1982, Geologic map and cross sections, Ridge Basin, southern California: Society of Economic Paleontologists and Mineralogists, Pacific Station, 2 plates, 5p.

DICKINSON, W. R., AND SNYDER, W. S., 1979, Geometry of triple junctions related to San Andreas transform: Journal of Geophysical Research, v. 84, B2, p. 561–572.

DONNELLY, J. M., McLAUGHLIN, R. J., GOFF, F. E., AND HEARN, B. C., 1976, Active faulting in the Geysers-Clear Lake area, northern California: Geological Society of America Abstracts with Programs, v. 8., p. 369–370.

EATON, J. E., 1939, Ridge Basin, California: American Association of Petroleum Geologists Bulletin, v. 23, p. 517–558.

ENSLEY, R. A., AND VEROSUB, K. L., 1982a, Biostratigraphy and magnetostratigraphy of southern Ridge Basin, central Transverse Ranges, California, *in* Crowell, J. C., and Link, M. H., eds., Geologic History of Ridge Basin, Southern California: Society of Economic Paleontologists and Mineralogists, Pacific Section, p. 13–24.

ENSLEY, R. A., AND VEROSUB, K. L., 1982b, A magnetostratigraphy study of the sediments of the Ridge Basin, southern California and its tectonic and sedimentologic implications: Earth and Plantary Science Letters, v. 59, p. 192–207.

FORESTER, R. M., AND BROUWERS, E. M., 1982, Paleoenvironmental implications of some estuarine and non-marine ostracodes from Ridge Basin, southern California, *in* Crowell, J. C., and Link, M. H., eds., Geologic History of Ridge Basin, Southern California: Society of Economic Paleontologists and Mineralogists, Pacific Section, p. 239–246.

GEE, D. G., 1975, A tectonic model for the central part of the Scandinavian Caledonides: American Journal of Science, v. 275, p. 468–515.

———, 1978, Nappe displacement in the Scandinavian Caledonides: Tectonophysics, v. 47, p. 393–410.

GLOPPEN, T. G., AND STEEL, R. J., 1981, The deposits, internal structure and geometry in six alluvial fan-fan delta bodies (Devonian-Norway)—a study in the significance of bedding sequence in conglomerates, *in* Ethridge, R. G., and Flores, R. M., eds, Recent and Ancient Nonmarine Depositional Environments: Models for Exploration: Society of Economic Paleontologists and Mineralogists Special Publication No. 31, p. 49–69.

HARDING, T. P., 1974, Petroleum traps associated with wrench faults: American Association of Petroleum Geologists Bulletin, v. 58, p. 1290–1304.

———, 1976, Predicting productive trends related to wrench faults: World Oil, v. 182, p. 64–69.

HARPER, A. S., AND OSBORNE, R. H., 1982, Fluviolacustrine facies of the Fisher Springs Sandstone Member, Mio-Pliocene Ridge Route Formation, Ridge Basin, southern California, *in* Crowell, J. C., and Link, M. H., eds., Geologic History of Ridge Basin, Southern California: Society of Economic Paleontologists and Mineralogists, Pacific Section, p. 105–114.

HEMPTON, M. R., AND DUNNE, L. A., 1984, Sedimentation in pull-apart basins: Active examples in eastern Turkey: Journal of Geology, v. 92, p. 513–530.

HEMPTON, M. R., DUNNE, L. A., AND DEWEY, J. F., 1983, Sedimentation in an active strike-slip basin, southeastern Turkey: Journal of Geology, v. 91, p. 401–412.

HERD, D. G., 1978, Intracontinental plate boundary east of Cape Mendocino, California: Geology, v. 6, p. 721–725.

HERD, D. G., AND HELLEY, E. J., 1977, Faults with Quaternary displacement, northwestern San Francisco Bay Region, California: United States Geological Survey Miscellaneous Field Studies Map MF-818, Scale 1:125,000.

HOEG, O. E., 1936, Norges fossile flora: Naturen, v. 60, p. 53–96.

HOLLYWOOD, J. M., AND OSBORNE, R. H., 1982, Sedimentology of the Frenchman Flat Sandstone Member of the Ridge Route Formation, Ridge Basin, southern California, *in* Crowell, J. C., and Link, M. H., eds., Geologic History of Ridge Basin, Southern California: Society of Economic Paleontologists and Mineralogists, Pacific Section, p. 115–126.

HOWELL, D. G., CROUCH, J. K., GREEN, H. G., McCULLOCH, D. S., AND VEDDER, J. G., 1980, Basin development along the late Mesozoic and Cainozoic California margin: A plate tectonic margin of subduction, oblique subduction and transform tectonics, *in* Ballance, P. F., and Reading, H. G., eds., 1980, Sedimentation in Oblique-Slip Mobile Zones: International Association of Sedimentologists Special Publication No. 4, p. 43–62.

IRVINE, P. H., 1977, The Posey Canyon Shale—a Pliocene lacustrine despoit of the Ridge Basin, southern California [unpubl. M.A. thesis]: Berkeley, University of California, 87 p.

JARVIK, ERIK, 1949, On the middle Devonian crossopterygians from the Hornelen field in western Norway: Universitet i Bergen Aarbok 1948, Naturvidenskap Raekke, No. 8, 48 p.

KIAER, J., 1918, Fiskerester fra the devoniske sandsten paa Norges vestkyst: Bergens Museum Aarbok, Naturvidenskap Raekke, No. 7, p. 1–17.

KOLDERUP, C. F., 1904, Vestlandets devonisk lagraekker: Naturen, p. 270–276.

———, 1915a, Vestlandets devonfelter og deres plantefossiler: Naturen, p. 217–232.

———, 1915b, Das vorkommen der pflanzenreste: Bergens Museum Aarbok 1914–1915, Naturvidenskap Raekke, No. 9, p. 1–11.

———, 1916, Vestnorges devon: Forhandlingen ved 16 Skandinaviske Naturforskermotet, p. 499–507.

———, 1927, Hornelen devonfelt: Bergens Museum Aarbok 1926, Naturvidenskap Raekke, No. 6, p. 1–56.

LARSEN, V., AND STEEL, R. J., 1978, The sedimentary history of a debris flow-dominated, Devonian alluvial fan—a study of textural inversion: Sedimentology, v. 25, p. 37–59.

LINK, M. H., 1982a, Stratigraphic nomenclature and age of Miocene strata, Ridge Basin, southern California, *in* Crowell, J. C., and Link, M. H., eds., Geologic History of Ridge Basin, Southern California: Society of Economic Paleontologists and Mineralogists, Pacific Section, p. 5–12.

———, 1982b, Petrography and geochemistry of sedimentary rocks, Ridge Basin, southern California, *in* Crowell, J. C., and Link, M. H., eds., Geologic History of Ridge Basin, Southern California: Society of Economic Paleontologists and Mineralogists, Pacific Section, p. 159–180.

———, 1982c, Provenance, paleocurrents, and paleogeography of Ridge Basin, southern California, *in* Crowell, J. C., and Link, M. H., eds., Geologic History of Ridge Basin, Southern California: Society of Economic Paleontologists and Mineralogists, Pacific Station, p. 265–276.

———, 1982d, Slope and turbidite facies of the Miocene Castaic Formation and the lower part of the Marple Canyon Sandstone Member, Ridge Route Formation, Ridge Basin, southern California, *in* Crowell, J. C., and Link, M. H., eds., Geologic History of Ridge Basin, Southern California: Society of Economic Paleontologists and Mineralogists, Pacific Section, p. 79–88.

———, 1982e, Introduction to the facies of the Ridge Route Formation, Ridge Basin, southern California, *in* Crowell, J. C., and Link, M. H., eds., Geologic History of Ridge Basin, Southern California: Society of Economic Paleontologists and Mineralogists, Pacific Section, p. 99–104.

———, 1982f, Sedimentology of the upper Miocene Piru Gorge Sandstone Member of the Ridge Route Formation, Ridge Basin, southern California, *in* Crowell, J. C., and Link, M. H., eds., Geologic History of Ridge Basin, Southern California: Society of Economic Paleontologists and Mineralogists, Pacific Section, p. 127–134.

———, 1983, Sedimentation, tectonics, and offset of Miocene-Pliocene Ridge Basin, California, *in* Andersen, D. W., and Rymer, M. J., eds., Tectonics and Sedimentation Along Faults of the San Andreas System: Society of Economic Paleontologists and Mineralogists, Pacific Section, p. 17–31.

———, 1984, Fluvial facies of the Miocene Ridge Route Formation, Ridge Basin, California, *in* Nilsen, T. H., ed., Fluvial Sedimentation and Related Tectonic Framework, Western North America: Special issue of Sedimentary Geology, v. 38, p. 263–285.

LINK, M. H., AND OSBORNE, R. H., 1978, Lacustrine facies in the Pliocene Ridge Basin Group: Ridge Basin, California, *in* Matter, Albert, and Tucker, M. E., eds., Modern and Ancient Lake Sediments: International Association of Sedimentologists Special Publication No. 2, p. 169–187.

LINK, M. H., OSBORNE, R. H., AND AWRAMIK, STANLEY, 1978, Lacustrine stromatolites and associated sediments of the Pliocene Ridge Route Formation, Ridge Basin, California: Journal of Sedimentary Petrology, v. 48, p. 143–178.

LINK, M. H., AND SMITH, P. R., 1982, Organic geochemistry of Ridge Basin, southern California, *in* Crowell, J. C., and Link, M. H., eds., Geologic History of Ridge Basin, Southern California: Society of Economic Paleontologists and Mineralogists, Pacific Section, p. 191–198.

MCDOUGALL, KRISTIN, 1982, Microfossil assemblages from the Castaic Formation, southern California, *in* Crowell, J. C., and Link, M. H., eds., Geologic History of Ridge Basin, Southern California: Society of Economic Paleontologists and Mineralogists, Pacific Section, p. 219–228.

MCLAUGHLIN, R. J., 1978, Preliminary geologic map and structural sections of the central Mayacmas Mountains and the Geysers steam field, Sonoma, Lake and Mendocino Counties, California: United States Geological Survey Open-File Map 78–389, scale 1:24,000, two sheets.

———, 1981, Tectonic setting of pre-Tertiary rocks and its relation to geothermal resources in the Geysers-Clear Lake area, *in* McLaughlin, R. J., and Donnelly-Nolan, J. M., eds., Research in the Geysers-Clear Lake Geothermal Area, Northern California: United States Geological Survey Professional Paper 1141, p. 3–23.

MCLAUGHLIN, R. J., AND NILSEN, T. H., 1980, Tectonic framework of the Neogene Little Sulphur Creek basins, Sonoma County, California: Geological Society of America Abstracts with Programs, v. 12, p. 119.

MCLAUGHLIN, R. J., AND NILSEN, T. H., 1982, Neogene non-marine sedimentation in small pull-apart basins of the San Andreas fault system, Sonoma County, California: Sedimentology, v. 29, p. 865–876.

MOODY, J. D., 1973, Petroleum exploration aspects of wrench fault tectonics: American Association of Petroleum Geologists Bulletin, v. 57, p. 449–476.

MOORE, D. G., 1969, Reflection profiling studies of the California continental borderland: Structure and Quaternary turbidite basins: Geological Society of America Special Paper No. 107, 142 p.

NATHORST, A. G., 1915, Zur Devonflora des westlichen Norwegens: Bergens Museum Aarbok 1914–1915, Naturvidenskap Raekke, No. 9, p. 12–23.

NILSEN, T. H., 1967, Old Red sedimentation in the Solund district, western Norway, *in* Oswald, D. H., ed., International Symposium on the Devonian System: Alberta Society of Petroleum Geologists, v. 2, p. 1101–1115.

———, 1973, Devonian (Old Red Sandstone) sedimentation and tectonics of Norway, *in* Pitcher, M. G., ed., Arctic Geology: American Association of Petroleum Geologists Memoir 19, p. 471–481.

NILSEN, T. H., MCLAUGHLIN, R. J., GRIFFIN, E. A., AND ZUFFA, G. G., 1980, Fluvial and lacustrine sedimentology of the Neogene Little Sulphur Creek basins, Sonoma County, California: Geological Society of America Abstracts with Programs, v. 12, p. 45.

PALMER, A. R., compiler, 1983, The Decade of North American Geology 1983 geologic time scale: Geology, v. 11, p. 503–504.

PAMPEYAN, E. H., HARSH, P. W., AND COAKLEY, J. M., 1981, Preliminary map showing recently active breaks along the Maacama fault zone between Hopland and Laytonville, Mendocino County, California: United States Geological Survey Miscellaneous Field Studies Map MF-1217, Scale 1:24,000, two sheets.

POLLARD, J. E., STEEL, R. J., AND UNDERSRUD, EINAR, 1982, Facies sequences and trace fossils in lacustrine/fan delta deposits, Hornelen Basin (M. Devonian), western Norway: Sedimentary Geology, v. 32, p. 63–87.

QUENNELL, A. M., 1958, The structural and geomorphic evolution of the Dead Sea Rift: Quarterly Journal of Geological Society of London, v. 114, p. 1–24.

RAMIREZ, V. R., 1983, Hungry Valley Formation: Evidence for 220 kilometers of post Miocene offset on the San Andreas fault, *in* Andersen, D. W., and Rymer, M. J., eds., Tectonics and Sedimentation Along Faults of the San Andreas System: Society of Economic Paleontologists and Mineralogists, Pacific Section, p. 33–44.

READING, H. G., 1980, Characteristics and recognition of strike-slip fault systems, *in* Ballance, P. F., and Reading, H. G., eds., 1980, Sedimentation in Oblique-Slip Mobile Zones: International Association of Sedimentologists Special Publication No. 4, p. 7–26.

RODGERS, D. A., 1980, Analysis of pull-apart basin development produced by *en echelon* strike-slip faults, *in* Ballance, P. F., and Reading, H. G., eds., 1980, Sedimentation in Oblique-Slip Mobile Zones: International Association of Sedimentologists Special Publication No. 4, p. 27–41.

SCHUBERT, CARLOS, 1980, Late Cenozoic pull-apart basins, Bocono fault zone, Venezuelan Andes: Journal of Structural Geology, v. 2, p. 463–468.

———, 1982, Origin of Cariaco basin, southern Caribbean Sea: Marine Geology, v. 47, p. 345–360.

SMITH, P. R., 1982a, Sedimentology and diagenesis of the Miocene Peace Valley Formation, Ridge Basin, southern California, *in* Crowell, J. C., and Link, M. H., eds., Geologic History of Ridge Basin, Southern California: Society of Economic Paleontologists and Mineralogists, Pacific Section, p. 151–158.

———, 1982b, Paleolimnology of Miocene lakes in Ridge Basin, southern California, *in* Crowell, J. C., and Link, M. H., eds. Geologic History of Ridge Basin, Southern California: Society of Economic Paleontologists and Mineralogists, Pacific Section, p. 259–264.

SMITH, P. R., HARPER, A. S., AND WOOD, M. F., 1982, Non-marine trace fossils in the Mio-Pliocene Ridge Basin Group, southern California, *in* Crowell, J. C., and Link, M. H., eds., Geologic History of Ridge Basin, Southern California: Society of Economic Paleontologists and Mineralogists, Pacific Section, p. 253–258.

SQUIRES, R. L., 1979, Middle Pliocene dragonfly nymphs, Ridge basin, Transverse Ranges, California: Journal of Paleontology, v. 53, p. 446–452.

STANTON, R. J., JR., 1982, Molluscan paleoecology and depositional environment of the Miocene Castaic Formation, Ridge Basin, southern California, *in* Crowell, J. C., and Link, M. H., eds., Geologic History of Ridge Basin, Southern California: Society of Economic Paleontologists and Mineralogists, Pacific Section, p. 211–218.

STEEL, R. J., 1976, Devonian basins of western Norway—sedimentary response to tectonism and to varying tectonic context: Tectonophysics, v. 36, p. 207–224.

STEEL, R. J., AND AASHEIM, S. M., 1978, Alluvial sand desposition in a rapidly subsiding basin (Devonian, Norway), *in* Miall, A. D., ed., Fluvial Sedimentology: Canadian Society of Petroleum Geologists Memoir 5, p. 385–412.

STEEL, R. J., AND GLOPPEN, T. G., 1980, Late Caledonian (Devonian) basin formation, western Norway: Signs of strike-slip tectonics during infilling, *in* Ballance, P. F., and Reading, H. G., eds., 1980, Sedimentation in Oblique-Slip Mobile Zones: International Association of Sedimentologists Special Publication No. 4, p. 79–103.

STEEL, R. J., MAEHLE, S., NILSEN, H., ROE, S. L., AND SPINNANGR, AANON, 1977, Coarsening-upward cycles in the alluvium of Hornelen Basin (Devonian) Norway: Sedimentary response to tectonic events: Geological Society of America Bulletin, v. 88, p. 1124–1134.

STEEL, R. J., MAEHLE, S., NILSEN, H., ROE, S. L., AND SPINNANGR, AANON, 1979, Reply to discussion by H. F. Garner: Geological Society of America Bulletin, v. 90, Part 1, p. 122–124.

STITT, L. T., 1982, Structure and late Miocene paleogeography of southern Ridge Basin, southern California, *in* Crowell, J. C., and Link, M. H., eds., Geologic History of Ridge Basin, Southern California: Society of Economic Paleontologists and Mineralogists, Pacific Section, p. 43–52.

WELTON, B. J., AND LINK, M. H., 1982, Vertebrate paleontology of Ridge Basin, southern California, *in* Crowell, J. C., and Link, M. H., eds., Geologic History of Ridge Basin, Southern California: Society of Economic Paleontologists and Mineralogists, Pacific Section, p. 205–210.

WOOD, M. F., AND OSBORNE, R. H., 1982, Sedimentology of the Mio-Pliocene Apple Canyon Sandstone Member, Ridge Route Formation, Ridge Basin, southern California, *in* Crowell, J. C., and Link, M. H., eds., Geologic History of Ridge Basin, Southern California: Society of Economic Paleontologists and Mineralogists, Pacific Section, p. 135–142.

YOUNG, D. R., 1982, Miocene molluscan paleontology of the Ridge Basin Group, Ridge Basin, southern California, *in* Crowell, J. C., and Link, M. H., eds., Geologic History of Ridge Basin, Southern California: Society of Economic Paleontologists and Mineralogists, Pacific Section, p. 219–228.

WALKER LAKE BASIN, NEVADA: AN EXAMPLE OF LATE TERTIARY (?) TO RECENT SEDIMENTATION IN A BASIN ADJACENT TO AN ACTIVE STRIKE-SLIP FAULT

MARTIN H. LINK AND MICHAEL T. ROBERTS

*Cities Service Oil and Gas Corporation, Exploration and
Production Research, P.O. Box 3908, Tulsa, Oklahoma 74102;*

AND

MARK S. NEWTON

*Los Angeles City College, 855 N. Vermont Avenue,
Los Angeles, California 90029*

ABSTRACT: Walker Lake sedimentary basin is a fault-controlled continental basin related to strike-slip faulting on the western side of the Basin and Range Province of Nevada. The Walker Lake Basin is contained within a triangular crustal block bounded by normal- to oblique-slip faults on the west, left-lateral faults on the south, and right-lateral strike-slip faults on the east (Walker Lane shear zone).

Modern Walker Lake is roughly one fourth the surface area and the water depth of its Pleistocene precursor. Carbon-rich (up to 2.5% total organic carbon) and uranium-rich sediments are currently accumulating in the deeper saline and anoxic parts of Walker Lake. If these conditions were to continue, significant potential hydrocarbon source rocks and uranium-bearing beds could accumulate.

Walker Lake Basin is being infilled by axially fed, sand-rich fluvial-deltaic deposits; side-fed, coarse-grained alluvial-fan/fan-delta deposits; and central fine-grained lacustrine deposits. Waves, wind, and lake-level fluctuations have caused reworking of the lower parts of fan-delta surfaces and the front (windward side) of the Walker River delta. Carbonate deposits, which include beach-rock horizons, stromatolites, oncolites, caliche, and tufas, locally form along the shorelines and spring areas of this predominantly coarse-grained clastic system.

INTRODUCTION

Families of generally northwest-striking right-lateral faults and east-northeast-striking left-lateral faults occur along the western margin of the Basin and Range Province in California and Nevada (e.g., Wright, 1976). Many of these faults have the characteristics of continental transform faults and appear to bound regions of differential crustal extension. These fault zones are well documented from field and earthquake studies (e.g., Gianella and Callaghan, 1934; Noble and Wright, 1954; Slemmons, 1957; Longwell, 1960; Nielsen, 1965; Shawe, 1965; Wright, 1976; Stewart, 1980). One group of right-slip faults forms the Walker Lane shear zone (Locke et al., 1940), which extends along the western margin of Nevada and into adjacent parts of California (Pease, 1969). The zone is locally as much as 80 km wide and 650 km long, extending from near Las Vegas, Nevada, on the south to the Honey Lake region of northeastern California on the north (Fig. 1). The Walker Lane appears to consist of numerous, discontinuous right-slip faults locally arrayed en echelon or in step-like patterns. Topographically, the Walker Lane is a northwest-trending zone containing discontinuous ranges of diverse orientation. It appears to be the boundary between a region of typical north- to northeast-trending ranges of the Basin and Range Province to the northeast, and north- to northwest-trending ranges including the Sierra Nevada to the southwest. It also appears to be a boundary between major regions of differential extension (e.g., Wright, 1976; Zoback and Zoback, 1980).

The magnitude and timing of the fault displacements within the Walker Lane vary, but individual fault displacements within the zone are on the order of 10 to 16 km (Nielson, 1965; Hardyman et al., 1975; Hardyman, 1978; Ekren et al., 1980). Total displacement across the zone may be as much as 130 to 190 km (Albers, 1967; Stewart, 1980). De-

formation for the entire zone has been postulated to have begun as early as late Early Jurassic or Middle Jurassic (Albers, 1964, 1967; Speed, 1978). Much of the deformation may date from the late Tertiary (early to middle Miocene; Nielsen, 1965; Wright, 1976; Ekren et al., 1980), and may be genetically related to Basin and Range tectonism.

The Walker Lake sedimentary basin is one of the best exposed of the basins that occur in or directly adjacent to the Walker Lane. The Walker Lake region (1° by 2° 1:250,000 scale quadrangle) has been extensively mapped and studied during the past 10 years by the U.S. Geological Survey in cooperation with the Nevada Bureau of Mines and Geology (Stewart and Carlson, 1978; Stewart et al., 1984). Detailed maps of the bedrock geology, recent faults, and surficial geology at various scales are available (Ross, 1961; Bingler, 1978; Ekren et al., 1980; Hardyman, 1980; Stewart and Johannesen, 1981a, b; Dohrenwend, 1982, 1983) and are discussed in this paper.

The purposes of this paper are (1) to describe and interpret the sedimentary deposits in Walker Lake sedimentary basin; and (2) to outline the tectonic setting of the basin and relate it to other strike-slip basins. Relatively few modern strike-slip related basins have been studied and discussed in terms of the interplay between tectonics and sedimentation (Crowell, 1974; Lewis, 1980; Hempton et al., 1983). From studies of such modern strike-slip-related basins, the geometry, drainage, and geomorphologic characteristics of ancient strike-slip basins can be inferred.

WALKER LAKE DRAINAGE BASIN

Walker Lake is in northern Mineral County, Nevada, about 10 km north of the City of Hawthorne and is the second largest lake wholly in Nevada (Figs. 2, 3). The lake trends north-northwest and is 26.2 km long and 7.7 km wide. The Walker River drainage basin occupies approximately 10,000

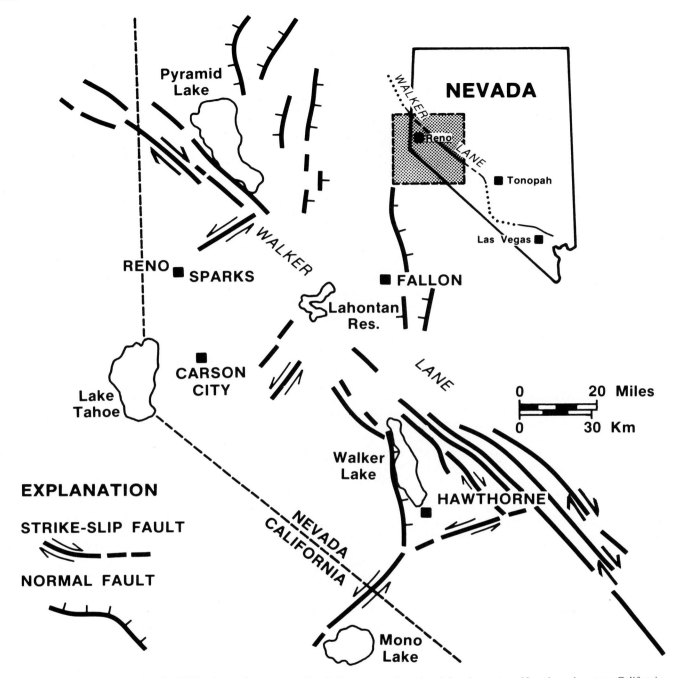

FIG. 1.—Index map showing the Walker Lane shear zone, other fault zones, and modern lakes in western Nevada and eastern California.

km² in western Nevada and adjacent parts of California. It is essentially a closed hydrologic system that drains from the central Sierra Nevada near Bridgeport and Topaz lakes in California and terminates in Walker Lake (Fig. 2). Walker Lake occurs in an asymmetrical sedimentary basin bounded by the faulted, steep front of the Wassuk Range on the west and the more gentle slopes of the Gillis Range on the east (Figs. 1, 2). Mount Grant, in the Wassuk Range, is 3,445 m (11,303 ft) high, whereas the maximum elevation in the Gillis Range is about 2,286 m (7,500 ft).

The lake level, which is currently about 1,212 m (3,976 ft) elevation, has been falling at an average rate of 0.61 m/yr since 1862, owing to ground-water withdrawal for irrigation (Rush, 1970). The maximum present depth of Walker Lake is 31.5 m, and its salinity, currently about 8.61‰, has been increasing. Chemical data for Walker Lake water reported in 1884 (Clarke, 1924) and 1966 (Rush, 1970) indicate that the lake is saturated with carbonate and precipitation could occur (Table 1).

The climate of the Walker Lake Basin region ranges from

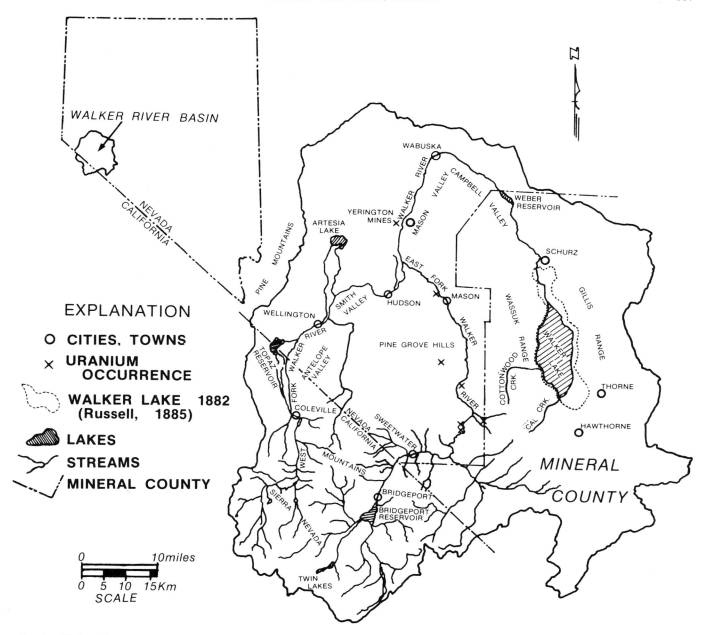

FIG. 2.—Walker River drainage basin showing major rivers, creeks, lakes, uranium occurrences, county boundaries, and towns in California and Nevada (after U.S. Department Agriculture, 1969).

arid to semiarid. Average precipitation ranges from 10 cm annually in the valley to 28 cm annually in the surrounding mountain ranges and Sierra Nevada to the west (Rush, 1970). Evaporation exceeds precipitation in the area. Shadscale, a kind of salt bush (*Atriplex canesens*), desert grasses, and other desert shrubs characterize the vegetation of the basin floor and slope. Mixed Juniper, Piñon Pine forests, and Mountain Mahogany trees are found higher in the ranges. Cottonwood trees and willows are common in the basin along the tributary streams, springs, and Walker River (Ross, 1961).

Walker Lake is one of the many sites once covered by Pleistocene Lake Lahontan (Russell, 1885; Morrison, 1964). Lake Lahontan and younger strandlines are visible along the margins of Walker Lake at elevations as high as 1,340 m (4,400 ft) and are common at 1,310 m (4,300 ft), which is usually accepted as the high-stand level for Lake Lahontan (U.S. Department of Agriculture, 1969). Middle to early Pleistocene (pre-Lake Lahontan) deposits are found on the east side of Walker Lake Basin, making this basin one of the earliest sites of Pleistocene lacustrine deposition in the area (Mifflin and Wheat, 1979). The various shoreline, la-

FIG. 3.—Lake-level maps for Walker Lake showing the Pleistocene Lake Lahontan maximum level, 1882 shoreline mapped by Russell (1885), and 1956 shoreline (Walker Lake, Nevada/California 1:250,000-scale map; after U.S. Department Agriculture, 1969).

custrine, and alluvial-fan/fan-delta deposits of Walker Lake Basin have been mapped and their relative ages determined by Dohrenwend (1982).

The size of Walker Lake has decreased dramatically, from a surface area of 880 km^2 or greater when it was part of Pleistocene Lake Lahontan, to 320 km^2 in 1882 (Russell, 1885), to 220 km^2 in 1956 (Fig. 3). The more recent reduction in lake size is related to man's agricultural influence. The reduction in the surface area of the lake by a factor of four since the Wisconsinan Lake Lahontan must be due, in part, to climatic changes. In the past thirty years, lake level has dropped an additional 4.9 m. The Pleistocene Lake Lahontan spill point for the Walker Lake drainage basin was northeast of Wabuska (Fig. 3), where it period-

TABLE 1.—CHEMICAL DATA FOR WALKER LAKE WATER

| | Concentration, ppm in year | | |
	1882 (Clark, 1924)	1966 (Rush, 1970)	1979 (Newton, unpubl.)
Cl$^-$	595	2020	—
SO$_4^{-2}$	532	1930	—
CO$_3^{-2}$	(total)	486	—
HCO$_3^-$	434	1640	—
Na$^+$	871	3040	—
K$^+$	—	160	—
Ca^{+2}	22.5	4.2	7.42
Mg^{+2}	39	124	130
Sr^{+2}	—	—	3.20
SiO$_2$	8	0.5	—
Total Salinity ‰	2.5	9.4	—
Ph	8.3 (est.)	9.3	9.3

ically drained into the Carson Sink area. During such times, and at least once in the past 10,000 years, Walker Lake was completely desiccated (Benson, 1978).

STRUCTURAL RELATIONSHIPS

The general geology and tectonic setting of the Walker Lake area are discussed by Ross (1961), Ekren et al. (1980), and Stewart et al. (1982). A late Cenozoic fault map and tectonic summary has been compiled by Dohrenwend (1983) for the Walker Lake area. The faults and their relationship to volcanism are discussed by Hardyman (1978), Speed and Coghill (1979), and Ekren et al. (1980). Geologic maps of individual quadrangles in the area include those by Bingler (1978), Ekren and Byers (1978a, b, c, d), Hardyman (1980), Stewart and Johannesen (1981a, b), and Stewart et al. (1981). The major faults and the distribution of volcanic rocks in the area are summarized in Figure 4.

The Walker Lake sedimentary basin is a fault-controlled basin related to normal faulting along the Wassuk Range front and to right-slip faulting along the northwest-trending Walker Lane shear zone. The age of this basin is at least early Pleistocene to Holocene based on the age of lacustrine deposits found within it (Morrison, 1964; Mifflin and Wheat, 1979). This basin is superimposed on a larger late Oligocene to early Miocene trough associated with Walker Lane and having roughly the same trend (Ekren et al., 1980). This trough is filled with volcanic and volcano-clastic rocks as thick as about 2,000 m thick and traceable for up to 400 km.

The modern basin appears to be in a half-graben bounded on its west side by a high-angle normal-fault zone along the steep eastern side of Wassuk Range. This fault zone includes a series of surface breaks at or near the range front. Normal faults of the zone cut both modern and older alluvial fans (Figs. 5A, B). These faults, like other range frontal zones in the area, are all high-angle zones of dip slip, strike north to northwest, are composed of bifurcating fault traces with recent vertical offset of 100 m or less, and range from 35 to 105 km in length (Dohrenwend, 1983). The Wassuk frontal fault zone has topographic expression of about 2,300 m and has also been interpreted as an oblique-slip fault with 5 to 10 km of right slip (Ekren et al., 1980). The Wassuk frontal fault zone either merges into or terminates at the northwest-trending Walker Lane right-lateral fault system (Figs. 4, 5B) to the north.

The Walker Lane shear zone in this area consists of at least five mappable, subparallel right-lateral faults, which form major topographic lineaments in the mountain ranges (Gillis and Gabbs Valley Ranges, Pilot Mountains, Garfield Hills, and Soda Spring Valley) adjacent to Walker Lake Basin. Total displacement along these right-slip faults is variously interpreted to be as little as 19 km since the late Miocene (Nielsen, 1965) or as much as 46 km (Hardyman et al., 1975). Major subalkalic ash flows, tuffs, and lava flows of intermediate composition, which in the Walker Lake Basin area range in age from 22 to 7 Ma, are cut by the strike-slip faults of the Walker Lane. Individual offsets along the faults are on the order 10 km or less, and total strike separation is 30 km (Ekren et al., 1980). These faults are

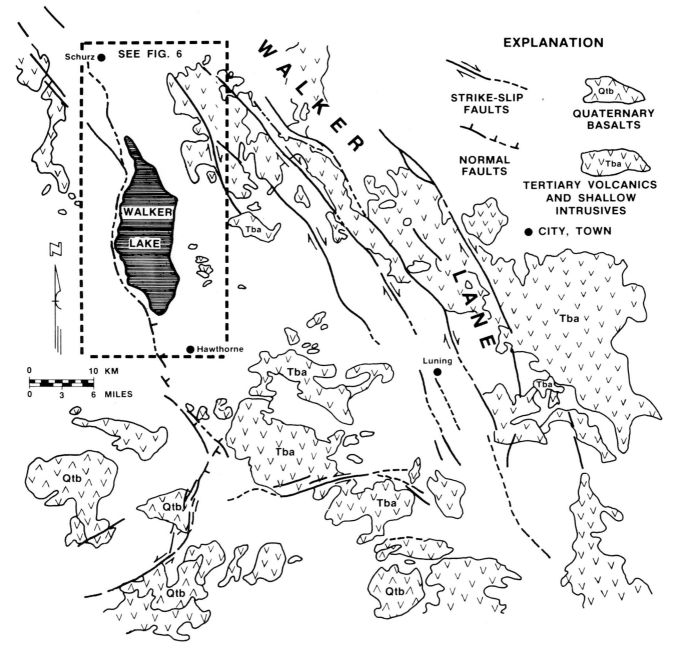

FIG. 4.—Generalized tectonic map and distribution of volcanic rocks for Walker Lake Basin. Note the triangular-shaped basin bounded by normal faults on the west side, left-slip fault on southern margin, and zone of subparallel right-slip faults of the Walker Lane on the east side of the basin.

relatively straight, continuous traces up to 20 km long with subsidiary faults that branch and splay from the main fault traces (Dohrenwend, 1983). The faults are considered to have had Quaternary slip, but there is little evidence of late Quaternary to historic activity (Howard et al., 1978; Dohrenwend, 1983).

At the southern end of Walker Lake sedimentary basin, the Wassuk frontal fault zone either merges with or is terminated by an east-northeast trending left-lateral fault system (Figs. 1, 4). This left-lateral fault zone may extend discontinuously westward to the Mono Lake area and eastward to the Walker Lane (Fig. 1). The fault system, known as the Mono Lake Basin—Excelsior Mountain trend (Dohrenwend, 1983), is expressed by a broad zone of northeast-to east-trending topography at high angles to the Basin and Range and Walker Lane trends. The zone is composed of (1) short, closely spaced, subparallel to bifurcating, high-angle dip-slip faults that cut late Tertiary and Quaternary volcanic rocks; and (2) left-lateral oblique-slip faults that are evenly spaced, high angle, and traceable for 5 to 10 km

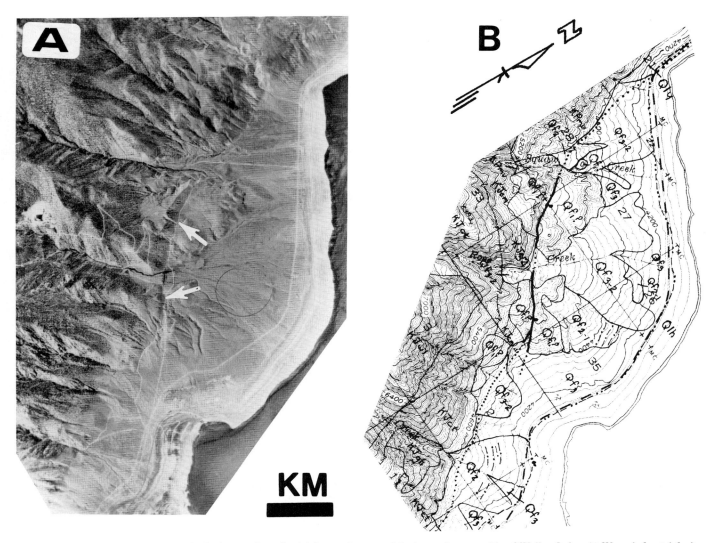

FIG. 5.—Aerial photograph and geological map of an alluvial fan cut by normal faults on the west side of Walker Lake: A) Wassuk frontal faults at southwestern end of Walker Lake (arrow indicates graben formed across prominent alluvial fan); and B) geological map with Quaternary surficial deposits (Dohrenwend, 1982). Abbreviations: Qlh, historic lacustrine deposits (upper Holocene); Qly, younger pluvial lake deposits (Holocene and upper Pleistocene); Qf₃, younger alluvial-fan deposits; Qf₃₋₂, Qf₂, Qf₂₋₁, intermediate alluvial-fan deposits; Qf₁₂, older alluvial-fan deposits; KJwl, KJgh, KJct, KJcm, KPmv, undifferentiated Cretaceous, Jurassic to Permian igneous-metamorphic rocks of the Wassuk Range; -----, shoreline terraces.

(Dohrenwend, 1983). The amount of strike slip along these faults is uncertain (Speed and Coghill, 1979).

Walker Lake sedimentary basin lies within a triangular crustal block bounded by the Wassuk frontal fault, the Walker Lane, and the Mono Lake fault systems (Fig. 1). The block is on the western margin of the Basin and Range Province, and extensive late Tertiary and Quaternary volcanic rocks occur adjacent to and within the block, suggesting major crustal extension. Extension leading to formation of the basin is interpreted to be the result of west-northwestward movement of the crustal block west of the Wassuk frontal fault (Sierra Nevada block of Wright, 1976). The movement is interpreted to have been accommodated also by strike-slip faulting on the south and east sides of the basin. The extension led to block faulting, subsidence, and volcanism within and adjacent to the block. The strike-slip boundaries

may have resulted from differential extension between the Walker Lake block and bounding crustal blocks. The timing and exact kinematics of the proposed fault movements have not been completely worked out. Similar triangular zones of extension bounded by strike-slip zones have been described by Wright (1976), especially in the Death Valley region south of Walker Lake. In our model, the Walker Lane and the left-lateral Mono Lake zone are viewed as analogous in role to the Death Valley—Furnace Creek right-lateral and the Garlock left-lateral fault zones. The Wassuk Range front, normal to the right-lateral oblique-slip fault zone, is viewed as analogous to the east-dipping normal faults and right-lateral faults along the Sierra Nevada block front of the Owens Valley region. These extension-related strike-slip fault zones contrast with the compressive or transpressional zones of the San Andreas (Reading, 1980).

Lithofacies

Walker Lake Basin is characterized by modern alluvial-fan/fan-delta, fluvial-deltaic, shoreline/dune, and lacustrine depositional environments. Deposition in similar environments in the basin is known to have occurred at least as early as the early(?) Pleistocene. The surface distribution of these sediments has been mapped by Dohrenwend (1982) and we have studied them in the field and from aerial photographs. The western side of Walker Lake Basin contains short, steep alluvial fans and fan deltas derived from the Wassuk Range. The eastern side of the basin contains lower-gradient, more areally extensive alluvial fans and fan deltas flanking the Gillis Range (Table 2, Figs. 5, 6, 7). The Walker River enters the basin from the north where it forms a delta and is a major source of fine-grained sediment for the basin. Shoreline-parallel sandflats and mudflats are laterally adjacent to the delta. Prominent older strandlines, beach ridges, sandflats, and mudflats occur along the southern margin of Walker Lake. The eastern shoreline is characterized by strandline plains on the lower parts of the alluvial fans, reworked sands in the form of eolian dunes and interdune deposits, and extensive submerged sandflats sloping to a depth of 26 m.

Alluvial-Fan/Fan-Delta Deposits

Numerous steep, narrow fans occur on the western side of the basin, whereas fewer, more gently sloping, wider fans occur on the eastern side of the basin (Figs. 5, 8). Sixteen alluvial fans measured on the western side of the basin range from 0.6 to 5.6 km wide and from 0.6 to 4.8 km long. They average 2.4 km wide and 1.9 km long (Table 2). The five fans that make up the eastern side of the basin range from 2.4 to 13.6 km wide and 3.2 to 8.8 km long. These fans average 6.7 km wide and 5.3 km long. The lower part of the fans extend into the lake, forming fan deltas. The slope of the fans on the western side of the basin range from 2°3′ to 14°25′ and averages 6°11′, whereas the fans on the eastern side ranges from 1°44′ to 4°11′ averaging 3°. The calculated average slopes of the alluvial fans are high because the lower parts of some of the fans are submerged and the surface of the lake was used as the lower limit of the fan in the measurement. The largest single fan measured extends out of Gillis Canyon on the northeastern side of the basin. It is over 13.6 km long and 8.8 km wide, and has a gradient of 1°53′ (Fig. 8C). The observed fan asymmetry is probably due to a coarser average grain size for the western fans and is related to the structural asymmetry of the basin.

The upper parts of the alluvial fans are characterized by one or several incised active channel(s), which radiate downfan into numerous smaller bifurcating channels, and numerous older channels (Fig. 5). The internal characteristics of these alluvial-fan deposits can be seen in several quarries and stream cuts where they consist of poorly sorted, crudely bedded, pebble to boulder conglomerates (Fig. 9). The upper alluvial fan is characterized by braided-stream, sheet-flood, and debris-flow deposits. The braided-stream deposits contain channelized and non-channelized gravel and sand beds up to 2 m thick. The gravels are clast-supported and locally cross-bedded with imbricated clasts dipping toward the mountain front (Figs. 9A, B). Large boulders are common, especially in the western fans. The active channels are of low relief, about 1 m deep and 10 to 20 m wide. Sand, silt, and organic debris make up the matrix material. Sheet-flood deposits are several centimeters to a meter thick and consist of sands and gravels in massive to laminated beds that are clast supported. These deposits are relatively flat-lying and sheet-like, and they are finer grained than the channel gravels. The channel and sheet-flood coarse-grained sediments are estimated to make up about 80% of the upper fan surface. The debris-flow deposits are distinctively brown, poorly sorted, and contain clasts up to boulder size together with logs that are matrix-supported (Fig. 9B). These deposits are estimated to make up about 20% of the upper fan and are interbedded with the fluvial gravels.

The middle to lower parts of some of these alluvial fans locally enter the lake where they form fan deltas (Fig. 8B).

TABLE 2.—SIZE AND SLOPE OF ALLUVIAL FANS AND FAN DELTAS FOR THE WALKER LAKE AREA

	West Side Fans[1]								East Side Fans[1]						
	Width		Length		Height		Slope		Width		Length		Height		Slope
	km	miles	km	miles	m	feet			km	miles	km	miles	m	feet	
1.	4	2.5	4.6	2.9	281	920	3°25′	1.	8	5.0	5.6	3.5	366	1200′	3°12′
2.	1.6	1.0	1.6	1.0	171	560	5°45′	2.	2.4	1.5	3.2	2.0	307	1007′	4°02′
3.	5.6	3.5	2.9	1.8	374	1227	7°25′	3.	3.2	2.0	3.2	2.0	368	1207′	4°11′
4.	1.6	1.0	1.3	0.8	160	524	7°08′	4.	6.4	4.0	5.6	3.5	244	800′	1°44′
5.	2.4	1.5	1.3	0.8	190	624	8°30′	5.	13.6	8.5	8.8	5.5	185	607′	1°53′
6.	0.6	0.4	1.0	0.4	160	524	7°37′	*Averages*							
7.	1.3	0.8	1.2	0.75	160	524	7°37′	5	6.7	4.2	5.3	3.3	294	964	3°
8.	1.9	1.2	1.3	0.8	160	524	7°08′								
9.	1.3	0.8	1.1	0.7	160	524	8°05′								
10.	0.8	0.5	1.1	0.7	152	500	7°44′								
11.	1.2	0.75	1.6	1.0	92	300	3°14′								
12.	2.8	1.75	2.8	1.75	122	400	2°30′								
13.	2.8	1.75	1.6	1.0	110	360	3°53′								
14.	5.3	3.3	4.8	3.0	268	880	3°10′								
15.	1.9	1.2	1.3	0.8	79	260	3°31′								
16.	2.4	1.5	2	1.25	92	300	2°′5′								
Averages															
16	2.4	1.5	1.9	1.2	170	559	6°11′								

[1]Measurements originally made in feet and miles using aerial photographs and topographic maps.

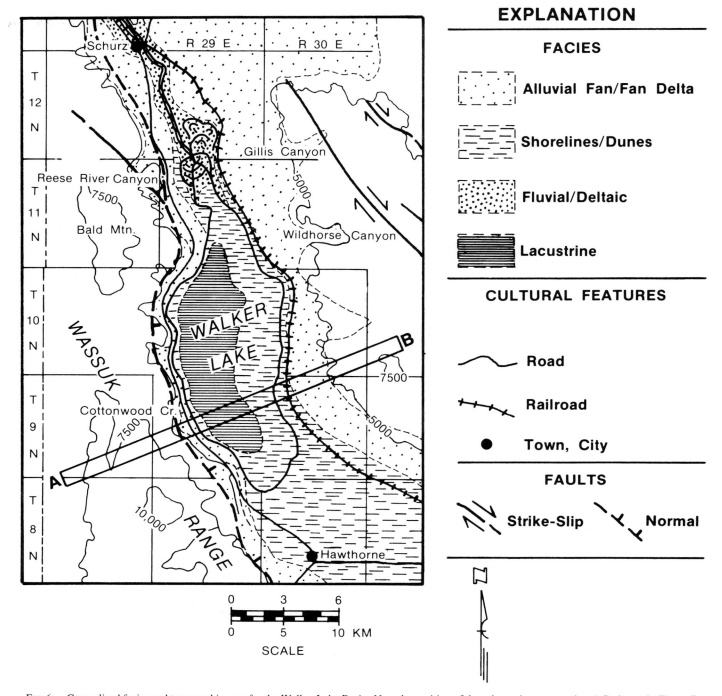

FIG. 6.—Generalized facies and topographic map for the Walker Lake Basin. Note the position of the schematic cross section A-B shown in Figure 7.

Several older Holocene(?) fan deltas are well exposed along the northwestern part of the basin where they are incised by Walker River (Fig. 8B). These fan deltas are much finer grained than the upper alluvial-fan surfaces and consist of medium- to coarse-grained sand interbedded with silt, mud, and gravel. Very distinctive Gilbert delta foresets are locally well preserved and dip 25° to 30° toward the lake (Figs. 9C, D). By measuring the vertical distance between the topset

and bottomset beds, water depths are inferred to have been between 7.6 and 10.7 m where the fan deltas prograded into the lake. Topset beds of these fan deltas include carbonate-cemented gravels and stromatolitic layers.

The alluvial-fan/fan-delta deposits were derived locally from the ranges bounding the basin and clast compositions reflect the rock types exposed in these ranges. Poorly sorted conglomerate is the dominant deposit in these systems, ex-

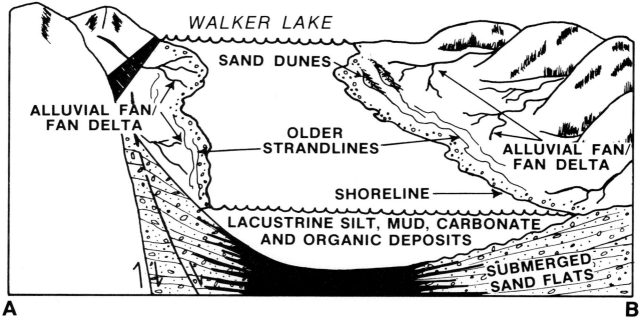

FIG. 7.—Schematic cross section A-B showing the facies and topographic relationships in Walker Lake Basin. See Figure 6 for location.

cept where shoreline processes have reworked the fans and in the foresets of the fan deltas. Shoreline processes have concentrated the coarser-grained sediment near shore and the finer-grained sediment offshore, where they intermix with distal fine-grained sediments from the Walker River.

Fluvial/Delta Deposits

In the area of study, the Walker River is a meandering river with a valley that is about 0.8 to 2.4 km wide (Figs. 2, 10). The river valley is confined to the lowest part of the sedimentary basin by the alluvial fans along both sides of the basin. Well-defined river terraces, oxbow lakes, and accretionary point-bar ridge-and-scour features are developed in the upper reaches of the river near Schurz (Figs. 2, 3). Near the southern end of the Walker River, extensive crevasse-splay lobes occur along the flanks of the river (Fig. 10D). Vegetated levees, sand-rich chute bars and longitudinal bars (islands) occur along the river's course (Figs. 10A, B). Oxbow lakes, swamps, and older abandoned channels and crevasse splays are laterally adjacent to the active channel (Figs. 10C, D).

The bed load in the river channel is generally medium- to fine-grained sand with some gravel, whereas the levees are finer grained. Most of the silt- and mud-size material transported by the river is deposited on the flood plain, in the delta plain, or transported to the deeper part of the lake. The Walker River delta environments are controlled by

the amount of fluvial sediment input, lake-level fluctuations, and wave and wind reworking. It is a fluvially dominated delta and has produced digitate, bifurcating distributary channels with channel-mouth bars. Local reworking by waves, wind, and lake-level fluctuations has distributed some of the sand and finer-grained sediment to the south along the lake shoreline in the form submerged sandflats and shoreline deposits. The development of several spits and the hooking of older sandy beach ridges toward the north (Figs. 11, 12) suggest that the dominant wind and wave directions are northward up the axis of the basin.

The Walker River and delta provides fill to the basin axially. The coarser-grained sediments are trapped in the fluvial and deltaic distributary channels, channel-mouth bars, and locally in crevasse-splay lobes. Fine-grained sediment derived from the delta front is transported to the deeper parts of the lake or reworked along the shorelines. If lake level remains stationary or continues to drop, the fluvial-deltaic system would continue to prograde down the axis of the basin and eventually fill the basin.

Shoreline/Dune Deposits

Shoreline deposits around Walker Lake are related to lake-level fluctuations, local wave and wind processes, spring waters and the carbonate saturation of the lake (Figs. 11, 12). The deposits include gravel or sand beaches and beach ridges, wave-cut terraces, carbonate-cemented beach rock

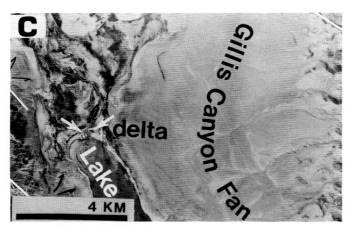

Fig. 8.—Aerial photographs of the alluvial fans and fan deltas adjacent to Walker Lake: A) the smaller, steeper fan deltas of the west side of the basin; B) small fan delta on the northwest side of the basin (note the area in the box of close-up photographs shown in Figures 9C and 9D); and C) the larger, more gently dipping alluvial fan (Gillis Canyon fan) on the northeast side of the basin. Note the inferred margins of the basin (line), the size and distribution of the fans on both sides of the basin, the position of the lake, delta, and river, and the concentric shorelines that ring the lower third to one half of the fans.

and associated stromatolites, oncolites, tufa deposits, submerged sandflats, and localized eolian sand dune and interdune deposits.

The gravel and sand beach ridges are the most prominent shoreline features rimming Walker Lake (Figs. 11, 12). Locally, 5 to 15 ridges of various ages are present. They are formed across the lower parts of the alluvial fans and are found adjacent to the Walker River delta in the north. Individual ridges are as high as 1 m and traceable on aerial photographs for several kilometers. They typically terminate by merging with other ridges. Many of these ridges are on terraced surfaces formed by wave erosion. The ridges consists mainly of pebble and cobble gravel on the alluvial-fan surfaces, and sand at the northern and southern ends of the lake. The clasts in the ridges are imbricated and together with crude internal bedding dip both in an on- and offshore direction. Spits and channel reentrants occur within the ridges. Similar features have been described by Russell (1885) and Gilbert (1890) for the Lake Lahontan and Bonneville basins, and the ridges are common features in lacustrine basins characterized by marked lake-level fluctuations.

The beach-rock deposits are carbonate-cemented hardgrounds containing both biochemical (stromatolites, oncolites, caliches, and tufa encrustations) and detrital grains (Fig. 13). These white to blue-gray carbonate deposits form distinctive rings around the basin and have been used to infer the highest stands of Lake Lahontan (between 1,310 and 1,340 m, or 4,300 to 4,400 ft) throughout most of Nevada (Russell, 1885; Morrison 1964). Associated with these carbonate deposits are molluscs, vertebrate bones (mostly birds and fish), and ostracodes. Walker Lake appears to be saturated with respect to calcium carbonate and the lake contains abundant and diverse green and blue-green algae and diatoms. The stromatolites, associated biota, and deposits are discussed by Osborne et al. (1982). It is thought that the biota directly cause carbonate precipitation in the lake. The carbonate beach-rock deposits are flat-lying to steeply dipping (basinward) hardground pavements, and consist of carbonate-cemented detrital gravels, sands, skeletal material, stromatolites, oncolites, and tufas (Fig. 13). Caliche also occurs as a cement in alluvial-fan deposits (Fig. 13D) and appears to be related to spring and groundwater deposition. Carbonate-cemented gravels or sands and *in situ* stromatolite horizons form many of the topset and upper parts of foresets in the fan-delta deposits. These carbonate deposits may have formed during submergence of the fan-delta deposits or may represent lake fluctuations. Such carbonate deposition in the nearshore lacustrine environment reflects the arid climate and chemistry of the lake waters. Many of the older carbonate deposits have been reworked and clasts of these deposits can be found along the western shoreline of the lake today. In addition to carbonate cementation of gravels and stromatolite formation, spring-derived tufa precipitation has been a locally important shoreline process. Radiocarbon dating of the stromatolites gives ages of 18,000 yr (pre-Lahontan), 11,000 to 13,000 yr (Lahontan), 1,600 yr, (post-Lahontan), and contemporary ages (Newton, unpubl. data).

Submerged sandflats occur on the north, east, and south sides of the basin and extend to water depths as great as 26 m. The sandflats consist of moderately well sorted fine- to medium-grained, rippled sand with abundant ostracod

FIG. 9.—Alluvial-fan and fan-delta deposits: A) in a quarry, flat-lying sheet-flood deposits (F), inclined (//), and locally channeled (--) braided gravels of the upper fan; B) mud-supported debris-flow deposit with boulder-size clasts in the upper fan; C) fan-delta bottomset (b) and foreset (f) beds; and D) close-up of fan-delta foreset beds shown in box in 9C with dips of 25° to 30° toward the basin. Inferred water depth for the fan delta is 7.6 to 10.7 m (25 to 35 feet), based on measurements of the vertical distance between the topset and bottomset beds.

shells. Extensive areas of rippled sand waves were observed in shallow water (1 m water depth). These sands occur on the submerged lower parts of the alluvial fans/fan deltas on the eastern side of the basin and are reworked into sheet sands on the northern and southern edges of the basin by shoreline and wave processes. The sands were chiefly derived from the fan deltas and river delta by wave reworking and longshore processes. These sands constitute an important part of the surficial deposits of Walker Lake Basin and have not been widely recognized in the ancient sedimentary record.

Eolian dunes and interdune areas occur on the east, north, and southeast sides of the basin. The dunes are small, rilled to irregular-shaped, 1 to 3 m high and tens of meters long. They consist of well-sorted fine- to medium-grained sand and are separated by vegetated interdune areas that were locally wet when observed. The dunes are localized in low-lying areas that face the prevailing wind direction in re-entrants between coalescing alluvial fans. These dune sands appear to be wind-reworked remnants of the submerged sandflat sheet sands exposed during the last three decades by the 4.9 m drop in lake level. Older eolian dunes occur

at about 1,250 m (4,100 ft), or the 1882 shoreline position, and probably represent reworking of Lake Lahontan deposits.

Lacustrine Deposits

Walker Lake has a surface area of about 220 km^2 and is about 31.5 m deep. The lake covers the lower parts of the basin-margin alluvial fans and fan deltas to a depth of about 26 m (Fig. 14). The lower fan sediments consist mainly of gravel, medium- to fine-grained sand, silt, and mud. A relatively flat central bottom where lacustrine sediments are deposited makes up about half of the lake's basinal area. It is about 1 to 6 km wide and 12 to 15 km long. The deep lacustrine sediment consists mainly of silt, clay, ostracod fragments, carbonate, and organic material. The mud is olive-brown to black, jelly-like in appearance, saline, and rich in carbonate. It is fetid with an H$_2$S odor and has an organic content estimated to be 1.2 to 2.5% by weight (Table 3). The carbonate content in the sediment is estimated to be greater than 10%. The carbonate mineral phase is hydrocalcite. It may have originated as (1) a thin flocculate precipitated in a turbid, wave-agitated nearshore setting, and/

FIG. 10.—Walker River fluvial/deltaic system: A) vegetated levees and the meandering Walker River where it enters Walker Lake; B) longitudinal chute bars of the Walker River (note the river's meandering course and the ripple-marked surfaces of the bars); C) aerial photograph of the Walker River delta (note the river-mouth bars, indicated by arrows, and the reworked linear shorelines); and D) aerial photograph of Walker River. Note the more recent crevasse splays (small arrows), older river course, and older, large crevasse splays (large arrows).

or (2) a biochemical precipitate related to phytoplankton conditioning of the water. The organic material is primarily derived from lake algae, with a minor contribution from fishes, diatoms, and ostracodes, and plant material from the surrounding land area. Both pelagic and benthic algae and diatoms live in the lake.

Cores taken in the lacustrine muds of the lake show an increase in salinity with depth (Benson and Leach, 1979). The high salinities of the bottom water and saturated nature of the lake water suggest that evaporites may also be present in the subsurface. Many of the nearby basins have extensive evaporites which formed during times of desiccation (Papke, 1976; Mifflin and Wheat, 1979). Benson (1978) reports several periods of desiccation in the Walker Lake Basin and at least one in the past 10,000 years.

In general, the mean surface-sediment particle-size diameter in the lake is highly correlated with water depth (Osborne et al., 1982). These lacustrine deposits are coarse grained in shallower water and fine grained offshore. Samples collected in water depths from 4.0 to 8.2 m have mean diameters that range from fine to very fine sand, respectively, whereas those collected from 25.0 to 31.5 m have mean diameters from medium silt to fine silt (Table 3).

Likewise, the weight percentages of clay increase from near zero in shoal areas to 15.9% in water 31.5 m deep (Table 3). The organic content also increases with depth, ranging from 0 to 2.5% by weight. Carbonate- and organic-rich sediments accumulate in the deepest part of the basin.

QUATERNARY STRATIGRAPHY

The age of the most recent sedimentary fill in Walker Lake Basin is Pleistocene to Holocene. These deposits are contemporaneous with Quaternary volanic deposits and overlie older Miocene and Pliocene volcanic and volcanic clastic rocks in the area (Fig. 4). The Quaternary surficial and lacustrine deposits can be grouped into four age sequences: (1) Pre-Lahontan (upper Pleistocene); (2) Lahontan (11,000 to 13,000 years old; uppermost Pleistocene); (3) post-Lahontan (Holocene); and (4) historical (Fig. 15). Their stratigraphic relationships are based on stratal superposition, onlap relations from maps and aerial photographs, and radiocarbon dating of interbedded carbonate deposits.

The pre-Lahontan deposits occur over a wide area on the western flank of the Gillis Range and along a narrow shoreline belt on the eastern side of the Wassuk Range. The pre-

FIG. 11.—Shoreline features in the Walker Lake Basin: A) concentric ridges and terraces on the northeast side of the basin; B) vertically stacked wave-cut terraces on the side of a fan delta on the northwest side of the basin (note the carbonate beach rock of the upper terraced surface); and C) wave-cut terraces and steeply dipping carbonate beach rock (b) on the west side of the basin.

Lahontan alluvial-fan and pediment deposits of the Gillis Range are prominent between elevations of 1,250 and 1,525 m (4,100 and 5,000 ft) (Fig. 15). They appear dark on aerial photos and on the ground owing to the development of desert varnish, soil, and concentrations of vegetation. In contrast to the younger alluvial sediments, the pre-Lahontan deposits are locally dissected by fluvial channels and covered by thin sheet-flood deposits. The second occurrence of these deposits is exposed above the western shoreline of Walker Lake between the 1979 shoreline at 1,208 m (3,964 ft) and 1,250 m (4,100 ft), about 3 km south-southeast of Copper Canyon (Fig. 15). The deposits consists of alluvial-fan, fan-delta, and shoreline facies that are arranged in a series of wave-cut terraces. These terraces were formed in pre-Lahontan time and are marked by three large stromatolite or tufa horizons at elevations of 1,208, 1,221, and 1,229 m (3,964, 4,005, and 4,032 ft) with some individual heads as large as 2 m wide and 1 m high. The strata consists of angular cobbles, gravel, and sands interbedded in the carbonate deposits and well-defined foreset beds dipping toward the lake. Two carbonate samples analyzed gave radiocarbon ages of 18,060 ± 570 yr.

Lahontan-aged deposits occur around the basin between 1,340 m (4,400 ft) and the present lake level. They consist of shoreline facies (gravels and sands, sandspits, and sandbars), alluvial-fan and fan-delta facies, and carbonate-cemented pediments. The most extensive of the deposits are the shoreline sands that probably originated as part of mobile sandflat-sheet deposits formed as lake level dropped. The Lahontan age assignment of these deposits is made because these sands show lateral continuity with prominent sandspits and bars of Lahontan age at 1,340 m (4,400 ft) on the southeastern shore (Mifflin and Wheat, 1979). The carbonate-cemented fanglomerates and gravels of the alluvial-fan and fan-delta(?) facies form a prominent pediment on the western shoreline between elevations of 1,250 and 1,310 m (4,100 to 4,300 ft). The deposit is up to 15 m thick and extremely durable, resembling concrete pavement. Cementation is inferred to have occurred during rising lake level and submergence of alluvial fan and fan deltas during the high-stand phase of Lake Lahontan. Calcium-rich groundwaters percolating through the extremely porous fan material may have reacted with the alkaline lake waters to produce this thick calcrete deposit. Summer water temperatures measured in still shoreline pools are as high as 40° C. This extreme shoreline environment is rich in algae and the water is greatly oversaturated (15 to 20 times) with respect to $CaCO_3$. Shoreline cementation is a highly

FIG. 12.—Shoreline deposits: A) modern gravel (G) ridge on the northwest side of Walker Lake, with carbonate mud flats (m) to the left of the ridge; B) quarried gravel-ridge deposits with imbricated and cross-bedded units dipping back toward the mountains; C) sand ridges (S) with a carbonate mud flat (m) on the landward side, and a rippled, partially submerged sandflat (I) on the lakeward side; and D) complex sand ridges with small recurved spits on the west side of Walker Lake suggesting longshore transport.

significant process in this basin, and may be an indicator of exact paleowater line. Walker Lake became the southernmost extension of Lake Lahontan about 11,000 to 13,000 years ago (Morrison and Frye, 1965) when water depths exceeded the 1,308 m (4,291 ft) elevation of the Adrian sill.

Post-Lahontan deposits occur on all sides of the basin between 1,310 and 1,340 m (4,100 to 4,400 ft) and can be separated from older deposits by their less dissected, less vegetated, and lighter-colored appearance. The post-Lahontan deposits consist of Holocene pediment surfaces on alluvial-fan, fan-delta, and shoreline deposits that are interbedded with stromatolite and tufa horizons, lacustrine and possible buried evaporite deposits (Fig. 15). Down-cutting channels, 3 m or more deep, flights of wave-cut terraces, and thin, uniform light-colored carbonate coatings suggest a gradual falling of lake level. The white carbonate deposit, present on all suitable substrates, is up to 0.5 m thick and is described by Osborne et al. (1982). It consists of stromatolites, tufa encrustations, oncolites, and mollusks. Radiocarbon dating of samples of these deposits gives ages of 1,620 ± 90 yr.

Alluvial fans and fan deltas of post-Lahontan age are locally quarried and well exposed on the west side of basin

where distinctive Gilbert delta foreset and bottomset beds can be observed (Fig. 9C). These deposits have wave-cut terraces and locally are incised by the Walker River in the northern part of the basin. Along the Walker River, post-Lahontan lacustrine mudstones are seen interbedded with the bottomset beds of the fan delta (Fig. 9C).

Benson (1978) and Benson and Leach (1979) found in cores and regional drainage patterns evidence for a major desiccation event in Walker Lake between 9,050 to 6,400 years ago. The basin did not receive significant discharge from the Walker River again until 5,000 years ago. This desiccation interval coincides with the supposed "altithermal period" and may correspond to the "upper salt" unit of Searles Lake (Smith, 1979). There is evidence that the course of the Walker River has flucuated in the past, sometimes emptying into the Carson Sink and causing desiccation of the Walker Lake Basin (Morrison and Davis, 1984). Benson and Leach (1979) suggested that a major evaporite deposit may be buried under Walker Lake because a 36-m core taken in the lake was saturated with NaCl at a sediment depth of 10.4 m. They suggested that a sharp halocline near the bottom of Walker Lake may also be due to the presence of a buried evaporite section.

The historical deposits in Walker Lake Basin are shown

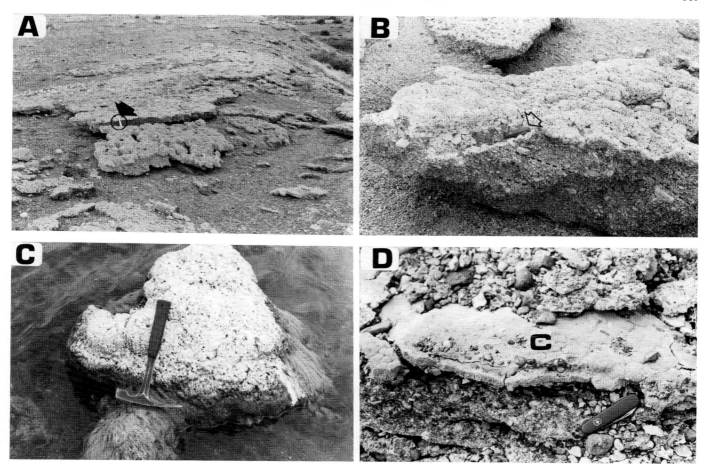

FIG. 13.—Some carbonate deposits in Walker Lake Basin: A) carbonate-cemented hardground pavement (beach rock; knife for scale; B) imbricated carbonate-cemented gravels and sands (knife for scale; C) stromatolite-encrusted clast covered by the green algae *Cladophora;* and D) carbonate-cemented gravels and caliche (C) horizon in an alluvial-fan deposit.

in Figure 6 and have been discussed above. Radiocarbon dating of historical carbonate deposits give contemporary ages. These carbonate rinds coat many of the carbonate deposits mentioned above and care must therefore be taken in sampling for isotopic dating. Oxygen and carbon isotope analyses of the historical carbonate samples, and of water from Walker Lake and rivers flowing into the basin indicate that the carbonates and CO_2 in equilibrium with the lake water have $\delta^{13}C$ values just above $+3\%_o$ (PDB). Carbon dioxide $\delta^{18}O$ values in equilibrium with the lake water at $25°$ C are $+2\%_o$ and the carbonate samples have $\delta^{18}O$ values ranging from -0.2 to $-1.8\%_o$. The spread of $\delta^{18}O$ values in the carbonate samples is interpreted to be due to differences in their temperature of formation, suggesting paleotemperature varied from $27°$ to $34°$ C. These temperatures are similar to those occurring along the lake shoreline in the summer months and suggest *in situ* contemporaneous carbonate deposition in the shallow, shoreline part of Walker Lake.

POTENTIAL ECONOMIC DEPOSITS

The 10,000 km² drainage area of Walker Lake Basin includes numerous uranium, silver, lead, copper, and iron sulfide deposits associated with igneous and metamorphic rocks. Uranium, as well as several other heavy elements, is extremely mobile in the acidic stream water that drains the Sierra Nevada and after a short residence time in a lake it can be concentrated by humic acid and aqueous sulfides in the organic-rich sediments of the lake (Benson and Leach, 1979). It is estimated that in the past 2 m.y., the amount of uranium alone transported to Walker Lake may be on the order of 4×10^8 kg (Benson and Leach, 1979). This suggests that closed basin termini may be sites for significant accumulations of minerals such as uranium. Ancient examples of lacustrine uraniferous deposits occur in the Miocene Artillery Peak–Date Creek Basin in Arizona (Otton, 1977; Sherborne et al., 1979), the Triassic-Jurassic Newark Basin in New Jersey and Pennsylvania (Turner-Peterson, 1977), and the Jurassic Morrison Formation in Utah (Peterson 1977). Salts and other economic brine deposits are common in nearby basins in Nevada and California (Papke, 1976).

Organic-rich sediments (1.5 to 2.5% TOC by weight) are currently accumulating in the deeper parts of Walker Lake, forming an area equivalent to about one third to one half of the lake's surface area. Most of the organic material ap-

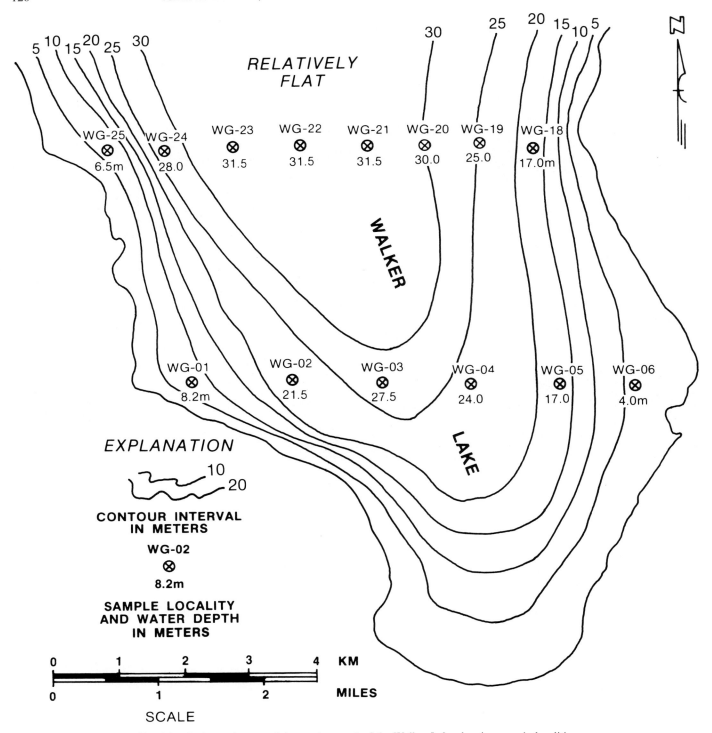

Fig. 14.—Bathymetric map of the southern end of the Walker Lake showing sample localities.

pears to be algal. The deep lake-bottom environments (deeper than 26 m) appear to be anoxic, saline, and contain two agents (humic acids and aqueous sulfides) capable of fixing uranium and other minerals. The preservation of such organic-rich lacustrine muds is likely if these conditions prevail. These types of lake deposits could make excellent source beds for hydrocarbons if they are not subsequently oxidized. Many ancient lake deposits contain excellent source rocks, notably the Eocene Green River Formation of the western United States, the Tertiary deposits of onshore China, and lacustrine deposits of Brazil, Angola, and Cabinda (Fouch, 1982).

TABLE 3.—SEDIMENT ANALYSES[1] FROM WALKER LAKE[2]

Grab Sample	Water Depth Meters	Mean Grain Size mm	Sorting[3]	Clay%[4]	TOC%[4]
WG-01	8.2	0.0960	1.45	0.8	1.0
WG-02	21.5	0.0820	1.81	1.6	1.0
WG-03	27.5	0.0069	2.02	11.9	1.75
WG-04	24.0	0.0433	2.03	2.8	0.4
WG-05	17.0	0.0310	2.03	2.3	0.6
WG-06	4.0	0.1750	2.17	4.2	0.2
WG-18	17.0	0.0420	1.79	3.5	—
WG-19	25.0	0.0140	2.63	7.8	—
WG-20	30.0	0.0073	2.03	10.3	1.7
WG-21	31.5	0.0096	2.37	14.0	1.2
WG-22	31.5	0.0093	2.51	15.9	2.5
WG-23	31.5	0.0068	2.12	15.9	2.5
WG-24	28.0	0.0094	2.49	11.3	2.5
WG-25	6.5	0.0974	1.05	0	—

[1]Analyses by M. S. Newton
[2]See Figure 14 for location of sample
[3]Sorting terminology after Folk, 1980: 1.0–2.0 poorly sorted, >2.0 very poorly sorted
[4]Clay % and TOC (Total organic carbon) % by weight are based on pipette analyses and acetone/distilled water washing of samples to remove organic acids.

DISCUSSION

Walker Lake sedimentary basin is a moderate-sized (20 by 60 km) basin associated with strike-slip faults related to Basin and Range extension. It is also associated with a center of late Tertiary to Recent volcanism. The basin lies in a crustal block of triangular shape bounded by normal-, right- and left-slip faults, and is down-dropped to the northwest. The basin is considered to be comparable to the classic Death Valley—Furnace Creek "pull-apart" basin of Burchfiel and Stewart (1966). Both basins have similar tectonic elements, facies distributions, and associated mineral deposits (Stewart, 1967; Hardie et al., 1978). Both Walker Lake and Death Valley basins are visualized to have formed under similar regional tectonic regimes, with the strike-slip faults and associated volcanism being a result of extensional tectonics. Numerous basins within and adjacent to the Walker Lane, such as Smith Valley, Carson Sink, Antelope Valley, and Pyramid Lake, may have formed in a similar manner and contain comparable facies.

The Great Basin "pull-apart" basins and extension-related strike-slip faults contrast in a marked way with the San Andreas fault system which delineates a major boundary between the Pacific and North American plates. The San Andreas fault zone is a wide zone of anatomosing strike-slip faults, which differ from those in the western Great Basin by being mostly compressional (transpressional) and marked by high-angle reverse faults and thrust faults associated with the strike-slip deformation. Both true pull-apart (Crowell, 1974; Mann et al., 1983) and fault-wedge basins (Crowell, 1974) occur as end members along and within the San Andreas transform margin. Parts of the Salton Trough, Los Angeles, and Santa Maria basins are examples of localized pull-apart basins, whereas Ridge Basin, San Timoteo Badlands, Elsinore Basin, San Francisco Bay, and Tomales-Bodega Bay areas are examples of fault-wedge basins. These basins are all essentially half grabens and similar in facies and facies distribution to the extensional basins that occur in the western Great Basin.

Walker Lake sedimentary basin and Ridge Basin, California, contain nearly identical sedimentary facies patterns despite the difference in detail of their tectonic settings (see Nilsen and McLaughlin, 1985 this volume). The Ridge Basin is a Miocene to Pleistocene half-graben filled with 12,000 m of marine and nonmarine sedimentary rocks (Crowell, 1974; Link and Osborn, 1978; Crowell and Link, 1982) and is the best exposed strike-slip basin in the San Andreas fault zone. The Violin Breccia in Ridge Basin is comparable to the narrow belt of alluvial fans/fan deltas flanking the Wassuk Range on the west side of Walker Lake. The Ridge Route Formation in Ridge Basin is similar in facies to the broad alluvial fans/fan deltas and shoreline facies on the east side of the Walker Lake Basin, flanking the Gillis Range. The lacustrine part of Walker Lake Basin is analogous to the Peace Valley beds in Ridge Basin. Also, the apparent tilting and onlapping relationships of the basins toward the north-northwest part of the wedge-shaped crustal block is similar. Major differences between the two basins are size (Ridge Basin is smaller: 10 by 40 km), patterns of volcanism (Ridge Basin contains only a few thin tuff horizons), fault relationships (Ridge Basin is more compressional with a master right-slip fault, the San Gabriel fault, and oblique-slip to high-angle reverse and thrust faults bounding the basin), and mineralization (none in Ridge Basin). Both basins appear to have good lacustrine organic-rich deposits (Crowell and Link, 1982).

Numerous other strike-slip basins have been identified around the world and are discussed in Crowell (1974), Ballance and Reading (1980), Hempton et al. (1983), Mann et al. (1983), and this volume. All these basins contain similar sedimentary patterns even though some are marine and others nonmarine. They are characterized by abrupt facies changes related to active tectonics, and generally form long, linear basins bounded by one or several strike-slip faults. These modern and ancient strike-slip basins contain thick, asymmetric sedimentary sequences that exhibit coarse-grained marginal facies on both sides of the basin and finer-grained facies at the center of the basin. The sediment dispersal directions in these basins are from (1) the margins of basins toward the basin center, usually in the form of alluvial fans and fan deltas, and (2) more longitudinal and down the axis of the basin in the form of fluvial, fluvial deltaic, and lacustrine or marine deposits. Axial turbidites may be present in the deeper marine or lacustrine basins.

SUMMARY AND CONCLUSIONS

Walker Lake is centered over a late Tertiary(?) to Recent sedimentary basin that is genetically related to the strike-slip faulting of the northwest-trending Walker Lane shear zone in the western Basin and Range Province, Nevada. The Walker Lake Basin lies on the northwest side of a fault-bounded, triangular crustal block. The block is bounded on its west side by north-striking high-angle normal- to oblique-slip faults along the Wassuk Range front. These normal faults may merge northward into a series of right-slip faults composing the Walker Lane, which forms the northeast boundary of the crustal block. The Wassuk Range frontal

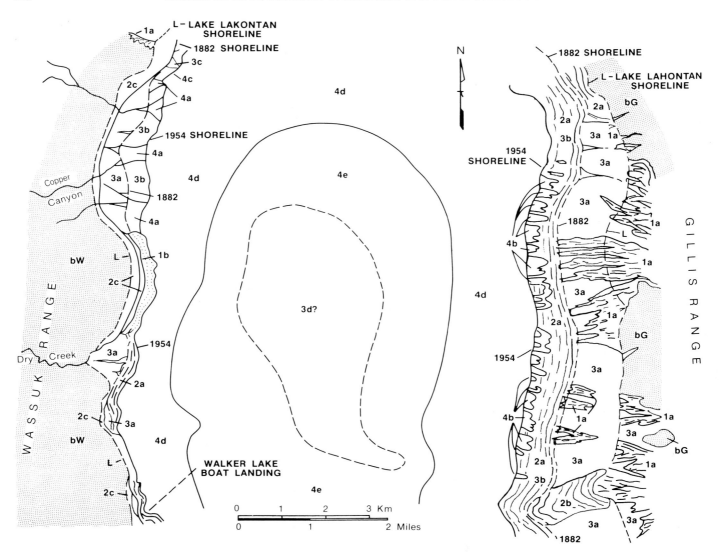

FIG. 15.—Outcrop map of the upper Quaternary stratigraphy for the central part of the Walker Lake sedimentary basin. See below for explanation.

Explanation

age	deposit/locality	lake stage
1. Pre-Lahontan Pleistocene (Wisconsinan)	1.a Dissected alluvial fans and pediment veneers modified by Lahontan shoreline; west flank of the Gillis Range.	
	1.b Wave-cut, terraced alluvial fans and fan deltas with large (2m) *in situ* lithoidal tufa or stromatolite deposits; east flank of the Wassuk Range below 4,100 feet and 3.5 km south-southeast of Copper Canyon along Walker Lake shoreline. Radiocarbon ages of 18,000 yr.	Low stand, dropping to a low of 3,950 ft.
2. Lahontan Pleistocene (Wisconsinan)	2.a Shoreline and sandflat-sheet sands; on west flank of the Gillis Range below 4,100 ft.	High stand, rising to a peak at Lake Lahontan high stand of 4,380 ft.
	2.b Same as 2.a, except it extends up to 4,400 ft elevation and forms spits at the Lahontan shoreline; west flank of the Gillis Range.	
	2.c Tufa-cemented fanglomerate capping faulted pediment remnants. Cementation is Lahontan but deposit may be part of 1.b; east flank of the Wassuk Range below 4,400 ft. Radiocarbon ages of 11,000 to 13,000 yr (Morrison and Frye, 1965).	
3. Post-Lahontan Holocene	3.a Holocene fan deltas, alluvial fans and pediment deposits. In places, incised by 3- to 6-m-deep channels. Lower margin is 1882 shoreline at 4,083 ft. Widespread on flanks of both the Gillis and Wassuk Ranges.	Several low stands, and possibly associated with complete desiccation.
	3.b Holocene stromatolites and tufas. Thin carbonate rinds (a few centimeters) uniformly coat-	

faults appear to merge or terminate southward into an east-northeast-striking, left-lateral fault zone, which forms the southern boundary of the crustal block. The Walker Lake Basin is interpreted to be a half graben formed by regional extension accommodated by the right- and left-lateral strike-slip zones. Initiation of this structural framework may have begun in middle to late Miocene as suggested by thick volcanic deposits near the projected junction of the right- and left-slip fault zones at the southeastern corner of the block.

The Walker Lake drainage basin occupies about 10,000 km^2 in western Nevada and parts of California. It is essentially a closed hydrologic system which drains from the crest of the Sierra Nevada in California and terminates in Walker Lake. North-northwest trending Walker Lake is 26.2 km long, 7.7 km wide, and is up to 31.5 m deep. The products of humic-acid weathering of Sierran granitic rocks are transported by Walker River to the lake, and the subsequent evaporation of this runoff in this semiarid basin, results in a saline and alkaline lake. Lake Lahontan (Wisconsinan) shorelines rim Walker Lake, and other Pleistocene deposits indicate former lake depths greater than 130 m and a surface area of 880 km^2, roughly four times larger than the present lake area. The lake has completely desiccated at least once in post-Pleistocene time.

Modern depositional settings of the basin include: alluvial fan/fan deltas, fluvial deltas, shorelines/dunes and lacustrine environments. The structurally asymmetric basin has a narrow belt of steep alluvial fans along the western shoreline (Wassuk Range) and gentle, more areally extensive fans along the eastern shoreline (Gillis Range) of the modern lake. The alluvial fans consists of poorly sorted conglomerate, sandstone, and minor mudstone and are transitional down-dip into fan deltas with well-developed Gilbert-delta foreset bedding. Walker River enters the lake from the north, where it forms a delta. The river is a major sediment source for the basin. Sand is concentrated in the fluvial and deltaic environments and locally is reworked along the shoreline around the delta and its dunes. The alluvial-fan, fan-delta, and much of the shoreline deposits consist largely of coarse-grained, locally derived materials from the flanking mountains. Carbonate-cemented horizons, stromatolites, oncolites, caliche, and tufas have formed along the shorelines. Silt, mud, carbonate and organic-rich sediments accumulate in the lake toward the center of the basin. Extensive submerged sandflats occur on the north, east, and south sides of the basin and formed by reworking of alluvial-fan, fan-delta, and delta sediments. Small localized sand dunes occur on the east side of the basin, in low areas, and in reentrants between coalescing alluvial fans. These Recent environments of deposition are representative of older deposits in the basin.

Lacustrine sediments in Walker Lake Basin have relatively high concentrations of carbonate, organic matter, and uranium. Humic acids and aqueous sulfides promote the concentration of uranium and other sulfates in this saline, alkaline lake that has locally anoxic bottom conditions. Under proper conditions such sediments could become significant hydrocarbon source beds and uranium deposits.

---

ing cobbles on wave-cut terraced alluvial fans and fan deltas below 4,150 ft. Deposit continuous and uniform on all bouldery surfaces on both sides of lake except where washed out by currently active fans (4.a). Radiocarbon ages of 1,600 yr.

3.c Holocene beach rock. Carbonate-cemented beach and delta foresets and shoreline deposits.

3.d Possible evaporite body that would occupy central basin below 3,861 ft, covered by 4.e.

4. Historical Holocene	4.a Active alluvial fans and fan deltas, on both flanks of the basin. No vegetation. Thin tufa coatings (2 mm rind) giving contemporary radiocarbon ages.	Several low stands,
	4.b Sand dunes. East shoreline, Rilled remnants of sandflat-sheet (4.c) exposed during the last 3 decades by 4.9-m drop in lake stage. Dunes above 4,100 ft are probably older eolian Lake Lahontan-aged deposits.	Starting in late 1800s, has been dropping rapidly from 4,083 ft to 3,960 ft in 1984.
	4.c Active shoreline deposits that consist of several zones, including a narrow shoreline on the west side of lake with beach foresets, gravels at the waterline, algae-covered cobbles and rippled sand lakeward.	
	4.d Lacustrine sandflat-sheet sands with rippled sand that extend from 1 to 26 m depth. Air photos show subaqueous sand waves with the shallower part being involved in longshore transport.	
	4.e Lacustrine silt and mud that are rich in organics and carbonate and in water deeper than 26 m. Asymmetrical sedimentary fill that is concentrated more on the west side of the basin.	

Other Symbols   bW—bedrock of Wassuk Range (Mesozoic plutonic and metavolcanic rocks.)
            bG—bedrock of Gillis Range (Tertiary volcanic and sedimentary rocks.)
            1882—shoreline of Walker Lake in 1882 (4,083 ft.)
            1954—shoreline of Walker Lake in 1954 (date of air photos and available topographic maps; 4,081 ft.)
            L—shoreline of Lake Lahontan high stand (4,380 ft.)

124   *MARTIN H. LINK, MICHAEL T. ROBERTS, AND MARK S. NEWTON*

ACKNOWLEDGMENTS

We thank K. T. Biddle, N. Christie-Blick, S. Y. Johnson, J. H. Stewart and R. P. Wright for kindly reviewing this paper. G. R. Licari, and R. H. Osborne are thanked for discussions and initial field work in Walker Lake Basin. A. S. Harper, D. Smith, and S. Wong helped with some of the original field work. E. Grossman helped with isotopic analyses used here, and S. Stine provided some unpublished data on Mono Lake. Radiocarbon dating was done at the University of California, Riverside, and at Simon Fraser University, British Columbia. Sediment and isotopic analyses were done at University of Southern California. Funding, typing, drafting, and logistical help were provided by Cities Service Research, Tulsa, Oklahoma. Maggie Draughon typed and helped edit the manuscript.

REFERENCES

ALBERS, J. P., 1964, Jurassic "oroclinal" folding and related strike-slip faulting in the western United States Cordillera (abs.): Geological Society of America Special Paper 76, p. 4.

———, 1967, Belt of sigmoidal bending and right-lateral faulting in the western Great Basin: Geological Society of America Bulletin, v. 78, p. 143–156.

BALLANCE, P. F., AND READING, H. G., eds., 1980, Sedimentation in oblique-slip mobile zones: International Association Sedimentologists Special Publication No. 4, 265 p.

BENSON, L. V., 1978, Fluctuations in the level of glacial Lake Lahontan during the last 40,000 years: Quaternary Research, v. 9, p. 300–318.

BENSON, L. V., AND LEACH, D. L., 1979, Uraniun transport in the Walker River Basin, California and Nevada: Journal of Geochemical Exploration, v. 11, p. 227–248.

BINGLER, E. L., 1978, Geologic map of the Schurz quadrangle: Nevada Bureau of Mines and Geology Map 60, Scale 1:48,000.

BURCHFIEL, B. C., AND STEWART, J. H., 1966, "Pull-apart" origin of the central segment of Death Valley, California: Geological Society of America Bulletin, v. 77, p. 439–442.

CLARKE, F. W., 1924, The data of geochemistry: 4th ed., United States Geological Survey Bulletin 770, 841 p.

CROWELL, J. C., 1974, Origin of late Cenozoic basins in southern California, in Dickinson, W. R., ed., Tectonics and Sedimentation: Society of Economic Paleontologists and Mineralogists Special Publication No. 22, p. 190–204.

CROWELL, J. C., AND LINK, M. H., eds., 1982, Geologic History of Ridge Basin Southern California: Society of Economic Paleontologists and Mineralogists, Pacific Section, 304 p.

DOHRENWEND, J. C., 1982, Surficial geology, Walker Lake 1° by 2° quadrangle Nevada-California: United States Geological Survey Miscellaneous Field Studies Map MF-1382 C, Scale 1:250,000.

———, 1983, Late Cenozoic fault map: United States Geological Survey Miscellaneous Field Studies Map MF-1382 D, Scale 1:250,000.

EKREN, E. B., AND BYERS, F. M., Jr., 1978a, Preliminary geologic map of the Luning NE quadrangle, Mineral and Nye Counties, Nevada: United States Geological Survey Open-File Report 78-915, Scale 1:48,000.

———, 1978b, Preliminary geologic map of the Luning NW quadrangle, Mineral County, Nevada: United States Geological Survey Open-File Report 78-916, Scale 1:48,000.

———, 1978c, Preliminary geologic map of the Luning SW quadrangle, Mineral County, Nevada: United States Geological Survey Open-File Report 78-917, Scale 1:48,000.

———, 1978d, Preliminary geologic map of the Luning SE quadrangle, Mineral and Nye Counties, Nevada: United States Geological Survey Open-File Report 78-918, Scale 1:48,000.

EKREN, E. B., BYERS, F. M., Jr., HARDYMAN, R. F., MARVIN, R. F., AND SILBERMAN, M. L., 1980, Stratigraphy, preliminary petrology, and some structural features of Tertiary volcanic rocks in the Gabbs Valley and Gillis Range, Mineral County, Nevada: United States Geological Survey Bulletin 1464, 54 p.

FOLK, R. L., 1980, Petrology of Sedimentary Rocks: Austin, Texas, Hemphill publication, 182 p.

FOUCH, T. D., 1982, Character of ancient petroliferous lake basins of the world (abs.): American Association of Petroleum Geologists Bulletin, v. 66, p. 1680–1681.

GIANELLA, V. P., AND CALLAGHAN, E., 1934, Cedar Mountain, Nevada, earthquake of December 20, 1932: Seismological Society of America Bulletin, v. 24, p. 345–377.

GILBERT, G. K., 1890, Lake Bonneville: United States Geological Survey Monographs, v. 1, 438 p.

HARDIE, L. A., SMOOT, J. P., AND EUGSTER, H. P., 1978, Saline lakes and their deposits: a sedimentological approach, in Matter, A. and Tucker M. E., eds., Modern and Ancient Lakes Sediments: International Association of Sedimentologists Special Publication No. 2, p. 7–41.

HARDYMAN, R. F., 1978, Volcanic stratigraphy and structural geology of Gillis Canyon quadrangle, northern Gillis Range, Mineral County, Nevada [unpubl. Ph.D dissertation]: Reno, University of Nevada, 377 p.

———, 1980, Geologic map of the Gillis Canyon quadrangle, Mineral County, Nevada: United States Geological Survey Miscellaneous Investigation Series Map I-1237, Scale 1:48,000.

HARDYMAN, R. F., EKREN, E. B., AND BYERS, F. M., Jr., 1975, Cenozoic strike-slip, normal, and detachment faults in northern part of Walker Lane west-central Nevada (abs.): Geological Society of America Abstracts with Programs, v. 7, p. 1100.

HEMPTON, M. R., DUNNE, L. A., AND DEWEY, J. F., 1983, Sedimentation in an active strike-slip basin, southeastern Turkey: Journal of Geology, v. 91, p. 401–412.

HOWARD, K. A., AARON, J. M., BRABB, E. E., BROCK, M. R., GOWER, H. D., HUNT, S. J., MILTON, D. J., MUEHLBERGER, W. R., NAKATA, J. K., PLAFKER, G., PROWELL, D. C., WALLACE, R. E., AND WITKIND, I. J., 1978, Preliminary map of young faults in the United States as a guide to possible fault activity: United States Geological Survey Miscellaneous Field Studies Map MF-916, Sheet 1, Scale 1:5,000,000.

LEWIS, K. B., 1980, Quaternary sedimentation on the Hikurangi oblique-subduction and transform margin, New Zealand, in Ballance, P. F., and Reading, H. G., eds., Sedimentation in Oblique-Slip Mobile Zones: International Association of Sedimentologists Special Publication No. 4, p. 171–190.

LINK, M. H., AND OSBORNE, R. H., 1978, Lacustrine facies in the Pliocene Ridge Basin Group, Ridge Basin, California, in Matter, A., and Tucker, M. E., eds., Modern and Ancient Lake Sediments: International Association of Sedimentologists Special Publication No. 2, p. 167–187.

LOCKE, A., BILLINGSLEY, P., AND MAYO, E. B., 1940, Sierra Nevada tectonic pattern: Geological Society of America Bulletin, v. 51, p. 513–539.

LONGWELL, C. R., 1960, Possible explanation of diverse structural patterns in southern Nevada: American Journal of Science, v. 258-A, p. 192–203.

MANN, P., HEMPTON, M. R., BRADLEY, D. C., AND BURKE, K., 1983, Development of pull-apart basins: Journal of Geology, v. 91, p. 529–554.

MIFFLIN, M. D., AND WHEAT, M. M., 1979, Pluvial lakes and estimated pluvial climates of Nevada: Nevada Bureau of Mines and Geology Bulletin 94, 57 p.

MORRISON, R. B., 1964, Lake Lahontan: Geology of southern Carson Desert, Nevada: United States Geological Survey Professional Paper 401, 156 p.

MORRISON, R. B., AND DAVIS, J. D., 1984, Supplementary guidebook for fieldtrip 13. Quaternary stratigraphy and archeology of the Lake Lahontan area: A reassessment: Desert Research Institute, University of Nevada, Social Sciences Center Technical Report No. 41, 2nd edition, 50 p.

MORRISON, R. B., AND FRYE, J. C., 1965, Correlation of the middle and late Quaternary successions of the Lake Lahontan, Lake Bonneville, Rocky Mountain (Wasatch Range), southern Great Plains, and eastern Midwest areas: Nevada Bureau of Mines Report 9, 45 p.

NIELSEN, R. L., 1965, Right-lateral strike-slip faulting in the Walker Lake, west-central Nevada: Geological Society of America Bulletin, v. 76, p. 1301–1308.

NILSEN, T. H., AND McLAUGHLIN, R. J., 1985, Comparison of tectonic framework and depositional patterns of the Hornelen strike-slip basin of Norway and the Ridge and Little Sulphur Creek strike-slip basins of California, *in* Biddle, K. T., and Christie-Blick, N., eds., Strike-slip Deformation, Basin Formation, and Sedimentation: Society of Economic Paleontologists and Mineralogist's Special Publication No. 37, p. 79–103.

NOBLE, L. F., AND WRIGHT, L. A., 1954, Geology of the central and southern Death Valley region, California: California Division of Mines Bulletin 170, p. 145–160.

OSBORNE, R. H., LICARI, G. R., AND LINK, M. H., 1982, Modern lacustrine stromatolites, Walker Lake, Nevada: Sedimentary Geology, v. 32, p. 39–61.

OTTON, J. K., 1977, Geology of uraniferous Tertiary rocks in the Artillery Peak-Date Creek Basin, west-central Arizona, *in* Campbell, J. A., ed., Short Papers of the United States Geological Survey Uranium-Thorium Symposium, 1977: United States Geological Survey Circular 753, p. 35–36.

PAPKE, K. G., 1976, Evaporites and brines in Nevada playas: Nevada Bureau of Mines and Geology Bulletin 87, 35 p.

PEASE, R. W., 1969, Normal faulting and lateral shear in northeastern California: Geological Society of America Bulletin, v. 80, p. 715–720.

PETERSON, F., 1977, Uranium deposits related to depositional environments in the Morrison Formation (Upper Jurassic), Henry Mountain mineral belt of southern Utah, *in* Campbell, J. A., ed., Short Papers of the United States Geological Survey Uranium-Thorium Symposium, 1977: United States Geological Survey Circular 753, p. 45–47.

READING, H. G., 1980, Characteristics and recognition of strike-slip fault systems, *in* Ballance, P. F., and Reading, H. G., eds., Sedimentation in Oblique-Slip Mobile Zones: International Association of Sedimentologists Special Publication No. 4, p. 7–26.

ROSS, D. C., 1961, Geology and mineral deposits of Mineral County, Nevada: Nevada Bureau of Mines Bulletin 58, 98 p.

RUSH, F. E., 1970, Hydrologic regime of Walker Lake, Mineral County, Nevada: United States Geological Survey Hydrologic Atlas HA-415, Scale 1:62,500.

RUSSELL, I. C., 1885, Geological history of Lake Lahontan, a Quaternary lake of northwestern Nevada: United States Geological Survey Monographs, v. 2, 288 p.

SHAWE, D. R., 1965, Strike-slip control of basin-range structure indicated by historical faults in western Nevada: Geological Society of America Bulletin, v. 76, p. 1361–1378.

SHERBORNE, J. E., Jr., BUCKOVIC, W. A., DEWITT, D. B., HELLINGER, T. S., AND PAVLAK, S. J., 1979, Major uranium discovery in volcaniclastic sediments, Basin and Range Province, Yavapai County, Arizona: American Association of Petroleum Geologists Bulletin, v. 63, p. 621–646.

SLEMMONS, D. B., 1957, Geological effects of the Dixie Valley-Fairview Peak, Nevada earthquakes of December 16, 1954: Seismological Society of America Bulletin, v. 47, p. 353–375.

SMITH, G. I., 1979, Subsurface stratigraphy and geochemistry of late Quaternary evaporites, Searles Lake, California: United States Geological Survey Professional Paper 1043, 130 p.

SPEED, R. C., 1978, Paleogeographic and plate tectonic evolution of the early Mesozoic marine province of the western Great Basin, *in* Howell,

D. G., and McDougall, K. A., eds., Mesozoic Paleogeography of the Western United States: Pacific Coast Paleogeography Symposium 2: Society of Economic Paleontologists and Mineralogists, Pacific Section, p. 253–270.

SPEED, R. C., AND COGBILL, A. H., 1979, Candelaria and other left-oblique slip faults of the Candelaria region, Nevada—Summary: Geological Society of America Bulletin, v. 90, p. 149–163.

STEWART, J. H., 1967, Possible large right-lateral displacement along fault and shear zones in the Death Valley-Las Vegas area, California and Nevada: Geological Society of America Bulletin, v. 78, p. 131–142.

———, 1980, Geology of Nevada: Nevada Bureau of Mines and Geology Special Publication 4, 136 p.

STEWART, J. H., AND CARLSON, J. E., 1978, Geologic map of Nevada: United States Geological Survey Miscellaneous Field Studies Map MF-930, in cooperation with the Nevada Bureau of Mines and Geology, Scale 1:500,000, 2 parts.

STEWART, J. H., AND JOHANNESEN, D. C., 1981a, Geologic map of the Hawthorne quadrangle, Mineral County, Nevada, with surficial geology by Dohrenwend, J. C.: United States Geological Survey Miscellaneous Field Studies Map MF-1277, Scale 1:62,500.

———, 1981b, Geologic map of the Powell Mountain quadrangle, Mineral County, Nevada, with surficial geology by J. C. Dohrenwend: United States Geological Survey Miscellaneous Field Studies Map MF-1268, Scale 1:62,500.

STEWART, J. H., REYNOLDS, M. W., AND JOHANNESEN, D. C., 1981, Geologic map of the Mount Grant quadrangle, Lyon and Mineral Counties, Nevada, with surficial geology by J. C. Dohrenwend: United States Geological Survey Miscellaneous Field Studies Map MF-1278, Scale 1:62,500.

STEWART, J. H., CARLSON, J. E., AND JOHANNESEN, D. C., 1982, Geologic map, Walker Lake 1° by 2° quadrangle, California-Nevada: United States Geological Survey Miscellaneous Field Studies Map MF-1382A, Scale 1:250,000.

STEWART, J. H., CHAFFEE, M. A., DOHRENWEND, J. C., JOHN, D. A., KISTLER, R. W., KLEINHAMPL, F. J., MENZIE, W. D., PLOUFF, D., ROWAN, L. C., AND SILBERLING, N. J., 1984, The conterminous United States mineral appraisal program: Background information to accompany folio of geologic, geochemical, geophysical, and mineral resourcers maps of the Walker Lake 1° × 2° quadrangles, California and Nevada: United States Geological Survey Circular 927, 22 p.

TURNER-PETERSON, C. E., 1977, Uranium mineralization during early burial, Newark Basin, Pennsylvania-New Jersey, *in* Campbell, J. A., ed., Short Papers of the United States Geological Survey Uranium-Thorium Symposium, 1977: United States Geological Survey Circular 753, p. 3–4.

UNITED STATES DEPARTMENT OF AGRICULTURE, 1969, USDA Report on water and related land resources, central Lahontan Basin, Walker River sub-basin, Nevada-California: USDA Nevada River Basin Survey, Carson City, Nevada, 231 p.

WRIGHT, L. A., 1976, Late Cenozoic fault patterns and stress fields in the Great Basin and westward displacement of the Sierra Nevada block: Geology, v. 4, p. 489–494.

ZOBACK, M. L., AND ZOBACK, M., 1980, State of stress in the conterminous United States: Journal of Geophysical Research, v. 85, n. B11, p. 6113–6156.

CENOZOIC SEDIMENTATION AND SEQUENCE OF DEFORMATIONAL EVENTS AT THE SOUTHEASTERN END OF THE FURNACE CREEK STRIKE-SLIP FAULT ZONE, DEATH VALLEY REGION, CALIFORNIA

IBRAHIM CEMEN[1] AND L. A. WRIGHT
Department of Geosciences
Pennsylavania State University
University Park, Pennsylvania 16802;
R. E. DRAKE
Department of Geology and Geophysics
University of California
Berkeley, California 94720;
AND
F. C. JOHNSON
American Borate Company
Tecopa, California 92389

ABSTRACT: Evidence of Cenozoic deformation and sedimentation along the southeasternmost 40 km of the Furnace Creek strike-slip fault zone, in the southwestern Great Basin, is contained in two successions of sedimentary and volcanic rocks. Each indicates a stage in the development of the fault zone and associated basins; each is bracketed by K/Ar age determinations. The older succession, dated at 25 to no less than 14 Ma, occurs in tilted fault blocks of the bordering Funeral Mountains, and predates major crustal extension. The succession is about 1,300 m in maximum exposed thickness, contains remnants of two northeast-sloping alluvial fans, onlaps a Paleozoic basement to the southwest, and includes an unconformity that cuts out progressively older units southwestward. These features indicate that a topographic high persisted where younger formations of the fault-controlled, northwest trending Furnace Creek Basin now occur, and also early vertical movement on the fault zone.

The younger succession, comprising the Artist Drive, Furnace Creek, and Funeral Formations of the basin (McAllister, 1970), ranges in age from about 14 Ma to about 4 Ma. It records the subsidence and deformation of the basin, and was associated with right-lateral slip on northwest-striking faults and with basin-range extension expressed in part by abundant north-to northeast-striking normal faults. The three formations have a composite maximum thickness of about 3,600 m, but the combined thickness varies greatly from place to place, owing to contemporaneous faulting, to changes in depocenters related to the fault movements and to a general thinning southeastward. The sedimentation and faulting also were accompanied by abundant plutonism and volcanism in the Greenwater Range and Black Mountains southwest of the fault zone. The fact that the normal faulting is regional, affecting terranes on both sides of the fault zone, qualifies the Furnace Creek as a special type of divergent strike-slip fault zone.

The contemporaneous strike slip within the 40-km segment of the fault zone probably increases northwestward from zero to no more than a few kilometers, but continues to increase farther northwestward. In the area of Furnace Creek Wash, strike slip is indicated by low-angle grooves and fault mullions, west-northwest-trending en echelon folds, north-northeast-striking normal faults, and by evidence that west-northwest-directed extension was greater on the southwest side of the fault zone than on the northeast side. Vertical movements are recorded by marginal conglomerates in each of the three formations and by northwest-trending folds with features characteristic of drag folds and forced folds. That the fault zone may penetrate deeply into the crust is suggested by the abundance of basalt flows and sills in the three formations of the basin as compared with the bordering terranes.

INTRODUCTION

Since the earliest systematic investigations of the geology of the southwestern part of the Great Basin in the early 1900s, the crust of this region has been recognized as pervasively broken by innumerable normal- to oblique-slip faults associated with regional extension. The normal faults tend to be oriented northward to north-northeastward and to tilt Oligocene and some Miocene rock units as severely as older units. Many of these faults offset rock units as young as Quaternary. Consequently, the extension has been recognized as confined largely or wholly to late Cenozoic time and as continuing to the present.

Several through-going strike-slip fault zones have moved contemporaneously with the normal faults. These, together with the major normal faults, have resulted in a mosaic of fault-bounded ranges and intervening valleys. The valleys commonly delineate well-developed sedimentary basins (Fig. 1). The basin sediments provide chronologic evidence of the onset and duration of contemporaneous crustal extension at given localities. Some of the basins are clearly related to the zones of strike slip and are thus relevant to this volume. One of these, a linear basin associated with the northwest-trending Furnace Creek fault zone and here designated as the "Furnace Creek Basin" (FC in Fig. 1), is discussed in this paper.

Much or all of the lateral movement on the Furnace Creek fault zone (also commonly called the "Northern Death Valley—Furnace Creek fault zone") has been contemporaneous with regional extension. The Furnace Creek fault zone qualifies as a special type of divergent strike-slip zone (see Christie-Blick and Biddle, 1985; and Harding et al., 1985, both in this volume) because the crust on both sides of the fault has been extended. It may also record the reactivation of an earlier established discontinuity (Wright, 1976). In fact, evidence too abundant to review here, indicates that most of the major fault zones of the southwestern Great Basin have been guided by discontinuities established in Mesozoic or Precambrian time (Wright et al., 1974, 1976;

---

[1]Present Address: Department of Geology, Oklahoma State University, Stillwater, Oklahoma 74074

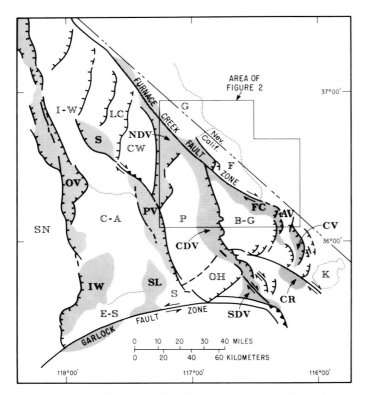

Fig. 1.—Generalized map of southwestern part of the Great Basin, California and Nevada, showing the ranges (light letters), the principal Cenozoic basins (heavy letters), and the major zones of Cenozoic faulting that define them. At most places along the throughgoing fault zones, movement is oblique: hachures, dominantly normal; arrows, dominantly strike slip; barbs dominantly reverse or thrust. Ranges: B-G, Black Mountains and Greenwater Range; C-A, Coso and Argus Ranges; CW, Cottonwood Mountains; E-S, El Paso Mountains and Spangler Hills; F, Funeral Mountains; G, Grapevine Mountains; I-W, Inyo and White Mountains; K, Kingston Range; LC, Last Chance Range; OH, Owlshead Range; P, Panamint Mountains; S, Slate Range; SN, Sierra Nevada. Basins: AV, Amargosa Valley; CDV, Central Death Valley; CR, China Ranch; CV, Chicago Valley; FC, Furnace Creek; IW, Indian Wells; NDV, Northern Death Valley; OV, Owens Valley; PV, Panamint Valley; S, Saline Valley; SL, Searles Lake Valley; SDV, southern Death Valley.

Williams et al., 1976; Wright, 1976) so that the orientation of an individual zone may indicate in only a general way the actual orientation of the late Cenozoic regional stress field. The observed grooves and mullions on exposed Cenozoic fault surfaces in the region are ordinarily oriented obliquely. Thus many of the "normal" and "strike-slip" faults of this text and the accompanying figures would be more accurately designated as oblique-slip, but dominated by either dip slip or strike slip.

The northwest-trending topographic depression defined by the Furance Creek fault zone terminates southeastward near Eagle Mountain (EM in Fig. 2) in the Amargosa Valley. There the fault zone seems to splay southward and to merge tangentially into normal faults (Fig. 2). From the Eagle Mountain area, it has been traced northwestward along Furnace Creek Wash and northern Death Valley, apparently terminating in Fish Lake Valley (beyond the area of Figure 2) on the Nevada side of the California-Nevada boundary.

In a general way, it forms the northeast edge, about 200 km long, of a large triangle. The Garlock fault zone with left-lateral slip forms the southern edge of the triangle and the dominantly normal Sierra Nevada fault zone forms the western edge (Fig. 1).

Apparently anomalous thickness differences in Precambrian and Paleozoic formations (Fig. 3) on opposite sides of the Furnace Creek fault zone, have been cited as evidence for right-lateral displacement measureable in tens of kilometers (Stewart, 1967; Stewart et al. 1968; Miller and Walsh, 1977; Poole and Sandberg, 1977). Evidence for an estimated 50 km of right-lateral offset of a Jurassic pluton, along a segment of the Furnace Creek Fault zone that lies about 150 km northwest of the Furnace Creek area (McKee, 1968), we find particularly convincing. However, although the right-lateral sense is undisputed, the magnitude and kinematics of the displacement remain conjectural. This is particularly true for the southeasternmost segment of the fault zone, including the part that coincides with the Furnace Creek Basin. Present considerations concerning the timing and magnitude of right-lateral displacement in the Furnace Creek Wash area center on two contrasting interpretations. Each is based on evidence that the fault zone does, indeed, terminate in the vicinity of Eagle Mountain, and that in this area, the percentage of crustal extension has been higher on the southwest side of the fault zone than on the northeast side (Wright and Troxel, 1970). The interpretations differ markedly, however, in the postulated magnitude and timing of displacement.

One interpretation holds that the displacement increases progressively northwestward to a maximum of about 50 km north of the northern end of Death Valley (Wright and Troxel, 1970). It is based upon the observation that for approximately 150 km northwestward from the apparent termination of the Furnace Creek fault zone, the crust on the southwest persistently contains indictors of marked crustal extension for which there are no counterparts on the northeast side. Evidence for crustal extension includes the field of Cenozoic plutons and units of the Central Death Valley volcanic field in the Black Mountains and Greenwater Range (B-G in Fig. 1), and several range-bounding normal faults. An interpretation involving laterally increasing strike slip permits the Furnace Creek Basin to have developed during the time of strike-slip deformation, but it requires no more than 10 km of right-lateral slip in the area of Furnace Creek Wash.

The other interpretation (Stewart, 1983) involves about 80 km of northwestward transport of the Panamint Mountains fault block (P in Fig. 1) along a low-angle detachment surface from an original location above the present Black Mountains and Greenwater Range. This interpretation is based largely on the observation that the Upper Proterozoic and Cambrian formations are thinner where exposed at the northern end of the Panamint Mountains than they are at a nearly opposite locality in the Funeral Mountains (F in Fig. 1) northeast of the fault zone. The 80 km of displacement is required to permit the matching of isopachs across the fault zone. It would also explain the absence of exposures of Upper Proterozoic and Paleozoic formations in much of the area occupied by the Black Mountains and Greenwater

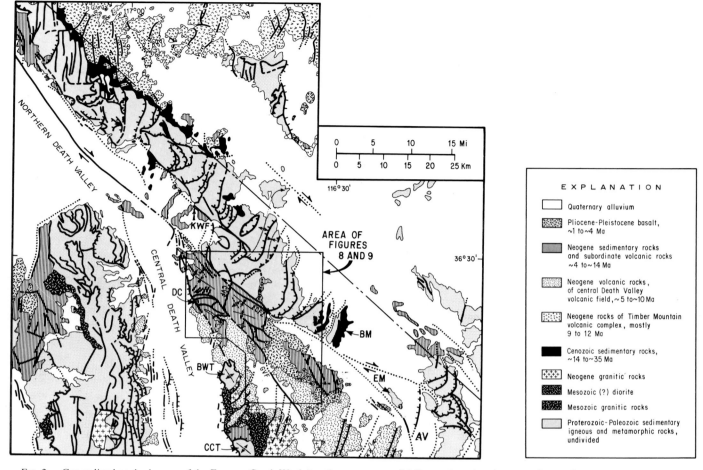

FIG. 2.—Generalized geologic map of the Furnace Creek Wash (northwestern part of delineated area) and surrounding region, showing in particular, the Neogene rock units of the Furnace Creek Basin, the Proterozoic-Paleozoic rocks of the Funeral Mountains on the northeast, the plutonic and volcanic rocks of the central Death Valley igneous field on the southwest and selected faults. Abbreviations: AV, Amargosa Valley; BM, Bat Mountain; BWT, Badwater turtleback; CCT, Copper Canyon turtleback; CVT, Cross-valley fault; DC, Desolation Canyon; EM, Eagle Mountain; FCFZ, Furnace Creek fault zone; GVF, Grand View fault zone, KWF, Keane Wonder fault. See Figure 1 for location.

Range. Because the two ranges are now extensively underlain by an autochthonous Upper Cenozoic sedimentary and volcanic cover, of which the formations of the Furnace Creek Basin are a part, the detachment and thus most of the movement on the Furnace Creek segment of the fault zone would predate the oldest basinal units. Although a detailed evaluation of the two interpretations lies beyond the objectives of this paper, several features of the Cenozoic geology of the area of the Furnace Creek basin point in favor of the first interpretation and are noted at the appropriate places below.

This paper is intended primarily as a summary of Cenozoic sedimentation along and near the southeasternmost 40-km segment of the fault zone, with special reference to the kinematics and chronology of successive phases in the history of the basin-related faulting. Cemen, and to a lesser extent Johnson and Wright, made the field observations; Drake provided K/Ar age determinations; and each author contributed to the interpretations. Wright and Cemen prepared the text and illustrations, with the exception of Figure

7, prepared by Johnson. For basic geologic data, we have drawn heavily upon the published and open-file geologic maps of J. F. McAllister (1970, 1971, 1973).

PREVIOUS INVESTIGATIONS

The sedimentary and layered volcanic rocks that define the Furnace Creek Basin consist, in upward succession, of the Artist Drive, Furnace Creek, and Funeral Formations (Fig. 4). These were named and first described by Thayer (cited by Noble, 1941). They were later delineated, in generalized fashion, on a geologic map by Noble and Wright (1954), who also observed that the basin is largely fault-controlled and lies between two contrasting terranes (Fig. 2). The terrane northeast of the basin, forms the backbone of the southern part of the present Funeral Mountains and consists chiefly of sedimentary rocks (now designated as miogeoclinal) that range in age from Late Proterozoic to Mississippian (Fig. 3). The terrane to the southwest, the Black Mountains and Greenwater Range, is underlain mostly

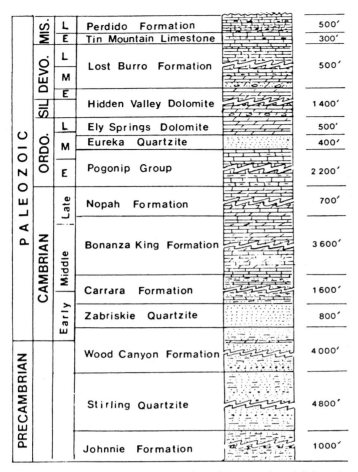

PALEOZOIC					Perdido Formation		500'
	MIS.		L		Perdido Formation		500'
			E		Tin Mountain Limestone		300'
	DEVO.		L		Lost Burro Formation		500'
			M				
	SIL.		E		Hidden Valley Dolomite		1400'
	ORDO.		L		Ely Springs Dolomite		500'
			M		Eureka Quartzite		400'
			E		Pogonip Group		2200'
	CAMBRIAN	Late			Nopah Formation		700'
					Bonanza King Formation		3600'
		Middle			Carrara Formation		1600'
		Early			Zabriskie Quartzite		800'
PRECAMBRIAN					Wood Canyon Formation		4000'
					Stirling Quartzite		4800'
					Johnnie Formation		1000'

FIG. 3.—Generalized columnar section of Proterozoic and Paleozoic stratigraphic units exposed in the southeastern and central parts of the Funeral Mountains (modified from Cemen et al., 1982).

by Tertiary volcanic rocks and by plutonic rocks. The plutonic rocks are now known to be largely of Tertiary age. Owing to the occurrence of borate deposits in the Furnace Creek Formation, the area of the Furnace Creek trough was mapped in detail by McAllister (1970, 1973) and by F. C. Johnson and others (unpublished maps in files of the American Borate Company) with emphasis upon the borate-bearing strata. The southern Funerals were also mapped by McAllister (1971, 1976a).

Also noted early in the investigations of Death Valley geology is another accumulation of Cenozoic sedimentary and volcanic rocks exposed in patches within and marginal to the Funeral and Grapevine Mountains (Stock and Bode, 1935; Noble and Wright, 1954; Reynolds, 1976). These rock units are most abundantly exposed along the northeast sides of these mountains, and presumably represent remnants of a much more continuous cover. They consist of fluvial and lacustrine strata, and subordinate layers of volcanic ash. In the Grapevines and northern Funerals, the lower part consists of the lower Oligocene Titus Canyon Formation of Stock and Bode (1935). Three of the patches are exposed at Bat Mountain and vicinity at the southern end of the Funerals (BM in Fig. 2, and Fig. 4). They were tentatively

correlated by Noble and Wright (1954) with the Titus Canyon and Artist Drive Formations, and were mapped in detail by Denny and Drewes (1965), McAllister (1971), and Cemen et al. (1982).

The upper 300+ m of the Bat Mountain section was found to consist of a remnant of an originally northeast-sloping alluvial fan. Its source area of Precambrian and Paleozoic rocks once lay to the southwest where the Cenozoic rocks of the Furnace Creek trough are now exposed.

Investigations of the chronology of the Tertiary rocks at Bat Mountain and also of the Artist Drive Formation of the Furnace Creek Basin (Cemen et al., 1982; Cemen, 1983) showed that the two are separated in time as well as space. Volcanic ash beds near the base of the Bat Mountain section yielded a K/Ar age of about 25 Ma, whereas ash beds near the base of the Artist Drive were dated at about 14 Ma. Meanwhile K/Ar age determinations obtained from volcanic rocks of the Greenwater Range and Black Mountains proved to range between 12 Ma and 4 Ma (Fleck, 1970; McAllister, 1973; Wright et al., 1981), thus confirming a contemporaneity with the formations of the Furnace Creek Basin already suggested in the mapping of the two geographically separate accumulations.

CENOZOIC STRATIGRAPHIC AND STRUCTURAL FEATURES AT BAT MOUNTAIN, SOUTHEASTERN FUNERAL MOUNTAINS

The Cenozoic sedimentary and volcanic rocks exposed at Bat Mountain lie immediately northeast of a hidden trace of a major fault of the Furnace Creek zone (Fig. 2). They record a post-25 Ma succession that predates the formation of the Furnace Creek Basin, and they probably record the inception of movement on the Furnace Creek fault zone in Cenozoic time. The rock units also predate the major extension expressed in the Funeral Mountains, as they have participated in the moderate tilting of fault blocks. In most places they dip about as steeply as the underlying Paleozoic units (Fig. 5), suggesting that they record most of the Tertiary tilting to affect this region.

The Cenozoic section at Bat Mountain is about 1,300 m in maximum exposed thickness and may be much thicker, as the youngest strata are hidden beneath alluvium. It is divisible into two parts, separated by an unconformity, above which is a conspicuous, persistent, and distinctive unit of algal limestone (Figs. 4, 5).

Both the upper and lower parts of the section contain evidence of a topographic high to the southwest that predates the Furnace Creek Basin by as much as 10 m.y. It extended across an area now occupied by formations of the basin and probably into the site of the Greenwater Range. The paleotopographic high appears to have been underlain by the same formations that are now exposed in the southern Funeral Mountains, and was strongly uplifted just prior to the formation of the basin.

At the base of the Cenozoic section at Bat Mountain are remnants of an alluvial fan, composed of angular clasts (locally derived from the underlying Cambrian and Devonian strata) in a silty matrix. A southerly source is indicated by a northward increase in the matrix-to-clast ratio, combined with the observation that the conglomerate wedges out

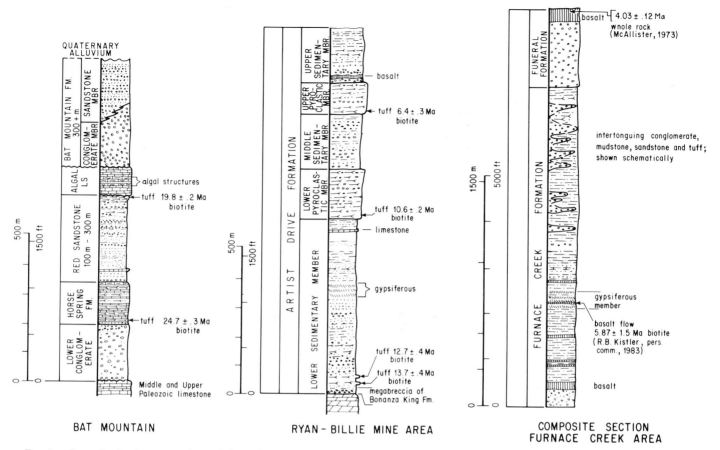

FIG. 4.—Generalized columnar sections of the sedimentary and volcanic successions exposed at Bat Mountain and vicinity and in the Furnace Creek Basin. Unless otherwise indicated radiometric age determination are by R. E. Drake. Age determination of basalt flow in lower part of Furnace Creek Formation is by P. E. Damon.

southward. The fan is the earliest evidence of Cenozoic uplift to the south (Fig. 6A). It is overlain by a predominantly lacustrine unit correlative, in a lithostratigraphic sense, with the Horse Spring Formation of southern Nevada. It is composed of interlayered limestone, siltstone, and tuff, a tuff layer low in the unit yielding a K/Ar age of about 25 Ma (Fig. 4). This unit also wedges out southward, in large part by onlap against the paleotopographic high (Fig. 6B). An overlying unit of red sandstone, with features characteristic of fluvial deposition (especially in the form of fining-upward lenses and unimodal current-direction indicators), also thins southward. A tuff layer in the more southerly exposures of the sandstone has been dated by the K/Ar method at 20 Ma. The thinning is caused by onlap and by a southward truncation of progressively older strata at an unconformity above the red sandstone. The unconformity provides evidence for active uplift of the topographic high (Fig. 6B).

An early episode of faulting was first detected by Denny and Drewes (1965), and later substantiated by the mapping of Cemen (Fig. 5). This episode is indicated by the fact that from place to place and on opposite sides of a given fault the algal limestone above the unconformity rests on rock units of different ages, and by the corollary observa-

tion that some faults that slightly offset the limestone offset the underlying rocks by much more. Palinspastic restoration of these faults, including rotation of the limestone to the horizontal show that they were once steeply dipping normal faults, defining horsts and grabens that, in general, post-date the Cenozoic units beneath the unconformity and pre-date the unconformity (Fig. 6C). We thus interpret these faults to have moved almost contemporaneously with the emplacement of the 20 Ma tuff exposed on the south side of Bat Mountain and immediately beneath the algal limestone. If so, they are the oldest extension-related features yet detected in the Death Valley region.

The unconformity and the overlying algal limestone record an interval of tectonic quiescence and low relief in the area surrounding the southern Funerals (Fig. 6D). The limestone contains abundant algal structures unique in the Bat Mountain section. It must have extended into and perhaps across the area now occupied by the formations of the Furnace Creek Basin, because in the more southerly exposures of the limestone, it is overlain by a conglomerate composed almost entirely of angular clasts identical with the limestone (Fig. 6E). This unit underlies a conglomerate (Fig. 4) composed of angular clasts, derived only from Upper Proterozoic and Paleozoic formations, and also con-

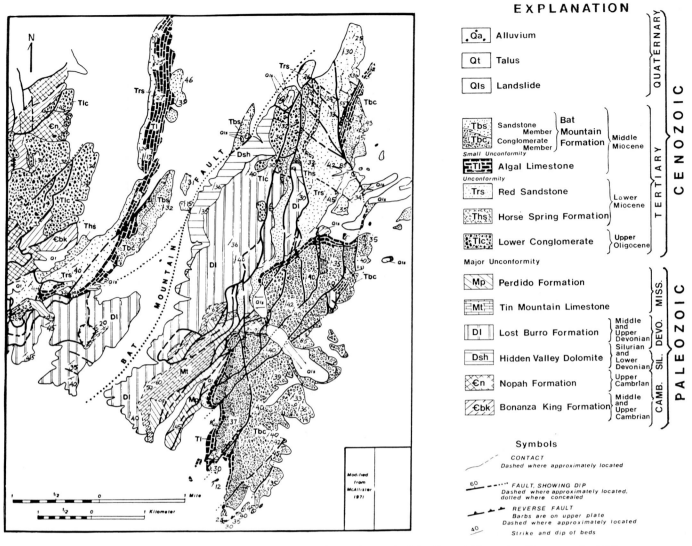

FIG. 5.—Geologic map of Bat Mountain and vicinity at the southern end of the Funeral Mountains (Cemen et al., 1982).

taining a subordinate sand- to clay-size matrix. The clasts tend to coarsen southward and to show an imbricate arrangement with a persistent southward dip. In general, this conglomerate records the unroofing of a miogeoclinal section, as the successively younger parts are composed of clasts of progressively older formations. Clasts as old as the Upper Proterozoic Johnnie Formation (Fig. 3) were noted high in the conglomerate. In the southeastern part of Bat Mountain, the exposed part of the conglomerate is about 300 m thick, its upper part being hidden beneath alluvium.

The distribution and textural characteristics of the conglomerate are diagnostic of a large alluvial fan, which, when rotated about a horizontal axis to its original position, would have sloped northeastward. The conglomerate thus provides evidence for rapid uplift and erosion of the same topographic high indicated by features in the older Cenozoic units, and thus for pronounced vertical movement at the site of the Furnace Creek fault zone (Fig. 6F). No later lateral movement is required as the indicated source area contained the same formations that now underlie the conglomerate.

The existence and composition of this fossil fan at Bat Mountain, together with the indicated southerly source area, also place constraints on the possibility that the block of the present Panamint Mountains once occupied the area southwest of the southern Funerals (Stewart, 1983). This interpretation would place the essentially complete section of Upper Proterozoic to Upper Paleozoic rocks, now preserved at Tucki Mountain at the north end of the Panamint Mountains, at a location from which these formations would have been already eroded during the formation of the fan.

STRATIGRAPHIC AND STRUCTURAL FEATURES
OF THE FURNACE CREEK BASIN

Regional Setting

The Furnace Creek Basin coincides with a topographic low about 50 km long, and is defined by the southeasternmost segment of the Furnace Creek fault zone. The low is traceable southeastward up Furnace Creek Wash across a drainage divide to the vicinity of Eagle Mountain in the Amargosa Valley (EM in Fig. 2). Near the confluence of Furnace Creek Wash and Death Valley, the formations exposed in the wash are truncated and dropped downward on

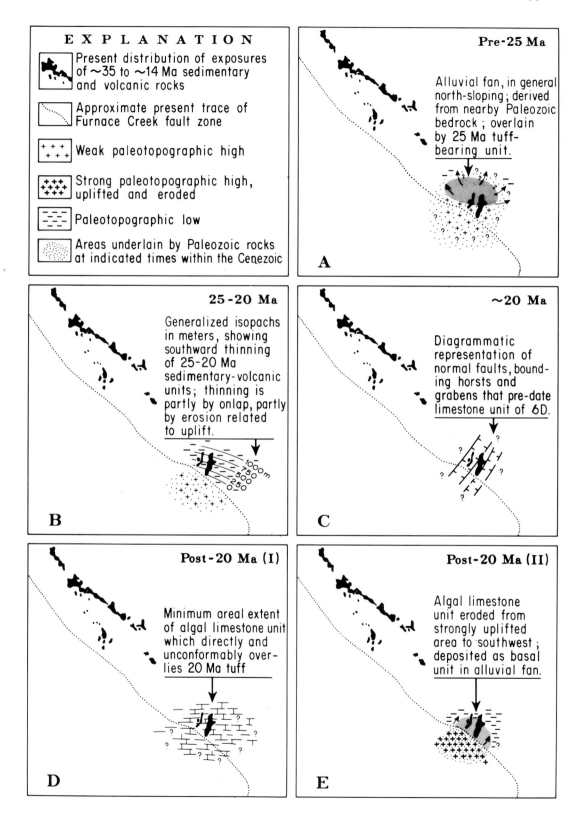

EXPLANATION

Present distribution of exposures of ~35 to ~14 Ma sedimentary and volcanic rocks

Approximate present trace of Furnace Creek fault zone

Weak paleotopographic high

Strong paleotopographic high, uplifted and eroded

Paleotopographic low

Areas underlain by Paleozoic rocks at indicated times within the Cenozoic

A — Pre-25 Ma
Alluvial fan, in general north-sloping; derived from nearby Paleozoic bedrock; overlain by 25 Ma tuff-bearing unit.

B — 25-20 Ma
Generalized isopachs in meters, showing southward thinning of 25-20 Ma sedimentary-volcanic units; thinning is partly by onlap, partly by erosion related to uplift.

C — ~20 Ma
Diagrammatic representation of normal faults, bounding horsts and grabens that pre-date limestone unit of 6D.

D — Post-20 Ma (I)
Minimum areal extent of algal limestone unit which directly and unconformably overlies 20 Ma tuff

E — Post-20 Ma (II)
Algal limestone unit eroded from strongly uplifted area to southwest; deposited as basal unit in alluvial fan.

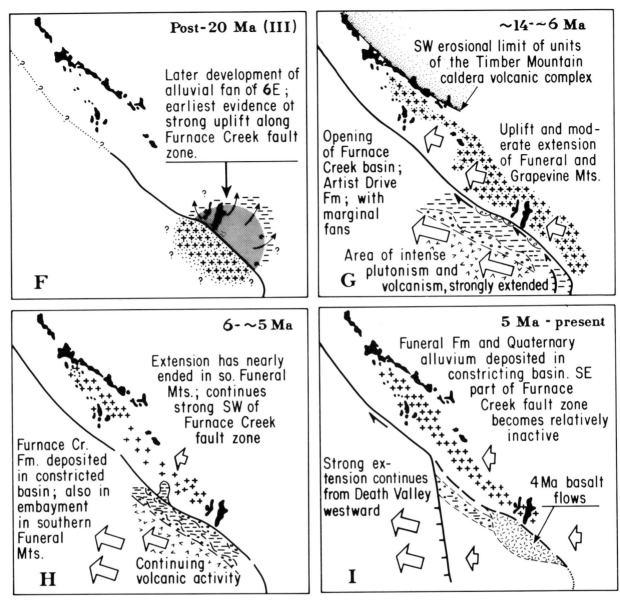

Fig. 6—A–I.—Cartoons showing an interpretation of the sequence of Cenozoic events along and near the southeasternmost part of the Furnace Creek fault zone. A palinspastic restoration of the patches of the pre-14-Ma Cenozoic rocks northeast of the fault zone would require a northwest-southeast foreshortening probably of 5 to 10%.

the west by the zone of normal faults that bounds the Black Mountains on the west (Fig. 2). Consequently, the three formations that define the basin must extend farther to the northwest and must be viewed as part of a more complex system of basinal deposits now largely hidden beneath Quaternary rocks of the floor of Death Valley (Fig. 1). That the basinal deposits are best exposed in Furnace Creek Wash and at the northern end of the Black Mountains, is attributable, in large part, to the headward erosion of Furnace Creek. But only a partial three-dimensional view of the sedimentary-volcanic fill is provided; the known exposures of the Artist Drive Formation (Fig. 4) are confined to the southwestern and central part of the basin and a veneer of

Quaternary alluvium commonly covers the younger formations.

Structural Features

The relatively through-going, northwest-striking faults that consistently define the Furnace Creek fault zone are represented, in the Furnace Creek Wash area, mainly by the fault contacts between the Tertiary formations and the older rock units of the Funeral Mountains. These faults bring the Paleozoic formations of the Funerals into contact with strata of the Funeral Formation and strata high in the Furnace Creek Formation (Fig. 4). If, in the subsurface, the Artist Drive Formation extends to the bounding faults and maintains its full measured thickness of about 1,200 m, a vertical displacement in excess of that figure is indicated. Direct evidence for lateral displacement on these faults is less obvious and consists mainly of exposures of fault surfaces that show low-angle to horizontal grooves and fault mullions. A branch of the main fault zone is joined by north-striking and west-dipping normal faults to another northwest-striking fault (the Keane Wonder fault; KWF in Fig. 2) along the southwestern margin of the northern Funeral Mountains. These faults thus form a right-stepping sigmoidal curve in the basin margin and provide evidence that some of the right-lateral motion on the Furnace Creek fault zone has been transferred to the Keane Wonder fault via the normal faults. The geometry and sense of offset along these basin-bounding faults are compatible with a pull-apart origin for the basin.

The Furnace Creek Basin also contains the northernmost exposures of the Grand View fault (GVF in Fig. 2), which strikes 5° to 20° more northerly than the main fault zone, and has been traced for at least 30 km southeastward from the Billie Mine area (Figs. 7, 8, 9). It, too, shows evidence of pronounced vertical movement (McAllister, 1970), but where exposed, displays obliquely oriented grooves with a dominant strike-slip component. It apparently terminates in the northwestern part of the basin. In that area, it is joined by a normal fault (down to the northwest), which extends northeastward apparently to connect with a principal bounding fault. The normal fault, here referred to by the

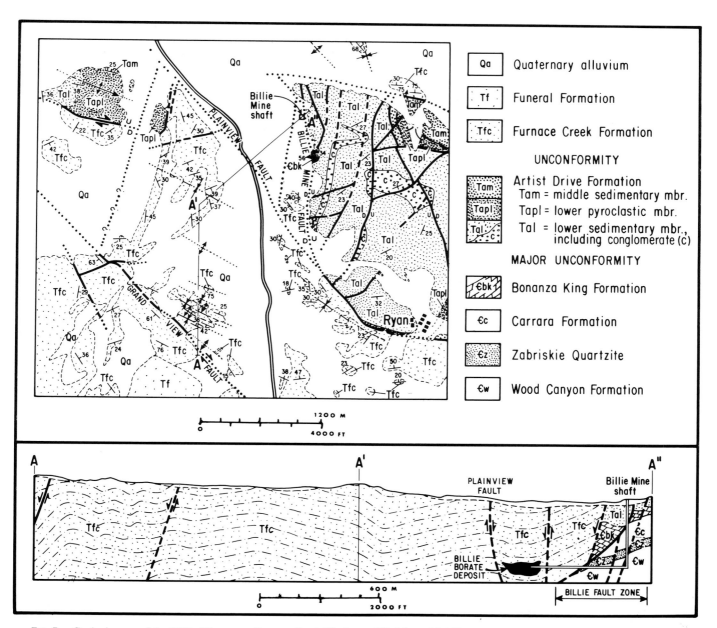

FIG. 7.—Geologic map of the Billie Mine area, Furnace Creek Wash, modified from McAllister (1970). See Figure 9, below, for location.

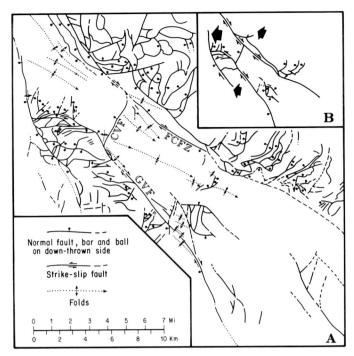

FIG. 8.—A) Tectonic map of the area of the Furnace Creek Basin modified and interpreted from data on the maps and columnar section of McAllister (1970, 1971, 1973). B) Kinematic interpretation of the features of 8A. Abbreviations: FCFZ, Furnace Creek fault zone; CVF Cross-valley fault; GVF, Grand View fault zone. See Figure 2 for location.

long-used name "Cross-valley fault" (CVF in Fig. 8), displaces Quaternary alluvium and brings strata of the Funeral Formation on the northwest into contact with strata of the Furnace Creek Formation and Quaternary alluvium. The result is another right-stepping fault pattern and an associated structural basin, compatible with right-lateral movement on the Grand View fault and in a pull-apart setting.

Beyond the juncture of the Grand View and the Cross-valley faults are numerous arcuate west- to southwest-striking faults (McAllister, 1970) that we interpret as splays of the Grand View fault and thus related to its termination (Fig. 8A). These splays displace members of the Artist Drive and Furnace Creek Formations in a manner that can be explained by dip slip, left slip, or left-oblique slip. We strongly suspect the last and suggest the kinematic interpretation shown in Figure 8B.

The Furnace Creek Wash segment of the basin is synclinal. Along the south side of the wash, the various units of the Artist Drive and Furnace Creek Formations dip northeastward. Northwest of the Cross-valley fault, the Funeral Formation also dips northeastward. In exposures along and near the Furnace Creek fault zone, the Tertiary strata generally dip southwestward. The syncline and smaller folds within it were originally believed to be compressional (Noble and Wright, 1954), but because the entire block of the Black Mountains has been tilted eastward to northeastward along the frontal fault zone, we attribute the southwest limb of the syncline to this tilting. Also, the northeast limb of

the syncline seems best interpreted as the result of down-dip drag along the high-angle bounding faults.

Smaller folds of various sizes characterize the internal part of the basin (Fig. 8A). The traces of the axial planes of some of these smaller folds are parallel with the trend of the basin, and with the traces of the northwest-striking faults; others trend 10° to 20° more westerly, an orientation compatible with right-lateral movement on the faults. Most of them probably qualify as forced folds, representing passive responses of the cover to movements in the underlying basement. A continuation of fold development into post-Funeral time is shown by the involvement in the folding of the Funeral Formation, in some places by about the same degree as the underlying Furnace Creek strata.

Basement Rocks of the Furnace Creek Basin

The base of the Tertiary fill of the Furnace Creek Basin is exposed at two, and perhaps three, localities. One is in Desolation Canyon along the Black Mountain front (DC in Fig. 2, and Fig. 9), and about 10 km south of the northeastern edge of the basin; another is at the Billie Mine (Figs. 7, 9), about 5 km from the northeastern edge; the third is at Eagle Mountain at the southeastern end of the basin (EM in Fig. 2). At each locality Tertiary sedimentary rocks rest upon the Cambrian Bonanza King Formation (Fig. 3). The main shaft at the Billie Mine penetrates units of the underlying Carrara Formation, Zabriskie Quartzite, and Wood Canyon Formation (Fig. 7). At Desolation Canyon and at the Billie Mine, the overlying strata mark the base of the Artist Drive Formation. At Eagle Mountain, only about 200 m of Tertiary strata are exposed, the higher strata being hidden beneath Quaternary alluvium. As the exposed beds consist of conglomerate, sandstone, and siltstone, which resemble the lower part of the Artist Drive elsewhere, and are unlike any of the Tertiary units at Bat Mountain, we tentatively correlate them with the Artist Drive. Thus, the underlying rocks at each of the three localities provide evidence that the Artist Drive Formation was deposited, largely or wholly upon a basement of Cambrian and Proterozoic rocks. They also provide evidence that the bedrock beneath the Artist Drive could not have been tectonically denuded of these formations by the westward transport of the Panamint Mountain block, and a further constraint is placed on that model. Indeed, remnants of the miogeoclinal cover are still preserved at various localities in the Black Mountains and Greenwater Range, including discontinuous exposures that overlie the Precambrian basement complex at each of the so-called turtleback folds along the west face of the Black Mountains. Two of these folds, the Badwaters turtleback (BWT) and Copper Canyon turtleback (CCT), are shown in Figure 2.

Artist Drive Formation

The Artist Drive Formation, named by Thayer (cited by Noble, 1941), divided into members and mapped by McAllister (1970, 1973), and measured in detail by Cemen (1983), is best and most continuously exposed on the Black Mountain front along the southwestern limb of the large

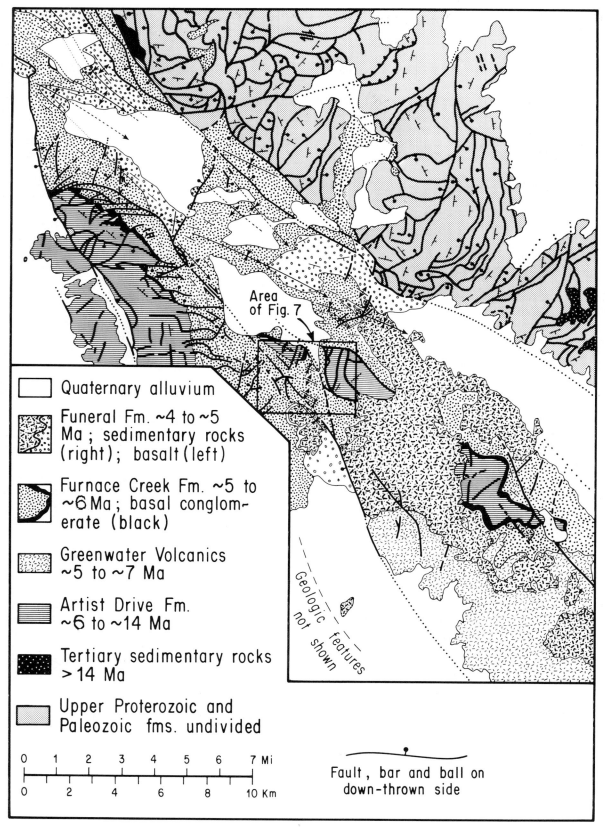

Quaternary alluvium

Funeral Fm. ~4 to ~5 Ma; sedimentary rocks (right); basalt (left)

Furnace Creek Fm. ~5 to ~6 Ma; basal conglomerate (black)

Greenwater Volcanics ~5 to ~7 Ma

Artist Drive Fm. ~6 to ~14 Ma

Tertiary sedimentary rocks > 14 Ma

Upper Proterozoic and Paleozoic fms. undivided

Area of Fig. 7

Geologic features not shown

Fault, bar and ball on down-thrown side

FIG. 9.—Generalized geologic map of the central and northwestern parts of the Furnace Creek Basin and vicinity, showing distribution of the Artist Drive, Furnace Creek, and Funeral Formations. See Figure 2 for location.

syncline (Fig. 9). It is also well exposed and more accessible in the vicinity of the Billie Mine on the northeast or opposite side of the Grand View fault (Fig. 7). At both localities, the formation consists of a lower, middle, and upper sedimentary member separated by two pyroclastic members (Fig. 4). At each locality the members have a combined thickness of about 1,300 m, but the Artist Drive of the Black Mountains front contains, in addition, abundant bodies of basalt and felsite, in the form of sills, flows, dikes and volcanic necks, whereas the section near the Billie Mine does not. Radiometric age determinations of tuff layers (Fig. 4) indicate that the five members were deposited through an interval extending from about 14 Ma to about 6 Ma. In the more southeasterly parts of the basin, the Artist Drive is poorly and discontinuously exposed, but consists of the same members recognized in Furnace Creek Wash area.

The three sedimentary members display variable lithology (claystone, siltstone, sandstone, conglomerate and subordinate tuff), textures, bedforms, and patterns of distribution that indicate rapid erosion of the southwestern margin of the basin and deposition in fluvial and lacustrine environments in the central part. The sandstones typically consist of well-developed fining-upward lenses with cross-bedding oriented north to northwest. We attribute these features to streams flowing from the southwestern margin of the basin and northward parallel with the axis of the basin. The pyroclastic members consist of poorly stratified tuff breccia and subordinate proportions of pumice lapilli and rock fragments derived from the Upper Proterozoic and Cambrian formations of the region.

Reconstructions of the development of the Furnace Creek Basin in Artist Drive time (Fig. 6G) are hindered by the limitation of the known exposures of the Artist Drive Formation to the southwestern side and central part of the basin. Its considerable thickness and lateral persistence along the basin together with the absence of the Artist Drive northeast of the bounding faults strongly suggest that the formation originally extended to these faults and not beyond, and that it is present in the subsurface for the full width of the basin.

Although the manner by which the Artist Drive terminates southwestward has yet to be well documented, reconnaissance mapping along the Black Mountain front suggests that it intertongues with volcanic rocks in that direction. At least some of the equivalent volcanic section apparently terminates against the faulted surface of the Badwater turtleback (BWT in Fig. 2).

Regardless of the exact nature of the southwestern margin of the basin in its early stages, episodic uplift there is indicated by the composition and textural features of sandstones and conglomerates in each of the sedimentary members. These members contain fragments of plutonic and volcanic rocks that could only have been derived from the area of the Black Mountains and Greenwater Range. Fragments of granitic rock, in the lower as well as the higher conglomeratic units, indicate that some of the acidic plutons in the source area had been emplaced and unroofed by 14 Ma (Fig. 6G).

The sandstones and conglomerates of the sedimentary members, as well as the tuffs of the pyroclastic members, contain fragments of the Upper Proterozoic and Cambrian formations of the region. Preliminary examination of thin sections of the sandstones has indicated that they ordinarily contain 20% or more each of quartz and feldspar and qualify either as volcanic arkoses or calcareous arkoses depending upon the predominating rock fragments.

In exposures near the Billie Mine (Fig. 7) and along the Black Mountains front, the lower sedimentary member contains a conglomerate subunit with textural features diagnostic of the distal parts of a north- to northeast-sloping alluvial fan or apron (Cemen, 1983). Nearly all of its clasts were derived from the Upper Proterozoic and Cambrian formations. Near the Billie Mine it is as much as 60 m thick, but it thins and fines northward to disappear within a kilometer of its most southerly exposures. The orientation of imbricated clasts also indicates a northward paleoslope.

Therefore, the clast content, paleoslope indicators, textural variability and thickness trends of the Artist Drive provide evidence that (1) between 14 and 6 Ma ago, the Upper Proterozoic and Cambrian formations were extensively exposed well to the south of the boundary of the basin, and (2) much of this miogeoclinal cover was eroded and carried northward to the basin, or was incorporated as fragments in tuffs during that interval. If correctly interpreted, this evidence also points against the possibility (Stewart, 1983) that these formations were entirely removed from the area of the present Black Mountains and Greenwater Range before the Furnace Creek Basin took form.

Furnace Creek Formation

The Furnace Creek Formation overlies the Artist Drive Formation, and in a general way, records a continuation of the stratigraphic patterns established in Artist Drive time, in that lacustrine and fluvial sediments dominate in the central part of the basin and intertongue with conglomerates derived from contrasting source areas to the northeast and southwest (Fig 6H; McAllister, 1970). The Furnace Creek, however, shows a much less persistent stratigraphy and a greater variability in thickness from place to place, along the length of the basin. The time interval represented by the Furnace Creek Formation has been moderately well bracketed by the 6.4 ± .3 Ma age of a tuff layer in the upper pyroclastic member of the Artist Drive Formation and the 4.03 ± .12 Ma age of a flow in the basalt of the Funeral Formation east of Ryan (Fig. 4). Where exposed within the Furnace Creek Basin (Fig. 9), the basal strata of the Furnace Creek Formation ordinarily rest concordantly upon the Artist Drive Formation, but in the vicinity of the Billie Mine (Fig. 7), they rest upon the middle members of the Artist Drive. At several places along the northeastern margin of the basin, the Furnace Creek rests depositionally upon the Paleozoic formations of the Funeral Mountains (Fig. 9). Moreover, lacustrine strata, identical with those of the Furnace Creek Formation, and unlike any in the Artist Drive or Funeral Formations, are exposed in a large amphitheater in the Funeral Mountains northeast of the Furnace Creek fault zone (Fig. 9). These strata and associated sedimentary breccias extend across the normal faults in the Paleozoic rocks without offset. If correlation with the Furnace Creek Formation is valid, this relationship indicates that most of the displacement along the normal faults in the Funeral Mountains occurred in pre-Furnace Creek time, in striking contrast with the displacement of Furnace Creek strata on the various faults of the Furnace Creek zone and the area to the southwest of it.

At the northwestern end of the basin, the Furnace Creek Formation was estimated by McAllister (1970) to be about 2,100 m thick. However, at a locality east of the Billie Mine, and at another about 5 km to the east, units of the Funeral Formation rest with strong angular discordance upon the Artist Drive Formation, and the Furnace Creek is missing altogether. At the Lila C Mine in the southwestern part of the basin, and south of the southern edge of Figure 9, it is about 500 m thick. The local absence of the Furnace Creek in the central part of the basin is probably an effect of pre-Funeral erosion of a major northwest-trending anticline, but much of the thickness variability in the Furnace Creek is probably attributable to contemporaneous faulting. Thus interpreted, the Cross-valley and Billie faults (Figs. 7, 8) define the margins of subordinate sedimentary as well as structural basins within the larger Furnace Creek Basin.

Where the base of the Furnace Creek Formation is exposed, its lowest unit ordinarily is a conglomerate. McAllister (1970) observed that the basal conglomerate is continuously exposed for about 8 km along the southwest side of Furnace Creek Wash (Fig. 9), thinning to disappearance southeastward from a maximum thickness of at least 110 m. Along the northeast side of the wash, the basal strata of the Furnace Creek are conglomeratic in the more northwesterly exposures and finer grained to the southeast. In the southeastern part of the Furnace Creek basin, the basal conglomerate is also present (McAllister, 1973). The intertonguing conglomerates and finer-grained sedimentary units (mostly lacustrine mudstones and sandstones), compose most of the remainder of the Furnace Creek Formation.

Of special significance with respect to the development of the basin are a number of observations originally made by McAllister (1970, 1976b). He noted that a large proportion of the clasts in the more southwesterly exposures of the basal Furnace Creek conglomerates, as well as in those higher in the formation, have been eroded from the Artist Drive Formation and transported northwestward. On the other hand, the clasts in the more northeasterly of the Furnace Creek conglomerates consist entirely of Paleozoic rocks derived from the area of the Funerals, providing evidence of vertical movement on the Furnace Creek fault zone and a basin margin that coincided with the fault zone.

McAllister also observed the high degree of textural variability in the Furnace Creek sedimentary units above the basal conglomerate, chiefly in lateral facies changes between conglomeratic and the finer-grained lacustrine strata. As indicated by McAllister, these facies changes commonly indicate the margins of alluvial fans. They also suggest to us closer and more localized source areas (during a time of

accelerated subsidence) than the source areas of the basal conglomerate of the Furnace Creek Formation and the conglomerates of the Artist Drive Formation.

A third observation concerns the volcanic material within and associated with the sedimentary rocks of the Furnace Creek Formation, specifically (1) the high proportion of volcanic ash in the form of discrete layers of tuff and as fragments in the siltstones and sandstones, (2) the abundant occurrence of basalt in flows and sills at various stratigraphic positions, and (3) the lateral gradation, west of Ryan (Figs. 7, 9), of sedimentary units into an accumulation of basalt and felsic volcanic rocks. The tuffaceous material in the rock units of the Furnace Creek Basin can be most logically attributed to volcanic activity in the central Death Valley volcanic field. The basalt associated with the Furnace Creek, as well as the basalt in the underlying Artist Drive and overlying Funeral Formation, forms a much higher proportion of the volcanic part of the succession than it does of the volcanic pile southwest of the trough. Thus the spatial relation between basaltic volcanism and the extension-related Furnace Creek fault zone suggests that the fault zone penetrates much or all of the crust, providing conduits for the continuing introduction of mantle-derived material.

Funeral Formation

The Funeral Formation records the most recent stages in the development of the Furnace Creek Basin (Fig. 6I). Its occurrences there, as mapped and described by McAllister (1970, 1973), underlie two segments of the basin separated by a 5-km segment in which the Funeral Formation is missing altogether (Fig. 9). Although the sedimentary part of the Funeral in each area has been commonly referred to as fanglomerate, it shows a considerable variation in lithology. Also variable is the degree of angular discordance between the Funeral Formation and underlying rock units.

Northwest of the Cross-valley fault (Fig. 8, 9) strata assigned to the Funeral Formation are confined to the central part of the basin, overlie the Furnace Creek Formation conformably or with slight angular discordance, and are as much as 700 m thick (Fig. 9). The Funeral Formation consists of a lower part of pebbly sandstone and mudstone and a conglomeratic upper part. The conglomerates contain abundant material eroded from the underlying Furnace Creek Formation, as well as from pre-Furnace Creek units. Thus the Funeral, when deposited, probably did not extend significantly beyond its present exposures, even though it is folded to about the same degree as the Furnace Creek.

In the southeastern part of the basin, the Funeral Formation is represented mainly by the 4 Ma basalt flows (Fig. 9), totaling as much as 160 m in exposed thickness, and containing subordinate interlayers of conglomerate and sandstone. This occurrence of the Funeral Formation remains essentially horizontal, and rests upon various members of the Furnace Creek and Artist Drive Formations with pronounced angular unconformity. If the two occurrences of the Funeral Formation represent essentially contemporaneous strata, they indicate that deformation and possibly sedimentation continued in the northwestern part of the basin, when the southeastern part was being eroded to a surface of low relief and then covered by the basalt flows and associated sediments.

DISCUSSION

Among the basins associated with strike-slip faults, the Furnace Creek Basin is atypical because it and the bounding Furnace Creek fault zone have formed in a region undergoing pervasive extension. The basin also is unusual in that it separates the two unlike terranes and evolved contemporaneously with the emplacement of the plutons and volcanic rocks that dominate the terrane on the southwest side. This kind of geologic setting is apparently not atypical of basinal deposits in the Great Basin, however, as R. G. Bohannon (personal commun., 1984) has recorded a similar setting in the Lake Mead area of southern Nevada.

The significance of the Furnace Creek Basin within the Cenozoic stratigraphic-tectonic framework of the Death Valley region remains incompletely understood, but when considered on the scales of Figure 2 and 9, the basin formation seems best attributed to a combination of (1) subsidence marginal to a volcanic field and contemporaneous with the volcanism, and (2) confinement between the northeast side of a major, northeast-tilting fault block and the high-angle faults of the Furnace Creek fault zone. As the outer limits of the basin are not defined by en echelon strike-slip faults, it does not strictly qualify as a pull-apart basin. A pull-apart origin, however, can be ascribed to relatively later secondary basins within the larger feature. The secondary basins would thus have formed as a result of interplay between the northwest-striking high-angle faults, northeast-striking normal faults, and by a right-stepping relation between the two.

Although no direct evidence of the magnitude of lateral slip in the Furnace Creek area has been detected, lateral slip is indicated by the low-angle striae and fault mullions, the en echelon folds, and the large flexure in the Proterozoic and Paleozoic strata along the northeast side of the Furnace Creek fault zone. A right-lateral sense of movement is indicated by the orientations of the en echelon folds and marginal flexures, and by the observation that the crust southwest of the fault zone has continued to extend northwestward after the cessation of most of the normal faulting in the southern Funeral Mountains.

We attribute the localization of the igneous activity southwest of the Furnace Creek fault zone to a greater degree of extension there than on the northeast side. This interpretation is compatible with evidence that the fault zone terminates at the southeastern end of the basin and has displaced the crust in a right-lateral sense. If the plutonic activity indicated by the granitic clasts in the lower part of the Artist Drive Formation was, indeed, favored by an early phase of the extension, some of the right-lateral movement could have occurred in pre-Artist Drive time. The proximity to the volcanic field also explains the unusually high proportion of volcanic ash and pyroclastic material associated with the sedimentary units of the Artist Drive and Furnace Creek Formations.

The emplacement of the plutons, the succeeding volcanism and the related crustal heating and cooling must have influenced, in some manner, the vertical movements along the southeastern part of the Furnace Creek fault zone and the Grand View fault. However, the chronology of the igneous events is still too incompletely documented to relate them to specific events recorded in the stratigraphy and structural features of either the Bat Mountain area or the Furnace Creek Basin.

The uplift recorded in the younger of the two fossil alluvial fans at Bat Mountain may well have accompanied the early phase of plutonism indicated by the clasts of granitic rock low in the Artist Drive Formation. But the uplift indicated by the pre-25 Ma fan predates by about 15 m.y. the nearest of the exposed plutons. The uplift may also have been the consequence of the initiation of lateral movement at the terminus of the Furnace Creek fault zone, as could be predicted from the theoretical models of Chinnery (1965).

From the physical stratigraphy, structural setting and chronology of the Tertiary formations of the Furnace Creek trough, several conclusions can be drawn: (1) In Artist Drive time, Upper Proterozoic and Paleozoic rock units were extensively exposed in the area of the Black Mountains and Greenwater Range to become incorporated as clasts in the northeast-sloping alluvial-fan and stream-deposits and in the pyroclastic units preserved in the present exposures of the Artist Drive Formation. (2) The basin contracted northeastward with time as shown by the abundance of recycled, southwesterly derived clastic material in both the Furnace Creek and Funeral Formations. (3) The rate of subsidence in the northwestern part of the trough increased with time, the 1,300 m of Artist Drive being deposited in ~8 m.y. and the 2,100 m of Furnace Creek in ~1 m.y. (4) The internal stratigraphy of the sedimentary fill, as recorded in each of the three formations, became less uniform with time, probably as a result of the increased effect of contemporaneous faulting as the subsidence rate increased.

ACKNOWLEDGMENTS

We thank Eugene G. Williams and Bennie W. Troxel for helpful critiques of our field work and interpretations, and we especially acknowledge the co-operation of James F. McAllister, who through his publications and field excursions has shared with us his intimate knowledge of the Furnace Creek area. We also thank K. T. Biddle, R. G. Bohannon, N. Christie-Blick, and E. L. Miller for critical and thoughtful reviews of the manuscript. Robert J. Texter drafted most of the illustrations. The project was supported by NSF Grants EAR-7927092 and EAR-8206627 to Wright.

REFERENCES

CEMEN, IBRAHIM, 1983, Stratigraphy, geochronology and structure of the selected areas of the northern Death Valley region, eastern California-western Nevada, and implications concerning Cenozoic tectonics of the region [unpubl. Ph.D. thesis]: University Park, Pennsylvania, Pennsylvania State University, 235 p.

CEMEN, IBRAHIM, DRAKE, R. E., AND WRIGHT, L. A., 1982, Stratigraphy and chronology of the Tertiary sedimentary and volcanic units at the southeastern end of the Funeral Mountains, Death Valley region, California, in Cooper, J. D., Troxel, B. W., and Wright, L. A., eds., Geology of Selected Areas in the San Bernardino Mountains, Western

Mojave Desert, and Southern Great Basin, California: Guidebook, field trip no. 9, 78th Annual Meeting Cordilleran section, Geological Society of America, p. 77–86.

CHRISTIE-BLICK, NICHOLAS, AND BIDDLE, K. T., 1985, Deformation and basin formation along strike-slip faults, *in* Biddle, K. T., and Christie-Blick, N., Strike-Slip Deformation, Basin formation, and Sedimentation: Society of Economic Paleontologists and Mineralogists Special Publication No. 37, p. 1–34.

CHINNERY, M. A., 1965, The vertical displacements associated with trans-current faulting: Journal of Geophysical Research, v. 70, p. 4627–4632.

DENNY, C. S., AND DREWES, HARALD, 1965, Geology of the Ash Meadows quadrangle, Nevada-California: United States Geological Survey Bulletin 1181-L, p. 1–56.

FLECK, R. J., 1970, Age and tectonic significance of volcanic rocks, Death Valley area, California: Geological Society of America Bulletin, v. 81, p. 2807–2816.

HARDING, T. P., VIERBUCHEN, R. C., AND CHRISTIE-BLICK, NICHOLAS, 1985, Structural styles, plate-tectonic settings and hydrocarbon traps of divergent (transtensional) wrench faults, *in* Biddle, K. T., and Christie-Blick, N., eds., Strike-Slip Deformation, Basin Formation, and Sedimentation: Society of Economic Paleontologists and Mineralogists Special Publication No. 37, p. 51–77.

MCALLISTER, J. F., 1970, Geology of the Furnace Creek borate area, Death Valley, Inyo County, California: California Division of Mines and Geology Map Sheet 14.

———, 1971, Preliminary geologic map of the Funeral Mountains in the Ryan quadrangle, Death Valley region, Inyo County, California: United States Geological Survey Open-File Report, Scale 1:62,500.

———, 1973, Geologic map and sections of the Amargosa Valley borate area—southeast continuation of the Furnace Creek area—Inyo County, California: United States Geological Survey Miscellaneous Geologic Investigations Map I–782.

———, 1976a, Geologic maps and sections of a strip from Pyramid Peak to the southeast end of the Funeral Mountains, Ryan Quadrangle, California, *in* Troxel, B. W., and Wright, L. A., eds., Geologic Features Death Valley, California: California Division of Mines and Geology Special Report 106, p. 63–65.

———, 1976b, Columnar sections of the main part of the Furnace Creek Formation of Pliocene (Clarendonian and Hemphillian) age across Twenty Mule Canyon, Furnace Creek borate area, Death Valley, California: United States Geological Survey Open-File Report 76–261.

MCKEE, E. H., 1968, Age and rate of movement of the northern part of the Death Valley-Furnace Creek fault zone, California: Geological Society of America Bulletin, v. 79, p. 509–512.

MILLER, R. H., AND WALSH, C. A., 1977, Depositional environments of Upper Ordovician through Lower Devonian rocks in the southern Great Basin, *in* Stewart, J. H., Stevens, C. H., and Fritch, A. E., eds., Paleozoic Paleogeography of the Western United States: Society of Economic Paleontologists and Mineralogists, Pacific Section, Pacific Coast Paleogeography Symposium 1, p. 165–180.

NOBLE, L. F., 1941, Structural features of the Virgin Spring area, Death

Valley, California: Geological Society of America Bulletin, v. 52, p. 941–1000.

NOBLE, L. F., AND WRIGHT, L. A., 1954, Geology of the central and southern Death Valley region, California, *in* Jahns, R. H., ed., Geology of Southern California: California Division of Mines and Geology Bulletin 170, Chap. II, contribution 4, p. 143–160.

POOLE, F. G., AND SANDBERG, C. A., 1977, Mississippian paleogeography and tectonics of the western United States, *in* Stewart, J. H., Stevens, C. H., and Fritsch, A. E., eds., Paleozoic Paleogeography of the Western United States: Society of Economic Paleontologists and Mineralogists, Pacific Section, Pacific Coast Paleogeography Symposium 1, p. 67–85.

REYNOLDS, M. W., 1976, Geology of the Grapevine Mountains, Death Valley, California: a summary, *in* Troxel, B. W., and Wright, L. A., eds., Geologic Features Death Valley: California Division of Mines and Geology Special Report 106, p. 19–25.

STEWART, J. H., 1967, Possible large right-lateral displacement along faults and shear zones in the Death Valley-Las Vegas area, California and Nevada: Geological Society of America Bulletin, v. 78, p. 131–142.

———, 1983, Extensional tectonics in the Death Valley area, California: transport of the Panamint Range structural block 80 km northwestward: Geology, v. 11, p. 153–157.

STEWART, J. H., ALBERS, J. P., AND POOLE, F. G., 1968, Summary of regional evidence for right-lateral displacement in the western Great Basin: Geological Society of America Bulletin, v. 79, p. 1407–1414.

STOCK, CHESTER, AND BODE, F. D., 1935, Occurrence of lower Oligocene mammal-bearing beds near Death Valley, California: National Academy of Sciences Proceedings, v. 21, p. 571–579.

WILLIAMS, E. G., WRIGHT, L. A., AND TROXEL, B. W., 1976, The Noonday dolomite and equivalent stratigraphic units, southern Death Valley region, California, *in* Troxel, B. W., and Wright, L. A., eds., Geologic Features Death Valley, California: California Division of Mines and Geology Special Report 106, p. 45–49.

WRIGHT, L. A., 1976, Late Cenozoic fault patterns and stress fields in the Great Basin and westward displacement of the Sierra Nevada block, discussion: Geology, v. 4, p. 489–494.

WRIGHT, L. A., AND TROXEL, B. W., 1970, Summary of regional evidence for right-lateral displacement in the western Great Basin: discussion: Geological Society of America Bulletin, v. 81, p. 2167–2174.

WRIGHT, L. A., OTTON, J. K., AND TROXEL, B. W., 1974, Turtleback surfaces of Death Valley viewed as phenomena of extensional tectonics: Geology, v. 2, p. 53–54.

WRIGHT, L. A., WILLIAMS, E. G., ROBERTS, M. T., AND DIEHL, P. E., 1976, Precambrian sedimentary environments of the Death Valley region, eastern California, *in* Troxel, B. W., and Wright, L. A., eds., Geologic Features Death Valley, California Division of Mines and Geology Special Report 106, p. 7–15.

WRIGHT, L. A., TROXEL, B. W., BURCHFIEL, B. C., CHAPMAN, R. H., AND LABOTKA, T. C., 1981, Geologic cross section from the Sierra Nevada to the Las Vegas Valley, eastern California to southern Nevada: Geological Society of America Map and Chart Series MC-28M.

THE DEAD SEA RIFT: IMPACT OF CLIMATE AND TECTONISM ON PLEISTOCENE AND HOLOCENE SEDIMENTATION

WARREN MANSPEIZER

Department of Geology, Rutgers University, Newark, New Jersey 07102

ABSTRACT: The Dead Sea Rift, a classic strike-slip lineament, occurs along a transform plate boundary that connects the Red Sea, where sea-floor spreading is occurring, with the Taurus Mountains, where there is plate convergence. The total strike slip along the transform since the Miocene is 105 km with the last 40 km having occurred in Plio-Pleistocene time, that is when the present Dead Sea Basin began to subside. The Dead Sea Basin, an active rhomb-shaped graben, is located within the Dead Sea transform. It is bounded on the east and west by two nearly vertical, north-striking normal faults, on the south by listric(?) normal faults, that dip steeply to the north and on the north by a gently sloping, south-facing flexure in the basement. Displacement along a zone of en echelon strike-slip faults formed 3 sedimentary basins whose depocenters migrated northward with time. Deposition began in the early Miocene, when a thick succession of continental red beds (the Hazeva Formation) were deposited in a basin south of the modern Dead Sea. Deposition migrated northward in the Pliocene, as an arm of the Mediterranean flowed south through the transform into the newly evolving Dead Sea Basin; there, marine clastic sediment and evaporites, including the Sedom Salt, were laid down. The youngest rift sequence, of Plio-Pleistocene to Holocene age, fills the northern basin of the Dead Sea with over 3,500 m of lacustrine evaporites and fluvial-deltaic clastic sediments.

Only the youngest syn-rift sequence crops out along the west bank of the northern basin, where it was mapped at a scale of 1:50,000 for a distance of 50 km, and is divisible into three stratigraphic units. From oldest to youngest these are the Samra Formation (debris-fan gravel); the Lisan Formation (fan-delta gravel, and lacustrine limestone and marl); and a unit of deltaic sands and beach gravels. Sedimentation along the rift is controlled by the interaction of rift tectonics and climate. Whereas tectonic activity creates the rift, the resulting rift morphology significantly modifies the climate. Thus it may be observed that as moist air from the Mediterrean Sea rises over the shoulders of the rift, it cools adiabatically yielding as much as 800–1,000 mm of rain per year. This rainfall contributes to high-discharge ephemeral streams that transport huge quantities of coarse clastic sediment eastward onto the narrow shelf of the Dead Sea; and to the drainage basin of the Jordan River, a perennial stream, that carries mud and fine-grained clastic sediment along the axis of the rift, where it constructs a large delta at the head of the Dead Sea. On the other hand, as the air descends into the basin, it warms adiabatically, evaporating more than 2,000 mm of water per year, thereby causing a concomitant drop in the Dead Sea level, precipitation of evaporites, and the reworking of shelf sediment into deeper water. In the absence of recurrent syn-depositonal faulting in the stratigraphic record, such actualistic models convincingly explain Pleistocene-Holocene sedimentary patterns along the Dead Sea.

And His feet shall stand in that day upon the mount
of Olives.
Which is before Jerusalem on the east,
And the mount of Olives shall be cleft in the midst
thereof Toward the east and toward the west;
So that there shall be a very great valley;
And half of the mountain shall remove toward the
north, And half of it toward the south.

ZECHARIAH XIV:4

INTRODUCTION

The Dead Sea Rift extends for about 1,000 km along a transform plate boundary that connects the Red Sea, where sea-floor spreading is occurring, to the Taurus-Zagros subduction zone, where there is plate convergence (Fig. 1; Freund, 1965; Freund et al., 1970). Breakup of the Arabo-African platform began in the Miocene (about 20 Ma; Garfunkel, 1981), resulting in left-lateral displacement of about 105 km along the transform (Quennel, 1959; Freund and Garfunkel, 1976; Bartov et al., 1980; Hatcher et al., 1981). The present structures of the rift, including the Dead Sea, formed during the last 40 km of slip, probably during Pliocene-Pleistocene time (Garfunkel, 1981). The rift is a conspicuous linear depression, marked by zones of rhomb-shaped grabens, en echelon strike-slip faults, pull-apart basins, and push-up structures (See Schulman and Bartov, 1978; Garfunkel, 1981). It comprises several morpho-tectonic provinces, including the Gulf of Elat (Aqaba), the Arava Valley, the Dead Sea, the Jordan Valley, and the Lebanon segment (Fig. 1). The Dead Sea Basin, a component of the larger Dead Sea Rift, is the primary concern of this paper;

it is an asymmetric, rhomb-shaped graben, bounded by high-angle normal faults (See Schulman and Bartov, 1978).

Sedimentation along the rift is controlled largely by the dynamics of rift morphology and climate. From the Pliocene to the Pleistocene, a series of small debris flows, alluvial fans and fan-deltas built across the rift into Lake Lisan (the precursor of the Dead Sea), which occupied the full width and much of the length of the Dead Sea Rift (Sneh, 1979). As the lake level dropped during the late Pleistocene and Holocene, beginning about 18,000 years ago, streams draining the shoulders of the rift were incised across these older fans, and built younger fans and deltas farther into the Dead Sea. The Jordan River, flowing south along the axis of the rift, also deposited a thick Holocene deltaic wedge that was incised, and partially reworked, as the lake level dropped and the northern shoreline of the Dead Sea retreated to the south.

The essence of this paper is to describe this fan-delta complex as it crops out for 50 km along the western side fault along the Dead Sea. This paper presents new lithofacies data about these fans, examines current meteorological data from the rift, and relates the development of these fan deltas to the interaction (as observed today) between rift tectonics and climates. The paper also outlines and reviews current published information on the tectonic history of the Dead Sea Rift.

STRATIGRAPHY

Vertical displacement along the transform zone produced a thick rift sequence in three distinct sedimentary basins whose depocenters migrated northward with time (Fig. 2;

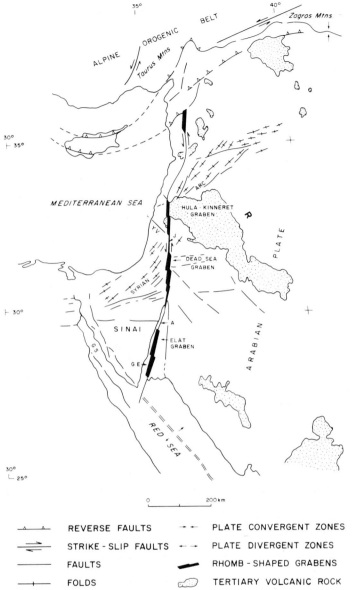

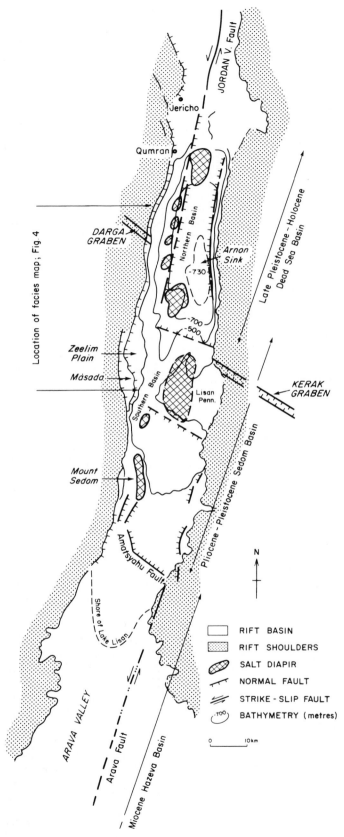

FIG. 1.—Regional tectonic setting of the Dead Sea Transform (after Garfunkel et al., 1981). Abbreviations: YV, Yisreel Valley; J, Jordan Valley; A, Arava Valley; G.S., Gulf of Suez; G.E., Gulf of Elat (Aqaba).

Zak and Freund, 1981). Deposition began in the earliest Miocene with the Hazeva Formation, a 2,000-m-thick continental fluvial red-bed sequence that filled a deep basin in the Arava Valley, south of the Dead Sea (Figs. 2, 3). The present Dead Sea Basin, however, began to subside only in the Pliocene. At that time, it was transgressed by an arm of the Mediterranean Sea that flowed south through the Yisreel and Jordan Valleys (Fig. 1), depositing over 4,000 m

FIG. 2.—Geologic sketch map of the Dead Sea region showing tectonic elements, physiographic features, northward migration of depocenters with time, and basin asymmetry expressed by bathymetry (modified from Zak and Freund, 1981; Garfunkel et al., 1981; Neev and Hall, 1979).

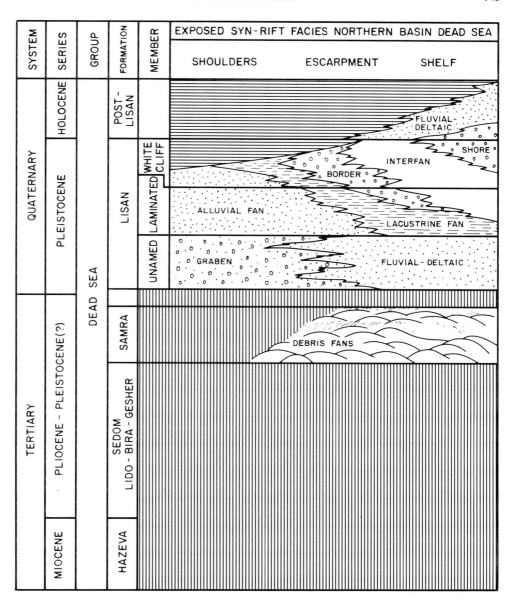

FIG. 3.—Plio-Pleistocene to Holocene syn-rift facies and formal stratigraphic units, along the west bank, northern basin of the Dead Sea (modified from Manspeizer, 1978).

of marine- to saline- and brackish-water sediments of the Bira-Lido-Gesher Formations, and rock salt of the Sedom Formation, in the southern basin of the Dead Sea (Fig. 3; Zak and Freund, 1981). The Sedom Formation is the source of the salt diapirs in the Dead Sea (see Fig. 2). The youngest (Plio-Pleistocene to Holocene) sediment of the rift fills the northern Dead Sea Basin with over 3,500 m of lacustrine evaporites and carbonates that are interbedded with fluvial-deltaic clastic sediments. The uppermost part of this sequence crops out along the western margin of the Dead Sea, where it is assigned to three stratigraphic units (Fig. 3): (1) the Samra Formation, a succession of Pliocene-Pleistocene(?), fresh- to brackish-water lacustrine debris-flow and fluvial-deltaic gravels, deposited during the earliest subsidence of the northern basin; (2) the Lisan Formation, a characteristically varved upper Pleistocene lacustrine carbonate sequence composed of interlaminated aragonite and calcite with interbeds of gravel; and (3) a young post-Lisan formation of Holocene fluvial-deltaic sands and shore gravels.

Samra Formation

The oldest syn-rift sequence found in this study crops out along the west bank of the Dead Sea, from Qumran to Ein Gedi, where it unconformably overlies Cretaceous carbonates (Figs. 3, 4). Typically it is cut by high-angle normal faults and is unconformably overlain by flat-lying, non-faulted Lisan strata. Although its age is poorly known, the formation is lithologically similar to, and occupies the same stratigraphic position as the Pliocene-Pleistocene(?) Samra Formation, as described by Begin (1975) from the Jericho

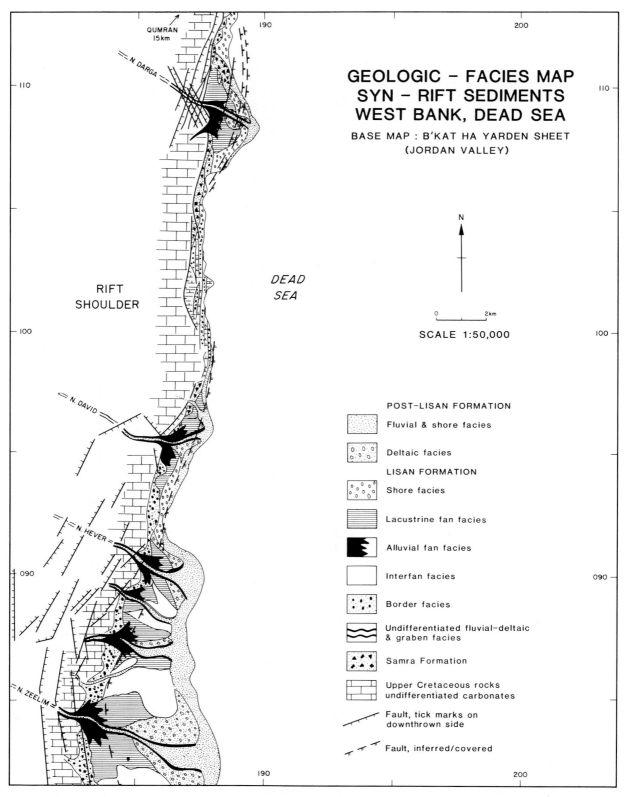

FIG. 4.—Geologic-facies map of syn-rift sediments, west bank, Dead Sea. Base map, B'kat Ha Yorden (Jordan Valley) Sheet, Scale 1:50,000, with Israel mapping grid. Fault patterns in the southern sector of the map drawn from Raz (1983).

area. It is thus tentatively correlated with that formation.

In the mapped area, the formation consists of overlapping mounds or pods of non-stratified and non-sorted gravels, some of which contain clasts as large as boulders. Large mounds are about 75 m wide and 25 m high, and small ones, about 3 to 5 m wide and 1 m high. The crests of these mounds are eroded and commonly are overlain by poorly sorted trough cross-stratified beds that thicken along the flanks of the mounds (Fig. 5). In longitudinal profile, the mounds resemble tear drops, pinching out toward the source on the west, and flaring out downcurrent to the east. Their flanks are inclined at about 25–30° to the north and south, whereas the downcurrent axis plunges about 35° due east, toward the Dead Sea. They are composed largely of well-indurated angular to subrounded carbonate clasts, averaging about 2–7 cm in diameter, and supported by a matrix of sand and fine-grained gravel. Larger clasts, measuring about 30–60 cm in diameter, are commonly composed of pebbly conglomerates, manifesting several generations of reworking. The formation also displays slump structures, reverse grading, and cyclic bedding.

The Samra Formation was deposited along a tectonically active margin. Faulting was episodic, beginning with the earliest record of subsidence in the northern basin. From Qumran to Nahal Darga (Figs. 2, 4) the formation accumulated along the margins of a moderately deep, and perhaps brackish-water lake; it consists of a succession of terrestrial to subaqueous lacustrine debris fans that were partially reworked by braided streams as the lake level dropped. South of Nahal Darga (along the depositional strike), the mounds pinch out and the formation consists almost exclusively of 100–150 m of poorly sorted gravel, trough cross-stratified on a large scale, and transported down the paleoslope to the east and southeast. North of the Dead Sea, near Jericho (Fig. 2), the Samra Formation consist of fluvial-deltaic calcarenites and conglomerates, with interbeds of siltstone, oolitic limestone, and marine brackish-water fossils (Begin, 1975). Near Jericho, the sediment was transported to the northeast, away from the Dead Sea, and into brackish-water lakes that may have been connected with the Mediterranean Sea during the Pliocene (Begin, 1975).

Lisan Formation

Resting unconformably on the Samra Formation is a younger fan complex, the Lisan Formation. Lartet (1869, cited in Begin et al., 1974) named the formation for the finely laminated white (aragonitic) and dark (calcitic) sediments that are about 40 m thick along the Dead Sea. The formation formed in Lake Lisan (the precursor of the Dead Sea) during the Wurm Stage of the late Pleistocene, from 60,000 to 18,000 years ago (Begin et al., 1974). At its maximum, the level of Lake Lisan was about 200 m above the present Dead Sea (−398 m) or 180 m below Mediterranean sea level (Neev and Hall, 1979). The rising waters flooded adjacent canyons, as well as the floor of the rift valley, from the Galilee on the north to the Arava Valley on the south, for a distance in excess of 200 km (Fig. 1). As the level of Lake Lisan rose, alluvial fans emerged from the narrow canyons, and prograded as fan-deltas (Holmes,

FIG. 5.—Cross-sectional view through a debris fan of the Samra Formation, along the Coast Road, south of Qumran. The fan consist of non-stratified and unsorted boulders (up to −11 phi) in a sandy gravel matrix; the upper contact of the fan is an erosion surface overlain by well-sorted, cross-stratified gravels that are inclined to the north (right). The photograph also shows an overlying younger debris fan, and the border fault (upper left).

1965) directly into the shallow waters of the lake and across slightly older fluvial-deltaic sands and gravels (Figs. 3, 4). The distal parts of these fans commonly interfinger with finely laminated aragonite and calcite, and with well-rounded beach gravels. These time-transgressive facies relationships are evident along the banks of present-day streams, which have been incised across older fan deltas (Fig. 6).

In contrast to prograding deltas, which become coarser grained upward, this transgressive sequence fines upward, and is characterized by fining-upward intervals of sand, mud, and aragonite. Locally, gypsum is common, as are mud cracks, root casts, ripple-drift cross-lamination, and minor unconformities with channel-lag deposits; collectively these structures suggest that deposition occurred on a shallow shelf alternately subject to episodes of transgression and regression.

Begin et al. (1974) presented a detailed treatment of the mineralogy, micropaleontology, and stratigraphy of this

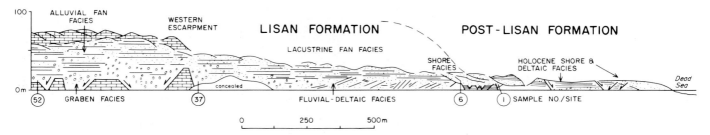

Fig. 6.—Facies distribution (exposed along a stream gorge) through the Lisan and post-Lisan fan-delta complex, Nahal Darga. Section drawn from field measurements and from a panorama of almost 50 photographs. Grain-size data were obtained from 52 stations along the stream cut. Only 4 sample sites are shown at the base of the diagram; the remaining sites may be determined by using a spacing grid of 1 sample/35 m. See Figure 8 for grain-size analyses and sampling procedures.

formation, particularly as it occurs north of the Dead Sea. They subdivided the unit (Fig. 3) into a lower member composed principally of varve-like white aragonite laminae with alternating dark detrital laminae; and an upper member that contains a higher chemical-to-detrital ratio with substantially more gypsum, aragonite, and diatomite. For the purpose of mapping these fans, the writer expanded the informal usage of the Lisan Formation, as defined by Lartet (1869), to include both the fan-delta complex and the underlying fluvial-deltaic sands. The relationship between facies and formal stratigraphic units is shown in Figure 3.

Facies Mapping

The Pleistocene and Holocene fan-delta complex was mapped (Manspeizer, 1978; Fig. 4) on a scale of 1:50,000 for a distance of 50 km, from Qumran, where the Dead Sea Scrolls were first found, to Masada, where the zealots held out against the Romans from 66 to 73 A.D. (Fig. 2). The subaerial parts of each fan-delta are small (averaging only 2–6 km^2) but the subaqueous portions occupy an area on the order of 100–200 km^2. Each fan system is delta-shaped in map view, convex upward in north-south section, and slopes about 10° to 15° towards the Dead Sea (Fig. 6). The catchment area of a typical fan is only 200–300 km^2 and occurs in a semi-arid terrain. In comparison, the drainage basin of the Jordan Delta is about 17,000 km^2, and occupies a temperate climate that in places receives rainfall in excess of 1,000 mm/yr. The fan-delta and associated facies, described below, are named for their inferred environment of deposition.

Graben facies.—High-angle normal faults, striking N50°W, cut the north-trending escarpment of the Dead Sea into rhomb-shaped grabens about 200 m wide and 800 m long (Fig. 7). The graben fill consists of very poorly sorted, non-stratified gravels composed of angular pebbles, cobbles and boulders with apparent mean diameters typically from −5.5 to −6.0 phi (Figs. 6, 8). Locally, the gravels interfinger with thin lenses of well-stratified and moderately sorted sandstones. These coarse-grained clastic sediments were deposited in steep canyons by short-duration debris and avalanche flows.

Fluvial-deltaic facies.—Stream deposits of the delta platform occur east of the escarpment and consist of reworked graben fill (Fig. 9). The channel deposits are poorly sorted, and coarse grained with apparent mean diameters between −6.5 phi near the border fault zone, and −5.0 phi, 2 km east of the border fault (Fig. 8). Bedding is obscure near the escarpment, but 1.0 to 1.5 km to the east, cross-bedding is conspicuous. Cut-and-fill structures are common, and most abundant where the gravels are in contact with lenses of gray marl and calcareous sandstones that typically have ripple-drift cross-stratification and trough cross-bedding. The facies is coeval with the graben facies and is slightly disconformable beneath the overlying lacustrine fan facies (Figs. 6, 9). One puzzling aspect of the facies is that cross-stratification is inclined toward the border fault (Fig. 9), suggesting that the facies may predate the basin, and may thus be as old as the Miocene Hazeva Formation.

Alluvial fan facies.—This facies includes the subaerial and proximal parts of the fan-deltas. Consequently it embraces both the shoulders of the rift, and the area east of the border fault (Figs. 4, 6, 7). It consists of poorly sorted cobble and boulder gravel, interbedded with calcarenites containing ripple marks, mud cracks, and burrows, and arranged in fining-upward sequences. Thin layers of thinly laminated aragonite and marl are common near the top of the section. The gravels are massively bedded (10–20 m thick), and pinch out by interfingering with calcarenites and marls having shallow-water structures. This facies is similar to the underlying graben deposits, but it contains proportionately less gravel, and more marl and calcarenite, and it is distinctively bedded (Fig. 7).

Lacustrine fan facies.—The lacustrine facies includes the shallow-water, and medial-to-distal parts of the fan-delta complex (Figs. 3, 4). It consists mainly of fine-grained gravels with interbeds of marl and calcarenites. The ratio of gravel to marl and calcarenite is about 4:1 and 3:1 at distances of about 1 km and 2 km, respectively, from the border fault. The gravels are typically massively bedded, well rounded, moderately well sorted, and have mean diameters of about −3 phi. Channeling and large-scale cross-bedding are common. The marls and sands are light gray, rhythmically bedded and finely laminated (1–5 mm); locally, they are iron-strained, cross-laminated and ripple-marked. Thinly laminated aragonite and thin lenses of coarsely crystalline gypsum occur sporadically near the upper half of this facies.

Interfan facies.—Varve-like white and dark laminae characterize this facies. The white laminae are composed of aragonite and locally of gypsum, whereas the dark laminae are marls composed mainly of calcite with varying amounts of quartz, dolomite, clay minerals (predominantly kaolinite with montmorillonite), and some gypsum and halite (Begin et al., 1974). The marl and aragonite are thinly laminated (average, 1–4 mm) and thinly bedded (average, 5–10 cm), whereas the gypsum is coarsely crystalline, and occurs in lenses 1–2 cm thick and in beds 20–30 cm thick. Both the marl and aragonite beds commonly show minor scour structures, lateral thinning and thickening of laminae, contorted bedding, and spectacular convolute beds (Fig. 10).

This facies is best developed on broad platforms (e.g., the large Zeelim plain, northeast of Masada; Fig. 2) where it interfingers with shallow-water sediments of the lacustrine fan and conglomerates of the border facies (Fig. 4). Ripple marks, mud cracks, and grain imbrication are locally common within the facies. Most noteworthy is the thick and repetitive sequence of thinly laminated marl and aragonite with numerous layers of mud cracks that are preserved with casts of gypsum (Fig. 11). The shallow-water origin of this facies is supported by the presence of gypsum, which, according to Friedman and Sanders (1978), is ubiquitous in the surface waters of the modern Dead Sea and on the bottom where the water is shallow, and where dissolved oxygen is present. In deeper water, gypsum is converted to calcite. The occurrence of cyclical, uniformly thin chalk laminae suggest that water fluctuations were controlled largely by seasonal climate rather than tectonics.

Border facies.—The presence of conglomerates and of cyclic arrangements of facies typifies this facies. The cycles, well displayed on lithologic logs from the Ein Gedi No. 2 well, consist largely of well-rounded and well-sorted gravels, fining upward into cross-laminated sandstones containing current-ripple and climbing-ripple stratification, which in turn give way to greenish marls with thin, discontinuous lenses of aragonite (Fig. 12). The facies also includes graded bedding and fining-upward sequences several meters thick. The border facies is a shallow-water deposit that accumulated adjacent to the western escarpment and exhibits considerable evidence for reworking.

FIG. 7.—The floor and wall (left of center) of a small graben along the shoulder of the Dead Sea Basin at Nahal Darga. Here the graben facies consists of unsorted boulder gravels overlain (background) by well-stratified gravels of the alluvial-fan facies. Scale: large boulder, right of center, is about −10 phi.

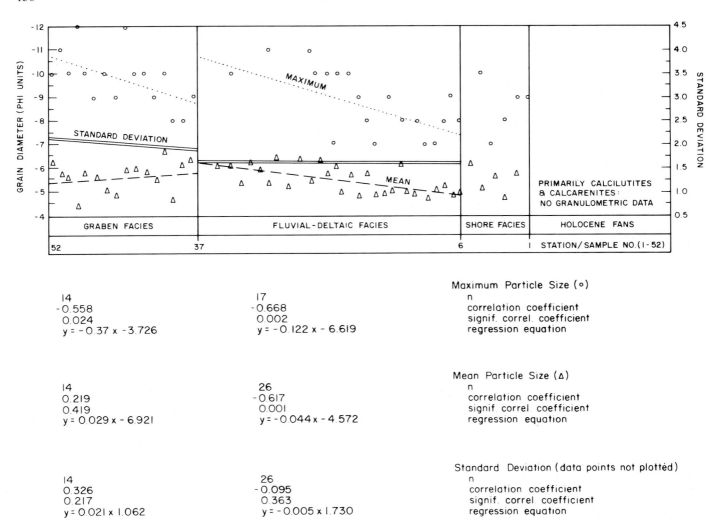

FIG. 8.—Regression analyses of grain-size data. Nahal Darga fan-delta (see Figs. 4, 6 for location), showing best-fit slopes for maximum (·····) and mean (-----) grain diameters, and standard deviation or grain sortings (==). Data control points are not plotted for grain sorting. The size data were collected in the field, using phi-rulers and frames to measure 100 particle diameters per station, which were set about 35 m apart (see Fig. 6 for location of stations 1, 6, 37, and 52). Phi arithmetic mean and standard deviation (method of moments; Folk, 1974) were computed from the field data.

Shore facies.—The beach facies is marked by well-rounded and moderately well-sorted gravels with large-scale tabular cross beds and slump structures, including interbeds of aragonite and ripple-marked sandstones.

Following deposition of these gravels and a fall in lake level, large-scale slides occurred in the gravel and throughout the Lisan clays. These slides superficially resemble listric normal faults and display as much as 20 m of vertical displacement. Arkin (1980), however, has shown that the clays of the Lisan Formation are structurally weak, having an open "house of cards" fabric, and a compressive strength only of about 4–7 kg/cm³. He suggests that the strength of the clay is due to the bonding of electrolytes within the open fabric of the clay, and that failure occurred when ground water leached the electrolytes following regression of Lake Lisan.

Post-Lisan Formation

Towards the end of the Pleistocene, as the climate became more arid, and the lake level began to fall (Neev and Hall, 1979), streams built successively younger Holocene fans farther out onto the shallow shelf, while beach gravels were laid down along the shore. A remarkably well-preserved vertical sequence of 28 regressional beach terraces along the western escarpment marks the high-water level of Lake Lisan and its intermittent retreat (Bowman, 1971).

As the lake reached the point of almost complete desiccation, evaporites were precipitated in the deeper parts of the basin. During the following pluvial period, streams carrying very fine grained sediment south along the axis of the rift deposited the thick deltaic wedge of the Jordan River Delta, a body of sediment about 13 km long, 8 km wide, and 400 m thick (Neev and Hall, 1979). Holocene fan-deltas that prograded across the western margin of the basin are typically coarser grained than the Jordan River Delta and finer grained than the underlying late Pleistocene fans. These young fans along the western margin of the rift consist primarily of alternating layers of calcarenite, marl, shale, and siltstone with an occasional gravel bed. The calcarenites commonly are well sorted, fine grained, convolutely bedded, ripple-marked and cross-stratified. Starved ripple marks, cut-and-fill channels, mud cracks, and uncommon aragonite laminae also occur in the sequence, as do moderately well-sorted shore gravels.

STRUCTURE

The Dead Sea Basin developed as an asymmetric, rhomb-shaped graben as a result of extension within a left-slip regime. Deformation within the basin is concentrated along oblique normal faults; north and south of the basin, however, it occurs along the major north-striking en echelon strike-slip faults (see Garfunkel et al., 1981; Reches and Hoexter, 1981).

The western border fault of the Dead Sea Basin forms a slightly arcuate, fairly continuous zone, several kilometers wide, of high-angle normal faults, fault blocks, and flexures (Fig. 4; Neev and Emery, 1967; Roth, 1970; Begin, 1975; Raz, 1983). The faults commonly dip from 60° to 80° eastward, exhibit only dip-slip slickensides, and cut the Samra fanglomerates. The fault contacts between these conglomerates and the older Cretaceous dolomites are strongly sheared, and the stratigraphic sections are repeated by high-angle normal faults, which typically step down towards the Dead Sea (Fig. 4). Near Masada, however (Fig. 2), high-angle normal faults step down away from the rift. Relief is on the order of 1,000 m, and the slopes are almost vertical. Near the top of the escarpment, shallow lacustrine algal tufa (Neev and Emery, 1967) drapes the border fault zone without discernible displacement, indicating that faults have been inactive there since the late Pleistocene, during the high stand of Lake Lisan. Elsewhere along the escarpment, flat-lying fan-delta beds of the Lisan Formation cross the border fault, and are not displaced (Fig. 6).

Several kilometers east of the escarpment, high-angle normal faults striking subparallel to the border fault cut

FIG. 9.—Coarse-grained fluvial-deltaic facies, in which low-angle foreset slopes are inclined towards the border fault of the Dead Sea Basin, where the facies is overlain by stratified gravels and marls of the lacustrine-fan facies. Almost 70 m of vertical section is exposed in this view, which is located about 300 m west of sample site 6, shown on Figure 6.

F ɪɢ . 10.—Contorted bedding in laminated aragonite and marl of the interfan facies, Lisan Formation, along Nahal Perazim. Note the three contorted intervals with abruptly truncated tops, asymmetric folds, and low- to high-angle thrust faults, separated by flat-lying units of marl and aragonite.

Holocene strata on the inner shelf, and along the outer shelf-slope margin of the modern Dead Sea (Fig. 4). Stratigraphic displacement along individual faults of the inner shelf measures only a few centimeters, but on the outer shelf-slope margin, displacement may be considerably greater (see Neev and Hall, 1976, Fig. 3a).

The floor of the Dead Sea (Neev and Hall, 1979) and the marginal rift highlands (Figs. 2, 4) are also cut by a family of high-angle, west-northwest-striking listric(?) normal faults. Examples are (1) the Amatsyahu fault zone (northern Arava Valley), which bounds the southern Dead Sea Basin with a family of faults that step down to the north; (2) the set of normal faults that cut across the Dead Sea trough, forming the horst of the Lisan Penninsula; (3) faults of the Kerak Graben in Jordan; and (4) the Wadie Darga faults of the west bank, which obliquely intersect the north-south border fault zone, forming small northwest-trending grabens on the western shoulder of the rift (Fig. 4). These oblique-striking faults cut the graben fill, indicating that they have been active since the onset of Lisan deposition. Active west-northwest-striking faults with as

much as 10 m of offset also have been recorded on the floor of the Dead Sea, where they were shown to coincide with epicenters of earthquakes by Ben-Avraham et al. (1984). However, no large oblique down-to-the-south normal faults have been reported from the northern end of the basin. Instead that part of the basin seems to be dominated by a basement flexure beneath the Jordan River delta (Figs. 13, 14).

In contrast to the western margin of the basin, with its broad shelf and vast prism of coalescing subaqueous fans, the eastern margin is dominated by an active transform plate boundary, a narrow shelf with no aqueous fans, and a spectacular bathymetric low, the Arnon Sink, as deep as 730 m below the Mediterranean sea level (Figs. 2, 14; see also Neev and Hall, 1979, Fig. 12). Isopachs drawn on the Lisan Formation and overlying sediments on the floor of the Dead Sea by Neev and Hall (1976, Figs. 3a, 3b, 3c, 6, 7, 8) consistently show them thickening and dipping toward the east, where they terminate abruptly against the transform plate boundary. Moreover, significantly older bedrock of Early Cretaceous, Triassic, Paleozoic and Precambrian age

FIG 11.—Alternating layers of thinly laminated white aragonite and equally thin laminae of gray marl (1–2 mm thick), with gypsiferous casts of mud cracks. Note (near pencil) an occasional horizon of imbricated and cross-stratified aragonite clasts. Gypsum is common in the section as small crystals on selected bedding planes. As many as 25 thin cycles of aragonite and marl are present, attesting to very shallow water deposition or to extreme fluctuations of Lake Lisan at this time. Outcrop is located on the broad Zeelim platform (see Fig. 4).

and seismic activity within that region involve several generations of tension fractures localized along pre-existing zones of weakness. Regional mapping of tectonic structures by Eyal and Reches (1983) generally support these findings, and show that the rift is characterized by east-northeast extension and north-northeast compression.

TECTONIC FRAMEWORK

As in most transforms, the Dead Sea Rift consists of a series of major en echelon strike-slip faults, rather than a single master fault. The major north-striking faults along the transform are arranged in a left-stepping en echelon pattern, and are joined together by a system of grabens (e.g., the Gulf of Elat, Dead Sea, and Hula-Kinneret Basins; Fig. 1; K. Arbenz, written commun., 1984). As structural features, these grabens develop primarily by extension along structures perpendicular to the major north-south strike-slip faults, so that for the Dead Sea, extension opens the basin primarily from south-to-north. This is compatible with the findings of Zak and Freund (1981), who suggest that the Dead Sea depositional basin migrated northward along the transform from the Miocene to the present. In the absence of upwelling mantle material, extension along the length of the transform must lead to crustal thinning and subsidence, most probably through listric normal faulting in the upper brittle crust and ductile attenuation in the lower crust (Fig. 13; K. Arbenz, written commun., 1984). While minor east-west extension may occur in the basin, the presence of thick Plio-Pleistocene sections along the border fault, shows that the present margins have been a major depositional site for a long time. The Dead Sea Basin has thus developed as an asymmetric, rhomb-shaped graben bounded by nearly vertical north-striking normal faults, less steeply dipping west-

FIG. 12.—Cyclic sedimentation in the border conglomerates of the Lisan Formation, Zeelim fan-delta. Each sequence begins with rounded and reworked cobbles and boulders, and abruptly fines upward into gravels and calcarenites with prominent ripple-drift cross-lamination. These sediments were deposited along the border fault, and in moderately shallow water.

crops out along the eastern shoulder, whereas only Upper Cretaceous bedrock is exposed along the western shoulder. Whereas these observations indicate that the rate of subsidence and sedimentation within the basin is higher along the transform plate boundary, they also show that the amount of vertical displacement, and the rates of uplift and erosion are similarly greater along the eastern margin of the basin.

These observations indicate that the eastern and western margins of the Dead Sea Basin are at present both dominated by high-angle faults with prominent normal slip and a secondary component of strike slip (Fig. 14). Behat and Rabinovitch (1983) even assert that the Dead Sea Rift was initiated through extensional tectonics and only subsequently involved strike-slip deformation. Detailed mapping of more than 2,700 fractures in a small (3.5 km²) area along the southern shoulder of the Dead Sea, near Masada (Fig. 2), led Arkin et al. (1981) to conclude that fault movement

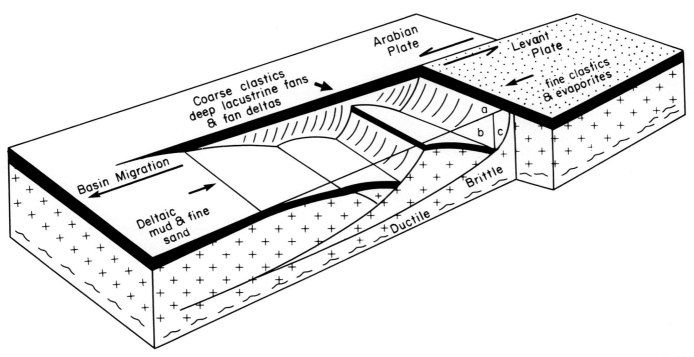

FIG. 13.—Schematic pull-apart model for the Dead Sea Basin, viewed to the southeast, modified from K. Arbenz (written commun., 1984). The nearly vertical walls are characterized by oblique slip (c), involving a component of dip slip (b) that exceeds the component of strike slip (a). Listric normal faults occur at the southern margin (right), and a basement flexure with minor normal faults at the northern one (left). The basin migrates to the north. The diagram does not illustrate the rhomb shape of the basin or its lateral asymmetry. The distribution of modern facies is shown around the basin margins.

northwest-striking listric(?) normal faults, and a south-facing basement flexure with minor normal faults (Fig. 13). The evolution of the Dead Sea Basin, from one dominated by strike slip to one characterized by extension, may thus reflect the concavity of the listric normal faults, with concomitant changes in its dip-slip and strike-slip components (Fig. 13).

These conclusions are similar to the findings of Choukroune et al. (1978) concerning transform faults in the FAMOUS study of the Mid-Atlantic Ridge. They report that (1) the main topographic expression of the transform fault, the transform valley, is mostly the result of normal faulting; (2) the main vertical motion between two plates takes place in the outer scarp area; and (3) the main dynamic pattern over the shoulders and outer scarps of the transform is quite probably due to vertical motion and normal faulting.

CLIMATE

Here I examine current meteorological data from the Dead Sea region, and analyze their relation to Holocene sedimentation within the rift. Actualistic models derived from the modern Dead Sea Rift are applied to the late Pleistocene. In the absence of evidence for repetitive syn-depositional faulting, facies in the Lisan Formation must reflect climatic, rather than tectonic, cycles.

The modern Dead Sea Rift embraces an extraordinary array of depositional environments, varying from paludal swamps in the Hula Valley to continental sabkhas in the Arava Valley, and coral reefs with mangrove swamps in the Gulf of Elat (Aqaba). This complex environmental assemblage is primarily due to the interaction of tectonism and climate. Tectonism plays the dominant role because it controls the distribution of basins and source terranes, the topography of the rift, and the drifting of plates as they pass through different climatic zones. Climate, which is significantly influenced by rift topography, controls evaporation and rainfall, and thereby strongly influences patterns of sediment transport and deposition.

Evaporation and rainfall in the Dead Sea Rift are mainly controlled by two climatic factors: (1) the global atmospheric pressure belts and atmospheric circulation; and (2) the orographic effect due to the uplift and descent of air as it flows over the shoulders of the rift. On a global scale, evaporation and rainfall are mainly controlled by the general circulation of the atmosphere, so that today the great deserts of the world coincide with subtropical high-pressure cells, and the equatorial rainforests are located beneath the equatorial low-pressure cells. During the summer months, the Dead Sea Rift lies within the subtropical high-pressure cell, centered in northern Lebanon and Syria. As dry air descends from the upper troposphere of the high-pressure cell, it warms adiabatically (due to an increase in atmospheric pressure), to bring hot, dry, continental tropical air to the Levant from the east. As the high-pressure cell is displaced southward during the winter, the region falls under the influence of middle latitude cyclones centered over Cyprus. Storms typically track from west to east across the northern part of the rift as alternating waves of low- and high-pressure cells. As the warm, moist air from the Mediterranean spirals upward, counterclockwise in a low-pressure cell, it cools adiabatically (due to a decrease of atmospheric pressure), bringing rainfall to the Levant from the west. The rift depositional environments are also affected by (1) 'khamsin' winds, which blow hot, dry continental tropical air masses from Asia toward low-pressure systems centered over Lybia and Egypt; (2) a 'sharov,' or high-pressure cell centered above the rift; and (3) an occasional low-pressure cell centered above the rift, bringing unseasonably warm air from the Red Sea (Orni and Efrat, 1971).

The rift basins, because of their special characteristics, also create unusual climates that cannot be explained through a general circulation model alone. Rifts typically have high-relief margins, extend for long distances across both lines of latitude and prevailing wind systems, and like the Dead Sea, East African, and San Andreas basins, lie adjacent to large water bodies. The orographic factor is clearly evident from the course of isohyets as they cross the Dead Sea Rift (Fig. 15; see Ashbel, 1948). Note that rainfall increases from less than 500 mm/yr on the Mediterranean coast to as much as 800 mm/yr on the eastern shoulder of the rift, and then drops abruptly to about 50–100 mm/yr on the shores of the Dead Sea, only to increase once again to over

800 mm/yr on the Moab Plateau in Jordan, and to over 1,000 mm/yr on the northern shoulders of the rift (Fig. 15). Evaporation, while increasing from about 1,400 mm/yr in the Galilee to 2,700 mm/yr in the Gulf of Elat, rises abruptly from 1,000 mm/yr in the Judean Hills to over 2,000 mm/yr in the Dead Sea (Fig. 15).

These rainfall and evaporation changes are due to the expansion and contraction of air as it passes through different altitudes. As warm, moist air from the Mediterranean Sea is uplifted over the western shoulder of the rift, it yields rain water for low-gradient perennial streams that flow westward away from the rift axis, and for high-discharge ephemeral streams that transport large quantities of coarse clastic sediments onto the floor of the rift, where fan-deltas prograde into the Dead Sea. The Jordan River, the only perennial stream feeding the Dead Sea, flows along the axis of the rift building a large delta of very fine grained clastic sediment at the head of the Dead Sea. The orographic effect on mean annual precipitation is on the order of 30 mm for each 100 m increase in elevation; on the other hand, the impact of longitude (distance from the sea) is only on the order of 1.5 to 3.1 mm/km, and the impact of latitude is about 1.1 to 1.9 mm/km (Diskin, 1970). Conversely, as the air descends onto the leeward side of the rift, it warms adiabatically, thereby increasing the rate of evaporation of

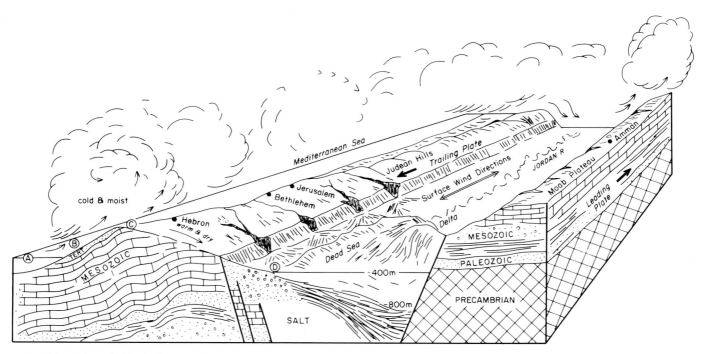

FIG. 14.—Schematic block diagram of the Dead Sea region, viewed to the north, showing generalized rift topography, tectonic framework, depositional domains, and meteorological patterns. A parcel of air at A (along the Mediterranean coast) having a temperature of 20° C, a dew point of 14° C and a dry adiabatic rate of 1° C/100 m, becomes saturated at B (600 m above sea level), At C, its temperature drops to 13° C with rain falling over the Judean Hills. As the air descends to the Dead Sea floor, it warms to 23° C, providing a heat sink for the precipitation of evaporites. Winds aloft blow from west to east, although surface winds within the rift blow from the north or south, along its axis. Perennial streams flow west towards the Mediterranean coast, while high-discharge ephemeral streams carry a large load of coarse clastic sediments via fan-deltas toward the Dead Sea. The Jordan River, a perennial stream whose drainage basin lies in a humid region, builds a large mud and fine-grained clastic delta along the axis of the basin. Tectonic asymmetry is shown by bedrock of different ages along rift shoulders, basin bathymetry, and thickening of depositional prism to the east.

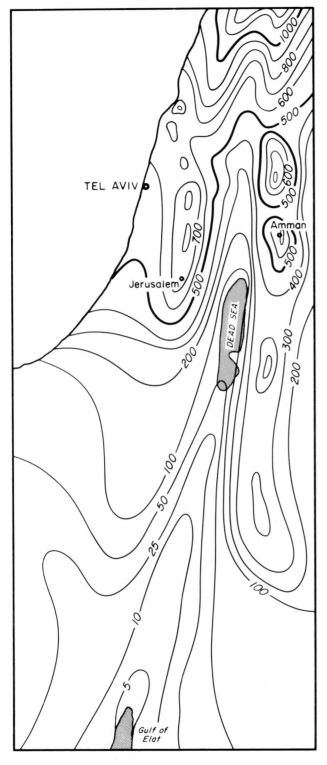

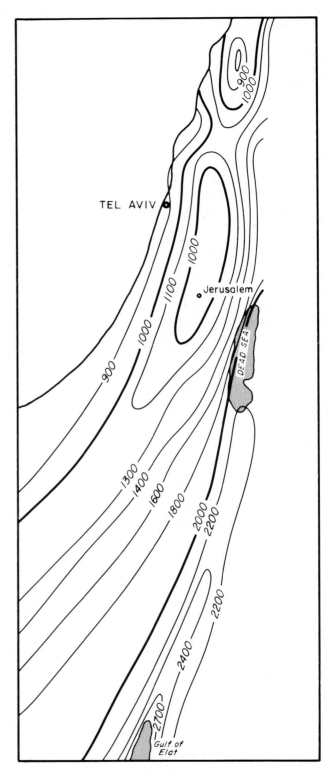

(A) ANNUAL RAINFALL (mm)   **(B)** ANNUAL EVAPORATION (mm)

Fig. 15.—A) Annual rainfall, and B) evaporation over the Dead Sea Rift (modified from Asbel, 1948). Note irregular contour intervals.

sea water and precipitation of gypsum and halite along the shallow shelf of the modern Dead Sea (see discussion by Friedman, 1965).

To illustrate this principle, consider a parcel of air along the Mediterranean coast with a temperature of 20° C and a dew-point temperature of 14° C. Condensation begins along the coast about 600 m above mean sea level, as cooling procedes along the dry adiabatic gradient of 1° C/100 m. From 600–800 m, condensation proceeds along the wet adiabatic gradient of 0.5° C/100 m, and is accompanied by moderate rainfall. Temperature at the Biblical cities of Jerusalem, Bethlehem, and Hebron (Fig. 15) is about 13° C, and the air is saturated. As the air descends to the floor of the rift, it warms at the dry adiabatic rate, less the latent heat of evaporation. The new temperature at the Dead Sea level is about 23° C (i.e., 800 m at 1° C/100 m, and 400 m at 0.5° C/100 m).

A more rigorous treatment of the adiabatic process is not warranted here, since a large number of environmental variables have been left unanswered (see Orni and Efrat, 1971). Temperatures, for example, are higher at the southern end of the Dead Sea because the northern basin is deeper and wider, thus setting into motion sea breezes which moderate the noonday heat. While the principal wind direction above the rift is from west to east, winds within the rift blow from the north and/or south (Fig. 14). Because the rift resembles a narrow corridor, closed on its sides and open at its end, surface wind directions other than north and south are rare. Changes in wind direction reflect changes in pressure gradient due to differential heating and cooling of land and water. High temperatures and strong winds contribute to the higher evaporation rate, which would be even higher were it not for the salinity of the water, the higher than usual barometric pressure, and the light mist hovering over the Dead Sea. If the waters in the Dead Sea were fresh, the annual rate of evaporation would be in excess of 3,000 mm. Moreover, the orographic factor produces a low average relative humidity of only 57%, and a very high barometric pressure of 800 mm (or 32 in) of mercury, which enriches the oxygen content above the Dead Sea by 6–8%, thus increasing precipitation of evaporites.

During the late Pleistocene, from 60,000 to 18,000 years ago, the rift was occupied by Lake Lisan, a fresh-water lake that extended for 225 km from the Sea of Galilee on the north to the Arava Valley on the south (Begin et al., 1974). At its highest, the lake level rose to about 180 m below the present-day Mediterranean sea level, while the lake floor was 400 m below that datum. The present lake floor, which according to Neev and Emery (1967) was down-faulted after deposition of the Lisan Formation, is at 800 m below mean sea level. Relief along the margins of the rift was slightly lower than today; thus the amount of adiabatic warming and cooling was not substantially different from that of the present. Although the location of the semi-permanent high- and low-pressure cells affecting the Levant was about the same, the pressure gradient was weaker than today (see Gates, 1982, Fig. 2.10). This had the effect of slightly decreasing the wind speed without changing its direction.

Winter monsoons, blowing from the east, brought cold air to the rift from the Eurasian continent, while summer winds from the west brought warm moist air from the Mediterranean Sea. Middle latitude cyclonic storms tracked west to east across the rift, bringing additional rain to the catchment basin and snow to the higher elevations of the Golan to the north. As seasons alternated between wet and dry, and cold to hot, land-locked Lake Lisan rapidly transgressed and regressed the shallow shelf, first innundating and later exposing both marginal fans and deltas. Geologic and atmospheric conditions during the late Pleistocene were such that variations of temperature, pressure, rainfall, wind direction and/or moisture content alone can explain the facies changes in the stratigraphic record. Recent observations (December, 1984) by the writer in Lakes Tanganyika and Turkana of East Africa clearly attest to the profound impact of monsoonal winds on geologic processes in rift-related lakes.

CONCLUSIONS

As a rhomb-shaped graben within a strike-slip regime, the Dead Sea Basin has characteristics similar to other extensional basins, namely high depositional rates, asymmetry of both basin fill and bathymetry, abrupt lateral facies changes, and margins bounded by normal faults. It differs, however, from other extensional basins in that the Dead Sea depositional center has migrated with time. Whereas continental red beds of the Miocene Hazeva Formation were deposited in a basin south of the Dead Sea, Pliocene evaporites and marine clastic sediments were laid down to the north in the newly subsiding Dead Sea Basin. By the Plio-Pleistocene, the depositional center had shifted farther north to the site of the modern Dead Sea, where over 3,500 m of clastic sediments and evaporites have accumulated.

Although the location and geometry of the Dead Sea depocenter is controlled by faulting, its basin facies are governed primarily by cyclical changes in lake level, which fluctuates principally with climatic variations rather than with tectonic pulses. Thus the Lisan Formation, which records multiple episodes of both very small and large-scale cyclic sedimentation, reflecting changes in the lake level, is without substantial evidence in the rock record of recurrent syndepositional faulting. In addition to creating the Dead Sea graben, strike-slip faulting also initiated the vertical tectonics (Fig. 13) that produced a rift system of high relief. The consequence of this was to create (in an otherwise arid climate, like the Middle East) a region of moderate to heavy rainfall along the rift shoulders, and a zone of excessive evaporation along its floor, through adiabatic expansion and contraction of rising and subsiding air masses (Fig. 14).

The Dead Sea Rift may thus serve as a guide for examining the tectonic and paleoclimatologic effects associated with the early stages of continental rifting and seafloor spreading. Further studies of climatic controls on deposition and sediment evolution in the Dead Sea Rift can help us understand and predict sediment distribution in other paleo-rift systems such as the Triassic-Jurassic rifts of North America and North Africa, the Cretaceous rift of central Africa; and the Jurassic-Cretaceous rifts in the South Atlantic.

ACKNOWLEDGMENTS

It is with considerable pleasure that I express my indebtedness to the faculty and staff at the Geological Survey of Israel and of the Hebrew University of Jerusalem; most particularly I wish to thank J. Arkin, Y. Bartov, the late R. Freund, Z. Garfunkel, D. Neev, and E. Zohar for introducing me to the challenges of the rift, and for their field assistance, stimulating and useful discussions, and general kindness shown to me and to my family. Careful and penetrating reviews by Kevin T. Biddle and John Van Wagoner and the editorial suggestions of N. Christie-Blick are greatly appreciated, as is the generous assistance given to me by K. Arbenz.

REFERENCES

ARKIN, Y., 1980, Underconsolidated sensitive clay in the Lisan Formation, Sedom southern Dead Sea Basin: The mineral industry in Israel—the fourth decade of the state: Geological Survey of Israel and Ministry of Energy and Infrastructure, 5th Conference on Mineral Engineering, p. 6–12.

ARKIN, Y., GILAT, A., AND AGNON, A., 1981, Mediterranean – Dead Sea Project: Geotechnical Survey of Area Proposed for an Underground Power Station: Israel Geological Survey Report MM/3/81, 15 p.

ASBEL, D., 1948, Bio-Climatic Atlas of Israel and the Near East: Meteorological Department, Hebrew University, Jerusalem, Israel, p. 41 and 51.

BAHAT, D. AND RABINOVITCH, A., 1983, The Initiation of the Dead Sea Rift: Journal of Geology, v. 91, p. 307–322.

BARTOV, Y., STEINITZ, G., EYAL, M., AND EYAL, Y., 1980, Sinistral movement along the Gulf of Aqaba—its age and relation to opening of the Red Sea: Nature, v. 285, p. 220–221.

BEGIN, Z. B., 1975, Paleocurrents in the Plio-Pleistocene Samra Formation (Jericho region, Israel) and their tectonic implication: Sedimentary Geology, v. 14, p. 191–218.

BEGIN, Z. B., EHRLICH, A., AND NATHAN, Y., 1974, Lake Lisan, the Pleistocene precursor of the Dead Sea: Geological Survey of Israel Bulletin No. 63, 30 p.

BEN-AVRAHAM, Z., BEYTH, M., ROTSTEIN, Y., AND EITAM, Y., 1984, Recent faults in the Lynch Straits-southern Dead Sea (abs.): Israel Geological Society, Annual Meeting, 1984, Proceedings, p. 19.

BOWMAN, D., 1971, Geomorphology of the shore terraces of the Late Pleistocene Lisan Lake (Israel): Palaeogeography, Palaeoclimatology, Palaeoecology, v. 9, p. 183–209.

CHOUKROUNE, P., FRANCHETEAU, J., AND LE PICHON, X., 1978, In situ structural observations along Transform Fault A in the FAMOUS area, Mid-Atlantic Ridge: Geological Society of America Bulletin, v. 89, p. 1013–1029.

DISKIN, M., 1970, Factors affecting variations of mean annual rainfall in Israel: International Association of Scientific Hydrology Bulletin, v. 15, No. 4, p. 41–49.

EYAL, Y. AND RECHES, Z., 1983, Tectonic analysis of the Dead Sea Rift region since the Late Cretaceous based on mesostructures: Tectonics, v. 2, p. 167–185.

FOLK, R. L., 1974, Petrology of Sedimentary Rocks: Austin, Texas, Hemphill Publication Company, 182 p.

FREUND, R., 1965, A model of the structural development of Israel and adjacent areas since Upper Cretaceous times: Geological Magazine, v. 102, p. 189–205.

FREUND, R., AND GARFUNKEL, Z., 1976, Guidebook to the Dead Sea Rift, Department of Geology, Hebrew University, Jerusalem, 27 p.

FREUND, R., GARFUNKEL, Z., ZAK, I., GOLDBERG, M., WEISSBROD, T., AND DERIN, B., 1970, The shear along the Dead Sea rift: Philosophical Transactions of Royal Society of London, Series A, v. 267, p. 107–130.

FRIEDMAN, G. M., 1965, On the origin of aragonite in the Dead Sea: Israel Journal of Earth Sciences, v. 14, p. 79–85.

FRIEDMAN, G. M., AND SANDERS, J. E., 1978, Principles of Sedimentology: John Wiley and Sons, New York, 792 p.

GARFUNKEL, Z., 1981, Internal structure of the Dead Sea leaky transform (Rift) in relation to plate kinematics: Tectonophysics, v. 80, p. 81–108.

GARFUNKEL, Z., ZAK, I., AND FREUND, R., 1981, Active faulting in the Dead Sea Rift: Tectonophysics, v. 80, p. 1–26.

GATES, W., 1982, Paleoclimatic Modeling—A review with reference to problems and prospects for the pre-Pleistocene, in Climate in Earth History: National Research Council Studies in Geophysics, p. 26–42.

HATCHER, R. D., REGAN, R. D., AND ABU-AJAMIEH, M., 1981, Sinistral strike-slip motion on the Dead Sea Rift: confirmation from new magnetic data: Geology, v. 9, p. 458–462.

HOLMES, A., 1965, Principles of Physical Geology: London, Thomas Nelson and Sons, 2nd edition, 1288 p.

LARTET, L., 1869; Essai sur la géologie de la Palestine et des contrées avoisinantes, telles que l'Egypte et l'Arabie, Comprenant les observations recueillies dans le cours de l'expédition du Duc de Luynes a la Mer Morte [thesis]: Paris, Victor Masson et fils, 109 p.

MANSPEIZER, W., 1978, Tectonics and sedimentation along the rift valley: The Dead Sea: Marginal Sedimentary Facies: 10th International Congress on Sedimentology, Jerusalem, International Association of Sedimentologists, Guidebook, p. 51–55 and p. 74–80.

NEEV, D., AND EMERY K. O., 1967, The Dead Sea: Depositional Processes and Environments of Evaporites: Ministry of Development, Geological Survey of Israel, Bulletin No. 41, 147 p.

NEEV, D., AND HALL, J. K., 1976, The Dead Sea Geophysical Survey, Final Report No. 2, Seismic results and interpretation: Report MGD, 6/76, Geological Survey of Israel, 21 p.

NEEV, D., AND HALL, J. K., 1979, Geophysical Investigations in the Dead Sea: Sedimentary Geology, v. 23, p. 209–238.

ORNI, EFRAIM, AND EFRAT, E., 1971, Geography of Israel: Philadelphia, The Jewish Publishing Society of America, 3rd edition, 551 p.

QUENNELL, A. M., 1959, Tectonics of the Dead Sea rift: Asociacion de Servicios Geologicos Africanos, 20th International Geological Congress, Mexico, 1956, p. 385–405.

RAZ, E., 1983, En Gedi: Structural Map (Preliminary Edition): Geological Survey of Israel, Scale 1:50,000.

RECHES, Z., AND HOEXTER, D., 1981, Holocene seismic and tectonic activity in the Dead Sea area: Tectonophysics, v. 80, p. 235–254.

ROTH, I., 1970, Geologic Map: Wadi el Qilt: Geological Survey, Israel, Scale 1:50,000.

SCHULMAN, N., AND BARTOV, Y., 1978, Tectonics and sedimentation along the rift valley, excursion Y2: 10th International Congress on Sedimentology, Jerusalem, International Association of Sedimentologists, Guidebook, p. 37–94.

SNEH, AMIHAI, 1979, Late Pleistocene fan-deltas along the Dead Sea Rift: Journal of Sedimentary Petrology, v. 49, p. 541–551.

STANHILL, G., 1981, Evaporation from the Dead Sea: Division of Agricultural Meteorology, Bet Dagan, Israel, Report No. 2, p. 41–51.

ZAK, I., AND FREUND, R., 1981, Asymmetry and basin migration in the Dead Sea Rift: Tectonophysics, v. 80, p. 27–38.

A RELEASING SOLITARY OVERSTEP MODEL FOR THE LATE JURASSIC-EARLY CRETACEOUS (WEALDIAN) SORIA STRIKE-SLIP BASIN (NORTHERN SPAIN)

MICHEL GUIRAUD[1] AND MICHEL SEGURET

Géologie Structurale,
USTL, Pl. Bataillon,
34060 Montpellier, France

ABSTRACT: The Soria Basin is a rhomb graben with borders that trend N60°E and N50°W. It was formed during the Late Jurassic – Early Cretaceous (Wealdian), when as much as 8 km of fluvially dominated deltaic strata accumulated in it. This sedimentary fill has been divided into five cyclothems, of which the lower four are discussed in this paper. Within the basin, a N50°W-trending, 50-km wide, syn-sedimentary syncline developed in the basin fill. This syncline was related to extensional tectonics, and to the formation of a half graben in Paleozoic basement overlain by competent Jurassic and incompetent Triassic strata.

Within the basin fill, the extensional deformation produced microstructures (stylolites, calcite tension gashes, and quartz dikes) with a coherent basin-wide pattern. The depocenter migrated with time from the southeastern corner of the basin during deposition of cyclothems I and II, to the northeastern corner during formation of cyclothems III and IV. High heat flow, related to crustal thinning in the area of greatest subsidence, led to metamorphism within the sediments. Conditions of metamorphism were a maximum temperature of 420° C, a temperature gradient of 100–150° C/km and pressures between 1–3 kb. At the same time, compressional deformation (N30°W-trending folds and associated cleavage) was induced along the southeastern margin of the basin, and erosion (uplift) occurred outside the basin.

Our interpretation of the geometry, sedimentation, tectonics, and thermal evolution of the Soria Basin is based on mathematical models and microtectonic analogue models of a releasing solitary overstep. In such models, stress/strain deviations and accumulations predict vertical motions (subsidence and uplift), and the geometry of structures in areas of extension (secondary normal faults, tension gashes) and in areas of compression (folds, cleavage). Most of the field data collected inside and outside the basin are consistent with a model of a releasing overstep along N60°E-striking sinistral strike-slip faults.

The proposed releasing overstep model differs from classical models of strike-slip basins by (1) taking into account stress/strain related to basin development, (2) explaining migration of the depocenter with time, and (3) predicting the geometry of secondary faults.

INTRODUCTION

The major types of basins associated with horizontal motion along strike-slip faults have received a good deal of attention in the last decade (Crowell, 1974a, b, 1976; Steel and Gloppen, 1980; Aydin and Nur, 1982; Burke et al., 1983; Mann et al., 1983). The term pull-apart basin has gained general acceptance for a depression produced by extension at a discontinuity or step along a strike-slip fault (Mann et al., 1983). We propose and discuss in this contribution a releasing overstep model for the development of the Late Jurassic-Early Cretaceous (Wealdian) Soria Basin of northern Spain.

The Soria Basin is characterized by (1) a well-preserved original geometry—the basin was little deformed by Eocene-to-Miocene Alpine orogenesis; (2) a simple rhomb shape with dimensions of 70 by 50 km and an area of about 3,500 km[2]; (3) as much as 8 km of fluvially dominated deltaic deposits; (4) metamorphism of the sedimentary fill during basin development; and (5) reduced thicknesses of Upper Jurassic-Lower Cretaceous deposits (0–500 m) or erosion of pre-basin deposits outside the basin boundaries. All of these characteristics are consistent with a strike-slip origin for the basin (Crowell, 1974a; Reading, 1980).

Our interpretation of the Soria Basin (Guiraud, 1983) is based on field analyses including (1) the geometry of depositional sequences; (2) facies evolution; (3) deformation associated with the basin formation; and (4) thermal evolution of the sediments. In order to interpret the field data, we refer to mathematical and microtectonic analogue models of strain/stress distribution in a releasing overstep discontinuity along strike-slip faults. In such an overstep, the far-field stress is disturbed in direction and magnitude. We establish the relationship between the strain observed inside and outside the basin and the proposed kinematic model of a releasing overstep. This model is characterized by simultaneous extension inside the basin and compression outside the basin.

THE SORIA BASIN

The Soria Basin is located in the northwestern Iberic Range, a little-deformed segment of the Alpine belt in Spain. To the north and south, the basin is bounded by the Oligocene-Miocene conglomerates and sandstones of the Ebro and Duero Basins (Fig. 1). To the west and east, the basin is bordered by Paleozoic basement and Jurassic carbonates of the Sierra Demanda and Sierra Montcayo. To the north, the margin of the basin is well defined by a thrust that juxtaposes Mesozoic sedimentary rocks above Cenozoic clastic deposits of the Ebro Basin. To the south, the Cenozoic conglomerates and sandstones of the Duero Basin partially overlap the Mesozoic formations.

This basin was first investigated by Sanchez Lozano and Palacios (1885), who described lagoonal and continental rocks of the area. Saenz Garcia (1932) defined three Upper Jurassic to Lower Cretaceous formations (Purbeckian, Wealdian, Urgonian). In subsequent studies, Beutler (1966) subdivided the so-called Wealdian deposits into five lithological units by means of successions of carbonate and clastic formations. A recent study by Salomon (1982, 1983) has permitted further lithostratigraphic subdivision and an interpretation of sedimentary environments and basin evolution during Late Jurassic to Early Cretaceous time.

The buried faults bounding the basin are well docu-

---

[1]Present Address: Department of Geology, Private Mail Bag 2084, University of Jos, Nigeria

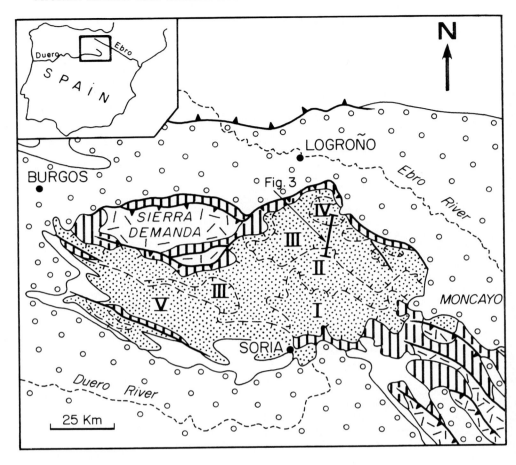

FIG. 1.—Geological setting and location of the Soria Basin, northern Spain.

mented by localized extension of the Upper Jurassic to Lower Cretaceous deposits (Salomon, 1982, 1983). Bounded by N50°E- to N60°E-striking and N50°W-striking faults, the Soria Basin exhibits a characteristic rhomboidal shape. These structural trends are related to Mesozoic and Alpine reactivation of Late Paleozoic crustal faults (Arthaud and Matte, 1977). Upper Jurassic to Lower Cretaceous sedimentary rocks are about 3 km thick in the southern part of the basin (Beutler, 1966; Salomon, 1982, 1983), and as thick as 6 km in the northern part (this study).

STRATIGRAPHY OF THE BASIN

Framework

The Upper Jurassic to Lower Cretaceous succession consists of five cyclothems, of which the lower four are discussed in this paper. Individual cyclothem are about 1,500 to 3,500 m thick, and bounded by well-defined erosion surfaces. Each cyclothem is composed of a basal coarse-grained clastic interval of fluvial origin that fines upward and is overlain by lacustrine to restricted shallow-marine carbonate rocks. The boundaries between cyclothems correspond to the progradation of clastic deposits over underlying carbonates. The major stratigraphic divisions or cyclothems, based upon these depositional breaks, are summarized in Figure 2. Three major complete sedimentary cyclothems and the basal part of a fourth have been recognized. Cyclothems III and IV are described in detail below because they are especially important in reconstructing the evolution of the basin.

Age of the Basin Deposits

The age of the rocks is uncertain because of the paucity of diagnostic faunas. Figure 2 includes ages proposed by Brenner (1976) and Salomon (1982). The two different dating schemes are based respectively on the study of charophyte and ostracode faunas by Brenner and of ostracodes by Salomon. Late Barremian-Early Aptian charophyta (*Chlypeastor lusitanicus* Grambast-Fessard), sampled in the

upper part of the Enciso Formation support the age designation of Brenner (Grambast-Fessard, personal commun., 1983).

Sedimentary Facies and Environments of Cyclothems III and IV

In the northern part of the basin, cyclothems III and IV are 5,200 m thick along the Yanguas-Enciso section (Fig. 3). Here the main facies of the uppermost part of cyclothem II (Valdeprado Formation) is represented by carbonates containing algal laminae, flat-pebble conglomerates, and desiccation cracks. Gypsum pseudomorphs and collapse breccia have also been observed. According to Salomon (1982), this facies accumulated in a sabkha environment. The contact between the sandstones of the Yanguas Formation and the underlying carbonates, an important discontinuity, is sharp but lacks topographic relief along the 70 km mapped.

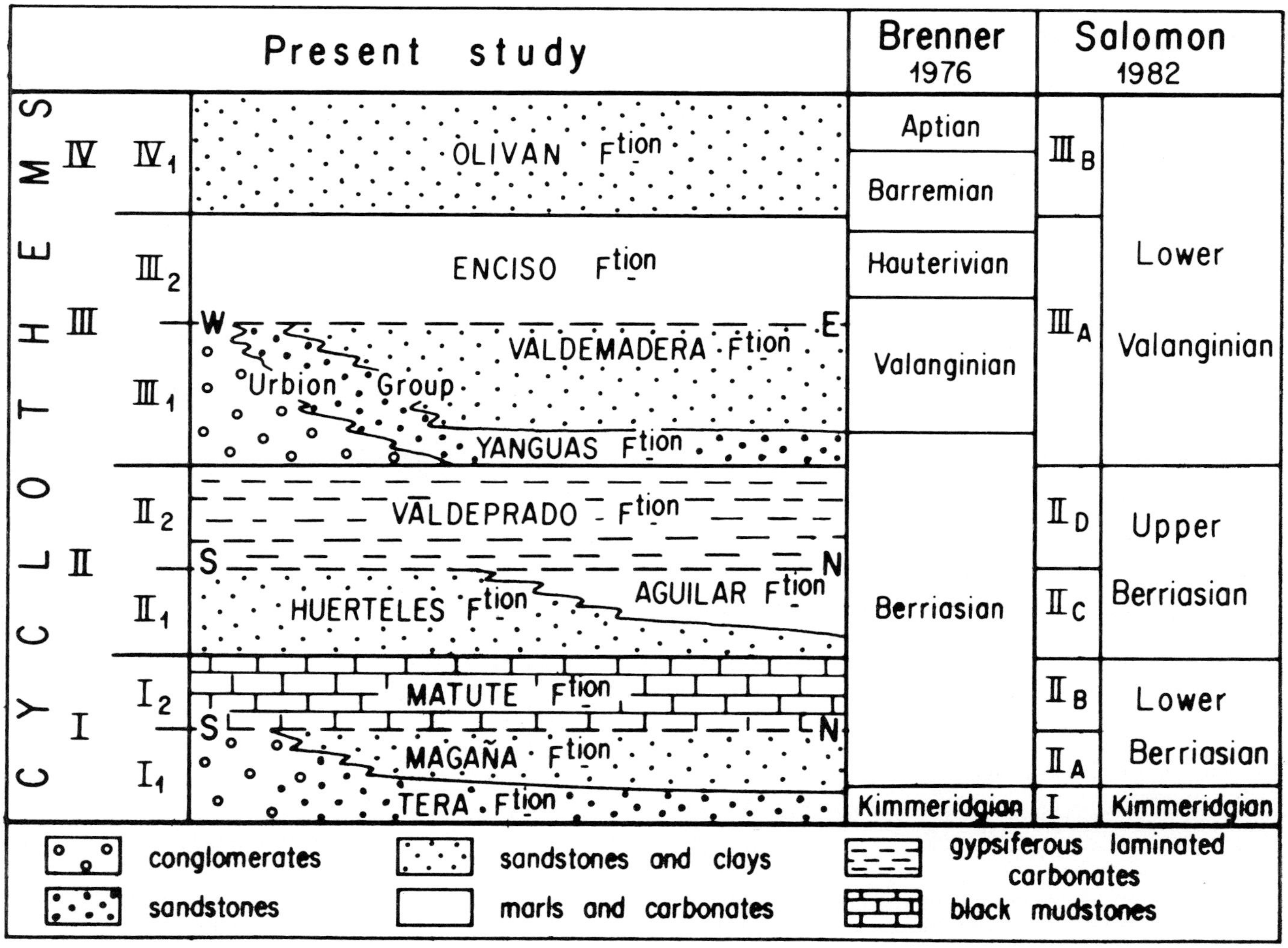

FIG. 2.—Stratigraphic subdivisions of the Upper Jurassic-Lower Cretaceous deposits of the Soria Basin.

Facies of the Urbion Group (Yanguas and Valdemadera Formations).—The Yanguas Formation is a 120-m-thick sandy body that includes 7 m of siltstones at the base. The sandstone is composed of a unique sedimentary facies characterized by curved internal erosional surfaces with lateral extents of 1 to 15 m. The Valdemadera Formation, defined by 2,200 m of medium- to very fine grained sandstones and grey or red claystones, consists of three facies. Facies A (see Fig. 3) makes up 70% to 80% of the unit. It includes grey and green claystones with abundant pyrite crystals. Thin subvertical carbonate filaments resembling root traces are present in interbedded siltstone layers. Facies B includes fine-grained sandstones and siltstones within thin, non-channelized, laterally extensive units. The typical vertical succession of internal sedimentary structures within a single bed includes planar laminae and current ripples commonly grading up into symmetrical ripples. Facies C consists of medium-grained, cross-bedded sandstone bodies, 2 to 6 m thick and 100 to 200 m long, with fining-upward sequences and evidence of lateral accretion. Claystone plugs are commonly observed capping the uppermost accretionary sandstone bed.

The sedimentary facies of the Yanguas Formation is similar to facies A2 of the Barcena Formation, described by Pujalte (1981), in the Upper Jurassic to Lower Cretaceous rocks of the northern Santander area. It can be related to deposits of braided fluvial channels. The associations of structures within the Valdemadera Formation, the lack of marine fossils, and the indications of subaerial conditions suggest a fluviatile environment. Facies A represents flood-plain deposits, facies B corresponds to overbank deposits, and facies C represents point-bar deposits of meandering channels. The predominance of flood-plain deposits and low dips of crossbedding related to lateral accretion suggests high-sinuosity rivers with channel depths ranging from 2 to 6 m.

Facies of the Enciso Formation.—The Enciso Formation is 1,450 m thick and involves the cyclic alternation of two sedimentary facies, a marl-carbonate facies and a silty to fine-grained facies. The extensive marl-carbonate facies includes three sub-facies. Sub-facies E_1 (see Figure 3) is an ostracode-bearing packstone. Sub-facies E_2 consists of a dark grey-green, highly bioturbated ostracode-bearing marl. Sub-facies E_3 is a black dolomicrite, including a vertical succession from a basal dark-weathering, structureless mudstone to a light-weathering mudstone with gypsum or anhydrite pseudomorphs, flat-pebble conglomerates, and desiccation cracks, generally filled with ostracod-bearing packstone. Numerous dinosaur foot prints have been found on the top of the carbonate beds. The silty and fine-grained facies is very similar to facies B of the Valdemadera Formation, described above.

An oligo-meso saline (0 to 9%) environment has been estimated using bore dosage (G. Chennaux, personal commun., 1983) and ostracod ecology (J. Lefevre, personal commun., 1983). However, the interpretation of the Enciso Formation environment remains dubious. A very low energy, low-salinity environment including fluviatile overbank deposits favors an interdistributary bay environment,

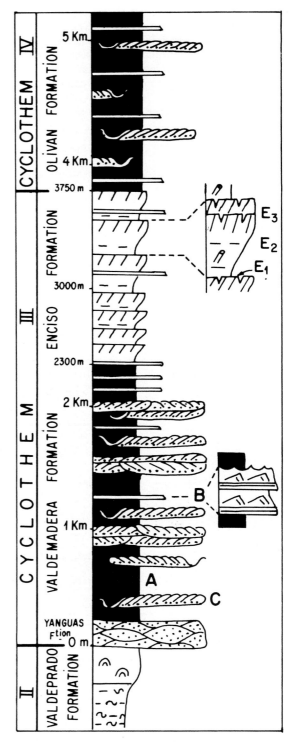

Fig. 3.—Stratigraphy of the Yanguas-Enciso section (see location, Fig. 1). Facies A, B, C, E1, E2, and E3 are described in the text.

but there is no evidence for channel-mouth bar deposits. The thickness (1,450 m) of the formation implies a long interval of constant environmental conditions with pronounced subsidence. We conclude that the Enciso Forma-

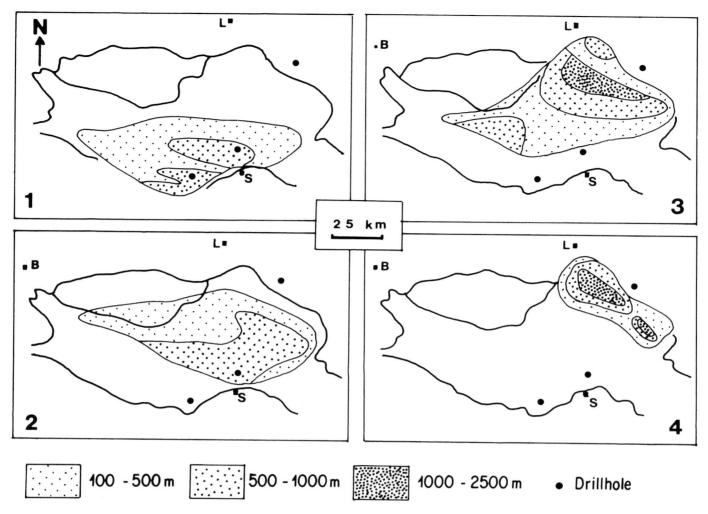

FIG. 4.—Isopach maps for cyclothems I, II, III, and IV in the Soria Basin.

tion represents the distal deposits of a fluvially dominated deltaic system deposited in restricted shallow-marine water. *Facies of the Olivan Formation.*—The Olivan Formation is much like the Valdemadera Formation and consists of repetitions of similar facies, described above as facies A, B, and C. However, the extensive facies A of the Olivan Formation consists of red silty clays with many mottled layers and root traces. Trees in living position have been found along with some dark lignitic beds (Tischer, 1966). The environment of deposition was fluvial, including floodplain, overbank and meandering-channel settings.

Depocenter Migration

The isopach maps shown in Figure 4 are drawn from measured sections and drill-hole data; Salomon's data (1982) are used for the southern area. These maps characterize the shape of the basin during four stages of deposition and the progressive northwards migration of depocenter through time. The basin is triangular in plan view and asymmetric in a roughly N50°E cross-sectional view. The thickness of cyclothems I and II is greatest in the southeastern corner of the basin (Salomon, 1982). No sediments of this age remain in the northwestern part. During the deposition of cyclothems III and IV, the thickest stratigraphic section was deposited in the northeastern corner of the basin. No deposits of the uppermost part of cyclothem III (Enciso Formation) or cyclothem IV (Olivan Formation) remain in the southern part of the basin.

The total stratigraphic thickness of the Soria Basin cyclothems is 8 km and is related to depocenter migration with time. This value is probably greater than the true basin depth, as previously shown by Steel and Gloppen (1980) for the Hornelen Basin.

STRUCTURAL GEOLOGY OF THE BASIN

Geological mapping at scales of 1:30,000 and 1:50,000 and microtectonic analysis have proven essential. On the basis of this work, we here document the association of

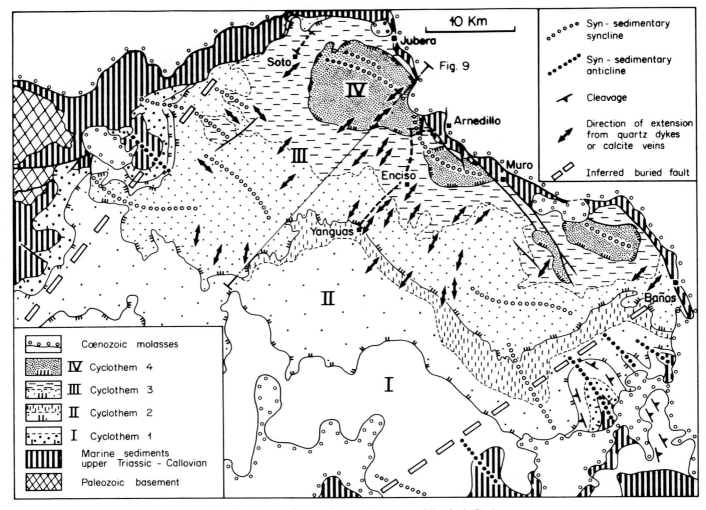

FIG. 5.—Structural map of the northern part of the Soria Basin.

extensional syn-sedimentary structures within the basin and compressional syn-depositional structures along the basin's southeastern margin.

A Syn-sedimentary Syncline

Upper Jurassic to Lower Cretaceous strata of the northern Soria Basin are folded in a 20 km-wide asymmetric syncline that trends N45°W (Fig. 5). The southern flank of the syncline dips at 20–30° to the north, and the northern flank dips at 60–80° to the south. Bed by bed mapping and cross-section correlations show a general thinning of the deposits from south to north. This thinning is well documented by comparing the total thickness of cyclothems III and IV along the southern flank (5,200 m in the Yanguas-Enciso section) to the total thickness along the northern flank (700 m to 1,000 m). Some divergent onlaps and intraformational angular unconformities (which die out laterally into the bedding) occur on the northern flank of the syncline. A detailed analysis of depocenter migration during the deposition of

cyclothems III and IV reveals the superposition of the depocenter with the syn-sedimentary synclinal axis and the progressive northward migration of both the depocenter and the syncline.

Extensional Deformation Along the Northern Margin and Within the Soria Basin

Along the northern margin of the basin (near Jubera, Muro de Aguas, and Baños de Fitero; Fig. 5), Lower Cretaceous rocks of the Urbion Group unconformably overlie marine Triassic to Oxfordian strata preserved in grabens of Callovian to Oxfordian age (Fig. 6). Relief on the contact is locally as great as 30 m. The Urbion Group conglomerates and sandstones of the Jubera section are themselves deformed by syn-sedimentary normal faults. A significant reduction in the thickness of the clastic sediments is evident on the horsts, but the detailed geometry of the Early Cretaceous horsts and grabens could not be precisely defined.

Within the Soria Basin (Soto, Enciso, and Cornago areas),

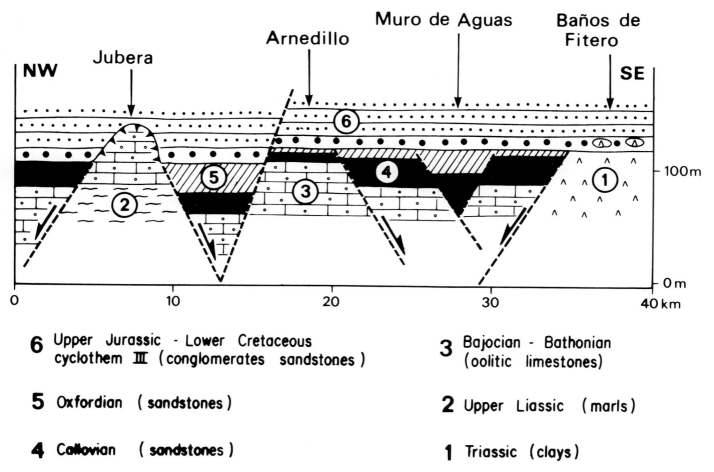

FIG. 6.—Relationships between continental Upper Jurassic to Lower Cretaceous sediments and marine Jurassic deposits along the northeastern border of the Soria Basin.

some of the north-dipping carbonate layers of the Enciso Formation are characterized by a set of tilted, extensional, brittle microstructures that clearly predate Cenozoic compressive microstructures. The N40°W- to N45°W-trending tension gashes are invariably perpendicular to stratification (Fig. 7) and are commonly arranged en echelon to potential normal faults. Fibrous calcite crystals, orthogonal to the tension-gash walls, define a N50°E extension direction. Stylolitic joints occur along bedding planes with stylolitic columns normal to bedding. The increase in density and length of columns near the tension gashes demonstrate the genetic relationship between the two microstructures.

The syn-depositional character of the syncline suggests that the tilting of the strata is also syn-sedimentary. Consequently, the extensional microstructures indicate an extensional phase of Early Cretaceous age. The calcite tension gashes define a basin-wide extension direction close to N50°E (see Fig. 5). This result is consistent with the observation of N45°W-striking syn-sedimentary normal faults in the Soto, Jubera, and Enciso sections.

Sandstone layers of the clastic Valdemadera and Olivan Formations contain 0.1- to 0.6-m-wide dikes infilled by automorphic and fibrous quartz (chlorite and siderite occur

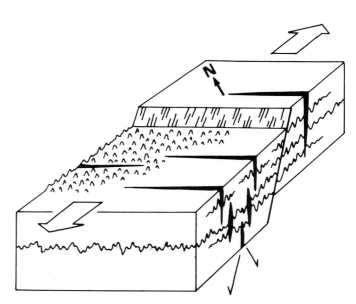

FIG. 7.—Structures formed during the Cretaceous extensional phase in carbonate rocks of the Enciso Formation. These are N45°W-striking normal faults, tension gashes, and stylolites (see the text for explanation).

locally on the walls). Quartz dikes are close to normal to the tilted bedding and generally strike in the same direction as the stratification. On a basin scale (see Fig. 5), they suggest an extension direction close to N40°–N50°E in agreement with that suggested by the calcite tension gashes.

Our interpretation of the syn-sedimentary syncline is as follows. To the north, the Soria Basin is bounded by a N50°W-striking thrust fault. We assume that this thrust resulted from the reactivation of an Early Cretaceous normal fault generated by northeast-southwest extension. We speculate that the Soria syncline developed over a half-graben within the basement (Fig. 8). During formation of the half-graben, marine Jurassic sediments (competent layer) were detached along the gypsum and claystones of the Upper Triassic section causing deformation synchronous with deposition of the Upper Jurassic to Lower Cretaceous deposits. On a north-south section, the progressive northwards migration of the depocenter with the syn-sedimentary syncline axis during sedimentation of cyclothems III and IV involved progressive basement block-faulting (Fig. 9). Note that this syn-depositional displacement confers a geometry similar to sigmoidal oblique progradational sequences observed on seismic lines from other basins. However, the sedimentological data presented here and other unpublished data show that all incremental deposits (time lines) were originally horizontal. Consequently, this geometry results from syn-sedimentary deformation and not from progradation.

Compressive Deformation Along the Southeastern Basin Margin

Folds, several kilometers long, occur along the southeastern margin of the basin (Fig. 5). Fold axes trend N50°W to N30°W and commonly have significant plunges. Fold limbs dip at 20° to 60°. Aerial photogeological study and field mapping show several progressive unconformities related to a significant thinning of cyclothems I and II toward the anticlines, and thickening toward the synclines. These folds were thus initiated during Late Jurassic to Early Cretaceous time. A steeply dipping cleavage (60° to 90°) is well developed. Cleavage refraction was noted between slaty cleavage in argillaceous beds and fracture cleavage in sandstone and carbonate layers. Geometrical relationships between bedding and convergent cleavage fans show that development of the cleavage occurred during the folding episode. The vertical cleavage corresponds to axial planes of the large-scale folds. The present structural data and the

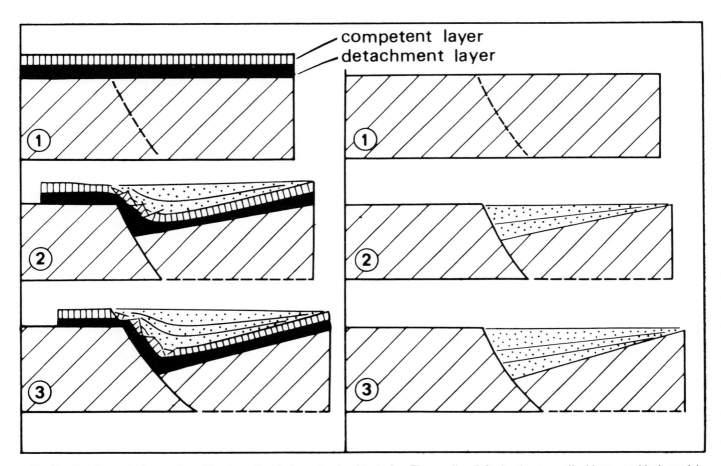

Fig. 8.—Development of a syn-depositional syncline during extensional tectonics. The syncline (left) develops on a tilted basement block overlain by a competent-incompetent layer; the half-graben basin (right) develops on a tilted block with no detachment layer (see text for explanation).

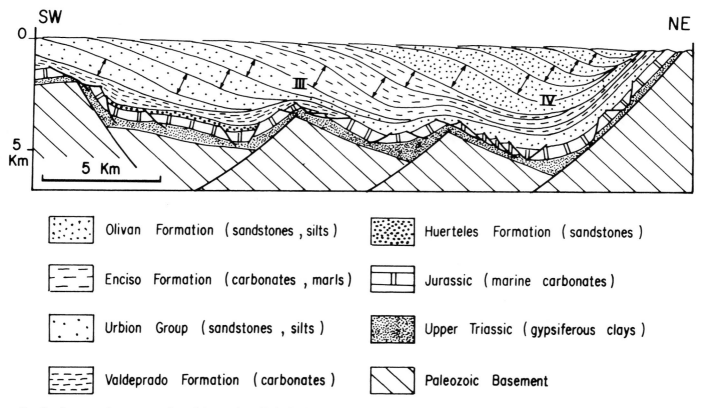

FIG. 9.—Interpretative cross section of the northern Soria Basin during Late Cretaceous time. The depocenter is shown for each Upper Jurassic-Lower Cretaceous sedimentary increment (see location on Fig. 5).

thickness reduction of basin deposits are quite consistent with the formation of a compressive southeastern margin during the genesis of the Soria Basin.

METAMORPHIC HISTORY

Petrology and Pressure/Temperature Conditions

By classical petrological and geochemical studies two isometamorphic zones may be defined (Fig. 10) in the northern part of the Soria Basin. These are an outer chlorite-pyrophyllite zone, defined by the following paragenesis: chlorite + phengite + pyrophyllite; and an inner chloritoid zone, defined by the following mineral association: phengite + chloritoid + chlorite. From thin-section study and microprobe analyses we can demonstrate three metamorphic reactions for the pelitic sequence.

Reaction 1: kaolinite + quartz $\leftrightarrows$ pyrophyllite + H_2O (Althaus, 1966; Thompson, 1970). Such a reaction is suggested by the absence of the kaolinite in the chlorite-pyrophyllite zone.

Reaction 2: chlorite + Al silicate $\leftrightarrows$ chloritoid + quartz + H_2O (Hoscheck, 1969). The occurrence of chlorite and chloritoid in the same thin section allows us to use this reaction.

Reaction 3: chloritoid + quartz $\leftrightarrows$ staurolite + almandine garnet + H_2O (Rao and Johannes, 1969). This experimental reaction was not observed in thin section. However, it can be used as upper limit of the metamorphism, knowing that chloritoid still occurs.

According to these previous works for pressures lower than 8 kb, the temperature of the described reactions R_1, R_2, and R_3 can be defined as follows:

$$T_{R1} = 350°–400° \text{ C}; \quad T_{R2} = 400°–420° \text{ C};$$
$$T_{R3} = 520°–580° \text{ C}.$$

We therefore infer that the maximum temperature of sediments in the Soria Basin during metamorphism was greater than or equal to 420° C and less than 520° C. The pressure may be settled using the stability curves of phengite in a pressure-temperature diagram established by Velde (1967). Each phengite composition can be represented by its number of Si^{4+} ions detected by microprobe analysis. The presence of phengite, alkali feldspar (orthoclase), and quartz in the thin sections studied, allows us to use the Velde method. Three phengites (A, B, C) of the chloritoid zone ($420° \leq T < 520°$) are plotted in Figure 11. The transfers of their corresponding number of Si^{4+} define a pressure on order of

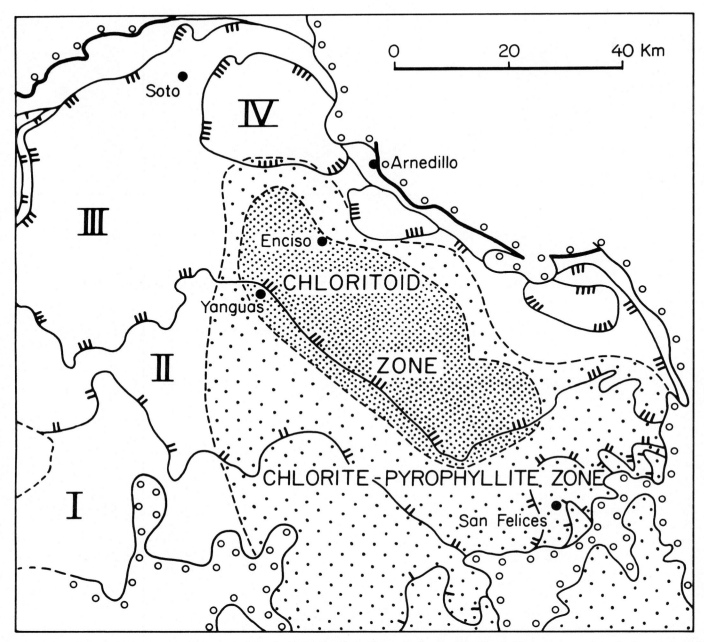

FIG. 10.—Map of the isometamorphic zones of the Soria Basin. As in Figure 5, the stratigraphic divisions correspond to the boundaries of cyclothems I, II, III, and IV.

0.5 kb to 4 kb, and a geothermal gradient on the order of 100°–150° C/km. The conditions of stability of the Soria Basin metamorphism (see Fig. 12) are defined as follows:

420° C ≤ maximum temperature < 520° C
0.5 kb < pressure < 4 kb
temperature gradient: 100°–150° C/km

Age of the Metamorphism in the Soria Basin

The isometamorphic zones are well located in the deepest part of the basin. In contrast, no metamorphism can be observed outside the basin (Sierra Demanda and Sierra Montcayo). The thin-section study of the Upper Jurassic to Lower Cretaceous metasediments sampled along the southeastern margin of the basin shows the relation between crystallization and deformation. The metamorphic phengites and chlorites are clearly syn- to post-kinematic with respect to the slaty cleavage described above from the Late Jurassic to Early Cretaceous rocks, which indicates that the Soria Basin metamorphism could be of Late Jurassic to Early Cretaceous age. Radiometric ages are required for accurate dating of the metamorphic event and ^{39}Ar–^{40}Ar dating of muscovites is in progress.

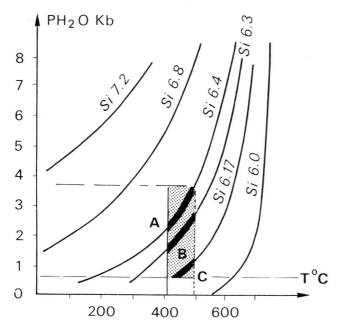

FIG. 11.—Metamorphic pressure/temperature conditions in the Soria Basin. The corresponding number of Si^{4+} ions of three phengites, A, B, and C of the chloritoid zone ($420°\,C \leq T < 520°\,C$) define pressures between 0.5 and 4 kb. Experimental data from Velde (1967).

Origin of the Metamorphism

The characteristics of the Soria Basin metamorphism (low pressure, high geothermal gradient) are similar to the pressure-temperature conditions prevailing in other modern and ancient pull-apart basins. Muffler and White (1969) describe low-grade metamorphic rocks of greenschist facies in the Salton Trough (southeastern California) where the temperature reaches 360° C at a depth of 2 km. Elders et al. (1972) show a progressive change to greenschist-facies mineralogy at a depth of only 1 km in the Imperial Valley pull-apart basin south of the Salton Sea. Temperatures in excess of 350° C exist at depths of less than 2 km and the thermal gradient here is about 100° C/km. A model of crustal thinning and associated magma generation during growth of that basin has been proposed by Fuis et al. (1984) on the basis of a seismic refraction survey. In their model, no granitic crust remains below the basin: unmetamorphosed Cenozoic sediments grade down into a "basement" composed of Cenozoic metasediments resting directly upon a "sub-basement" thought to be a mafic intrusive complex. Temperatures higher than 300° C are inferred at depths ranging from 4.8 to 7.6 km corresponding to the transition zone between unmetamorphosed sediments and metasedimentary basement.

These different data, together with the interpretation of the Salton Sea area, may be used in an interpretation of the evolution of the Soria Basin. The temperature gradient within Soria Basin is on the order of 100° C/km and the depth of the 420° C isograd may be estimated at 4–6 km. So the thermal behavior of the Soria Basin is similar to that of the Salton Trough. The Soria Basin is characterized by significant crustal extension (Fig. 9) and thinning, and it is likely

that the inferred high heat flow is related to such crustal thinning, although no intrusive or volcanic activity has been detected.

ANALOGUE MODELS FOR THE FORMATION OF THE SORIA BASIN

With regard to its rhomboidal shape, sediment thickness (a cumulative stratigraphic thickness as great as 8 km; 3 to 5 km at any given location), and high geothermal gradient, the Soria Basin presents the major characteristics of a typical pull-apart basin. Unfortunately, the original faults bordering the basin are buried or reactivated, and the strike-slip and dip-slip margins cannot be directly observed. For 20 years, pull-apart basins along strike-slip faults have received consideration from experimentalists and theoreticians, who have added pull-apart experimental or computed models to those based on geophysical and field data. Consequently, in order to integrate field data into a general scheme of basin evolution, mathematical and microtectonic

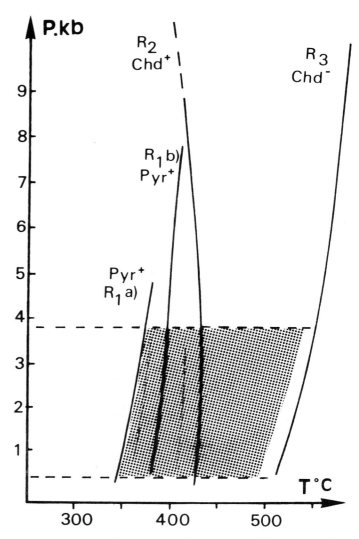

FIG. 12.—Pressure/temperature stability domain of the metamorphism in the Soria Basin. Reactions R1, R2, and R3 are detailed in the text. Pressure conditions are defined by Figure 11.

analogue models are used. The present investigation was restricted to the following observations: (1) the geometry of the basin, displaying the general structural trends of N45°W and N50°E to N60°E; (2) the location of the basin depocenter and its migration with time; (3) the N50°E direction of basin opening; and (4) the association of localized extensional and compressional structures during Late Jurassic to Early Cretaceous time. The experimental models relating to a releasing solitary overstep enabled us to interpret the field data and to present a possible scheme for the development of the Soria Basin.

The Mathematical Models

The results of three mathematical models are summarized (Rodgers, 1980; Segall and Pollard, 1980; Liu Xiaohan, 1983). These models have the common characteristics of delineating the strain or stress patterns of a releasing solitary overstep. In each case, σ_1 represents the maximum compressive stress, and σ_3 represents the minimum compressive stress. All the mathematical models are derived from the theory of elastic deformation by integrating the equations of Chinnery (1961, 1963). They are developed from a homogeneous, linear, elastic material, and thus their

application in the case of an already fractured crust and with finite displacement along the faults remains dubious. In addition to the theoretical models, we briefly present results from microtectonic examples (Rispoli, 1981; Liu Xiaohan, 1983).

Rodgers (1980) calculated the vertical displacement of an originally horizontal surface. A basin is initiated inside a releasing overstep and uplifts develop outside the ends of the master faults (Fig. 13A). The magnitudes of uplift and depression depend on geometry of the step. As in the geological study of the Soria Basin, this theoretical result shows the synchronous association of compressional and extensional structures. A possible development of the pull-apart basin is simulated by an increase in offset along the master faults (Fig. 13B). In this way, the secondary normal faults move in a direction opposite to that of the propagating master faults.

Segall and Pollard (1980) calculated the stress configuration at a releasing overstep assuming a uniform far-field applied stress. Inside the releasing overstep, the mean stress of $1/2 (\sigma_1 + \sigma_3)$ is reduced to 0.6 times the value of the applied stress field. The minimum compressive stress, σ_3, is deflected and tends to be parallel to the strike-slip fault. Outside the step near the tips of the strike-slip faults, the

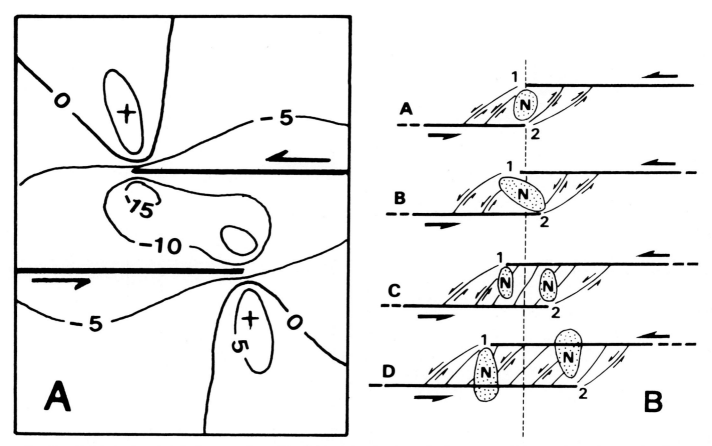

FIG. 13.—A) Calculated vertical displacement of an original horizontal surface in a releasing overstep. Values are in centimeters; the offset on each master strike-slip fault (heavy lines) is 1 m (Rodgers, 1980). B) Possible development of a pull-apart basin (Rodgers, 1980). The numbers refer to the location of the ends of the master faults; the labelled "N" zones are zones of normal faulting.

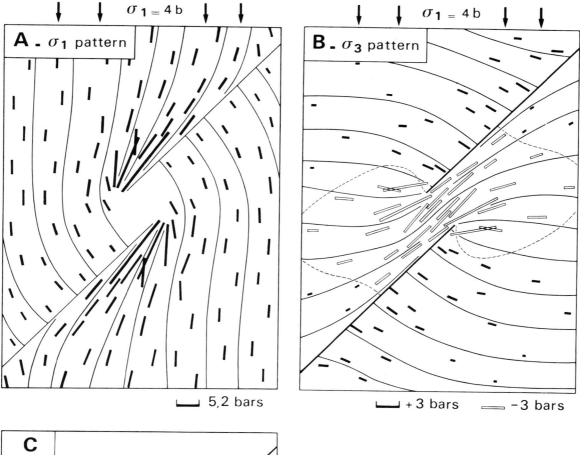

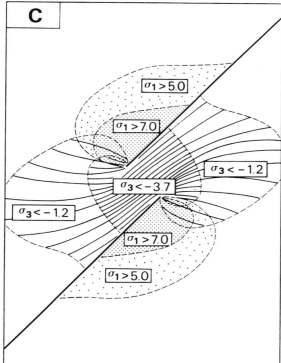

FIG. 14.—Mathematical model of stress distribution in a releasing overstep (Liu Xiahoan, 1983): A) Directions and values of maximum compressive stress, σ_1; B) Directions and positive or negative values of minimum stress, σ_3; C) Distribution of high values of the maximum principal stress (stippled area), and of large negative values of the minimum principal stress (hachured area).

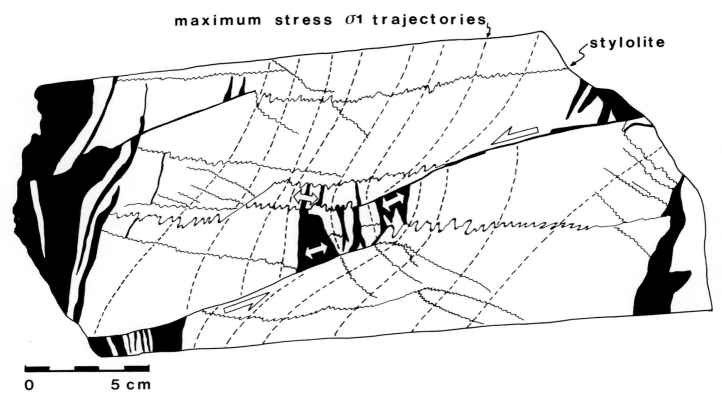

FIG. 15.—Analogue microtectonic model of strain/stress distribution in a releasing solitary overstep developed in micritic carbonate (Liu Xiahoan, 1983).

mean stress increases up to 1.4 times the value of the applied stress field.

Liu Xiaohan (1983) calculated the stress patterns for different types of fault association. To reduce calculations, the finite-element method was used. Considering a releasing overstep, the maximum stress, σ_1 (4 bars), is applied along the two horizontal boundaries in Figure 14. The two vertical boundaries in Figure 14 are free. The discontinuities strike at 45° to the boundaries and to σ_1. The values and directions of stress are calculated for each point shown in Figure 14. The stress trajectories and the areas of stress accumulation are well defined. Figure 14B displays the trajectories and values of the minimum stress (σ_3). Inside the step, σ_3 is parallel to the faults and has high negative absolute values ($\sigma_3 < -3.7$ bars; Fig. 14C). Therefore, a significant induced tensional stress characterizes this zone. Outside the ends of the faults, the maximum stress, σ_1 (Fig. 14A), is deflected and tends to be parallel to the strike-slip faults. It increases to a value of almost twice the far-field values ($\sigma_1 > 7$ bars, Fig. 14C).

The Microtectonic Models

The analysis of natural deformation observed at metric, centimetric, and microscopic scales provides information

FIG. 16.—Releasing solitary overstep basin model: a, area of induced tensile stress (basin). The density of stipping approximates the magnitude of subsidence. The secondary normal faults are perpendicular to their branching master faults, and are curved. b, area of concentration of compressive stress (shortening, uplift, source of sediment).

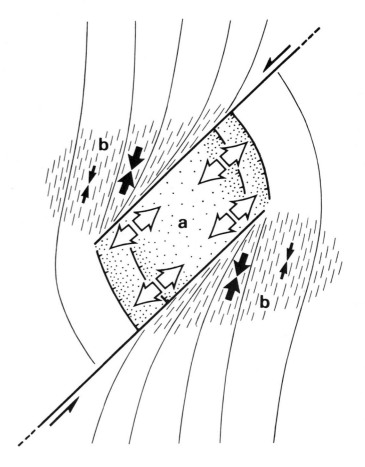

on strain patterns in a solitary overstep between strike-slip microfaults.

Ripoli (1981) describes the spatial distribution of compressional and extensional microstructures associated with a strike-slip microfault. The high concentration of stylolites identified on one side of the fault tip determines the concentration of maximum stress, σ_1. The direction of stylolitic peaks indicates the direction of σ_1, which tends to be parallel to the strike-slip microfault. At the opposite side of the fault tip, the presence of tension gashes reveals a concentration of minimum stress, σ_3.

From sections of hand specimens, Liu Xiaohan (1983) studied a releasing solitary overstep between two strike-slip microfaults (Fig. 15). The fault tips are localized where microfaults stop being linear and extend into a stylolitic plane; however, the motion along a microfault is partially absorbed in stylolitic dissolution at the end of the microfault, and the fault tips are not closely defined. The deviation and

concentration of maximum stress, σ_1, are characterized by the direction and length of stylolitic peaks. Outside the step and fault tip, stylolitic peaks determine a concentration and deviation of the maximum stress, σ_1. Inside the step the curved tension gashes open normal to their branching strike-slip fault at some distance from the tip, with a maximum aperture against the fault. The arrangement of tension gashes is comparable to the disposition of normal faults characterized by Rodger's mathematical model (Fig. 13B).

Model of a Releasing Solitary Overstep

Although several reservations about their applications can be made, the mathematical and microtectonic models supply corresponding information on strain-stress patterns and secondary fault geometry in a releasing solitary overstep (Fig. 16). Outside the step and near the tips of the strike-slip faults, the maximum stress, σ_1, tends to be parallel to

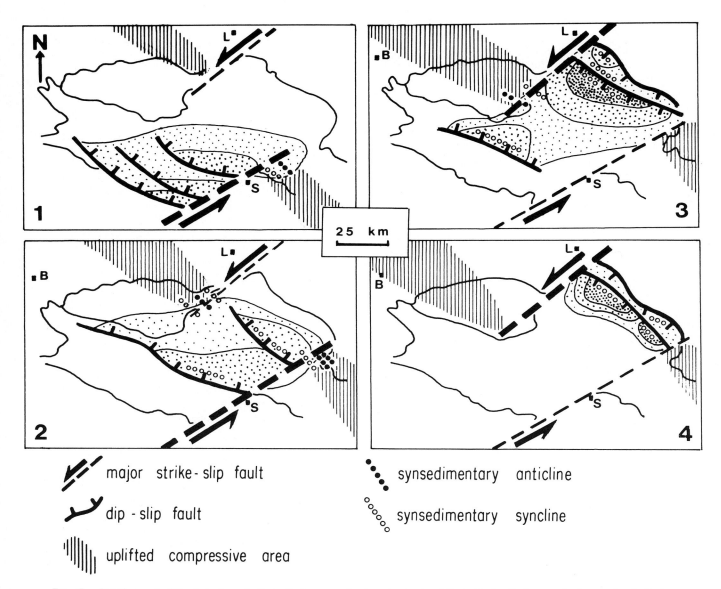

FIG. 17.—Development of the Late Jurassic-Early Cretaceous releasing step basin of Soria Basin. B, Burgos; L, Longroño; S, Soria.

the major faults, and it significantly increases in magnitude. Inside the step, σ_1 tends to be normal to the strike-slip faults and reaches low values; the minimum stress, σ_3, is parallel to the major faults and approaches a negative value. Although the stress conditions are compressive at the boundaries, the step area represents a zone of induced tensile stress. The curved tensile fractures (tension gashes and normal faults) are approximately normal to the corresponding branching strike-slip faults. As the propagating strike-slip faults develop, two zones of maximal aperture (depocenter, tension gashes) form inside the step (see Fig. 16). When the tips of the strike-slip faults migrate, the tensile fracturing migrates in the opposite sense, and so the basin lengthens.

Correspondence Between Field and Model Data

The main structural data collected in the Soria Basin are consistent with a basin that opened in a releasing solitary overstep between two sinistral N50°E- to N60°E-striking strike-slip faults. Inside the basin, the calcite and quartz tension gashes opened in a N50°E direction (Figs. 8, 9) and characterize the Late Jurassic to Early Cretaceous extensional phase as identified within the basin. From these data, the N50°E- to N60°E-trending margins can be defined as strike-slip margins.

The folds described above along the southeastern margin could be interpreted as compressional structures localized at the outside end of the tip of a major, buried N60°E-striking strike-slip fault. Moreover, the direction of the N30°W-trending syn-depositional folds of this zone is in agreement with the deviaiton of the maximum compressive stress anticipated by the model if the N50°E- to N60°E-trending strike-slip margin had sinistral motion only.

EVOLUTION OF THE SORIA BASIN

A model for the tectonic-sedimentary evolution of the Soria Basin can now be proposed (Fig. 17). According to the relationship between the syn-depositional syncline defined within the basin and the formation of the basement half-graben, the axes of the different syn-sedimentary synclines are used to locate basement-involved normal faults. During the sedimentation of the cyclothem I (Fig. 17), the southeastern margin of the basin acted as a sinistral strike-slip fault. Some N45°W structural trends operated as north-facing normal faults, thus localizing the depocenter. According to the location of the coarsest deposits of the Tera and Magana Formations (Fig. 2), the main source of detrital material was located along the southeastern strike-slip margin. During deposition of cyclothem II (Fig. 17), the depocenter tended to migrate northwards. Figure 17 shows the displacement of syn-depositional compressive folds that can probably be associated with the northward migration of the active N50°E- to N60°E-striking strike-slip fault tip. During sedimentation of cyclothem III (Fig. 17), the depocenter was localized in the northeastern part of the basin, related to the activation of N45°W-striking south-facing normal faults. The alluvial-fan conglomerates and braided-stream coarse-grained sandstones deposited along the northwestern basin margin may argue for pronounced uplift

of the Sierra Demanda (Fig. 1) that became a source of clastic material. These observations are consistent with the activation of the northwestern strike-slip margin. Cyclothem IV deposits are located in the northeastern corner of the basin. The lack of outcrop of proximal facies may indicate that the margins of the depositional basin were farther north.

During genesis of the Soria Basin, the lateral motion along the two major strike-slip faults was asymmetric. The southeastern strike-slip fault was first the most active (Fig. 17), at variance with the later activity of the northwestern strike-slip fault (Fig. 17). The total width of the Soria Basin results from this evolution.

CONCLUSIONS

The main characteristics of the Soria Basin are more adequately interpreted as a releasing solitary overstep model rather than other pull-apart or strike-slip basin models. The model developed in this contribution differs from the classic pull-apart model (Crowell, 1974a) in some interesting ways. The genesis of a releasing solitary overstep is characterized by the temporal association of extensional tectonics inside the step and compressional tectonics outside the step. Inside the step, the associated normal faults are secondary faults. They are normal to their branching strike-slip faults and are curved. Associated with the forward migration of each tip of the major strike-slip faults, the normal faults are successively initiated in an opposite sense. Consequently, the basin length increases. The migration of the depocenter with time limits the deposition of youngest sediments near the dip-slip and strike-slip margins of the basin and not in the center of the basin as in Crowell's pull-apart model (Crowell, 1974a, Fig. 8). Outside the basin, the two compressive areas (Sierra Demanda, Sierra Montcayo; Fig. 1) could form the highest topographic zones and generate the coarsest material supplied to the basin. The model predicts the location of uplifted source areas. This information could be useful in reservoir prediction using models of sediment dispersal related to the pattern of drainage (Ballance, 1980).

ACKNOWLEDGMENTS

This work was made possible by financial support from Elf Aquitaine. We are grateful for the useful reviews and critical comments made by K. T. Biddle, N. Christie-Blick, and two anonymous reviewers. We also thank Liu Xiaohan for allowing us to use his unpublished data in figures 14 and 15.

REFERENCES

ALTHAUS, E., 1966, Die bildung von Pyrophyllit und Andalousit zwischen 2000 und 7000 bars H$_2$O-Druck: Naturwissenschafsen, v. 53, p. 105–106.
ARTHAUD, F., AND MATTE, P., 1977, Late Paleozoic strike-slip faulting in southern Europe and northern Africa: Result of a right lateral shear zone between the Appalachians and the Urals: Geological Society of America Bulletin, v. 88, p. 1305–1320.
AYDIN, A., AND NUR, A., 1982, Evolution of pull-apart basins and their scale independence: Tectonics, v. 1, p. 91–105.

BALLANCE, P. F., 1980, Models of sediment distribution in non-marine and shallow marine environments, *in* Ballance, P. F. and Reading, H. G., eds., Sedimentation in Oblique-Slip Mobile Zones: International Association of Sedimentologists Special Publication No. 4, p. 229–236.

BEUTLER, A., 1966, Geologische untersuchungen in Wealden and Utrillas. Schichten im Westteil der Sierra de Los Cameros (Nordwestlich Iberisch Kettin): Geologisches Jahrbuch Beihefte, v. 44, p. 103–121.

BRENNER, P., 1976, Ostracoden und Charophyten des Spacrischen Wealdien: Paleontographica Abteilung A, v. 152, p. 113–201.

BURKE, K., MANN, P., AND KIDD, W., 1982, What is a ramp-valley?: 11th International Congress on Sedimentology, Hamilton, Ontario, Canada, International Assocication of Sedimentologists, Abstracts of Papers, p. 40.

CHINNERY, M. A., 1961, The deformation of the ground around surface faults: Seismological Society of America Bulletin, v. 51, p. 355–372.

––––––., 1963, The stress changes that accompany strike-slip faulting: Seismological Society of America Bulletin, v. 53, p. 921–932.

CROWELL, J. C., 1974a, Origin of Late Cenozoic basins in southern California, *in* Dickinson, W. R., ed., Tectonics and Sedimentation: Society of Economic Paleontologists and Mineralogists Special Publication No. 22, p. 190–204.

––––––., 1974b, Sedimentation along the San Andreas Fault, California, *in* Dott, R. H., Jr., and Shaver, R. H., eds., Modern and Ancient Geosynclinal Sedimentation: Society of Economic Paleontologists and Mineralogists Special Publication No. 19, p. 292–303.

––––––., 1976, Implications of crustal stretching and shortening of coastal Ventura basin, California, *in* Howell, D. G., ed., Aspects of the Geologic History of the California Continental Borderland: American Association of Petroleum Geologists, Pacific Section, Miscellaneous Publication 24, p. 365–382.

ELDERS, W. A., REX, R. W., MEIDAV, T., ROBINSON, P. T., AND BIEHLER, S., 1972, Crustal spreading in southern California: Science, v. 178, p. 15–24.

FUIS, G. S., MOONEY, W. D., HEALY, J. H., MCMECHAN, G. A., AND LUTTER, W. J., 1984, A seismic refraction survey of the Imperial Valley region, California: Journal of Geophysical Research, v. 89, p. 1165–1189.

GUIRAUD, M., 1983, Evolution tectono-sédimentaire du bassin Wealdien (Crétacé inferieur) en relais de décrochements de Longroño-Soria (NW Espagne) [unpubl. 3 eme cycle Thesis]: Montpellier, Université des Sciences et Techniques du Languedoc, 183 p.

HOSCHECK, G., 1969, The stability of staurolite and chloritoid and their significance in metamorphism of pelitic rocks: Contributions to Mineralogy and Petrology, v. 22, p. 208–232.

LIU XIAOHAN, 1983, Perturbations de contraintes liées aux structures cassantes dans les calcaires fins du Languedoc. Observations et simulations mathématiques [unpubl. 3 eme cycle thesis]: Montpellier, Université des Sciences et Techniques du Languedoc, 130 p.

MANN, P., HEMPTON, M. R., BRADLEY, D. C., AND BURKE, K., 1983,

Development of pull-apart basins: Journal of Geology, v. 91, p. 529–554.

MUFFLER, L. J. P., AND WHITE, D. E., 1969, Active metamorphism of Upper Cenozoic sediments in the Salton Sea geothermal field and Salton Trough, southeastern California: Geological Society of America Bulletin, v. 80, p. 157–181.

PUJALTE, V., 1981, Sedimentary succession and paleoenvironments within a fault-controlled basin: the "Wealden" of the Santander area, northern Spain: Sedimentary Geology, v. 28, p. 293–325.

RAO, B. B., AND JOHANNES, W., 1969, Further data on the stability of staurolite + quartz and related assemblages: Neus Jahrbuch Mineralogie, v. 10, p. 437–447.

READING, H. G., 1980, Characteristics and recognition of strike-slip fault systems, *in* Ballance, P. F., and Reading, H. G., eds., Sedimentation in Oblique-Slip Mobile Zones: International Association of Sedimentologists Special Publication No. 4, p. 7–26.

RODGERS, D. A., 1980, Analysis of pull-apart basin development produced by en echelon strike-slip faults, *in* Ballance, P. F., and Reading, H. G., eds., Sedimentation in Oblique-Slip Mobile Zones: International Association of Sedimentologists Special Publication No. 4, p. 27–41.

RISPOLI, R., 1981, Stress fields about strike-slip faults inferred from stylolites and tension gashes: Tectonophysics, v. 75, p. T29–T36.

SAENZ GARCIA, G., 1932, Notas para el estudio de la facies Wealdica Espanola: Association Española para el progresso de las ciencias, p. 59–76.

SALOMON, J., 1982, Les formations continentales du Jurassique supérieur-Crétacé inférieur en Espagne du Nord (Chaîne Cantabrique et NW ibérique): Mémoire Geologique Université Dijon, N.6, 228 p.

SALOMON, J., 1983, Les phases "fossé" dans l'histoire du bassin de Soria (Espagne du Nord) au Jurassique supérieur-Crétacé inférieur: Bulletin Centres Recherche Exploration Production Elf-Aquitaine, v. 7, p. 399–407.

SANCHEZ LOZANO, R., AND PALACIOS, P., 1885, La formacion Wealdense en las provincias de Soria y Logroño: Bolotin Comision Mapa Geologica España, v. 12, p. 109–140.

SEGALL, P. AND POLLARD, D. O., 1980, Mechanics of discontinuous faults: Journal of Geophysical Research, v. 85, p. 4337–4350.

STEEL, R., AND GLOPPEN, T. G., 1980, Late Caledonian (Devonian) basin formation, western Norway: signs of strike-slip tectonics during infilling, *in* Ballance, P. F., and Reading, H. G., eds., Sedimentation in Oblique-Slip Mobile Zones: International Association of Sedimentologists Special Publication No. 4, p. 79–103.

THOMPSON, A. B., 1970, A note on the kaolinite-pyrophyllite equilibrium: American Journal of Science, v. 268, p. 454–458.

TISCHER, G., 1966, Über die Wealden-Ablagerung and die Tektonik der östlichen de los Cameros in den nordwestlichen Iberischen Ketten (Spanien): Geologisches Jahrbuch Biehefte, v. 44, p. 123–164.

VELDE, B., 1967, Si^{4+} contents of natural phengites: Contributions to Mineralogy and Petrology v. 14, p. 250–258.

THE PIAUI BASIN: RIFTING AND WRENCHING IN AN EQUATORIAL ATLANTIC TRANSFORM BASIN

PEDRO V. ZALAN

PETROBRAS/Depex, Av. Chile 65, Rio de Janeiro, RJ 20035, Brazil;

ERIC P. NELSON AND JOHN E. WARME

Department of Geology, Colorado School of Mines, Golden, Colorado 80401;

AND

THOMAS L. DAVIS

Department of Geophysics, Colorado School of Mines, Golden, Colorado 80401

ABSTRACT: The east-trending Piaui Basin off the northern Brazilian continental margin is an Atlantic-type rifted basin. The stratigraphic and structural framework of the basin is interpreted as recording wrenching during separation of South America and Africa along the equatorial Romanche Fracture Zone. Superposition of rifting and wrenching resulted in a diverse set of structures that are uncommon in Atlantic-type basins.

The basin was initiated during Aptian time. The structural grain of the adjacent Precambrian Parnaiba Platform probably influenced the orientations of normal faults which strike between northeast and east-northeast. Dominantly non-marine siliciclastic sediments were deposited during the rift stage, prior to the development of a mid-Aptian (112 Ma) regional unconformity. Oblique, northeast-southwest separation of South America and Africa between mid-Aptian and early Cenomanian time, was accompanied by the deposition of thick transitional marine and marine clastic sediments.

During the middle Cenomanian, the direction of sea-floor spreading between South America and Africa changed to east-west, leading to convergent wrenching between asperities of the two continents along the Romanche Fracture Zone. Rift faults are thought to have been reactivated as oblique- and strike-slip faults; other faults are interpreted as synthetic (N70° to N75°W) and antithetic (north to N20°E) strike-slip faults. Associated structures include flower structures, en echelon folds, and shale ridges (N20°E). Wrenching created a 200 km by 50 km uplifted transpressive belt (Atlantic High) in the Piaui Basin, where erosion occurred from the Late Cretaceous to the Eocene. Oligocene-Miocene shallow-marine sediments cover the Cretaceous rocks unconformably over most of the basin.

INTRODUCTION

Most Gondwanaland reconstructions show a tight fit of Africa and South America where the Romanche Fracture Zone projects into northeastern Brazil between the Barreirinhas and Piaui-Ceará Basins (Fig. 1; Rabinowitz and LaBrecque, 1979; Martin, et al., 1981; Pindell, 1985). Because this segment of the Brazilian margin was parallel to interplate flow lines during opening of the equatorial Atlantic (Fig. 2), schemes of early opening involve some degree of wrenching between Africa and South America along this segment (and possibly along the western extension of the St. Paul Fracture Zone; see Fig. 4 below; and Pindell, 1985, Fig. 7). In addition, reconstructions call on earlier and more rapid opening of the south Atlantic relative to the equatorial Atlantic (Fig. 2).

The variation in the history of opening and subsequent tectonism along the Brazilian margin is recorded in the stratigraphy and structure of the basins along the margin (see Ojeda, 1982, Fig. 2; Asmus and Baisch, 1983, Fig. 1). Along the southeastern margin of Brazil, the post-breakup sedimentary section is generally thicker and there is less structural relief on basement fault blocks than along the northeastern margin of Brazil. Whereas the stratigraphy and structure in basins off southeastern Brazil are quite similar, in basins off northeastern Brazil, both stratigraphy and structure vary from one basin to another. In addition, tectonic structures (other than those of extensional origin) and related angular unconformities have been found only in the basins along the northeastern margin (Barreirinhas and Piaui-Ceará Basins; Ojeda, 1982; Zalan, 1983, 1984). These features, together with the kinematics of continental break-up and the opening of the equatorial and south Atlantic (Fig. 2), indicate that the northeastern margin of Brazil, particularly the east-west segment, experienced a tectonic history fundamentally different from that of the southeastern margin. This history involved development of a transform margin. According to Scrutton (1982), such margins are characterized by a narrow transition between continental and oceanic crust, limited thinning of the continental crust, and a steep continental slope. In detail, however, the basins of northeastern Brazil are quite variable: some were structurally inverted during wrenching; others remained on structurally low blocks and accumulated thicker sections. Thus northeastern Brazil has characteristics of both passive margins and transform margins similar to the southern California borderland. Here we present structural and stratigraphic data from the Piaui Basin that support an interpretation involving wrenching.

The east-trending Piaui Basin is situated offshore northeastern Brazil (Fig. 1). Although it may be considered a subdivision of the Ceará Basin (Piaui-Camocim sub-basin) to the east, we regard it as a separate basin because of its large size (approximately 12,000 km²) and its distinctive geology. Sedimentation in the Piaui Basin began in the Cretaceous, and its origin and development are related to the opening of the equatorial Atlantic Ocean. Preliminary studies of seismic sections (Miura and Barbosa, 1972) revealed the presence of folds and reverse faults in the Piaui Basin, as well as in the eastern part of the adjacent Barreirinhas Basin to the west. Such structures are unusual in Atlantic-type margins, and are described further in this paper.

DATA BASE AND METHODS

Most of the available information on the Piaui Basin is geophysical. It includes 80 migrated seismic lines, representing 3,800 km of section with an average spacing of

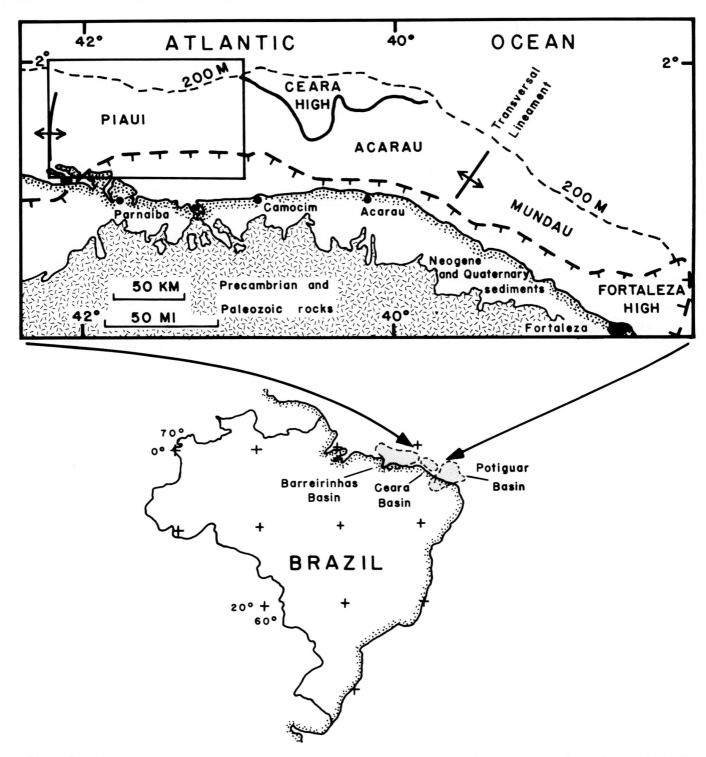

FIG. 1.—Location map, showing sub-basins (Piaui, Acarau, and Mundau) and main structural elements in Ceará (or Piaui-Ceará) Basin. Box outlines area of Figures 13, 14.

about 5 km, and one aeromagnetic survey. A Landsat image covering the adjacent onshore geology was used to interpret the structural fabric of the Piaui Basin. Data from three wells in the basin and from two wells on the basin margin were also available, but we had access to only four cores.

Standard methods were used in this study to analyze the seismic sections, drill cuttings, and down-hole log re-

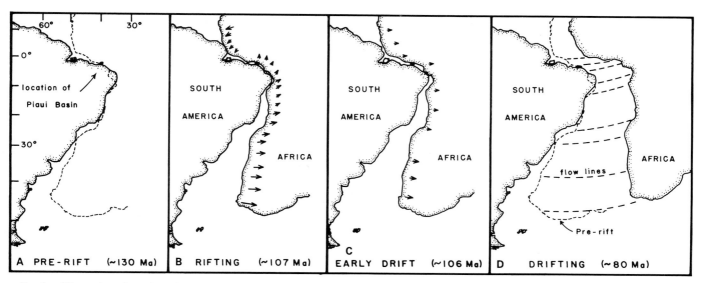

FIG. 2.—Kinematics of opening of South Atlantic (after Rabinowitz and LaBrecque, 1979). Arrows show motion of Africa with respect to South America. From left to right: A) Pre-rift configuration (note apparent overlap at site of present Niger delta). B) Rift phase (rotation pole at 2.5° S, 45.0° W; note compression in far north, and extension near Piaui Basin and in far south). C) Early drift or wrench stage (note wrenching near Piaui Basin). D) Late drift or passive margin phase (shown only up to 80 Ma).

sponses (see Fig. 7 below). Biostratigraphic and paleoenvironmental data were provided from the analysis of cuttings by paleontologists at PETROBRAS (the Brazilian national oil company). Cuttings were obtained for every 3 m drilled, but most paleontological data were derived from analyses of composite samples representing intervals of 15 m. Ages were determined from composite assemblages of foraminifera and palynomorphs. The pre-Albian Cretaceous stages in this region cannot be directly correlated with the international stages because of a lack of marine strata (and fauna) in this interval. Thus the Brazilian Alagoas Stage is used here and is roughly equivalent to Aptian.

Cuttings, electric logs, and the limited core material were studied in detail at the Rio de Janeiro laboratories of PETROBRAS. These data were integrated with the seismic profiles in the correlation of unconformities and major rock units, and in the interpretation of depositional environments. Comparisons with the seismic expressions of stratigraphic sequences and structures in other Brazilian Atlantic-margin basins, as interpreted by PETROBRAS, assisted this analysis.

PREVIOUS WORK

The northern Brazilian margin gained attention when Miura and Barbosa (1972) first described folds and reverse faults in the Piaui and Barreirinhas Basins. They attributed these structures to shortening produced by transcurrent movement between South America and Africa along the Romanche Fracture Zone, probably during Coniacian and Santonian time. Subsequent papers refer only briefly to these structures (Asmus and Porto, 1972; Bryan et al., 1972; Rabinowitz and LaBrecque, 1979). However, these early papers generally lack the subsurface data that have since been obtained from drilling and from closely spaced seismic profiles on the continental shelf.

Geological and geophysical data are available from the African counterpart basins along Ghana and the Ivory Coast, and from the Atlantic floor between the two continents. Kumar and Ladd (1974) suggested that a change in the pole of rotation between the South American and African plates caused changes in the configuration of oceanic fracture zones, thus forming the compressional features. Sibuet and Mascle (1978) and Wilson and Williams (1979) also adopted the concept of large-scale geometric reorganization of lithospheric plates as an explanation of some equatorial Atlantic-margin geological features. Rabinowitz and LaBrecque (1979) calculated revised poles of rotation for South America and Africa, and indicated that continental separation along the Brazilian margin near the Piaui Basin involved initial normal separation, followed by east-west separation, in turn followed by more oblique separation (Fig. 2).

Figueiredo et al. (1983) studied the structural features in the Barreirinhas Basin (Fig. 1), and concluded that they are related to dextral strike slip along the Romanche Fracture Zone during the late Aptian. The Piaui Basin contains many similar features, and our results suggest that they are also a result of wrenching. However, some high-angle faults in the Piaui Basin are interpreted to have formed during the inital rifting stage.

REGIONAL GEOLOGY

The Piaui Basin trends east-west for 150 km, and is roughly parallel to the coasts of the Brazilian states of Ceará, Piaui, and Maranhao. It is bounded on the south by shallow basement, which extends to the north far beyond the 200 m isobath. Four large structures are present within, or bound, the basin: the Parnaiba Platform, the Tutoia High, the Atlantic High, and the Ceará High (Fig. 3).

The Parnaiba Platform is a northward bulge of shallow basement that extends offshore and marks the southern edge

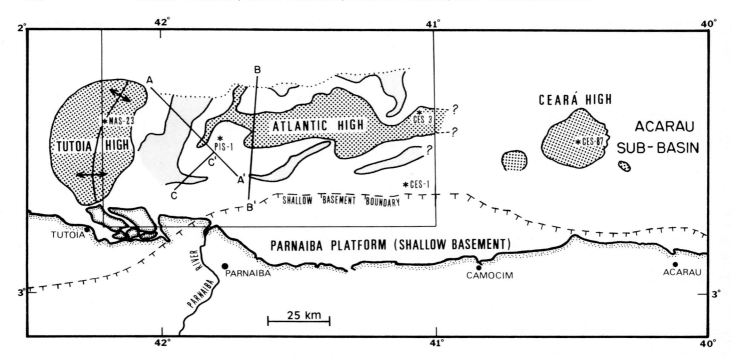

FIG. 3.—Index map of Piaui Basin showing location of study area (rectangle), major tectonic elements, well locations, and seismic profile lines (A, B, C). Large synformal depressions are shown in light pattern; antiformal highs are shown in dark pattern. The rift sequence is indistinguishable on seismic profiles north of the dotted line. Position of Tutoia High from Teixeira et al. (unpubl. studies, 1982).

of the Piaui Basin. This platform crops out on the coast as Precambrian (undifferentiated) gneisses, migmatites and granites, Proterozoic metasedimentary and metavolcanic rocks, and Paleozoic-Mesozoic sedimentary rocks of the cratonic Parnaiba Basin.

The Tutoia High is a large north-northeast-trending anticlinal structure that forms the boundary between the Piaui Basin and the Barreirinhas Basin. Based on data from one well (MAS-23), Teixeira et al. (unpubl. studies, 1982) interpreted the Tutoia High to be an inversion structure, in which a previous depocenter was uplifted as a result of transpression. The Atlantic High is a major structure that trends east-west in the north-central part of the basin, and is also thought to consist of basin sediments inverted by wrenching.

The Ceará High separates the Piaui Basin from the Acarau

sub-basin of the Ceará Basin to the east. Data presented by Miura and Barbosa (1972) and by Bryan et al. (1972) suggest that this high is composed of Precambrian crystalline rock. If so, it may have been uplifted during the wrenching phase of the continental separation. However, data from a recent well (CES-87), together with our reinterpretation of seismic data, suggest that the Ceará High may be an acidic volcanic/intrusive complex, possibly genetically related to the strike-slip tectonism.

The Piaui Basin is situated at the westward extension of the Romanche Fracture Zone (Fig. 4). This fracture zone dominates the equatorial megashear zone, which formed at about 100 to 90 Ma as a result of sea-floor spreading between South America and Africa (Bonatti and Crane, 1984). The Romanche Fracture Zone crosses the entire Atlantic Ocean from the Gulf of Guinea, where it forms the Ivory

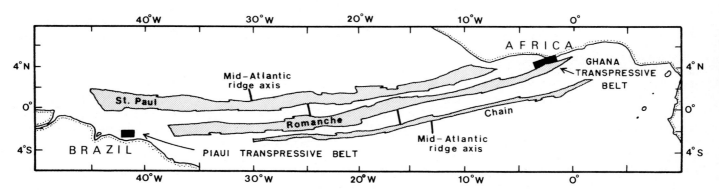

FIG. 4.—Relation of Piaui Basin to equatorial Megashear Zone (from Bonatti and Crane, 1984). Note position of Piaui Basin and Ghana transpressional belt on opposite ends of Romanche Fracture Zone.

Coast-Ghana Ridge (Delteil et al., 1974), to 37°30′ W on the continental rise off northeastern Brazil. Although the primary expression of a fracture zone is in oceanic crust, its location may have been influenced by major pre-existing zones of weakness in the continental crust (Sykes, 1978). The continuation of the Romanche Fracture Zone under the sedimentary succession of the Piaui Basin is inferred from wrench-related deformation throughout the basin. The Ceara and Atlantic Highs have previously been considered expressions of the continental extension of this oceanic fracture zone (Miura and Barbosa, 1972); our results reinforce that hypothesis.

STRATIGRAPHY

The overall stratigraphic framework of the study area (Fig. 5) was obtained by regional correlation based on seismic data. Strata in the Piaui Basin are predominantly siliciclastic and range from approximately Aptian (early Alagoas) to Quaternary. Three major tectono-stratigraphic sequences are present, separated by two regional unconformities (Figs. 5, 6). A tectono-stratigraphic sequence is defined informally here as a relatively conformable sequence corresponding to a single tectonic phase in the evolution of a basin.

The interpretation of the depositional environments represented in the wells was based on the integration of three major sources of information: paleontological data, lithology and sedimentological characteristics of cutting samples, and electrical log responses (see Fig. 7). The microfauna present in the cuttings were used to define continental, transitional, and marine environments. For example, environments inferred to be continental are characterized by the absence of marine faunas, especially foraminifera, and

by the abundance of palynomorphs and continental ostracodes (Zalan, 1983). Paleobathymetric information was also obtained from the marine faunas, using especially the characteristics of forams and ratios of planktonic to benthic types. Rock type, grain size, sorting, roundness, composition, and color were also described from the cuttings.

Rift Sequence

The rift sequence is the deepest, oldest, and probably the thickest unit present in the basin. Paleontological determinations, mainly palynological, from two wells (CES-3, MAS-23; Fig. 3) indicate an early Alagoas age for the upper part of the sequence. The great thickness of sediments present below the bottom of the wells (especially CES-3 and PIS-1; Fig. 5) suggests that older sediments (Neocomian) should be included in this sequence.

Rift-sequence rocks in the southernmost part of the basin (well CES-1) consist of conglomerates and red beds typical of faulted-margin settings (Figs. 5, 6). Along the east-west axis of the basin, this sequence is represented by intercalated mudstones and fining-upward sandstones, interpreted as meandering fluvial deposits, and by highly radioactive shales, interpreted as lacustrine deposits (wells MAS-23 and PIS-1; Fig. 7B). Basalts are present in PIS-1 (Fig. 7A). Well CES-3 encountered a thick interval of medium-grained sandstones with a blocky log response, interpreted as representing a braided fluvial system, between two intervals of intercalated shales and fine- to very fine grained fining-upward sandstones, interpreted as representing meandering fluvial systems (Fig. 5). The stratigraphic and paleogeographic development of the rift sequence is shown schematically in Figure 8.

The top of the rift sequence is a basinwide surface that

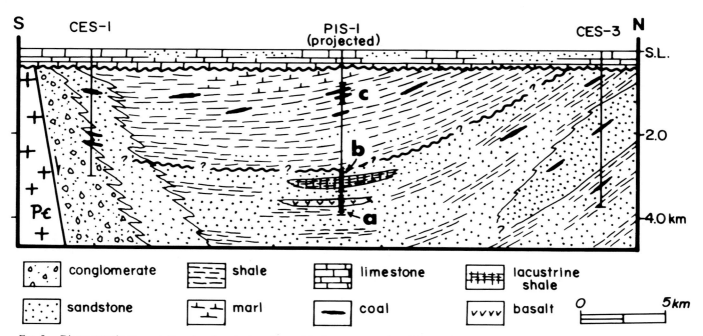

FIG. 5.—Diagrammatic cross section showing stratigraphic and facies reconstruction. The uplifted Parnaiba Platform (south) contributed sediments to alluvial fans and fan-deltas; the Atlantic High shed coarse siliciclastic sediment from the north. The center of the trough contains sediments interpreted in the PIS-1 well (Fig. 7) as representing rift (a, b), drift (c), and late-drift sequences.

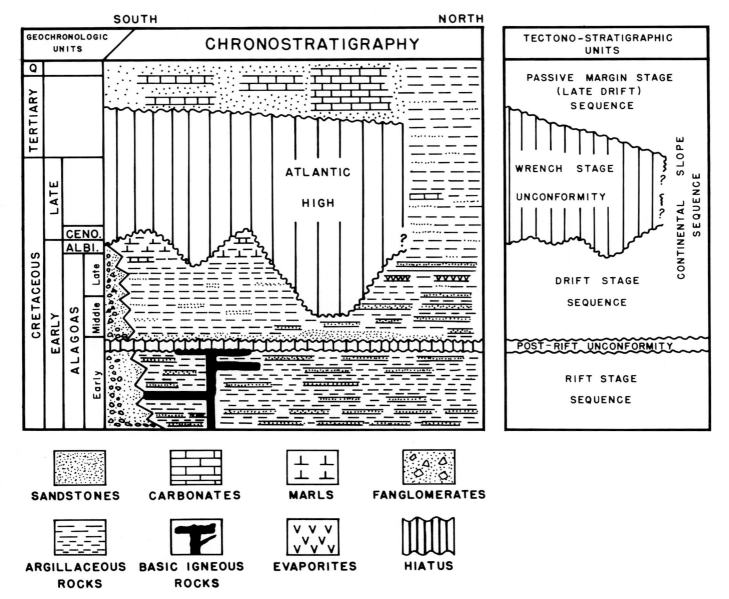

FIG. 6.—Schematic N-S chronostratigraphic section of the Piaui Basin. Geochronologic time scale is not linear. Column on right summarizes interpreted tectono-stratigraphic sequences and events.

separates high-amplitude continuous and discontinuous reflections of the rift sequence from the seismically featureless basal part of the overlying drift sequence (Figs. 9, 10, 11). This surface, interpreted as a post-rift unconformity, is overlain and underlain by sediments of the Alagoas stage and may be related to the major worldwide sea-level fall that occurred at about 112 Ma (see Vail et al., 1977).

Drift Sequence

The drift sequence is the intermediate tectono-stratigraphic sequence (Figs. 5, 6) and encompasses all strata above the post-rift unconformity and beneath the unconformably overlying lower Tertiary rocks. Paleontological data indicate an age from middle/late Alagoas to Cenomanian.

Based on the interpretation of seismic profiles, well-logs (see Fig. 7C), and cuttings from the few available wells, the lower part of the sequence probably represents deltaic, shoreline, and tidal-flat environments in the west (well PIS-1) and continental deposits in the east (CES-1; Fig. 3). The upper part of the drift sequence is entirely marine. The distribution of facies suggests that the drift sequence records the gradual invasion of marine water from the west and northwest (Fig. 8). This may be due to local basin geometry or due to diachronous continental separation.

The drift sequence is composed of an extremely thick shale section in the western half of the Piaui Basin (2,100 m in well PIS-1), and of well-stratified shelf- and shoreline-related sandstones in the eastern half (Fig. 5). The thick shale sequence is interpreted as prodeltaic/basinal deposits, equivalent to the Arpoador Formation of the Canarias Group

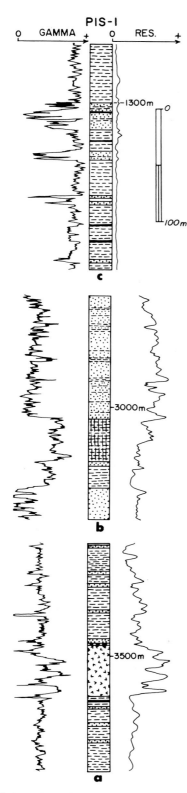

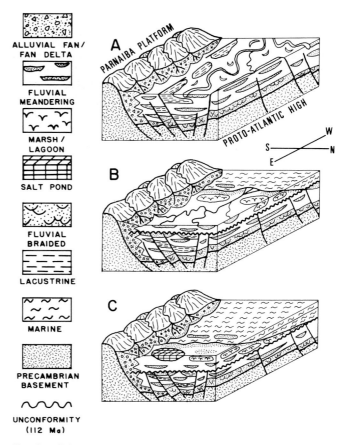

FIG. 8.—Schematic illustrations of stratigraphic and paleogeographic evolution of the Piaui Basin prior to wrenching. A) Rift stage. B) Early drift stage, after development of the proposed regional post-rift unconformity. C) Drift stage showing marine incursion from the west.

FIG. 7.—Lithology determined from cuttings, and depositional environments interpreted from paleontological analysis and log responses in well PIS-1. A) Early rift basalts and fluvial meander-belt deposits. B) Syn-rift lacustrine shales, with gas shows, analyzed by PETROBRAS as potential source rocks, and fluvial (braided?) facies. C) Drift-sequence prodelta marine and delta plain sediments. See Figure 5 for legend and position in well.

in the Barreirinhas Basin (Ojeda, 1982). This interval played an important role during the wrenching stage when several diapiric shale ridges were formed. Local salt accumulations in the upper part of the Alagoas stage are inferred from the presence of several collapse structures interpreted from the seismic data (see Fig. 14 below). Albian and Cenomanian sediments in the upper part of the drift sequence characteristically represent open-marine conditions, reflecting a eustatic rise at this time (Vail et al., 1977). These sediments consist of marls and limestones (well PIS-1) or shales (well CES-1), and are the equivalents of the Preguicas and Peria Formations, respectively, of the Ceju Group in the Barreirinhas Basin (unpublished studies of PETROBRAS, 1982; Ojeda, 1982). Rocks as old as Aptian and Albian were removed by erosion from the Atlantic High.

The drift sequence was deposited during an interval of passive subsidence between the rift stage and the wrench stage. There is little evidence in the Piaui Basin for major syn-depositional deformation such as localized internal unconformities or thickness changes.

Tertiary Sequence (Late Drift or Passive-Margin Sequence)

Within the basin the drift sequence is unconformably overlain by Oligocene and younger flat-lying sediments

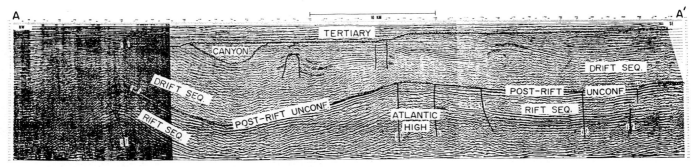

FIG. 9.—Migrated seismic section oriented northwest-southeast from the western part of the basin (see Fig. 3 and Figs. 13, 14).

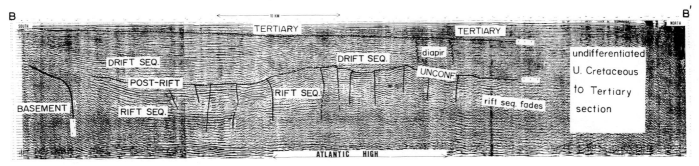

FIG. 10.—Migrated seismic section oriented north-south from the central part of the basin (see Fig. 3 and Figs. 13, 14).

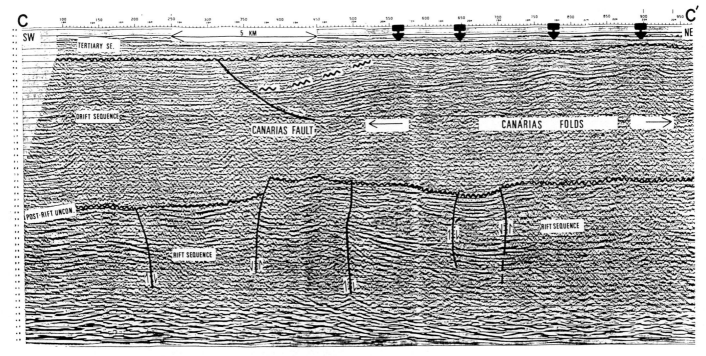

FIG. 11.—Migrated seismic section oriented southwest-northeast from the southwestern part of the basin, and passing through the Canarias folds and Canarias fault (see Fig. 3, and Figs. 13, 14 below).

composed predominantly of coarse-grained sandstones (interpreted to represent fan-deltas), interlayered with calcarenites and calcirudites (Figs. 5, 6). To the north, near the shelf break, a complete sequence of argillaceous beds represents a prograding slope sequence that includes most or all of the Upper Cretaceous and Tertiary.

STRUCTURES

Each of the tectono-stratigraphic sequences described above displays a characteristic set of structural features, most of which probably formed during a widespread wrenching stage that affected both the rift and drift sequences. However, the structures in the rift sequence are strikingly different from those in the drift sequence because each responded differently during deformation. Figures 9, 10, and 11 show examples of these contrasting structural patterns on seismic profiles. Figure 12 illustrates some individual structural features on short seismic lines. A brief discussion of the structural framework of each sequence follows.

Rift Sequence

The most characteristic structures of the rift sequence are high-angle faults (with either normal or reverse separation), and intervening major or minor folds. Figure 13 shows the distribution of high-angle faults and folds mapped at the top of the rift sequence. Flower structures appear to be common (Fig. 12A), and are locally associated with folds. On seismic sections, the rift sequence appears generally in the lower half of the section, extending downward from 1.8 to 2.9 seconds two-way travel time. The basal part is generally devoid of significant reflections, and the upper part shows strong, continuous reflections, especially in the central and southwest part of the basin area. It is in this uppermost interval where most of the high-angle faults splay upward to form what appear to be flower structures. Owing to the lack of continuous reflections in the drift sequence, many of these faults seem to die out or terminate at the post-rift unconformity. However, some splay upwards into the drift sequence, and a few may reach as far as the unconformity at the top of the drift sequence.

Many of the high-angle faults have characteristics recently interpreted to indicate strike slip: near-vertical attitudes, flower structures, changes from normal to reverse separation along strike, upthrown block switching sides along strike, reversal or change in fault throw with depth, abrupt change in seismic character across the fault, and abrupt change in thickness of seismic facies across the fault (Harding, 1985).

Drift Sequence

The most characteristic structures of the drift sequence are interpreted as diapiric shale ridges and folds (Figs. 12B, 14). Faults are relatively rare, and only a few were identified and mapped. They are present either as reverse faults at the termination of shale ridges, or as splays of the major fault zones in the rift sequence. Some flower structures are present within the drift sequence. Some seismic sections contain areas where dipping reflections terminate downwards against sub-horizontal reflections, possibly indicating localized thrusting. A large canyon fill (12 km by 5 km) was mapped at the top of the drift sequence (Figs. 9, 14). The Canarias fault, one of the few faults mapped within the drift sequence, is shown on Figures 11 and 14. It is a listric normal growth fault and is oriented perpendicular to the trend of the folds and shale ridges (Fig. 14).

Tertiary Sequence

The Tertiary sequence was not subjected to significant tectonism, and is for the most part structureless. The most prominent structures are growth faults and rollover anticlines in the northwestern part of the study area (Fig. 14).

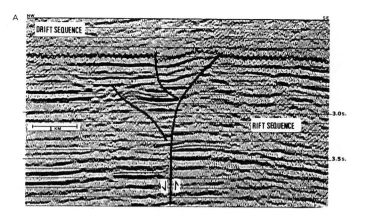

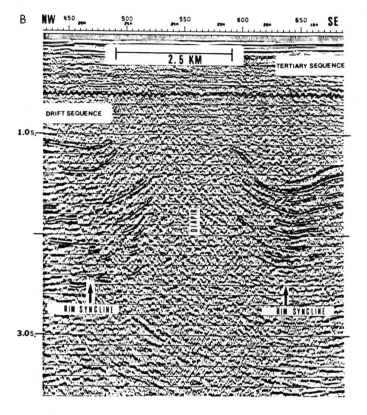

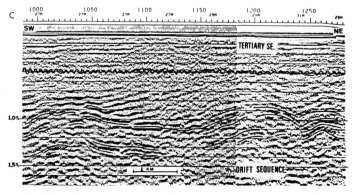

FIG. 12.—Examples of structural features seen in seismic profiles in Piaui Basin: A) flower structure; B) shale ridge, C) folds within drift sequence.

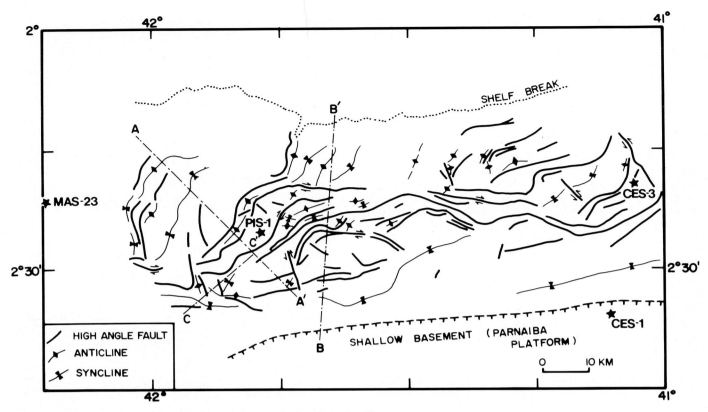

FIG. 13.—Map of major structures (high-angle faults and folds) within the rift sequence.

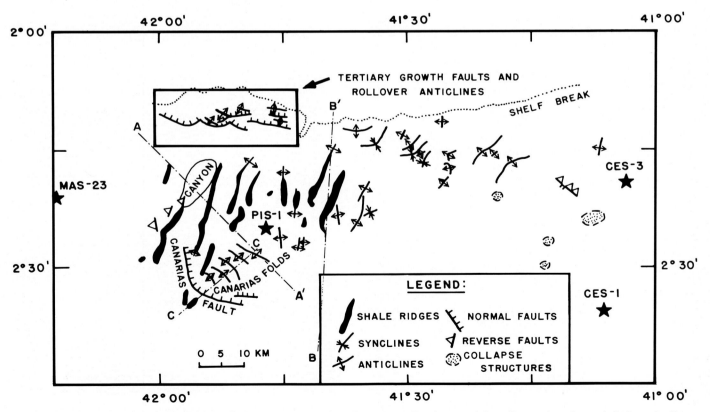

FIG. 14.—Map of major structures (shale ridges and folds) within the drift sequence. Box in upper left outlines region of growth faults and rollover anticlines in the Tertiary passive-margin sequence.

We interpret these structures to be related to recent deltaic sedimentation at the mouth of the Parnaiba River. In a few places, the Tertiary sequence has been affected by minor reactivation of faults and diapirs in the drift sequence.

Discussion

There are many factors accounting for the different structural styles described above; they include the age, tectonic history, and lithological composition of each sequence. The rift sequence is the oldest and deepest, and is composed mostly of continental sediments affected by syn-sedimentary faulting. During wrenching the behavior of this sequence was relatively brittle; it was more indurated than the overlying drift sequence. Much of the deformation accompanied oblique- and strike-slip reactivation of the pre-existing network of faults, although some new faults and folds may have developed.

The following characteristics of the drift sequence influenced deformation during the wrenching stage: (1) During deformation, the sequence consisted of newly deposited, poorly indurated sediments. (2) Shales dominated the western half of the basin. (3) Evaporites were deposited in the eastern part. (4) Shallow burial depths prevailed during deformation. (5) The sequence contained few pre-existing faults. These factors resulted in a relatively ductile behavior in the drift sequence during wrenching. Similar variable structural response by stratigraphic units during wrenching was reported in the North Sea by Glennie and Boegner (1981), where the Permian Zechstein halite responded to movement on underlying high-angle faults by plastic flow and diapirism.

STRUCTURAL ANALYSIS

The orientations of structural features (Figs. 13, 14) were analyzed statistically to investigate possible regional patterns of strain. In this approach, similar to that of Glennie and Boegner (1981), the cumulative lengths of faults or the axial traces of folds are plotted as a function of orientation. Structures used in this analysis are (1) high-angle faults in the rift sequence, and (2) folds and shale ridges in the drift sequence, and folds in the rift sequence.

High-angle Faults

Most high-angle faults were mapped only if they cross two or more seismic sections, and appear to be confined to the rift sequence (Figs. 9, 10, 11, 13, 14). Figure 15 shows the statistical distribution of the orientations of such faults in the rift sequence. Although faults of almost all orientations are present, three main modes are obvious. Most of the faults strike between north-northeast and due east (N25°E − S85°E). Faults striking between N70°W and N45°W, and between N20°W and due north form two other distinct groups. When all these data are plotted separately into an eastern domain and a western domain (bounded approximately by 41°40′ W), most of the faults striking northwest to northwest are present in the eastern domain (Figs. 13, 15).

Because most of the high-angle faults are confined to the rift sequence, it is likely that their origin is most closely related to the early stages of continental breakup. In addition, the main trend of the rift sequence faults (east-northeast to due east) is approximately parallel to the dominant orientation of lineaments in the Precambrian basement, as interpreted from Landsat and aeromagnetic analyses. These data show that the structural grain of the adjacent Precambrian terrain is dominated by major lineaments oriented between N35°E and N80°E (Zalan, 1984). This correlation suggests that the high-angle faults were controlled by the pre-existing anisotropy within the continental basement.

Although the time, or times, when the high-angle faults formed is not well constrained, we tentatively suggest that the other two main groups (striking west-northwest and north-northwest) may have developed during the wrenching stage. This suggestion is based on the fact that their mean orientations are subparallel to the trends of synthetic and antithetic faults that would hypothetically form in association with a major wrench system parallel to the Romanche Fracture Zone (Fig. 16). Secondary faults with this configuration have been produced in modelled wrench systems (e.g., Tchalenko, 1970; Wilcox et al., 1973), and have been observed in many natural wrench systems (e.g., Wilcox et al., 1973). In addition, the orientation of the listric normal Canarias fault (perpendicular to compressional structures) is consistent with such a strain pattern. However, because the two secondary groups are concentrated in separate geographic domains (Fig. 15), they may require some other explanation.

Compressional Structures

Figure 17 shows the distribution of trends of axial traces of folds and shale ridges. Most of these features trend northeast to north-northeast, with modal values of N20°E (shale ridges, rift sequence folds) and N35°E (drift sequence folds). The similarity in the trends of the shale ridges and folds suggests (1) that the shale ridges may be folds whose amplitudes have been enhanced by diapirism; and (2) that the ridges and folds developed at approximately the same time and in the same strain field, and/or that their orientation was controlled by some pre-existing anisotropy in the underlying basement. A secondary group of folds in the drift sequence, the Canarias folds (Figs. 11, 14, 17), are transverse (N60°W) to the other compressional features. Their origin is unknown, but they may be related to the Canarias fault.

In summary, most of the high-angle faults fall into three main sets on the basis of orientation. Because the dominant, east-northeast-striking set is mainly present in the rift sequence, and is parallel to the main grain in the continental basement, we suggest that these faults were formed during continental rifting. However, many may continue up into the drift sequence suggesting reactivation during wrenching. Faults striking between west-northwest and northwest are subparallel to hypothetical secondary faults associated with a proposed wrench system along the trend of the Romanche Fracture Zone. Although the time of post-rift faulting and folding is poorly constrained, we suggest that they developed in response to transpression during the early drifting stage.

Any interpretation of fault and fold trends in terms of a dynamic-kinematic model must be based on (1) a knowl-

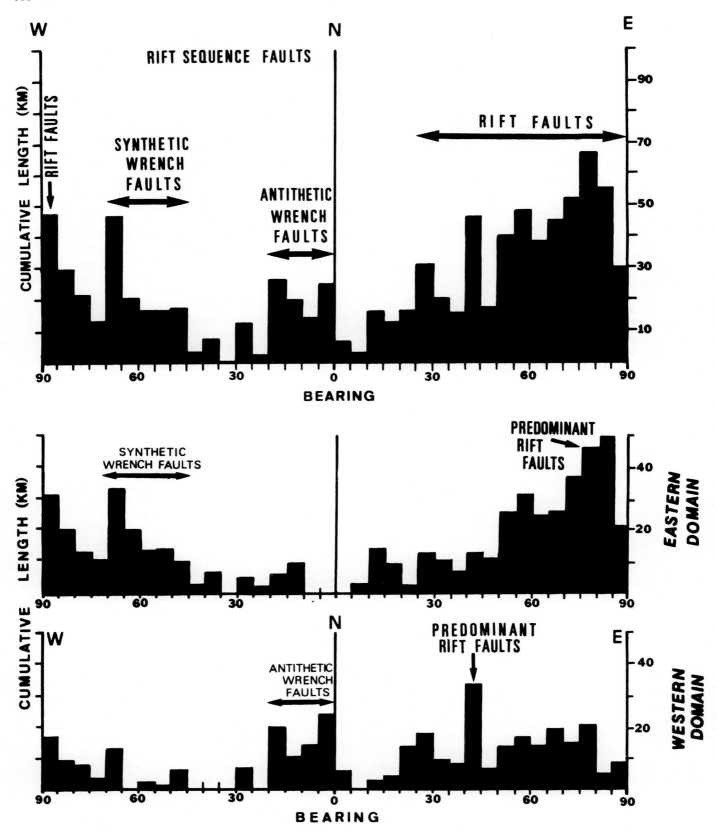

FIG. 15.—The orientation of faults at the top of the rift sequence showing an interpretation of the tectonic origin of each modal group. In the lower two diagrams, the faults are divided into eastern and western domains.

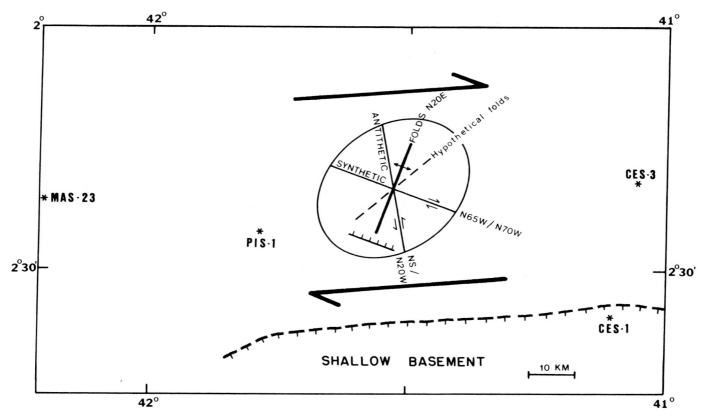

FIG. 16.—Comparison of the main fault and fold trends with the hypothetical orientation of faults and folds for right-lateral shear parallel to the Romanche Fracture Zone.

edge of timing relationships, and (2) an assumption that pre-existing anisotropies have not controlled the orientation of structures. Although we have little control on these factors, our model is consistent with the overall rift-drift tectonic history of the Brazilian margin, and it fits the hypothetical strain pattern expected for a rifted and wrenched setting (Fig. 16).

The Atlantic High

The Atlantic High is a prominent east-northeast-trending feature in the Piaui Basin, the areal expression of which was mapped from seismic sections where the top of the rift sequence was shallower than 2.0 seconds two-way travel time (Figs. 3, 10). This high is composed of several uplifted blocks and contains anticlines, synclines, and both reverse and normal faults. These structural features are more numerous and pronounced than elsewhere in the Piaui Basin. On seismic sections, the Atlantic High appears as an elevated platform cut by several high-angle faults (Figs. 9, 10). The trend and position of the Atlantic High correspond roughly to the westward extension of the Romanche Fracture Zone, and we suggest that wrenching along the fracture zone may be responsible for structures in the Atlantic High.

The Ceará High

The areal distribution of the Ceará High, shown in Figure 3, was determined by analysis of seismic and aeromagnetic data. The Ceará High is seismically reflection-free, and roughly circular in map view. Two similar but smaller highs were mapped in the same area. On the aeromagnetic map, the zones are coaxial with a pronounced, elliptically shaped, negative magnetic anomaly. Although the precise boundaries of this anomaly are obscure, the western and eastern boundaries appear to differ. To the west, the high is thought to be faulted against dipping sedimentary rocks. To the east, the reflection-free high passes laterally into what appear to be horizontal, undisturbed sedimentary rocks of the Acarau sub-basin.

Well CES-87, drilled on top of the Ceará High, penetrated 400 m of Tertiary platform sediments (quartzose sands and calcarenites), and entered a section of conglomerate and breccia interbedded with volcanic rock (total depth, 1,600 m). A core of the breccia is composed predominantly of fragments of granite, porphyritic granite, and rhyolite in a feldspathic sand matrix. The presence of rhyolite and granite in the breccia, the featureless character on the seismic sections, and the low negative magnetic anomaly (possibly indicating depletion in magnetite), suggest that the Ceará High is an acidic volcanic/intrusive complex. Because the Ceará High is aligned with the Atlantic High and the Romanche Fracture Zone, this proposed magmatic complex may be genetically related to wrench tectonism. We note, however, that igneous rocks are rare in transpressive belts (Reading, 1980), and their significance is not understood.

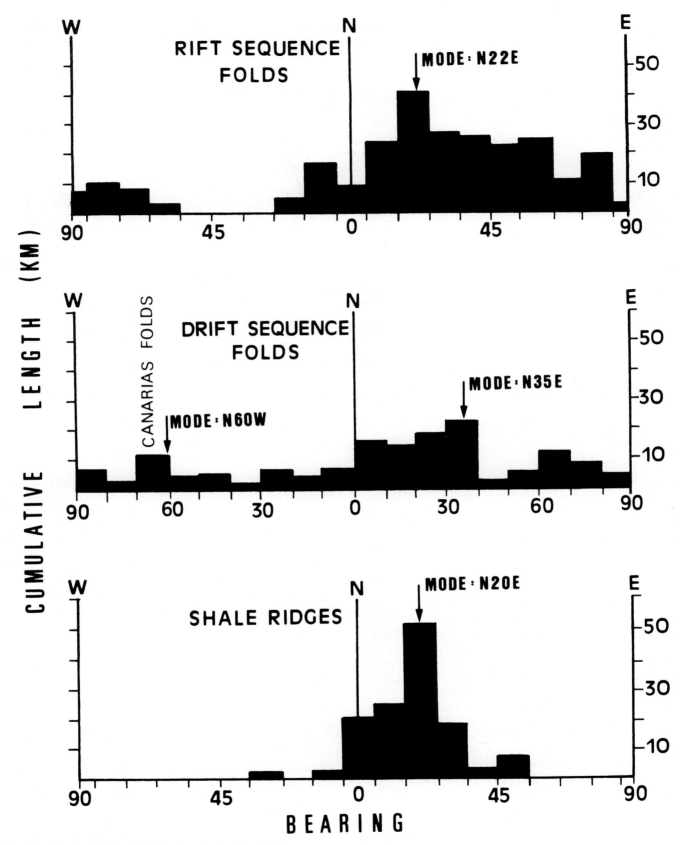

FIG. 17.—Trends of folds and shale ridges in the Piaui Basin.

GEOLOGICAL EVOLUTION OF THE PIAUI BASIN–CONCLUSIONS

We have divided the geological evolution of the Piaui Basin into four stages (rift, drift, wrench, and late drift or passive-margin stages). The rift and wrench stages were marked by tectonic activity; the drift and late drift stages were tectonically quiescent intervals characterized by sedimentation typical of passive margins. The following outline of the geologic history is somewhat speculative owing to the limited constraints on the timing of specific events, and the lack of definitive evidence for lateral offset along major faults.

Rift Stage (approximately 130-112 Ma)

During the rifting stage, Africa separated from South America at a high angle to the continental margin along the northern Brazilian margin (Fig. 2A; Rabinowitz and LaBrecque, 1979). The orientation of normal faults formed during rifting was influenced by the N35°E- to N80°E-trending Precambrian structural grain. The western half of the Piaui Basin is dominated by N40°E- to N45°E-striking normal faults, whereas the eastern part is dominated by N75°E- to N85°E-striking faults, forming a distribution of extensional faults that parallels the edge of the Parnaiba Platform (Fig. 13). Sedimentation during the rifting stage was essentially clastic and predominantly continental in nature (fluvial, alluvial fan, and lacustrine depositional environments), and was accompanied locally by the extrusion of basalts. The first marine invasions may have come from the west, and took place sometime near the end of this stage (as indicated by well PIS-1). The end of the rifting stage documented here, and suggested by Rabinowitz and LaBrecque (1977) to have occurred at 107–111 Ma, coincided with an Aptian worldwide sea-level fall (112 Ma; Vail et al., 1977).

Drift Stage (Approximately 112–98 Ma)

During the drifting stage, Africa continued to separate from South America at a high angle (Fig. 2B; Rabinowitz and LaBrecque, 1979), and the equatorial Atlantic was already open, as indicated by widespread marine deposits in the upper portion of the drift sequence. The age of the oldest oceanic crust in this region is unknown because oceanic magnetic anomalies are not observed in the area. This is because oceanic crust at the margins of the equatorial Atlantic formed at a north-south spreading center near the paleomagnetic equator and is part of the Cretaceous Quiet Zone (Cande and Rabinowitz, 1979). Thus it is not clear when sea-floor spreading began.

Rifting may have propagated eastward and was active in the eastern Ceará and Potiguar Basins to the east (see Fig. 1). Sedimentary environments were transitional to marine during the worldwide Aptian through early Cenomanian sea-level rise. At least 3,000 m of clastic sediments were deposited in near-shore and prodeltaic environments during the late Alagoas Stage. Albian and Cenomanian sediments characteristically represent more open marine environments. Anomalously thick Albian and Cenomanian deposits in the adjacent Acarau Basin suggest that some transtensional wrenching may have already occurred during this time.

Wrench Stage (Approximately 98–95 Ma)

During late Albian to middle Cenomanian time, major transform movements began along the oceanic Romanche Fracture Zone, probably due to a shift in the spreading direction from northeast-southwest to east-west. Rabinowitz and LaBrecque (1979) suggested that a change in the pole of rotation between Africa and South America occurred around 106 ±5 Ma and resulted in east-west transform motion along this segment of the Brazilian margin. A slightly later pole shift, uncertainties in the time of inception of wrenching, or uncertainties in the absolute time scale probably account for the slightly different estimates for the age of onset of wrenching. The continents were probably close enough to allow interaction between major irregularities (such as the Parnaiba Platform bulge) along their rifted margins, and many rift faults (initially normal) were probably reactivated as oblique- and strike-slip faults. Normal offset on these faults may have been greatly diminished and even inverted through reverse faulting. Synthetic and antithetic (based only on orientation) strike-slip faults developed along pre-existing rift faults. Flower structures, en echelon folds, and shale ridges also formed at this time. By the end of this tectonism, most of the area had been subject to uplift as a result of transpression, particularly along the Atlantic High. Wrenching ceased when the continents had separated sufficiently so that irregularities along their margins could no longer interact. By this time, a transpressive belt 200 km long and 50 km wide had emerged in the Piaui Basin, and remained emergent until mid-Tertiary time.

Passive Margin or Late Drifting Stage (Late Cenomanian-Present)

From Late Cretaceous to Eocene time (approximately 60 m.y.) the transpressive belt was eroded and finally submerged (owing to gradual subsidence of the continental crust below it). Steckler and Watts (1982) calculated that, over a 60-m.y.-interval, a rifted margin should generally subside 3–4 km beneath the continental shelf. This figure does not match our interpretation of the Piaui Basin, where erosion occurred from the Cenomanian to the Oligocene. However, Scrutton (1982) indicated that transform continental margins typically display limited thinning of the continental crust, a factor that would decrease the expected subsidence. In addition, the variable tectonic history of basins along the northeastern Brazilian margin suggests that wrenching may have caused some basins to subside rapidly, while others were being eroded. Oligocene-Miocene shallow-marine shelf sediments were deposited over most of the Piaui Basin. Only north of the Atlantic High was sedimentation apparently continuous throughout the Late Cretaceous and Tertiary in the form of thick continental-slope sequences.

The stratigraphic and structural data presented for the Piaui Basin, although limited, may be useful as a predictive model in other basins that underwent a similar convergent wrenching stage in an overall context of rifting and continental separation.

ACKNOWLEDGMENTS

The authors are indebted to PETROBRAS for permission to publish this paper. This work is a summary of P. V.

Zalan's dissertation research at the Colorado School of Mines, funded by a grant from PETROBRAS. The manuscript benefitted greatly from reviews by J. D. Haun, K. T. Biddle, M. A. Uliana, and F. L. Wehr, and from the editorial suggestions of N. Christie-Blick.

REFERENCES

ASMUS, H. E., AND BAISCH, P. R., 1983, Geological evolution of the Brazilian continental margin: Episodes, v. 1983, p. 3–9.

ASMUS, H. E., AND PORTO, R., 1972, Classificacao das bacias sedimentares brasileiras segundo a tectonica de placas: 26th Brazilian Geological Congress, Proceedings, v. 2, p. 67–90.

BONATTI, E., AND CRANE, K., 1984, Oceanic fracture zones: Scientific American, v. 250, No. 5, p. 40–51.

BRYAN, G. M., KUMAR, N., AND CASTRO, P. J. M., 1972, The north Brazilian ridge and the extension of equatorial fracture zones into the continents: 26th Brazilian Geological Congress, Proceedings, v. 2, p. 133–144.

CANDE, S., AND RABINOWITZ, P. D., 1979, Magnetic anomalies bordering the continental margins of Brazil: Offshore Brazil Map Series, American Association of Petroleum Geologists, Tulsa, Okla.

DELTEIL, J-R., VALERY, P., MONTADERT, L., FONDEUR, C., PATRIAT, P., AND MASCLE, J., 1974, Continental margin in the northern part of the Gulf of Guinea, in Burk, C. A., and Drake, C. L., eds., Geology of Continental Margins: New York, Springer, p. 297–311.

FIGUEIREDO, A. M. F., CARMINATTI, M., FILHO, J. A. P., AND TEIXEIRA, L., 1983, Barreirinhas basin, an equatorial Atlantic transform basin (abs.): American Association of Petroleum Geologists Bulletin, v. 67, p. 449.

GLENNIE, K. W., AND BOEGNER, P. L. E., 1981, Sole pit inversion tectonics, in Illing, L. V., and Hobson, G. D., eds., Petroleum Geology of the Continental Shelf of Northwest Europe: London, Institute of Petroleum, p. 110–120.

HARDING, T. P., 1985, Seismic characteristics and identification of negative flower structures, positive flower structures, and positive structural inversion: American Association of Petroleum Geologists Bulletin, v. 69, p. 582–600.

KUMAR, N., AND LADD, J. W., 1974, Origin of compressional structures on the shelf off Brazil and Ghana formed during the opening of the equatorial Atlantic: Geological Society of America Abstracts with Programs, v. 6, p. 835.

MARTIN, A. K., HARTNADY, C. J. H., AND GOODLAD, S. W., 1981, A revised fit of South America and south central Africa: Earth and Planetary Science Letters, v. 54, p. 293–305.

MIURA, K., AND BARBOSA, J. C., 1972, Geologia da plataforma continental do Maranhao, Piaui, Ceara e Rio Grande do Norte: 26th Brazilian Geological Congress, Proceedings, v. 2, p. 57–66.

OJEDA, H. A. O., 1982, Structural framework, stratigraphy, and evolution of Brazilian marginal basins: American Association of Petroleum

Geologists Bulletin, v. 66, p. 732–749.

PINDELL, J. L., 1985, Alleghenian reconstruction and subsequent evolution of the Gulf of Mexico, Bahamas, and Proto-Caribbean: Tectonics, v. 4, p. 1–39.

RABINOWITZ, P. D., AND LABRECQUE, J., 1979, The Mesozoic South Atlantic ocean and evolution of its continental margins: Journal of Geophysical Research, v. 84, p. 5973–6002.

READING, H. G., 1980, Characteristics and recognition of strike-slip fault systems, in Ballance, P. F., and Reading, H. G., eds., Sedimentation in Oblique-Slip Mobile Zones: International Association Sedimentologists Special Publication No. 4, p. 7–26.

SCRUTTON, R. A., 1982, Crustal structure and development of sheared passive continental margins, in Scrutton, R. A., ed., Dynamics of Passive Margins: American Geophysical Union Geodynamics Series, v.6, p. 133–140.

SIBUET, J. C., AND MASCLE, J., 1978, Plate kinematic implications of Atlantic equatorial fracture zone trends: Journal of Geophysical Research, v. 83, p. 3401–3421.

STECKLER, M. S., AND WATTS, A. B., 1982, Subsidence history and tectonic evolution of Atlantic-type continental margins, in Scrutton, R. A., ed., Dynamics of Passive Margins: American Geophysical Union Geodynamics Series, v. 6, p. 184–196.

SYKES, L. R., 1978, Interplate seismicity, reactivation of pre-existing zones of weakness, alkaline magmatism, and other tectonism postdating continental fragmentation: Journal of Geophysical Research, v. 16, p. 621–688.

TCHALENKO, J. S., 1970, Similarities between shear zones of different magnitude: Geological Society of America Bulletin, v. 81, p. 1625–1640.

TEIXEIRA, L., AMORIM, J., CARMINATTI, M., AND FIGUEIREDO, A. M. F., 1982, Projeto Barreirinhas-Reavaliacao da bacia Cretacea—Area terrestre e maritima: unpubl. PETROBRAS Internal Report, 58 p.

VAIL, P. R., MITCHUM, R. M., JR., AND THOMPSON, S., III, 1977, Seismic stratigraphy and global changes of sea level, part 4; global cycles of relative changes of sea level, in Payton, C. E., ed., Seismic Stratigraphy - Applications to Hydrocarbon Exploration: American Association of Petroleum Geologists Memoir 26, p. 83–97.

WILCOX, R. E., HARDING, T. P., AND SEELY, D. R., 1973, Basic wrench tectonics: American Association of Petroleum Geologists Bulletin, v. 57, p. 74–96.

WILSON, R. C. L., AND WILLIAMS, C. A., 1979, Oceanic transform structures and the development of Atlantic continental margin sedimentary basins—a review: Journal of the Geological Society of London, v. 136, p. 311–320.

ZALAN, P. V., 1983, Stratigraphy and petroleum potential of the Acarau and Piaui-Camocim sub-basins, Ceara basin, offshore northeastern Brazil [unpubl. masters thesis]: Golden, Colorado, Colorado School of Mines, 155 p.

———, 1984, Tectonics and sedimentation of the Piaui-Camocim sub-basin, Ceara basin, offshore northeastern Brazil [unpubl. Ph.D. thesis]: Golden, Colorado, Colorado School of Mines, 128 p.

THE NONACHO BASIN (EARLY PROTEROZOIC), NORTHWEST TERRITORIES, CANADA: SEDIMENTATION AND DEFORMATION IN A STRIKE-SLIP SETTING[1]

LAWRENCE B. ASPLER, AND J. A. DONALDSON

Department of Geology, Carleton University, Ottawa, Ontario, Canada, K1S 5B6

ABSTRACT: The Nonacho Group is a sequence of talus, alluvial-fan, braided-stream, pond, beach, fan-delta, and lacustrine rocks. Deposition occurred over an area 200 km by 60 km, in rhomb, wedge, and rectangular sub-basins. The sub-basins were separated during sedimentation by basement uplifts, as indicated by changes in clast composition, unit thickness, and facies that occur across fault-bounded basement inliers. Sinistral strike slip along near-vertical, north-northeast-striking faults that border the western margin of the Nonacho Basin is regarded as having controlled basin evolution. Syn-depositional strike slip is suggested by an extreme longitudinal cumulative thickness (>40 km) in conjunction with a uniform, lower greenschist facies metamorphic grade. A sinistral sense is indicated by (1) the location of the basin where there is a left-hand stepover in the regional fault system; (2) the probable southward sequential development of the basin as the basin floor moved, like a conveyor belt, northwards along faults concentrated on the western margin; and (3) the inferred tectonic emplacement from south of the basin of source rocks for conglomerates stratigraphically high in the sequence. The western margin of the basin was bordered by high-gradient alluvial fans and braid-plains, which in turn were located on the edge of a relatively deep lake. The eastern margin was characterized by a low-gradient alluvial plain with isolated ponds. This inferred paleotopography is consistent with active faulting along the western margin during sedimentation. Post-depositional sinistral faulting is indicated by traces of a regionally penetrative cleavage (and axial surfaces of related folds) that trend at an angle of 20°–40° to the average strike of the faults. Cleavage traces define sigmoidal patterns of heterogeneous sinistral shear. Zones of high strain on the western side of the basin display a lower fabric-to-fault angle than zones of low strain in the central and eastern parts of the basin. A gently plunging displacement vector is confirmed by near-horizontal stretching lineations (pebbles and quartz fibers), which reflect the finite extension direction adjacent to shear zones.

INTRODUCTION

The Nonacho Basin was the site of Early Proterozoic terrigenous clastic sedimentation in the northwestern part of the Canadian Shield (Fig. 1). It consists of a number of sub-basins distributed over an area 200 km by 60 km. Between 1978 and 1983, eleven months of field work were spent updating the pioneering studies of Henderson (1937) and McGlynn (1966, 1970, 1971), focussing on basic mapping of the basin, defining stratigraphic units, establishing stratigraphic and structural patterns, devising paleogeographic models, and evaluating the significance of the basin for the evolution of the northwestern Churchill Province (Aspler 1985a,b).

The coarse continental clastic rocks that are typical of strike-slip basins can accumulate in a variety of tectonic settings (Miall, 1981). Presented herein are stratigraphic, sedimentologic, and structural data suggesting that initiation, growth, and deformation of the Nonacho Basin were a consequence of sinistral strike-slip faulting.

STRATIGRAPHIC FRAMEWORK

Basement

The basement of the Nonacho Group consists of amphibolite- to granulite-grade granitoid gneisses, metasedimentary and metavolcanic rocks, and less metamorphosed granitoid rocks. K/Ar studies have yielded ages as old as 2,420 Ma (Burwash and Baadsgaard, 1962); a Rb/Sr study gave an age of 2,524 ± 27 Ma (Baadsgaard and Godfrey, 1972). Anatectic Aphebian granites have been dated at 1,938 ± 29 Ma (Rb/Sr whole rock; Nielsen et al., 1981), 1,944 ± 16 Ma, and 1,900 ± 40 Ma (U/Pb zircon; Bostock, 1982).

Depositional Setting

Stratigraphic units of the Nonacho Group (Fig. 2) consist of a limited suite of facies that are repeated through time and space. Radical thickness changes of individual stratigraphic units (Figs. 2, 3) are a reflection of both the terrestrial depositional system and the rugged basement topography that can be generated in a strike-slip setting (see Fig. 8 of Spörli, 1980). The principal deposits (and inferred depositional environments) include: a breccia facies (talus slope); a conglomerate facies (proximal alluvial fan); an interbedded conglomerate-sandstone facies (distal alluvial fan); a sandstone and pebbly sandstone facies (alluvial plain); an interbedded sandstone-mudstone facies (pond); and a mixed facies (perennial lake with beach, beach cliff and fan-delta marginal sub-facies). These various facies are described and interpreted below.

Megacyclicity

As is evident from Figure 2, deposition in the Nonacho Basin was cyclic on a scale of thousands to tens of thousands of meters. One fining-upward megasequence extends from talus-slope breccias and proximal alluvial-fan conglomerates at the base of Unit A, through distal alluvial-fan interbedded conglomerates and sandstones at the top of Unit A, to alluvial-plain sandstones at the top of Unit B. This sequence probably represents fault-induced basin initiation, followed by backwasting of source areas during tectonic quiescence. The abrupt (but conformable) contact of lacustrine pelites of Unit C with underlying fluvial sandstones of Unit B signals a change in depositional regime, and marks the onset of a coarsening-upward sequence in which distal lacustrine rocks pass up-section to marginal lacustrine rocks, which are in turn overlain by fluvial sandstones of Unit D. The abrupt transition is probably due to lakes on the western (high-gradient) side of the basin abutting against the fluvial facies without well-developed transitional environments (see below). The abruptness is prob-

[1]Contribution 09-84 of the Ottawa-Carleton Centre for Geoscience Studies.

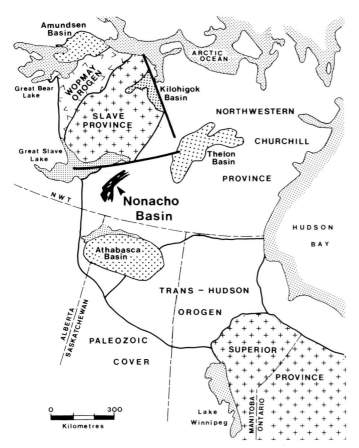

Fig. 1.—Location of the Nonacho Basin relative to elements of the Canadian Shield.

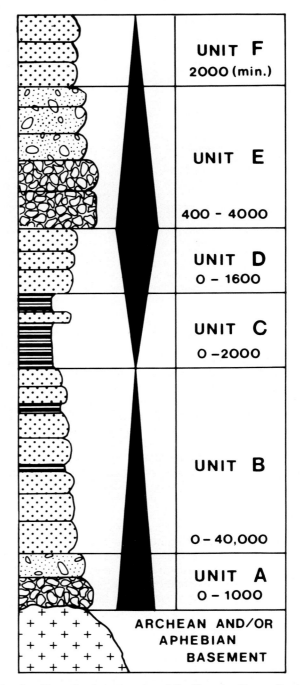

Fig. 2.—Generalized stratigraphy of the Nonacho Group. Symbols: crosses, basement; open circles, conglomerate; open circles and stipple, interbedded conglomerate and sandstone; stipple, sandstone and pebbly sandstone; bars, interbedded sandstone, siltstone, and mudstone. Black wedges represent fining-upward and coarsening-upward megasequences. Thicknesses are in meters.

ably also a consequence of relatively sudden, fault-induced basin deepening, which inhibited the development of a complete transgressive sequence. Renewed basement uplift is supported by the appearance of intraformational sandstone clasts in Unit C and Unit E, and the local development of conglomeratic beds consisting of angular mono-mictic granite fragments in both units. In contrast to Unit A conglomerates, which contain almost exclusively clasts of granitoid rocks, amphibolite, paragneiss and milky quartz, Unit E conglomerates contain a diverse assemblage of clast types, including varicolored phyric and aphyric, felsic to mafic volcanic rocks; quartz arenite; and intraformational sandstone and conglomerate. As discussed below, the appearance of this polymictic clast suite in stratigraphically high units is thought to reflect tectonic emplacement of source rocks during sedimentation.

The third and final megasequence is represented by the fining-upward trend of Unit E conglomerates and interbedded conglomerates and sandstones, and overlying Unit F sandstones. This is considered to indicate a time of relative tectonic quiescence as the basin "recovered" from the second cycle.

Asymmetry of Unit Development and Distribution

Several aspects of the stratigraphy and sedimentology of the Nonacho Group vary laterally in cross sections parallel to and transverse to the basin (Fig. 3). Features in transverse cross section are discussed first. The unconformity between the Nonacho Group and the basement is generally obscured by faulting on the west; most exposures where the unconformity is preserved are on the east. Unit A is generally thicker on the western side of the basin (as thick as

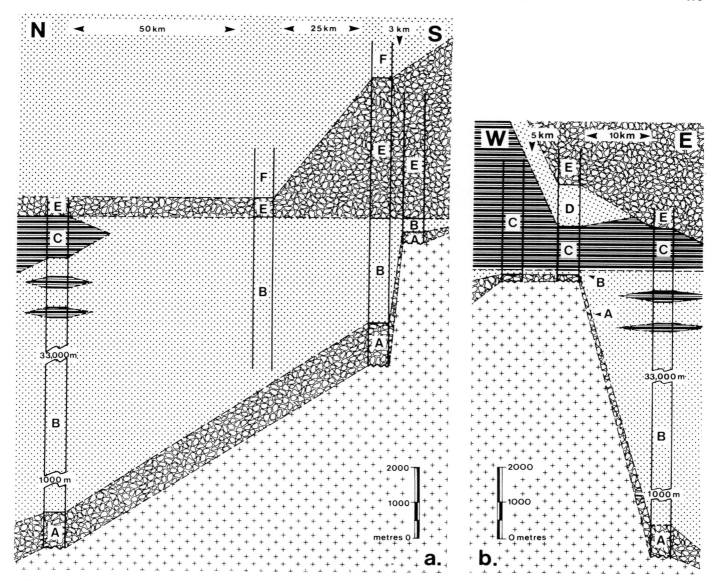

FIG. 3.—Generalized stratigraphic relations of the Nonacho Group: A) North to south; B) west to east. Thickness estimates for the composite sections are based on mapping at a scale of 1:125,000. Symbols: open circles, conglomerate and interbedded conglomerate and sandstone; stipple, sandstone and pebbly sandstone; bars, interbedded sandstone, siltstone, and mudstone.

1,000 m) relative to the eastern side, where the unit is locally absent. Unit B is generally thinner on the western side (as thin as 150 m) relative to the eastern side (as thick as 40,000 m). A full suite of sub-facies in Unit C, representative of a perennial deep-water lake, is developed only on the western side of the basin; on the east, local pond deposits occur. Conglomeratic rocks near the top of Unit C on the western side of the basin (conglomerate-pelite subfacies) contain monomictic granitic and intraformational sandstone fragments, whereas to the east, equivalent rocks contain a diverse clast suite. Sparse conglomerate beds in polymictic Unit E, containing up to 90% angular granitoid clasts, are exposed only in the western and northwestern parts of the basin.

Intraformational sandstone clasts occur in conglomerates of Unit A, but only in the extreme southern part of the basin. The first appearance of caliche in the south is in Unit A, whereas in the north, caliche does not appear until the top of Unit C. An estimate of at least 40,000 meters was obtained for the cumulative longitudinal thickness of Unit B by measuring down-plunge along hinge lines of folds cored by the unit in the main basin. It should be emphasized that this cumulative longitudinal thickness does not represent the true thickness of the section at any one locality. Unit B thins dramatically southwards at the expense of Unit E. Unit C is absent in the southern part of the main basin. Unit E conglomerates in the southern part of the basin commonly consist of up to 20% intraformational sandstone clasts, in contrast to exposures in the north where such clasts generally constitute less than a few percent. Beds with 50–90% quartz arenite clasts are developed only on the southern and eastern sides of the basin.

FACIES ANALYSIS

Breccia Facies

Description.—The breccia facies is confined to the lower 10 m of Unit A, above the basal unconformity. A zonation is commonly developed, which consists of the following: pre-Nonacho Group fractures in basement filled with coarse sandstone and locally derived breccia fragments; angular blocks rotated slightly out of position from *in situ* basement; and locally derived breccia (Fig. 4A). The breccia is clast-supported, with clasts as large as 2 m, and the matrix is generally coarse-grained sandstone.

Interpretation.—The breccia is interpreted as a talus deposit on the basis of clast angularity, clast-supported framework, predominance of clasts derived from immediately adjacent basement, and the nature of the transition between basement and breccia.

Conglomerate Facies

Description.—Thickly bedded, non-stratified conglomerates (Fig. 4B) are dominant in the lower parts of Units A and E. Clasts as large as 1 m occur in a matrix of coarse-grained sandstone; frameworks are both clast-supported and matrix-supported. Bedding is locally manifested by variations in clast size or clast/matrix ratio, or by crude parallel stratification defined by aligned clasts. Sandstone interbeds are rare, generally less than a few tens of centimeters thick, and spaced more than 5 m apart; they have a wedge or sheet geometry, and are parallel-stratified. Fining-upward sequences of pebble-cobble conglomerate to granule conglomerate to parallel-stratified sandstone to centimeter-scale mud drapes occur locally. Millimeter- to centimeter-scale mudstone-carbonate laminites occur as isolated lenses, generally less than a few meters thick, within basal Unit A conglomerates. Features indicative of clastic deposition of the carbonates are lacking. Bedding-parallel zones in which coarse calcite cement has encrusted and replaced clasts occur locally.

Interpretation.—The conglomerates are interpreted as proximal alluvial-fan deposits. Low clast diversity, low roundness, and the presence of large clasts imply active nearby sources. The general absence of stratification and preferred clast orientation, and the poor sorting of the gravel mode suggest disequilibrium, non-tractional depositional processes. The coarse-grained sandstone matrix of matrix-supported conglomerates indicates deposition by low-viscosity flash floods and not high-viscosity debris flows (see Rodine and Johnson, 1976; Bull, 1977; Eriksson, 1978; Hampton, 1979; Allen, 1981). Flash floods were probably a consequence of the lack of land plants to impede direct runoff (see Schumm, 1967; Cotter, 1978). Clast-supported conglomerates with a crude stratification probably formed by the migration of diffuse gravel sheets and/or low-relief bars for which downstream migration was too rapid for the development of angle-of-repose slip faces (see Hein and Walker, 1977). Discontinuous sandstone layers with parallel stratification are probably upper-flow-regime flash-flood sheets, similar to those described by McKee et al. (1967), Rust (1978, 1984), Winston (1978), Turnbridge (1981, 1983), Vos and Tankard (1981), Stear (1983), and Ballance (1984).

The laminated mudstone-carbonate sub-facies is interpreted to represent ephemeral-pond deposition, mainly on the basis of the fine-scale interlamination of the constituent rock types and their restricted occurrence within the alluvial-fan facies. Typically, ephemeral ponds occupy the central parts of closed intramontane basins, beyond the distal fringes of associated alluvial fans (Eugster and Hardie, 1975; Nilsen, 1982). In the present example, because the mudstone-carbonate laminites occur as lenses within the coarse-grained fan facies, a different model is favored. The ponds are considered to have originated near fan apices, within interfan areas and in hollows where the groundwater table intersected fan surfaces (see McGowen and Groat, 1971; Elmore, 1983). Zones of coarse calcite cement are interpreted as caliche.

Interbedded Conglomerate-Sandstone Facies

Description.—In this facies, sheets of granule-pebble and pebble-cobble conglomerate alternate with sandstone on a decimeter to meter scale (Fig. 4C). The conglomerates range from clast- to matrix-supported, with a matrix of coarse-grained sandstone. The most common sedimentary structures are parallel stratification and normal grading. Planar cross-stratification and inverse grading occur locally. In the sandstones, heavy mineral layers commonly define parallel and low-angle stratification; upper flow-regime sedimentation is indicated by local current lineation and bed thicknesses up to 1 m (see Harms et al., 1982). Fining-upward sequences, which consist of a scour surface, overlain by conglomerate, and parallel-stratified or cross-stratified sandstone, are common.

Interpretation.—A change up-section in Units A and E from proximal to distal alluvial-fan sedimentation is indicated in the conglomerate-sandstone facies by a decrease in grain size and an increase in the number and thickness of sandstone interbeds. The sheet-like geometry of the deposits, the coarse-grained sandstone matrix of the conglomerates, local small-scale scours, and well-developed parallel stratification in the sandstones suggest that floodwaters, at least partly channelized in upper fan reaches, may have dispersed as unconfined, shallow, areally extensive, high-energy sheets in lower parts of the fan (see Hardie et al., 1978; Winston, 1978).

Conglomerates with normal grading probably represent floods that decelerated sufficiently slowly to enable sorting. Those conglomerates with inverse grading may represent the effects of high dispersive pressures and grain-flow behavior. Clast-supported conglomerates with a crude stratification probably represent diffuse gravel sheets or longitudinal bars; the local planar cross-stratified conglomerates may represent transverse bars. In the sandstones, the predominance of parallel stratification and the paucity of lower flow-regime dune and ripple cross-stratification may indicate rapid deceleration of flash floods without reworking as a result of bedform migration (see Boothroyd and Ashley, 1975; Eriksson, 1978; Turnbridge, 1981, 1983).

Sandstone and Pebbly Sandstone Facies

Description.—This facies consists of about 90% medium- to coarse-grained arkose. Sedimentary structures include

FIG. 4.—Sedimentary facies of the Nonacho Group. A) Breccia consisting of foliated granitic clasts, interpreted as a talus deposit (breccia facies, basal Unit A). B) Proximal alluvial-fan deposits (conglomerate facies, Unit A), consisting of massive, clast-supported boulder-cobble conglomerate with coarse-grained sandstone matrix. Note large clast with folded fabric. C) Sheet alternations of parallel-stratified sandstone and conglomerate; distal alluvial-fan deposit (interbedded conglomerate-sandstone facies, Unit E). D) Fluvial parallel-stratified sandstone cut by low-angle stratification (with pebbles), and in turn cut by a single pebble layer (sandstone and pebbly sandstone facies, Unit B). E) Stepped outcrop consisting of fluvial fining-upward sequences of parallel-stratified sandstone to mudstone (sandstone and pebbly sandstone facies, Unit B).

FIG. 5.—Sedimentary facies of the Nonacho Group. A) Tabular set of cross-stratified sandstone between beds of parallel-stratified sandstone (sandstone and pebbly sandstone facies, Unit D). B) Distal lacustrine fining-upward sequences on a centimeter scale, indicated by arrows (siltstone

parallel stratification (with current lineation); low-angle stratification; planar and trough cross-stratification (up to about 1 m thick); reactivation surfaces; small-scale scours; and channels. Heavy-mineral bands are remarkably well developed (Figs. 4D, 5A). Locally, decimeter-scale mudstone beds (commonly brecciated) separate otherwise massive sandstones. Decimeter- to meter-scale fining-upward sequences of scour surface (± mudchip breccias) to parallel-stratified sandstones to mudstones occur locally (Fig. 4E). Pebbles form single-pebble or decimeter-scale layers at the bases of fining-upward sequences as well as discrete beds. Generally, as much as 50% of the framework is dispersed in a coarse-grained sandstone matrix. In single-pebble layers, individual pebbles are isolated, occur in clusters, or are uniformly distributed. In bedding-plane views, pebbles are commonly distributed in trains aligned parallel to paleocurrents.

Interpretation.—The sandstone and pebbly sandstone facies is interpreted to have been deposited in extensive alluvial plains on which flash-flood processes were predominant. The alluvial plains are visualized as having been fed by tributaries leading from the alluvial fans that deposited Units A and E. Parallel-stratified sandstones probably represent flood sheets; tabular planar and trough cross-stratification probably formed by the migration of straight- and sinuous-crested bars. Simple fining-upward sequences of parallel-stratified sandstones to mudstones reflect rapid flood deceleration. Pebbly sandstones in which the framework is supported by a coarse-grained sandstone matrix support a flash-flood interpretation. The linear pattern of single-pebble trains in bedding-plane view probably resulted from helical flow with current-parallel vortices.

Interbedded Sandstone-Mudstone Facies

Description.—This pelite-rich facies occurs (1) as isolated lenses enclosed within rocks of the sandstone and pebbly sandstone facies of Unit B; (2) as the Unit C equivalent on the eastern side of the basin; and (3) as the Unit F equivalent in the MacInnis sub-basin (Fig. 6). The facies consists of rhythmic interlayers of sandstone and mudstone on two scales. First, meter-scale mudstone-rich beds alternate with meter-scale beds of sandstone. Second, within the mudstone-rich layers, sandstones and mudstones alternate on a centimeter- to decimeter-scale. Contacts between the rock types are sharp. The interlayering is rhythmic, but fining-upward sequences are not developed.

The mudstones are structureless, lacking the delicate lamination and fine cross-stratification shown by pelitic rocks of the mixed facies (described below). Sandstones within the mudstones are generally massive, but commonly display parallel stratification and planar or trough cross-stratification. Symmetric dune bedforms are locally preserved.

Some of the centimeter-scale layers are wavy and discontinuous.

The meter-scale sandstone beds commonly display large-scale planar or trough cross-stratification. Parallel-stratified heavy-mineral bands in horizontal and low-angle sets are common. Mudchip breccias, mudcracks, and mud saucers occur locally.

Interpretation.—The interbedded sandstone-mudstone facies is interpreted to represent deposition in ponds that occupied subtle topographic depressions in interfluves of the alluvial plains (see Seni, 1980). Pelitic rocks are negligible in associated alluvial-fan and alluvial-plain deposits. Apparently fluvial processes were efficient in washing out fine-grained sediment. The localization of isolated pelite-rich rocks seems to demand the existence of a standing body of water capable of trapping the fine-grained material. The ponds were probably ephemeral; standing-water deposition alternated with fluvial and possibly eolian deposition. The term "pond" is used here to denote small standing bodies of water which, even though perennial over the short term (years or decades), may be ephemeral over periods of longer duration. Although strict definition of the term "playa lake" is without climatic or genetic inference (Reeves, 1978), the term "pond" is preferred because of the common usage of "playa lake" for saline lakes, alkali lakes or inland sabkhas. Because there is no evidence of elevated salinities, the relatively short-lived nature of the ponds probably does not have any paleoclimatic significance. Rather, the low areas are envisaged as having been periodically filled and re-established as a result of lateral migration of the principal sites of fluvial deposition.

The centimeter- to decimeter-scale sandstone layers within the mudstones are difficult to interpret in terms of specific depositional processes. The most likely agents include subaerial sheet floods (during dry intervals), sheet floods entering the ponds (during high-water stands) and wind drifts, both as sheets migrating into ponds and as adhesions of sand on damp sediment surfaces (see Glennie, 1970; Adams and Patton, 1979; Stewart and Walker, 1980). Although sandy sheetwash deposits, which form when unconfined floods spread over an alluvial plain and enter a lake, generally include fining-upward sequences and structures such as basal scours and graded beds (Hardie et al., 1978; Winston, 1978; Hubert and Hyde, 1982; Smoot, 1983; Ballance, 1984; Grover, 1984), some examples do not (Hesse and Reading, 1978; Sneh, 1979). Thin discontinuous wavy sandstone laminae may have formed owing to an inadequate supply of wind-blown sand drifting on a recently emerged pond surface.

The meter-scale sandstone interbeds are interpreted as subaerial deposits on the basis of locally developed mud-saucers and mudcracks. The development of decimeter-scale planar and trough cross-stratification suggests fluvial bed-

---

sub-facies, Unit C). C) Proximal lacustrine fining-upward sequence of coarse-grained sandstone to ripple-drift cross-stratified sandstone to siltstone to mudstone (graded sandstone-siltstone sub-facies, Unit C). D) Fining-upward sequence consisting of granule-conglomerate to cross-stratified very fine-grained sandstone to siltstone (conglomeratic pelite sub-facies, Unit C). E) Reactivation surface in coarse-grained sandstone defined by heavy mineral bands (graded-sandstone-siltstone sub-facies, Unit C). F) Sandstone protolith (background) passes along strike to intraformational sandstone breccia. (Intraformational sandstone breccia sub-facies, Unit C).

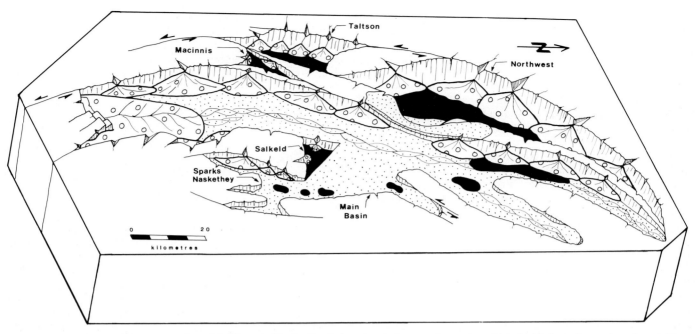

FIG. 6.—Paleogeographic reconstruction of the Nonacho Basin. Symbols: open circles, conglomerate and interbedded conglomerate and sandstone; stipple, sandstone, and pebbly sandstone; black, interbedded sandstone, siltstone, and mudstone.

form migration. Mudchip breccias at the base of sandstone layers are probably a product of the incorporation of desiccated muds during flooding. The common occurrence of thick parallel-stratified heavy-mineral bands suggests that flash-flooding was significant. This is supported by the local development of overturned cross-stratification. Delicate mudsaucers, likely to be preserved intact only when buried by eolian sands (Glennie, 1970), suggest an eolian contribution.

Mixed Facies

Description.—The mixed facies includes an assemblage of sub-facies which constitutes Unit C on the western side of the basin. Within an overall coarsening-upward trend, the sub-facies are arranged in 100-m-scale coarsening-upward sequences.

The siltstone sub-facies occurs primarily at the base of Unit C. It consists of fining-upward sequences of sandstone to siltstone to mudstone (Fig. 5B). Siltstone is the dominant rock type. The scale of interlayering ranges from centimeter-thick siltstone to mudstone couplets, to more fully developed decimeter-scale sequences. Lateral continuity of beds, even thin ones, is marked. Complete Bouma cycles were not observed; the most common sequence is an erosional surface overlain by massive or graded siltstone (or fine-grained sandstone) grading to mudstone. However, a lower parallel-stratified sandstone or ripple-drift cross-stratified sandstone component is locally developed. Mudstone rip-ups occur locally at the base of some sequences. Indications of subaerial exposure are lacking. Carbonates occur locally as one- to two-meter beds of millimeter- to centimeter-scale calcite or dolomite layers intercalated with

mudstone, and as centimeter-scale layers at the tops of siliciclastic Bouma-like sequences.

Upsection, fining-upward sequences of the graded sandstone-siltstone sub-facies tend to be more proximal in aspect (Fig. 5C). Sandstone layers are generally thicker, parallel stratification and ripple-drift cross stratification are common, and there are local mudcracks, pebbles and granules, wave ripples, lenticular and flaser bedding, and variegated redbeds (drab sandstones and red mudstones). In addition, meter-scale interbeds of coarse-grained sandstone with large bedform-related sedimentary structures appear. These sandstones are sheet-like, and the most common sedimentary structure is parallel stratification in horizontal or low-angle sets. The contacts of amalgamated sandstone beds are commonly low-angle truncation surfaces. Planar and trough cross-beds and reactivation surfaces are locally developed (Fig. 5E). Contacts between the meter-scale sandstones and the fining-upward sequences described above are sharp.

Two sub-facies occur as discontinuous lenses that interfinger with rocks of the graded sandstone-siltstone sub-facies near the top of Unit C. The conglomeratic pelite sub-facies contains monomictic granitoid and intraformational sandstone clasts (1) at the base of centimeter to decimeter-scale Bouma-like fining-upward sequences (Fig. 5D), (2) at the base of decimeter- to meter-scale granule-pebble conglomerate to sandstone (and rarely red mudstone) fining-upward sequences, and (3) as massive, meter-scale sheets of clast-supported granule to pebble (and locally cobble) conglomerate with a coarse-grained sandstone matrix.

The second sub-facies consists of clast-supported, non-stratified intraformational sandstone breccias (Fig. 5F). The clasts, up to boulder size, are set in a matrix of sandstone.

Relict scarps, in which *in situ* sandstone protolith passes along strike into breccia, are locally preserved. Fragment and scarp boundaries are sharp, indicating a high degree of induration. Breccia beds from 10 cm to more than 5 m thick are interbedded with undisturbed sandstones. The latter are identical to those which make up the breccia fragments. Sandstone and conglomerate beds with subrounded to well-rounded sandstone clasts are locally developed.

Interpretation.—The mixed facies is interpreted to have been deposited in an open, deep-water, perennial lake. In the Precambrian, distinguishing lake from marine environments is necessarily indirect. Regional aspects which indicate lacustrine deposition are enclosure of sub-basins by basement highs during sedimentation (see below), and stratigraphic position between units of fluvial sediments. The facies model is based on studies of present-day siliciclastic lakes formed mainly in temperate, humid climates such as in Alaska, Canada, and the Swiss Alps (Houbolt and Jonker, 1968; Gilbert, 1975; Gustavson, 1975; Sturm and Matter, 1978; Lambert and Hsü, 1979; Pharo and Carmack, 1979; Gilbert and Shaw, 1981). Represented are an open deep-water siliciclastic facies in which carbonate sedimentation occurred locally, a shallow-water facies, and a variety of marginal environments, which alternated along what was probably an extensive shoreline (including beaches, fan-deltas and beach cliffs). Coarsening-upward sequences of deep-water to marginal facies probably reflect sudden (fault induced?) deepening and infilling of the lake.

Rocks of the siltstone sub-facies are interpreted to be distal lacustrine turbidites on the basis of classical features such as extreme lateral continuity of evenly bedded, silt-dominant sandstone-siltstone interlayers, and the arrangement of the layers in Bouma-type fining-upward sequences with sharp erosive bases. Calcite and dolomite laminites are interpreted as chemical precipitates formed within the deep-water facies. They may represent sites where background siliciclastic sedimentation was sufficiently low to allow relatively undiluted accumulation of carbonate. The mudstone intercalations may represent interruptions by turbidity currents (see Figs. 22 and 28 of Dean and Fouch, 1983). Carbonate rocks do not axiomatically imply warm tropical conditions (Leonard et al. 1981; Walter and Bauld, 1983). Chemically precipitated rhythmites are an integral component of the deep parts of modern, predominantly siliciclastic lakes formed in cold wet climates (Müller and Wagner, 1978; Schäfer and Stapf, 1978; Sturm and Matter, 1978; Lambert and Hsü, 1979; Dean, 1981).

Fining-upward sequences in the graded sandstone-siltstone sub-facies are considered to be shallow-water equivalents of those in the siltstone sub-facies on the basis of (1) a thicker sandstone component (generally decimeter- rather than centimeter-scale), (2) more fully developed Bouma sequences, (3) locally developed mudcracks and wave-ripples, (4) local pebbly beds, and (5) a locally developed red coloration in fine-grained segments. Lenticular and flaser bedding may be wave-generated, and independent of tidal processes (de Raaf et al., 1977); their local development supports a shallow-water environment (at least one above storm wave-base).

The distinction between beach and fluvial deposits on the basis of lithology and sedimentary structures alone is difficult (Clifton, 1973). Although a fluvial component cannot be ruled out entirely, a beach origin for the meter-thick sandstone interbeds in the graded sandstone-siltstone sub-facies is favored by the following: (1) the intimate interbedding of the sandstones with the turbidites described above, (2) an absence of internal fluvial-type fining-upward sequences, and (3) the transition up-section to sandstones that more closely resemble fluvial deposits. Parallel-stratified heavy mineral bands may be from swash-backwash action on the beach face or upper-flow-regime transport in the surf zone (Thompson, 1937; Clifton, 1969); the low-angle discordances and stratification may be swash cross-stratification (Harms et al., 1982); and the planar cross-stratification may represent the migration of a variety of beach bars (Hoyt, 1962; Hine, 1979; Hunter et al., 1979).

The conglomeratic pelite sub-facies is considered to represent progradation of fan-deltas away from syn-sedimentary basement uplifts and into the lake. Mixed centimeter-scale couplets and decimeter-scale fining-upward sequences (with pebbly bases) are considered to represent prodelta to delta-front environments. Sequences that consist of channelized pebbly sandstone to parallel- or cross-stratified sandstone to red mudstone are interpreted to be lower fan-delta-plain deposits. The non-stratified meter-scale sheets of conglomerate are considered to represent proximal fan-delta-plain deposits (see Wescott and Ethridge, 1980, 1983; Dutton, 1982). Fan growth appears to have been restricted to specific sites in response to syn-sedimentary uplift of basement blocks. This is suggested by local development of the facies, basement-derived clasts of essentially one rock type, and abundance of intraformational sandstone clasts.

The intraformational sandstone breccias are inferred to be talus deposits of penecontemporaneously cemented sandstone, which accumulated at the foot of beach cliffs. Talus deposition is inferred on the basis of position adjacent to scarps of protolith rocks, clast size (up to 1 m), clast shape (many tabular slabs), clast roundness (many angular), clast composition (entirely intraformational), clast-supported framework, lack of stratification, evidence of autobrecciation, and evidence of only minor movement of blocks out of alignment with the *in situ* protolith. Sandstones and conglomerates with sandstone clasts probably were derived by reworking of breccia fragments by wave action. Similar deposits have been described from Scotland by Lawson (1976).

PALEOGEOGRAPHIC SYNTHESIS

Paleogeographic Model

Stratigraphic and sedimentologic observations imply an active western margin along which relatively continuous faulting was responsible for maintaining a rugged high-gradient paleotopography, and a relatively passive eastern margin of gentle gradient (Fig. 6). The contrast in preservation of the unconformity is compatible with faulting having been more intense and long-lived on the west. The greater thickness of Unit A on the west is regarded as a consequence of protracted faulting creating relief sufficient for the accumulation of a thick conglomeratic fill. The fact that

Unit B is relatively thin in the west, and that a full suite of facies is developed in Unit C are indications of the difference in paleotopography from west to east: on the west, high-gradient alluvial fans and braid-plains were located on the edge of a relatively deep permanent lake; to the east, the broad alluvial plain was characterized by a low gradient and local ponds. The changes in thickness and facies coincide with a major fault and a nearly continuous basement high, suggesting that movement along the fault was responsible for maintaining a longitudinal basement divide, which segmented the main basin during sedimentation. Monomictic granitic and intraformational sandstone clasts in the conglomeratic pelite sub-facies of Unit C, and conglomerate beds with monomictic granitic clasts in normally polymictic Unit E, are also considered to reflect more active syn-sedimentary faulting on the western side of the basin than on the eastern side.

Evidence for Syn-sedimentary Strike Slip

The extreme cumulative sedimentary thickness measured longitudinally through the Nonacho Basin (> 40 km), coupled with a uniformly low grade of metamorphism, is characteristic of strike-slip basins (see Crowell, 1974a,b; 1982a,b). Sites of sedimentation were physically moved along strike-slip faults so that there was longitudinal stacking of depositional units, like shingles on a roof. Extreme longitudinal thicknesses result, but at any one site, thicknesses are not sufficiently great to induce significant burial metamorphism. We cannot demonstrate with certainty the sense of strike slip during sedimentation, lacking definitive data such as clasts that can be matched with a distinctive point source (Crowell 1952, 1982b) or successive longitudinal offsetting of alluvial fans (Steel and Gloppen, 1980). However, a number of indirect features indicate sinistral movement.

First, the faults that border and transect the Nonacho Basin are part of a system that extends for about 450 km, from the northern tip of the basin to the Paleozoic cover south of Lake Athabasca (Fig. 7). Strike slip in the basement has been documented by Bostock (1984) and Culshaw (1984). Near-horizontal stretching lineations were observed in basement rocks bordering the basin in the present study. The time of initiation of the faults is unknown, but we can specify first movement as being before Nonacho Group sedimentation, because basal Nonacho conglomerates contain clasts of mylonitized granitic rocks and folded paragneiss near basement exposures where these rocks occur *in situ* (Fig. 4B). The basin is developed at a double bend in the Yatsore-Hill Island and Salkeld faults and at a left stepover with the MacInnis fault. Positioning of the basin where

these structural changes occur is probably not a coincidence; such geometry is consistent with sinistral strike slip (see Rodgers, 1980; Aydin and Nur, 1982; Mann et al.,

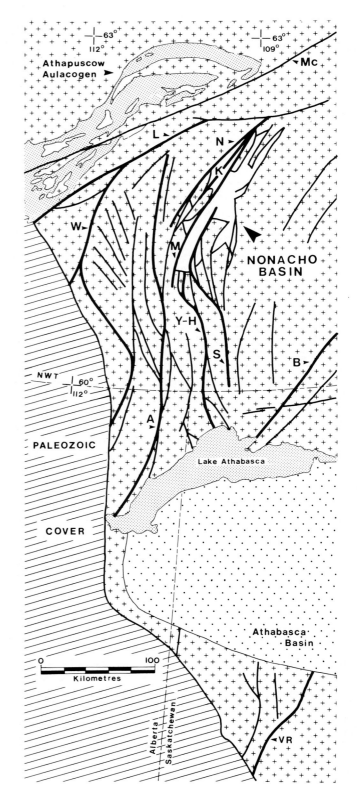

FIG. 7.—The Nonacho Basin with respect to shear zones in the northwestern Churchill Province. The basin is located at a double bend in the Yatsore-Hill Island (Y-H) and Salkeld (S) faults, and at a stepover between the Yatsore-Hill Island fault and the MacInnis (M), King (K) and Noman (N) faults. Other major faults: A, Allen; B, Black Bay; L, La Loche; Mc, McDonald; VR, Virgin River; W, Warren. Basement shears after Godfrey (1958), Reinhardt (1969), MacDonald and Broughton (1980), Bostock (1982, 1984), and Culshaw (1984).

1983). Northward thinning of Unit E conglomerates and thickening of Unit B sandstones away from the southern margin of the basin is consistent with the stepover having been active during sedimentation. It is uncertain how closely the present southern margin of the basin coincides with the margin at the time of sedimentation, but the thickly bedded, cobble-boulder conglomerates of Unit E in the south suggest nearby sources.

A second feature consistent with sinistral strike slip on the western margin is the abundance of intraformational sandstone clasts in conglomerates in the southern part of the basin. The basin formed in the south after deposition, lithification, uplift and cannibalization of previously deposited sediments in the north. In addition, the Nonacho Group was deposited mainly in a cold, wet paleoclimate, with local caliche indicating brief dry periods (Aspler and Donaldson, 1984). Although we appreciate the uncertainty of this assumption, if the first appearance of caliche can be used as an approximate time marker, then the observation that caliche occurs low in the section in the south but high in the section in the north, implies that deposition of several thousands of meters of sediment already had taken place in the north before sedimentation started in the south, confirming southward-younging of lithostratigraphic units.

Third, conglomerates provide a sampling of the basement rocks exposed during sedimentation, and therefore progressive stripping of basement should be reflected by a reversed stratigraphy in clast compositions (see Miall, 1970). The Nonacho conglomerates do not reflect such erosion into progressively deeper structural levels, because high-grade clasts are dominant low in the section (Unit A), and undeformed, unmetamorphosed porphyries, tuffs and quartz arenites occur high in the section (Unit E). To explain why the source rocks to Unit E conglomerates were not exposed initially, and how they became exposed in proximal sites without progressive stripping, we suggest tectonic emplacement during sedimentation (see Fig. 8 of Bluck, 1980). The direction of tectonic transport is considered to have been from the south for three reasons: (1) Unit E thickens southward; (2) proximal-fan facies pass laterally into distal-fan facies toward the north; and (3) beds rich in quartz arenite clasts occur only in the southern and eastern parts of the main basin. This is consistent with the (now eroded) source rocks having been emplaced from an area south of the basin as a result of sinistral motion along the Yatsore-Hill Island and Salkeld faults (Fig. 7). Potential sources for the volcanic clasts have not yet been discovered in the basement terrane. The alternative that the diversity results from the tapping of distant source areas via the development of a well-integrated tributary system is rejected because (1) the nature of the conglomerate facies at the base of the unit suggests that sources were close; (2) the admixture of hard and soft clast types suggests that the concentration is not a transport-distance effect; (3) the restriction of beds rich in quartz arenite cobbles and boulders to particular areas suggests deposition immediately adjacent to specific sources; and (4) the polymictic conglomerates appear above a basin-wide coarsening-upward megasequence (with retreat and integration of source areas, fining-upward sequences generally result; Heward 1978).

Evidence for Separate Sub-basins

The main basin and sub-basins were probably separate, at least in part, during sedimentation. This conclusion is based on changes in thickness, facies, and clast composition across major faults and across large basement inliers. The faults are thought to have controlled the geometry of the basin throughout its history.

As discussed above, the thickening of Unit B, and the disappearance of a full suite of facies in Unit C east of the MacInnis fault, is taken to indicate that the northwest part of the basin was separated from the main basin by a ridge of basement bounded on the east by the MacInnis fault (Figs. 6, 7). For similar reasons, the Taltson Outlier probably was separated from the main basin. It is uncertain if it was separated from the basin farther north, but the present-day basement block between the two suggests that this may have been so. The MacInnis Outlier is considered to have evolved independently because (1) sedimentation started with deposition of Unit E; (2) Unit E conglomerates are enriched in milky-quartz pebbles (locally 90%), in contrast to those in areas across intervening basement blocks; and (3) the interbedded sandstone-mudstone facies is developed in Unit F only in this area. The absence of stratigraphically lower units and the position of the outlier within a narrow block of basement that extends for almost the full length of the basin suggest that the sub-basin formed late, as a result of partial internal segmentation of the block.

Within the main basin, an abrupt appearance of pelitic rocks of Unit C (absent southward) is spatially related to termination of the Salkeld fault and a 90° bend in the basin border. This geometry is consistent with pulling apart of the main basin from the adjacent basement block along sinistral faults to the west. The area may never have been entirely separated from the main basin; rather, the abrupt and local appearance of the pelites probably reflects an isolated topographic low within the main basin generated by the pull-apart.

The Sparks-Naskethey sub-basin is regarded as having been separated from the main basin to the west by the large basement block between the two because of (1) thinning of Unit B across the block (400 m on the west, 2,100 m on the east), (2) development of Unit C on the east (Unit C is absent on the west), and (3) the high content of quartz arenite clasts in Unit E conglomerates on the east relative to exposures of Unit E immediately west of the basement block.

DEFORMATION

Steeply dipping north-northeast-striking faults concentrated on the western side of the basin are considered to have been the main loci of left-lateral strike slip (Fig. 8). Cross faults that bring basement into contact with Nonacho sediments at high angles to the main north-northeast trends probably had a more significant dip-slip component. The first post-depositional set of structures (D_1) consists of a sporadically developed but regionally distributed widely spaced cleavage, and related local folds whose presence is inferred on the basis of changes in facing direction of later (D_3) folds (see Kehlenbeck, 1984). During D_2, fault-bounded blocks rotated about gently plunging, east-trending axes.

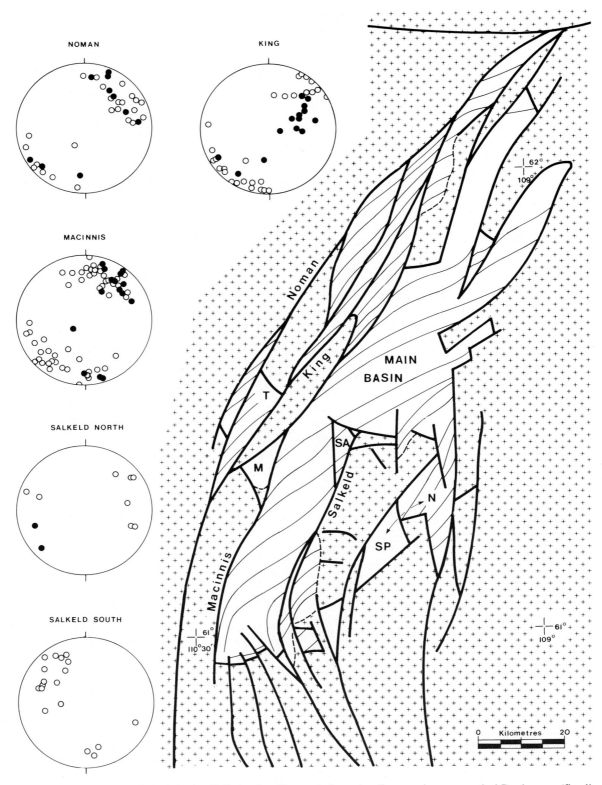

FIG. 8.—Nonacho faults, cleavage, and sub-basins. Foliation form lines are of a regionally pervasive near-vertical D$_3$ cleavage (fine lines) related to major near-vertical faults (bold lines). Lower hemisphere equal-area projections for faults: open circles, fiber-growths; dots, stretched-pebble lineations. Sinistral strike slip is indicated by the near-horizontal lineations and by the angular relationship between faults and cleavage (see text). Sub-basins: M, MacInnis; SA, Salkeld; SP-N, Sparks-Naskethey; T, Taltson.

FIG. 9.—Structures related to D₃ deformation. A) Gently plunging stretching lineation in monomictic conglomerate, adjacent to Noman fault (view is of near-vertical surface; black line indicates horizontal). B) Near-horizontal quartz fibers in near-vertical S₃ surface (adjacent to King fault). C) View to the southeast of a steep southeast-dipping fault surface (minor fault in King fault zone). Fiber-lineations (parallel to pen) are near-horizontal. The stepped pattern indicates sinistral shear (see Ramsay and Huber, 1983). D) Wavy fiber growths (pencil is parallel to the horizontal) along King fault. E) View of near-horizontal surface of pelitic rocks (Unit C) adjacent to King fault. Nearly vertical compositional layering is parallel to S₃. Nearly vertical discrete shears cutting across the layering from upper right to lower left at about 30° are similar to shear bands and indicate sinistral slip (see White et al., 1980; Platt and Vissers, 1980; Simpson, 1984; Weijermars and Rondeel, 1984). F) Tight fold in pelitic rocks (Unit C) adjacent to minor fault within King fault zone. Angle between trace of nearly-vertical axial surface and trace of nearly-vertical fault suggests sinistral shear.

This rotation was inferred through an analysis of pre-D_3 overturning of bedding, and the mismatch in pre-D_3 orientations of D_1 cleavage from block to block (Aspler, 1985a). Although rotations about vertical axes commonly are emphasized in splaying strike-slip systems (Freund, 1974), rotations about gently plunging axes are implicit in the depression associated with fault divergence, and uplift associated with fault convergence (Kingma, 1958; Lensen, 1958; Crowell, 1974a,b).

Figure 8 summarizes the variation in trend of the regionally penetrative, near-vertical D_3 cleavage (and related folds). Observations worthy of emphasis are: (1) the cleavage trends at angles of 20°–40° clockwise relative to the faults; (2) the cleavage is sigmoidal in blocks between faults (i.e., cleavage-fault angles are low adjacent to the faults and increase away from the faults); and (3) in zones west of MacInnis fault, the cleavage-fault angle is less than in zones such as the main basin. The configuration is analogous to a regional-scale version of "C" (shear) and "S" (plane of flattening) fabrics of Berthé et al. (1979). The cleavage is inferred to be related to the faults because, accompanying the changes in angle towards the faults is an increase in cleavage (and fold) intensity. This leads to the following suggestions: (1) the cleavage developed concurrently with strike-slip faulting; (2) the orientation of the cleavage relative to the faults indicates sinistral motion; (3) the sigmoidal pattern is a reflection of heterogeneous strain (Ramsay and Graham, 1970), with an increase in rotation (thus a decrease in cleavage-fault angle) towards the faults; and (4) the decreased angle west of MacInnis fault is an indication that this area is a zone of higher shear strain (see Davies, 1984). A low-angle displacement vector along the faults is confirmed by gently plunging pebble lineations and quartz fiber growths within S_3 (Figs. 8, 9) which are inferred to indicate the extension direction. Plunge reversals through the horizontal indicate changes in the dip-slip component, suggesting movement analogous to porpoises diving in and out of the water.

Post-D_3 structures include local open folds which crumple previously steeply dipping beds about a near-horizontal axis, and a conjugate set of strike-slip faults. The latter have the same orientation and sense (northwest-striking left-lateral faults, northeast-striking right-lateral faults) as those generated during terminal collision in Wopmay Orogen, and described by Hoffman and St.-Onge (1981) and Hoffman et al. (1984). In contrast to the major north-northeast-striking faults described above, breccia zones are common along the surfaces of these conjugate faults, indicating brittle-deformation. Although fault reactivation may have occurred, the conjugate faults generally cross-cut the north-northeast-striking faults, and no genetic relationship between the two can be demonstrated.

SUMMARY

The Nonacho Basin shares several traits with other terrigenous basins, including a thick fill of talus, alluvial-fan, fluvial, lacustrine, pond, fan-delta, and beach deposits; faults that were active during sedimentation; a paucity of volcanic rocks; and lower greenschist grade of metamorphism. Sinistral strike-slip faults on the western side of the basin (linked by cross-faults that probably had a more significant dip-slip component) are considered responsible for the development and maintenance of a system of rhomb, wedge, and rectangular sub-basins and intervening basement uplifts. Abrupt changes in clast composition, unit thickness, and facies that are coincident with fault contacts and basement inliers suggest that the sub-basins were at least partly independent during sedimentation.

The inferred paleotopography of the basin is asymmetric, with a high-gradient, active western margin characterized by abrupt transitions from alluvial fans to deep-water perennial lakes; and low-gradient, relatively passive eastern and southern margins characterized by gradual transitions and isolated ponds (Fig. 6). The asymmetry is similar to other strike-slip basins (Steel and Gloppen, 1980; Link, 1982; McLaughlin and Nilsen, 1982; Hempton et al., 1983; Dunne and Hempton, 1984; Hempton and Dunne, 1984). Indicators of strike slip during sedimentation include: the position of the basin in a releasing stepover in the regional fault network; extreme longitudinal stratigraphic thickness, inferred southward diachroneity of the basin fill, and the tectonic emplacement of source rocks to stratigraphically high conglomerates. Angular relationships between faults and related cleavage and folds indicate post-depositional heterogeneous sinistral shear.

ACKNOWLEDGMENTS

We benefited from discussions with R. L. Brown, W. K. Fyson, J. C. McGlynn, and S. M. Roscoe. E. J. Hurdle assisted in the field from 1978 to 1981. C. Journeay and E. J. Hurdle drafted the figures. K. T. Biddle, N. Christie-Blick, S. K. Hanmer, P. F. Hoffman, and R. Tirrul commented on an early version of the manuscript. Biddle, Christie-Blick and an anonymous reviewer provided helpful suggestions for preparation of the final draft. Field work was supported by the Department of Indian Affairs and Northern Development. We thank NSERC for a post-graduate scholarship to LBA and Operating Grant A5536 to JAD. W. A. Padgham, Chief Geologist, D.I.A.N.D. (Yellowknife) is thanked for his continued logistic, financial, and moral support.

REFERENCES

ADAMS, J., AND PATTON, J., 1979, Sebkha-dune deposition in the Lyons Formation (Permian) northern Front Range, Colorado: Mountain Geologist, v. 16, p. 47–57.

ALLEN, P. A., 1981, Sediments and processes on a small stream-flow dominated, Devonian alluvial fan, Shetland Islands: Sedimentary Geology, v. 29, p. 31–66.

ASPLER, L. B., 1985a, Geology of Nonacho Basin (Early Proterozoic) NWT [unpubl. Ph.D. thesis]: Ottawa, Ontario, Carleton University, 385 p.

————, 1985b, Geological maps of Nonacho Basin (1:125,000 and 1:250,000): Department of Indian Affairs and Northern Development Open File, Yellowknife, Northwest Territories.

ASPLER, L. B., AND DONALDSON, J. A., 1984, Paleoclimatology of Nonacho Basin (Early Proterozoic) Northwest Territories, Canada: Geological Society of America Abstracts With Programs, v. 16, p. 433.

AYDIN, A., AND NUR, A., 1982, Evolution of pull-apart basins and their scale independence: Tectonics, v. 1, p. 91–105.

BAADSGAARD, H., AND GODFREY, J. D., 1972, Geochronology of the Canadian Shield in northeastern Alberta, II: Charles-Andrew-Colin Lakes area: Canadian Journal of Earth Sciences, v. 9, p. 863–881.

BALLANCE, P. F., 1984, Sheet-flow-dominated gravel fans of the nonmarine Middle Cenozoic Simmler Formation, central California: Sedimentary Geology, v. 38, p. 337–359.

BERTHÉ, D., CHOUKROUNE, P., AND JEGOUZO, P., 1979, Orthogneiss, mylonite and non coaxial deformation of granites: the example of the South Armorican Shear Zone: Journal of Structural Geology, v. 1, p. 31–42.

BLUCK, B. J., 1980, Evolution of a strike-slip fault-controlled basin, Upper Old Red Sandstone, Scotland, *in* Ballance, P. F., and Reading, H. G., eds., Sedimentation in Oblique-Slip Mobile Belts: International Association of Sedimentologists Special Publication No. 4, p. 63–78.

BOOTHROYD, J. C., AND ASHLEY, G. M., 1975, Process, bar morphology, and sedimentary structures on braided outwash fans, northeastern Gulf of Alaska, *in* Jopling, A. V., and McDonald, B. C., eds., Glaciofluvial and Glaciolacustrine Sedimentation: Society of Economic Paleontologists and Mineralogists Special Publication No. 23, p. 193–222.

BOSTOCK, H. H., 1982, Geology of the Fort Smith map area, District of Mackenzie, Northwest Territories: Geological Survey of Canada Open File 859, 53 p.

————, 1984, Preliminary geological reconnaissance of the Hill Island Lake and Taltson Lake Areas, District of Mackenzie, *in* Current Research Part A: Geological Survey of Canada Paper 84-1A, p. 165–170.

BULL, W. B., 1977, The alluvial fan environment: Progress In Physical Geography, v. 1, p. 222–270.

BURWASH, R. A., AND BAADSGAARD, H., 1962, Yellowknife-Nonacho age and structural relations: Royal Society of Canada Special Publication 4, p. 22–29.

CLIFTON, H. E., 1969, Beach lamination: nature and origin: Marine Geology, v. 7, p. 553–559.

————, 1973, Pebble segregation and bed lenticularity in wave-worked versus alluvial gravel: Sedimentology, v. 20, p. 173–187.

COTTER, E., 1978, The evolution of fluvial style, with special reference to the central Appalachian Paleozoic, *in* Miall, A. D., ed., Fluvial Sedimentology: Canadian Society of Petroleum Geologists Memoir 5, p. 361–384.

CROWELL, J. C., 1952, Probable large lateral displacement on San Gabriel Fault, southern California: American Association of Petroleum Geologists Bulletin, v. 36, p. 2026–2035.

————, 1974a, Sedimentation along the San Andreas Fault, California, *in* Dott, R. H., Jr., and Shaver, R. H., eds. Modern and Ancient Geosynclinal Sedimentation: Society of Economic Paleontologists and Mineralogists Special Publication No. 19, p. 292–303.

————, 1974b, Origin of late Cenozoic basins in southern California, *in* Dickinson, W. R., ed., Tectonics and Sedimentation: Society of Paleontologists and Economic Mineralogists Special Publication No. 22, p. 190–204.

————, 1982a, The tectonics of Ridge Basin, southern California, *in* Crowell, J. C., and Link, M. H., eds., Geologic History of Ridge Basin, Southern California: Society of Paleontologists and Mineralogists, Pacific Section, p. 25–41.

————, 1982b, The Violin Breccia, Ridge Basin, southern California, *in* Crowell, J. C., and Link, M. H., eds., Geologic History of Ridge Basin, Southern California: Society of Economic Paleontologists and Mineralogists, Pacific Section, p. 89–97.

CULSHAW, N. G., 1984, Rutledge Lake, Northwest Territories; a section across a shear belt within the Churchill Province, *in* Current Research Part A: Geological Survey of Canada Paper 84-1A, p. 331–338.

DAVIES, F. B., 1984, Strain analysis of wrench faults and collision tectonics of the Arabian-Nubian Shield: Journal of Geology, v. 92, p. 37–53.

DEAN, W. E., 1981, Carbonate minerals and organic matter in sediments of modern north temperate hard-water lakes, *in* Etheridge, F. G., and Flores, R. M., eds., Recent and Ancient Nonmarine Depositional Environments: Models for Exploration: Society of Economic Paleontologists and Mineralogists Special Publication No. 31, p. 213–231.

DEAN, W. E., AND FOUCH, T. D., 1983, Lacustrine environment, *in* Scholle, P. A., Bebout, D. G., and Moore, C. H., eds., Carbonate Depositional Environments: American Association of Petroleum Geologists Memoir 33, p. 97–130.

DE RAAF, J. F. M., BOERSMA, J. R., AND VAN GELDER, A., 1977, Wave generated structures and sequences from a shallow marine succession,

Lower Carboniferous, County Cork, Ireland: Sedimentology, v. 24, p. 451–483.

DUNNE, L. A., AND HEMPTON, M. R., 1984, Deltaic sedimentation in the Lake Hazar pull-apart basin, south-eastern Turkey: Sedimentology, v. 31, p. 401–412.

DUTTON, S. P., 1982, Pennsylvanian fan-delta and carbonate deposition, Mobeetie Field, Texas Panhandle: American Association of Petroleum Geologists Bulletin, v. 66, p. 389–407.

ELMORE, R. D., 1983, Precambrian non-marine stromatolites in alluvial fan deposits, the Copper Harbor Conglomerate, Upper Michigan: Sedimentology, v. 30, p. 829–842.

ERIKSSON, K. E., 1978, Alluvial and destructive beach facies from the Archean Moodies Group, Barberton Mountain Land, South Africa and Swaziland, *in* Miall, A. D., ed., Fluvial Sedimentology: Canadian Society of Petroleum Geologists Memoir 5, p. 287–312.

EUGSTER, H. P., AND HARDIE, L. A., 1975, Sedimentation in an ancient playa-lake complex: the Wilkins Peak Member of the Green River Formation of Wyoming: Geological Society of America Bulletin, v. 86, p. 319–324.

FREUND, R., 1974, Kinematics of transform and transcurrent faults: Tectonophysics, v. 21, p. 93–134.

GILBERT, R., 1975, Sedimentation in Lillooet Lake, British Columbia: Canadian Journal of Earth Sciences, v. 12, p. 1697–1711.

GILBERT, R., AND SHAW, J., 1981, Sedimentation in proglacial Sunwapta Lake, Alberta: Canadian Journal of Earth Sciences, v. 18, p. 81–93.

GLENNIE, K. W., 1970, Desert sedimentary environments: Amsterdam, Elsevier, 222 p.

GODFREY, J. D., 1958. Aerial photographic interpretation of Precambrian structures north of Lake Athabasca: Research Council of Alberta, Geological Division, Bulletin 1, 9 p.

GROVER, J. A., 1984, Petrology, depositional environments and structural development of the Mineta Formation, Teran Basin, Cochise County, Arizona: Sedimentary Geology, v. 38, p. 87–105.

GUSTAVSON, T. C., 1975, Sedimentation and physical limnology in proglacial Malaspina Lake, southeastern Alaska, *in* Jopling, A. V., and McDonald, B. C., eds., Glaciofluvial and Glaciolacustrine Sedimentation: Society of Economic Paleontologists and Mineralogists Special Publication No. 23, p. 249–263.

HAMPTON, M. A., 1979, Buoyancy in debris flows: Journal of Sedimentary Petrology, v. 49, p. 753–758.

HARDIE, L. A., SMOOT, J. P., AND EUGSTER, H. P., 1978, Saline lakes and their deposits: a sedimentological approach, *in* Matter, A., and Tucker, M. E., eds., Modern and Ancient Lake Sediments: International Association of Sedimentologists Special Publication No. 2, p. 7–42.

HARMS, J. C., SOUTHARD, J. B., AND WALKER, R. G., 1982, Structures and sequences in clastic rocks: Society of Economic Paleontologists and Mineralogists Short Course Notes 9.

HEIN, F. J., AND WALKER, R. G., 1977, Bar evolution and development of stratification in the gravelly, braided Kicking Horse River, British Columbia: Canadian Journal of Earth Sciences, v. 14, p. 562–570.

HEMPTON, M. R., AND DUNNE, L. A., 1984, Sedimentation in pull-apart basins: active examples in eastern Turkey: Journal of Geology, v. 92, p. 513–530.

HEMPTON, M. R., DUNNE, L. A., AND DEWEY, J. F., 1983, Sedimentation in an active strike-slip basin, southeastern Turkey: Journal of Geology, v. 91, p. 401–412.

HENDERSON, J. F., 1937, Nonacho Lake area, Northwest Territories: Geological Survey of Canada Paper 37-2, 22 p.

HESSE, R., AND READING, H. G., 1978, Subaqueous clastic fissure eruptions and other examples of sedimentary transposition in the lacustrine Horton Bluff Formation (Mississippian), Nova Scotia, Canada, *in* Matter, A., and Tucker, M. E., eds., Modern and Ancient Lake Sediments: International Association of Sedimentologists Special Publication No. 2, p. 239–256.

HEWARD, A. P., 1978, Alluvial fan sequence and megasequence models: with examples from Westphalian D-Stephanian B coalfields, northern Spain, *in* Miall, A. D., ed., Fluvial Sedimentology: Canadian Society of Petroleum Geologists Memoir 5, p. 669–702.

HINE, A. C., 1979, Mechanisms of berm development and resulting beach growth along a barrier spit complex: Sedimentology, v. 26, p. 333–351.

HOFFMAN, P. F., AND ST.-ONGE, M. R., 1981, Contemporaneous thrusting and conjugate faulting during second collision in Wopmay Orogen: implications for the subsurface structure of post-orogenic outliers, *in* Current Research Part A: Geological Survey of Canada Paper 81-1A, p. 251–257.

HOFFMAN, P. F., TIRRUL, R., GROTZINGER, J. P., LUCAS, S. B., AND ER-IKSSON, K. A., 1984, The externides of Wopmay Orogen, Takijuq Lake and Kikerk Lake map areas, District of Mackenzie, *in* Current Research Part A: Geological Survey of Canada Paper 84-1A, p. 383–395.

HOUBOLT, J. J. H. C., AND JONKER, J. B. M., 1968, Recent sediments in the eastern part of the Lake of Geneva (Lac Leman): Geologie en Mijnbouw, v. 47, p. 131–148.

HOYT, J. H., 1962, High-angle beach stratification, Sapelo Island, Georgia: Journal of Sedimentary Petrology, v. 32, p. 309–311.

HUBERT, J. F., AND HYDE, M. G., 1982, Sheet-flow deposits of graded beds and mudstones on an alluvial sandflat-playa system: Upper Triassic Blomidon redbeds, St. Mary's Bay, Nova Scotia: Sedimentology, v. 29, p. 457–474.

HUNTER, R. E., CLIFTON, H. E., AND PHILIPS, R. L., 1979, Depositional processes, sedimentary structures, and predicted vertical sequences in barred nearshore systems, southern Oregon Coast: Journal of Sedimentary Petrology, v. 49, p. 711–726.

KELLENBECK, M. M., Use of stratigraphic and structural-facing directions to delineate the geometry of refolded folds near Thunder Bay, Ontario: Geoscience Canada, v. 11, p. 23–32.

KINGMA, J. T., 1958, Possible origin of piercement structures, local unconformities and secondary basins in the eastern geosyncline, New Zealand: New Zealand Journal of Geology and Geophysics, v. 1, p. 269–274.

LAMBERT, A., AND HSÜ, K. J., 1979, Non-annual cycles of varve-like sediment in Walensee, Switzerland: Sedimentology, v. 29, p. 453–462.

LAWSON, D. E., 1976, Sandstone-boulder conglomerates and a Torridonian cliffed shoreline between Gairloch and Stoer, northwest Scotland: Scottish Journal of Geology, v. 12, p. 67–88.

LENSEN, G. J., 1958, A method of horst and graben formation: Journal of Geology, v. 66, p. 579–587.

LEONARD, J. E., CAMERON, B., PILKEY, O. H., AND FRIEDMAN, G. M., 1981, Evaluation of cold-water carbonates as possible paleoclimatic indicators: Sedimentary Geology, v. 28, p. 1–28.

LINK, M. H., 1982, Slope and turbidite facies of the Miocene Castaic Formation and the lower part of the Marple Canyon Sandstone Member, Ridge Route Formation, Ridge Basin southern California, *in* Crowell, J. C., and Link, M. H., eds., Geologic History of Ridge Basin, Southern California: Society of Paleontologists and Mineralogists, Pacific Section, p. 79–87.

MACDONALD, R., AND BROUGHTON, P., 1980, Geological map of Saskatchewan, provisional edition, Scale 1:1,000,000: Saskatchewan Geological Survey.

MANN, P., HEMPTON, M. R., BRADLEY, D. C., AND BURKE, K., 1983, Development of pull-apart basins: Journal of Geology, v. 91, p. 529–554.

MCGLYNN, J. C., 1966, Thekulthili Lake area, *in* Report of Activities: Geological Survey of Canada Paper 66-1A, p. 32–33.

———, 1970, Study of the Nonacho Group of sedimentary rocks, Nonacho Lake, Taltson and Reliance areas, District of Mackenzie, *in* Report of Activities: Geological Survey of Canada Paper 70-1A, p. 154–155.

———, 1971, Stratigraphy, sedimentology and correlation of the Nonacho Group, District of Mackenzie, *in* Report of Activities: Geological Survey of Canada Paper 71-1A, p. 140–142.

MCGOWEN, J. H., AND GROAT, C. G., 1971, Van Horn Sandstone, West Texas: an alluvial fan model for mineral exploration: Bureau of Economic Geology, The University of Texas at Austin Report of Investigations 72, 57 p.

MCKEE, E. D., CROSBY, E. J., AND BERRYHILL, H. L., 1967. Flood deposits, Bijou Creek, Colorado, June 1965: Journal of Sedimentary Petrology, v. 37, p. 829–851.

MCLAUGHLIN, R. J., AND NILSEN, T. H., 1982, Neogene nonmarine sedimentation and tectonics in small pull-apart basins of the San Andreas Fault System, Sonoma County, California: Sedimentology, v. 29, p. 865–876.

MIALL, A. D., 1970, Devonian alluvial fans, Prince of Wales Island, Arctic Canada: Journal of Sedimentary Petrology, v. 40, p. 556–571.

———, 1981, Alluvial sedimentary basins: tectonic setting and basin architecture, *in* Miall, A. D., ed., Sedimentation and Tectonics in Alluvial Basins: Geological Association of Canada Special Paper 23, p. 1–33.

MÜLLER, G., AND WAGNER, F., 1978, Holocene carbonate evolution in Lake Balaton (Hungary): a response to climate and impact of man, *in* Matter, A., and Tucker, M. E., eds., Modern and Ancient Lake Sediments: International Association of Sedimentologists Special Publication No. 2, p. 55–80.

NIELSEN, P. A., LANGENBERG, C. W., BAADSGAARD, H., AND GODFREY, J. D., 1981, Precambrian metamorphic conditions and crustal evolution, northeastern Alberta, Canada: Precambrian Research, v. 16, p. 171–193.

NILSEN, T. H., 1982, Alluvial fan deposits, *in* Scholle, P. A., and Spearing, D., eds., Sandstone Depositional Environments: American Association of Petroleum Geologists Memoir 31, p. 49–86.

PHARO, C. H., AND CARMACK, E. C., 1979, Sedimentation processes in a short residence-time intermontane lake, Kamloops Lake, British Columbia: Sedimentology, v. 26, p. 523–541.

PLATT, J. P., AND VISSERS, R. L. M., 1980, Extensional structures in anisotropic rocks: Journal of Structural Geology, v. 2, p. 397–410.

RAMSAY, J. G., AND GRAHAM, R. H., 1970, Strain variations in shear belts: Canadian Journal of Earth Sciences, v. 7, p. 786–813.

RAMSAY, J. G., AND HUBER, M. I., 1983, The Techniques of Modern Structural Geology, Volume 1: Strain Analysis: London, Academic Press, 307 p.

REEVES, C. C., JR., Economic significance of playa lake deposits, *in* Matter, A., and Tucker, M. E., eds., Modern and Ancient Lake Sediments: International Association of Sedimentologists Special Publication No. 2, p. 277–290.

REINHARDT, E. W., 1969, Geology of the Precambrian rocks of Thubun Lakes Map-Area in relationship to the McDonald Fault system: Geological Survey of Canada Paper 69-21, 29 p.

RODINE, J. D., AND JOHNSON, A. M., 1976, The ability of debris, heavily freighted with coarse clastic materials, to flow on gentle slopes: Sedimentology, v. 23, p. 213–234.

RODGERS, D. A., 1980, Analysis of pull-apart basin development produced by *en echelon* strike-slip faults, *in* Ballance, P. F., and Reading, H. G., eds., Sedimentation in Oblique-Slip Mobile Zones: International Association of Sedimentologists Special Publication No. 4, p. 27–41.

RUST, B. R., 1978, Depositional models for braided alluvium, *in* Miall, A. D., ed., Fluvial Sedimentology: Canadian Society of Petroleum Geologists Memoir 5, p. 605–626.

———, 1984, Proximal braidplain deposits in the Middle Devonian Malbaie Formation of eastern Gaspe, Quebec, Canada: Sedimentology, v. 31, p. 675–695.

SCHÄFER, A., AND STAPF, K. R. G., 1978, Permian Saar-Nahe Basin and Recent Lake Constance (Germany): two environments of lacustrine algal carbonates, *in* Matter, A., and Tucker, M. E., eds., Modern and Ancient Lake Sediments: International Association of Sedimentologists Special Publication No. 2, p. 81–106.

SCHUMM, S. A., 1967, Speculations concerning paleohydrological controls of terrestrial sedimentation: Geological Society of America Bulletin, v. 79, p. 1573–1588.

SENI, S. J., 1980, Sand body geometry and depositional systems, Ogallala Formation, Texas: Bureau of Economic Geology, the University of Texas at Austin, Report of Investigation 105, 36 p.

SIMPSON, C., 1984, Borrego Springs-Santa Rosa Mylonite Zone: a Late Cretaceous west-directed thrust in southern California: Geology, v. 12, p. 8–11.

SMOOT, J. P., 1983, Depositional subenvironments in an arid closed basin; the Wilkins Peak Member of the Green River Formation (Eocene), Wyoming, USA: Sedimentology, v. 30, p. 801–828.

SNEH, A., 1979, Late Pleistocene fan-deltas along the Dead Sea Rift: Journal of Sedimentary Petrology, v. 49, p. 541–552.

SPÖRLI, K. D., 1980, New Zealand and oblique-slip margins: tectonic development up to and during the Cainozoic, *in* Ballance, P. F., and Reading, H. G., eds., Sedimentation in Oblique-Slip Mobile Zones: International Association of Sedimentologists Special Publication No. 4, p. 147–170.

STEAR, W. M., 1983, Morphologic characteristics of ephemeral stream channel and overbank splay sandstone bodies in the Permian Lower

Beaufort Group, Karoo Basin, South Africa, *in* Collinson, J. D., and Lewin, J., eds., Modern and Ancient Fluvial Systems: International Association of Sedimentologists Special Publication No. 6, p. 405–420.

STEEL, R. J., AND GLOPPEN, T. G., 1980, Late-Caledonian (Devonian) basin formation, western Norway: signs of strike-slip tectonics during infilling, *in* Ballance, P. F., and Reading, H. G., eds., Sedimentation in Oblique-Slip Mobile Zones: International Association of Sedimentologists Special Publication No. 4, p. 79–103.

STEWART, W. D., AND WALKER, R. G., 1980, Eolian coastal dune deposits and surrounding marine sandstones, Rocky Mountain Supergroup (Lower Pennsylvanian), southeastern British Columbia: Canadian Journal of Earth Sciences, v. 17, p. 1125–1140.

STURM, M., AND MATTER, A., 1978, Turbidites and varves in Lake Brienz (Switzerland): deposition of clastic detritus by density currents, *in* Matter, A., and Tucker, M. E., eds., Modern and Ancient Lake Sediments: International Association of Sedimentologists Special Publication No. 2, p. 145–168.

THOMPSON, W. O., 1937, Original structures of beaches, bars and dunes: Geological Society of America Bulletin, v. 48, p. 723–752.

TURNBRIDGE, I. P., 1981, Sandy high-energy flood sedimentation: some criteria for recognition, with an example from the Devonian of S. W. England: Sedimentary Geology, v. 28, p. 79–95.

———, 1983, Alluvial fan sedimentation of the Horsehoe Park Flood, Colorado, USA, July 15, 1982: Sedimentary Geology, v. 36, p. 15–23.

VOS, R. G., AND TANKARD, A. J., 1981, Braided fluvial sedimentation in the Lower Paleozoic Cape Basin, South Africa: Sedimentary Geology, v. 29, p. 171–193.

WALTER, M. R., AND BAULD, J., 1983, The association of sulphate evaporites, stromatolitic carbonates and glacial sediments: examples from the Proterozoic of Australia and the Cainozoic of Antarctica: Precambrian Research, v. 21, p. 129–148.

WEIJERMARS, R., AND RONDEEL, H. E., 1984, Shear band foliation as an indicator of sense of shear: field observations in central Spain: Geology, v. 12, p. 603–606.

WESCOTT, W. A., AND ETHRIDGE, F. G., 1980, Fan-delta sedimentology and tectonic setting: American Association of Petroleum Geologists Bulletin, v. 64, p. 374–399.

WESCOTT, W. A., AND ETHRIDGE, F. G., 1983, Eocene fan-delta—submarine fan deposition in the Wagwater Trough, east-central Jamaica: Sedimentology, v. 30, p. 235–247.

WHITE, S. H., BURROWS, S. E., CARRERAS, J., SHAW, N. D., AND HUMPHREYS, F. J., 1980, On mylonites in ductile shear zones: Journal of Structural Geology, v. 2, p. 175–187.

WINSTON, D., 1978, Fluvial systems of the Precambrian Belt Supergroup, Montana and Idaho, U.S.A., *in* Miall, A. D., ed., Fluvial Sedimentology: Canadian Society of Petroleum Geologists Memoir 5, p. 343–360.

NEOTECTONICS OF A STRIKE-SLIP RESTRAINING BEND SYSTEM, JAMAICA

PAUL MANN,

Institute of Geophysics, University of Texas at Austin, Austin, Texas 78751;

GRENVILLE DRAPER

Department of Geology, Florida International University, Miami Florida 33199;

AND

KEVIN BURKE

Lunar and Planetary Institute, 3303 NASA Road 1, Houston, Texas 77058; and
Department of Geosciences, University of Houston/Central Campus, Houston, Texas 77004

ABSTRACT: The pattern of faulting and fault-related deformation in Cretaceous to Neogene rocks, together with the distribution of Neogene sediments, suggest that Jamaica is the site of two complex, right-stepping restraining bends along the strike-slip plate boundary between the North American and Caribbean plates. Block convergence at the eastern bend between the Plantain Garden and Duanvale fault zones is manifest by topographic uplift (>2 km), rapid erosion, and northwest-southeast shortening of Cretaceous and Paleogene metamorphic, volcanic and sedimentary rocks in the Blue Mountains and Wagwater Belt. Limited data on the age of faulting in Jamaica suggest that deformation and uplift related to bends in the faults probably began in the middle to late Miocene and was roughly contemporaneous with initial strike slip along the eastern extension of the Plantain Garden fault zone in southern Hispaniola. Uplift and deformation at the western bend is less prominent, and for the most part involves carbonate rocks that are cut by numerous west-facing fault scarps thought to be formed by east-dipping high-angle reverse faults. Both bends appear to have nucleated on northwest-striking normal faults that bounded Paleogene rifts. Maps of historic and recorded earthquakes on Jamaica indicate a close spatial association between the two restraining bends and the largest-magnitude events. In Jamaica, as in other active and ancient strike-slip zones, it is unclear how observed compressional deformation relates to the following three mechanisms: (1) *restraining bend development* or interaction of two parallel, overstepping strike-slip faults; (2) *simple shear* adjacent to a single strike-slip fault; or (3) *end effects* caused by termination of a single strike-slip fault.

INTRODUCTION

Fieldwork along active strike-slip boundaries has shown that strike-slip faults parallel to the direction of relative plate motion are characterized by relatively narrow fault zones, whereas faults striking oblique to the direction of plate motion form wider zones of either convergence (transpression) or divergence (transtension; Harland, 1971). Convergent areas, generally referred to as restraining bends (Crowell, 1974) or push-ups, are characterized by thrusting and mountains. Divergent or releasing fault bends develop into extensional pull-apart basins (Mann et al., 1983) that are typically submarine or at least topographically low.

Restraining bends are found along most active strike-slip fault systems and occur at a variety of scales. General characteristics of major bends (that is, master-fault separation or steps of roughly 100 km) have been described in the Transverse Ranges of southern California (Crowell, 1979), the Southern Alps of New Zealand (Walcott and Cresswell, 1979), the Lebanon Ranges of Lebanon (Hancock and Atiya, 1979), and the several mountain ranges which make up the island of Hispaniola (Mann et al., 1984). The complexity of strike-slip zones, particularly ancient inactive ones, makes it difficult to distinguish restraining bend structure from: (1) compressional deformation related to simple shear adjacent to the strike-slip fault, or (2) compressional deformation at the termination of a single strike-slip fault (see Aydin and Page, 1984, for a discussion of complexities along the San Andreas fault). Data from active areas provide an instantaneous picture of fault interactions, which can act as models for interpreting ancient strike-slip systems.

We present here an integrated tectonic, sedimentary, and seismic description of a Neogene restraining bend system that occurs on the island of Jamaica (Fig. 1). The Jamaican bend system is considerably smaller and appears to have a less complicated origin and tectonic development than the larger restraining bends mentioned above. We hope that documentation of the structural and sedimentary characteristics of the Jamaican bends may prove useful in identifying and interpreting ancient areas of strike-slip convergence including those suggested in such diverse settings as accretionary prisms (Karig, 1979), and collisional zones (Woodcock and Robertson, 1982; Brun and Burg, 1982).

GEOLOGIC AND TECTONIC SETTING OF JAMAICA

The island of Jamaica, measuring approximately 225 km by 100 km, forms an emergent part of the Nicaragua Rise along the northern edge of the Caribbean plate (Fig. 1). The Nicaragua Rise is a broad submarine swell of intermediate crustal thickness (~22 km) that extends from Honduras in Central America to southern Haiti on the island of Hispaniola (Arden, 1975). Jamaica lies entirely within a 200-km-wide seismic zone of left-lateral strike-slip deformation between the North American and Caribbean plates (Burke et al., 1980; Fig. 1). Two throughgoing strike-slip fault systems interrupted by push-up and pull-apart segments have been identified in the plate boundary zone: the Septentrional-Oriente-Swan-Motagua fault system passes to the north of Jamaica and bounds the Cayman Trough pull-apart structure (Holcombe et al., 1973), while the Enriquillo-Plantain Garden-Swan fault system extends from the Dominican Republic in central Hispaniola, through Jamaica, and appears to merge with the Swan fault zone along the southern edge of the Cayman Trough (Mann et al., 1983; Fig. 1). The total rate of relative plate motion is at least 2 cm/yr (present rate of spreading in the Cayman Trough, Macdonald and Holcombe, 1978) and perhaps as fast as 4 cm/yr (derived from the configuration of the seismic zone in the northeastern Caribbean, Sykes et al., 1982). Compilations of field and seismic data suggest that, as a general rule, strike-slip faults in this part of the Caribbean parallel

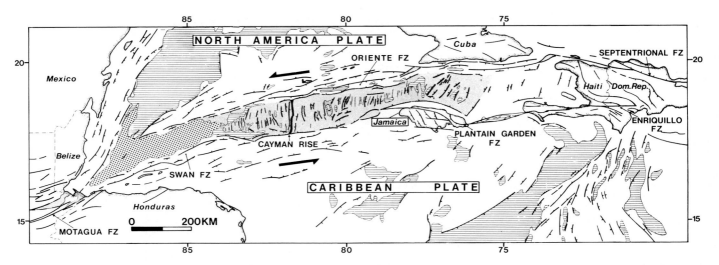

FIG. 1.—Recent fault and sediment map of the North American-Caribbean plate boundary zone (modified from Case and Holcombe, 1980; Mann et al., 1983). Dot pattern represents submarine-fan complex and horizontal lines indicate distal turbidites. Note position of Jamaica at compressional right step or restraining bend between the left-lateral Enriquillo-Plantain Garden fault zone and faults along the southern margin of the Cayman Trough pull-apart. Fine stipple pattern indicates Neogene pull-apart basins produced at extensional left steps between left-lateral strike-slip faults within the plate boundary zone.

predicted east-trending interplate slip lines (Jordan, 1975), whereas reverse faults commonly strike northwest, and are associated with uplifted land areas such as Jamaica (Burke et al., 1980) and Hispaniola (Mann et al., 1984). In this paper, high-angle reverse faults and low-angle thrusts are both referred to as reverse faults for the sake of simplicity. Normal faults typically strike northeast and are associated with offshore sedimentary basins such as the Yallahs Basin south of Jamaica (Burke, 1967).

Jamaica consists of widely exposed Cenozoic carbonate rocks unconformably overlying a basement composed mostly of Albian to Maastrichtian volcanic, volcaniclastic, and plutonic rocks (Fig. 2). These basement rocks are typical of those found in arcs built entirely on oceanic crust, and they have been interpreted as an intra-oceanic arc complex, active throughout most of the Cretaceous (Roobol, 1972; Horsfield and Roobol, 1974; Draper, 1979; Grippi and Burke, 1980). The Cretaceous rocks are locally exposed in a num-

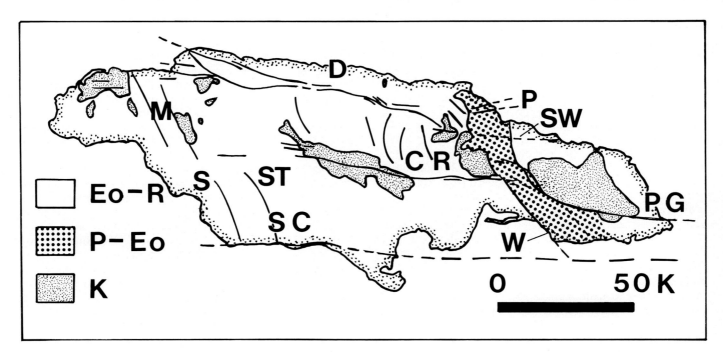

FIG. 2.—Major late Neogene faults of Jamaica and schematic geological map. Ages of rocks: K, Cretaceous; P-Eo, Paleocene-Eocene sediments of Wagwater Belt; Eo-R, Eocene to Recent limestones. Fault abbreviations: D, Duanvale; P, Peyton's Cove; SW, Seafield-Whitehall; CR, Crawle River; PG, Plantain Garden; W, Wagwater; SC, South Coast; ST, Spur Tree; S, Santa Cruz; M, Montpelier-Newmarket.

ber of elongate inliers that have been uplifted along north-west-striking, reverse faults (Fig. 2). In addition to arc rocks, a Paleogene graben, the Wagwater graben, has been uplifted by reverse faulting along its former, northwest-striking boundary faults (Fig. 2). The deformed sedimentary and volcanic rocks of the reactivated Wagwater graben are here referred to as the Wagwater Belt.

In this paper, we describe the predominately compressional structures developed within the Cretaceous inliers of eastern Jamaica (Blue Mountain and Sunning Hill inliers), the Paleogene Wagwater Belt, and the widely exposed Cenozoic carbonate rocks of central and western Jamaica. We relate these structures to restraining bends or right steps in Neogene strike-slip faults.

STRUCTURE OF THE BLUE MOUNTAINS

Fault Pattern and General Structure

As the nearly east-striking Plantain Garden fault zone veers to the northwest in southeastern Jamaica, it splays into a complex zone of generally northwest-striking faults along the southwestern corner of the Blue Mountains (Figs. 2, 3, 4). This fault zone reaches a maximum width of 7–8 km and generally separates arc rocks exposed in the Blue

Mountains from rift-related rocks of Paleocene and Eocene age that were deposited in the Wagwater graben (Fig. 2). The highest peaks in Jamaica, which are over 2 km in elevation, are found immediately to the north of this region in the northwest-trending Main Ridge of the Blue Mountains (black triangles in Fig. 4).

The two morphologically most prominent fault zones within this region are the Yallahs and Blue Mountain fault zones, which occupy prominent linear valleys and are easily visible on radar imagery (Wadge and Dixon, 1984; Figs. 3, 4). The Yallahs fault zone is convex southward, and describes an arc that generally separates a thick (~6.8 km) sedimentary sequence of Paleogene volcanic and clastic rocks, deposited in the Wagwater Belt to the west, from a thinner (0.1–0.5 km) sequence of clastic rocks, deposited along the edges of the graben. The apparent northward extension of the Yallahs fault zone, the Silver Hill fault zone (Fig. 7, below), continues almost to the coast of northern Jamaica, and like the Yallahs fault zone, separates thick rift sediments from much thinner graben-edge deposits. The Yallahs and Silver Hill faults therefore appear to represent reactivated normal faults that formed the eastern edge of the Paleogene graben (Green, 1977).

The morphologically prominent Blue Mountain fault zone

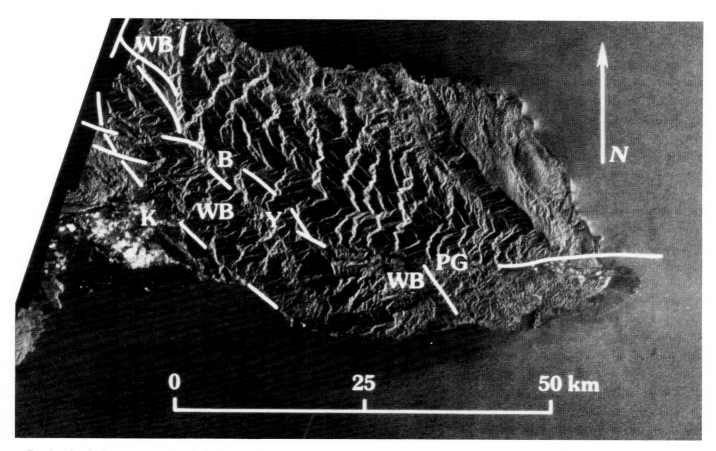

FIG. 3.—Synthetic aperature radar (SAR) image of eastern Jamaica (processed by T. Dixon, Jet Propulsion Lab). Radar illumination direction is from east-southeast. White lines emphasize major fault zones: PG, Plantain Garden; Y, Yallahs; B, Blue Mountain. WB indicates area of Wagwater Belt, a reactivated Paleogene graben. Steep topography of the central and southern Wagwater Belt and Blue Mountains results in strong distortion or "layover" effects. Note narrowing of the Wagwater Belt from south to north. Metropolitan Kingston is indicated by the letter K.

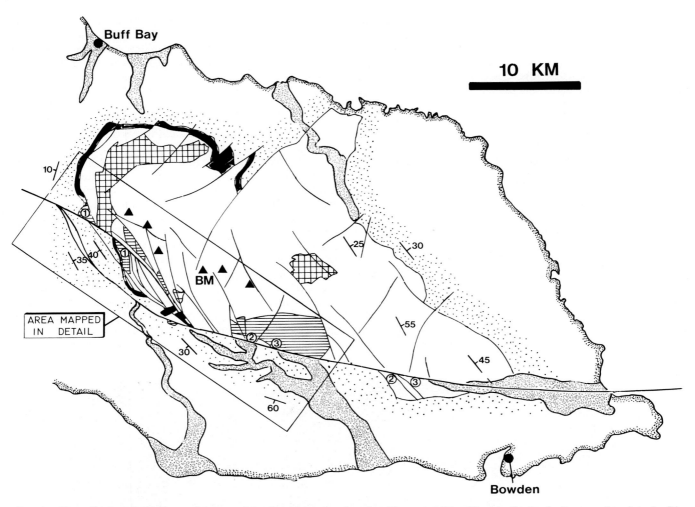

FIG. 4.—Generalized geological map of the restraining bend in eastern Jamaica. The east-striking Plantain Garden fault zone splays into the Blue Mountain and Yallahs fault zones southwest of Blue Mountain Peak (BM; elevation, 2,256 m). Details are shown in Figs. 5 and 6. Explanation of map patterns and symbols: gravel pattern, fluvial and fan-delta sediment shed off the Blue Mountains domal uplift; dotted pattern, Paleogene rocks (mostly Richmond Formation); checkered pattern, late Cretaceous plutons; horizontal lined pattern, blueschists, greenschists, and amphibolites; grey, upper Paleocene Chepstow Limestone; black, serpentinite; unpatterned area of Blue Mountains, mostly unmetamorphosed volcaniclastic sedimentary and volcanic rocks; black triangles, peaks of the Main Ridge of the Blue Mountains. Circled numbers refer to strike-slip offsets along the Blue Mountain and Plantain Garden fault zones, discussed in text.

is convex northwards (Fig. 3, 4), and generally separates thin graben-edge clastic rocks, unconformable on Cretaceous arc basement, from other Cretaceous arc rocks that constitute the Main Ridge of the Blue Mountains. The Cretaceous rocks found near the Blue Mountain and Yallahs fault zones include fine-grained, low-grade blueschist, greenschist, mafic to quartzo-feldspathic amphibolite, and unmetamorphosed granodiorite (Draper et al., 1976; Draper 1979). These rocks appear to represent the deeper structural levels of a Cretaceous arc-trench complex (Draper, 1979) that has been locally upthrust along the Plantain Garden, Yallahs, and Blue Mountain fault zones (Fig. 4). The deformation of this arc-trench complex is discussed in more detail below. To the northeast of the Blue Mountains, unmetamorphosed Cretaceous rocks, largely granodiorite, and volcanic and volcaniclastic rocks, are generally tilted towards the northeast (Fig. 4) and appear to represent the

higher structural levels of a volcanic arc. Eocene to Miocene limestones in the eastern Blue Mountains are tilted towards the northeast at as much as 30° (McFarlane, 1977; Wadge and Draper, 1978).

The eastward extension of the Yallahs-Blue Mountain fault zones, the Plantain Garden fault zone, forms a remarkably straight boundary between the Cretaceous rocks of the Blue Mountains and predominantly alluvial deposits to the south. The escarpment north of the fault rises steeply to elevations of 600 m. Two types of morphological feature common to active strike-slip faults are present: offset drainages (the Banana River and an unnamed river draining Round Hill), and faceted spurs (the segment between Hillside and Cross Pass). However, the sinistral offset on the drainages is variable, less than 200 m, and therefore, unreliable for determining the magnitude of strike slip.

There are several apparent left-lateral offsets along the

fault between northwest-striking Cretaceous rocks south of the fault and the Blue Mountains (Fig. 4). An apparent offset of about 10–12 km is suggested by the lithologic similarity of Campanian rocks north and south of the fault (offset no. 2 in Fig. 4). In addition, a 100-m-wide Cretaceous syenite dike, an unusual rock type in Jamaica, appears to have been left-laterally offset by 12.8 km (offset no. 3 in Fig. 4; Draper, 1979), although reconnaissance mapping indicates that the dike in the southern area is approximately parallel to the Plantain Garden fault zone. The exact magnitude of offset is therefore suspect. Wadge and Eva (1978) have pointed out another apparent offset (also about 10 km, and not shown in Fig. 4) of a northwest-trending gravity and magnetic high that is possibly associated with ultramafic rocks.

As the Plantain Garden fault zone curves northward into the Yallahs and Blue Mountain fault zones, apparent left-lateral offset decreases. Green (1977) pointed out a maximum apparent offset of about 3.5 km along the Blue Mountain fault zone on the basis of the close correspondence of several formations and complex structures on opposite sides of the fault (offset no. 1 in Fig. 4). Traced westward into the Wagwater Belt, the offset appears to diminish in rocks of approximately the same age to about 1.5 km (Green, 1977).

The decrease of offset westwards is attributed to splaying of the Plantain Garden fault zones into several faults that cut across the south central part of the Wagwater Belt (Figs. 4, 5). Displacement on the Plantain Garden fault is thus accommodated by smaller offsets on the Yallahs, Blue Mountain, and other splay faults.

A maximum positive free-air gravity anomaly exceeding 300 mgal and a maximum positive Bouguer anomaly exceeding 155 mgal have been measured in the Blue Mountains area (Wadge et al., 1983), and these observations are consistent with pronounced vertical displacements accompanying horizontal movements on the Plantain Garden, Yallahs, and Blue Mountain fault zones. The free-air anomaly is centered on the highest part of the Blue Mountains (BM in Fig. 4). The Bouguer anomaly is displaced to the east and situated over the metamorphic rocks north of the Plaintain Garden fault zone. Gravity modeling indicates considerable relief (>5 km) on Cretaceous arc rocks north of the Plantain Garden fault zone (Wadge et al., 1983). This relief is greater than that observed in the northern Blue Mountains (<4 km), consistent with an asymmetric (up-to-south) domal structure (Fig. 4).

Fault-Zone Structure, Southwest Blue Mountains

The structure of the Blue Mountain and Yallahs fault zones was mapped at a scale of 1:12,500 over an accessible and deeply eroded area of the southwest Blue Mountains (Figs. 4, 5; Kemp, 1971; Draper, 1979; Mann, 1983). Major river valleys parallel the Plantain Garden fault zone, and no cross-sectional exposures of the fault have been found. In the Clydesdale area, the Blue Mountain and Yallahs fault zones both consist of thin (10–500 m), fault-bounded slices of Cretaceous arc rocks (Figs. 4, 5). In the Yallahs fault zone, a 5-km-long slice of Cretaceous volcanic rocks forms a

faulted inlier within the Paleogene graben and graben-edge clastic rocks of the Wagwater Belt. In the Blue Mountain fault zone, a 9-km-long slice of Cretaceous marble, volcanic, and volcaniclastic rocks is faulted against a variety of other Cretaceous rocks (Figs. 4, 5). The slices are generally coherent, but cataclastically deformed in places. In the case of the Yallahs fault zone, mylonitic impure limestone, with strong foliation apparently due to localized strain, occurs in relatively planar zones along the edges of the structural slice (section D, Fig. 7 below). Clastic rocks dip steeply along both faults bounding the slice. The straightness of the bounding fault contacts in map view suggests steeply dipping fault planes that uplift the slice along both edges. A large component of vertical displacement is suggested by vertically striated fault planes with local gouge zones. The symmetry of deformation on both sides of the slice does not suggest a thrust slice preferentially overthrust in one direction as suggested in the older literature (Zans et al., 1962). Instead, the block appears to be a "pop-up" structure (Butler, 1982, Fig. 16) between steeply and inwardly dipping reverse faults with perhaps a significant component of strike slip.

A different structure appears at the southeastern convergence of the Blue Mountain and Yallahs fault zones (Figs. 4, 5). Serpentinite is exposed in a roughly triangular area (black in Fig. 5). Fault contacts between the serpentinite and Eocene sedimentary rocks (Richmond Formation) indicate a post-Eocene, probably Neogene, diapiric emplacement of the serpentinite. This structure has been interpreted by Draper (1979) as resulting from dilation produced by a northwestward motion of the Blue Mountain block as it abutted against the Blue Mountain and Yallahs fault zones. Mobile serpentinite derived from ultramafic rock of the Cretaceous trench complex then filled in the floor of the resulting "sphenochasm." It is also possible that sediment once filled in the upper part of this structure, but that the sediment was eroded during Plio-Pleistocene uplift of the region.

Structure in the Metamorphic Rocks

Elongate outcrops of regionally metamorphosed, fault-bounded schists with a strong foliation occur adjacent to the Plantain Garden, Yallahs, and Blue Mountain fault zones (Kemp, 1971; Draper et al., 1976; Draper, 1979, Fig. 5). With one minor exception, these are the only outcrops of metamorphic rocks on Jamaica. The close spatial association of the metamorphic rocks with the faults suggests that the former were uplifted as a result of compression at the Neogene restraining bend (Fig. 5). The rocks were initially deformed and metamorphosed in the lower part of an Early Cretaceous accretionary complex at depths of greater than 20 km (Draper, 1978, 1979), but early fabrics defined by oriented minerals were subsequently folded in Neogene time (Fig. 5).

Folding of the schists about northwest-trending axes is more intense adjacent to the Blue Mountain and Yallahs fault zones than north of the Plantain Garden fault zone (Fig. 5). Outcrops are also more elongate adjacent to the Blue Mountain and Yallahs fault zones, reflecting a dense

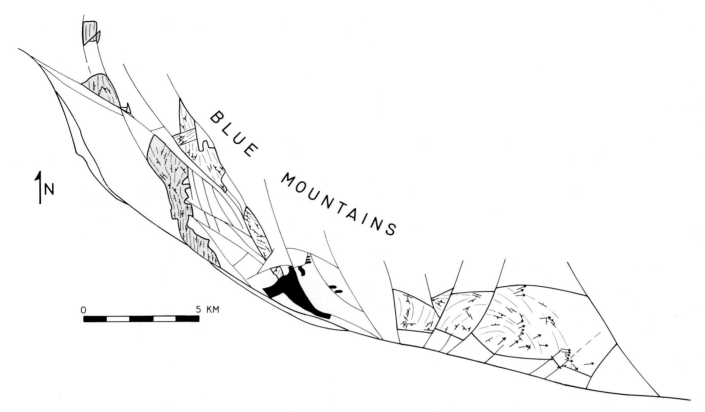

FIG. 5.—Outcrop pattern and structure of the metamorphic rocks of the southern Blue Mountains (modified from Draper, 1979). Arrows are lineations defined by preferred orientations of acicular minerals (especially crossite and tremolite). Strike and dip of foliation is defined by chlorite and compositional banding. Unshaded area with structure symbols represents blueschists and associated low-grade greenschists and amphibolites, and includes structurally disrupted basalt, gabbro, and peridotite (collectively known as the Mt. Hibernia Schists). Shaded area with structure symbols represents banded mafic to quartzo-feldspathic schists (collectively known as the Westphalia Schists). Black area represents serpentinite. Mapped faults juxtaposing different rock types are indicated by heavy lines.

network of predominantly northwest-striking faults bounding the outcrops. Minor northeast-trending folds are enigmatic, and it is not clear if they are synchronous with the northwest-trending folds, or the result of an earlier phase of deformation (Draper, 1979).

Several north- to northeast-striking faults disrupt the elongate eastern outcrop of the Mt. Hibernia Schists (unshaded in Fig. 5). These faults appear to be normal faults which accommodate belt-parallel extension of the schist block. This extension is also reflected immediately to the west in the northeasterly orientation of the serpentinite diapir (Fig. 5), and far to the south, in the northeast-striking faults bounding the Yallahs Basin (Fig. 11 below).

Structure of Sedimentary Rocks, Southern Blue Mountains

Sedimentary rocks of Cretaceous to Eocene age are folded on a large scale only in the southeasternmost Blue Mountains (Wadge and Draper, 1978; Draper, 1979). Stereographic plots for most areas along the Plantain Garden fault zone do not display convincing girdles indicative of regional cylindrical folds, although a few outcrop-scale folds have been observed (Wadge and Draper, 1978).

Structure of Sedimentary Rocks, Wagwater Belt

The structure of Paleogene sedimentary rocks deposited within and adjacent to the Wagwater graben, extensively studied by Armour-Brown (1966, and unpubl. data), Green (1977), Draper (1979), and Mann (1983), is summarized in Figure 6 by equal-area plots of poles to bedding, and in Figure 7 by a generalized map and four cross sections. The Wagwater Belt is important for determining the magnitude of strike-slip deformation in Jamaica because the belt was undeformed prior to the onset of strike-slip faulting in the Neogene.

Strike-slip deformation of the Wagwater Belt varies considerably along strike. The northern Wagwater Belt is only slightly deformed, and the upper unit of the graben, the Richmond Formation, is widely exposed (sections A and B, Fig. 7)). Upright, en echelon folds are thought to be due to incipient left-lateral shearing along the Peyton's Cove fault zone (PC in Fig. 7; Mann and Burke, 1980). The Richmond Formation generally dips to the northeast, except in the south, where it is truncated by the Seafield-Whitehall fault zone (SW in Fig. 7), and strikes east parallel with the fault.

South of this fault zone, deformation is more intense, particularly in the area of overlap between the Plantain Garden and Cavaliers fault zones (PG and C, and section D, Fig. 7). The structure of this central part of the Wagwater Belt consists of an elongate northwest-trending dome whose northern end is sharply defined by the Seafield-Whitehall fault zone (SW in Fig. 7), but whose southern end appears to plunge beneath younger rocks. Older rocks exposed in the core of the dome consist of red conglomerate and sandstone of the Wagwater Formation, and interbedded volcanic rocks of the Newcastle Formation (Green, 1977; Fig. 7). Most of the mapped faults within the Wagwater and Silver Hill fault zones are reverse faults, although normal faults occur in the crest of a hanging-wall anticline associated with overthrusting along the Wagwater fault (sections C and D, Fig. 7). Adjacent to the Yallahs fault zone, the rocks are complexly deformed by steeply dipping faults, most of which appear to be reverse faults, although correlation of offset units is difficult (section D, Fig. 7). Slices of Cretaceous arc rocks have been locally upthrust along both the Wagwater and Yallah fault zones. Folds related to the numerous thrusts are as young as Pliocene (Green, 1977; Fig. 8).

Folding in the southern Wagwater Belt is clearly visible only in the well-bedded Richmond Formation. Folds are generally meter-scale chevrons or kink bands with well-defined hinge zones and interlimb angles of about 60° to 110° (Fig. 9). Hinge zones are commonly fractured with displacements of centimeters to decimeters. Pervasive folding is obvious in the southern Wagwater Belt because dips commonly change abruptly in road and stream sections. Equal-area plots of poles to bedding indicate tilting and folding about northwest- to north-trending axes, and that folding is better developed south of the eastern extension of the Cavaliers fault than north of it (Fig. 6).

In summary, structural data suggest that the Wagwater Belt is most intensely deformed in the southern and central parts, where the Wagwater Belt is narrowest (Fig. 2).

STRUCTURE OF CENTRAL AND WESTERN JAMAICA

Two morphologically prominent sets of lineaments are readily apparent on synoptic radar imagery of the dominantly karst terrane of central and western Jamaica: (1) a northwest-trending set of mostly unnamed faults that form

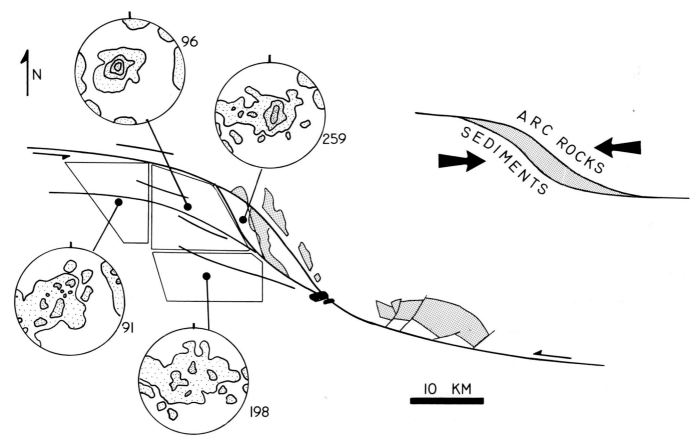

FIG. 6.—Fold patterns in the Wagwater Belt and southern Blue Mountains demonstrated by equal-area lower-hemisphere projections of poles to bedding, and contoured using the methods of Starkey (1977). Contour interval is 4 times random density distribution. Number of observations used for construction shown to right of each diagram. Areas of metamorphic rocks are shown in grey. Inset shows diagrammatically the location of metamorphic rocks upthrust within the restraining bend, and juxtaposed with higher-level arc rocks of the Blue Mountains and with sediments of the Wagwater Belt.

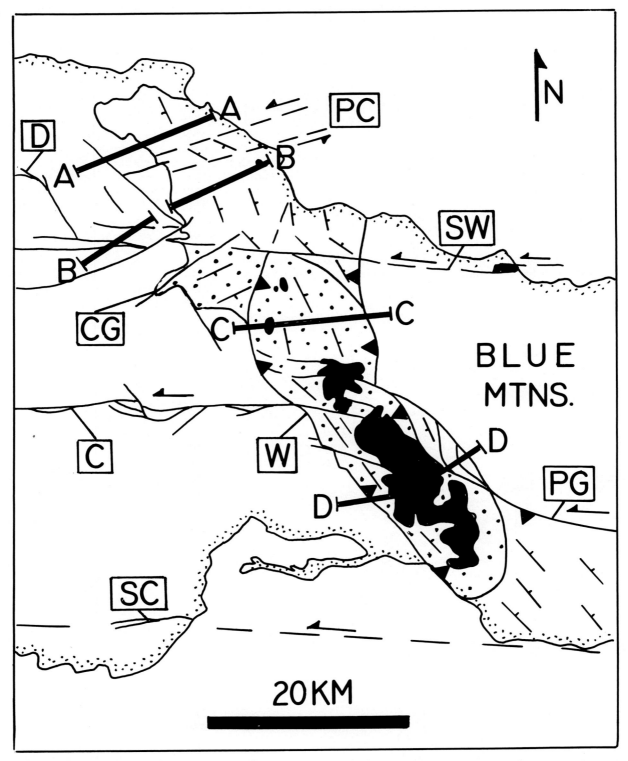

Fig. 7A.—Generalized structural map of the Wagwater Belt. Cross sections A, B, C and D are shown in Figure 7B. Strikes of bedding in Wagwater Belt are indicated. Black areas represent outcrop of volcanic rocks related to Paleogene rifting. Stippled area is outcrop of Wagwater Formation within a large dome developed in a right-stepping restraining bend between the Plantain Garden fault zone (PG) in the southeast and the Cavaliers (C) and Duanvale (D) fault zones in the northwest. Unpatterned areas with strike-lines represent outcrop of Richmond Formation, which is stratigraphically higher than the Wagwater Formation and has not been extensively uplifted by strike-slip deformation. Other faults affecting rocks of the Wagwater Belt are the Cuffy Gulley (CG), Wagwater (W), Peyton's Cover (PC), and Seafield-Whitehall (SW) faults.

WAGWATER BELT — JAMAICA

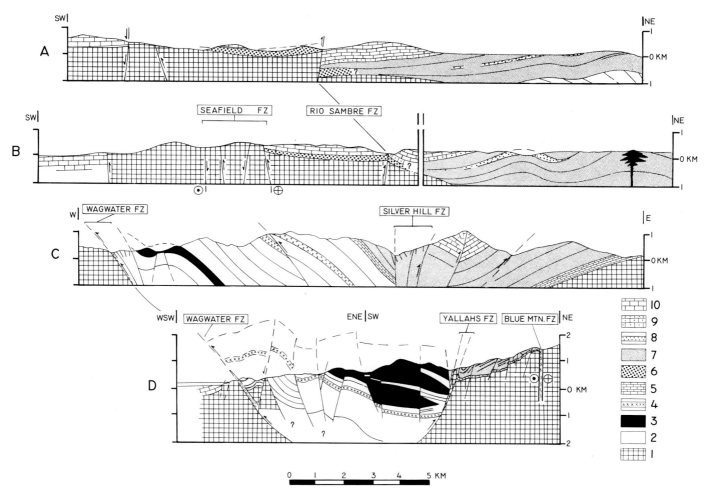

FIG. 7B.—Cross sections of the Wagwater Belt: Sections A and B, southwestern half of each section is from mapping of the Benbow Cretaceous Inlier by Burke et al. (1968), and northeastern half is from Mann (1983); Section C is from unpublished data of A. Armour-Brown (used with permission of the author) and Black et al. (1972); Section D is from Green (1977) and Mann (1983). Shortening of southern Wagwater Belt is related to development in post-middle Miocene time of a restraining bend. Numbered lithologic symbols: 1, Cretaceous metamorphic, igneous and sedimentary rocks; 2, Paleocene to lower Eocene Wagwater Formation; 3, Paleocene to lower Eocene Newcastle volcanics; 4, Pencar River Member of Wagwater Formation; 5, upper Paleocene Chepstow Formation; 6, Paleocene(?) Pembroke Hall Formation; 7, lower Eocene Richmond Formation; 8, Albany Member of Richmond Formation; 9, uppermost lower Eocene Langley Member of Richmond Formation; 10, middle Eocene Yellow Limestone Formation.

west-facing, arcuate scarps, and (2) an east-trending set that is concentrated in three complex braided zones of scarps within the Duanvale, Crawle River-Cavaliers, and South Coast fault zones (Wadge and Dixon, 1984, Figs. 2, 10).

The highest mountains in central and western Jamaica (about 1 km) are generally bounded on at least one side by prominent scarps of the northwest-trending fault set. The prominence of the fault scarps suggests active displacement. For example, the Spur Tree fault zone in southwestern Jamaica (ST in Fig. 2) is only 50 km long, yet forms a continuous 800-m-high west-facing scarp. To the west, the Santa Cruz fault forms a similar continuous west-facing scarp with over 600 m of relief (S in Fig. 2). Together with less prominent escarpments in the Montpelier-Newmarket fault zone (M in Fig. 2), the Spur Tree and

Santa Cruz fault zones define local areas of high ground (Santa Cruz and May Day Mountains and the western part of "Cockpit Country") between the Duanvale fault zone and the western onshore exposure of the South Coast fault zone (D and SC in Fig. 2). None of these escarpments appears to be conspicuously offset by the Crawle River fault zone (CR in Fig. 2), which is therefore thought not to be an active throughgoing feature.

A second area of prominent northwest-trending fault scarps and high ground (Dry Harbour Mountains and eastern part of "Cockpit Country") occurs within the overlap area of the Duanvale and Crawle River fault zones (Figs. 2, 10). These fault scarps, convex to the west, are of lower relief (100–200 m) than those in western Jamaica (Fig. 10). Detailed analysis of radar imagery by Wadge and Dixon (1984)

against northwest-striking faults near the western edge of the northern Wagwater Belt (Fig. 10). On the basis of this observation and the lack of offset piercing points, Horsfield (1974) argued against major strike slip on the Duanvale fault zone. However, Wadge and Dixon (1984) have suggested about 3 km of apparent left-lateral displacement at the western end of the fault.

The Crawle River-Cavaliers fault zone (CR and C in Fig. 10) bisects the largest Cretaceous inlier of central Jamaica and is followed by a number of river valleys, but there are no major escarpments along the fault. The Crawle River-Cavalier fault zone appears to terminate eastward against several northwest-striking faults, some of which slightly offset the Wagwater fault zone and appear to represent the complex western continuations of the Blue Mountain and Yallahs fault zones (Fig. 2).

The South Coast fault zone (SC in Fig. 10) defines the southern coast of Jamaica and forms a steep coastal escarpment as high as 600 m. Burke (1967) has suggested on the basis of seafloor morphology that the fault may con-

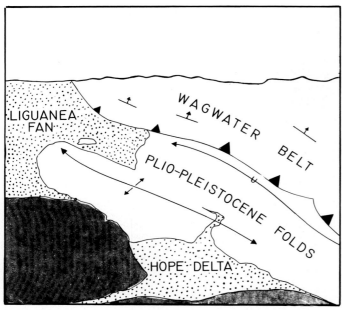

FIG. 8.—Oblique aerial view and interpretation of folds in Pliocene carbonate rocks adjacent to the Wagwater fault zone near Kingston (photo by S. Tyndale-Biscoe). Approximate distance across base of photograph is 7 km.

reveals a scissor-like sense of vertical displacement (east side up in the south and west side up in the north) for several of these faults.

The northwest-striking faults of this area curve into and terminate along the east-trending Duanvale fault zone, a complex anastamosing zone of discontinuous scarps downthrown generally to the north (Figs. 2, 10). Some fault segments near the eastern end of the Duanvale fault zone are morphologically indistinct and a few appear to terminate

FIG. 9.—Kink folding of well-bedded, lower Eocene turbidites of the Richmond Formation in the southwestern Blue Mountains near Clydesdale. Strike-slip deformation is of Neogene age.

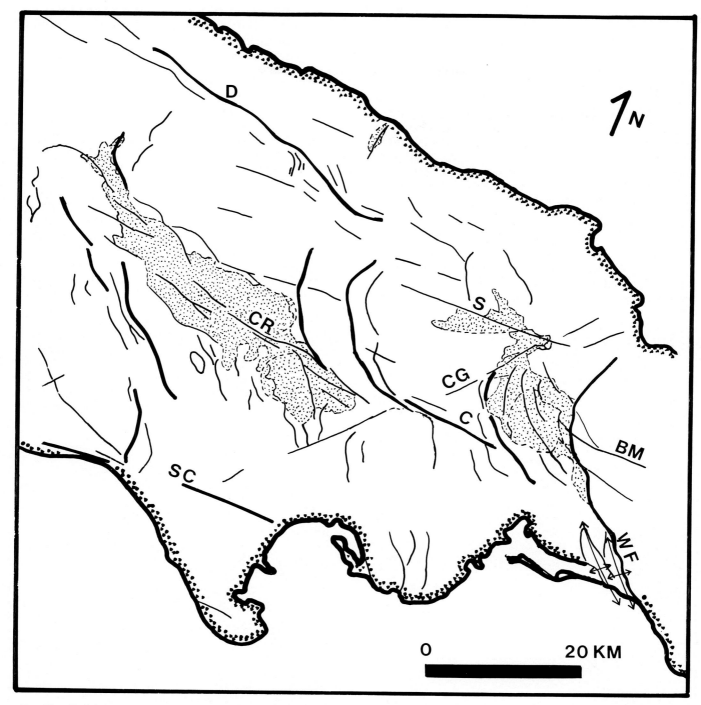

FIG. 10.—Fault interpretation of central Jamaica using SEASAT image provided and processed by T. H. Dixon (entire image published by Wadge and Dixon, 1984). Faults abbreviations: D, Duanvale; S, Seafield; CR, Crawle River; C, Cavaliers; SC, South Coast; CG, Cuffy Gulley; BM, Blue Mountain; WF, Wagwater. Shaded areas indicate outcrop of Cretaceous rocks apparent on radar image. Heavy dark lines are conspicuous scarps. Note crescentic scarps, probably reverse faults, linking the Cavaliers-Crawle River fault zone to the Duanvale fault zone.

tinue eastward beneath the sea as far as the Yallahs Basin (Fig. 2).

Few structural studies have been undertaken in central and western Jamaica, and so the structure of faults there is poorly known. The lack of obvious stratification in many exposures of the carbonates and the fact that fault zones are commonly occupied by river valleys has hindered structural investigation. Other than the scissor-like sense of vertical displacements on faults southwest of the Duanvale fault zone, little is known of the attitude of the fault planes forming the prominent northwest-trending set of escarpments. Burke et al. (1980) suggested that the northwest-striking faults are high-angle reverse faults for the following reasons: (1) they strike subparallel with the Wagwater fault zone, which is a known reverse fault; (2) they are of comparable age to the Wagwater fault zone (3 Ma or less; Robinson, 1971); (3) they cut west-verging syncline-anticline pairs in the same way as the Wagwater fault zone (see Fig. 3b of Burke et al., 1980); (4) high ground east-northeast of most of the faults coincides with young anticlines, compatible with active upthrusting; and (5) the curvature of the outcrop patterns of the faults is generally appropriate for thrusts dipping at about 60° to the east-northeast, which is approximately the dip of the Wagwater fault zone.

AGE OF DEFORMATION, UPLIFT, AND FAULT-CONTROLLED SEDIMENTATION

The age of compressional deformation and uplift in Jamaica can be inferred from the age of the youngest rocks deformed, the age of major unconformities, and the age of clastic rocks eroded off the central parts of the island and now cropping out in coastal areas.

Some of the deformation in the Wagwater Belt must have taken place after the deposition of the Richmond Formation in the early Eocene. In the southern Wagwater Belt, carbonate rocks as young as early Miocene commonly dip as steeply as 40°, suggesting still younger deformation. Lower Miocene carbonate rocks in central and western Jamaica

(Fig. 2) are sinistrally displaced as much as 10 km by east-striking strike-slip faults such as the Cavaliers fault (Burke et al., 1980), and they are folded along northwest-trending axes. The most obvious manifestation of folding in central Jamaica is the large anticlinal structure of the Central Inlier, which has been breached to expose Cretaceous rocks over a wide area (Fig. 2).

A widespread unconformity of early Miocene age in the carbonate section of central Jamaica may date the onset of compressional deformation and uplift in this part of the island (Eva and McFarlane, 1979). Green (1977) suggested that uplift and strike-slip faulting in eastern Jamaica is marked by a late-middle Miocene unconformity.

Along the northern coast near Buff Bay (Fig. 4), an abrupt facies change between white chalky limestones and grey or brown marls of late Miocene age is well established (Blow, 1969), and may reflect the early erosion and exposure of Cretaceous and Paleogene clastic rocks in the Blue Mountains. Carbonate rocks of late Miocene age around the periphery of the Blue Mountains and of late Miocene to Pliocene age along the southern coast become increasingly conglomeratic upwards in the section (Horsfield, 1974), again suggesting early erosion of the Blue Mountains.

Differential uplift of eastern Jamaica explains the lack of carbonate rocks and the exposure of the deep structural levels of Cretaceous arc rocks in the Blue Mountains (Fig. 4). Much of the eroded detritus is probably to be found either within the half dozen large fan-delta complexes that currently radiate outward from the center of the Blue Moun-

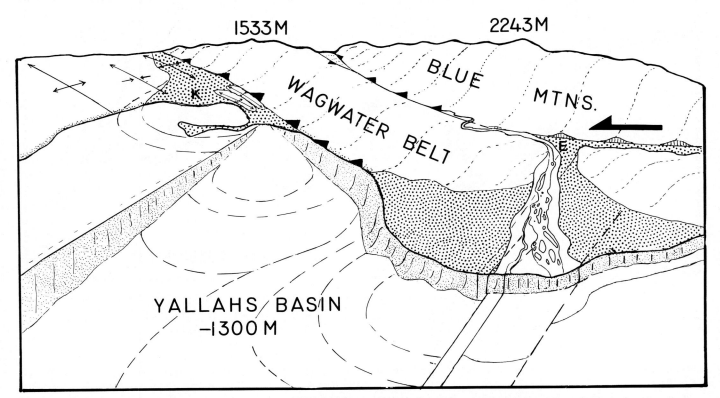

FIG. 11.—Schematic drawing illustrating the general pattern of sedimentation related to the uplift of the restraining bend in eastern Jamaica. Stipple represents fluvial and fan-delta deposits derived from the Blue Mountain uplift. K, Kingston; E, Easington. Approximate distance across the lower part of sketched area is 20 km (modified from Burke, 1967; Wescott and Ethridge, 1980).

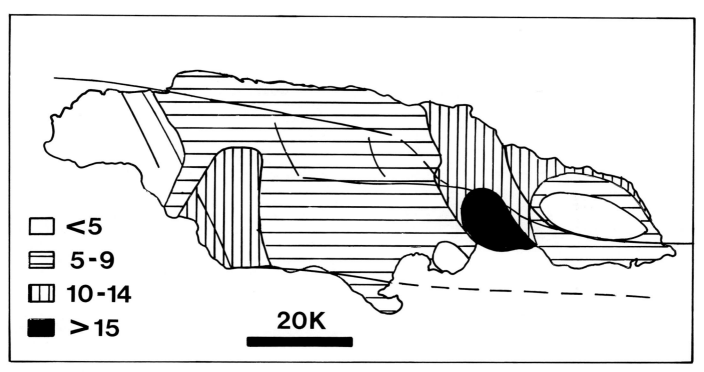

FIG. 12.—Geographical distribution of the number of times per century that intensities of modified Mercalli VI or greater have been reported in Jamaica from 1880 to 1960 (from Shepherd and Aspinall, 1980).

tains or within offshore fault-bounded depressions like the Yallahs Basin (Burke, 1967; Wescott and Ethridge, 1980, 1982; Fig. 11). Actively subsiding basins like the Yallahs (water depth, 1,300 m) and smaller basins in the Jamaica Passage appear to trap most of the sediment derived from eastern Jamaica before it has a chance to travel far along the Caribbean seafloor (Fig. 11).

SEISMICITY OF JAMAICA

There are no known large historical or modern earthquakes associated with the Yallahs or Blue Mountain fault zones, although it seems possible that the great earthquake of 1692 ($MM = X$), which triggered large landslides in various parts of eastern Jamaica, may have resulted from slip on one or more of those faults (earthquake intensities all refer to the Modified Mercalli Scale and are stated as MM followed by Roman numerals). Shepherd and Aspinall (1980) have compiled historical records from the period 1880–1960 and have drawn a contour map of the number of times per century that intensities $MM = VI$ or greater were reported in Jamaica. Their map (reproduced as Fig. 12) indicates two north-northwest-trending belts of greatest historical felt seismicity ($MM = X$). Part of the apparently high frequency of earthquakes in eastern Jamaica may reflect the distribution of populated areas like Kingston as well as greater accelerations of seismic waves in alluvium beneath these settled areas.

In western Jamaica, a northwest-trending belt of greatest historical felt intensities ($MM = X$) follows the Spur Tree and Santa Cruz fault zones (ST and S in Fig. 2). Earth-

quakes recorded by five or more seismograph stations during this century cluster in two east-trending bands roughly parallel to the Duanvale and South Coast fault zones (D and SC in Fig. 2). The close association of historical earthquakes with northwest-striking faults, which are known or inferred to be thrust faults, suggests continued fault activity to the present.

RESTRAINING BEND INTERPRETATIONS OF JAMAICA

Two tectonic models for the development of reverse faults and uplift in Jamaica have been published previously, but neither model has emphasized fault interaction (i.e., restraining bend formation) between right-stepping, left-lateral strike-slip faults. Burke et al. (1980) suggested, mainly from a compilation of the results of field geologic studies, that Jamaica is a broad zone of Neogene east-west strike-slip motion, distributed on three main fault zones: (1) the Plantain Garden-Cavaliers-Crawle River fault zone; (2) the Duanvale fault zone; and (3) the South Coast fault zone (Fig. 2). Cumulative strike slip of about 40 km identified on these faults forms the basis of their interpretation. Burke et al. (1980) further suggested that northwest-striking faults are thrust or reverse faults whereas northeast-striking faults are normal faults, and that both types are secondary features related to strike-slip deformation. (Fig. 13A.)

Wadge and Dixon (1984) modified the simple shear model of Burke et al. (1980) mainly on the basis of a detailed analysis of radar imagery. They proposed two alternative models: In one, hypothetical east-west convergence across Jamaica initially produced north-trending folds that subse-

A. SIMPLE SHEAR

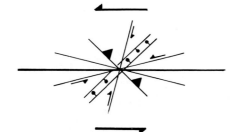

B. FAULT TERMINATION

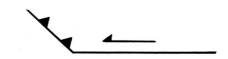

C. FAULT INTERACTION

Fig. 13.—Models for second-order compressional structures in strike-slip zones. See text for discussion.

quently experienced anticlockwise rotation during displacement along the Plantain Garden-Cavaliers-Crawle River fault zone. In the other model, initial northeast-southwest convergence, perhaps related to convergent relative plate motion, produced northwest-trending folds that were subsequently reactivated by strike slip along the Duanvale and South Coast fault zones.

Here we outline a simpler model, shown schematically in Figure 13C, which emphasizes the generation of compressional structures at discrete stepovers or restraining bends between interacting strike-slip faults rather than along the entire length of individual strike-slip faults (see Burke et al., 1980; Fig. 13A). The topography of Jamaica, locations of both recorded earthquakes and those with greatest felt intensities over the last 100 years, the pattern of Neogene deformation (scarps, mapped faults, folds), and Neogene uplift history are all consistent with the idea of two right-stepping restraining-bend fault segments that relay left-lateral strike slip from two throughgoing fault zones along the southern part of the island (the Enriquillo-Plantain Garden and South Coast fault zones) to a single throughgoing fault zone, the Duanvale fault zone, along the northern part of the island (Fig. 2). Transfer of displacement at the eastern bend between the Enriquillo-Plantain Garden and Duanvale

fault zones occurs by (1) upthrusting and apparent left-lateral displacement along the southern edge of the Blue Mountains (Blue Mountain fault zone); (2) uplift and doming of the Blue Mountains to elevations over 2 km; (3) reverse and left-lateral strike-slip faulting along the Yallahs-Silver Hill fault zones; (4) reverse displacement on the Wagwater fault zone; (5) uplift and doming of the Wagwater Belt; (6) apparent left-lateral displacement on the Crawle River-Cavaliers fault zone; and (7) reverse and scissor faulting on northwest-striking arcuate faults situated between the Crawle River-Cavaliers and Duanvale fault zones. The Duanvale fault zone continues westward offshore as a bathymetric trough interrupted by one pull-apart basin, and it appears to merge with the southern edge of the Cayman Trough near the Mid-Cayman Spreading Center (Case and Holcombe, 1980, Fig. 1). Transfer of displacement at the western bend between the South Coast and Duanvale fault zones occurs by (1) thrusting along the Spur Tree, Santa Cruz and Montpelier-Newmarket fault zones; and (2) local uplift with exposure of Cretaceous rocks (Fig. 2).

The two belts of greatest earthquake intensity parallel the two bends (Fig. 12). The isoseismic map, compiled after the 1957 Montego Bay earthquake (Robinson et al., 1958), shows a pronounced parallelism of the greater intensity contours with the Montpelier-Newmarket-Santa Cruz fault zones and suggests recent activity on the western bend.

Limited data on the age of faulting in Jamaica suggest that most faulting and accompanying folding began in the middle to late Miocene, only slightly before Pliocene initiation of strike slip along the Enriquillo-Plantain Garden fault zone in southern Hispaniola (Fig. 1). However, only the Plantain Garden fault zone of eastern Jamaica is physically continuous with the single post-Miocene throughgoing strike-slip fault system of southern Hispaniola. The Duanvale-Seafield-Whitehall fault zone has not yet been identified beneath the sea east of Jamaica (Fig. 2). An eastern submarine extension of the South Coast fault zone has been postulated on the basis of limited bathymetric and seismic data, but this fault is not known offshore to the west. Mapping of submarine faults should be given priority in future work.

The Duanvale-Seafield-Whitehall fault zone is apparently slightly older than the Enriquillo-Plantain Garden-Crawle River-Cavaliers-Duanvale system. The latter nucleated on the northwest-trending Yallahs-Silver Hill fault zone that once formed the eastern normal-fault boundary of the Paleogene Wagwater graben (Fig. 2). The Blue Mountain fault zone appears slightly younger than the reactivated Yallahs reverse fault and is interpreted as a "bypass" fault that reduces the asperity at the bend. The western bend similarly nucleated on northwest-striking normal faults bounding the Paleogene Montpelier-Newmarket trough that is known only from boreholes (Meyerhoff and Krieg, 1977). It appears to be at an early stage of development because uplifts are relatively small and bypass faults are not obvious.

DISCUSSION

The interpretation of structures and related sedimentary deposits within broad strike-slip plate boundary zones like

Jamaica is often difficult because of the similar effects of three different mechanisms for compressional deformation: *simple shear* adjacent to a single strike-slip fault, *termination* of a single strike-slip fault, and *interaction* of two strike-slip faults to form restraining bends (Fig. 13, Aydin and Page, 1984). The purpose of this discussion is briefly to review our understanding of these different mechanisms and to point out some of the problems in interpreting the Jamaican data.

Faults occurring within strike-slip zones can be classified on the basis of either their *relative size* (first-order faults are larger, second-order are smaller) or their *time relationships* (primary faults develop first; secondary faults, later; Groshong and Rodgers, 1978). Interpretations of second-order compressional structures (i.e., folds and thrusts) in strike-slip zones can be divided into several groups: (1) Second-order folds and thrusts are primary features that develop in originally unfaulted rock prior to the formation by simple shear of a throughgoing strike-slip fault (Tchalenko, 1970; Fig. 13A). (2) Second-order folds and thrusts are secondary features caused by slip along the strike-slip fault (Moody and Hill, 1956; Fig. 13A). (3) Second-order folds and thrusts form at the ends of strike-slip faults (Chinnery, 1966; Fig. 13B). (4) Second-order folds and thrusts form at localized bends or compressional stepovers where two strike-slip faults interact (Segall and Pollard, 1980; Fig. 13C). All of the hypotheses except the second are based on both field and experimental observations as well as physical theory (Groshong and Rodgers, 1978). Hypothesis 2 is based mainly on field observations and is now favored by few geologists. However, all of the hypotheses predict similar patterns of second-order structures adjacent to and within strike-slip zones (Groshong and Rodgers, 1978; Aydin and Page, 1984).

In this paper, we have attempted to demonstrate the importance of compressional fault interaction or bend formation (hypothesis 4, Fig. 13C) in Jamaica. The fundamental observation supporting the bend hypothesis for Jamaica is that a major left-lateral strike-slip fault strikes into the mountainous island from the southeast corner while another major strike-slip fault strikes into the island from the northwest corner (Fig. 2). However, in the absence of more detailed seismic and structural studies it is difficult to distinguish the relative roles of simple shear and fault termination (Fig. 13A, B) in the present pattern of bend deformation or in its development over the past several million years. For example, it seems likely that simple shear generated second-order folds and thrusts as primary features prior to the formation of the throughgoing east-west strike-slip faults such as the Duanvale and Crawle River (Figs. 2, 13A). Following propagation of throughgoing strike-slip faults, large folds such as the Central Cretaceous Inlier (Fig. 10) may have formed as compressional structures at strike-slip fault terminations (Fig. 13B). Later (the present stage), strike-slip faults such as the Duanvale and Crawle River may have interacted and generated folds at a restraining bend (Fig. 13C). Therefore, the restraining bend systems described in Jamaica probably represent the culmination of a complex structural development involving superposition of all of the mechanisms shown in Figure 13.

The details of this structural development await more detailed structural and sedimentary studies.

ACKNOWLEDGMENTS

We thank Raymond Wright, Arthur Geddes and their colleagues at the Geological Survery of Jamaica and the faculty of the Geology Department of the University of the West Indies for their continuing assistance. Special thanks to Ashlyn Armour-Brown for allowing access to his unpublished field data, to Tim Dixon for donating SAR radar imagery (used in Fig. 3), and to Conliffe Simpson for field assistance. Linda Handley kindly helped prepare the manuscript. This manuscript benefited from helpful reviews by Kevin Biddle, Nick Christie-Blick, Carlos Dengo, and Geoff Wadge. Mann and Burke were supported by NSF-EAR 7803319 and NASA-NAG5155, and Draper was supported by the University of the West Indies. University of Texas Institute for Geophysics contribution 606 and Lunar and Planetary Institute contribution 557.

REFERENCES

ARDEN, D. D., JR, 1975, Geology of Jamaica and the Nicaragua Rise, *in* Nairn, A. E. M., and Stehli, F. H., eds., The Ocean Basins and Margins: New York, Plenum, v. 3, p. 617–661.

ARMOUR-BROWN, A., 1966, *in* Annual Report of the Geology Survey Department for the year ending 31st March, 1966: Geological Survey Department, Kingston, Jamaica, p. 9–10.

AYDIN, A., AND PAGE, B. M., 1984, Diverse Pliocene-Quaternery tectonics in a transform environment: San Francisco Bay Region, California: Geological Society of America Bulletin, v. 95, p. 1303–1317.

BLACK, C. D. G., GREEN, G. W., AND NAWROCKI, P. E., 1972, Preliminary investigations of the Castleton copper prospect: Geological Survey Department, Kingston, Jamaica, Economic Geology Report Number 4, 26 p.

BLOW, W. H., 1969, Late middle Eocene to Recent planktonic foraminiferal biostratigraphy, *in* Bronnimann, P., and Renz, H. H., eds., Proceedings International Conference on Planktonic Microfossils, v. 1, p. 199–421.

BRUN, J. P., AND BURG, J. P., 1982, Combined thrusting and wrenching in the Ibero-Armorican arc: a corner effect during continental collision: Earth and Planetary Science Letters, v. 61, p. 332–339.

BURKE, K., 1967, The Yallahs Basin: a sedimentary basin southeast of Kingston, Jamaica: Marine Geology, v. 5, p. 45–60.

BURKE, K., ROBINSON, E., AND COATES, A. G., 1968, The geology of the Benbow Inlier, Jamaica: Transactions of 4th Caribbean Geological Conference, Trinidad, p. 299–307.

BURKE, K., GRIPPI, J., AND ŞENGÖR, A. M. C., 1980, Neogene structures in Jamaica and the tectonic style of the Northern Caribbean Plate Boundary Zone: Journal of Geology, v. 88, p. 375–386.

BUTLER, R. W. H., 1982, The terminology of structures in thrust belts: Journal of Structural Geology, v. 4, p. 239–245.

CASE, J. E., AND HOLCOMBE, T. L., 1980, Geologic-tectonic map of the Caribbean region: United States Geological Survey Miscellaneous Investigation Series Map I-1100.

CHINNERY, M. A., 1966, Secondary faulting, II, geological aspects: Canadian Journal of Earth Sciences, v. 3, p. 175–190.

CROWELL, J. C., 1974, Sedimentation along the San Andreas Fault, California, *in* Dott, R. H., Jr., and Shaver, R. H., eds., Modern and Ancient Geosynclinal Sedimentation: Society of Economic Paleontologists and Mineralogists Special Publication No. 19, p. 292–303.

———, 1979, The San Andreas Fault through time: Journal of Geological Society London, v. 136, p. 293–302.

DRAPER, G., 1978, Coaxial progressive pure shear and deformation associated with subduction: Nature, v. 275, p. 735–736.

———, 1979, Tectonics of the regionally metamorphosed rocks of eastern Jamaica [unpubl. Ph.D. thesis]: Kingston, University of the West Indies, 277 p.

DRAPER, G., HARDING, R. R., HORSFIELD, W. T., KEMP, A. W., AND TRESHAM, A. E., 1976, Low grade metamorphic belt in Jamaica and its tectonic implications: Geological Society of America Bulletin, v. 87, p. 1283–1290.

EVA, A. N., AND MCFARLANE, N., 1979, Tertiary to early Quaternary facies relationships in Jamaica (abs): 4th Latin American Geological Congress, Port of Spain, Trinidad and Tobago.

GREEN, G. W., 1977, Structure and stratigraphy of the Wagwater Belt, Jamaica: Overseas Geology and Mineral Resources, No. 48, 21 p.

GRIPPI, J., AND BURKE, K., 1980, Submarine-canyon complex among Cretaceous island-arc sediments, western Jamaica: Geological Society of America Bulletin, v. 91, p. 179–184.

GROSHONG, R. H., JR., AND RODGERS, D. A., 1978, Left-lateral strike-slip fault model, in Structural Style of the Arbuckle Region: Geological Society of America South-Central Field Trip Guide No. 3.

HANCOCK, P. L., AND ATIYA, M. S., 1979, Tectonic significance of mesofracture systems associated with the Lebanese segment of the Dead Sea transform fault: Journal of Structural Geology, v. 1, p. 143–153.

HARLAND, W. B., 1971, Tectonic transpression in Caledonian Spitsbergen: Geological Magazine, v. 108, p. 27–42.

HOLCOMBE, T. L., VOGT, P. R., MATTHEWS, J. E., AND MURCHISON, R. R., 1973, Evidence for seafloor spreading in the Cayman Trough: Earth and Planetary Science Letters, v. 20, p. 357–371.

HORSFIELD, W. T., 1974, Major faults in Jamaica: Journal of Geological Society of Jamaica, v. 14, p. 1–15.

HORSFIELD, W. T., AND ROOBOL, M. J., 1974, A tectonic model for the evolution of Jamaica: Journal of Geological Society of Jamaica, v. 14, p. 31–38.

JORDAN, T. H., 1975, The present-day motions of the Caribbean plate: Journal of Geophysical Research, v. 80, p. 4433–4439.

KARIG, D. E., 1979, Material transport within accretionary prisms and the "knocker" problem: Journal of Geology, v. 88, p. 27–39.

KEMP, A. W., 1971, The geology of the southwestern flank of the Blue Mountains, Jamaica [unpubl. Ph.D. dissertation]: Kingston, University of the West Indies, 307 p.

MACDONALD, K. C., AND HOLCOMBE, T. L., 1978, Inversion of magnetic anomalies and seafloor spreading in the Cayman Trough: Earth and Planetary Science Letters, v. 40, p. 407–414.

MANN, P., 1983, Cenozoic tectonics of the Caribbean: structural and stratigraphic studies in Jamaica and Hispaniola (Parts 1 and 2) [unpubl. Ph.D. thesis]: Albany, State University of New York at Albany, 688 p.

MANN, P., AND BURKE, K., 1980, Neogene wrench faulting in the Wagwater Belt, Jamaica: Transactions of the 9th Caribbean Geological Conference, Santo Domingo, Dominican Republic, p. 95–97.

MANN, P., BURKE, K., AND MATUMOTO, T., 1984, Neotectonics of Hispaniola: plate motion, sedimentation, and seismicity at a restraining bend: Earth and Planetary Science Letters, v. 70, p. 311–324.

MANN, P., HEMPTON, M. R., BRADLEY, D. C., AND BURKE, K., 1983, Development of pull-apart basins: Journal of Geology, v. 91, p. 529–554.

MCFARLANE, N., 1977, 1:250,000 Geological map of Jamaica, second edition: Mines and Geology Division, Kingston, Jamaica.

MEYERHOFF, A. A., AND KRIEG, E. A., 1977, Future Jamaican exploration justified: Oil and Gas Journal, Aug. 29, p. 79–85.

MOODY, J. D., AND HILL, M. J., 1956, Wrench-fault tectonics: Geological Society of America Bulletin, v. 67, p. 1207–1346.

ROBINSON, E., 1971, Observations on the geology of the Jamaican bauxite: Journal of Geological Society of Jamaica, Bauxite Special Issue, p. 3–9.

ROBINSON, E., VERSEY, H. R., AND WILLIAMS, J. B., 1958, The Jamaica earthquake of March 1, 1957: Transactions of the 2nd Caribbean Geology Conference, Mayaguez, Puerto Rico, p. 50–57.

ROOBOL, M. J., 1972, The volcanic geology of Jamaica: Transactions of the 6th Caribbean Geological Conference, Maragarita, Venezuela, p. 100–107.

SEGALL, P., AND POLLARD, D. D., 1980, Mechanics of discontinuous faults: Journal of Geophysical Research, v. 85, p. 4337–4350.

SHEPHERD, J. B., AND ASPINALL, W. P., 1980, Seismicity and seismic intensities in Jamaica, West Indies: a problem in risk assessment: Earthquake Engineering and Structural Dynamics, v. 8, p. 315–335.

STARKEY, J., 1977, The contouring of orientation data represented in spherical projection: Canadian Journal of Earth Science, v. 14, p. 268–277.

SYKES, L. R., MCCANN, W. R., AND KAFKA, A. L., 1982, Motion of Caribbean plate during the last 7 million years and implications for earlier Cenozoic movements: Journal of Geophysic Research, v. 87, p. 10656–10676.

TCHALENKO, J. S., 1970, Similarities between shear zones of different magnitudes: Geological Society of America Bulletin, v. 81, p. 1625–1640.

WADGE, G., AND DIXON, T. H., 1984, A geological interpretation of SEASAT-SAR imagery of Jamaica: Journal of Geology, v. 92, p. 561–581.

WADGE, G., AND DRAPER, G., 1978, Structural geology of the southeastern Blue Mountains, Jamaica: Geologie Mijnbouw, v. 57, 347–352.

WADGE, G., DRAPER, G., AND ROBINSON, E., 1983, Gravity anomalies in the Blue Mountains, eastern Jamaica: Transactions of the 9th Caribbean Geological Conference, Santo Domingo, August 1980, p. 467–474.

WADGE, G., AND EVA, A. N., 1978, The geology and tectonic significance of the Sunning Hill Inlier: Journal of the Geological Society of Jamaica, v. 17, p. 1–15.

WALCOTT, R. I., AND CRESSWELL, M. M., eds., 1979, The origin of the Southern Alps: Royal Society of New Zealand, Bulletin 18, 147 p.

WESTCOTT, W. A., AND ETHRIDGE, F. G., 1980, Fan-delta sedimentology and tectonic setting—Yallahs fan delta, southeast Jamaica: American Association of Petroleum Geologists Bulletin, v. 64, p. 374–399.

WOODCOCK, N. H., AND ROBERTSON, A. H. F., 1982, Wrench and thrust tectonics along a Mesozoic-Cenozoic complex, SW Turkey: Journal of the Geological Society of London, v. 139, part 2, p. 147–163.

ZANS, V. A., CHUBB, L. J., VERSEY, H. R., WILLIAMS, J. B., ROBINSON, E., AND COOKE, D. L., 1962, Synopsis of the geology of Jamaica: Jamaica Geological Survey Department Bulletin, no. 4, 72 p.

STRIKE-SLIP FAULTING AND RELATED BASIN FORMATION IN ZONES OF TECTONIC ESCAPE: TURKEY AS A CASE STUDY[1]

A. M. C. ŞENGÖR, AND NACİ GÖRÜR

İ.T.Ü. Maden Fakültesi
Jeoloji Bölümü, Teşvikiye
Istanbul, Turkey;

AND

FUAT ŞAROĞLU
Maden Tetkik ve Arama Enstitüsü
Temel Araştırmalar Dairesi
Ankara, Turkey

ABSTRACT: Strike slip on various scales and on faults of diverse orientations is one of the most prominent modes of deformation in continental convergence zones. Extreme heterogeneity and low shear strength of continental rocks are responsible for creating complex "escape routes" from nodes of constriction along irregular collision fronts toward free faces formed by subduction zones. The origin of this process is poorly understood. The two main models ascribe tectonic escape to buoyancy forces resulting from differences in crustal thickness generated by collision and to forces applied to the boundaries of the escaping wedges. Escape tectonics also creates a complicated geological signature, whose recognition in fossil examples may be difficult. In this paper we examine the Neogene to present tectonic escape-dominated evolution of Turkey both to test the models devised to account for tectonic escape and to develop criteria by which fossil escape systems may be recognized.

Since the late Serravallian (~12 Ma), the tectonics of Turkey has been dominated by the westward escape of an Anatolian block ("scholle") from the east Anatolian convergent zone onto the oceanic lithosphere of the Eastern Mediterranean Sea, mainly along the North and East Anatolian strike-slip faults. This tectonic regime generated four distinct neotectonic provinces: (1) The East Anatolian contractional province, located mainly east of where the North and East Anatolian faults meet, and characterized by roughly north-south shortening; (2) the weakly active North Turkish province situated north of the North Anatolian fault, and characterized by limited east-west shortening; (3) the West Anatolian extensional province characterized by north-south extension; and (4) the Central Anatolian "ova" province characterized by northeast-southwest shortening and northwest-southeast extension. Large, roughly equant, complex basins ("ovas") form peculiar structural elements of the Central Anatolian province. The two latter provinces are located within the westerly-moving Anatolian scholle.

A number of pull-apart and fault-wedge basins have formed along the North and East Anatolian fault zones in addition to several other "incompatibility basins," arising from space problems where these faults interfere with each other and with other large-scale structures. Incompatibility basins seem to have the most complicated structural history. The pull-apart basins are located on either primary or secondary releasing bends along the North and East Anatolian faults. The secondary type is related to the intersection of east-trending zones of high convergent strain with the North and East Anatolian fault zones.

The tectonic escape regime in and around Turkey was not caused by buoyancy forces resulting from crustal thickness differences, but such forces may have been maintaining it. A knowledge of the geology of escape-related basins is critical both for our understanding of the nature of tectonic escape, and for its recognition in the geological record. We believe that the present tectonic scheme of Turkey constitutes an excellent guide for understanding the causes and consequences of escape, and for the recognition of its fossil representatives.

INTRODUCTION

Although the famous Swiss geologist Arnold Escher von der Linth was perhaps the first to recognize a strike-slip fault in the Säntis mountains south of Wildkirchli in the canton of Appenzell in the 1850's (Suess, 1885, p. 153–154), the year 1985 marks the centenary of the initial appreciation of their widespread occurrence and significance in the structural evolution of the lithosphere (Suess, 1885, p. 153–164), and the quinquagenary of the initial realization that strike-slip motion on large scales is capable of generating basin complexes of considerable size and complexity (Becker, 1934; Lotze, 1936, following earlier hints by Kossmat, 1926, and von Seidlitz, 1931). The growing awareness of the importance of strike-slip zones since then (e.g., Ketin, 1948; Moody and Hill, 1956; Carey, 1958; Wilson, 1963) finally culminated in the advent of the the-

ory of plate tectonics that requires that one of the three kinds of boundaries dividing the lithosphere into mobile plates have strike-slip displacement. Thus, with plate tectonics, strike-slip zones have come to be regarded as one of the three kinds of "mobile belts" (Wilson, 1965; Reading, 1980), in which over 95% of the tectonic deformations of the lithosphere have been concentrated.

Isacks et al. (1968) noted that the present-day mobile belts, the plate boundaries, defined by seismically active regions, are narrow and cleanly delineated in the oceans, whereas in continental areas they are wide and diffuse, regardless of the type of boundary. The greatest width of such zones is present in Asia today, where ongoing post-collisional convergence between India and the main body of the continent is being converted into strain.

The difference in behavior between continental and oceanic lithosphere results from the buoyancy and low shear strength of continental rocks (McKenzie, 1969). McKenzie (1969) indicated that lithosphere carrying a normal thickness of continental crust cannot be subducted continuously because it would not sink into the denser asthenosphere. Thus when two continents are apposed across a suture, continuing con-

---

[1]This paper is dedicated to Professor Sırrı Erinç, the Turkish nestor of geomorphology and Quaternary geology, on the occasion of his retirement from active teaching and in grateful recognition of his fundamental contributions to our understanding of the young tectonics of Turkey.

vergence must be converted into intracontinental strain. McKenzie (1972) argued that after collision begins at points representing continental promontories, further convergence may be accommodated by thickening of continental crust and/or by the formation of small plates to consume the oceanic lithosphere remaining in embayments. This process of sideways expulsion of lithospheric wedges from zones of convergence is one that had been already noted by Argand (1920, Fig. 1), although its first explicit statement seems to have been made by Cloos (1928, p. 303–311; esp. footnote 2, p. 309–310) on the example of the Mesozoic-Cenozoic geometry and kinematics of the block mosaic ("*Schollen*") in Central Europe (see esp. his Fig. 16), albeit in a fundamentally different framework from that of plate tectonics.

Molnar and Tapponnier (1975) later pointed out that the Cenozoic tectonics of Asia represents a much grander example of continental wedges fleeing from loci of convergence, in this particular instance the locus of convergence being between India and Asia (Fig. 1). They used the slip-line field theory and modelled Asia as a plastic-rigid material indented by a rigid die represented by India (Tapponnier and Molnar, 1976). They compared the slip lines with the large active strike-slip faults of central and southeastern Asia such as the Altin Dagh, Kang Ting, and the Red River faults (Fig. 1).

Although Molnar and Tapponnier's (1975) paper helped to focus attention on processes that deform the continents during collision, and especially on large strike-slip faults that accommodate the convergence, the validity of their application of the slip-line field theory to continental deformation has been questioned because it utilized an instantaneous, two-dimensional model developed for homogeneous media to account for processes affecting three-dimensional heterogeneous objects over tens of millions of years and because it included the incorrect assumption that faults are located along slip-lines.

Two main different approaches were subsequently tried to find an appropriate dynamic explanation for the kinematic picture described by Molnar and Tapponnier (1975). The one by England and McKenzie (1982, 1983) was to model the continental lithosphere as a thin viscous sheet moving on an inviscid fluid. They concluded that crustal thickness differences formed by flow within the continental lithosphere may be the driving agents of the sideways motions observed in Asia (Fig. 1). Conversely, Tapponnier et al. (1982, and in press) believe, on the basis of indentation experiments, that forces applied at the boundaries of moving lithospheric wedges are the cause of tectonic escape. As long as the rheological properties of lithospheric rocks at given depths remain conjectural, the only way to test the models proposed to account for the sideways expulsion of continental material during convergence, called "tectonic escape" by Burke and Şengör (in press), is to undertake a detailed examination of the geology of regions where this process is now occurring, with special reference to the temporal relationships between different structures that constitute a tectonic escape system and to the evolution of associated strain geometry.

In addition to the light it sheds on our understanding of the behavior of continental lithosphere, the process of tectonic escape is important also from the viewpoint of the complexity of its geological signature. Figure 1 shows the great variety of the structures it generates. They range from contractional through strike-slip to extensional structures connected to each other through a complex array of discontinuities. Not only do discrete pieces of the lithosphere, termed scholles (Dewey and Şengör, 1979), move coherently along strike-slip faults, but they also deform internally producing large bulk strains related in a very complicated fashion to the velocities of the bounding major plates (McKenzie and Jackson, 1983). Different kinds of basins of various sizes and locations may form as a result of these strains. These basins are perhaps the most important escape-related structures not only because they may contain significant hydrocarbon reserves but also because they commonly preserve the only stratigraphic record of events associated with escape and thus allow us to study the critical temporal aspects of the process.

In such escape systems, scholle boundaries frequently change position and character, as do the fault systems within them, largely because of incompatibility problems generated within the overall convergent regime. Tectonic escape is thus likely to dominate the "orogenic result" because not only does it overprint earlier structures in complex ways, but it is also generally governed by fast relative-motion rates (Dewey and Şengör, 1979). This would make the interpretation of events that pre-dated escape extremely laborious; if tectonic escape goes unnoticed, attempts at reconstruction of the evolution of mountain belts, especially the construction of balanced cross sections across major orogens, are unlikely to be realistic.

Both to test the models proposed to explain the dynamics of tectonic escape and to learn more about the geology of this process we have initiated a case study in Turkey (Fig. 1, boxed area), where post-collisional convergence is currently underway and where the widest possible range of collision-related deformation styles and associated structures are observed. Also, we have a better knowledge of stratigraphy in Turkey than anywhere else where tectonic escape is now occurring. This allows us to document the detailed temporal relationships among a variety of structures, most of which are now seismically active.

This paper is a preliminary progress report. In the following sections we first summarize the post-Oligocene tectonic evolution and paleogeographic development of Turkey to establish a framework within which we then describe the evolution of various escape-related structures, especially the strike-slip-related basins. Much of our data is summarized in Figures 2 through 6, and these figures form the basis of our narrative. We have given the references to these data mainly in the figure captions so as not to interrupt the flow of the text unnecessarily.

SUMMARY OF THE NEOTECTONIC EVOLUTION OF TURKEY

Turkey forms the better-developed western part of the asymmetric tectonic escape system caused by post-colli-

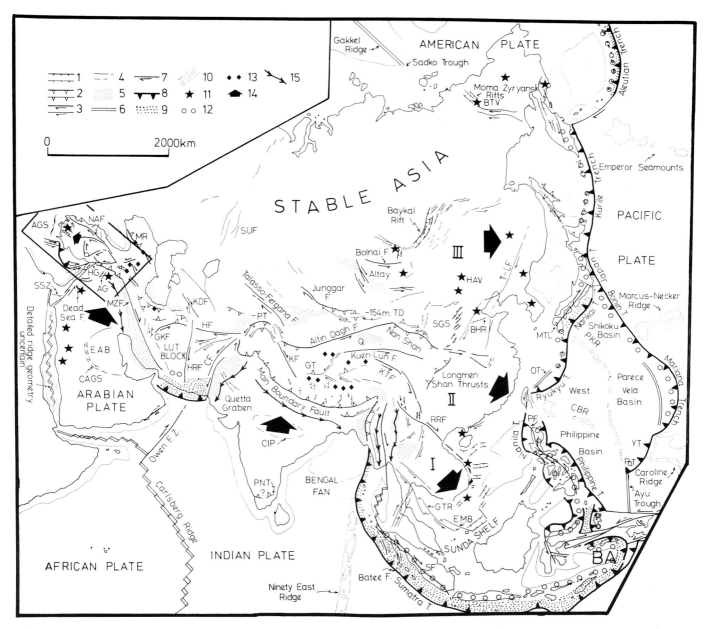

FIG. 1.—Neotectonic map of Asia showing most of the major active structures that have evolved since the beginning of the Cenozoic era, but mostly after the India-Asia collision during the Eocene; an insignificantly small number of the structures shown may now be inactive (after Şengör, in press).

Key to legend: 1, first- and second-order normal fault; 2, first- and second-order thrust fault; 3, first- and second-order strike-slip fault; 4, unspecified and/or suspected fault; 5, area of folding (commonly superficial); 6, sea-floor spreading center; 7, major transform fault; 8, active subduction zone; 9, active subduction-accretion complex; 10, aseismic submarine ridge; 11, intra-plate volcanism; 12, subduction volcanism; 13, "Tibet-type" collisional volcanism; 14, direction of approximate movement relative to stable Asia of various continental fragments in Asia; 15, major course of drainage.

Key to lettering: AG, Akçakale graben; AGS, Aegean graben system; BA, Banda Sea; BHR, Bo Hai rift; BTV, Balagan-Tas volcano; CAGS, Central Arabian graben system; CBR, Central Basin Ridge; CF, Chaman fault; CIP, Central Indian Plateau; EAB, East Arabian block; EMB, East Malaya Basin; GKF, Great Kavir fault; GT, Gerze thrust; GTR, Gulf of Thailand rift; HAV, Hsing An fissure volcanos; HF, Herat fault; HG, Hatay graben; HRF, Harirud fault; KDF, Kopet Dagh fault; KF, Karakorum fault; KTF, Kang Ting fault; MR, Main Range of the Greater Caucasus; MTL, Median tectonic line; MZF, Main Zagros fault; NAH, North Anatolian fault; OT, Okinawa trough; PaT, Palau trench; PF, Philippine fault; PKR, Palau-Kyushyu ridge; PNT, Palni-Nilgiri Hills thrust; PT, Pamir thrust; Q, Qaidam basin; RRF, Red River fault; SGS, Shansi graben system; SF, Sumatra fault; SSZ, Sinai shear zone; SUF, South Ural seismic faults; TD, Turfan depression; T-LF, Tan Lu fault; YT, Yap trench.

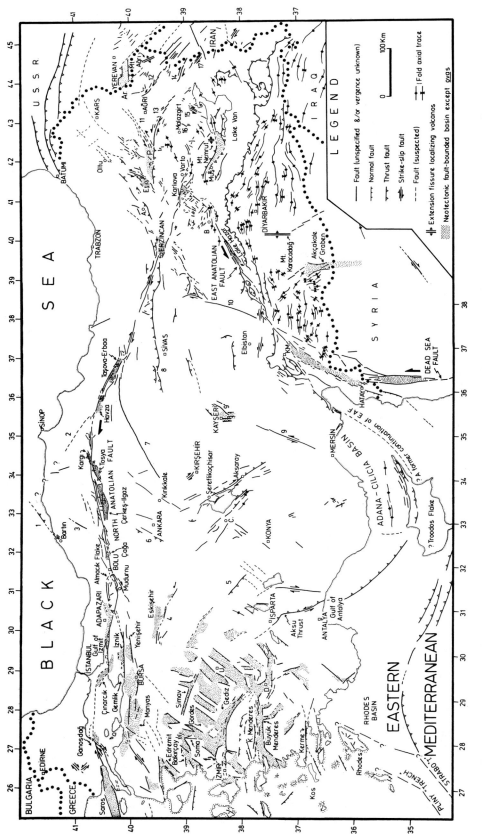

FIG. 2.—Simplified map of Turkey showing all the major neotectonic structures that have been mapped in the field or established from seismic reflection profiling in marine areas. The Çınarçık Basin and the conjectured former continuation of the East Anatolian fault have been mapped from bathymetry alone. The distribution of the large, roughly equant Neogene depressions in Central Anatolia ("ovas") is omitted because the sedimentary fill of these basins commonly overflows their tectonic boundaries, largely concealing the structures and rendering the basin contours irrelevant for the purposes of this map.

Key to numbers: 1, Bartın fault and its submarine continuation; 2, Conjectured; ?incipient Havza-İnebolu fault; 3, Safranbolu fault; 4, Eskişehir fault; 5, Sultandağ fault; 6, Ankara fault; 7, Kırıkkale-Erbaa fault; 8, Sivas fault; 9, Ecemiş fault; 10, Malatya fault; 11, Kağızman fault; 12, Balık Gölü fault; 13, Tutak fault; 14, Çaldıran fault; 15, Süphan fault; 16, Malazgirt fault; 17, Hasantimur Gölü fault. Key to lettering: A, Aşkale; Ar, Ararat Basin; B, Bingöl; D, Denizli, E, Erzurum; KM, Kahramanmaraş; P, Pasinler Basin; U, Uşak.

The map is compiled and synthesized, in addition to our own observations, from the following: Paréjas and Pamir (1939); Zeschke (1954); M.T.A. (1961–1964); Burshtar and Tolmachevskiy (1965); Allen (1969); Erol (1969, 1982b); Ketin (1968, 1969, 1985); Arpat and Şaroğlu (1972, 1975); Angelier (1973); Tokay (1973); Konuk (1974); Seymen (1975); Ben-Menahem et al. (1976); Brinkmann (1976); McKenzie (1976); Arpat et al. (1977); Berckhemer (1977); Letouzey et al. (1977); Tchalenko (1977); Toksöz et al. (1977); Atalay (1978); Biju-Duval et al. (1978); Evans et al. (1978); Şengör (1978); Sidorenko (1978); Wong et al. (1978); Aktimur (1979); Aktimur et al. (1979); Dumont et al. (1979); Le Pichon et al. (1979); Şaroğlu and Güner (1979, 1981); Tatar (1978); Angelier et al. (1981); Jongsma and Mascle (1981); Barka (1981, 1983a, b; 1985); Yılmaz et al. (1982); Erkal (1983); Özgül et al. (1983); Şengör et al. (1983); Bilgin (1984); Güner (1984); Okay (1984); Boray and Şaroğlu (in press). We have also used the unpublished active and potentially active fault map of Turkey, Scale: 1:2,000,000 prepared as a part of the Mineral Research and Exploration Institute (M.T.A.) project "Map of Active Faults and the Neotectonics of Turkey" and the 1:250,000-Scale unpublished geological map of southeastern Turkey by Turkish Petroleum Co. geologists. M.Ş. Abdüsselamoğlu, İ. Ketin, O. Sungurlu and Y. Yılmaz also contributed unpublished data.

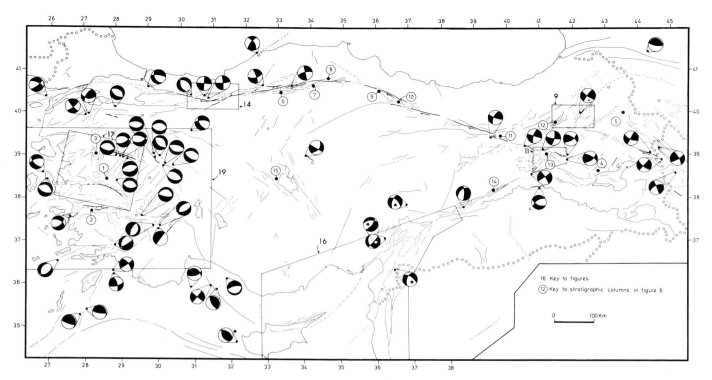

FIG. 3.—A neotectonic map of Turkey drawn from Figure 2 and showing all the reliable focal mechanism solutions (after McKenzie, 1978; Jackson and McKenzie, 1984; four solutions from Büyükaşıkoğlu, 1979, identified with asterisks, and one from Toksöz et al., 1983, identified with a cross); the locations of the stratigraphic columns shown in Figure 6, and the locations of Figures 9, 11, 14, 16, 17, and 19. Heavy black lines represent ground breaks during earthquakes, summarized in McKenzie (1978) and Jackson and McKenzie (1984). Note added in proof: The fault-plane solution of the Lice earthquake in SE Turkey is shown as a normal fault. It should be a thrust solution.

sional convergence of the Arabian platform and Asia (Fig. 1). The active tectonic scheme of Turkey is characterized by a great variety of structure families, the largest of which belongs to strike-slip fault systems (Fig. 2). Available high-quality focal-mechanism solutions of earthquakes (Fig. 3) also substantiate the view that major strike-slip faults in Turkey predominate over other kinds of active structures.

Because it is not possible to gain a satisfactory understanding of the active strike-slip systems in Turkey without considering their broader context (Şengör, 1979; Şengör et al., 1982), we present here the history of evolution of the active tectonics of Turkey and surrounding regions. The tectonic evolution and corresponding paleogeography of the Eastern Mediterranean region since early Miocene time are shown schematically in Figures 4 and 5. The stratigraphic evolution of selected basins is summarized in Figure 6. Figure 7 is a map of the neotectonic provinces in Turkey.

At this point it is useful to indicate what we mean by neotectonics. Şengör (1980, 1982) defined a neotectonic period as the time that elapsed since the last major whole-sale tectonic reorganization in a region of interest. He argued that the middle Miocene collision of Arabia with Anatolia introduced such drastic changes in the tectonic evolution of all of Turkey that it forms a convenient landmark to separate the country's neotectonic development from its paleotectonic development (Fig. 6A). Erol (1981, 1982a)

pointed out that this division was also in good agreement with the geomorphological development.

One of the most persistent problems in the study of the neotectonic evolution of the eastern Mediterranean area and surrounding regions has been the lack of detailed correlations among the Mediterranean (marine), Paratethyan, and terrestrial sequences. In this paper, we follow the most recent correlation scheme of Rögl and Steininger (1983) as revised by F. Steininger (personal commun., 1985) on the basis of Berggren et al. (in press) for the Mediterranean-Paratethyan correlations; whereas we follow Erol (1981, 1982a) for the correlation of terrestrial sequences, erosion surfaces, and terrace systems with each other and with the Mediterranean and Paratethyan sequences. The time scales used by Rögl and Steininger (1983) and Erol (1981, 1982a) do not differ significantly from one another, and the existing differences are probably not greater than the inherent uncertainties of the geomorphological correlation methods employed by Erol.

Neotectonic Evolution and Coeval Paleogeographic Development

During the early Miocene (Aquitanian through Langhian), continuing intracontinental convergence associated with the Paleocene-Eocene collision along the Izmir-Ankara-Erzincan and Inner Tauride sutures (1 and 2 respectively in

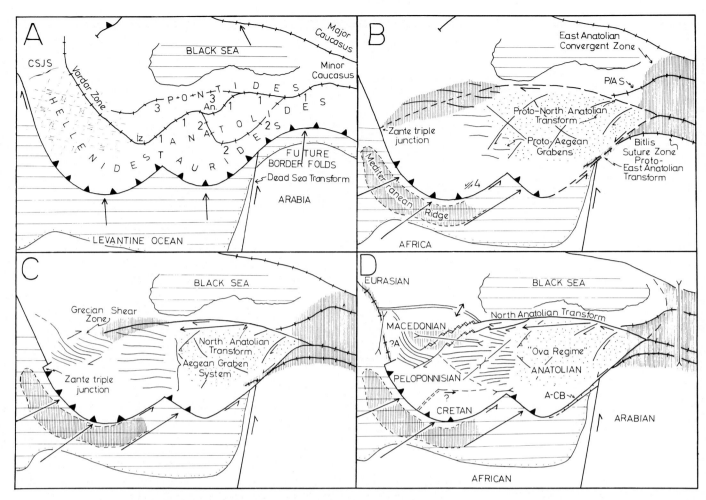

Fig. 4.—Tectonic evolution of Turkey and surrounding areas since the early Miocene (modified from Şengör and Canıtez, 1982). A, early Miocene; B, middle to late Miocene; C, Pliocene; D, Present.

Key to ornament: Ladder pattern, Alpide sutures and/or zones of intracontinental convergent high strain; lines with black triangles, subduction zones (triangles on the upper plate); double lines, zones of extensional high strain; simple lines with half arrows along them, strike-slip faults; full arrows, direction and approximate rate of relative motion of plates and scholles; closely spaced vertical ruling, regions of diffuse convergent strain; sparse stippling, regions of diffuse extension; widely spaced horizontal ruling, oceanic lithosphere.

In A, CSJS represents a hypothetical conjugate shear joint system in the Hellenides generated during the paleotectonic episode, which is believed to have localized the Grecian Shear Zone and prevented a straight continuation of the North Anatolion fault towards the west, to the Ionian Sea.

Key to numbers: 1, İzmir-Ankara-Erzincan suture; 2, Inner Tauride suture; 3, Intra-Pontide suture; 4, Ierapetra strike-slip fault. Key to lettering: A, Albanian Scholle, An, Ankara; A-CB, Adana/Cilicia Basin; P/AS, Pontide/Anatolide suture.

Fig. 4A) characterized much of the present area of Anatolia (Şengör and Yılmaz, 1981). The Menderes Massif in the west continued its uplift (Şengör et al., 1984) and probably represented a high region not dissimilar in character to the present Gurla Mandata or Guntschu uplifts of the Himalaya (Gansser 1964, 1977). In Crete (Wachendorf et al., 1980), in the Lycian Taurus (Gutnic et al., 1979), and in northern Cyprus, which was then still attached to Turkey (Ducloz, 1972), south-vergent nappe movements continued until the (?early) Serravallian, while north-directed back-thrusting characterized wide areas in the Pontides (Şengör and Yılmaz, 1981), indicating the continuing north-south shortening in western and central Turkey. Corresponding with this tectonic picture, the paleogeographic panorama (Fig. 5A) was

one of an eastward-sloping highland represented by western and central Anatolia and a lowland with a rolling topography in large parts of the Pontides locally covered by terrestrial sediments. Occasional marine invasions from the east and south inundated only small areas in central Anatolia and limited evaporite deposits were formed (Erol, 1969; Fig. 6A, col. 15). Sparse volcanic activity, mainly of calc-alkalic type, took place in northwestern Anatolia and in the Galatean Massif (west of 1 in Fig. 5A; Şengör and Yılmaz, 1981).

The land area in western and central Turkey was surrounded by shallow seas in the north, east, and south. Especially in the south and east, the deposition of widespread reefal limestones (Altınlı, 1966; Şengör and Kidd, 1979;

Şaroğlu and Güner, 1981, for eastern Turkey; Gutnic et al., 1979, for southern Turkey) indicates a tropical climate, which is substantiated by studies on the terrestrial sediments that accumulated around the periphery of the western and central Anatolian highland (Erol, 1981). The actual elevation of this highland is difficult to assess, but minimum and maximum figures may be surmised for its western part, the Menderes Massif, following a reasoning outlined below. Akkök (1981, 1983) argued that the latest Eocene metamorphic mineral assemblages in the central part of the Menderes Massif near Derbent indicate a paleopressure of about 5–6 kb corresponding with a burial depth of 15–20 km. In that particular locality we do not know when exactly the overlying rocks were removed but the earliest postmetamorphic sediments are of middle Miocene age (Fig. 5 A,B). If the overlying rocks were still there during the early Miocene (Fig. 5A), we should add that 15–20 km of crustal thickness to a thickness of 50 km obtained by multiplying the present crustal thickness under western Turkey (±35 km, Le Pichon and Angelier, 1981) by a conservative fac-

tor of 1.5 to remove the effects of the post-Serravallian extension (see Şengör, 1982). This would give a crustal thickness between 65 km and 70 km and would represent a maximum possible thickness. The latter of these two values is similar to the thickness of the crust beneath Tibet (Hirn et al., 1984) and the Altiplano (Ocola et al., 1971) indicating an average elevation with respect to the present sea level of some 4 to 5 km. A minimum crustal thickness is obtained by assuming that all of the 15–20 km overburden had been eroded completely by early Miocene time. In that case we reverse the post-Serravallian extension and obtain a crustal thickness of about 50 km, which would mean an elevation with respect to the present sea level of about 2 km. The real elevation was probably somewhere between the two extremes. In Figure 6B we assumed an early Miocene average height of western Turkey of about 3 km.

During the early Miocene, Arabia had not yet collided with Eurasia along the Bitlis suture zone in southeastern Turkey as indicated by the development of the Midyat limestones on the north Arabian shelf (Figs. 4A, 5A) and the

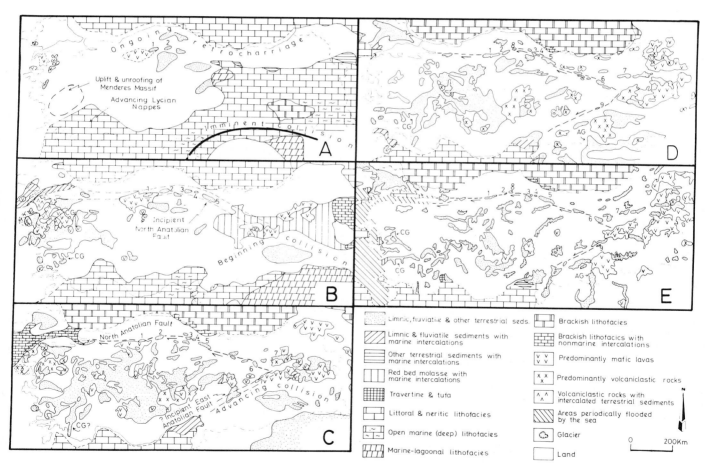

FIG. 5.—Non-palinspastic paleogeographic maps of Turkey for the early (A), middle (B), late Miocene (C), Pliocene (D), and Pleistocene (E); simplified and modified from Lüttig and Steffens (1976) on the basis of Kaya (1981), Barka (1981), Hempton et al. (1983), and our observations mainly in southern and southeastern Turkey. The thick black line in A in southeastern Turkey represents a zone along which palinspastic continuity was significantly disturbed owing to later subduction and strike slip. Key to numbers: 1, Çerkeş-Kurşunlu Basin; 2, Tosya Basin; 3, Havza Basin; 4, Lâdik Basin; 5, Taşova-Erbaa Basin; 6, Erzincan Basin; 7, Erzurum Basin; 8, Kargı Basin; 9, Lake Hazar Basin. Key to lettering: AG, Akçakale Graben; CG, Cross-graben.

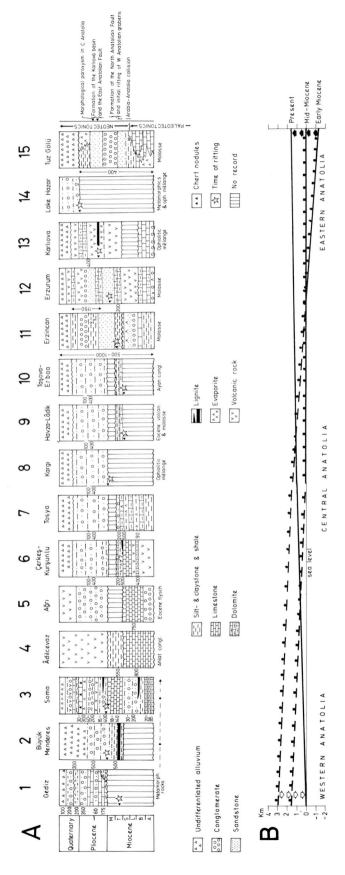

FIG. 6.—A. Correlation chart showing simplified post-Oligocene stratigraphic sections for selected sites in Turkey (for locations see Fig. 3): most local unconformities and disconformities not co-extensive with the basins could not be shown because of the chosen scale. On the right is a list of significant events in the neotectonic evolution of Turkey. Figures to left of each column give approximate thicknesses in meters. The sources for the individual columns are as follows:

(1) Erentöz and Ternek (1968), F. Şaroğlu (unpublished data); (2) Erentöz and Ternek (1968), Becker-Platen (1970); (3) Brinkmann et al. (1970). Nebert (1978); (4) and (5) Şengör and Yılmaz (1981); (6), (7), and (8) Lüttig and Steffens (1976), Barka (1981), Barka and Hancock (1984); (9) and (10) Irrlitz (1972), Barka (1981), Barka and Hancock (1984); (11) and (12) Irrlitz (1972), Lüttig and Steffens (1976); (13) Nakoman (1968), F. Şaroğlu (unpublished data); (14) Hempton et al. (1983), Dunne and Hempton (1984); (15) Erol (1969, 1981), N. Görür (unpublished data).

B. Schematic, east-west sections from Turkey showing the change in average elevation of the topographic surface since the early Miocene with respect to sea level. Because the maximum sea-level change since then is small (<300 m) with respect to the magnitudes of elevation change, a single sea level was taken as representative of the entire interval.

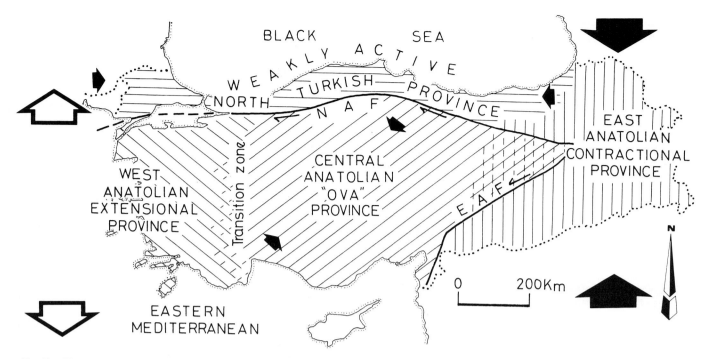

FIG. 7.—Neotectonic provinces of Turkey: Black arrows show shortening and white arrows extension directions. Arrow size is crudely proportional to magnitude of total shortening.

persistence of deep-water environments in extreme southeastern Turkey around Hakkâri (Fig. 5A; N. Görür, unpubl. data). The development of shallow-water carbonates throughout eastern Turkey also corroborates this view.

It is important to emphasize that the North and East Anatolian faults did not exist during the early Miocene. The Çerkeş-Kurşunlu (1 in Fig. 5A, loc. 6 in Fig. 6A) and the Tosya (2 in Fig. 5A, col. 7 in Fig. 6A) Basins, later to be incorporated into the North Anatolian fault zone, were developing as parts of post-collisional flysch-molasse basins along the Intra-Pontide suture (3 in Fig. 4A), and represented the last remnants of the previously much larger Mudurnu-Göynük basin complex of Saner (1980). Barka and Hancock (1984) interpreted the deposition of early Miocene sediments in these basins (Sivricek and Devren Formations) as having taken place during a modest compressional episode. This compression was coeval and parallel in orientation with the last back-thrusting events in the Pontides (Şengör and Yılmaz, 1981; Fig. 5A), and thus may be viewed as a part of the continuing north-south shortening of the Turkish orogen at this time.

During the middle to early-late Miocene (mainly Langhian-Serravallian), Arabia and Eurasia collided along the Bitlis suture in southeastern Turkey (Şengör and Yılmaz, 1981; Fig. 4B). This collision resulted in the uplift of the mountains along the suture, and the quiet shallow-marine depositional environments of eastern Turkey were converted into molasse basins in which terrestrial redbeds were deposited between intervals of marine incursions (Figs. 5B, cols. 4, 5 in Fig. 6A). It is important to emphasize the presence of

these marine incursions, the last of which occurred during the Serravallian (Gelati, 1975), because they show that substantial parts of eastern Turkey were close to sea level at least until after the Serravallian.

During the same time interval, there was no marine deposition in western or central Anatolia. It is likely that these regions were still fairly high, because after at least 10 m.y. of (?steady) extension and subsidence, western Turkey was still at least some 200 m higher during the last Pleistocene glaciation than now (de Planhol, 1953). The 1.5 km elevation shown in Figure 6B is believed to be a generous estimate of the elevation decrease since early Miocene time.

It is during the middle to early-late Miocene (late Serravallian-Tortonian) that there is abundant evidence of the initiation of westerly motion of the Anatolian scholle (Fig. 4B). The detailed stratigraphic and structural studies of Barka (1981) and Barka and Hancock (1984) of the basins aligned along the North Anatolian fault and the refinement of the correlations between Paratethyan and Mediterranean sequences (Rögl and Steininger, 1983, as revised by F. Steininger, personal commun., 1985) have enabled us to establish the time of formation of the North Anatolian fault zone as late Serravallian. At this time, the lowermost parts of the lower Pontus Formation of Irrlitz (1972) began to accumulate in basins (3, 4, 5 in Fig. 5B, and cols. 9, 11 in Fig. 6A) along a broad shear zone concurrently with synsedimentary faulting (Barka and Hancock, 1984). At the same time, a broad furrow formed along the northern coast of the future Sea of Marmara and connected with the northern Aegean Sea via the Saros graben (Fig. 5B). This furrow localized limnic and fluviatile sedimentation with repeated marine incursions coming from the northern Aegean Sea. The Aegean Sea area itself was first inundated during the Tortonian from the south via a narrow strait that formed between the Cyclades and the Anatolian landmass (see Fig. 8, Dermitzakis and Papanikolaoy, 1981). The furrow in Turkey reached as far east as Istanbul, but did not yet connect with the terrestrial basin of Yalova farther to the southeast, which marked the western termination of the broad shear zone of the incipient North Anatolian fault. The right-stepping primary offset of the marine furrow from the incipient North Anatolian fault zone is interesting, because its location coincides with the present Çınarcık pull-apart basin. The Ergene Basin in Thrace also began its neotectonic episode of subsidence during the Tortonian (Fig. 5C).

The Tortonian is also the time when the Aegean and the west Turkish grabens, at least the major ones such as the Büyük Menderes (Becker-Platen, 1970; Fig. 6A, col. 2), Gediz (Fig. 6A, col. 1) and associated smaller grabens (such as the Soma Graben) that trend at high angles to the Gediz Graben (Fig. 2, Fig. 6A, col. 3), and the Saros Graben (Dermitzakis and Papanikolaoy, 1981), began to form (Figs. 4B, 5B). Marine invasion in the axial part of the future Aegean Sea at the same time shows that the uplands occupying this region during the early Miocene had already subsided locally beneath sea level as a result of the beginning extensional tectonics (Meulenkamp, 1982; Şengör, 1982).

Another piece of important evidence showing the beginning westerly motion of the Anatolian scholle comes from the Ierapetra half graben in eastern Crete (Fig. 4B, loc. 4). This northeast-striking, left-lateral strike-slip-controlled structure formed during the late Serravallian on top of and disrupting the Eocene-?early Miocene nappe pile in this part of the island (Fortuin and Peters, 1984). It indicates that the previously convergent eastern half of the Hellenic trench system switched to strike-slip faulting most likely in response to the westerly motion of the Anatolian scholle.

Erol (1981) showed that during the middle Miocene (±Langhian to early Serravallian), much of the land surface in Anatolia formed an undulating topography occupied by broad freshwater lakes separated by low and wide swells. Later, during the Tortonian, this gentle topography began to be broken up by numerous fault systems; the lake basins narrowed, and became surrounded by extensive pediments. Meulenkamp (1982) showed that a similar evolution also took place in the southern Aegean. This shows that as the Anatolian scholle began its westerly movement, it also started to be disrupted internally.

Changes in the spatial distribution of subduction-related volcanoes both in the Aegean area and in Anatolia at the end of the Serravallian also appear to indicate the onset of the westerly motion of Anatolia. The pre-Tortonian volcanism associated with the southernmost subduction zone in Turkey was distributed uniformly, although sparsely, along the entire length of the future Anatolian scholle. Beginning with the Tortonian, it not only increased in intensity but also became concentrated in two areas: the southern Aegean and southeastern central Anatolia, behind the Hellenic

and the Cyprus subduction zones, respectively (Lüttig and Steffens, 1976; Innocenti et al., 1981).

Dewey and Şengör (1979) argued that the active zone of north-south extension reaching from Thrace to Yugoslavia probably had formed as a consequence of the westward-moving Anatolian scholle's trying to rip away eastern Thrace and Macedonia from the areas north of them (Fig. 4B). The onset of subsidence in the Ergene Basin during the Tortonian corroborates this view and indicates that Anatolia's attempt to tear away Macedonia and eastern Thrace had begun by Tortonian time.

During the late Miocene (mainly Messinian), the present tectonic regime of Turkey was well established (Fig. 4C). From this time onward marine waters never again invaded any significant portion of the Turkish landmass (see Figs. 5C, D, E). Barka and Hancock (1984) noted that at this time the broad shear zone developed earlier in the place of the future North Anatolian fault was transformed into a narrow fault zone. Its geomorphological expression, however, first appeared later during the early Pliocene (Figs. 4C, 5C). Barka and Hancock's (1984) inference is based on the distribution of mesofaults within the Pontus Formation which was deposited during late Serravallian to earliest Pliocene time (see Şengör et al., 1983; Barka and Hancock, 1984; F. Steininger, personal commun., 1985). Mesofaults related to the North Anatolian fault zone, which are confined to the lower Pontus Formation (late Serravallian-Tortonian), occur in a relatively broad belt (10–15 km) across which shear was distributed. Those that affect also the upper Pontus Formation (Messinian-earliest Pliocene) are confined to a narrow zone (2–3 km) astride the main trace of the fault.

Most of the strike-slip basins along the North Anatolian fault were also established by Messinian time (localities 1–6 in Fig. 5C; cols. 6–11 in Fig. 6A). This progressive increase in the number of the strike-slip basins along the course of the North Anatolian fault through time is very similar to the formation of Neogene basins along the San Andreas fault zone in western North America (Blake et al., 1978, Fig. 2; compare with our Fig. 5). At the western end of the fault zone, the former marine furrow finally joined the fault zone via a northwest-trending depression, the Çınarcık Basin.

Graben complexes continued to subside and increased in number during the Messinian. Böger (1978) pointed out that normal faulting began on the island of Kos during the Messinian. The Kerme graben, in which the island of Kos is located, is perhaps also of the same age (Şengör, 1982). The rapid subsidence of the graben floors localized sedimentation and greatly restricted its areal distribution (Fig. 5C). Although freshwater lakes had occupied considerable areas in central Turkey during the Tortonian, their areal extent diminished in the Messinian, both because of increased relief owing to increased faulting and because of the arid climate that developed along with the Messinian desiccation of the Mediterranean.

In eastern Turkey, terrestrial sediments were accumulating in east-trending basins. In this area, dominantly calc-alkalic volcanism of high K-type increased in intensity between the Serravallian and early Pliocene (Innocenti et al.,

1980), and following the enlargement of topographically high areas, it also began spreading laterally (compare Fig. 5B with 5C) (Innocenti et al., 1981).

During the Pliocene (Fig. 4C, 5D), the area of sedimentation in Turkey decreased somewhat compared with the late Miocene and the northeast- and northwest-trending topographic prominences in central Anatolia became more pronounced. Along the North Anatolian fault, the existing basins enlarged and new ones formed. Erol (1981) noted that during the early Pliocene, the first indications of continuous deep and narrow depressions along the North and the East Anatolian fault zones could be recognized. The sea retreated from the western end of the North Anatolian fault zone, although elsewhere in the future Aegean Sea, it considerably enlarged its area, indicating the continuing subsidence of this extensional region (Dermitzakis and Papanikolaoy, 1981).

There is no unequivocal evidence along the course of the East Anatolian fault to date its inception with any precision. The beginning subsidence of the Karlıova Basin (Fig. 6A, col. 13) at its eastern end, which evolved under the direct influence of the fault, indicates, however, that the fault had already formed during the Pliocene (Nakoman, 1968; Fig. 6A, col. 13).

During the Pleistocene (Figs. 4D, 5E), Turkey acquired its present geography. The sea returned to the western termination of the North Anatolian fault, but this time along two strands instead of the previous one. It also invaded the coastal areas of Turkey along the Aegean Sea and extended into some of the grabens that continued their growth (e.g., Erinç, 1955a,b; Bilgin, 1969b) only to be pushed back by delta progradation during historic times.

Figure 6B illustrates, in schematic fashion, the evolution of average topographic elevations in Turkey during the late Cenozoic along a generalized east-west cross-section. The most conspicuous aspect of these cross-sections is the reversal of elevation of the topographic surface around a pivot in central Anatolia. We emphasize that the initiation of the westerly flight of the Anatolian scholle began at a time when the topographic surface was still sloping to the east. The topographic surface was probably nearly horizontal during the early Pliocene and began sloping westwards towards the end of this epoch. Even now it does not dip westwards as strongly as it did eastwards during early Miocene time.

Neotectonic Provinces of Turkey

As a result of the tectonic evolution outlined in this section, four major "neotectonic provinces" have formed in Turkey (Şengör and Dyer, 1979; Şengör, 1980). Each of these is characterized by a distinctive set of structures and a specific form of generalized finite strain (Fig. 7). In each province all the structures function together to accomplish the overall deformation of the area. The actual type of structure, such as folds, thrust faults, strike-slip faults, normal faults, or extensional fissures, seems dependent on a variety of factors such as the presence and orientation of anisotropies in the rocks, crustal thickness, and compati-

bility with nearby structures. Under the influence of continuing deformation and the factors listed above, a variety of structures formed and evolved in each of these provinces. Within the same deformation field, changes in these factors also bring about changes in the orientation and types of structures. Thus, progressive deformation gives rise to a complex array of structures thay may have variable life spans and complex overprinting relationships even in simple deformation fields similar to the cases illustrated by Flinn (1962).

Neotectonic structures also have complex relationships with older, paleotectonic structures. Many of the former appear to have been controlled by the latter ones. Following Şengör (1982), we classify the relations between paleotectonic and neotectonic structures in Turkey under three different categories.

(1) Resurrected structures: These form by reactivating a paleotectonic structure in its former role during the neotectonic episode. For example, if a compressional paleotectonic basin continues to function as a compressional basin during the neotectonic period it would be called a resurrected structure.

(2) Replacement structures: These neotectonic structures coincide in space with paleotectonic structures, but do not perform the same function as the latter. The new structure "replaces" the old one. A thrust fault reactivating a former normal fault or a strike-slip fault nucleating on a rotated suture are examples of replacement structures.

(3) Revolutionary structures: These structures are independent of the pre-existing fabric, and they are probably the rarest neotectonic structures in Turkey.

Revolutionary structures are often easy to recognize. Problems frequently arise on the other hand in the identification of resurrected and replacement structures, particularly the former. If paleo- and neotectonic periods in the history of the development of resurrected and replacement structures are not distinguished, these structures may lead to completely erroneous results regarding the time of establishment of the neotectonic regime in a given area. Also the ratios of the number of each of these three kinds of structures to the total number of all structures in a given area may constitute a guide to the relationships between the paleo- and neotectonic regimes in that area.

The distinguishing characteristics of the four neotectonic provinces of Turkey, defined on the basis of the form of the generalized finite strain, as expressed by all three kinds of neotectonic structures, may be summarized as follows (Figs. 2, 3, 7): (1) The east Anatolian contractional province is characterized by roughly north-south shortening, now accommodated dominantly by folding and thrusting along its southern and northern boundaries, and by strike-slip faulting in its central part. (2) The north Turkish province is characterized by very weak active tectonism. Available meager evidence indicates that it may be shortening east-west (Şengör et al., 1983). (3) The west Anatolian extensional province is dominated by generally east-trending grabens that have been evolving under a north-south exten-sional regime. It is generally believed that the amount of extension since the Tortonian in western Turkey may have been about 30%. (4) The central Anatolian "ova" province is the quietest both seismically and structurally after the north Turkish province. Most of its neotectonic features lie buried under extensive pluvial lake sediments. Large, roughly equant extensional basins bounded by more than two oblique-throw faults, called ovas (see Şengör 1979, 1980), are its characteristic structural forms. Over a narrow transition zone (Fig. 7), they gradually pass into and connect with the gra-bens of western Turkey (Fig. 2). A limited amount of northeast-southwest shortening with corresponding north-west-southeast extension may control the structural evolu-tion of this province, but the lack of detailed information about its buried structures makes an analysis difficult.

In addition to these four major provinces, the North and East Anatolian faults form nearly autonomous boundary re-gions between some of the provinces. The dominantly strike-slip-related structures along these large fault zones cannot be considered to belong to any of the four neotectonic prov-inces of Turkey.

In the following sections we describe the strike-slip sys-tems and associated basin types in each of the neotectonic provinces of Turkey as well as in the North and East An-atolian fault zones. The poorly-understood ova province is not treated because of lack of information.

BASINS OF THE EAST ANATOLIAN CONTRACTIONAL PROVINCE

The East Anatolian contractional province represents the vice from which the Anatolian scholle is escaping west-wards (Fig. 7). It acquired its individuality after the for-mation of the North and the East Anatolian faults. Pro-gressive north-south shortening has characterized its evolution during the neotectonic period, and generated a rich assort-ment of structures forming a complex but discontinuous network of high strain zones (Fig. 2).

Although the northeast- and southeast-striking conjugate strike-slip fault systems dominate the active tectonics of eastern Turkey, the most spectacular structures associated with the north-south convergence are the east-trending ramp basins. In the absence of detailed studies, the faults bound-ing such basins have generally been described as normal faults (e.g., Oswald, 1912; Sür, 1964; Atalay, 1983), al-though in 1953, Erinç interpreted the main southern boun-dry fault of the Lake Van depression as a steep thrust. Re-cent field observations (e.g., Şaroğlu and Güner, 1981), combined with seismic reflection profiling, both on land and on Lake Van (Wong et al., 1978) have shown, how-ever, that the east-trending basins in eastern Turkey are ramp basins of compressional origin, indicating the north-south shortening of the plateau (Fig. 8; Wong et al., 1978).

Şaroğlu and Güner (1981) argued that at the beginning of the neotectonic period, eastern Turkey represented a low-lying peneplain locally interrupted by east-trending depres-sions filled with sandstones, mudstones, and intercalated andesitic basalt flows, tuffs, and agglomerates. They showed that these basins occupied wider areas during the late Mio-

cene-Pliocene than during the Pleistocene and that their margins had not been faulted before the late Pliocene, or at least not to the extent that they were later. Şaroğlu and Güner (1981) concluded from this that during the late Mio-cene, the land surface of eastern Turkey had been warped into large and broad articlines and synclines perhaps owing to the buckling of the lower Miocene shallow-water lime-stone layer above the softer flysch and mélange of the East Anatolian accretionary complex (Şengör and Yılmaz, 1981). The resulting shallow synclines formed broad basins in which mainly clastic terrestrial sediments accumulated. Because folding was faster than erosion, these broad synclinal basins also formed wide-bottomed valleys in which meandering streams became established. With continued shortening the simple folds were disrupted by thrusts along the flanks of the synclines and thus turned into ramp basins.

The best-known ramp structure in eastern Turkey is that of the Muş - Lake Van Basin complex (Fig. 2) separated by the volcanic edifice of Mt. Nemrut since the middle Pleistocene into the dry, sediment filled Muş Basin and the lake-filled Van depression (Erinç, 1953; Şaroğlu and Güner, 1981; Güner, 1984). Figure 8 is a north-south seismic re-flection profile across of the Muş Basin. This profile, when interpreted in the light of the surface observations, shows the compressional nature of the Muş Basin and the large bounding thrust faults. Another noteworthy feature on this profile is the thin wedge of sediment on the northern margin of the depression, disrupted by splays of the major northern boundary thrust of the basin. This wedge thickens south-wards, that is towards the basin, and extends farther north than the northern boundary thrust, thus corroborating the views of Şaroğlu and Güner (1981) about the evolution of the east Anatolian ramp basins. Similar contractional faults disrupting originally gentle, very large wavelength (~10 km) folds have also been detected by seismic reflection profiling in the Pasinler Basin (P in Fig. 2), in which only the south-ern margin seems to have been disrupted by major thrust faults (O. Sungurlu, written commun., 1984). Shallow seismic reflection profiles in Lake Van show that contrac-tional structures affect even the youngest sediments of the lake (Wong et al., 1978).

Ramp basins such as the Muş - Lake Van Basin and the Pasinler Basin have variable relations, not everywhere well-exposed, with the prominent but discontinuous strike-slip systems of eastern Turkey. For example the left-lateral, northeast-striking Kağızman and the right-lateral north-west-striking Tutak fault systems (Fig. 2) intersect just east of the Pasinler Basin, and together accommodate north-south shortening (Fig. 2). West of the Pasinler Basin is a north-northeast-trending system of strike-slip faults with a slight component of east-west extension. The Erzurum Basin (E in Fig. 2), which developed during Tortonian-Messinian time (Irrlitz, 1972; as revised by Rögl and Steininger, 1983), is a complex pull-apart basin that formed along a right-step-ping offset on this fault system (Atalay, 1978).

In contrast to the Pasinler half-ramp basin, the Muş-Lake Van ramp basin forms one part of a long east-trending zone of compressional high strain that extends from east of the town of Bingöl (B in Fig. 2) nearly to the Iranian border,

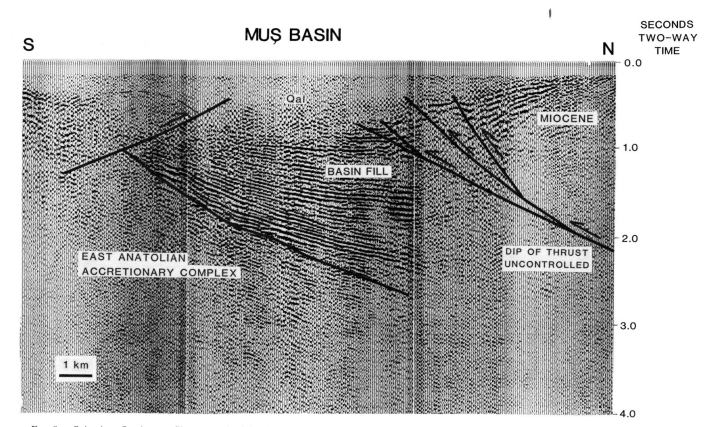

FIG. 8.—Seismic reflection profile across the Muş Basin. The northward-thrust metamorphic rocks of the Bitlis Massif in the south are probably the cause of a velocity pull-up creating the apparently significant northerly dip of the basement under the basin. To the north, the abrupt truncation of the prominent basin-fill reflections seems to be an artifact of the sharp reduction in data quality, perhaps because of deformation under the thrust sheets forming the northern frame of the basin (seismic line courtesy of O. Sungurlu).

where it connects, much like the Pasinler Basin, with two fault systems with northeasterly and northwesterly strikes (Fig. 2). The northwest-striking Salmas fault experienced 4 m of right-lateral offset during the May 6, 1930, Salmas earthquake, together with 5 m of dip slip (north side down; Tchalenko, 1977).

The area between the Pasinler and the Muş - Lake Van ramp basins is characterized by irregularly distributed and discontinuous folds, strike-slip faults, and extensional fissures (Fig. 2). The folds south of Varto parallel the trace of the August 19, 1966, Varto-Üstükran earthquake fault (Ambraseys and Zatopek, 1968; Fig. 2), which showed predominant right-lateral motion at the surface, but whose fault-plane solution (McKenzie, 1972) contains an appreciable thrusting component (Fig. 3). This zone of convergent high strain connects westwards with the North and East Anatolian faults, and eastwards diffuses into the area of distributed convergent strain between the Pasinler and the Muş - Lake Van ramp basins (Fig. 2).

North of the Pasinler half-ramp basin, between Oltu and Kars (Fig. 2), a number of faults and fault-controlled geomorphological features (mainly stream valleys) define a northeast- and northwest-striking conjugate pattern, although detailed observations in this area to establish ages

and directions of movement on these structures are unfortunately not available.

The character and the overall geometric and kinematic pattern of the East Anatolian contractional province are repeated within it at smaller scales. This is best illustrated by the population of faults to the east of Erzurum, a number of which, located between the villages of Kızlarkale in the south and Çimli and Balabantaş in the north, were reactivated by the Narman-Horasan earthquake (October 30, 1983; Toksöz et al., 1983; Fig. 9). The most salient feature of Figure 9A is the abundance of northeast-, north-northeast-, north-northwest-, and west-northwest-striking strike-slip faults, with only one north-striking extensional feature that formed during the Narman-Horasan earthquake (Fig. 9B). A number of west-northwest-striking faults have also been observed to the immediate southwest of the conjugate strike-slip fault system of Çimenli, but their nature could not be determined. The discontinuous character of these structures and their complex relationships with one another are clearly visible in Figure 9A. Yet they all work in such a way as to accomplish north-south shortening with corresponding east-west extension (active structures that thicken the crust are not present within the area of Fig. 9A).

Figure 9B shows an active example of structure com-

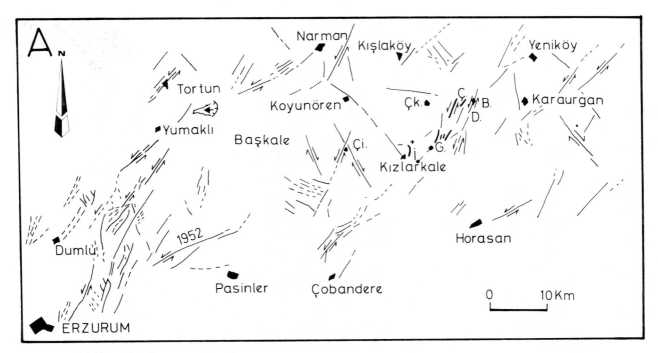

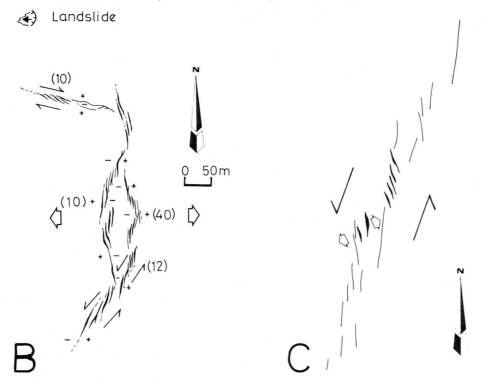

FIG. 9.—A) Active fault pattern to the east of Erzurum (for location see Fig. 3) and the surface breaks generated during the October 30, 1983, Narman-Horasan earthquake (after Barka et al., 1983).

B) Surface break consisting of an extensional mini-graben (in the center) and a right-lateral (in the north) and a left-lateral strike-slip fault (in the south) that formed during the Narman-Horasan earthquake just to the east of the village of Kızlarkale. The topographic slope of the location of this break precludes the possibility of its being the expression of an incipient land-slide (after Özgül et al., 1983).

C) Surface break consisting of a series of en echelon extensional and shear fractures that formed during the same earthquake to the north of the village of Gerek. The sketch was drawn without a scale (after Özgül et al., 1983).

patibility on a still smaller scale. The two sides of the surface break illustrated here moved away from one another during the Narman-Horasan earthquake as shown by the broad arrows. The motion was accommodated by a small graben in the central part of the break, which connected with strike-slip systems at both ends. The northern system, with west-northwest strike, is characterized predominantly by strike slip (with perhaps minor convergence, as suggested by the depression it encloses), whereas the southern system with north-northeast strike, exhibits a considerable component of east-west extension (Özgül et al., 1983). We see here that it is the orientation of the structures in an overall north-south contractional system that determines their character. Local factors, however, can also influence direction of local movement. For instance, the "Riedel with Riedel" structures illustrated in Figure 9C, which also originated during the Narman-Horasan earthquake, have an associated extension orientation rotated slightly counterclockwise toward the east-northeast under the influence of the strike-slip system in which they formed.

The overall pattern of structures present in the East Anatolian contractional province is presented in a schematic summary diagram in Figure 10. This figure shows the relationships of individual structures to one another and to the overall pattern of strain in the province. The figure also summarizes the various kinds of strike-slip related basins that formed in this environment.

The most prominent feature of this diagram is the conjugate strike-slip fault systems dividing the area into rhombohedral blocks that move east or west, away from the zone of convergence. If an east-trending zone of higher convergent strain than the surrounding areas intersects one of these conjugate wrench faults, the portion of the wrench fault located within the zone of high convergent strain begins to rotate towards an east-west orientation, with an ever increasing component of convergence across it. If this convergence is taken up by one or more uniformly vergent thrust(s), a kind of foredeep (4 in Fig. 10) forms in front of them. The asymmetric Pasinler Basin may well have formed through such a mechanism, connecting the Kağızman strike-slip system with another one represented by the discontinuous fault segments north of Karlıova (Fig. 2). Pasinler-type basins may in time turn into ramp basins by the generation of another thrust on their originally unfaulted side, or alternatively, they may initiate as ramp basins. The Ararat Basin of the Armenian S.S.R. (Ar in Fig. 2), containing more than 2 km of folded and faulted Neogene sediment (Burshtar and Tolmachevskiy, 1965), may be an example of such a ramp basin.

Similar compressional basins may originate if two parallel but non-overlapping wrench faults are connected by a system of uniformly vergent or convergent thrusts (3 in Fig. 10) along a restraining bend (Crowell, 1974). Ideally, basins of type 4, as illustrated in Figure 10, are never quite perpendicular to the orientation of the axis of maximum shortening, whereas those of type 3 can be.

If two non-overlapping but parallel wrench faults are connected through a releasing bend (Crowell, 1974), typical pull-apart structures may develop (2, 6 in Fig. 10). The most conspicuous example of such a pull-apart structure in the East Anatolian convergent province appears, at the first sight, to be the Erzurum Basin (E in Fig. 2, Irrlitz, 1972; Atalay, 1978). However, the marked non-parallelism of its bounding faults, and its spindle shape with a long axis nearly parallel with the strike-slip system in which it originated, induce one to speculate whether the Erzurum Basin could have formed through a mechanism of bifurcation of extensional cracks propagating towards one another as shown by Bahat (1983). The geometry of the rhomb structures illustrated by Bahat (1983, Fig. 6b) is very similar to that of the Erzurum Basin. If this comparison is valid, the Erzurum Basin may be more like basin 5 than basin 2 in Figure 10. The component of extension indicated by the hybrid nature of surface breaks associated with the nearby Narman-Horasan earthquake may thus be responsible for the bifurcation and propagation of the fractures producing the rhomboid geometry, and the strike-slip component permits a pull-apart basin to develop. This example is presented here to emphasize the complexity of both local factors and the available physical mechanisms that may influence the formation of "pull-aparts" in regions of such complicated strike-slip fault development as eastern Turkey.

The existence of fissures that form parallel with the orientation of shortening in eastern Turkey and that localize volcanoes have long been speculated upon, but Şengör and Kidd (1979) were the first to show that the young volcano of Mt. Nemrud was located on a north-south orientated fissure that functioned as a magma conduit. Dewey et al. (in press) concluded, mainly on the basis of the distribution of the parasitic cones and using the principle outlined by Nakamura (1977), that at least three of the major volcanoes in eastern Turkey, (Nemrud, Süphan, and Ağrı), formed on major fissures whose orientations deviated from a north-south line at angles between zero and only about 20°. The Süphan fissure connects towards the north with the north-northeast-striking, left-lateral Süphan fault system (15 in Fig. 2), and probably occupies a small pull-apart structure. Mt. Ağrı may occupy a similar releasing bend, but in a right-lateral system (Fig. 2), such as that of 6 in Fig. 10.

If extensional fissures originate independently of strike-slip systems (as 1 in Fig. 10), they may propagate into such systems (as indicated to the south of 1 in Fig. 10), and unless very accurate timing can be obtained, the end result in such cases would be indistinguishable from cases where the strike-slip systems formed first.

An interesting feature of the extensional fissures in the eastern Turkish high plateau is that they have not developed into north-south graben complexes such as the ones described from Tibet (e.g., Chen Zhi-Ming, 1981; Tapponnier et al., 1981), and consequently do not form basins. This cannot readily be attributed to the crustal-thickness differences between eastern Turkey and Tibet, because sideways motion seems equally common in both plateaux. However, the formation of wrench-bounded wedges and absence of north-trending grabens in eastern Turkey may be due to a relatively narrow collision front (~500 km) that is convex towards the plateau; the Himalayan collision front is ~2,000 km long and concave toward the plateau. The difference in the strength of the basement rocks (in Tibet old, strong continental material; in eastern Turkey, weaker Mesozoic ac-

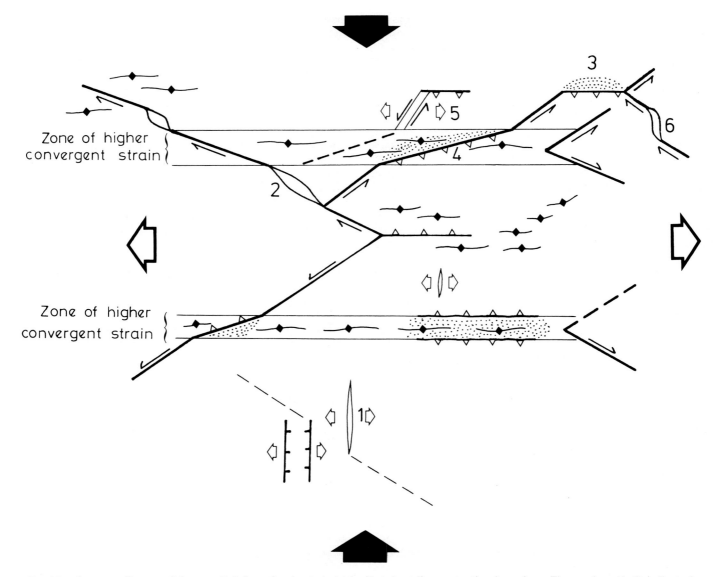

FIG. 10.—Summary diagram of the overall deformational pattern in the East Anatolian contractional province. The numbers (1–6) indicate the types of basins that form in this environment. See text for discussion.

cretionary complex material; see Şengör, 1984) enabling the Tibetan lithosphere to split along latitudinal fault systems, while allowing the basement of Turkey to squash much more easily, may be another reason for the absence of major, shortening-parallel grabens in the east Anatolian high plateau.

In addition to north-south oriented fissures localizing volcanoes, such as in the enormous shield of Karacadağ (Erinç, 1971, Fig. 156; Şengör et al., 1978, Fig. 2), similarly orientated grabens (e.g., the Akçakale graben; Fig. 2) formed along the southern margin of the east Anatolian contractional province, across the Bitlis suture on the north Arabian foreland. An interesting observation in the case of both the Karacadağ fissure and the Akçakale graben is that northwest-trending zones of discontinuous faulting emanate from the southern end of the former and from the northern

end of the latter (Fig. 2). These may be zones of incipient strike-slip faulting, as in the case of 1 in Figure 10.

In summary, we note that some basins in eastern Turkey form along releasing (2, 6 in Fig. 10) or restraining (3, 4, in Fig. 10) bends of strike-slip faults, which themselves may be primary or secondary features. Others originate by complex patterns of fracture propagation along oblique-slip faults. Yet others first form in isolation to link up later with strike-slip systems (1, Fig. 10). Most tend to maintain their orientation with respect to the regional orientation of the maximum shortening axis (1, 2, 3, 4 in Fig. 10); some, however, must rotate around vertical axes (4, 5 in Fig. 10). Although strike slip influences the orientation of small, local features in a few cases (Fig. 9C), most larger structures in the East Anatolian contractional province are the products of the overall north-south convergence. They all accommodate this shortening, some by thickening the crust and some others by elongating it perpendicular to shortening. It may thus be unproductive to ask whether the strike-slip faults formed the other structures or vice-versa, because they all seem to be parts of a larger system.

Because of extensive young volcanic cover and the sparsity of detailed field observations, the relationships between the paleotectonic and the neotectonic structures of the East Anatolian contractional province have not yet been established. One interesting question is whether any of the paleotectonic upper slope basins of the East Anatolian accretionary complex (Şengör and Yılmaz, 1981) have been "resurrected" in the form of neotectonic half or full ramp basins. Similarly, subduction-related paleotectonic strike-slip faults within the East Anatolian accretionary complex, like the Batee fault of the Sumatran accretionary complex (Fig. 1; Karig, 1980), may have been "resurrected" by neotectonic strike-slip faults of similar sense of offset.

The general pattern of deformation we describe here resembles greatly that present in Tibet (Şengör and Kidd, 1979; Chen Zhi-Ming, 1981; Rothery and Drury, 1984), although east-trending compressional structures are much more common in Turkey than in Tibet, and north-trending extensional ones much less common. The sparsity of young compressional structures in Tibet, however, may be simply an artifact of the sparsity of data. The general rhomb pattern, on the other hand, is very common on both plateaux. In fact, escape tectonics as a whole is only a special case of what Rothery and Drury (1984) so aptly termed rhombohedral block tectonics, where one "rhomb" attains very much larger dimensions than the rest. Thus, in the following section we describe the North and the East Anatolian strike-slip fault systems, which bound the largest of the westerly moving "rhombs" in the overall north-south intracontinental convergent regime.

STRIKE-SLIP BASINS ALONG THE NORTH AND EAST ANATOLIAN FAULT ZONES

In the following account we do not place much emphasis on the origin and evolution of obvious pull-apart structures, sag ponds, or wedge-basins (Crowell, 1974) along either of the fault zones, because not only have the theoretical aspects of such basins been treated recently by a number of authors (e.g., Crowell, 1974, 1976; Dewey, 1978; Rodgers, 1980; Aydın and Nur, 1982, 1985 this volume; Mann et al., 1983), but also their representatives along the North and the East Anatolian faults have been subjects of numerous discussions (e.g., Akkan, 1964; Seymen, 1975; Şengör, 1979; Barka, 1981, 1985; Allen, 1982; Aydın and Nur, 1982; Hempton et al., 1983; Barka and Hancock, 1984; Dunne and Hempton, 1984). We instead emphasize other types of strike-slip-related basins along the two fault zones that have so far received only slight attention or have never been discussed before, and discuss the pull-apart basins only when our views differ from those expressed by others.

The Karlıova "Triple Junction"

The Karlıova "triple junction" area is located where the North and East Anatolian faults meet, and the zone of higher convergent strain trends east-southeastwards from that point (Şengör, 1979; Fig. 2). We use "triple junction" in quotation marks, because although three zones of high strain meet at one "point" near Karlıova, none of the blocks they define is torsionally rigid, nor is deformation concentrated enough along any one of them to allow a simple, rigorous geometrical analysis of the region. Figure 11 shows a simplified map of the distribution of faults and folds in the vicinity of the triple junction, the mechanism proposed to explain its evolution, and the present topography of the Karlıova region.

As seen in Figure 11A, both the North and East Anatolian faults maintain their well-defined topographic expression until they come together some 3 km east of the village of Kargapazarı (Fig. 11B). East of that point we no longer find either of the prominent left-lateral systems, although a zone of shortening with a strike-slip component roughly parallel with the North Anatolian fault continues to the east of Varto (Fig. 2). Farther east, not even that remains and deformation becomes more diffuse. Microearthquake studies also show that no single fault zone similar to the North Anatolian fault continues into eastern Turkey (Fig. 3; Ateş, 1982).

The Karlıova triple junction resembles a west-facing equivalent of the eastern end of the Pasinler Basin where two strike-slip faults meet at the end of a zone of high convergent strain. However, the tectonic style of the wedge defined by the two faults in the case of Karlıova is fundamentally different from regions outside it. Within the wedge, there are conspicuously fewer neotectonic structures than east of it. Although a number of east-striking thrust faults north of Elbistan and Sivas (Fig. 2) indicate the presence of north-south shortening in the wedge, its magnitude seems much less than that in eastern Turkey. Moreover, at least some studied segments of the Malatya fault system (Fig. 2, loc. 10), and the strike-slip fault south of Elbistan have displacements compatible with roughly east-west shortening and corresponding north-south extension of the wedge (Şengör, 1979). These observation, when combined with others in Central Anatolia, indicate the presence of northeast-southwest-directed shortening with corresponding northwest-southeast extension in the wedge defined by the North and East Anatolian faults until the meridian of Antalya is reached (Figs. 2, 7). We therefore believe that the Anatolian scholle has not been shortening north-south at the same rate as has eastern Turkey and that this statement is valid even as far east as Karlıova.

Between Karlıova and Kargapazari, the present topography is one of a distinct depression bounded by the North and East Anatolian faults and prominences within the Anatolian scholle (Fig. 11B). Stratigraphic studies show that the topographic basin of Karlıova also coincides with a sedimentary basin of Pliocene age (Nakoman, 1968; Fig. 6A, col. 13).

Figure 11C illustrates the main elements of the tectonics of the Karlıova triple junction. The diagonally ruled regions (I and II) represent the East Anatolian contractional province and are here taken to be torsionally rigid for geometric simplification. Region III represents the Anatolian scholle and is also assumed to be torsionally rigid. The zone of

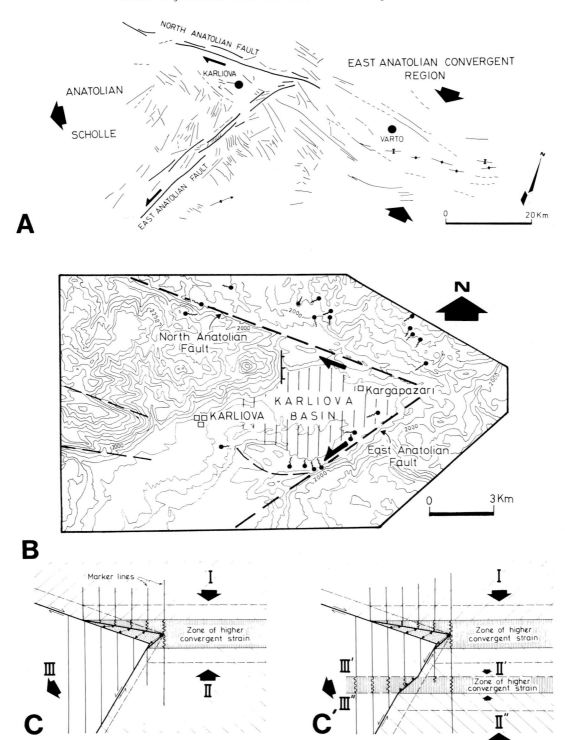

Fɪɢ. 11.—A) Fault pattern around the Karlıova "triple junction" (F. Şaroğlu, unpubl. data). See Figure 3 for location.

B) The topographic (active) basin of Karlıova. Note how well defined topographically both the North and the East Anatolian faults are. The small, tadpole-like objects are springs.

C) Simplified model of the kinematics of deformation around the Karlıova "triple junction." I, II, and III are assumed to be torsionally perfectly rigid blocks, whereas the zone of higher convergent strain is assumed to undergo only vertical plane strain (modified from Şengör, 1979). See text for discussion. C′) Same as C, but with two zones of higher convergent strain. The geometry simulates the observed geometry between Bingöl and Muş (for localities see Fig. 2), where an east-trending zone of higher convergent strain represented by folding and thrusting seems to induce bending in the trace of the East Anatolian fault.

higher convergent strain is a region in which homogeneous shortening is assumed to be taken up only by vertical plane strain. Figure 11C shows the result of 50% shortening in the zone of higher convergent strain. Block III moves as shown along the major strike-slip faults and a basin similar to the one around Karlıova forms. As north-south shortening proceeds, the basin fill is itself incorporated into the zone of higher convergent strain. Highly deformed and discontinuous patches of very young clastic rocks located just to the east of the Karlıova "triple junction" within the zone of convergence and resembling greatly the fill of the Karlıova Basin may be representatives of the fill material overtaken by contractional deformation. An interesting aspect of the kinematics of the model basin is that the direction of extension varies from one part of the basin to another, and structures forming in response to extension have correspondingly different orientations, rendering an understanding of the evolution of the basin very difficult. Although the model predicts the Karlıova Basin, it is difficult to test its detailed implications, because structures within the latter have not been mapped.

Figure 11C′ shows a simple relaxation of the condition of torsional rigidity of the blocks around the triple junction by introducing a second zone of east-trending higher convergent strain that cuts through both blocks II and III. This is essentially the same geometry as illustrated by the "zone of higher convergent strain" in Figure 10. As in that case, note here also that the strike-slip fault intersected by the zone of higher convergent strain begins to be curved as a result of rotation. If the zone itself is narrow, and strike slip continues on the major fault, the rotated segment will begin to acquire an ever-increasing component of convergence. This is actually the situation observed where the Muş - Lake Van-Bingöl zone of folds and thrusts intersects the East Anatolian fault and induces a change in orientation to create a restraining bend (Fig. 2; Arpat and Şaroğlu, 1972).

On the other hand, if the number and widths of zones of higher convergent strain are large, then the rotated portions of the original strike-slip fault function as strike-slip faults and the unrotated parts become sites of pull-apart basins. Simple homogeneous constriction of an extruding wedge during tectonic escape, as depicted in Figure 3D of Mann et al. (1983), thus cannot generate pull-apart structures. To do so, strain must be heterogeneous and be compartmentalized into transverse zones of differential strain as shown in Figures 10 and 11C′. In Figure 11C′ only a very simplified geometry is illustrated. In more complicated, and probably more realistic situations, zones II′ and II″ may experience different amounts of strain, and the corresponding strike-slip fault segments north and south of the zone of higher convergent strain between II′ and II″ may become non-parallel thus creating further complexities in their evolution.

Another aspect of the zones of higher convergent strain in eastern Turkey is the problem of the depth to which they may extend, an issue addressed by Dewey et al. (in press). At present, evidence is equivocal to support their thesis requiring that crustal flakes about 10 km thick determine the pattern of surface deformation in eastern Turkey. The geometry depicted in Figure 11C′ provides one possible test to check this hypothesis. If the seismicity of the fault segment south of the southern zone of higher convergent strain is offset at depth to give rise to an en echelon cross-sectional view (as, for example, in the San Andreas fault), the depth of the zone of offset would correspond with the depth to which the effects of greater convergence extend. Thus, very detailed studies to establish the precise locations of hypocenters along the East Anatolian fault may provide clues to the sub-surface pattern of deformation.

In summary, it seems clear that the style of deformation both in the East Anatolian contractional province and in the Central Anatolian "ova" province and the ways in which they may be related, critically influence not only the geometry and kinematics of the Karlıova triple junction region, but also those of the North and East Anatolian strike-slip faults.

Basins of the North Anatolian Fault

West of the Karlıova triple junction, hybrid fault-plane solutions of the East Anatolian convergent province are replaced by strike-slip solutions along the North and East Anatolian faults. The North Anatolian fault has been broken along most of its length during this century by catastrophic earthquakes (Ketin, 1948), whereas the East Anatolian fault has been more quiescent and only the Bingöl earthquakes of May 22, 1971, produced a significant surface break (Seymen and Aydın, 1972; McKenzie, 1976).

The Erzincan (Fig. 5, loc. 6; Fig. 6A, col. 11), Taşova-Erbaa or Reşadiye (Fig. 5, loc. 5; Fig. 6A, col. 10), and Havza-Lâdik (Fig. 5, locs. 3, 4; Fig. 6A, col. 9) Basins are pull-apart basins along the eastern, west-northwest-striking segment of the North Anatolian fault. We have searched for evidence on both sides of the eastern segment of North Anatolian fault for wide zones of higher convergent strain corresponding with the ideal strike-slip segments of the fault to see whether the mechanism of inhomogeneous wedge constriction discussed above could be applicable to the origin of these pull-aparts—in other words, to see whether these pull-apart basins were located on primary or secondary releasing bends. As Figure 2 shows, there is in fact such a zone stretching east-west from west of Sivas through Refahiye (Tatar, 1975) to Aşkale (Tatar, 1978). Just to the north of Sivas, Görür (unpubl. data) observed high-angle thrust faults overriding Pliocene sediments, and Tatar (1975) observed east-northeast-trending anticlines that fold Pliocene sediments near Refahiye (just northwest of Erzincan). From there the zone continues into the young thrust zone of Aşkale (Tatar, 1978). Barka (1983b) observed that the strike of the North Anatolian fault zone between Karlıova and Erzincan is about 125°, whereas between Erzincan and Erbaa it is only 105°; from Erbaa to Havza, the fault again follows a strike of about 120° (Fig. 2). The Erzincan, Taşova-Erbaa, and Havza-Lâdik pull-apart structures are located where the strike of the fault zone is more northerly. These observations are compatible with the hypothesis that the Sivas-Aşkale zone of higher convergent strain was responsible for the rotation of the Erzincan-Erbaa sector of the North Anatolian fault zone into a more easterly strike, and because this long portion continued to function as a strike-slip fault (Seymen, 1975; Tatar, 1978), the more northerly but shorter segments became sites of releasing bends and thus pull-apart basins. In connection with this hypothesis, we observe that there are no pull-apart structures between Havza and Adapazarı, where no intersecting zones of higher convergent strain interfere with the course of the fault.

The Çerkeş-Kurşunlu (or and Çerkeş-Ilgaz), Tosya, and the much younger Kargı Basin (Figs. 2, and cols. 6, 7, 8 in Fig. 6A) are ramp basins, the two former being inherited from compressional molasse basin remnants of the earlier Mudurnu-Göynük Basin (see Barka, 1983b, 1985; Barka and Hancock, 1984). Barka (1983b) suggested that the basins may have resulted from convergent strike slip, but this explantion seems unlikely in the view of the ideal strike-slip focal mechanism solutions for earthquakes between Tosya and Adapazarı (Fig. 3).

The western end of the North Anatolian fault zone in and around the Sea of Marmara is a region of complex geometry and abundant basins. Dewey and Şengör (1979) indicated that just east of Bolu (Fig. 2), the fault zone begins to bifurcate and that the tectonically controlled depression of Çağa (Erinç et al., 1961), located at the point of bifurcation, may be an extensional fault-wedge basin of the type described by Crowell (1974). Westwards, the northern strand continues north of the Almacık flake (see below), through the Adapazarı plain (Bilgin, 1984), and beneath the sea in the Gulf of Izmit (Fig. 2). All along this strand, the fault has a very straight and well-developed topographic expression. Between Adapazarı and the western end of the Gulf of Izmit, its trace resembles a narrow graben, with well-developed normal fault scarps especially along its southern side. Bilgin (1967) also noted the existence of locally faulted, north-facing monoclines along the southern coast of the Gulf of Izmit. Thick marine Pleistocene deposits of Tyrrhenian I age have been uplifted within the graben (Erinç, 1956). These sediments contain abundant evidence of syn-sedimentary meso-scale strike-slip and other kinds of faults (Şengör et al., 1982). Along the same strand on the western side of the Sea of Marmara, south of Ganosdağ (Fig. 2), the trace of the fault is still well-developed morphologically and broke as recently as in 1912 (Ambraseys, 1970). This trace then strikes westwards into the Saros graben (Fig. 2). Between the two segments of the northern strand on opposite sides of the sea of Marmara are two well-defined submarine deeps. The eastern one, the Çınarcık Basin, is here interpreted as a pull-apart (for its detailed bathymetry, see Gunnerson and Özturgut, 1974), whereas the western one is the compressional depression immediately east of the Ganosdağ (see Şengör, 1978; Fig. 2). The southern strand of the North Anatolian fault, which branches off east of Bolu, has a less continuous trace, but more seismicity than the northern strand (Crampin and Üçer, 1975). It seems, at least from the twentieth century record, that much of the current westerly motion of the Anatolian scholle is accommodated by the southern strand. Between the two strands there are a number of

east-trending nearly continuous normal-fault systems in the Sea of Marmara region, the most prominent of which is the Lake Iznik - Gulf of Gemlik graben complex (Fig. 2). We emphasize that the Gulf of Izmit and the Lake Iznik - Gulf of Gemlik graben complexes are parallel with the graben complexes of the west Anatolian extensional province.

Figure 12 is an attempt to explain the evolution of the northern and southern strands of the North Anatolian fault and the other neotectonic structures seen in and around the Sea of Marmara. The hypothesized geometry of the fault before the onset of the north-south extension of the Aegean area and western Turkey is illustrated in Figure 12A. Because of the cumulative amount of north-south extension increases from western Turkey towards the Aegean (Le Pichon and Angelier, 1981), we divided the western end of the fault into two domains, one with only 30% extension and the other with 50% extension, scaled respectively with the Sea of Marmara and with the northern Aegean Sea between the Gulf of Saros and the Khalkidiki Peninsula (Fig. 13). This division follows the estimates of McKenzie (1978) and Şengör (1978) for extension in the Aegean Sea and in western Turkey. The original, pre-extension curvature of the fault is schematized by using the pre-extension palinspastic reconstruction of the Aegean area by Le Pichon and Angelier (1981), and is probably a function of the pre-existing zones of weakness such as old, paleotectonic fracture patterns (Fig. 4A) of the Hellenides (Dewey and Şengör, 1979; Şengör, 1979; McKenzie and Jackson, 1983).

During extension, the original curvature of the fault is accentuated. If strike slip continues on the westernmost strand, the central strand acquires the character of a transtensional fault zone, generating narrow grabens such as the one forming the Gulf of Izmit (Fig. 12B). Eventually, north-south extension in both segments bends the fault sufficiently to preclude further strike slip, and a new strand forms to the south. At this stage the oblique-slip faults of the abandoned strand are left as narrow graben complexes aligned along the fossil strike-slip fault trace (Fig. 12C).

The extremely youthful topography and the discontinuous nature of its trace probably indicate that the southern strand of the North Anatolian fault is still in the process of breaking through. The northern strand originated during the middle to early-late Miocene in the Marmara area (Fig. 4), but the southern strand initiated during the Pleistocene. As the earthquake evidence shows the northern strand to have a strike-slip component (e.g., the 1912 Mürefte-Şarköy earthquake; Ambraseys, 1970; Allen, 1975), we believe that the western end of the North Anatolian fault zone is at present in a transition phase between stages B and C shown in Figure 12. In the light of this interpretation, the east-trending normal fault systems of the interior and the surrounding of the Sea of Marmara appear to be related to a diffuse zone of transcurrent crustal thinning. This zone has evolved as a result of the incompatibilities that arise from the interference of a regional extensional regime, in which the cumulative amount of extension increases from east to west, with the originally curved western end of the North Anatolian fault.

If the system shown in Fig. 12B episodically locks it may

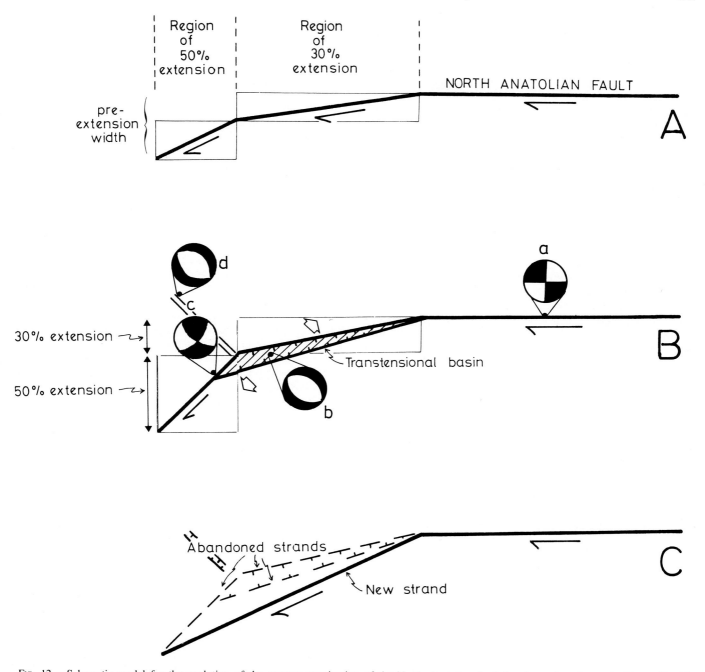

FIG. 12.—Schematic model for the evolution of the western termination of the North Anatolian fault for the region shown in Figure 13. A) Hypothetical course of the fault before the onset of Aegean extension, B) Possible geometry of the fault immediately after extension, C) Present geometry of the fault. Fault-plane solutions a, b, c, and d shown in B are theoretical. See text for discussion.

begin tearing the areas west of it from regions lying to the north along normal faults (labelled d). Such a zone exists in Thrace and has limited seismicity. Magnetotelluric measurements by Ilkışık (1980) disclosed a region of partial melting at a depth of 12 to 14 km, where the crustal thickness seems reduced to about 28–30 km. This zone of extension continues westwards through Bulgaria and separates a Macedonian scholle from the rest of the Balkans north of it, along which Macedonia seems to be in the process of separating from the Balkans (Dewey and Şengör, 1979).

There are two other aspects of basin evolution along the North Anatolian fault that we need to touch upon before we go on to the discussion of basin genesis in other parts of Turkey. One of these is the formation of large rotating flakes and the other is segments that show a reversed sense of offset along the course of the fault.

The generation of shallow flakes from flower structures along strike-slip fault zones (Dewey, 1982) is well known from field (Yeats, 1981), experimental (Gallo et al., 1980) and seismic studies (Anonymous, 1984). Because they move on large thrust faults, these flakes may form extensive basin complexes by isostatic loading around their periphery, as is partly the case in the Ventura Basin of California (Crowell, 1976; Blake et al., 1978; Yeats, 1983). No rotated flakes (or any other rotated structures) have been previously recognized along the North Anatolian fault. We here suggest that the structure, which we informally name the "Almacık flake" (Fig. 2), may represent one such object, which has rotated in a clockwise sense more than 110° since the initiation of faulting. Figure 14A shows a simplified geological map of the flake and the surrounding area. The flake consists of three sub-domains: a metamorphosed Paleozoic succession in the east, a central zone of large sheared slabs of upper Mesozoic ophiolites and oceanic sediments; and a metamorphic upper Cretaceous limestone-clastic succession in the west. Yılmaz et al. (1982) interpreted the ophiolitic zone as a portion of the latest Mesozoic-earliest Cenozoic Intra-Pontide suture zone between the Rhodope-Pontide fragment in the north and the Sakarya continent in the south (Şengör and Yılmaz, 1981). The metamorphosed Paleozoic succession of the eastern half of the Almacık flake belongs to the Rhodope-Pontide fragment, whereas the Cretaceous rocks represent a piece of the northern margin of the Sakarya continent.

In this region, most such tectonostratigraphic domains are aligned in roughly east-trending belts, (Fig. 4A; Şengör and Yılmaz, 1981), but within the flake, they are oriented northnortheast. Figure 14 shows the overall trace of the Intra-Pontide suture to emphasize the conspicuous difference between its strike and the strike of the segment contained within the Almacık flake. The mainly Eocene andesitic volcanic rocks located on the northern side of the flake provide an excellent opportunity to check the rotation hypothesis paleomagnetically.

The flake is surrounded at the surface by strike-slip and high-angle thrust faults (Fig. 14B; Abdüsselâmoğlu, 1959). We speculate that these thrust faults flatten at depth under the flake and then re-steepen to join the main strike-slip fault zone. We believe that the southern part of the Adapazarı Basin around Akyazı (Fig. 14A) may have subsided as a kind of foreland trough in front of the thrust faults that bound the flake along its northwestern border.

A peculiar feature of the southern, dominantly strike-slip border faults of the Almacık flake is the presence of at least two left-lateral strands within a family of right-lateral faults composing the North Anatolian fault zone (Fig. 15A). Hancock and Barka (1981) suggested that the central portion of the fault zone may have reversed its sense of motion during Pliocene - early Pleistocene time. Şengör et al. (1983) pointed out, however, that a large-scale reversal of motion along the North Anatolian fault would have important repercussions elsewhere within the neotectonic frame of Turkey for which no evidence was known. Şengör et al. (1983) showed that local reversals along individual fault strands within a strike-slip zone could occur without requiring any change in the overall movement along the fault zone as a whole (Fig. 15B). If graben pairs 1–4 and 2–3 are active alternately, the middle fault segment experiences a sense of motion opposite that of the main fault during the activity of grabens 2 and 3. As pull-apart structures can be much longer than they are wide, such graben-linking, inter-pull-apart strike-slip faults may be indistinguishable from parallel strands of the main fault zone of the type illustrated in Figure 15A.

In summary, basins along the North Anatolian fault zone include (1) classical, simple pull-apart structures located along releasing bends that may be either primary (as in the

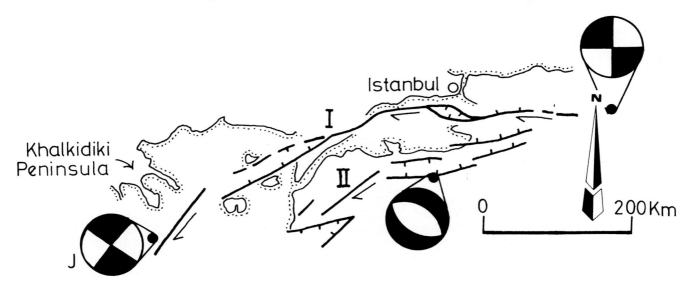

FIG. 13.—Map showing a simplified version of the geometry and kinematics at the western termination of the North Anatolian fault (after Şengör, 1978). I and II are northern and southern strands of the North Anatolian fault respectively. Fault plane solution J is from Jackson et al. (1983).

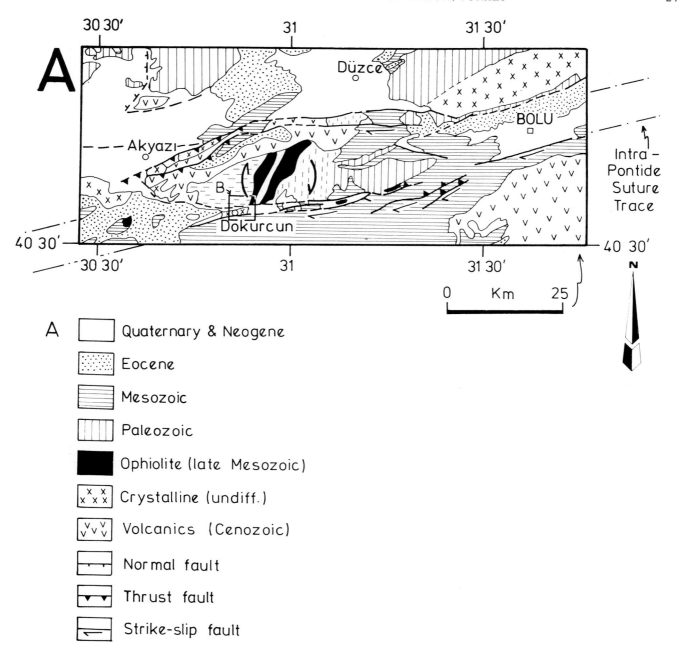

A
- ☐ Quaternary & Neogene
- ☷ Eocene
- ☰ Mesozoic
- ☷ Paleozoic
- ■ Ophiolite (late Mesozoic)
- x x / x x x Crystalline (undiff.)
- v v v / v v v Volcanics (Cenozoic)
- ⊢─⊢ Normal fault
- ▼─▼ Thrust fault
- ⊢──⊣ Strike-slip fault

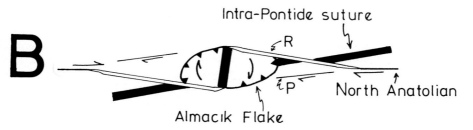

FIG. 14.—A) Geological map of the "Almacik flake" (simplified from M.T.A., 1961–1964; see Fig. 3 for location).
B) Interpretation of the "Almacik flake" as a rotated flake along the North Anatolian fault, formed through a mechanism termed "Riedel flaking" by Dewey (1982). R and P indicated opened Riedel and P shears. See text for discussion. Small box labelled B in A represents the area of Fig. 15A.

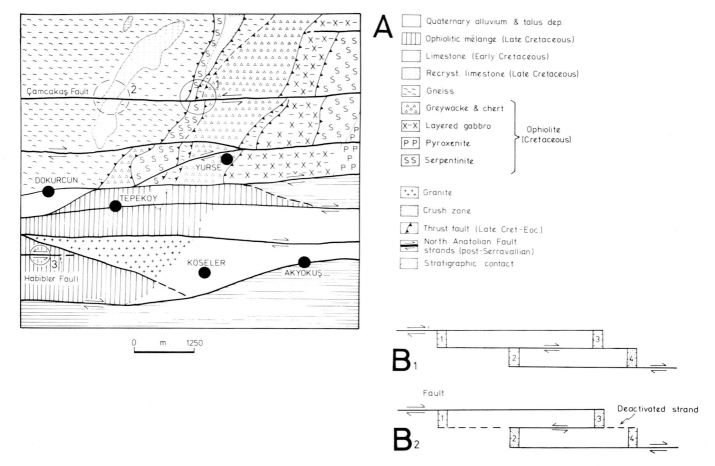

FIG. 15.—A) Geological map of the southern boundary of the "Almacık flake" near the town of Dokurcun, showing the evidence for left-lateral motion on the Çamcakaş strand (redrawn from Sipahioğlu, 1984). 1 and 2 are localities where left-lateral offsets were mapped.

B) A possible mechanism suggested by Şengör et al. (1983) to account for the observed local reversals along certain strands of the North Anatolian fault. 1 shows "normal," whereas 2 shows "reversed" motion along the middle strand, while movement on the main fault zone remains right-lateral throughout. Stippled areas represent active grabens.

example of the Çınarcık Basin) and secondary (as in the case of the Erzincan, Taşova–Erbaa, and Havza-Lâdik Basins); (2) compressional ramp basins; (3) long transtensional basins that arise because of incompatibilities induced by progressively increasing curvatures along the course of the fault; (4) fault-wedge basins where the fault strands bifurcate; and (5) foreland-type compressional basins that form around the periphery of rotating flakes.

Basins of the East Anatolian Fault

The strike-slip basins along the East Anatolian fault are in principle no different from those of the North Anatolian fault zone. In addition to the Karlıova Basin at its eastern end, the fault zone includes the Bingöl compressional basin (Arpat and Şaroğlu, 1972; Fig. 2) and the Lake Hazar pull-apart basin (Hempton et al., 1983; Figs. 2; 6A, col. 14).

The one peculiar basin type associated with the western end of the East Anatolian fault is the Adana/Cilicia Basin complex, which also includes the Gulf of Iskenderun Basin and the Hatay graben (Figs. 2, 16A). Although these basins

have long been studied because of their hydrocarbon potential (Ternek, 1953, 1957; Schmidt, 1961; Görür, 1977, Evans et al., 1978; Ketin et al., 1981; Yalçin and Görür, 1984), seismicity (Pinar, 1953; Büyükaşikoğlu, 1979; Jakson and McKenzie, 1984), and young volcanic activity (Bilgin, 1969a; Bilgin and Ercan, 1981), no satisfactory model has yet been proposed for their origin or present kinematics. We emphasize first that no evidence of an important rifting event is seen on land in the Burdigalian to the Pliocene sedimentary record of the Adana Basin. These rocks record the simple outward building out of a south-southwest-facing basin margin by submarine fan and delta progradation. However, rifting did occur in the submarine part of the Adana Basin between Cyprus and Turkey and in the separate subaerial Hatay graben (Fig. 16A).

Figure 16B puts the basins in a plate-tectonic context. A key feature is the Kahramanmaraş "quadruple junction," located at the corner of three continental blocks and one oceanic plate. If the southeast Taurus boundary thrust accommodating the relative motion between eastern Turkey and the Arabian plate is assumed to be absent, the vector

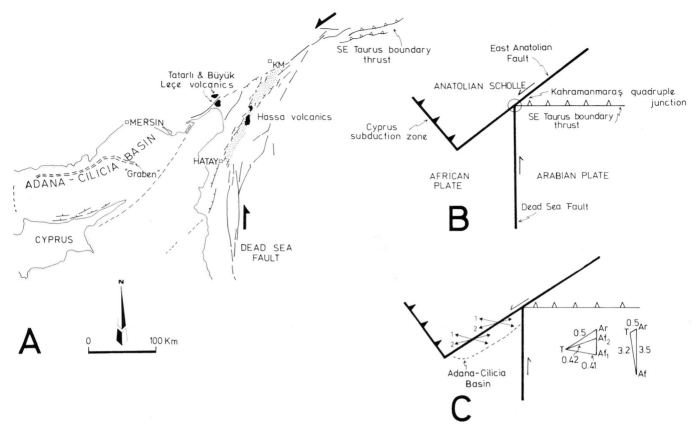

FIG. 16.—A) Tectonics of the Adana-Cilicia Basin (redrawn from Fig. 2 with data added from Bilgin 1969a; Evans et al., 1978; and Bilgin and Ercan, 1981). See Fig. 3 for location. KM is Kahramanmaraş.

B) Schematic large-scale tectonic elements of the Kahramanmaraş quadruple junction.

C) Relative motion vectors across the Cyprus-Kahramanmaraş segment of the southern boundary of the Anatolian scholle. The velocity triangle on the right shows the relative plate motion rates if there were no southeast Taurus boundary thrust. In this case the rate of extension across the Cyprus-Kahramanmaraş segment equals 3.2 cm/yr in a north-northwest to south-southeast orientation. Such a high rate of extension in this region is not observed today.

The velocity triangle on the left shows two different situations where 3.16 cm/yr (Ar-Af$_1$) and 3.4 cm/yr (Ar-Af$_2$) of the Arabia-Anatolia motion are absorbed by the southeast Taurus boundary thrust. In these cases, the rate of relative motion across the Cyprus-Kahramanmaraş segment amounts to 0.41 and 0.42 cm/yr, respectively. Double-headed arrows designated 2 and 1 give the orientation of these extensions respectively. Because no reliable fault plane solutions exist in this region, it is difficult to choose any one of these models; both seem compatible with the known surface geology.

diagram shown on the right side in Fig. 16C predicts the orientation and magnitude of the relative motion vector across the boundary between the Anatolian scholle and the African plate. It can be readily seen that this is an unreasonable assumption, as it leads to an unacceptably fast extension across the Cyprus-Kahramanmaraş segment of the Anatolia/Africa boundary, and also requires oblique extension across the Cyprus subduction zone, which is contradicted by observation (Jackson and McKenzie, 1984). If we assume that 3.16 cm/yr of the total 3.5 cm/yr of the Africa/ Arabia motion is absorbed by the southeast Taurus boundary thrust, then we get an Anatolia/Africa motion of 0.41 cm/yr shown by the TAf$_1$ vector. If we further increase the amount of motion absorbed by the boundary thrust to 3.4 cm/yr, the orientation of extension across the Anatolia/ Africa boundary switches to that shown by TAf$_2$, and the rate of extension grows to 0.42 cm/yr. In both cases, oblique extension is required across the Cyprus-Kahramanmaraş

segment of the Anatolia/Africa boundary, and oblique convergence across the Cyprus subduction zone. The presence of extensional structures in the submarine Adana/Cilicia Basin (Evans et al., 1978) and of Quaternary tholeiitic plateau basalts on its northeastern margin (Bilgin and Ercan, 1981; Fig. 16A) shows that the kinematic picture described may be appropriate for the post-Serravallian tectonics of the Cyprus-Kahramanmaraş segment of the Anatolia/Africa boundary.

The mechanism proposed here for the opening of the Adana/Cilicia Basin shows how incompatibility problems arising from the buoyancy of the continental lithosphere may generate complex basin types at strike-slip fault intersections. An interesting aspect of this mechanism is its dependence on the amount of absorption of the Arabia/Africa motion by the southeast Taurus boundary thrust. The same absorption also affects the rate of westerly motion of the Anatolian scholle. A high degree of absorption may be the explanation of the much slower (1–2 cm/yr) rate of movement of the Anatolian scholle with respect to Africa and Arabia than predicted (∼4 cm/yr: Dewey and Şengör, 1979).

STRIKE-SLIP SYSTEMS IN THE "WEAKLY ACTIVE" NORTH TURKISH
NEOTECTONIC PROVINCE

Şengör et al. (1983) showed that the neotectonic deformation pattern in northern Turkey north of the North Anatolian fault may be explained by a limited amount of east-

west shortening. Because the neotectonic movements in this province are exceedingly small no large basins associated with strike slip have formed here. Even the existence of strike slip on faults such as the Havza-İnebolu or Safranbolu faults (Fig. 2, locs. 2, 3 respectively) remains to be documented by field work.

The only well-documented active structure in this province is the Bartın fault (Fig. 2, loc. 1) on which the earthquake of September 3, 1968, occurred. Fault-plane solutions for this shock indicate right-lateral strike-slip with a

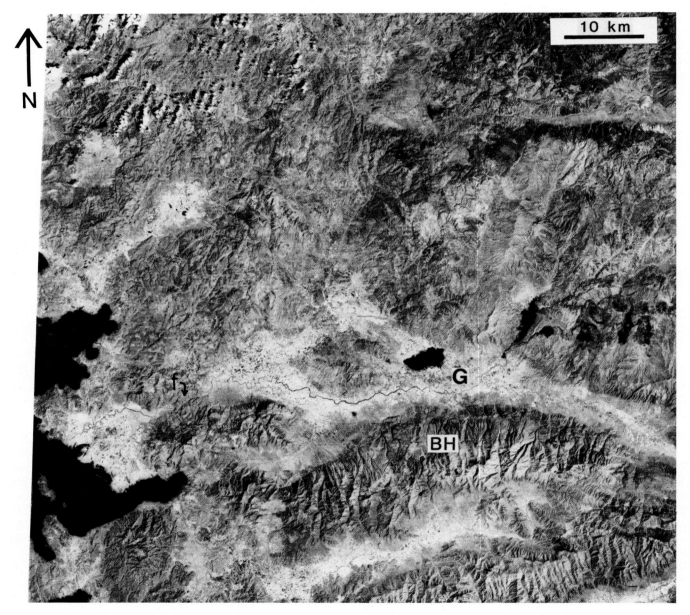

FIG. 17.—ERTS image of a part of western Turkey showing the Bozdağ horst (see Fig. 19), the Gediz, Küçük Menderes, Bakırçay and Simav grabens and the Manisa horst; in the extreme southeast only a small portion of the Büyük Menderes graben is visible (for location see Fig. 3). Notice the pronounced faulted southern margin of the asymmetric Gediz graben as opposed to its unfaulted northern margin (except in the extreme southeast of the graben), where the graben fill laps onto the metamorphic rocks of the Menderes Massif. Extending in a north-northeast direction from the graben are cross faults that bound several cross-grabens (cf. Fig. 19), although they are much less pronounced than the main graben. These cross-grabens and intervening cross-horsts form a pattern reminiscent of the key-board of a piano. BH-Bozdag Horst; G-Gediz graben.

strong component of east-west thrusting (McKenzie, 1972; Şengör et al., 1983). Letouzey et al. (1977) indicated the existence of a young thrust fault along strike from the Bartın earthquake fault on the basis of seismic reflection profiling supporting the thrust interpretation of the Bartın fault. Here we would like to point out the possibility of a connection between the Havza-İnebolu fault (Fig. 2, loc. 2) and the Bartın fault as shown in Figure 2, perhaps indicating the incipient formation of a new strand of the North Anatolian fault to push the western Pontides onto the oceanic lithosphere of the Black Sea. If this happens, a small escape system into the Black Sea would form closing at least its western part. Similar mechanisms may have operated in the past to obliterate Black Sea-like remnant oceanic pockets in older orogenic zones, as in the example of the eastern flysch basin of the eastern Carpathians (Burchfiel, 1980).

BASINS AND STRIKE-SLIP FAULTS IN THE WEST ANATOLIAN EXTENSIONAL PROVINCE

Western Turkey, together with its western prolongation the Aegean Sea, is one of the better-known regions in the world where diffuse extension is now occurring in a wide area along a number of sub-parallel normal faults bounding graben complexes. This extensional area was first recognized by Philippson (1910–1915) and since then has been the subject of numerous studies (see Şengör, 1982, for a summary). Dewey and Şengör, (1979) pointed out that the extension in the Aegean area and in western Turkey is the result of an east-west shortening being relieved by north-south extension and consequent lateral spreading of the continental material onto the easily subductable oceanic lithosphere of the eastern Mediterranean. This east-west shortening may be, they argued, the result of the obstruction of the Anatolian scholle's westerly motion owing to the southwesterly bend in the course of the North Anatolian fault zone in the Aegean and in Greece.

Although the morphologically most outstanding (Fig. 17) and the seismically most active structures (McKenzie, 1978) of the west Anatolian extensional province are the east-trending graben complexes, there is also a large number of north-northeast-, north-, and north-northwest-trending graben structures and isolated faults with apparently dominant dip slip in this region that seem associated with the major east-trending grabens (Fig. 2). Because the seismicity associated with these structures generally does not produce large-magnitude shocks yielding reliable fault-plane solutions and because they are not as prominent in the field as

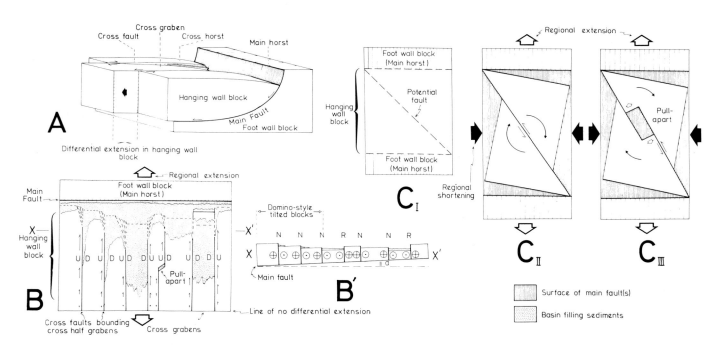

FIG 18.—A) Block diagram showing the origin of cross-grabens and cross-horsts as a consequence of differential extension in the hanging-wall block of a large listric normal fault, the differential extension being accommodated by oblique-throw faults. For simplicity, we have chosen a smoothly curved listric fault for illustration.

B) Sketch map showing key-board tectonics caused by differential extension in the hanging-wall block of the main listric fault. Note that both strike-slip and dip-slip displacements along the oblique-throw faults bounding cross-grabens and cross-horsts diminish towards a hypothetical line of no differential extension. The magnitude of the vertical component of motion varies, assuming other parameters to be equal, according to the magnitude of differential extension of the hanging-wall block. See text for discussion.

B′) Cross section showing the results of the main fault's having a purely cylindrical geometry and a plunging fold axis. If the fault surface is tilted as shown, all the blocks resting on it will be tilted also and all the originally vertical faults will acquire normal (N) and reverse (R) separation depending on which side is upthrown; (a) is the angle of plunge. See text for discussion.

C) Extension of the hanging-wall block by strike-slip faulting in cases where it is shortened sideways. Note that if shortening keeps pace with extension, the strike-slip fault (in C_II) and any pull-apart basins that may form along it (in C_III) must rotate towards the direction of extension. See text for discussion.

the east-west grabens (Fig. 17), they have so far received much less attention than the latter. But these structures may hold the key to our understanding of the deep structure and the overall style of deformation of western Turkey.

McKenzie (1978) and, following him, Şengör (1978) argued that the normal faults bounding the major east-west grabens in western Turkey may be listric and flatten at depth. J. A. Jackson (personal commun., 1982) has shown that some of the normal faults that delimit the Büyük Menderes graben in the north and that are located in Pliocene and Pleistocene sediments (Erinç, 1955a; Becker-Platen, 1970) have dips as low as 20° and they show evidence of considerable rotation around horizontal axes. Erinç (1970) and Şengör (1978, 1982) also pointed out that numerous other rotated fault blocks in western Turkey are associated with the major, east-striking normal faults. Moreover, by far the largest portion of the earthquake aftershocks that occur along the major normal faults are distributed over a large area away from these faults (Ergin et al., 1972), and do not necessarily correlate with the distribution of faults mapped at the surface, indicating the presence of other faults at depth or listric fault planes. In a recent seismological study of the normal faulting events associated with the Alaşehir (March 28, 1969) and Gediz (March 28, 1970) earthquakes, Eyidoğan and Jackson (1985) demonstrated that these two earthquakes were clearly multiple events, their seismograms indicating that at least two discrete sub-events were involved in producing the surface faulting. Eyidoğan and Jackson (1985) pointed out that the seismograms of these events also contain later, longer period signals that they viewed as representing source complexities. They modeled these later signals by sub-events with long time functions on very gently dipping, detachment-type faults. Eyidoğan and Jackson (1985) concluded that when, during an earthquake, the upper brittle crust ruptures through its entire thickness and thus high strain rates are imposed on the top part of the lower crust, the lower crust too ruptures by brittle failure. This failure occurs on very gently dipping discrete faults that otherwise deform by distributed creep. In the case of the Alaşehir and Gediz faults, the gently dipping faults of the lower crust are only the continuations of the more steeply dipping normal faults in the upper crust. The overall geometry of such a fault system is therefore roughly listric, but with an angular cross section with two or more straight segments meeting at corners instead of a smooth curved cross section (see Fig. 25a, Eyidoğan and Jackson, 1985).

All these indications converge to suggest the presence of large-scale listric normal faults in western Turkey. We accept this as a working hypothesis that will form the basis of the following discussion.

Figure 18A shows a block diagram illustrating such a listric normal fault. If the hanging wall block of this fault is extended by different amounts in different compartments, the differential extention is accommodated by detached strike-slip faults, depicted in Figure 18A as striking at 90° to the strike of the main listric fault for simplicity. Because the compartment, which is extended internally less than its neighboring blocks, moves farther away from the main horst, its tip bends down more with respect to its neighbours creating a roll-over anticline of tighter flexure. Depending on the radius of curvature of the listric normal fault this also leads to more pronounced tilting towards the main fault of the top surface of the less-extended compartment, and induces a dip-slip component on its bounding strike-slip faults. The dip-slip component is greatest near the main listric fault, and progressively decreases away from it (ignoring the thinning induced on the neighbouring blocks by their greater internal extension). In the field, one thus obtains the image of a T-shaped graben, where the top bar of the T corresponds with the main half graben, and its tail with the block subsided along the oblique-slip hinge faults forming a cross-graben. Although the tail of the T appears as a graben in the field, theoretically the net amount of extension across it is zero.

Figure 18B shows the map view of such a situation where the hanging wall block is divided into eight compartments all with different amounts of extension. An imaginary line of no differential extension is assumed to exist at the bottom of the figure. Notice how a number of half and full grabens form at the surface and strike at 90° to the trace of the main fault. Notice also that along the strike-slip faults, all kinds of strike-slip related structures, such as pull-apart basins, may form. Because the offsets of these faults change along their strike, the size and/or intensity of development of such structures also tends to vary, creating a complex structural picture. Moreover, if the basal listric fault is tilted in one direction, all the originally vertical oblique-slip hinge faults bounding the different compartments also experience the same tilting as shown in Figure 18B′. If the tilting is considerable ($a > 15°$), this will produce at the surface the image of a series of regularly tilted blocks that may variously be seen as a series of regularly dipping normal faults or reverse faults, or irregularly alternating normal and reverse faults. If the normal and reverse faults are then interpreted as extensional and contractional features respectively, it may lead to an erroneous and confusing picture.

The actual situation encountered north of the Gediz graben in western Turkey is extremely similar to the hypothetical picture illustrated in Figure 18B. Kaya (1979, 1981) has documented oblique slip on the north-northeast-striking faults north of the western end of the Gediz graben and also showed that they are hinge faults. A nearly mirror image of the same situation with respect to the Bozdağ Horst (Fig. 19) exists in the Büyük Menderes Graben, where its northern boundary fault represents the main listric normal fault. One fault-plane solution (of the July 7, 1955, Söke-Balat earthquake; Fig. 3), at the western end of the Büyük Menderes graben, where the northern main boundary fault turns towards the southwest, is hybrid showing extension combined with right-lateral strike slip, entirely compatible with the view that the northern boundary fault is mainly a large listric normal fault and turns into an oblique-slip extensional fault at its ends; the north-northeast elongation of the isoseismal lines associated with the same shock further supports this view (Öcal, 1958). The fault-plane solution displayed in our Figure 3 was taken from McKenzie (1972), which had been run in the mantle. Since then it turned out that nearly all the extensional earthquake events had taken place in the crust. If one reruns the same solution in the crust the effect is to increase the strike-slip component, thus bringing the solution into even closer agreement with our model (D. McKenzie, personal commun., 1985; see also McKenzie, 1972, p. 124–125).

According to the view advocated here, the southern main boundary fault of the Gediz graben (I in Fig. 19; the Alaşehir Fault of Eyidoğan and Jackson, 1885) and the northern main boundary fault of the Büyük Menderes graben (II in Fig. 19) are the two major listric normal faults of western Turkey (Fig. 19, cross section) and they probably underlie major portions of this part of the country. Areas north and south of these two large structures represent the hanging-wall blocks of the fault systems. These hanging-wall blocks are divided into a number of compartments with variable amounts of internal extension resembling the key-board of a piano in plan view. These compartments are separated from one another by oblique-slip hinge faults which bound half or full grabens and also apparently localize weak but widespread seismicity (Fig. 19).

A similar picture has been emerging from the Basin and Range province of the western United States in the last five years. Field mapping and deep-crustal seismic reflection profiling have disclosed there the existence of large-scale, low-angle normal faults with hanging-wall blocks that themselves undergo large extensional strains (e.g. Wernicke, 1981; Allmendinger et al., 1983). Hunt and Mabey (1966) and Stewart (1983) have shown, in the Panamint Range of the Death Valley area, that hanging-wall blocks may move for considerable distances on such low-angle faults along "thin-skinned" strike-slip faults of the kind we believe to exist in western Turkey.

Extreme caution should be exercised, however, when one wishes to obtain amounts and directions of regional extension from such thin-skinned ("detached") strike-slip faults. An extreme case where neither the orientation, nor the amount of extension can be obtained from such structures is illustrated in Figures $18C_I$ through $18C_{III}$. If the hanging-wall block in Figure $18C_I$ is shortened sideways concurrently with extension, it may break along a strike-slip fault shown in Figure $18C_{II}$. The two halves will then rotate as shown and two half-grabens with varying amounts of extension may form along the northern and the southern ends of the hanging wall. As extension proceeds not only the two halves of the hanging-wall block, but also the strike-slip fault separating them will rotate towards the extension direction. If a pull-apart (Fig. $18C_{III}$), or any other kind of strike-slip-related structure should form along that fault it too will rotate. If the pinning of the two halves of the hanging-wall block to the bounding horsts is not complete, then the offset along the strike-slip fault can give us no idea (neither a minimum nor a maximum) of the amount of the regional extension responsible for its formation. Nor can the orientation of the fault at any one time during the extension yield any useful information about the orientation of the extension. If conjugate strike-slip faults become active alternately they can rotate each other up to a theoretical limit of 45° or 60° (depending on the original angle they make with the orientation of the regional extension) away from the orientation of the regional extension. Paleomagnetic measurements on Neogene volcanic rocks around the

Izmir region by Kissel et al. (1985) for example have revealed a wide scatter. When their sample stations are plotted on a neotectonic map of the area, individual fault-bounded blocks are seen to contain measurements consistent with each other (Fig. 19). They show that some blocks rotated clockwise whereas others rotated anticlockwise since at least the Pliocene. Kissel et al. (1985) interpret the anticlockwise rotations as expression of the increasing curvature of the Aegean arc while they do not comment on the cause of the clockwise rotations. We suggest that the clockwise rotations may have occurred through a mechanism similar to the one described in Fig. 18C. An east-northeast-striking young (right-lateral?) strike-slip fault just to the northeast of Izmir (f in Fig. 2) may have acquired its present strike through such a rotation. In fact the block north of that fault does appear to have undergone a small degree of clockwise rotation since the eruption of the mafic rocks (middle Miocene or younger?) on which paleomagnetic measurements of Kissel et al. (1985) were made (Fig. 19). The evidence for east-west (or northeast-southwest) shortening just to the east of the west Anatolian extensional province, the hypothesis that the Aegean and west Turkish region itself may be shortening in a generalized east-west direction (Dewey and Şengör, 1979; Le Pichon and Angelier, 1981), and the preliminary paleomagnetic observations by Kissel et al. (1985) invite one to reflect on the possibility of the widespread occurrence of such rotations of large blocks around vertical axes in western Turkey. Because sedimentary fills overflowing the tectonic boundaries of basins commonly hide very large structures, their recognition may become extremely difficult.

One final question that we must answer in defense of our hypothesis is why the structures in western Anatolia developed in the particular pattern that we see today. At this point only a possibility can be indicated, because studies to establish the relationships between the neotectonic and paleotectonic structures in western Anatolia are neither as numerous nor as detailed as would be required to give an accurate answer.

Figure 20 shows the strike- and trendlines of the penetrative and semi-penetrative paleotectonic structures in the basement of western Anatolia, which are likely to control the nucleation of shallow structures in the lithosphere. The parallelism they display with the orientation of the neotectonic structures (compare with Figure 19) is truly remarkable. We suggest that the old structures may have controlled the location of the younger ones, which is in agreement with the hypothesis advanced by McKenzie (1972). If graben complexes also formed during the paleotectonic contractional episode in western Turkey parallel with the orientation of shortening, some of these may have been inherited by the neotectonic episode as resurrected cross-grabens. Some anomalously large Neogene thicknesses in the cross-grabens, such as the one near Gördes (Fig. 2) which approaches 3,000 m or graben-fill sequences dating from the early Miocene (Kaya, 1981), may be indicative of the existence of resurrected structures in western Turkey. The boundary faults of such grabens are likely to be only replacement structures in the strict sense of the word, because oblique-slip faults of the neotectonic episode would "re-

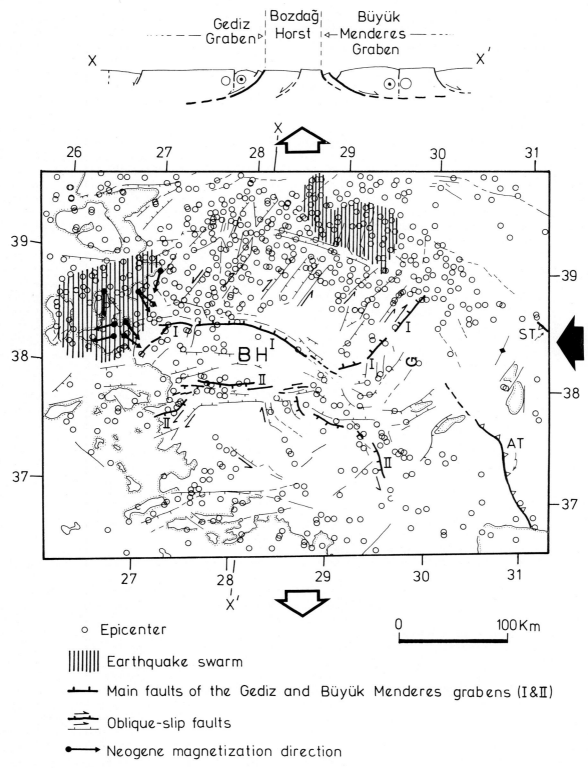

FIG. 19.—Map showing the distribution of epicenters of earthquakes that occurred in the western Anatolian graben system between 1976 and 1980 (after Üçer et al., 1981), and the major neotectonic structures. Strike-slip components on many of the cross-faults that extend north-north-eastwards from the Gediz graben and south from the Büyük Menderes graben are conjectural (except near the coast, where strike slip has been demonstrated either by field mapping or by focal-mechanism solutions). Notice the concentration of epicenters north of the main fault of the Gediz graben. BH is the Bozdağ Horst. Thin arrows on the map show Neogene paleomagnetic inclinations and declinations after Kissel et al. (1985). Large white arrows show the generalized direction of extension. The cross section at the top shows the hypothetical extension at depth of the main faults of the Gediz and the Büyük Menderes grabens. See Figure 3 for location.

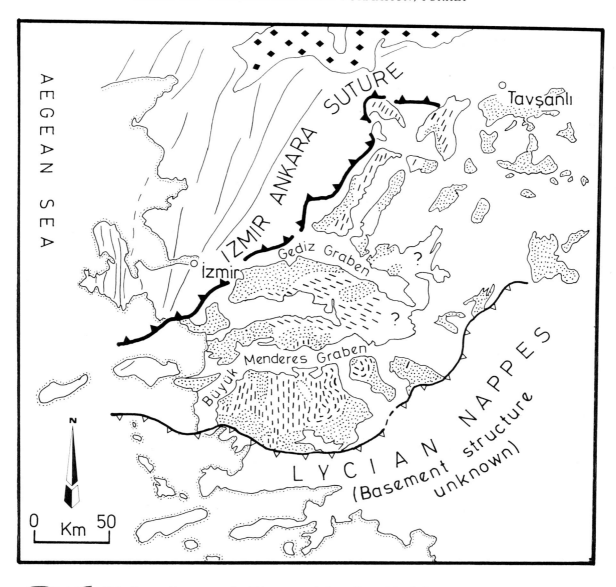

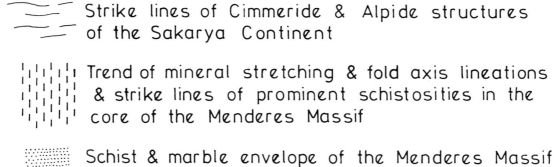

FIG. 20.—Sketch map of the strike- and trendlines of penetrative and semi-penetrative mesoscopic paleotectonic structures in the basement of the western Anatolian graben system that may have affected the nucleation of the later neotectonic structures (the map was compiled and simplified from Brinkmann, 1976; Jacobshagen and Skala, 1977; and Şengör et al., 1984). A comparison of this map with Figures 2 and 17 shows the remarkable parallelism between paleotectonic and neotectonic structures.

place" the pure normal faults of the paleotectonic episode, thus creating a complicated kinematic history along such faults. One must be careful not to try to infer the evolution of the neotectonic regime by using the sedimentary fill of such resurrected cross-grabens, for they would indicate a longer history than the actual duration of the neotectonic episode, because their resurrection may have gone on without the generation of any break in the sedimentary evolution of their fill. However, in the cross-grabens north of Izmir, along the coast, Kaya (1981) documented a subsidence history that began during the early Miocene and continued to the present but with a marked break in tectonic activity during the 16.2 to 11 Ma interval. Clearly, the events predating this interval belong to the paleotectonic regime whereas those after it represent neotectonic activity resurrecting the paleotectonic basin.

In conclusion, the strike-slip faults of the West Anatolian extensional province appear to be subordinate, shallow ("thin-skinned" or "detached") features accommodating differential extension between hanging-wall-block compartments that move on enormous low-angle normal faults. Despite their secondary role, they generate a variety of basin types (T grabens, thin-skinned pull-apart basins, etc.) and may lead to very serious errors in the interpretation of the kinematic history of the region if not correctly recognized. Many of the neotectonic structures of western Anatolian extensional province seem to be replacement structures with respect to the penetrative and semi-penetrative paleotectonic (Alpide and Pan-African) structures of the basement, and some may be resurrected Tibet-type grabens that originally formed during the last phases of the paleotectonic evolution of western Turkey.

DISCUSSION AND CONCLUSIONS

As the discussions in the preceding paragraphs and a comparison of Figures 1 and 2 show, the geology of tectonic escape is extremely complicated at all scales. In Turkey, in an area smaller than 1 million km^2, we have noted the existence of a high plateau characterized by a diffuse convergent regime, dominated by strike-slip faults. The plateau takes up the convergence between the European and the Arabian plates and is grossly divided into rhombohedral sub-domains with east-trending long axes. From the plateau, two prominent strike-slip fault systems extend westwards and bound a large piece of continental material squeezed out of the convergent area, the Anatolian scholle. We have observed a complex relationship between these faults and the convergent regime farther east; we noticed that in places north-south convergence extends into the continental wedge across the faults and induces strike changes onto the faults creating pull-apart basins and/or compressional segments, depending on whether segments preserving their original strike or the rotated ones continue to function as strike-slip faults. We have concluded that simple homogeneous constriction of the escaping wedge can generate neither pull-apart basins nor push-up ridges along the strike-slip boundaries of the wedge. To form releasing or constraining bends along these strike-slip boundaries, the constriction must be inhomogeneous and be divided into

compartments trending normal to the direction of convergence.

If the motion of the wedge away from the zone of constriction is checked by another continental object, the wedge will try the next easiest route to continue its escape. In the process it may disintegrate internally. We have interpreted the west Anatolian extensional province, where the distal end of the escaping wedge (with respect to the convergent area) is extending in a north-south direction, to continue the escape onto the eastern Mediterranean oceanic lithosphere away from the obstruction of the Balkan peninsula. We realized that this extension not only generates extremely complex, thin-skinned basin complexes associated with similarly thin-skinned strike-slip faults, but it also interferes with the North Anatolian fault creating long and narrow transtensional basins parallel with the trace of the fault.

Where strike-slip faults interfere with one another within the continental lithosphere, the buoyancy and low shear strength of the latter give rise to problems of incompatibility and "holes" open that appear as basins at the surface. The Karlıova and Adana/Cilicia-type basins are characterized by extremely complicated kinematics both in space and in time. The Karlıova-type basins are progressively destroyed also by the mechanisms that create them, by incorporating them into the zone of convergence. Recognition of their fossil analogues would therefore be difficult, which may explain why none has been reported so far.

Strike-slip faults branching off the boundaries of the escaping wedge have proven to be discontinuous features. They do not cleanly segment the Anatolian scholle into discrete blocks, but instead form only one type of member of a complex set of structures that also includes thrust and normal faults.

When we set out to study the neotectonics of Turkey and the surrounding regions, our main objectives were to devise tests to check the models proposed to account for tectonic escape, and to assemble criteria to recognize its fossil representatives. In the following final sections we discuss the conclusions of this study as they relate to these problems.

Causes of Escape

This study has documented that when the North Anatolian fault zone appeared for the first time during the (?late) Serravallian-Tortonian interval, eastern Turkey was either still below sea level or only just above it. This low region occupied a large area extending into northeastern Iran (Berberian and King, 1981), and northeastwards beyond the greater Caucasus (Vinogradov, 1967), where only two isolated small ranges belonging to the Main Range of the greater Caucasus formed "low mountainous topography" providing clastic sediment to the surrounding shallow seas. By contrast, western and central Turkey were fairly high. In western Turkey elevations reached about 1.5 km. Because recent studies in Crete also indicate that the previously subduction-related tectonism of the Pliny-Strabo segment of the Hellenic arc switched to a strike-slip regime during late Serravallian time (Fortuin and Peters, 1984), and that the Aegean region began dissolving into discrete blocks (Meulenkamp, 1982), we believe that the Anatolian scholle must have begun its westerly escape by the late Serravallian time. Therefore this escape occurred away from an area of low topography (almost at sea level) in the convergent zone of eastern Turkey and towards an area of high topography (1.5 km) in western Turkey and the Aegean, implying that at least in this particular instance buoyancy forces arising from crustal thickness differences were probably not the cause of the initiation of the escape.

An interesting point concerns the evidence indicating the much later initiation of the East Anatolian fault than most other escape-related structures in Turkey. Initial escape may have been accommodated by oblique slip on the Bitlis suture zone. Only after the Bitlis suture zone was consolidated by convergent strain and (at depth) metamorphism, and the elevation of eastern Turkey started to increase, a new fault formed to the north of and in places disrupting the suture and began to accommodate the westerly motion of Turkey. This is supported by the observation that the cumulative offset of the North Anatolian fault since the early Pliocene, that is the time of initiation of the East Anatolian fault (not the total offset of the North Anatolian fault) is about 25 km (Barka and Hancock, 1984), which is similar to the 22 km observed along the East Anatolian fault. Thus the lower magnitude of total offset along the East Anatolian fault (22 km versus 85 ± 5 km) seems not to result from any rotational movement of the Anatolian scholle as a whole, as claimed by Rotstein (1984), but from the younger age of the East Anatolian fault.

In conclusion the origin of the westerly escape of the Anatolian scholle seems to be the forces applied to its boundaries or perhaps other forces applied to its lower surface by the mantle, and not the buoyancy forces arising from crustal thickness differences. It is a different matter to maintain the escape. Kasapoğlu and Toksöz (1984), for example, found that the forces applied to the boundaries of the Anatolian scholle were inadequate to maintain its present westerly motion at the observed rates. We believe that the gravitational potential of the East Turkish high plateau or forces applied to the Anatolian scholle from beneath may very well supply the extra force needed.

Geological Signature of Escape

As Burke and Şengör (in press) pointed out, both a zone of convergence to expel and a free-face to receive a wedge of buoyant material are necessary conditions for tectonic escape to occur. The subsequent closure of the oceanic area where the free-face is located generates a suture younger than that located in front of the original zone of constriction. Thus the obvious places in the geological record to start searching for escape regimes are where sutures progressively young in one direction.

In well-developed escape regimes, escape occurs from an area of high topography formed due to convergence towards a lowland generally created by extensional tectonics. Therefore regions characterised by extensive terrains of high temperature/low pressure metamorphic rocks, extensive structural imbrication complicated by the presence of crisscrossing, steep, ductile shear zones, and abundant calc-al-

kalic high-K granitic and alkalic intrusions (representing the eroded roots of Tibetan- or East Anatolian-type convergent high plateaux) passing sideways, parallel with sutures younging in same direction, into regions where large sedimentary basins are nested on former zones of distributed extension, are good candidates to be products of continental escape. Between these two regions of contrasting tectonic development, namely the compressional and the extensional areas, a series of disconnected basins of similar age may line up marking the course of former strike-slip faults. Burchfiel (1980), Burchfiel and Royden (1982), and Royden et al. (1982) have documented a fossil escape system in the eastern Alpine/Carpathian/Pannonian region very similar to the one we have described above and having all the fingerprints mentioned. Basins with very complex strain histories may mark places where large, intracontinental strike-slip faults interfered in the past to generate incompatibility problems. The detection of such spots may lead the geologist to the discovery of the associated fossil strike-slip faults.

All these features indicative of escape may be compressed across the regional strike one or more times. In such cases the recognition of former high plateaux from evidence contained within them will be even more difficult, for most of their structures will be rotated into parallelism with the overall trend of the convergent zone. The extensional zone and the basins associated with the strike-slip faults may be easier to recognize. Many of the Devonian and Permo-Triassic clastic sedimentary rocks in the area of the Altaids in Central Asia (Şengör, in press) may have accumulated in escape-related basins associated with early and late Paleozoic continental collisions related to the evolution of the Altaids. Most of these basin complexes have been compressed out of recognition during the later Cimmeride and Alpide collisions farther to the south (Şengör, 1984).

Pull-apart basins and the long and narrow transtensional basins of the Sea of Marmara type may appear as ultramafic/granulitic associations in nappes representing deep continental crust/upper mantle cross sections. The cross sections they represent would be very thin (8–15 km) and show rapid passages into basinal-type supracrustal rocks. Kornprobst and Vielzeuf (1984) argued that such zones of crustal thinning associated with strike-slip regimes may be recognized in the fossil record by analogy with present-day fracture zones, and A. Nicolas (personal commun., 1984) showed that detailed fabric studies in the metamorphic basements of such fossil basins may yield the sense and orientation of strike slip. Isotopic dating of such basin-floor remnants would yield the age of extension that created the basin floor and therefore the age of escape.

The lesson the neotectonics of Turkey teaches about the causes and evolution of tectonic escape is, therefore, that they are complex and that more than one factor is involved in the origin and development of escape systems. Models emphasizing only one factor are unlikely to be realistic, and complex models, compatible with the geology of as large a number as possible of escape systems, both active and fossil, must be devised to account for their formation and maintenance.

Similarly, there is no single criterion to recognize fossil

escape systems in the geological record. Instead, an assemblage of structures with temporal/spatial relationships similar to the ones we have described above must be mapped in detail to be able to identify fossil escape systems with confidence. Mann and Bradley (1984) pointed out that hydrocarbon exploration in strike-slip zones requires awareness of several distinct basin types, traditionally defined on the basis of bounding fault geometry, which include pull-aparts, fault-wedge basins, fault-angle basins, fault flank basins, and ramp valleys. This study has shown that although such a classification covers all possible geometries of strike-slip related basins, their kinematic evolution may greatly vary within one class rendering the above classification misleading. Incompatibility effects seem to dominate the causes of basin formation in complexly faulted areas, of the kind commonly generated in zones of tectonic escape.

Sophisticated techniques such as deep crustal seismic reflection profiling or high-precision paleomagnetic studies are helpful in studies of tectonic escape, but they can only be exploited fully if used in conjunction with stratigraphic/structural/petrological studies tied to detailed field mapping.

ACKNOWLEDGMENTS

This study forms a part of an integrated research program on the tectonic evolution of Turkey and surrounding regions undertaken at the Istanbul Technical University with the aid of a number of institutions in and outside Turkey. Drs. Aykut Barka and Paul Hancock provided us with pre-prints of a number of their papers on the North Anatolian fault. Profs. N. Canıtez, K. Ergin, S. Erinç, O. Erol, İ. Ketin, Drs. R. Akkök, A. Boray, A. Okay, N. Özgül, Y. Yılmaz, and Messrs. E. Arpat, O. Sungurlu, Y. Güner, and C. Cooper helped us by discussions and by providing unpublished information. We are also grateful to Profs. B. C. Burchfiel, K. Burke, J. F. Dewey and to Drs. R. Armijo, J. A. Jackson, D. P. McKenzie, L. H. Royden and P. Tapponnier for discussions on the behavior and evolution of the continental lithosphere and the geology of tectonic escape. K. Biddle and N. Christie-Blick are thanked for having invited this contribution to be read at the San Antonio symposium. They also helped to improve this manuscript by their excellent reviews. We thank Mr. İzzet Türkkal for the photographic reproductions and the Sürat Daktilo Company for preparing the typescript.

REFERENCES

ABDÜSSELÂMOĞLU, M. Ş., 1959, Almacıkdağı ile Mudurnu ve Göynük Civarının Jeolojisi (Geology of Almacıkdağı and the surroundings of Mudurnu and Göynük): İstanbul Üniv., Fen Fak. Monogr., v. 14, 94 p.

AKKAN, E., 1964, Erzincan ovası ve çevresinin jeomorfolojisi (Geomorphology of the Erzincan Plain and its Surroundings): Ankara Üniversitesi Dil ve Tarih-Coğrafya Fakültesi Yayınları, No. 153, 104 p.

AKKÖK, R., 1981, Menderes Masifinin gnayslarında ve şistlerinde metamorfizma koşulları, Alaşehir-Manisa (Metamorphic conditions of the gneisses and schists of the Menderes Massif, Alaşehir-Manisa): Türkiye Jeoloji Kurumu Bülteni, v. 24, p. 11–20.

———, 1983, Structural and Metamorphic evolution of the northern part of the Menderes Massif: New data from the Derbent area and their implication for the tectonics of the Massif: Journal of Geology, v. 91, p. 342–350.

AKTİMUR, S., 1979, Malatya-Sivas dolayının uzaktan algılama yöntemi ile çizgiselliklerinin incelenmesi (A study of the lineaments of the Malatya-Sivas region by remote sensing): Maden Tetkik ve Arama Enstitüsü, Derleme No: 6651, unpublished report.

AKTİMUR, T., İRFAN, Y., ÖKTEM, A., AND OLGUNER, A., 1979, Ozalp ve çevresinin yerbilim verileri (Geoscience data of Ozalp and Surrounding regions): Maden Tetkik ve Arama Enstitüsü, unpublished report.

ALLEN, C. R., 1969, Active faulting in northern Turkey: California Institute of Technology, Division of Geological Sciences Contribution No. 1577, 32 p.

———, 1975, Geological criteria for evaluating seismicity: Geological Society of America Bulletin, v. 86, p. 1041–1057.

———, 1982, Comparisons between the North Anatolian Fault of Turkey and the San Andreas Fault of California, in Işıkara, A. M. and Vogel, A., eds., Multidisciplinary Approach to Earthquake Prediction: Wiesbaden, Friedrich Vieweg and Sohn, p. 67–85.

ALLMENDINGER, R. W., SHARP, J. W., VON TISH, D., SERPA, L., BROWN, L., KAUFMAN, S., AND OLIVER, J., 1983, Cenozoic and Mesozoic structure of the eastern Basin and Range province, Utah, from COCORP seismic-reflection data: Geology, v. 11, p. 532–536.

ALTINLI, E., 1966, Doğu ve Güneydoğu Anadolu'nun jeolojisi. Kısım I (Geology of eastern and southeastern Anatolia): Maden Tetkik ve Arama Enstitüsü Dergisi, v. 66, p. 35–74.

AMBRASEYS, N. N., 1970, Some characteristic features of the North Anatolian fault zone: Tectonophysics, v. 9, p. 143–165.

AMBRASEYS, N. N., AND ZATOPEK, A., 1968, The Varto-Ustukran (Anatolia) earthquake of 1966 August 19: A summary of a field report: Seismological Society of America Bulletin, v. 58, p. 47.

ANGELIER, J., 1973, Sur la néotectonique Egéenne: failles anté-tyrrhéniennes et post-tyrrhéniennes dans l'île de Karpathos (Dodécanése, Gréce): Comptes Rendus de la Société géologique de France, 7°sér., v. 15, p. 105–106.

ANGELIER, J., DUMONT, J. F., KARAMANDERESİ, H., POISSON, A., ŞİMŞEK, Ş., AND UYSAL, Ş., 1981, Analyses of fault mechanisms and expansion of south-western Anatolia since the late Miocene: Tectonophysics, v. 75, p. T.1–T.9.

ANONMYMOUS, 1984, A new paradigm for understanding southern San Andreas fault tectonics: Lamont Newsletter, Summer 1984, p. 6–7.

ARGAND, E., 1920, Plissements précurseurs et plissements tardifs des chaînes de montagnes: Actes de la Société helvetique des Sciences Naturelles, Neuchâtel, p. 1–27.

———, 1924, La tectonique de l'Asie: Report of the 13th International Geological Congress, v. 1, p. 171–372.

ARPAT, E., AND ŞAROĞLU, F., 1972, The East Anatolian Fault System: Thoughts on its development: Bulletin of the Mineral Research and Exploration Institute of Turkey, Ankara, v. 78, p. 33–39.

ARPAT, E., AND ŞAROĞLU, F., 1975, Türkiye'deki bazı önemli genç tektonik olaylar (On some important young tectonic events in Turkey): Türkiye Jeoloji Kurumu Bülteni, v. 18, p. 91–10.

ARPAT, E., ŞAROĞLU, F., AND İZ, H. B., 1977, 1976 Çaldıran depremi (Çaldıran Earthquake): Yeryuvarı ve İnsan, p. 29–41.

ATALAY, İ., 1978, Erzurum ovası ve çevresinin jeolojisi ve jeomorfolojisi (Geology and geomorphology of the Erzurum Plain and its surroundings): Atatürk Üniversitesi Yayınları, No. 543, 96 p.

———, 1983, Muş ovası ve çevresinin jeomorfolojisi ve toprak coğrafyası (Geomorphology and soil geography of the Muş Plain and its surroundings): Ege Üniversitesi Edebiyat Fakültesi Yayınları, n. 24, İzmir, 154 p.

ATEŞ, R., 1982, Earthquake activity on the North Anatolian Fault Zone, in Işıkara, A. M. and Vogel, A., eds., Multidisciplinary Approach to Earthquake Prediction: Wiesbaden, Friedrich Vieweg und Sohn, p. 95–113.

AYDIN, A., AND NUR, A., 1982, Evolution of pull-apart basins and their scale independence: Tectonics, v. 1, p. 91–106.

AYDIN, A., AND NUR, A., 1985, The types and role of stepovers in strike-slip tectonics, in Biddle, K. T., and Christie-Blick, N., eds., Strike-Slip Deformation, Basin Formation, and Sedimentation: Society of Economic Paleontologists and Mineralogists Special Publication No. 37, p. 1–34.

BAHAT, D., 1983, New aspects of rhomb structures: Journal of Structural Geology, v. 5, p. 591–601.

BARKA, A. A., 1981, Seismo-tectonic aspects of the North Anatolian fault zone [unpubl. Ph.D. thesis]: Bristol, University of Bristol, England, 335 p.

———, 1983a, Doğu Anadolu'da ve Marmara çevresinde gelecekte ola-bilecek bazı büyük depremlerin olasılı episantr alanları (Likely epicen-tral areas of some possible future large magnitude earthquakes in east-ern Turkey and in Marmara region): Yeryuvarı ve İnsan, v. 8, p.30–33.

———, 1983b, Büyük magnitüdlü depremlerin episantr alanlarını önceden belirleyebilecek bazı jeolojik veriler (Geological evidence to predict the epicentral areas of large-magnitude earthquakes): Türkiye Jeoloji Ku-rumu Bülteni, v. 26, p. 21–30.

———, 1985, Kuzey Anadolu fay zonundaki bazı Neojen-Kuvaterner havzalarının jeolojisi ve tektonik evrimi (Geology and tectonic evolu-tion of some of the Neogene-Quaternary basins in the North Anatolian fault zone): in Ketin Simpozyumu Kitabı, Ankara, Türkiye Jeoloji Ku-rumu, p. 209–227.

BARKA, A. A., AND HANCOCK, P. L., 1984, Neotectonic deformation pat-terns in the convex-northwards arc of the North Anatolian fault zone, in Dixon, J. E., and Robertson, A. H. F., eds., The Geological Evo-lution of the Eastern Mediterranean: Geological Society of London Special Publication No. 17, p. 763–774.

BARKA, A. A., ŞAROĞLU, F., AND GÜNER, Y., 1983 Horasan-Narman de-premi ve bu depremin Doğu Anadolu neotektoniğindeki yeri (Earth-quake of Horasan-Narman and its place in the neotectonics of eastern Turkey): Yeryuvarı ve İnsan, v. 8, p. 16–21.

BECKER, H., 1934, Die Beziehungen zwischen Felsengebirge und Großem Becken im westlichen Nordamerika: Zeitschrift der deutschen geolo-gischen Gesellschaft, v. 86, p. 115–120.

BECKER-PLATEN, J. D., 1970, Lithostratigraphisce Untersuchungen im Känozoikum Südwest-Anatoliens (Türkei): Beihefte zum geologischen Jahrbuch, v. 97, 244 p.

BEN-MENAHEM, A., NUR, A., AND VERED, M., 1976, Tectonics, seismic-ity and structure of the Afro-Eurasian junction—the breaking of an incoherent plate: Physics of the Earth and Planetary Interiors, v. 12, p. 1–50.

BERBERIAN, M., AND KING, G. C. P., 1981, Towards a paleogeography and tectonic evolution of Iran: Canadian Journal of Earth Sciences, v. 18, p. 210–265.

BERCKHEMER, H., 1977, Some aspects of the evolution of marginal seas deduced from observations in the Aegean region, in Biju-Duval, B., and Montadert, L., eds., Structural History of the Mediterranean Ba-sins: Paris, Editions Technip, p. 303–313.

BERGGREN, W. A., KENT, D. V., FLYNN, J. J., AND VAN COUVERING, J. A., in press, Cenozoic geochronology, in Snelling, N. J., ed., Geo-chronology and the Geologic Time Scale: Geological Society of Lon-don Special Paper.

BİLGİN, T., 1967, Samanlı Dağları (Samanlı Mountains): İstanbul Üniversitesi Yayınları, No. 1294, 196 p.

———, 1969a, Ceyhan doğusunda volkanik şekiller ile Hassa Leçesi (Volcanic landforms east of Ceyhan and the Leçe of Hassa): İstanbul Üniversitesi Yayınları, No. 1494, p. 5–26.

———, 1969b, Biga yarımadası güneybatı kısmının jeomorfolojisi (Geo-morphology of the southwest part of the Biga Peninsula): İstanbul Üniversitesi Yayınları, No. 1433, 273 p.

———, 1984, Adapazarı ve Sapanca oluğunun alüvyal morfolojisi ve Kuaternerdeki jeomorfolojik tekâmülü (Alluvial morphology of the Adapazarı-Sakarya trough and its geomorphological evolution during the Quaternary): İstanbul Üniversitesi Edebiyat Fakültesi Yayınları 199 p.

BİLGİN, A. Z., AND ERCAN, T., 1981, Cephan-Osmaniye yöresindeki Ku-vaterner bazaltlarının petrolojisi (Petrology of the Quarternary basalts in the Ceyhan-Osmaniye area): Türkiye Jeoloji Kurumu Bülteni, v. 24, p. 21–30.

BIJU-DUVAL, B., LETOUZEY, J., AND MONTADERT, L., 1978, Structure and evolution of the Mediterranean basins, in Hsü, K. J., Montadert, L., et al., Initial Reports of the Deep Sea Drilling Project, V. XLII, Part I, Washington, U.S. Government Printing Office, p. 951–984.

BLAKE, JR., M. C., CAMPBELL, R. H., DIBBLEE, T. W., JR., HOWELL, D. G., NILSEN, T. H., NORMARK, W. R., VEDDER, J. C., AND SILVER, E. A., 1978, Neogene basin formation in relation to plate-tectonic evo-lution of San Andreas Fault System, California: American Association of Petroleum Geologists Bulletin, v. 62, p. 344–372.

BORAY, A., AND ŞAROĞLU, F., in press, Evidence for east-west shortening in the apical region of the Angle of Isparta: Tectonophysics.

BÖGER, H., 1978, Sedimentary history and tectonic movements during the late Neogene, in Closs, H., Roeder, D., and Schmidt, K., eds., Alps, Appennines, Hellenides: Stuttgart, Schweizerbart'sche Verlags-buchhandlung, p. 510–512.

BRINKMANN, R., 1976, Geology of Turkey: Stuttgart, Enke, 158 p.

BRINKMANN, R., FEIST, R., MARR, W. U., NICKEL, E., SCHLIMM, W., AND WALTER, H. R., 1970, Geologie der Soma Dağları: Maden Tetkik ve Arama Enstitüsü Bülteni, v. 74, p. 7–23.

BURCHFIEL, B. C., 1980, Eastern European Alpine System and the Car-pathian orocline as an example of collision tectonics: Tectonophysics, v. 63, p. 31–61.

BURCHFIEL, B. C., AND ROYDEN, L., 1982, Carpathian foreland fold and thrust belt and its relation to Pannonian and other basins: American Association of Petroleum Geologists Bulletin, v. 66, p. 1179–1195.

BURKE, K., AND ŞENGÖR, A. M. C., in press, Tectonic escape in the evo-lution of the continental crust, in Barazangi, M., ed.: American Geo-physical Union Special Publication.

BURSHTAR, M. S., AND TOLMACHEVSKIY, A. A., 1965, Novyye dannyye o glubinnom stroyenii Araratskoy kotlaviny v Armyanskoy SSR. Dok-lady Akademii Nauk SSSR, v. 165, p. 1135–1138.

BÜYÜKAŞIKOĞLU, S., 1979, Sismolojik verilere göre Güney Anadolu ve Doğu Akdenizde Avrasya-Afrika levha sınırının özellikleri (Character-istics of the Africa-Eurasia plate boundary in southern Turkey and the eastern Mediterranean according to seismological data): Istanbul Tek-nik Üniversitesi, Maden Fakültesi Yayınları, İstanbul, 75 p.

CAREY, S. W., 1958, A tectonic approach to continental drift, in Carey, S. W., ed., Continental Drift, A Symposium: Geology Department, University of Tasmania, Hobart, p. 177–355.

CHEN ZHI-MING, 1981, Structural origin of lakes on the Xizang Plateau, in Geological and Ecological Studies of Qinghai-Xizang plateau: Sci-ence Press, Beijing, p. 1769–1776.

CLOOS, H., 1928, Bau and Bewegung der Gebirge in Nordamerika, Skan-dinavien und Mittleuropa: Berlin, Gebrüder Borntraeger, 87 p.

CRAMPIN, S., AND ÜÇER, S. B., 1975, The seismicity of the Marmara Sea region of Turkey: Geophysical Journal of Royal Astronomical Society, v. 40, p. 269–288.

CROWELL, J. C., 1974, Origin of late Cenozoic basins in southern Cali-fornia, in Dickinson, W. R., ed., Tectonics and Sedimentation: Society of Economic Paleontologists and Mineralogists Special Publication No. 22, p. 190–204.

———, 1976, Implications of crustal stretching and shortening of coastal Ventura basin, California: American Association of Petroleum Geol-ogists, Pacific Section, Miscellaneous Publication 24, p. 365–382.

DERMITZAKIS, M. D., AND PAPANIKOLAOY, 1981, Paleogeography and geo-dynamics of the Aegean region during the Neogene: Proc. VIIth. In-ternational Congress on Mediterranean Neogene, Athens, 1979, p. 245–289.

DEWEY, J. F., 1978, Origin of long transform-short ridge systems: Geo-logical Society of America Abstracts with Programs, v. 10, p. 388.

———, 1982, Plate tectonics and the evolution of the British Isles: Jour-nal of the Geological Society of London, v. 139, p. 371–414.

DEWEY, J. F., HEMPTON, M. R., KIDD, W. S. F., ŞAROĞLU, F., AND ŞENGÖR, A. M. C., in press, Shortening of continental lithosphere: the neotec-tonics of eastern Anatolia—a young collision zone: Geological Society of London Special Publication on Collision Tectonics.

DEWEY, J. F., AND ŞENGÖR, A. M. C., 1979, Aegean and surrounding regions: Complex multiplate and continuum tectonics in a convergent zone: Geological Society of America Bulletin, Part I, v. 90, p. 84–92.

DUCLOZ, C., 1972, The geology of the Pellapais-Kythrea area of the cen-tral Kyrenia Range: Geological Survey Department Cyprus, No. 6, 75 p.

DUMONT, J. F., UYSAL, Ş., SİMŞEK, Ş., KARAMANDERESİ, İ. H., AND LE-TOUZEY, J., 1979, Güneybatı Anadolu'daki grabenlerin oluşumu (On the origin of the grabens in southwestern Anatolia): Maden Tetkik ve Arama Enstitüsü Dergisi, v. 92, p. 7–16.

DUNNE, L. A., AND HEMPTON, M. R., 1984, Deltaic sedimentation in the Lake Hazar pull-apart basin, south-eastern Turkey: Sedimentology, v. 31, p. 401–412.

ENGLAND, P., AND MCKENZIE, D. P., 1982, A thin viscous sheet model for continental deformation: Geophysical Journal of the Royal Astronomical Society, v. 70, p. 295–321.

———, 1983, Correction to: a thin viscous sheet model for continental deformation: Geophysical Journal of the Royal Astronomical Society, v. 73, p. 523–532.

ERENTÖZ, C., AND TERNEK, Z., 1968, Türkiye'deki termomineral kaynaklar ve jeotermik enerji etüdleri (Thermo-mineral springs and geothermal studies in Turkey): Maden Tetkik ve Arama Enstitüsü Dergisi, v. 70, p. 1–57.

ERGIN, K., UZ, Z., AND GÜÇLÜ, U., 1972, 28 Mart 1970 Gediz depremi art sarsıntılarının incelenmesi (Study of the aftershocks of the March 28th, 1970 Gediz earthquake): İstanbul Teknik Üniversitesi Arz Fiziği Enstitüsü Yayınları, No. 29, 50 p.

ERINÇ, S., 1953, Doğu Anadolu Coğrafyası (Geography of eastern Anatolia): İstanbul Üniversitesi Yayınları, No. 572, 124 p.

———, 1955a, Über die Entstehung und morphologische Bedeutung des Tmolosschutts: University of Istanbul Geographical Institute Review, No. 2, p. 57–72.

———, 1955b, Die morphologische Entwicklungsstadien der Küçükmenderes-Masse: University of Istanbul Geographical Institute Review, No. 2, p. 93–95.

———, 1956, Yalova civarında bahri Pleistosen depoları ve taraçaları (Marine Pleistocene deposits and terraces around Yalova): Türk Coğrafya Dergisi, v. 12, p. 188–190.

———, 1970, Kula ve Adala arasında genç volkan reliyefi (Young volcanic morphology between Kula and Adala): İstanbul Üniversitesi Coğrafya Enstitüsü Dergisi, v. 9, p. 7–31.

———, 1971, Jeomorfoloji II (Geomorphology II): İstanbul Üniversitesi Yayınları, No. 1628, 487 p.

ERINÇ, S., BILGIN, T., AND BENER, M., 1961, Çağa Depresyonu ve Boğazı (The Gorge and Depression of Çağa): İstanbul Üniversitesi Coğrafya Enstitüsü Dergisi, v. 6, p. 170–173.

ERKAL, T., 1983, Structure, sedimentology and geomorphology related to active faulting in the Gaziköy-Şarköy area, Thrace, Turkey [unpubl. M.Sc. thesis]: Bristol, England, University of Bristol, 238 p.

EROL, O., 1969, Tuzgölü havzasının jeolojisi ve jeomorfolojisi (Geology and geomorphology of the Tuzgölü Basin): TÜBİTAK: unpublished report.

———, 1981, Neotectonic and geomorphological evolution of Turkey: Zeitschrift für Geomorphologie, Neue Folge, Supplement Band 40, p. 193–211.

———, 1982a, Batı Anadolu genç tektoniğinin jeomorfolojik sonuçları, in Erol, O., and Oygür, V., eds., Batı Anadolu'nun genç tektoniği ve volkanizması paneli: Türkiye Jeoloji Kurumu, Ankara, p. 15–20.

———, 1982b, Türkiye jeomorfoloji haritası, 1/2.000.000: Maden Tetkik ve Arama Enstitüsü, Ankara.

EVANS, G., MORGAN, P., EVANS, W. E., EVANS, T. R., AND WOODSIDE, J. M., 1978, Faulting and halokinetics in the northeastern Mediterranean between Cyprus and Turkey: Geology, v. 6, p. 392–396.

EYİDOĞAN, H., AND JACKSON, J. A., 1985, A seismological study of normal faulting in the Demirci, Alaşehir and Gediz earthquakes of 1969–70 in western Turkey: implications for the nature and geometry of deformation in the continental crust: Geophysical Journal of the Royal Astronomical Society, v. 81, p. 569–607.

FLINN, D., 1962, On folding during three-dimensional progressive deformation: Quarterly Journal of the Geological Society of London, v. 118, p. 385–433.

FORTUIN, A. R., AND PETERS, J. M., 1984, The Prina Complex in eastern Crete and its relationship to possible Miocene strike-slip tectonics: Journal of Structural Geology, v. 6., p. 459–476.

GALLO, D. G., KIDD, W. S. F., SLOAN, H. S., AND ŞENGÖR, A. M. C., 1980, Large angular rotations of blocks along strike-slip zones as shallow décollement features (abs.): EOS, v. 61, p. 1120.

GANSSER, A., 1964, Geology of the Himalayas: London, Wiley-Interscience, 289 p.

———, 1977, The great suture zone between Himalaya and Tibet. A preliminary account: Colloques internationaux du Centre National des Recherches Scientifiques, No. 268, Ecologie at Géologie de l'Himalaya, p. 181–192.

GELATI, R., 1975, Miocene marine sequence from Lake Van, eastern Turkey: Rivista Italiana di Paleontologia e Stratigrafia, v. 81, p. 477–490.

GÖRÜR, N., 1977, Sedimentology of the Karaisalı Limestone and associated clastics (Miocene) of the north west flank of the Adana Basin, Turkey [unpubl. Ph.D. thesis]: London, University of London, 244 p.

GÜNER, Y., 1984, Nemrut yanardağının jeolojisi, jeomorfolojisi ve volkanizmasının evrimi (The geology, geomorphology and the evolution of the Nemrut Volcano): Jeomorfoloji Dergisi, v. 12, p. 23–65.

GUNNERSON, C. G., AND ÖZTURGUT, E., 1974, The Bosporus, in Degens, E. T., and Ross, D. A., eds., The Black Sea—Geology, Chemistry and Biology: American Association of Petroleum Geologists Memoir 20, p. 99–144.

GUTNIC, M., MONOD, O., POISSON, A., AND DUMONT, F. D., 1979, Géologie des Taurides occidentales (Turquie): Mémoires de la Société géologique de France, Nouvelle Serie, v. 58, 112 p.

HANCOCK, P. L., AND BARKA, A. A., 1981, Opposed shear senses inferred from neotectonic mesofracture systems in the North Anatolian fault zone: Journal of Structural Geology, v. 3, p. 383–392.

HEMPTON, M. R., DUNNE, L. A., AND DEWEY, J. F., 1983, Sedimentation in an active strike-slip basin, southeastern Turkey: Journal of Geology, v. 91, p. 401–412.

HIRN, A., NERCESSIAN, A., SAPIN, M., JOBERT, G., XU ZHONG XIN, GAD EN YUAN, LU DE YUAN, AND TENG JI WEN, 1984, Lhasa block and bordering sutures—a continuation of a 500-km Moho Traverse through Tibet: Nature, v. 307, p. 25–27.

HUNT, C. B., AND MABEY, D. R., 1966, General Geology of Death Valley, California—stratigraphy and structure: United States Geological Survey Professional Paper 494-A, p. 1–165.

İLKİŞIK, M., 1980, Trakya'da yerkabuğunun manyetotelürik yöntemle incelenmesi (A magnetotelluric study of the Earth's crust in Thrace): İstanbul Teknik Üniversitesi Maden Fakültesi Yayınları, 99 p.

INNOCENTI, F., MANETTI, P., MAZZUOLI, R., PASQUARÈ, G., AND VILLARI, L., 1981, Neogene and Quaternary volcanism in the Eastern Mediterranean. Time-space distribution and geotectonic implication, in Wezel, F. C., ed., Sedimentary Basins of Mediterranean Margins: Bologna, Tecnoprint, p. 369–385.

INNOCENTI, F., PISA, R., MAZZUOLI, C., PASQUARE, G., SERRI, G., AND VILLARI, L., 1980, Geology of the volcanic area north of Lake Van (Turkey): Geologische Rundschau, v. 69, p. 292–322.

IRRLITZ, W., 1972, Lithostratigraphie und tektonische Entwicklung des Neogens in Nordostanatolien: Beihefte zum Geologischen Jahrbuch, 120 p.

ISACKS, B., OLIVER, J., AND SYKES, L. R., 1968, Seismology and the new global tectonics: Journal of Geophysical Research, v. 73, p. 5855–5899.

JACKSON, J. AND MCKENZIE, D. P., 1984, Active tectonics of the Alpine-Himalayan Belt between western Turkey and Pakistan: Geophysical Journal of the Royal Astronomical Society, v. 77, p. 185–264.

JACKSON, J. A., KING, G. C. P., AND VITA-FINZI, C., 1983, Neotectonics of the Aegean: an alternative view: Earth and Planetary Science Letters, v. 61, p. 300–318.

JACOBSHAGEN, V., AND SKALA, W., 1977, Geologie der Nord-Sporaden und die Struktur-Prägung auf der Mittelägäischen Inselbrücke: Annales géologiques des Pays helleniques, v. 28, p. 233–274.

JONGSMA, D., AND MASCLE, J., 1981, Evidence for northward thrusting southwest of the Rhodes Basin: Nature, v. 293, p. 49–51.

KARIG, D. E., 1980, Material transport within accretionary prisms and the "knocker" problem: Journal of Geology, v. 88, p. 27–39.

KASAPOĞLU, K. E., AND TOKSÖZ, M. N., 1984, Consequence of collision of Arabian and Eurasian plates—finite element model: Tectonophysics v. 100, p. 71–95.

KAYA, O., 1979, Ortadoğu Ege çöküntüsünün Neojen stratigrafisi ve tektoniği (East-central Aegean depression: Neogene stratigraphy and tectonics): Türkiye Jeoloji Kurumu Bülteni, v. 22, p. 35–58.

———, 1981, Miocene reference section for the coastal parts of West Anatolia: Newsletter of Stratigraphy, v. 10, p. 164–191.

KETİN, İ., 1948, Über die tektonisch-mechanischen Folgerungen aus den grossen anatolischen Erdbeben des letzten Dezenniums: Geologische Rundschau, v. 36, p. 77–83.

———, 1968, Relations between general tectonic features and the main earthquake regions of Turkey: Bulletin of the Mineral Resources and Exploration Institute, Ankara, v. 71, p. 63–67.

———, 1969, Über die nordanatolische Horizontalverschiebung: Bulletin of the Mineral Research and Exploration Institute of Turkey, Ankara, v. 72, p. 1–28.

———, 1985, Türkiye'nin bindirmeli-naplı yapısında yeni gelişmeler ve bir örnek: Uludağ Masifi (New developments in the nappe tectonics of Turkey with the example of Uludağ Massif): Ketin Simpozyumu Kitabı, Ankara, Türkiye Jeoloji Kurumu, p. 19–36.

KETIN, I., GÖRÜR, N., AND AKKÖK, R., 1981, Petrol bölgelerimizin genel jeolojik durumları ve petrol olanakları hakkında görüşler (Geological setting of our petroleum districts and thoughts on their petroleum potential): Petrol İşleri Genel Müdürlüğü Dergisi, v. 25, p. 121–124.

KISSEL, C., LAJ, C., MERCIER, J. L., POISSON, A., SAVAŞÇIN, Y., AND SIMEAKIS, K., 1985, Tertiary rotational deformations in the Aegean domain (Abs.): Terra Cognita, v. 5, p. 139.

KONUK, Y. T., 1974, Geologie eines Abschnitts der Nordanatolischen Erdbebenlinie bei Kamil/Osmancık (Prov. Çorum; Türkei) [unpubl. Ph.D. thesis]: Bonn, Rheinischen Friedrich-Wilhelms-Universität 83 p.

KORNPROBST, J., AND VIELZEUF, D., 1984, Transcurrent crustal thinning: a mechanism for the uplift of deep continental crust/upper mantle associations, in Kornprobst, J., ed., Kimberlites II: The Mantle and Crust-Mantle Relationships: Amsterdam, Elsevier, p. 347–359.

KOSSMAT, F., 1926, Die Mediterranen Kettengebirge in ihrer Beziehung zum Gleichgewichtszustande der Erdrinde: Abhandlungen der mathematisch-physikalischen klasse der sächsischen Akademie der Wissenschaften, v. 38, p. 1–63.

LE PICHON, X., AND ANGELIER, J., 1981, The Aegean Sea: Philosophical Transactions of the Royal Society of London, Series A, v. 300, p. 357–372.

LE PICHON, X., AUBOUIN, J., LYBERIS, N., MONTI, S., RENARD, V., GOT, H., HSÜ, K., MART, Y., MASCLE, J., MATTHEWS, D., MITROPOULOS, D., TSOFLIAS, P., AND CHRONIS, G., 1979, From subduction to transform motion: A seabeam survey of the Hellenic trench system: Earth and Planetary Science Letters, v. 44, p. 441–450.

LETOUZEY, J., BIJU-DUVAL, B., DORKEL, A., GONNARD, R., KRISTCHEV, K., MONTADERT, L., AND SUNGURLU, O., 1977, The Black Sea: a marginal basin-Geophysical and geological data, in Biju-Duval, B. and Montadert, L., eds., Structural History of the Mediterranean Basins: Paris, Editions Technip, p. 363–376.

LOTZE, F., 1936, Zur Methodik der Forschungen über saxonische Tektonik: Geotektonische Forschungen, v. 1, p. 6–27.

LÜTTIG, G., AND STEFFENS, P., 1976, Explanatory notes for the paleogeographic atlas of Turkey from the Oligocene to the Pleistocene: Hannover, Bundesanstalt für Geowissenschaften und Rohstoffe, 64 p.

MANN, P., AND BRADLEY, D., 1984, Comparison of basin types in active and ancient strike-slip zones (abs.): American Association of Petroleum Geologists Bulletin, v. 68, p. 503.

MANN, P., HEMPTON, M. R., BRADLEY, D. C., AND BURKE, K., 1983, Development of pull-apart basins: Journal of Geology, v. 91, p. 529–554.

MCKENZIE, D. P., 1969, Speculations on the consequences and causes of plate motions: Geophysical Journal of the Royal Astronomical Society, v. 18, p. 1–32.

———, 1972, Active tectonics of the Mediterranean region: Geophysical Journal of the Royal Astronomical Society, v. 30, p. 109–185.

———, 1976, The East Anatolian Fault: A major structure in Eastern Turkey: Earth and Planetary Science Letters, v. 29, p. 189–193.

———, 1978, Active tectonics of the Alpine-Himalayan belt: The Aegean Sea and Surrounding regions: Geophysical Journal of the Royal Astronomical Society, v. 55, p. 217–254.

MCKENZIE, D. P., AND JACKSON, J., 1983, The relationship between strain rates, crustal thickening, palaeomagnetism, finite strain and fault movements within a deforming zone: Earth and Planetary Science Letters, v. 65, p. 182–202.

MEULENKAMP, J. E., 1982, On the pulsating evolution of the Mediterranean: Episodes, 1982, p. 13–16.

MOLNAR, P., AND TAPPONNIER, P., 1975, Cenozoic tectonics of Asia: effects of a continental collision: Science, v.189, p. 419–426.

MOODY, J. D., AND HILL, M. J., 1956, Wrench-fault tectonics: Geological Society of America Bulletin, v. 67, p. 1207–1246.

M.T.A. (Maden Tetkik ve Arama Enstitüsü), 1961–1964 Geological map of Turkey, scale 1/500.000, 18 sheets: Ankara, Maden Tetkik ve Arama Enstitüsü.

NAKAMURA, K., 1977, Volcanoes as possible indicators of tectonic stress orientation—principle and proposal: Journal of Volcanology and Geothermal Research, v. 2, p. 1–16.

NAKOMAN, E., 1968, Karlıova-Halifan linyitlerinin sporo-pollinik etüdleri (Sporo-palinologic study of the Karlıova-Halifan lignites): Türkiye Jeoloji Kurumu Bülteni, v. 11, p. 68–116.

NEBERT, K., 1978, Das braunkohlenführende Neogengebiet von Soma, West-anatolien: Bulletin of Mineral Research and Exploration Institute of Turkey, Ankara, v. 90, p. 20–72.

ÖCAL, N., 1958, 16 Temmuz 1955 Söke-Balat Zelzelesi (The Söke-Balat earthquake of July 16th, 1955): İstanbul Kandilli Rasathanesi Sismoloji Yayınları, n. 2, 8 p.

OCOLA, L. C., MEYER, R. P., AND ALDRICH, L. T., 1971, Gross crustal structure under Peru-Bolivia Altiplano: Earthquake Notes, v. 13, p. 33–48.

OKAY, A. I., 1984, The geology of the Agvanis metamorphic rocks and surrounding areas: Bulletin of Mineral Research and Exploration Institute of Turkey, Ankara, v. 99/100, p. 51–57.

OSWALD, F., 1912, Armenien: Handbuch der regionalen Geologie, v. 3, 40 p.

ÖZGÜL, N., SEYMEN, İ., AND ARPAT, E., 1983, 30 Ekim 1983 Horasan-Narman depreminin makrosismik ve tektonik özellikleri (Macroseismic and tectonic characteristics of the Horasan-Narman earthquake): Yeryuvarı ve İnsan, v. 8, p. 21–25.

PAREJAS, E., AND PAMIR, H. N., 1939, Le tremblement de terre du 19 avril 1938 en Anatolie centrale: İstanbul Üniversitesi Fen Fakültesi Mecmuası, v. 4, p. 183–193.

PHILIPPSON, A., 1910–1915, Reisen und Forschungen im westlichen Kleinasien: Ergänzungshefte 167, 172, 177, 180, 183 der Petermanns Mitteilungen, Gotha, Justus Perthes.

PINAR, N., 1953, La géologie du bassin d'Adana (Turquie) et le séisme du 22 Octobre 1952: Revue de la Faculte des Sciences de l'Universite d'Istanbul, sér. A., v. 18, p. 231–241.

DE PLANHOL, J., 1953, Les formes glaciaires du Sandras dag et la limite des negies éternelles quaternaires dans le SW de l'Anatolie: Comptes Rendus de la Société géologique de France, No. 13, p. 263–625.

READING, H. G., 1980, Characteristics and recognition of strike-slip fault systems, in Ballance, P. F., and Reading H.G., eds., Sedimentation in Oblique-Slip Mobile Zones: International Association of Sedimentologists Special Publication No. 4, p. 7–26.

RODGERS, D., 1980, Analysis of pull-apart basin development produced by en echelon strike-slip faults, in Ballance, P. F., and Reading, H. G., eds., Sedimentation in Oblique-Slip Mobile Zones: International Association of Sedimentologists Special Publication No. 4, p. 27–41.

RÖGL, F., AND STEININGER, F. F., 1983, Vom Zerfall der Tethys zu Mediterran und Paratethys: Annalen des Naturhistorischen Museums Wien, v. 85/A, p. 135–163.

ROTHERY, D. A., AND DRURY, S. A., 1984, The neotectonics of the Tibetan Plateau: Tectonics, v. 3, p. 19–26.

ROTSTEIN, Y., 1984, Counterclockwise rotation of the Anatolian Block: Tectonophysics, v. 108, p. 71–91.

ROYDEN, L. H., HORVATH, F., AND BURCHFIEL, B. C., 1982, Transform faulting, extension and subduction in the Carpathian Pannonian region: Geological Society of America Bulletin, v. 93, p. 717–725.

SANER, S., 1980, Batı Pontidler'in ve komşu havzaların oluşumlarının levha tektoniği kuramıyla açıklanması, Kuzeybatı Türkiye (Plate tectonic explanation of the origin of the western Pontides and neighboring basins): Maden Tetkik ve Arama Enstitüsü Dergisi, v. 93/94, p. 1–20.

ŞAROĞLU, F., AND GÜNER, Y., 1979, Tutak diri fayı, özellikleri ve Çaldıran fayı ile ilişkisi (The active Tutak fault, its characteristics and relations to the Çaldıran fault): Yeryuvarı ve İnsan, v. 4, p. 11–14.

ŞAROĞLU, F., AND GÜNER, Y., 1981, Doğu Anadolu'nun jeomorfolojik gelişimine etki eden öğeler, jeomorfoloji, tektonik, volkanizma ilişkileri (Factors affecting the geomorphological development of eastern Turkey—relations between geomorphology, tectonics, and volcanism): Türkiye Jeoloji Kurumu Bülteni, v. 24, p. 39–50.

SCHMIDT, G. C., 1961, Stratigraphic nomenclature for the Adana region petroleum district VII: Petroleum Administration Bulletin, Ankara, v. 6, p. 47–63.

SEIDLITZ, W., VON, 1931, Diskordanz und Orogenese der Gebirge am Mittelmeer: Berlin, Gebrüder Borntraeger, 651 p.

ŞENGÖR, A. M. C., 1978, Über die angebliche primäre vertikaltektonik im Ägäistraum: Neues Jahrbuch der Geologie und Paläontologie, Monatshefte, n. 11, p. 698–703.

———, 1979, The North Anatolian transform fault: its age, offset and tectonic significance: Journal of the Geological Society of London, v. 136, p. 269–282.

———, 1980, Türkiye'nin neotektoniğinin esasları (Fundamentals of the neotectonics of Turkey): Türkiye Jeoloji Kurumu Konferans Serisi, No. 2, 40 p.

——, 1982, Ege'nin neotektonik evrimini yöneten etkenler (Factors governing the neotectonic evolution of the Aegean): in Erol, O., and Oygür, V., eds., Batı Anadolu'nun genç tektoniği ve volkanizması paneli. Türkiye Jeoloji Kurumu, Ankara, p. 59–72.

——, 1984, The Cimmeride Orogenic System and the tectonics of Eurasia: Geological Society of America Special Paper 195, 82 p.

——, in press, Asia, in Fairbridge, W. R., ed., Encyclopaedia of World Regional Geology, v. 2 (Eastern Hemisphere): Stroudsburg, Pennsylvania, Dowden, Hutchinson and Ross.

ŞENGÖR, A. M. C., AND CANITEZ, N., 1982, The North Anatolian Fault: Alpine Mediterranean Geodynamics: American Geophysical Union Geodynamics Series, v. 7., p. 205–216.

ŞENGÖR, A. M. C., AND DYER, J., 1979, Neotectonic provinces of the Tethyan orogenic belt of the eastern Mediterranean: variations in tectonic style and magmatism in a collision zone (abs.): EOS, v. 60, p. 390.

ŞENGÖR, A. M. C., AND KIDD, W. S. F., 1979, Post-collisional tectonics of the Turkish-Iranian Plateau and a comparison with Tibet: Tectonophysics, v. 55, p. 361–376.

ŞENGÖR, A. M. C., AND YILMAZ, Y., 1981, Tethyan evolution of Turkey: a plate tectonic approach: Tectonophysics, v. 75, p. 181–241.

ŞENGÖR, A. M. C., BURKE, K., AND DEWEY, J. F., 1978, Rifts at high angles to orogenic belts: Tests for their origin and the upper Rhine Graben as an example: American Journal of Science, v. 278, p. 24–40.

ŞENGÖR, A. M. C., BURKE, K., AND DEWEY, J. F., 1982, Tectonics of the North Anatolian Transform Fault, in Işıkara, A. M., and Vogel, A., eds., Multidisciplinary approach to Earthquake Prediction: Wiesbaden, Friedrich, Vieweg and Sohn, p. 3–22.

ŞENGÖR, A. M. C., BÜYÜKAŞIKOĞLU, S., AND CANITEZ, N., 1983, Neotectonics of the Pontides: implications for 'incompatible' structures along the North Anatolian fault: Journal of Structural Geology, v. 5, p. 211–216.

ŞENGÖR, A. M. C., SATIR, M., AND AKKÖK, R., 1984, Timing of tectonic events in the Menderes Massif, western Turkey: Implications for tectonic evolution and evidence for Pan-African basement in Turkey: Tectonics, v. 3, p. 693–707.

SEYMEN, İ., 1975, Kelkit vadisi kesiminde Kuzey Anadolu Fay Zonunun tektonik özelliği (Tectonic characteristics of the North Anatolian Fault zone in the Kelkit Valley segment): İstanbul Teknik Üniversitesi Maden Fakültesi Yayınları, İstanbul, 198 p.

SEYMEN, İ., AND AYDIN, A., 1972, The Bingöl Earthquake fault and its relation to the North Anatolian Fault Zone: Bulletin of Mineral Research and Exploration Institute of Turkey, Ankara, v. 79, p. 1–8.

SIDORENKO, A. V., 1978, Karta razlomov territorii SSSR i sopredelnich stran, 1/2,500,000: Akademi Nauk, SSSR, Moscow.

SİPAHİOĞLU, S., 1984, Kuzey Anadolu Fay Zonunun Yapısı, Abant-Dokurcun arasındaki yerleşimi ve fizyografik-jeomorfolojik özellikleri (The structure of the North Anatolian fault zone, its setting between Abant and Dokurcun and physiographic-geomorphological characteristics): İstanbul Üniversitesi Mühendislik Fakültesi Jeofizik Mühendisliği Bölümü, 50 p.

SUESS, E., 1885, Das Antlitz der Erde: Vienna, Tempsky, 778 p.

STEWART, J. H., 1983, Extensional tectonics in the Death Valley area, California: Transport of the Panamint Range structural block 80 km northwestward: Geology, v. 11, p. 153–157.

SÜR, Ö., 1964, Pasinler ovası ve çevresinin jeomorfolojisi (Geomorphology of the Pasinler Plain and its surroundings): Ankara Üniversitesi Yayınları No. 154, 93 p.

TAPPONNIER, P., AND MOLNAR, P., 1976, Slip-line field theory and large-scale continental tectonics: Nature, v. 264, p. 319–324.

TAPPONNIER, P., PELTZER, G., AND ARMIJO, R., in press, On the mechanics of the collision between India and Asia: Geological Society of London Special Publication on Collision Tectonics.

TAPPONNIER, P., LEDAIN, A. Y., ARMIJO, R., AND COBBOLD, P., 1982, Propagating extrusion tectonics in Asia: new insights with experiments with plasticine: Geology, v. 10, p. 611–616.

TAPPONNIER, P., MERCIER, J. L., ARMIJO, R., TONGLIN, H., AND JI, Z., 1981, Field evidence for active normal faulting in Tibet: Nature, v. 294, p. 410–413.

TATAR, Y., 1975, Tectonic structures along the North Anatolian fault zone, north-east of Refahiye (Erzincan): Tectonophysics, v. 29, p. 401–410.

——, 1978, Kuzey Anadolu Fay Zonu'nun Erzincan-Refahiye arasındaki bölümü üzerinde tektonik incelemeler (Tectonic studies on the Erzincan-Refahiye segment of the North Anatolian Fault): Hacettepe Yerbilimleri, v. 1–2, p. 201–236.

TCHALENKO, J. S., 1977, A reconnaissance of the seismicity and tectonics at the northern border of the Arabian Plate (Lake Van region): Revue de géographie physique et de géologie dynamique, v. 19, p. 189–208.

TERNEK, Z., 1953, Mersin-Tarsus kuzey bölgesinin jeolojisi (Geology of the northern sector of the Mersin-Tarsus region): Maden Tetkik ve Arama enstitüsü Dergisi, Ankara, v. 44/45, p. 18–62.

——, 1957, The Lower Miocene (Burdigalian) formations of the Adana Basin, their relations with other formations, and oil possibilities: Bulletin of Mineral Research and Exploration Institute of Turkey, Ankara, v. 49, p. 60–80.

TOKAY, M., 1973, Kuzey Anadolu Fay Zonunun Gerede ile Ilgaz arasındaki kısmında jeolojik gözlemler (Geological observations on the North Anatolian fault zone between Gerede and Ilgaz): in Kuzey Anadolu Fayı ve Deprem Kuşağı Simpozyumu, Maden Tetkik ve Arama Enstitüsü, Ankara, p. 12–29.

TOKSÖZ, M. N., ARPAT, E., AND ŞAROĞLU, F., 1977, East Anatolian earthquake of 24 November 1976: Nature, v. 270, p. 423–425.

TOKSÖZ, M. N., GUENETTE, M., GÜLEN, L., KEOUGH, G., PULLI, J. J., SAV, H., AND OLGUNER, A., 1983, Narman-Horasan depresiminin kaynak mekanizması (Source mechanism of the Narman-Horasan earthquake): Yeryuvarı ve İnsan, v. 8, p. 47–52.

ÜÇER, B., EVANS, J. R., AND STAFF OF KANDİLLİ OBSERVATORY, 1981, Epicentres in western Turkey: Institute of Geological Sciences, Global Seismology Unit, Report No. 152, 131 p.

VINOGRADOV, A. P. (editor-in-chief), 1967, Atlas of the Lithological-Paleogeographical Maps of the USSR, Grossheim, V. A., and Khain, V. E., eds.: Academy of Sciences of the USSR, Moscow, v. 4.

WACHENDORF, H., GRALLA, P., KOLL, J., AND SCHUTZE, I., 1980, Gedynamik des mittelkretischen Deckenstapels (nördliches Dikti-Gebirge): Geotektonische Forschungen, v. 59, 72 p.

WERNICKE, B., 1981, Low-angle normal faults in the Basin and Range province: nappe tectonics in an extending orogen: Nature, v. 291, p. 645–648.

WILSON, J. T., 1963, Hypothesis of Earth's behaviour: Nature, v. 198, p. 925–929.

——, 1965, A new class of faults and their bearing on continental drift: Nature, v. 207, p. 343–347.

WONG, H. K., DEGENS, E. T., AND FINCKH, P., 1978, Structures in Modern Lake Van sediments as revealed by 3.5 KHz high resolution profiling, in Degens, E. T., and Kurtman, F., eds., The Geology of Lake Van: Maden Tetkik ve Arama Enstitüsü Yayını, Ankara, No. 169, p. 11–19.

YALÇIN, M. N., AND GÖRÜR, N., 1984, Sedimentological evolution of the Adana Basin, in Tekeli, O., and Göncüoğlu, M. C., eds., Geology of the Taurus Belt, Ankara, MTA, p. 165–172.

YEATS, R. S., 1981, Quarternary flake tectonics of the California Transverse Ranges: Geology, v. 9, p. 16–20.

——, 1983, Large-scale Quarternary detachments in Ventura Basin, southern California: Journal of Geophysical Research, v. 88, p. 569–583.

YILMAZ, Y., GÖZÜBOL, A. M., AND TÜYSÜZ, O., 1982, Geology of an area in and around the Northern Anatolian Transform Fault Zone between Bolu and Akyazı, in Işıkara, A. M., and Vogel, A., eds., Multidisciplinary approach to Earthquake Prediction: Wiesbaden, Friedrich Vieweg and Sohn, p. 45–66.

ZESCHKE, G., 1954, Der Simav-Graben und Seine Gesteine: Türkiye Jeoloji Kurumu Bülteni, v. 5, p. 179–189.

PERICOLLISIONAL STRIKE-SLIP FAULTS AND SYNOROGENIC BASINS, CANADIAN CORDILLERA

G. H. EISBACHER
Geologisches Institut, Universität Karlsruhe,
Kaiserstrasse 12, 7500 Karlsruhe 1,
Federal Republic of Germany

ABSTRACT: During large-scale convergence of oceanic and arc-type lithospheric fragments towards a cratonic promontory along western North America from Middle Jurassic through Paleogene time, non-subductable crust of the approaching Pacific realm was deflected dextrally northward or sinistrally southward from this 'reverse indenter' in the California-Nevada region. Paleontologic and paleomagnetic data suggest oblique dextral displacements on the order of 1,500 to 2,000 km for the accreted terranes in the western Cordillera of Canada. These dextral displacements were first concentrated along closing sutures (from Middle Jurassic to Early Cretaceous time); later they were also taken up by pericollisional fault zones, which propagated into the western parts of the Cordilleran thrust belt and involved the Coast Plutonic Complex (mid-Cretaceous to Paleocene).

A-subduction in the thrust belt and inferred B-subduction west of the Coast Plutonic Complex were thus accompanied by dextral displacements within the Omineca and Coast fault arrays respectively, imparting northwest-directed stretching fabrics onto ductile metamorphic or igneous rocks, and discrete fault strands on high-level crustal rocks.

The convergent strike-slip fault motions in the Canadian Cordillera created mainly sedimentary source areas rather than subsiding basins. Pericollisional basins that did receive clastic materials from zones of oblique convergence were (1) marginal basins in the process of closing, (2) relict or tectonically overloaded depressions on accreting terranes, (3) foreland basins created by thrust propagation in the miogeoclinal succession, and (4) small pull-apart or restraining bend depressions near high-angle strike-slip faults.

Basins in the accreted terrane complexes west of the Cordilleran thrust belt received most of their detrital material from exposed volcanic, plutonic, and oceanic sedimentary rocks; the predominantly turbiditic basin fill suffered repeated deformation, high sustained heat flow, and intrusive activity. The foreland basin to the east of the thrust belt, on the other hand, received most of its detrital input from carbonate and quartz-rich clastic rocks of the miogeocline and metamorphosed equivalents; the predominantly shallow-water clastic deposits of the foreland basin experienced considerably less deformation and thermal alteration than the varied sedimentary assemblages of the accreted belt.

INTRODUCTION

A system of lithospheric plates moving on the surface of the earth implies the existence of zones along which substantial transcurrent displacement occurs. Strike-slip faults generally develop along transform plate boundaries connecting zones of crustal divergence with zones of crustal convergence and in areas where thickness, geometry, and ductility variations in the lithosphere prevent pure orthogonal plate convergence or divergence. Thus strike-slip faults are rarely simple, and rarely accommodate just strike slip.

The presence of strike-slip faulting along subsiding sedimentary basins has been well documented along young transform faults near continental margins and extending intraplate settings such as those of the western United States, the Middle East, and New Zealand (Burchfiel and Stewart, 1966; Crowell, 1974; Ballance and Reading, 1980; Aydin and Nur, 1982; Mann et al., 1983). In these tectonic environments the concept of pull-apart basins has been documented to the fullest. In contrast, the role of strike-slip faults with respect to basin evolution in broadly convergent crustal regimes has attracted considerably less attention. This is not surprising: along accretionary sedimentary wedges of subduction zones or in the cratonic foreland of thrust belts, the combined effects of crustal thickening, thrust-sheet loading, and thrust progradation may be so dominant that the possible significance of transcurrent displacements is easily overlooked. Early-developed strike-slip faults can be rotated, overprinted, disrupted, or modified beyond recognition during subsequent crustal convergence; they may also be overridden by younger thrust faults. The influence of early strike-slip faults on basin geometry thus could be fundamental but nevertheless would be extremely difficult to document once the synorogenic clastic wedges have been deformed by continuing convergence. However, evidence is becoming abundant for strike-slip faulting along recently active convergent plate margins such as those of the Indonesian-West Pacific regions where transform, intraplate, and subduction-related oblique motion occurs along and across sedimentary troughs (Mitchell and McKerrow, 1975; Curray et al., 1978; Hamilton, 1979; Taponnier et al., 1982; Karig, 1983). Short-term variations in the direction of displacement vectors can induce dramatic changes in the sense of displacement on pre-existing major fault zones, and as a consequence, the regional fault pattern can be expected to include partly active, curved, and obliterated fault strands (Dewey, 1980). Tectonic relief, sedimentary source areas, and depositional sinks vary greatly along trend and only a few unique criteria besides the extreme variability of the sedimentary successions can be derived from these recent settings and applied unequivocally to the study of older orogenic belts (Ballance, 1980; Mann et al., 1983). The problem is compounded in ancient orogenic belts created by oblique convergence. There, strike-slip motion may be apparent from the fabric of mylonitic rocks (Nicolas et al., 1977). However, such areas have often been dramatically uplifted and eroded; the relationship between crustal shear belts and nearby synorogenic basins is thus generally tenuous.

Ongoing work in the Canadian Cordillera indicates that much of the late Mesozoic-Paleogene orogenic evolution was controlled intricately by major dextral accretion of crustal blocks along the western edge of the North American craton. Syntectonic sedimentary basins, major tectonic structures, and metamorphic-plutonic suites are sufficiently well preserved to test how strike-slip faults affected the basin development along the dextrally accreting crustal terranes and in the orogenic foreland. This requires first a step-by-step analysis of the oblique convergence as reflected in the basin fill and tectonic structures adjacent to the basins, and second a model of the crustal dynamics that contributed to this process.

OBLIQUE CONVERGENCE IN THE CANADIAN CORDILLERA

The Canadian Cordillera is the central segment of the North American Cordillera. Although there are similarities between the tectonic elements of the Canadian Cordillera and those of the adjacent United States, several tectonic-stratigraphic peculiarities impart a distinct identity to the Canadian segment. These elements are best discussed by proceeding from east to west.

The Cordilleran foreland is composed of Paleozoic and Mesozoic platform deposits that overlie the Precambrian crystalline basement rocks of the North American craton. Basement dips gently westward underneath the deformed Mesozoic, Paleozoic, and upper Proterozoic miogeoclinal successions of the Cordilleran thrust belt (Fig. 1). In the westernmost parts of the thrust belt, the total inferred thickness of the sedimentary prism is in excess of 10 to 15 km. Progressively older successions are exposed in east-verging folds or thrust sheets. Thrust faults merge with gently west-dipping master detachments above and eventually within the cratonic basement. Thrust sheets composed of mechanically competent Paleozoic carbonates also involve the synorogenic clastic deposits of the foreland. Tectonic foreshortening of the miogeoclinal strata by thrust faulting and folding attains its greatest value (100 to about 200 km) in the southern segment of the Canadian thrust belt; it decreases to about half this value (50 to 100 km) farther to the north (Bally et al., 1966; Roeder, 1967; Price, 1981; Thompson, 1981). In the western part of the thrust belt, Paleozoic miogeoclinal carbonate successions change facies into fine-grained slope or basin deposits. Here, both Paleozoic and Proterozoic rocks are exposed along the leading edge of thrust plates, but are more commonly exposed in the cores of tightly folded anticlines, broad synclines, and along oblique tears. The clastic successions generally display penetrative or spaced cleavage and the deeper subsurface structures are increasingly enigmatic towards the Rocky Mountain Trench, a major northwest-trending physiographic depression within the westernmost part of the thrust belt. Near the Rocky Mountain Trench, Proterozoic miogeoclinal sedimentary rocks first display high-grade regional metamorphism (kyanite-sillimanite grade) and slices of Precambrian crystalline basement re-emerge at the surface (Campbell et al., 1973; Simony et al., 1980; Evenchick, 1984; Gabrielse, 1985). The Rocky Mountain Trench also marks the transition from mainly northeast-verging folds and thrust faults to a mainly polyphase deformed, high-grade metamorphic orogenic core zone involving sedimentary rocks of the outer cratonic margin and the Precambrian gneissic

basement (Evenchick et al., 1984). A large part of the metamorphic core zone of the thrust belt is characterized by large-scale west-verging folds, which have been overprinted later by east-directed thrusts and high-angle faults (Brown, 1978; Mansy, 1980). The complex northwest-trending zone of structural divergence resulting from this overprint includes major fan-like anticlinoria that localized subsequent strands of strike-slip fault systems. Regional metamorphism and deformation dated by syn-kinematic to post-kinematic plutons and metamorphic minerals, occurred between about 170 and 140 Ma and again between about 70 and 40 Ma (Pigage, 1977; Parrish, 1979; Archibald et al., 1983). The western border of the metamorphic parts of the thrust belt is a major tectonic contact, the Teslin Suture. It separates deformed autochthonous and parautochthonous North American crust from crustal slices of displaced terranes that accreted against it (Monger and Price, 1979; Tempelman-Kluit, 1979).

Prior to the mid-Jurassic deformation that led to the eventual closure of the Teslin Suture, slivers of upper Paleozoic to lower Mesozoic chert, basinal clastics, mafic volcanics, granitoids, and ultramafic rocks were emplaced eastward over rocks of the miogeocline. These rocks, possibly remnants of a marginal basin, are now preserved as flat-lying or intensely deformed klippen on miogeoclinal successions east of the Teslin Suture ('Slide Mountain Terrane' in Fig. 1). The Teslin Suture itself is an extremely complex zone that locally was disrupted by later strike-slip faulting and partly obliterated by intrusion of granitoid plutons. In some areas (e.g., Yukon cataclastic belt), the Teslin Suture is a belt several tens of kilometers wide that consists of cataclastic rocks including autochthonous, allochthonous, and exotic successions (D. J. Tempelman-Kluit, 1979, and personal commun., 1984). In the region west of the Teslin Suture, but east of the Coast Plutonic Complex, the Cordillera is underlain by distinctly elongate upper Paleozoic to mid-Mesozoic crustal blocks, which presumably amalgamated with each other before they accreted and suffered further deformation along the western margin of North America (Monger et al., 1982). These blocks have been termed, from east to west, the 'Quesnel', 'Cache Creek', and 'Stikine' Terranes, and are shown as such in Figures 1 and 2. Along with the Slide Mountain Terrane, they are referred to collectively as the Intermontane terranes.

The Quesnel Terrane, west of the Teslin Suture, contains mainly lower Mesozoic volcanic rocks and fine-grained basinal deposits. The oceanic Cache Creek Terrane, in fault contact with the Quesnel Terrane, consists of upper Paleozoic-Triassic chert, volcanic rocks, carbonates, ophiolitic melange, plus the partly superimposed Upper Triassic to Lower Jurassic flysch facies of the Whitehorse Trough. The Stikine Terrane represents mainly a calcalkalic volcano-plutonic arc succession of early Mesozoic age superimposed on an older crustal block; radiometric and paleontologic ages indicate that arc-related magmatic activity lasted from late Triassic to middle Jurassic time (220 to 170 Ma). The coeval flysch deposits of the Whitehorse Trough, which overlie both Stikine and Cache Creek rocks, probably represent forearc deposits that are part of a subduction complex involving the subjacent Cache Creek, Quesnel, and

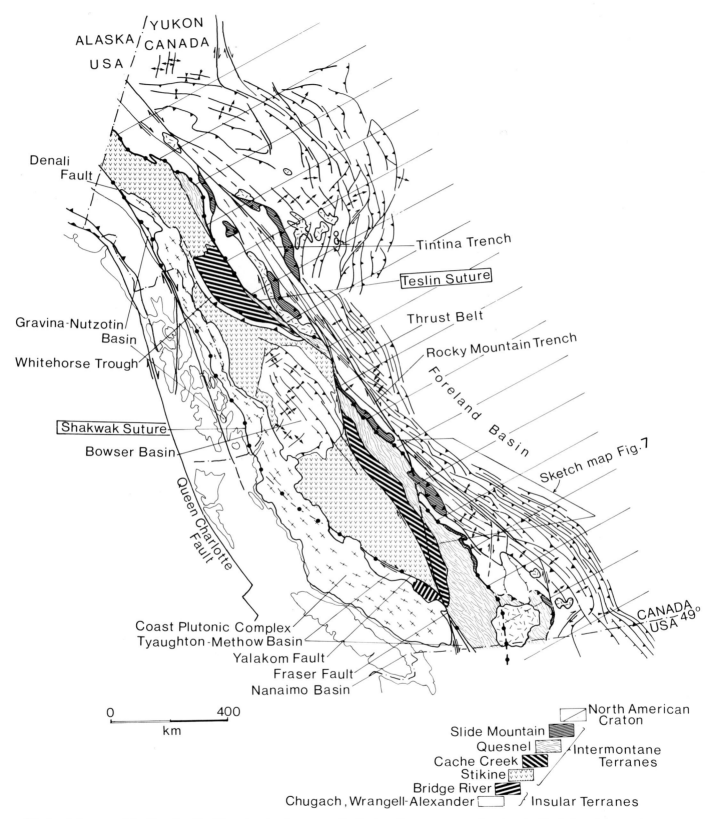

FIG. 1.—Index map of the Canadian Cordillera showing the present distribution of terranes accreted against the North American craton and major structures created within the thrust belt during late Mesozoic-Paleogene dextral convergence. The Teslin Suture separates the composite Intermontane terrane complex from the thrust belt; the Shakwak Suture separates the Intermontane terrane complex from the Insular terrane complex.

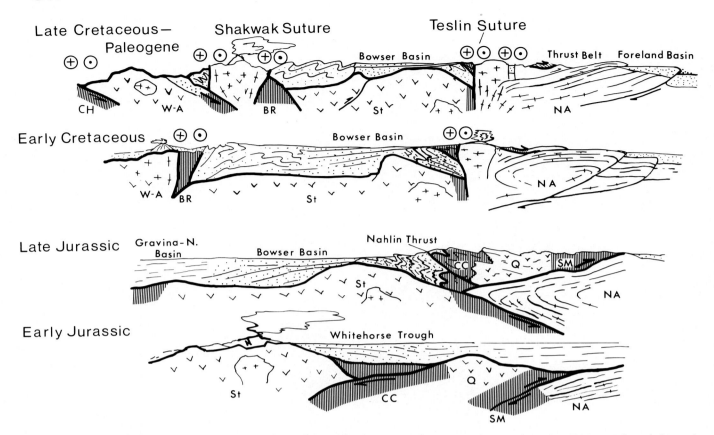

FIG. 2.—Schematic evolution of the pericollisional clastic basins and strike-slip faults in a southwest-northeast oriented transect through the north-central Canadian Cordillera. Vertical ruling indicates oceanic-type terranes and ophiolitic melanges (**SM**, Slide Mountain Terrane; **CC**, Cache Creek Terrane; **BR**, Bridge River Terrane; **CH**, Chugach Terrane). V-pattern indicates island arcs and oceanic plateaus (**Q**, Quesnel Terrane; **St**, Stikine Terrane; **W-A**, Wrangell-Alexander Terrane). **NA**, North American craton. Note that the Cretaceous-Paleogene dextral strike-slip faults developed along zones of complex crustal convergence characterized by multiple changes in structural vergence and subsequent plutonic activity.

Slide Mountain Terranes prior to their emplacement against North America (Tempelman-Kluit, 1979). Locally, slivers of this west- to southwest-dipping subduction complex display high-pressure low-temperature metamorphic mineral assemblages characteristic of subduction zones (Paterson, 1977; Erdmer and Helmstaedt, 1983). However, the eventual closure of the Whitehorse Trough and Teslin Suture does not seem to have been related to westward subduction, but to a new regime of crustal convergence (Eisbacher, 1974, 1981). Beginning in Middle Jurassic time, oblique convergence of the Intermontane terranes produced southwest-verging folds and thrust faults in both the Whitehorse Trough and its substratum; both emerged along a major southwest-directed thrust, the Nahlin thrust (Fig. 2). The thrust mass overrode the arc-derived flysch and probably also parts of the arc itself (Monger, 1977). This oceanic thrust plate overloaded the crust of the approaching arc, and as a consequence, a massive clastic wedge rich in chert and volcanic rock fragments prograded over the depressed Stikine Terrane forming the Bowser Basin molasse (Fig. 2). By the end of Bajocian time (about 176 Ma) all but traces of vol-

canic activity had ceased in the northern part of the Stikine Terrane (Tipper and Richards, 1976; Gabrielse et al., 1980; R. Anderson, personal commun., 1984). Clastic progradation from the upthrust oceanic Cache Creek Terrane persisted for at least 30 m.y., although detritus from the deformed and metamorphosed thrust belt did not yet reach the Bowser Basin.

On the west, the Intermontane terranes are bordered by another major but now partly cryptic suture zone, here termed the 'Shakwak Suture'. The Shakwak Suture marks a belt along which crustal units of the composite Paleozoic to Mesozoic Wrangell-and-Alexander Terrane and others, collectively referred to as the Insular terranes, impinged against the Intermontane terranes. Owing to widespread obliteration of this suture by the subsequent emplacement of the Coast Plutonic Complex, the tectonic history of the collision is not entirely clear. However, stratigraphic-structural evidence west and east of the Coast Plutonic Complex indicates that collision was preceded by closure of a marginal flysch trough partly underlain by oceanic crust (Berg et al., 1972; Eisbacher, 1976; Davis et al., 1978). It appears that the suture closed by dextral oblique movement in late Cretaceous time (about 70 to 60 Ma), but that intense heating, igneous activity, and strike-slip deformation continued along the entire length of the belt (Roddick and Hutchison, 1974; Eisbacher, 1976; Nokleberg et al., in press).

The emplacement of large quantities of quartz diorite and granodiorite in the Coast Plutonic Complex was probably the result of sustained easterly to northeasterly subduction of Pacific oceanic crust, first beneath the approaching Insular terranes and then beneath the entire accreted crustal assemblage of the western Cordillera. Radiometric ages in the Coast Plutonic Complex decrease in general from west to east and range between about 130 to 45 Ma; however, the bulk of the ages falls between 75 and 45 Ma (Roddick and Hutchison, 1974). Remnants of a Cretaceous-Paleogene sedimentary trench-forearc assemblage have been identified along the west coast of Canada and in the Chugach Terrane of southeast Alaska (Muller, 1977; Nilsen and Zuffa, 1982). Much of the Coast Plutonic Complex bears evidence of intermittent oblique shear. Pre-late Mesozoic crystalline basement and volcanic rocks, late Mesozoic clastic basins, and ultramafic rock bodies are dismembered into slices with northwesterly structural trends. Most late Mesozoic intrusive rocks of the Coast Plutonic Complex display a regional northwest-striking foliation and many show penetrative northwest-trending mineral lineations. However, discrete displacements have been documented only for young fault strands.

During Tertiary time, active convergence along the Pacific margin gradually shifted northwestward and recent deformation seems to be limited to the dextral Queen Charlotte fault, which connects spreading centers of the Juan de Fuca ridge in the south with a region of lithospheric subduction and crustal convergence in southern Alaska.

Paleontologic and paleomagnetic data collected in the accreted crustal blocks west of the Teslin Suture have further elucidated the evidence for large transcurrent motions during and after accretion. Exotic faunas incompatible with faunal assemblages found in coeval upper Paleozoic to lower Mesozoic rocks on the North American craton seem to indicate large relative displacements postdating the age of these rocks (Monger and Ross, 1971; Tipper, 1981; Tozer, 1982). Paleomagnetic studies of igneous rocks west of the Teslin Suture indicate that major motion of the displaced terranes must have been northward with respect to North America with a strong component parallel to the trend of the inferred sutures: a dextral post-Early Cretaceous shift of about 13° to 20° latitude has been proposed for the displacement of the Stikine Terrane, and northward motion of as much as 18° is inferred for rocks of the Wrangell Terrane (Jones et al., 1977; Irving et al., 1980). The southern part of the Coast Plutonic Complex may have shifted between 10° and 20° northward (E. Irving, personal commun., 1984). Substantial rotation of smaller pieces, possibly accompanied by shear within Intermontane terranes has also been implied from the paleomagnetic data (Irving, 1983). With respect to the North American craton, dextral displacements in excess of 1,000 to 1,500 km therefore have to be considered for the accreted terranes of the western Cordillera. The 100 to 200 km of crustal shortening registered in the deformed

rocks of the thrust belt thus represent only the dip-slip vector of convergence within a part of the outer cratonic margin—one order of magnitude smaller than the strike-slip vector during oblique convergence. Considerable overlap of strike-slip and dip-slip motion thus is to be expected along the closing Teslin Suture and within the higher-grade metamorphic domains of the western Thrust Belt. As the dextral oblique motion probably occurred contemporaneous with B-subduction along the convergent Pacific margin and A-subduction along the thrust belt, the steeply dipping subcrustal zone of mantle convergence could have played a major role in the initial development of the dextral shear pattern (Roeder, 1973). The shear between the two subduction zones seems to have been concentrated mainly near the collision belts fringing the Shakwak and Teslin Sutures. For ease of discussion, the two pericollisional fault arrays are referred to below as the Coast array (on the west) and the Omineca array (on the east).

PERICOLLISIONAL FAULT ARRAYS

The foregoing discussion illustrates that young and physiographically prominent fault strands might not necessarily be those with the largest cumulative displacements. During closure of crustal sutures, strike-slip faulting probably first occurred *along* the sutures and then spread into adjacent metamorphic-plutonic belts underlain by correspondingly weak crust. Displacements along a suture would be difficult to detect, simply because by definition there are few matching linear geologic features on either side of it, and early strike-slip faults may have been overthrust, rotated, and metamorphosed.

In the Canadian Cordillera, major discrete strike-slip faults were first recognized along the physiographic trenches of the *Omineca fault array*. Along the Tintina Trench, Roddick (1967) and Tempelman-Kluit (1971, 1979) suggested post-Early Cretaceous dextral offsets on the order of 400 to 500 km, a value derived from apparent strike separation of upper Proterozoic miogeoclinal clastic rocks, of allochthonous slices resting on miogeoclinal rocks, and of lower Cretaceous quartzites cut by the Tintina fault zone. Extensive regional mapping by Gabrielse (1985) along the northern Rocky Mountain Trench led to his proposal of dextral displacements ranging from about 750 km to possibly more than 900 km along the southern extension of the Tintina fault; in addition, some 300 km of dextral offsets may occur along fault strands splaying westward off the northern Rocky Mountain Trench. The estimates by Gabrielse (1985) are based on separation of lower Paleozoic shelf-slope transitions and displaced granitic rocks. Fault movements along both the Tintina and northern Rocky Mountain trenches have been bracketed broadly between mid-Jurassic and Paleogene time. However, clastic rocks found along fault strands near the physiographic trenches seem to be mainly mid-Cretaceous to Paleogene in age (Hughes and Long, 1980; Long, 1981).

In the *Coast fault array* the most prominent fault are the Denali fault and the Yalakom-Fraser fault. Although the Denali fault shows dextral offsets on the order of 350 to 400 km that were achieved mainly in Eocene time, in north-central Alaska the fault is still active; however, cumulative displacement in post-Eocene time appears to have been on the order of only a few tens of kilometers (Forbes et al., 1973; Eisbacher, 1976; Lanphere, 1978; Nokleberg et al., in press). The estimate of Paleogene displacement is based on apparent strike separation of the linear northern Coast Plutonic Complex and associated metamorphic rocks, and a possible match-up of disrupted parts of the Upper Jurassic-Lower Cretaceous Gravina-Nutzotin flysch belt. The southward continuation of the Denali fault into southeast

Alaska along the west side of the Coast Plutonic Complex is not entirely clear.

Along the Yalakom fault on the east side of the Coast Plutonic Complex, late Cretaceous displacements on the order of 150 to 160 km have been inferred for offset Mesozoic rock units. The Yalakom fault itself seems to have been offset by about 110 km in Paleogene time by the Fraser fault (Kleinspehn, 1982). This composite value of about 270 km of dextral shear is probably a minimum value. Other faults with possibly considerable dextral displacements are

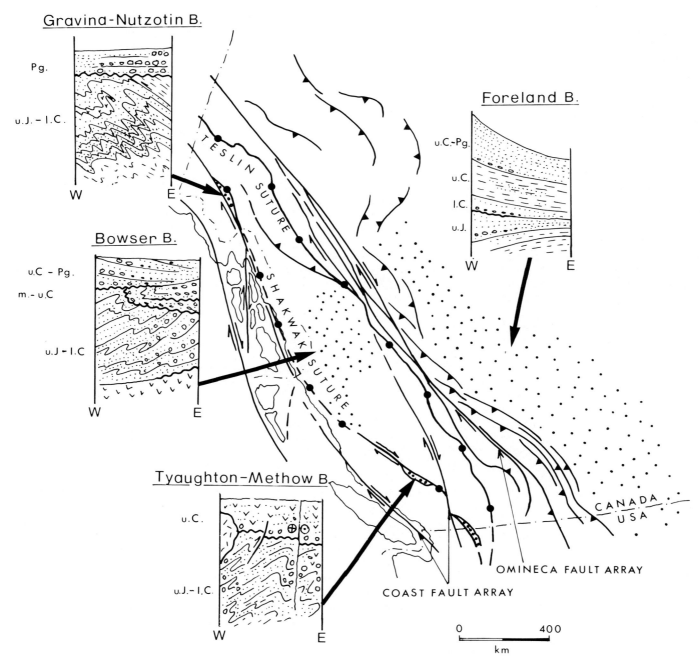

FIG. 3.—Location and structural style (schematic) of the major pericollisional basins in relation to sutures and dextral fault arrays of the Canadian Cordillera. Wavy lines in the insets indicate significant unconformities. **u.J.,** Upper Jurassic; **l.C., m.C.,** and **u.C.** are Lower, mid-, and Upper Cretaceous, respectively; **Pg.,** Paleogene.

probably hidden in the poorly exposed region between the Coast and Omineca fault arrays or have been obliterated by intrusion of the Coast Plutonic Complex.

PERICOLLISIONAL SEDIMENTARY BASINS

During oblique convergence in late Mesozoic to Paleogene time, several basins received clastic input reflecting the tectonic evolution of the adjacent regions: the large foreland basin east of the thrust belt; the pericollisional strike-slip basins of the Omineca fault array east of the Teslin Suture; the pericollisional Bowser Basin west of the Teslin Suture; the severely disrupted basins of the Coast fault array; and the poorly preserved forearc basins along the Pacific rim. Figure 3 shows the location of the major sedimentary basins that were filled and deformed during late Mesozoic-Paleogene accretion.

In general, the basins west of the Teslin Suture are filled with erosional products derived from volcanic-magmatic terranes trapped behind a gradually approaching B-subduction complex. The modal composition of the detrital sediments is predominantly subquartzose and dominated by volcanic detritus; this modal composition and high heat flow led to extensive alteration and cementation of sandstones (e.g., Read and Eisbacher, 1974). In contrast, the basins to the east of the Teslin Suture were filled predominantly with quartzose clastic sediments derived from the rising thrust belt and its sialic metamorphic infrastructure. This compositional distinction between the basins, however, becomes more blurred in the youngest basin fill.

Using the dextral displacements inferred from paleomagnetic data for a semiquantitative paleogeographic restoration of the synorogenic basins, a tectonic evolution can be reconstructed in terms of three broadly bracketed time intervals: (1) middle Jurassic to early Cretaceous; (2) mid-Cretaceous to Paleocene; and (3) Eocene.

MIDDLE JURASSIC-EARLY CRETACEOUS (170 TO 120 MA)

In middle and late Jurassic time, the Intermontane terranes converged against the North American craton and created major compressional structures on either side of the closing Teslin Suture (Fig. 4). Petrologic, structural, and radiometric data from the metamorphosed miogeoclinal succession east of the Teslin Suture indicate that deeply buried core-zone rocks of the thrust belt experienced rapid cooling between about 165 and 140 Ma, with uplift amounting locally to as much as 8 to 10 km. In the western foreland basin, prograding upper Jurassic quartzose clastic sediments containing detrital muscovite and volcanic pebbles were derived by unroofing of rising metamorphic culminations (M. McMechan, personal commun., 1984). In general, the uppermost Jurassic to lowermost Cretaceous clastic rocks of the foreland basin were derived from miogeoclinal rocks rich in quartz and chert and were deposited as alluvial, deltaic, and turbiditic successions locally more than 600 m thick in a northwest-plunging basin that subsided in front of the incipient thrust belt (Hamblin and Walker, 1979; Poulton, in press).

West of the Teslin Suture, relict oceanic sediments and the crustal substratum of the Whitehorse Trough were thrust

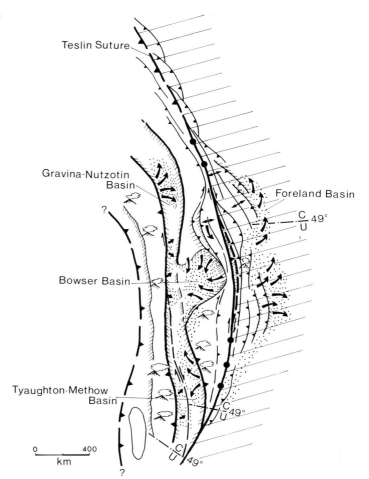

FIG. 4.—Paleogeographic model of the Middle Jurassic to Early Cretaceous setting of the clastic basins near the Teslin Suture. Diagonal ruling indicates the extent of North American continental crust locally overlapped in the west by ophiolitic thrust masses of the closing Teslin Suture, and characterized by uplift of metamorphic core zones. The accreted terranes west of the thrust belt are shown approximately 1,500 to 2,000 km south of their present position with respect to North America (49th parallel is used as displaced marker line). Strike-slip faulting and arc polarity west of the marginal Gravina-Nutzotin-Bowser-Tyaughton-Methow troughs are speculative. Areas of volcanic activity and generalized sediment dispersal are shown by cones and arrows, respectively.

southwestward over the extinct Stikine arc along the Nahlin thrust and subsidiary faults (Fig. 2). Thus overloaded by the thrust plate and possibly stretched along north- to northeast-striking faults, the Stikine Terrane subsided. An enormous quantity of chert and volcanic rock fragments was eroded from the thrust plate and carried towards the Bowser Basin. From Bathonian to earliest Cretaceous time, some 3,000 m of river-dominated deltaic deposits, prodelta shales, and turbidite fans accumulated in the basin (Eisbacher, 1981; Bustin and Moffat, 1983). However, no quartzose clastic sediments derived from east of the Teslin Suture seem to have reached the Bowser Basin until about Hauterivian time (about 125 Ma). For some 30 to 40 m.y., an elevated and tectonically active range composed of quartz-poor Intermontane rock types must have blocked retrogressive expansion of the Bowser drainage towards miogeoclinal realms located east of the Teslin Suture. It is also possible that residual elongate depressions (strike-slip fault valleys?) along the Teslin Suture trapped westward-transported clastic debris from the rising metamorphic core zones. The fill of these linear depressions would have been largely removed by later erosion. Scant evidence for such a scenario comes from scattered outcrops of chert-pebble conglomerates and related fine-grained deposits with abundant mica and quartz near the Teslin Suture. These coal-bearing nonmarine to paralic deposits are locally more than 1,000 m thick (e.g., Tantalus Formation in the Yukon) and could have been derived partly from miogeoclinal rocks. Unfortunately, the age of these deposits is poorly constrained, other than having been assigned latest Jurassic to early Cretaceous ages on the basis of plant fossils.

Within the northern Bowser Basin, paleocurrents are predominantly to the south-southwest, and turbidite facies probably extended westward into a deep basin that was later obliterated by the intrusion of the Coast Plutonic Complex (Eisbacher, 1981). Towards the southern part of the basin, paleocurrents flowed westward, carrying detritus derived from volcano-plutonic fault blocks located within the Stikine Terrane. Intraformational unconformities, sedimentary slide blocks, and syn-depositional folds parallel to the basin margins indicate that subsidence of the Bowser Basin was accompanied by regional deformation of its basement. The syn-depositional deformation led to overall shoaling and progradation of coarse uppermost Jurassic-lowermost Cretaceous nonmarine facies towards the center of the basin (Eisbacher, 1981).

On the western margin of the basin the tectonic setting of Bowser sedimentation is unclear. Nevertheless, middle to upper Jurassic turbidite fans seem to have fringed the southwestern parts of the Stikine Terrane as well. In the southwesternmost part, volcanolithic flysch occurs at the base of the Tyaughton-Methow Basin (Coates, 1974). The clastic sedimentary rocks of this trough are tectonically juxtaposed against a belt of ophiolitic melange (Bridge River Terrane), which might represent the remnants of a Triassic-early Jurassic (oceanic?) basin floor (Monger et al., 1982). Judging from granitoid rocks intruded during this time interval to the east of this trough, some east-directed oblique subduction may have affected the Intermontane terranes. A sedimentary succession partly coeval with the basal Tyaughton-Methow strata also occurs along the west side of the northern Coast Plutonic Complex. There, the upper Jurassic to lower Cretaceous Gravina-Nutzotin Basin was filled by volcanolithic turbidite successions more than 3,000 m thick that were derived mainly from the Insular terranes on the west (Berg et al., 1972; Eisbacher, 1976). The sedimentary trough itself probably represented an obliquely rifted back-arc basin located between the Intermontane terranes on the east and the Wrangell and Alexander Terranes of the approaching Insular complex on the west. The connection between the Tyaughton-Methow and Gravina-Nutzotin Basins, if it existed, was later obliterated by basin closure, dextral shear, and intrusive activity along the Shakwak Suture.

In summary, the middle Jurassic to earliest Cretaceous

tectonic setting as depicted in Figure 4, was dominated by convergence along the Teslin Suture, which, for a considerable time, acted as an effective border between the source areas for the foreland basin to the east and those for the Bowser Basin to the west. Although strike-slip faulting is difficult to demonstrate during this stage of convergence, it is possible that major oblique faults skirted the southern margin of the Bowser Basin and fringed the Gravina-Nutzotin and Tyaughton-Methow back-arc troughs.

MID-CRETACEOUS TO PALEOCENE (120 TO 58 MA)

In Hauterivian to Barremian time (about 130 to 120 Ma), large parts of the eastern Cordillera underwent pedimentation. Uplift of the metamorphic belt east of the Teslin Suture slowed considerably (Archibald et al., 1983) and there was conspicuously little igneous activity during this time interval (R. L. Armstrong, personal commun., 1983). In both the foreland basin and Bowser Basin, thin but laterally extensive sheets of pediment gravels composed of chert and vein quartz disconformably overlie older synorogenic clastic rocks (McLean, 1977; Schulteis and Mountjoy, 1978; Eisbacher, 1981). Following pedimentation, clastic sedimentation in the foreland continued in mid-Cretaceous (Albian) time in fluvial, paralic, and marine settings along a north-dipping paleoslope that faced a residual shale trough centered on the northern platform and adjacent miogeoclinal realms (Carmichael, 1982; Leckie and Walker, 1982). A variety of generally quartzose nonmarine sandstone facies, interlaced with coal and shale tongues, indicate low depositional gradients along the basin axis. Low regional gradients are also suggested perpendicular to the axis of the foreland by the occasional progradation of thin feldspathic sandstones of Cordilleran provenance across its entire width (Putnam and Pedskalny, 1983). However, in the southern parts of the foreland basin granite-pebble-bearing fluvial conglomerates and redbeds indicate steeper gradients near the active thrust belt. Feldspathic detritus in Albian clastic rocks was derived from the volcanic carapaces of a regionally extensive suite of granitic rocks, which by then began to intrude the metamorphic core of the western thrust belt. These plutons generally yield radiometric cooling ages ranging between 110 and 90 Ma (Gabrielse and Reesor, 1974). Some K-Ar dates from low-grade metamorphic rocks of the thrust belt also fall into this time interval, and probably indicate uplift. By Cenomanian time (about 95 Ma), the axis of the foreland basin began to shift eastward, and muscovite-rich quartzose sediments began to prograde southeastward. Subsidence of the foreland accelerated in Campanian to Paleocene time (about 80 to 60 Ma) and fluvial-deltaic sediments of varied Cordilleran provenance filled the depression in front of the southern thrust belt (Price and Mountjoy, 1970; Beaumont, 1981).

The emplacement of high-level granitoids east of the Teslin Suture between 110 and 90 Ma also heralds the onset of extensive dextral motion on the Omineca fault array, first along the Teslin Suture, and later on strands straddling the metamorphic highs of the western thrust belt. Muscovites in sheared plutonic rocks astride discrete faults are only a few million years younger than structurally undisturbed

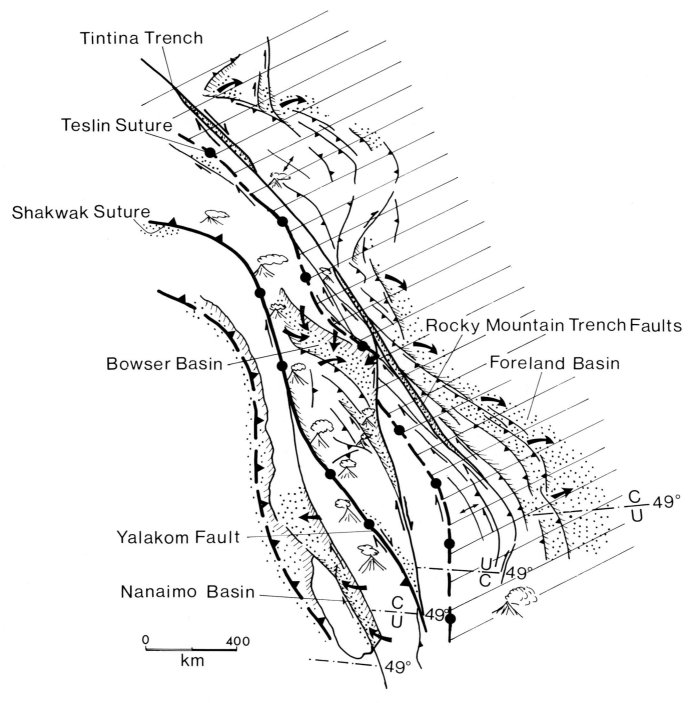

FIG. 5.—Paleogeographic model for the mid-Cretaceous to Paleocene setting of the clastic basins near the active Omineca fault array (e.g., Rocky Mountain Trench) east of the Teslin Suture. Dextral displacements also dominated the closing Shakwak Suture. Arrows indicate sediment dispersal from the partly volcanic source areas in the west and the rising thrust belt on the east.

phases of the same plutons (Gabrielse, 1985). West of the Teslin Suture, the transpression of the Intermontane terranes created a distinct crowding (syntaxis) of older thrusts and oblique strike-slip faults along the eastern margin of the Bowser Basin (Fig. 5). Truncation and rotation of older structures in this restraining bend of the Omineca fault ar-

ray not only created the subsiding structural re-entrant that favored entrapment of clastic sediments, but also accentuated the curvature of the Teslin Suture to such an extent that transcurrent motion began to shift onto straight fault strands east of the Teslin Suture (e.g., Rocky Mountain Trench). This eastward shift may have been aided by a pre-existing zone of structural divergence in the western thrust belt, where west-directed fold structures had been overprinted by east-directed structures (see Fig. 2). At depth, strike-slip faulting probably occurred along gently southwest-dipping surfaces, which originated as crustal thrusts but now were used as surfaces along which the brittle upper crust slid northwestward on a relatively wide ductile detachment zone in the middle crust (see below).

Beginning in about mid-Cretaceous time, braided and meandering streams carried varied detritus from both sides of the Teslin Suture towards the depressed eastern half of the Bowser Basin (Eisbacher, 1981). Some 500 to 1,000 m of nonmarine fluvial and lacustrine sediments, now partly derived from uplifted fault blocks along the Omineca fault array, were laid down unconformably on deformed older basin fill or on pediments along the eastern border of the basin. Similar suites of clastic sediments were deposited by anastomosing streams in the narrow fault valleys along the Omineca fault array (Eisbacher, 1981; Long, 1981). By Santonian-Campanian time (about 85 to 75 Ma), the Bowser Basin experienced further deformation along trends parallel to the rising Coast Plutonic Complex. The western margins of the basin were also intruded by numerous high-level granodiorite plutons associated with acidic to intermediate-composition volcanic activity. Eventually, deformation of the older deposits of the Bowser Basin restricted clastic sedimentation to a small peripheral trough located northeast of its center, and east-flowing rivers vigorously reworked subjacent strata (Eisbacher, 1981). Some 300 to 800 m of alluvial-fan conglomerates, sandstones, and ashfall tuffs accumulated in the elongate residual basin. K-Ar whole-rock ages obtained from tuffaceous beds in this succession led to the assumption that all the tuffaceous sediments were deposited in Eocene time (Eisbacher, 1981). This is probably not so, and the associated conglomeratic facies could be as old as 70 to 80 Ma. The radiometric ages probably reflect a late thermal and intrusive event that caused profound mineralogical alterations of sandstones (Read and Eisbacher 1974), and led to some of the abnormally high coal ranks known from other parts of the Bowser Basin. Upper Cretaceous-Paleogene ashfall tuffs also occur in the foreland basin deposits.

In the marginal basins along the Shakwak Suture, crustal convergence is evident from Hauterivian time onward. In the Gravina-Nutzotin Basin, the monotonous flysch succession was folded and intruded by ultramafic and granodiorite plutons between about 120 and 105 Ma. The youngest marine strata in the basin appear to be of Albian age (Berg et al., 1972; Eisbacher, 1976). In the Tyaughton-Methow basin to the south, several thousand meters of Hauterivian to lower Albian turbidite deposits continued to accumulate in a roughly northwest-trending basin. Paleocurrents indicate west- to northwest-dipping paleoslopes (Kleinspehn, 1982). The clastic succession probably represents a narrow fore-arc basin coeval with volcanic-plutonic activity on the east (Davis et al., 1978; Tennyson and Cole, 1978). However, other coarse facies, derived from local plutonic source areas in the intensely deformed and later intruded western parts of the basin, indicate the approach of the Insular terranes. The entire basin fill was subsequently cut by the northwest-striking Yalakom fault and the two halves of the basin were displaced dextrally by 150 to 160 km (Kleinspehn, 1982). A late Albian to Cenomanian nonmarine sedimentary and andesitic volcanic suite, indicating high-gradient depositional slopes and provenance of volcanic-sedimentary detritus from both east and west, probably heralds incipient closure and strike-slip deformation along the Shakwak Suture between about 100 and 85 Ma (Tennyson and Cole, 1978; Kleinspehn, 1982; Trexler and Bourgeois, 1983).

Following the closure of the Shakwak Suture, fault-controlled basins began to develop on the west side of the Coast Plutonic Complex between Santonian and Maastrichtian time (about 85 to 65 Ma). The best exposed of these is the Nanaimo Basin, which received a varied assemblage of turbiditic marine, coal-bearing paralic, and fluvial clastic sediments whose aggregate thickness is in excess of several thousand meters (Muller and Jeletzky, 1970). Facies and thickness changes along northwest-striking faults are abrupt; both paleocurrents and modal composition indicate that the bulk of the quartzose lithic or feldspathic sandstones were derived from the Coast Plutonic Complex to the east (Ward and Stanley, 1982). The basin thus probably represents part of a new fore-arc basin dominated by strike-slip faults; coeval distal trench deposits were later removed by continued dextral strike-slip faulting along the continental edge west of the accreted Vancouver Island block or were subducted underneath the Insular terranes and the growing Coast Plutonic Complex (Muller, 1977).

In summary, mid-Cretaceous to Paleogene tectonics were dominated by large-scale dextral strike-slip and thrust faulting in the pericollisional domain east of the Teslin Suture. Strike-slip and thrust faulting along the western thrust belt seem to have been intimately related to the same process of crustal convergence. Oblique dextral shear also closed marine basins along the Shakwak Suture and created some of the discrete fault strands of the Coast fault array. Intrusive activity and volcanism along the Shakwak Suture obliterated much of the earlier sedimentary record. Thus oblique subduction seems to have dominated the fore-arc region west of the rising Coast Plutonic Complex as well, although the magnitude of subduction is unknown.

EOCENE (58 TO 37 MA)

During early Eocene time, a large segment of the Canadian Cordillera between the Omineca and Coast fault arrays experienced a dramatic change in the tectonic setting. Most of the area underwent basin-and-range type extension along north-northeast-oriented grabens or half-grabens, which were filled with volcanic flows, pyroclastic, and minor clastic detritus derived from adjacent uplifted blocks. Regionally, the hitherto dominant northeast-directed convergence was replaced by west-northwest-directed extension (Fig. 6). In the western thrust belt the physiographic expression of longitudinal thrust faults and strike-slip systems was accentuated by listric extension faulting (Bally et al., 1966). Judging from the broad spectrum of stratigraphic ages reported for the generally small outcrop areas along the Omineca fault array (Long, 1981), nonmarine clastic sedimen-

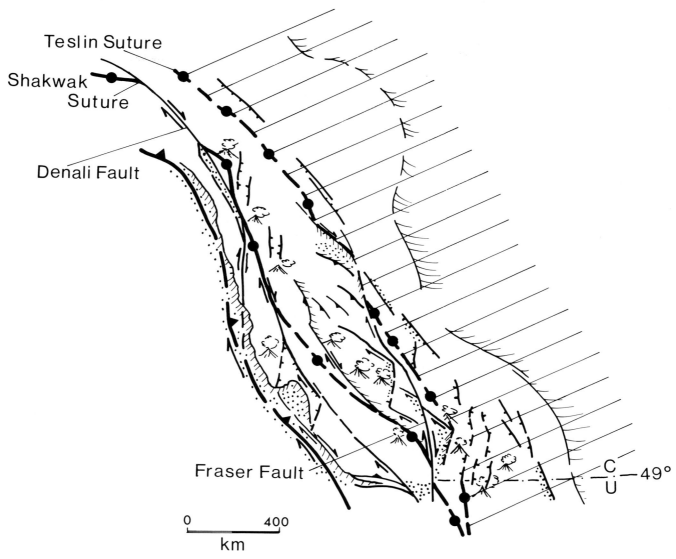

Fig. 6.—Paleogeographic model of the predominantly Eocene extensional volcanic-clastic basins. Active strike-slip faulting and oblique-dextral convergence continues along the Coast fault array east and west of the closed Shakwak Suture and along the Coast Plutonic Complex.

tation probably continued along previously or newly established fault blocks. The change of the regional stress pattern from one with northeast-directed maximum principal stresses to one with northeast-directed intermediate principal stresses effectively terminated thrust progradation in the foreland and marked the end of subsidence there. Large-scale uplift and erosion began to dominate.

However, in the region of the Coast Plutonic Complex north-northeast-directed compressive stresses still prevailed, and along the Coast fault array, dextral strike-slip faulting continued. Along the Denali fault, west of the Coast Plutonic Complex, some 300 to 400 km of dextral strike slip occurred, and along the Fraser fault, about 110 km of dextral strike slip has been inferred (Eisbacher, 1976; Kleinspehn, 1982). Along both fault zones small non-volcanic pull-apart basins are found. These and other less dis-

tinct dextral faults may interlace with post-compressional transform faults related to the west-northwest-oriented regional basin-and-range type extension (Ewing, 1980; Eisbacher, 1981), which may have led locally to dramatic uplift (Hollister, 1982).

The continued westward shift of active strike-slip faulting eventually involved much of the Cretaceous-Paleogene fore-arc along the Pacific margin and shifted most of the accretionary wedge towards southern Alaska (Nilsen and Zuffa, 1982).

In summary, the Eocene tectonic setting of the western Cordillera was dominated by uplift, basin-and-range type extension, and volcanic activity. Strike-slip faulting continued along the Coast fault array, but merged with extension-related transform faults of the southern Canadian Cordillera.

STRIKE-SLIP FAULTING AND THRUST BELT DEFORMATION

A most intriguing aspect of the large dextral convergence that took place in Cretaceous to Paleogene time (about 100 to 55 Ma) within the Canadian Cordillera is the relationship between strike-slip motion along northwest-trending Omineca fault array and contemporaneous thrust faulting of similar trend in the thrust belt. Both dextral strike-slip displacements on the Omineca fault array and orthogonal shortening in the thrust belt affected miogeoclinal strata and basement rocks. A lack of subsurface control throughout most of the western thrust belt, however, precludes a reliable definition of the basement-cover contact there. Also, the geometry of the largely thermally controlled brittle-ductile transition zone is poorly known. However, it appears that much of the dextral shear occurred within higher-grade metamorphic rocks in the area between the Teslin Suture and the Rocky Mountain Trench. The anastomosing fault pattern in general has a northwest trend but includes many connecting strands whose strike deviates from this general direction. Some of the strands may have originated as en echelon compressive structures; others may represent reactivated basement faults, and still others may be older strike-slip faults that were truncated and rotated by throughgoing younger strike-slip faults (Gabrielse, 1985). The Cretaceous to Paleogene fault-related clastic rocks, which in general are poorly exposed, are also disrupted by faults. Displaced pediments below some of the conglomeratic strata indicate substantial vertical motion along the faults. Radiometric dates reflecting uplift within the metamorphic rocks range from about 120 Ma to 45 Ma; however, there does not appear to be any systematic regional pattern. The fault-bounded clastic basins contain detritus derived from sedimentary strata, high-grade metamorphic rocks, and the plutonic rocks intruded into them. Most of the metamorphic rocks along the Omineca fault array display a strong penetrative stretching lineation (e.g., elongate pebbles) parallel to the northwesterly strike of the faults. Penetrative linear and planar elements are generally cut by faults and locally are overprinted by kink bands and younger folds. Such secondary features are commonly at considerable angles to the main faults (Eisbacher, 1972; Gabrielse, 1985).

Over great distances, thrust faults, folds, and cleavage within the thrust belt also trend obliquely into the easternmost strands of the Omineca fault array (Figs. 1 and 3). However, the structures in the thrust belt vary considerably as to their trend, and magnitude of displacement. Broadly, tectonic shortening decreases from south to north, as does the width of the exposed regionally metamorphosed rocks in the western thrust belt; the thickness of the miogeoclinal succession, on the other hand, increases towards the north, particularly the thickness of the Proterozoic section. With respect to strike-slip and thrust faulting, the Canadian thrust belt can be divided into three segments: a southern, a central, and a northern one.

In the southern thrust belt, regional structures within the sedimentary cover indicate shortening on the order of 100 to 200 km east of the Rocky Mountain Trench (Bally et al., 1966; Price, 1981). Along the Rocky Mountain Trench the strike-slip faults of the Omineca fault array seem to merge with thrust faults or relatively gently inclined faults of unknown magnitude and sense of displacement (e.g., Purcell Fault of Simony and Wind, 1970). In the central thrust belt, folds and thrust faults seem to have the same general trend as in the southern thrust belt. However, shortening across the belt seems to be substantially smaller, and the Omineca fault array appears to truncate thrust faults and folds of the thrust belt (Fig. 1). A change in trend at the releasing bend of the Omineca fault array coincides with the appearance of small clastic basins along the strike-slip faults, and this region also served as a major conduit for clastic sediments derived from the west and carried eastward into the foreland basin (Eisbacher et al., 1974). In the northern thrust belt, fold and fault structures diverge in a clockwise sense from the strike-slip faults of the northern Rocky Mountain Trench and, in a broad arc, rejoin the Tintina Trench near the Alaska border. Shortening in the thrust belt here probably increases from the southeast to the northwest; the sense of thrusting in general is to the north-northeast. The arcuate shape of the thrust belt, the emplacement of discordant plutons into cleaved strata near the Omineca fault array, and the common northeast- to north-directed dextral oblique thrusts within the thrust belt are probably controlled by older extensional structures or facies changes within the miogeoclinal succession. In the northernmost segments of the thrust belt, the Omineca fault array merges with, but also truncates, complex structures related to the emplacement of allochthonous terranes over the distal miogeocline (Tempelman-Kluit, 1979).

The transition from the strike-slip to the thrust regime in the western thrust belt is best displayed by structures exposed in the area near McBride, where the southern thrust-dominated segment changes to the strike-slip-dominated central segment (Fig. 7). In this area, the thrust belt narrows markedly towards the northwest, and the western edge of the foreland basin approaches to within 40 km of the Rocky Mountain Trench. The Rocky Mountain Trench in this region is a trough-shaped valley 4 to 5 km wide. It separates generally cleaved upper Proterozoic clastic rocks in east-verging folds and thrust faults from stratigraphically equivalent but polydeformed metamorphic rocks (Campbell et al., 1973; Ghent et al., 1977; Mountjoy, 1978). In the vicinity of McBride, the northeast wall of the Rocky Mountain Trench is characterized by a zone approximately 1 km wide of tightly deformed upper Proterozoic-Cambrian clastic rocks. This zone of intense deformation contains many rootless folds with vertical axial planes and mylonitic shear zones displaying subhorizontal stretching lineations. At several localities along the wall of the Rocky Mountain Trench, displacement indicators suggest dextral semi-penetrative shear.

Northeast of the shear belt, cleaved sedimentary rocks are overprinted by pervasive east-dipping crinkle lineations. Thrust faults become dominant in competent Paleozoic carbonate successions farther to the northeast (Mountjoy, 1978). West of the trench, the generally higher-grade metamorphic rocks display complex but northwest-trending penetrative planar and linear fabrics that are cut by faults (Campbell et al., 1973). The Omineca fault array undergoes a distinct change in trend from northwesterly to north-

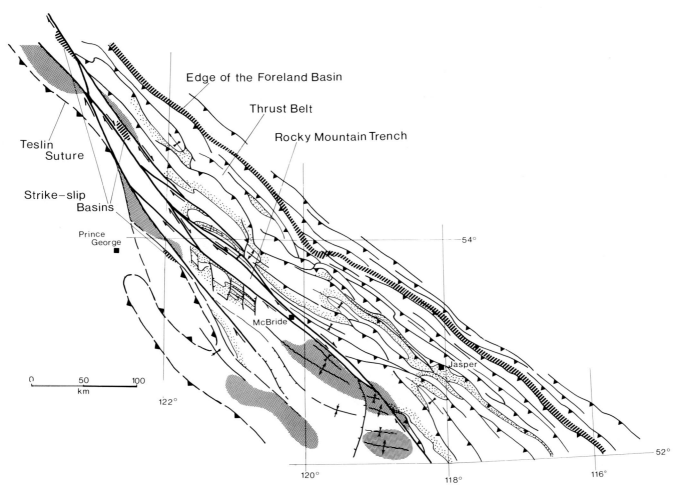

FIG. 7.—Simplified map of the transition thrust belt-Omineca fault array in the McBride area of British Columbia-Alberta (for location see Fig. 1). Wavy patterns indicate high-grade metamorphic complexes; dots indicate Upper Proterozoic-Cambrian boundary beds. Note the appearance of small strike-slip basins at the 'releasing bend' of the Omineca fault array.

northwesterly (Figs. 1 and 7). Remnants of upper Cretaceous to Paleocene nonmarine clastic basins appear first along the more northerly striking faults northwest of McBride. Radiometric ages obtained from metamorphic rocks along both sides of the trench suggest intermittent uplift of the fault blocks between about 120 and 55 Ma, followed by Eocene extensional faulting and plutonic activity. The pattern of thrust faults, the mylonitic dextral shear zones along the trench, the penetrative stretching lineations in and away from discrete shear zones, and the appearance of fault-bounded basins identify this area as one where northeast-southwest convergence in the thrust belt changes into semipenetrative dextral strike-slip deformation (releasing bend of Crowell, 1974).

From the tectonic evolution of the collision zone east of the Teslin Suture (Fig. 2), it appears that southwest-directed A-subduction of the cratonic basement probably created most of the west-dipping fabrics in the transition zone. Judging from high-grade metamorphic rocks exposed along the Rocky Mountain Trench (Simony et al., 1980), most of the fabrics probably developed below the brittle-ductile transition zone at mid-crustal levels. However, during westward under-

thrusting, these rocks reached a regime dominated by the considerably greater northwest-directed displacement vectors of the Omineca fault array. Thus strain in the underthrust rocks took on fabrics expressing the penetrative northwest-directed flow at deeper crustal levels and discrete dextral strike-slip faulting near the surface. Figure 8 is a schematic illustration of the proposed model. A zone of A-subduction characterized by relatively small southwest directed displacement vectors creates thrust faults in the sedimentary cover and upper basement levels, while the much larger northwest-directed displacements subject the tectonically thickened crust to ductile dextral shear. The model suggests that northwest-directed dextral displacements nucleated along west-dipping surfaces created by crustal wedging east of the closing Teslin Suture, but later also occurred along many gently dipping surfaces within the pericollisional realm of the thrust belt. In general, particles within underthrust rocks along the Omineca fault array would thus follow gently inclined paths (indicated schematically in Fig. 8). This in turn would result in the uneven regional metamorphic cooling patterns which vary from fault slice to fault slice.

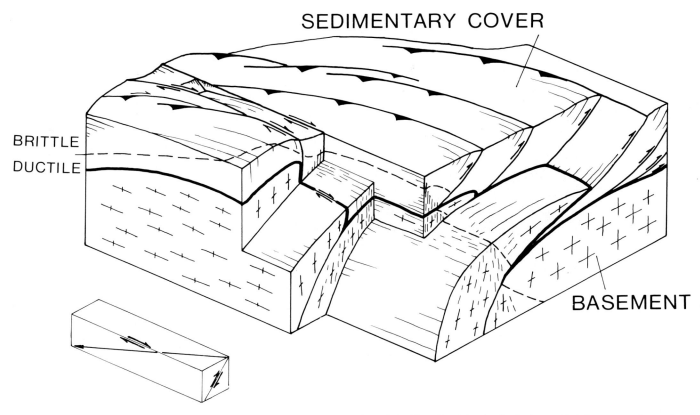

FIG. 8.—The relation of thrust faults to dextral shear zones in the western thrust belt. Slow A-subduction in the layered brittle crust on the northeast changes to more rapid dextral shear in the ductile parts of the deeper crust on the southwest. Southwest-dipping fabrics created by A-subduction are converted into diffuse zones of dextral flow or discrete faults. Relative magnitude of displacements by the two interfering processes and brittle-ductile transition is indicated schematically in the lower left corner of the diagram.

A 'REVERSE INDENTER' MODEL FOR CORDILLERAN STRIKE-SLIP FAULTING

In recent years several mechanisms have been proposed to explain syn- and post-collisional strike-slip deformation in convergent intraplate regions (Tapponnier and Molnar, 1976; Molnar and Tapponnier, 1977; Dewey and Şengör, 1979). Most of these models require that tectonically thickened continental crust behaves akin to a plastic slab yielding to a more rigid lithospheric indenter. A slip-line field diverging from the indenter defines the trend of vertical surfaces along which shear stresses attain a maximum value; slip lines that coincide with steeply dipping pre-existing zones of weakness in the upper crust (e.g., older sutures) yield preferentially during continued intraplate convergence. Rather than deforming along crustal thrust faults, discrete blocks (Tapponnier et al., 1982), or pervasively yielding wedges (Şengör and Kidd, 1979), escape laterally away from the areas of convergence towards extending or 'free' regions such as back-arc basins. Thrust faults, however, continue to be the main mode of deformation in zones of orthogonal convergence. On a regional scale, thrust faults thus can be expected to merge with oblique- and strike-slip faults. It is tempting to see how the 'indenter model' might be applied to the large displacements in the western Canadian Cordil-

lera. To do this, a broader perspective of late Mesozoic Cordilleran evolution has to be taken.

Data presented in the foregoing sections show that discrete or distributed strike-slip deformation along the Omineca and Coast fault arrays of the Canadian Cordillera is predominantly dextral. By their very magnitude, dextral displacements also dominate most of the fault pattern of Alaska. South of the Canadian segment of the North American Cordillera, accretionary tectonics in late Mesozoic time seems to reflect mainly orthogonal convergence of oceanic realms against the North American craton (see Burchfiel and Davis, 1975; Dickinson, 1976; Coney, 1978; Ernst, 1981). In a general scenario favored by many workers in the western United States, the late Mesozoic-Paleogene Franciscan melanges are considered to represent a sedimentary wedge, accreted as a subduction complex against the adjacent Great Valley forearc assemblage, and related to the Andean-type volcano-plutonic Sierra Nevada arc complex at the edge of the North American craton. The total amount of orthogonal subduction of oceanic lithosphere beneath North America, estimated roughly from the overall asymmetry in the age of the Pacific ocean basin, seems to exceed several thousand kilometers. An eastward propagation of crustal thrust faults, roughly coeval with subduction, involving miogeoclinal strata and basement, was probably facilitated by thermal weakening of the crust (Burchfiel and Davis, 1975). Thrust faults first nucleated near plutons of middle to late Jurassic age at the boundary between the arc and craton; subsequently, thrust faulting spread to deeper and more easterly located crustal zones (Allmendinger and Jordan, 1981). Although many geologic data support such a setting for the evolving Cordilleran orogen of the United States, Saleeby (1983) recently has sounded a note of caution and suggested that some of the Franciscan subduction process may have occurred contemporaneously with oblique-dextral motions. Oblique shear certainly seems to have been significant in Paleogene time (e.g., Alvarez et al., 1980), and transform dextral displacements dominated the Neogene California margin.

Moving farther south to the southwestern United States and Mexico, the amount of later Mesozoic crustal convergence seems to decrease, but geologic and paleomagnetic data indicate that in conjunction with convergence, significant sinistral faulting occurred. Some of the relevant data have been compiled recently by Dickinson (1983), Oldow (1984), and Urrutia-Fucugauchi (1984). Cretaceous sinistral displacement in excess of 500 km is plausible for the Nacimiento fault, a southeast-striking structural feature parallel to the Jurassic(?) Mojave-Sonora Megashear (Dickinson, 1983). Other faults striking east-southeasterly across Mexico probably also experienced significant sinistral displacements in Cretaceous time (Urrutia-Fucugauchi, 1984). Although it would go beyond this paper to discuss the broader implications, it seems possible that the dextral faults of Canada-Alaska and the sinistral faults of Mexico represent zones of lateral lithospheric escape from the area of orthogonal plate convergence in the western United States. Cretaceous crustal stretching in the Bering back-arc basin to the north (Fisher et al., 1982) and in the Gulf of Mexico (Dickinson, 1983) are certainly reminiscent of the crustal stretching along the free ends of the Anatolian strike-slip system of Turkey (Şengör, 1979; Şengör, 1985 this volume).

If these broad relationships have any plausibility, then the rigid indenter model can be simply reversed. A spur or blunt promontory of rigid continental lithosphere in the region of Nevada-California was exposed to easterly or northeasterly subduction of oceanic lithosphere. Convergence of thicker crustal elements within the oceanic realms (e.g., arcs and plateaus) produced a pattern of stress trajectories within the upper plate that diverged from the promontory (Fig. 9). More rigid fragments that impinged near the promontory were deflected by the reverse indenter either to the north or south along slip lines (arrows in Fig. 9) oriented at about 45° to the maximum horizontal stress ($S_{h\ max}$ in Fig. 9). The fact that several crustal fragments that have accreted north of the cratonic promontory have paleomagnetic signatures placing them originally south of it simply implies that convergence of the oceanic plate probably had a northeast-component; these crustal fragments still approached the area of the reverse indenter when they were deflected northward by it. Changes in the direction of principal convergence may have caused variations in the amount of crustal material deflected northwards or southwards. An active sub-

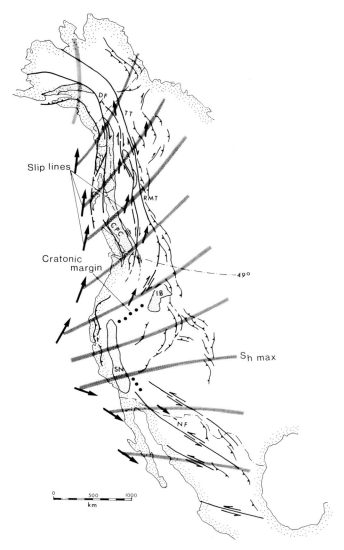

FIG. 9.—The 'reverse indenter' interpretation of the Cretaceous-Paleogene strike-slip system and related thrust belts of western North America: a relatively rigid cratonic promontory in the California-Nevada region concentrates horizontal compressive stress trajectories ($S_{h\ max}$) during convergence and subduction of Pacific crust. Slip on shear surfaces oriented at about 45° to the maximum horizontal stresses allows lateral escape of non-subductable crust (arrows). Shear is concentrated near pre-existing lithospheric weaknesses such as the Teslin and Shakwak Sutures and related metamorphic-plutonic zones of Canada. Dextral displacements decrease towards the 'stiff' cratonic lithosphere, but stretching may extend along trend with the dextral and sinistral faults towards the Bering Sea and the Gulf of Mexico respectively. **NF**, Nacimiento fault; **SN**, Sierra Nevada Batholith; **IB**, Idaho Batholith; **CPC**, Coast Plutonic Complex; **RMT**, Rocky Mountain Trench; **TT**, Tintina Trench; **DF**, Denali fault.

duction zone west of the Coast Plutonic Complex probably shifted northward during displacements along the pericollisional fault zones, as envisaged by Beck (1983). Crustal zones of weakness near the Teslin and Shakwak Sutures are therefore considered lateral splays of the composite Franciscan subduction complex that enclosed laterally escaping non-subductable lithospheric terranes.

CONCLUSIONS

In summary, the following conclusions can be drawn.

(1) The sedimentary fill of late Mesozoic-Paleogene clastic basins in the Canadian Cordillera reflects the gradual suturing and oblique convergence of oceanic and arc-type lithospheric fragments against the western edge of the relatively rigid North American craton.

(2) Pericollisional shortening of North American crust (A-subduction) and dextral convergence created the thrust belt along the miogeoclinal hinge. It became the principal source area for shallow-water orogenic clastic sediments of the foreland basin; the composition of these deposits is predominantly quartzose.

(3) Complex fault-controlled basins developed on and between accreting terranes west of the thrust belt. Their sedimentary fill reflects mainly subquartzose plutonic, volcanic, and sedimentary source areas of the accreted oceanic and arc complexes.

(4) Dextral Cretaceous-Paleogene strike-slip motion on the order of 1,000 to 2,000 km resulted from north-directed lateral escape of non-subductable crustal terranes from near a reverse indenter, a westward projecting promontory of continental lithosphere in the Nevada-California region.

(5) The pericollisional dextral Omineca and Coast fault arrays nucleated near zones of complex structural overprinting along the closing Teslin and Shakwak Sutures respectively. Subsequently, strike-slip faulting spread into adjacent domains of high-grade regional metamorphism. In the thrust belt, large-scale dextral shear affected planar fabrics created by A-subduction. Many gently dipping planar structures in high-grade core terranes of other orogens (e.g., Brevard fault in the Appalachians) may also have experienced strike-slip displacements.

(6) The large dextral shifts along the accreting crustal terranes resulted in repeated and intense deformation of turbiditic, deltaic, and fluvial successions deposited on them. Some of the basins were almost entirely disrupted or obliterated by ongoing dislocations and by intrusions related to B-subduction along the Pacific continental margin. Intrusive activity and enhanced heat flow within the accreted terranes also induced widespread zeolitization of the volcanic-rich sandstones.

(7) In general, clastic deposits located along strike-slip faults of convergent crustal settings form disrupted outcrop areas, are commonly overridden by oblique thrust faults and may be eroded soon after their deposition due to uplift of thickened crust. Detritus eroded from convergent oblique fault zones generally ends up in pericollisional basins, which are commonly related to closing sutures along accreting terranes or to thrust progradation in foreland basins.

ACKNOWLEDGMENTS

I would like to acknowledge stimulating discussions during preparation of this paper with D. J. Tempelman-Kluit. Earlier versions of the manuscript were critically read by K. T. Biddle, N. Christie-Blick, D. Seely, and C. C. Wielchowsky. As some of the notions put forth here are speculative, the responsibility for errors rests entirely with me. Ms. B. Vanlier processed the manuscript.

REFERENCES

ALLMENDINGER, R. W., AND JORDAN, T. E., 1981, Mesozoic evolution, hinterland of the Sevier orogenic belt: Geology, v. 9, p. 308–313.

ALVAREZ, W., KENT, D. V., PREMOLI-SILVA, I., SCHWEICKERT, R. A., AND LARSON, R. A., 1980, Franciscan Complex limestone deposited at 17° south paleolatitude: Geological Society American Bulletin, v. 91, p. 476–484.

ARCHIBALD, D. A., GLOVER, J. K., PRICE, R. A., FARRAR, E., AND CARMICHAEL, D. M., 1983, Geochronology and tectonic implications of magmatism and metamorphism, southern Kootenay Arc and neighbouring regions, southeastern British Columbia. Part I: Jurassic to mid-Cretaceous: Canadian Journal of Earth Sciences, v. 20, p. 1891–1913.

AYDIN, A., AND NUR, A., 1982, Evolution of pull-apart basins and their scale independence: Tectonics, v. 1, p. 91–105.

BALLANCE, P. F., 1980, Models of sediment distribution in nonmarine and shallow marine environments in oblique-slip fault zones, in Ballance, P. F. and Reading, H. G. eds., Sedimentation in Oblique-Slip Mobile Zones: International Association of Sedimentologists Special Publication No. 4, p. 229–236.

BALLANCE, P. F., AND READING, H. G., eds., 1980, Sedimentation in Oblique-Slip Mobile Zones: International Association of Sedimentologists Special Publication No. 4, 265 p.

BALLY, A. W., GORDY, P. L., AND STEWART, G. A., 1966, Structure, seismic data, and orogenic evolution of southern Canadian Rocky Mountains: Bulletin of Canadian Petroleum Geology, v. 14, p. 337–381.

BEAUMONT, C., 1981, Foreland basins: Geophysical Journal of Royal Astronomical Society, v. 65, p. 291–329.

BECK, M. E., JR., 1983, On the mechanism of tectonic transport in zones of oblique subduction: Tectonophysics, v. 93, p. 1–11.

BERG, H. C., JONES, D. L., AND RICHTER, D. H., 1972, Gravina-Nutzotin Belt—tectonic significance of an upper Mesozoic sedimentary and volcanic sequence in southern and southeastern Alaska: United States Geological Survey Professional Paper 800-D, p. D1–D24.

BROWN, R. L., 1978, Structural evolution of the southeast Canadian Cordillera: A new hypothesis: Tectonophysics, v. 48, p. 133–151.

BURCHFIEL, B. C., AND DAVIS, G. A., 1975, Nature and controls of Cordilleran orogenesis, western United States: extensions of an earlier synthesis: American Journal of Science, v. 275A, p. 363–396.

BURCHFIEL, B. C., AND STEWART, J. H., 1966, 'Pull-apart' origin of the central segment of Death Valley, California: Geological Society America Bulletin, v. 77, p. 439–442.

BUSTIN, R. M., AND MOFFAT, I., 1983, Groundhog coal field, central British Columbia: Reconnaissance stratigraphy and structure: Bulletin of Canadian Petroleum Geology, v. 31, p. 231–245.

CAMPBELL, R. B., MOUNTJOY, E. W., AND YOUNG, F. G., 1973, Geology of McBride map-area, British Columbia: Geological Survey of Canada Paper 72-35, 104 p.

CARMICHAEL, S. M. M., 1982, Depositional environment and paleocurrent trends in the Gates Member, Northeast Coalfield: Geological Fieldwork 1981: Ministry of Energy, Mines and Petroleum Resources, Paper 1982-1, p. 244–258.

COATES, J. A., 1974, Geology of the Manning Park area, British Columbia: Geological Survey of Canada Bulletin 238, 177 p.

CONEY, P. J., 1978, Mesozoic-Cenozoic Cordilleran plate tectonics: Geological Society of America Memoir 152, p. 33–50.

CROWELL, J. C., 1974, Origin of late Cenozoic basins in southern California, in Dott, R. H., Jr., and Shaver, R. H., eds., Modern and Ancient Geosynclinal Sedimentation: Society of Economic Paleontologists and Mineralogists Special Publication No. 19, p. 292–303.

CURRY, J. R., MOORE, D. G., LAWVER, L. A., EMMEL, F. J., RAITT, R. W., HENRY, M., AND KIECKHEFER, 1979, Tectonics of the Andaman Sea and Burma, in Watkins, J. S., Montadert, L., and Dickerson, P. W., eds., Geological and Geophysical Investigations of Continental Margins: American Association of Petroleum Geologists Memoir 29, p. 189–198.

DAVIS, G. A., MONGER, J. W. H., AND BURCHFIEL, B. C., 1978, Mesozoic construction of the Cordilleran 'collage', central British Columbia to central California, in Howell, D. G., and McDougall, K. A., eds.,

Mesozoic Paleogeography of the Western United States: Society of Economic Paleontologists and Mineralogists Pacific Coast Paleogeography Symposium 2, p. 1–32.

DEWEY, J. F., 1980, Episodicity, sequence, and style at convergent plate boundaries, *in* Strangway, D. W., ed., The Continental Crust and its Mineral Deposits: Geological Association of Canada Special Paper 20, p. 553–573.

DEWEY, J. F., AND ŞENGÖR, A. M. C., 1979, Aegean and surrounding regions: complex multiplate and continuum tectonics in a convergent zone: Geological Society of America Bulletin, v. 90, p. 84–92.

DICKINSON, W. R., 1976, Sedimentary basins developed during evolution of Mesozoic-Cenozoic arc-trench system in western North America: Canadian Journal of Earth Sciences, v. 13, p. 1268–1287.

———, 1983, Cretaceous sinistral strike slip along Nacimiento Fault in coastal California: American Association of Petroleum Geologists Bulletin, v. 67, p. 624–645.

EISBACHER, G. H., 1972, Tectonic overprinting near Ware, northern Rocky Mountain Trench: Canadian Journal of Earth Sciences, v. 9, p. 903–913.

———, 1974, Evolution of successor basins in the Canadian Cordillera, *in* Dott, R. H., Jr., and Shaver, R. H., eds., Modern and Ancient Geosynclinal Sedimentation: Society of Economic Paleontologists and Mineralogists Special Publication No. 19, p. 274–291.

———, 1976, Sedimentology of the Dezdeash flysch and its implications for strike-slip faulting along the Denali Fault, Yukon Territory and Alaska: Canadian Journal of Earth Sciences, v. 13, p. 1495–1513.

———, 1981, Late Mesozoic-Paleogene Bowser Basin Molasse and Cordilleran tectonics, western Canada, *in* Miall, A. D., ed., Sedimentation and Tectonics in Alluvial Basins: Geological Association of Canada Special Paper 23, p. 125–151.

EISBACHER, G. H., CARRIGY, M. A., AND CAMPBELL, R. B., 1974, Paleo-drainage pattern and late-orogenic basins of the Canadian Cordillera, *in* Dickinson, W. R., ed., Tectonics and Sedimentation: Society of Economic Paleontologists and Mineralogists Special Publication No. 22, p. 143–166.

ERDMER, P., AND HELMSTAEDT, H., 1983, Eclogite from central Yukon: a record of subduction at the western margin of ancient North America: Canadian Journal of Earth Sciences, v. 20, p. 1389–1408.

ERNST, W. G., ed., 1981, The Geotectonic Development of California, Rubey Volume 1: Los Angeles, California, Prentice Hall, 706 p.

EVENCHICK, C. A., 1984, Structure and stratigraphy in the hanging wall of the Sifton Fault, Sifton Ranges, northern British Columbia: Geological Survey of Canada Paper 84-1A, p. 105–108.

EVENCHICK, C. A., PARRISH, R. R., AND GABRIELSE, H., 1984, Precambrian gneiss and late Proterozoic sedimentation in north-central British Columbia: Geology, v. 12, p. 233–237.

EWING, T. E., 1980, Paleogene tectonic evolution of the Pacific Northwest: Journal of Geology, v. 88, p. 619–638.

FISHER, M. A., PATTON, W. W., JR., AND HOLMES, M. L., 1982, Geology of Norton Basin and continental shelf beneath northwestern Bering Sea, Alaska: American Association of Petroleum Geologists Bulletin, v. 66, p. 255–285.

FORBES, R. B., TURNER, D. L., STOUT, J., AND SMITH, T. E., 1973, Cenozoic offset along the Denali Fault, Alaska (abs.): Eos, v. 54, p. 495.

GABRIELSE, H., 1985, Major dextral transcurrent displacements along the northern Rocky Mountain Trench and related lineaments in north-central British Columbia: Geological Society America Bulletin, v. 96, p. 1–14.

GABRIELSE, H., AND REESOR, J. E., 1974, The nature and setting of granitic plutons in the central and eastern part of the Canadian Cordillera: Pacific Geology, v. 8, p. 109–138.

GABRIELSE, H., WANLESS, R. K., ARMSTRONG, R. L., AND ERDMAN, L. R., 1980, Isotopic dating of Early Jurassic volcanism and plutonism in north-central British Columbia: Geological Survey of Canada Paper 80-1A, p. 27–32.

GHENT, E. D., SIMONY, P. S., MITCHELL, W., PERRY, J., ROBBINS, D., AND WAGNER, J., 1977, Structure and metamorphism in southeast Canoe River area, British Columbia: Geological Survey of Canada Paper 77-1C, p. 13–17.

HAMBLIN, A. P., AND WALKER, R. G., 1979, Storm-dominated shallow marine deposits: the Fernie-Kootenay (Jurassic) transition, southern Rocky Mountains: Canadian Journal of Earth Sciences, v. 16, p. 1673–1690.

HAMILTON, W., 1979, Tectonics of the Indonesian region: U.S. Geological Survey Professional Paper 1078, 345 p.

HOLLISTER, L. S., 1982, Metamorphic evidence for rapid (2 mm/yr) uplift of a portion of the Central Gneiss Complex, Coast Mountains, B.C.: Canadian Mineralogist, v. 20, p. 319–332.

HUGHES, J. D., AND LONG, D. G. F., 1980, Geology and coal resource potential of early Tertiary strata along Tintina Trench, Yukon Territory: Geological Survey of Canada Paper 79-32, 21 p.

IRVING, E., 1983, Fragmentation and assembly of the continents, mid-Carboniferous to present: Geophysical Surveys, v. 5, p. 299–333.

IRVING, E., MONGER, J. W. H., AND JOLE, R. W., 1980, New paleomagnetic evidence for displaced terranes in British Columbia, *in* Strangway, D. W., ed., The Continental Crust and its Mineral Deposits: Geological Association of Canada Special Paper 20, p. 441–456.

JONES, D. L., SILBERLING, N. J., AND HILLHOUSE, J., 1977, Wrangellia—a displaced terrane in northwestern North America: Canadian Journal of Earth Sciences, v. 14, p. 2565–2577.

KARIG, D. E., 1983, Accreted terranes in the northern part of the Philippine Archipelago: Tectonics, v. 2, p. 211–236.

KLEINSPEHN, K. L., 1982, Cretaceous sedimentation and tectonics, Tyaughton-Methow Basin, southwestern British Columbia [unpubl. Ph.D. thesis]: Princeton New Jersey, Princeton University, 184 p.

LANPHERE, M., 1978, Displacement history of the Denali Fault system, Alaska and Canada: Canadian Journal of Earth Sciences, v. 15, p. 817–822.

LECKIE, D. A., AND WALKER, R. G., 1982, Storm- and tide-dominated shorelines in Cretaceous Moosebar-Lower Gates interval—outcrop equivalents of Deep Basin Gas Trap in western Canada: American Association of Petroleum Geologists Bulletin, v. 66, p. 138–157.

LONG, D. G. F., 1981, Dextral strike-slip faults in the Canadian Cordillera and depositional environments of related fresh-water intermontane coal basins, *in* Miall, A. D., ed., Sedimentation and Tectonics in Alluvial Basins: Geological Association of Canada Special Paper 23, p. 153–186.

MCLEAN, J. R., 1977, The Cadomin Formation: stratigraphy, sedimentology, and tectonic implications: Bulletin of Canadian Petroleum Geology, v. 25, p. 792–827.

MANN, P., HEMPTON, M. R., BRADLEY, D. C., AND BURKE, K., 1983, Development of pull-apart basins: Journal of Geology, v. 91, p. 529–554.

MANSY, J. L., 1980, La Cordillère canadienne au Nord et au centre de la Colombie Britannique (Canada): Revue de Géologie Dynamique et de Géographie Physique, v. 22, p. 233–254.

MITCHELL, A., AND MCKERROW, W., 1975, Analogous evolution of the Burma orogen and the Scottish Caldonides: Geological Society of America Bulletin, v. 86, p. 305–315.

MOLNAR, P., AND TAPPONNIER, P., 1977, Relation of the tectonics of eastern China to the India-Eurasia collision: application of slip-line field theory to large-scale continental tectonics: Geology, v. 5, p. 212–216.

MONGER, J. W. H., 1977, Upper Paleozoic rocks of the western Canadian Cordillera and their bearing on Cordilleran evolution: Canadian Journal of Earth Sciences, v. 14, p. 1832–1859.

MONGER, J. W. H., AND PRICE, R. A., 1979, Geodynamic evolution of the Canadian Cordillera—progress and problems: Canadian Journal of Earth Sciences, v. 16, p. 770–791.

MONGER, J. W. H., PRICE, R. A., AND TEMPELMAN-KLUIT, D. J., 1982, Tectonic accretion and the origin of the two major metamorphic and plutonic welts in the Canadian Cordillera: Geology, v. 10, p. 70–75.

MONGER, J. W. H., AND ROSS, C. A., 1971, Distribution of fusulinaceans in the Canadian Cordillera: Canadian Journal of Earth Sciences, v. 8, p. 259–278.

MOUNTJOY, E. W., 1978, Mount Robson: Geological Survey of Canada Map 1499.

MULLER, J. E., 1977, Evolution of the Pacific Margin, Vancouver Island and adjacent regions: Canadian Journal of Earth Sciences, v. 14, p. 2062–2085.

MULLER, J. E., AND JELETZKY, J. A., 1970, Geology of the Upper Cretaceous Nanaimo Group, Vancouver Island and Gulf Islands, British Columbia: Geological Survey of Canada Paper 69-25, 77 p.

NICOLAS, A., BOUCHEZ, J. L., BLAISE, J., AND POIRIER, J. P., 1977, Geological aspects of deformation in continental shear zones: Tectonophysics, v. 42, p. 55–73.

NILSEN, T. H., AND ZUFFA, G. G., 1982, The Chugach Terrane, a Cretaceous trench-fill deposit, southern Alaska, *in* Leggett, J. K., ed., Trench-Forearc Geology: Geological Society of London Special Publication No. 10, p. 213–227.

NOKLEBERG, W. J., JONES, D. L., AND SILBERLING, N. J., in press, Origin, migration, and accretion of the MacLaren and Wrangellia Terranes, Eastern Alaska Range, Alaska: Geological Society of America Bulletin.

OLDOW, J. S., 1984, Evolution of a late Mesozoic back-arc fold and thrust belt, northwestern Great Basin: Tectonophysics, v. 102, p. 245–274.

PARRISH, R. R., 1979, Geochronology and tectonics of the northern Wolverine Complex, British Columbia: Canadian Journal of Earth Sciences, v. 16, p. 1428–1438.

PATERSON, I. A., 1977, The geology and evolution of the Pinchi Fault Zone at Pinchi Lake, central British Columbia: Canadian Journal of Earth Sciences, v. 14, p. 1324–1342.

PIGAGE, L. C., 1977, Rb-Sr dates for granodiorite intrusions on the northeast margin of the Shuswap Metamorphic Complex, Cariboo Mountains, British Columbia: Canadian Journal of Earth Sciences, v. 14, p. 1690–1695.

POULTON, T. P., in press, The Jurassic of the Canadian Western Interior, from the 49° latitude to Beaufort Sea: *in* Canadian Society of Petroleum Geologists, Memoir 9.

PRICE, R. A., 1981, The Cordilleran foreland thrust and fold belt in the southern Canadian Rocky Mountains, *in* McClay, K. R., and Price, N. J., eds., Thrust and Nappe Tectonics: Geological Society of London Special Publication No. 9, p. 427–448.

PRICE, R. A., AND MOUNTJOY, E. W., 1970, Geologic structure of the Canadian Rocky Mountains between Bow and Athabasca rivers—a progress report, *in* Wheeler, J. O., ed., Structure of the Southern Canadian Cordillera: Geological Association of Canada Special Paper 6, p. 7–25.

PUTNAM, P. E., AND PEDSKALNY, M. A., 1983, Provenance of Clearwater Formation reservoir sandstones, Cold Lake, Alberta, with comments on feldspar composition: Bulletin of Canadian Petroleum Geology, v. 31, p. 148–160.

READ, P. B., AND EISBACHER, G. H., 1974, Regional zeolite alteration of the Sustut Group, north-central British Columbia: Canadian Mineralogist, v. 12, p. 527–541.

RODDICK, J. A., 1967, Tintina Trench: Journal of Geology, v. 75, p. 23–33.

RODDICK, J. A., AND HUTCHISON, W. W., 1974, Setting of the Coast Plutonic Complex, British Columbia: Pacific Geology, v. 8, p. 91–108.

ROEDER, D. H., 1967, Rocky Mountains—der geologische Aufbau des kanadischen Felsengebirges: Beitrage zur Regionalen Geologie der Erde, v. 5, 318 p.

————, 1973, Subduction and orogeny: Journal of Geophysical Research, v. 78, p. 5005–5024.

SALEEBY, J. B., 1983, Accretionary tectonics of the North American Cordillera: Annual Review of Earth and Planetary Sciences, v. 15, p. 45–73.

SCHULTHEIS, N., AND MOUNTJOY, E. W., 1978, Cadomin conglomerate of Alberta, a result of early Cretaceous uplift of the Main Ranges: Bulletin of Canadian Petroleum Geology, v. 26, p. 297–342.

ŞENGÖR, A. M. C., 1979, The North Anatolian transform fault: its age, offset and tectonic significance: Journal of Geological Society of London, v. 136, p. 269–282.

ŞENGÖR, A. M. C., AND KIDD, W. S. F., 1979, Post-collisional tectonics of the Turkish-Iranian Plateau and a comparison with Tibet: Tectonophysics, v. 55, p. 361–376.

ŞENGÖR, A. M. C., GÖRÜR, N., AND ŞAROĞLU, F., 1985, Strike-slip faulting and related basin formation in zones of tectonic escape: Turkey as a case study, *in* Biddle, K. T., and Christie-Blick, N., eds., Strike-Slip Deformation, Basin Formation, and Sedimentation: Society of Economic Paleontologists and Mineralogists Special Publication No. 37, p. 227–264.

SIMONY, P. S., GHENT, E. D., CRAW, D., MITCHELL, W., AND ROBBINS, D. B., 1980, Structural and metamorphic evolution of northeast flank of Shuswap Complex, southern Canoe River area, British Columbia, *in* Crittenden, M. D., Jr., Coney, P. J., and Davis, G. H., eds., Cordilleran Metamorphic Core Complexes: Geological Society of America Memoir 153, p. 445–461.

SIMONY, P. S., AND WIND, G., 1970, Structure of the Dogtooth Range and adjacent portions of the Rocky Mountain Trench: Geological Association of Canada Special Paper 6, p. 41–51.

TAPPONNIER, P., AND MOLNAR, P., 1976, Slip-line field theory and large-scale continental tectonics: Nature, v. 264, p. 319–324.

TAPPONNIER, P., PELTZER, G., LEDAIN, A. Y., ARMIJO, R., AND COBBOLD, P., 1982, Propagating extrusion tectonics in Asia: New insights from simple experiments with Plasticine: Geology, v. 10, p. 611–616.

TEMPELMAN-KLUIT, D. J., 1971, Stratigraphy and structure of the 'Keno Hill Quartzite' in Tombstone River-Upper Klondike River map-areas, Yukon Territory: Geological Survey of Canada Bulletin 180, 102 p.

————, 1979, Transported cataclasite, ophiolite, and granodiorite in Yukon: evidence of arc-continent collision: Geological Survey of Canada Paper 79-14, 27 p.

TENNYSON, M. E., AND COLE, M. R., 1978, Tectonic significance of Upper Mesozoic Methow-Pasayten Sequence, northeastern Cascade Range, Washington and British Columbia, *in* Howell, D. G., and McDougall, D. R., eds., Mesozoic Paleogeography of the Western United States: Society of Economic Paleontologists and Mineralogists Pacific Coast Paleogeography Symposium v. 2, p. 499–508.

THOMPSON, R. I., 1981, The nature and significance of large 'blind' thrusts within the northern Rocky Mountains of Canada, *in* McClay, K. R., and Price, N. J., eds., Thrust and Nappe Tectonics: Geological Society of London Special Publication No. 9, p. 449–462.

TIPPER, H. W., 1981, Offset of an upper Pliensbachian geographic zonation in the North American Cordillera by transcurrent movement: Canadian Journal of Earth Sciences, v. 18, p. 1788–1792.

TIPPER, H. W., AND RICHARDS, T. A., 1976, Jurassic stratigraphy and history of north-central British Columbia: Geological Survey of Canada Bulletin 170, 73 p.

TOZER, E. T., 1982, Marine Triassic faunas of North America: their significance for assessing plate and terrane movements: Geologische Rundschau, v. 71, p. 1077–1104.

TREXLER, J. H., AND BOURGEOIS, J., 1983, Stratigraphy and sedimentation of the Virginian Ridge Formation (Albian-Cenomanian)—a key to the history of the Methow Basin (abs.): Geological Association of Canada, Annual Meeting 1983, p. 69.

URRUTIA-FUCUGAUCHI, J., 1984, On the tectonic evolution of Mexico: paleomagnetic constraints, *in* Van der Voo, R., Scotese, C. R., and Bonhommet, N., eds., Plate Reconstruction from Paleozoic Paleomagnetism: American Geophysical Union Geodynamics Series, v. 12, p. 29–47.

WARD, P., AND STANLEY, K. O., 1982, The Haslam Formation: a late Santonian-early Campanian forearc basin deposit in the Insular Belt of southwestern British Columbia and adjacent Washington: Journal of Sedimentary Petrology, v. 52, p. 975–990.

EOCENE STRIKE-SLIP FAULTING AND NONMARINE BASIN FORMATION IN WASHINGTON

SAMUEL Y. JOHNSON[1]

Department of Geology, Washington State University, Pullman, Washington 99164-2812

ABSTRACT: Eocene right-lateral displacements occurred along several major fault zones in western and central Washington, including the Straight Creek fault, the Entiat-Leavenworth fault system, and probably the Puget fault, a covered north-trending structure in the Puget Lowland. Within this strike-slip framework, nonmarine sediments accumulated in the Chuckanut, Puget-Naches, Chiwaukum graben, and Swauk Basins to form some of the thickest (more than 6,000 m) alluvial sequences in North America. To varying degrees, the basins are characterized by (1) high sediment-accumulation rates, implying rapid subsidence; (2) abrupt local stratigraphic thickening and thinning; (3) intrabasinal and basin-margin unconformities; (4) abrupt facies changes; (5) fault-induced drainage re-organization; (6) intermittent internal drainage; and (7) interbedded and intrusive relationships with extension-related(?) volcanic rocks. Similar sandstone petrography between basins suggests a common sediment source with possible local or temporary connection between basins. Sedimentation and deformation throughout the province were diachronous, and the basins experienced rapid alternating subsidence and uplift. Orientations of folds and faults are consistent with regional right-lateral shear.

The 90-km-wide composite Chuckanut-Puget-Naches Basin probably formed as a pull-apart basin between the Straight Creek and Puget faults. The 50-km-wide Swauk Basin may have formed as a fault-wedge basin between the Straight Creek and Entiat-Leavenworth faults. These basins are large when compared to most modern and ancient basins controlled by strike-slip faults, suggesting that processes of strike-slip basin formation operate at a variety of scales. The 20-km-wide Chiwaukum graben probably formed as a pull-apart basin between the Entiat and Leavenworth faults.

Eocene strike-slip faulting was probably driven by oblique convergence of the Kula plate below North America. Although strike-slip basins in Washington occupied a forearc or possibly an intra-arc tectonic setting in this ancient continental margin, they differ significantly from typical forearc or intra-arc basins.

INTRODUCTION

In recent years, strike-slip faulting has been shown to be an important control on the formation, sedimentation style, and deformation of sedimentary basins (e.g., Crowell, 1974a, b; Ballance and Reading, 1980). Recognition of this control has provided a valuable tool in regional stratigraphic analysis and paleogeographic reconstruction of ancient continental margins. During the Eocene in western Washington, several basins formed in a strike-slip province associated with an oblique convergent margin. These basins contain as much as 6,000 to 9,000 meters of nonmarine strata and form some of the thickest alluvial sequences in North America. For the most part, the relationships between basin formation, sedimentation, deformation, and strike-slip faulting in western Washington have not been explored. The purpose of this paper is to describe briefly the history of these basins in the context of their tectonic setting. This task is hindered by limited sedimentologic data, a lack of conclusive data concerning fault timing and offset, post-Eocene uplift and erosion, and the presence of widespread, thick Oligocene and younger cover. This synthesis should therefore be viewed as a starting point for more detailed work in the future.

REGIONAL GEOLOGIC AND TECTONIC SETTING

The Cascades of western Washington (Fig. 1) are part of a linear mid- to late-Cenozoic volcanic arc associated with subduction of the Juan de Fuca plate below North America (Atwater, 1970). In Washington, this volcanic belt is floored by a diverse suite of structurally complex Precambrian(?) to Eocene basement (Misch, 1966; Tabor et al., 1980, 1982a, b). Exotic or allocthonous origins have been postulated for many of the older basement rocks (Misch, 1966; Vance et al., 1980; Whetten et al., 1980; Frizzell et al., 1982; Tabor et al., 1982c), but final stages of accretion and/or thrusting were completed before the Eocene. By Eocene time, these older rocks formed the continental framework of the Washington continental margin. This framework was cut by a major Late Cretaceous-early Tertiary dextral transcurrent fault network that in Washington includes the Straight Creek, Ross Lake, Chewack-Pasayten, Entiat-Leavenworth, and Puget(?) fault systems (Davis et al., 1978; Gresens, 1982; Johnson, 1984a). The Eocene nonmarine basins and strata that are the subject of this paper formed within this network and compose the youngest part of the basement for volcanic rocks of the Cascades. Differential late Tertiary uplift has brought the basement rocks to the surface in the northern Washington Cascades, whereas to the south, the basement disappears below a voluminous cover of Cenozoic volcanic rocks. Eocene nonmarine sedimentary rocks now form discontinuous outcrop belts that are preserved in structurally low areas on the eastern and western flanks of the northern and central Washington Cascades.

The Olympic Peninsula and Washington Coast Range to the west (Fig. 1) have a geologic history very different from that of the Cascades. Their basement is formed by the Crescent Formation (Cady, 1975), a thick sequence of upper Paleocene(?) to lower Eocene marine basaltic rocks interbedded with minor continent-derived clastic sedimentary rocks (Blue Mountain unit of Tabor and Cady, 1978a). The Crescent Formation is interpreted as a seamount province that formed near, and was quickly accreted to, the continental margin (Cady, 1975; Duncan, 1982; Wells et al., 1984). This basement terrane is overlain by middle Eocene and younger strata of mostly marine origin, and has been underthrust in the core of the Olympic Mountains by Eocene and younger subduction-zone deposits (Tabor and Cady, 1978a, b).

[1]Present Address: United States Geological Survey, MS-916, Box 25046, DFC, Denver, Colorado 80225

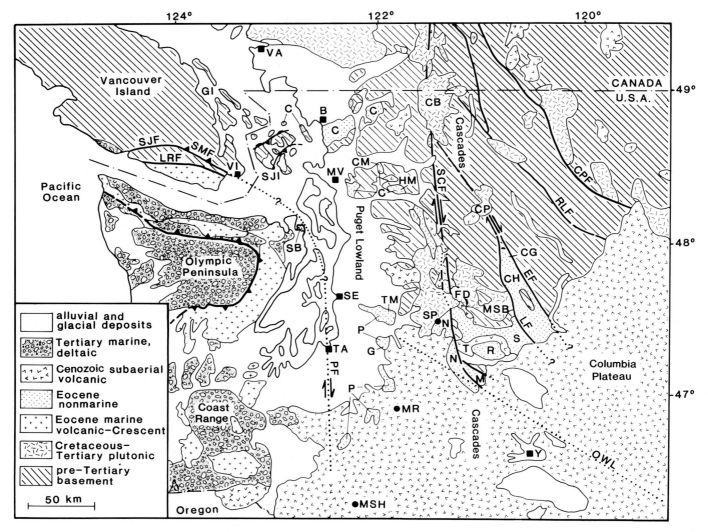

FIG. 1.—Generalized geologic map of western Washington and southwestern British Columbia. Abbreviations used on map: B, Bellingham; C, Chuckanut Formation; CB, Chilliwack batholith; CG, Chiwaukum graben; CM, Coal Mountain; CP, Cloudy Pass pluton; CPF, Chewack-Pasayten fault; CH, Chumstick Formation; EF, Entiat fault; FD, Foss River-Deception Pass graben; G, Green River section of Puget Group; GI, Gulf Islands; HM, Higgins Mountain; LF, Leavenworth fault; LRF, Leech River fault; M, Manastash River block; MR, Mount Rainier; MSB, Mount Stuart batholith; MSH, Mount Saint Helens; MV, Mount Vernon; N, Naches Formation; OWL, Olympic-Wallowa lineament; P, Puget Group and Raging River Formation; R, Roslyn Formation; RLF, Ross Lake fault; S, Swauk Formation; SB, Scow Bay unit; SCF, Straight Creek fault; SE, Seattle; SJF, San Juan fault; SJI, San Juan Islands; SMF, Survey Mountain fault; SP, Snoqualmie Pass; T, Teanaway Formation; TA, Tacoma; TM, Tiger Mountain; VA, Vancouver; VI, Victoria; Y, Yakima.

This paper concerns the Eocene geologic evolution of the continental portion of the Washington continental margin. Recent plate reconstructions of the Pacific Northwest (Duncan, 1982; Engebretson, 1982; Wells et al., 1984) indicate that during most of the Eocene, Washington was located north of a triple junction between the oceanic Kula and Farallon plates, and the North American plate (Fig. 2). Relative motions between plates (Engebretson, 1982) suggest that both the Kula-North American and the Farallon-North American plate boundaries were characterized by rapid convergence. The basaltic seamounts of the Crescent Formation are inferred to have been accreted from the Kula plate onto the North American plate during this interval (Duncan, 1982; Wells et al., 1984). Johnson (1984a) and Wells et al. (1984) have inferred that the oblique compo-

nent of Kula-North American convergence provided the driving force for Eocene strike-slip faulting and basin development in western Washington. Eocene subduction in the Pacific Northwest has been inferred to be shallow, and coupled with volcanism in the Challis volcanic "arc" (Armstrong, 1979; Dickinson, 1979; Hammond, 1979). This unusually broad diffuse volcanic belt extends southeastward through central British Columbia and into northeastern Washington, Idaho, and Montana (Fig. 2). Because of the diffuse character of volcanism, the western boundary of this "arc" has been difficult to define. Vance (1982) suggested that the Challis "arc" may have extended far to the west to include Eocene volcanic and plutonic rocks in western Washington. The Eocene basins to be discussed here (Fig. 2) therefore occupied an intra-arc or forearc tectonic setting. The transition from broad, diffuse volcanism in the Challis "arc" to more restricted volcanism along the linear Cascade trend began in the late Eocene, apparently as a response to plate reorganization and increased angle of subduction (Dickinson, 1979; Vance, 1979, 1982).

FAULTS

Three main strike-slip fault systems appear to control the distribution and geometry of Eocene nonmarine basins in western Washington. These include the Straight Creek fault (Vance, 1957; Misch, 1966, 1977), the Entiat fault (Laravie, 1976; Tabor et al., 1980, 1982a), and the inferred Puget fault (Johnson, 1984a) (Fig. 2). The Ross Lake and Chewack-Pasayten faults to the east (Fig. 1) are probably also major Late Cretaceous to early Tertiary strike-slip faults (Misch, 1977; Okulitch et al., 1977; Davis et al., 1978; Hoppe, 1982). These two faults do not appear to have greatly influenced Eocene basin development or sedimentation patterns in western Washington.

Straight Creek Fault

The Straight Creek fault (Fig. 1) extends from south to north for about 280 km through the Washington Cascades and into British Columbia before merging with the north-

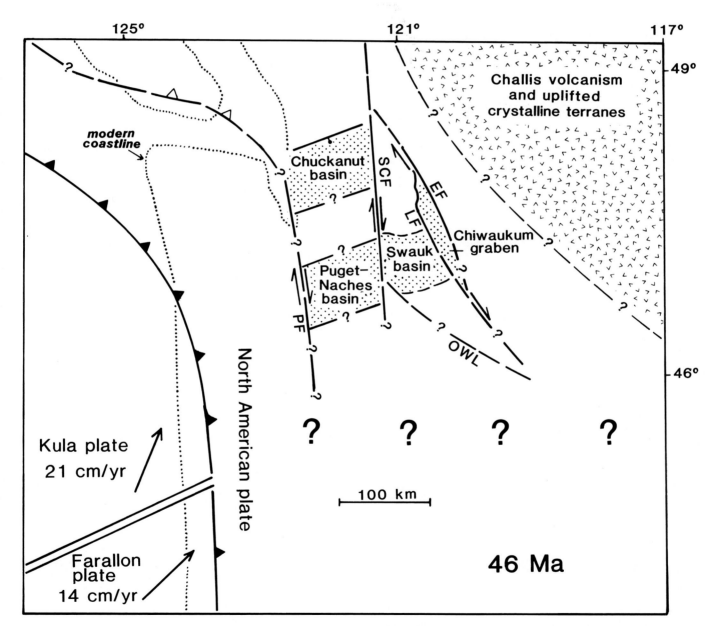

FIG. 2.—Schematic paleotectonic reconstruction for the Pacific Northwest at approximately 46 Ma. The reconstruction assumes no more than a few tens of kilometers of offset on the Straight Creek and Entiat fault systems after 46 Ma. Large question marks show area in which Eocene geology is largely concealed by younger volcanic and sedimentary rocks. Plate motion data from Engebretson (1982) and Wells et al. (1984). Basin formation and evolution are discussed in the text.

northwest-striking Fraser fault system (Vance, 1957; Misch, 1966, 1977; McTaggart and Thompson, 1967). At its southernmost exposure near Snoqualmie Pass, it breaks into several south-southeast-striking splays (Frizzell et al., 1984; Tabor et al., 1984) that parallel the Olympic-Wallowa lineament (Raisz, 1945). These splays disappear to the southeast below Miocene Columbia River Basalt and their southeastern extent is not known. Estimated offset on the Straight Creek fault based on displacement of metamorphic rocks and structures ranges from about 90 to 190 km (Misch, 1977; Okulitch et al., 1977; Frizzell, 1979; Vance and Miller, 1981; J. A. Vance, personal commun., 1984). To the north in Canada, however, cumulative offsets of 450 to 500 km have been postulated to explain offset geologic terranes on the Fraser fault system (Gabrielse and Dodds, 1977; Gabrielse et al., 1977). This fault is intruded by the Oligocene Chilliwack batholith (Misch, 1966). Eocene offset is indicated by localized intense deformation of Eocene strata along the fault (Ashleman, 1979; Vance and Miller, 1981; Tabor et al., 1982b, 1984), and the presence of slices and small basins of Eocene rocks within the fault zone (Milnes, 1976; Dotter, 1977; McDougall, 1980; Tabor et al., 1984). Frizzell (1979), Ewing (1980), and Vance (personal commun., 1984) have suggested that most or all dextral offset on the Straight Creek is Eocene.

Entiat and Leavenworth Faults

The Entiat fault zone (Fig. 1) extends north-northwest for about 160 km through the eastern Cascades and forms the eastern boundary of the Eocene Chiwaukum graben (Tabor et al., 1980, 1982a). The graben terminates to the north, where the Entiat fault merges with the Leavenworth fault, the western boundary of the graben. The southern extensions of the Entiat and Leavenworth faults, and the Chiwaukum graben, are covered by Miocene Columbia River Basalt. About 30 km north of the graben, the Entiat fault is intruded by the Miocene Cloudy Pass pluton. Still farther north, the fault probably merges with the Straight Creek fault, although the area of juncture is intruded by the Chilliwack batholith (J. A. Vance, personal commun., 1984). Locally the fault zone is up to 1,800 m wide and contains mylonites, ultramafic rocks, discontinuous elongate lenses of sheared rocks, and horizontal to vertical slickensides (Laravie, 1976; Tabor et al., 1980).

Gresens (1982a) suggested strike-slip offset on the Entiat fault on the basis of its linear trace, and contrasts in pre-Tertiary bedrock on opposite sides of the fault. He further suggested that the Chiwaukum graben is a pull-apart basin that opened in response to a transfer or stepover of dextral offset from the Leavenworth to the Entiat fault. The presence of sheared Eocene boulder conglomerate adjacent to both the Entiat and Leavenworth faults (Laravie, 1976; Tabor et al., 1980, 1982a) and the thick (6,000 to 9,000 m) Eocene fill of the Chiwaukum graben (Gresens, 1982b) indicate significant dip separation on each fault during the Eocene. However, marker units on opposite sides of the fault system needed to demonstrate and quantify strike-slip displacement have not yet been recognized. Coupling dextral strike-slip displacement on the Entiat-Leavenworth fault

and possibly structures farther east (Chewack-Pasayten fault?) with that of the Straight Creek fault might help explain significant differences between estimated offset along the Straight Creek fault system in Canada and that in Washington.

Puget Fault

Johnson et al. (1981, 1984) and Johnson (1984a,b) have inferred the presence of a major north-striking right-slip fault of Late Cretaceous to Eocene age buried below upper Eocene and younger deposits of the Puget Lowland. This inferred fault, here referred to as the Puget fault, is assumed to lie along a prominent gravity anomaly (Bonini et al., 1974), and may form a major structural boundary between the Eocene basaltic basement (Crescent Formation) of the Olympic Mountains and Washington Coast Range to the west, and pre-Tertiary basement terranes of the Cascades to the east. Paleomagnetic data for the Eocene basaltic rocks west of the inferred structure suggest a component (up to 350 km?) of northward transport (Bates et al., 1981, M. E. Beck, personal commun., 1983). Further indications of possible strike slip include a prominent petrographic mismatch of coeval Eocene sandstones on opposite sides of the inferred structure (Figs. 3, 4) and the nature of sedimentation and deformation in the zone between the inferred fault and the Straight Creek fault to the east (discussed below). At its northern end, the fault is inferred to bend to the northwest (MacLeod et al., 1977) and merge with the west-northwest-striking Survey Mountain and San Juan faults on southern Vancouver Island. Thrusting in the San Juan Islands (Brandon et al., 1983) and on the Survey Mountain fault may have a transpressional origin that is at least partly associated with this prominent bend. The amount of strike slip on this structure has not been determined, but is inferred to be significant. Cowan (1982) and Johnson (1984a) have each suggested that large tracts of pre-Tertiary rock once lay west of their present limits in western Washington and on southern Vancouver Island (i.e., west of the inferred Puget fault), and that these missing units may now reside in southern Alaska.

SEDIMENTARY BASINS

Major Eocene nonmarine basins in western Washington recognized in this paper include the Chuckanut, Puget-Naches, Swauk, and Chiwaukum graben Basins (Figs. 1, 2, 3). Each basin is areally distinct and has a unique sedimentary and deformational history. However, because of discontinuous exposure and incomplete preservation of basin fills, and the probability of temporary connections between basins, the original boundaries and dimensions of these basins are difficult to define (all discussed below). Eocene nonmarine rocks in western Washington also occur in the Manastash River block (Tabor et al., 1984) and other small fault-bounded blocks and basins within the Straight Creek fault zone (McDougall, 1980; Tabor et al., 1982b).

Chuckanut Basin

The Chuckanut Formation (the fill of the Chuckanut Basin) crops out in several discontinuous belts preserved in

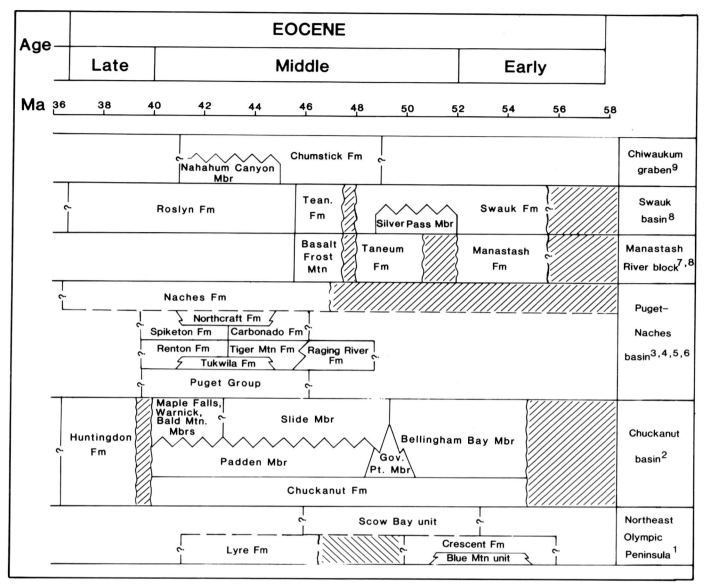

FIG. 3.—Correlation chart of Eocene strata in western Washington. Wavy lines and patterns indicate unconformities and intervals of hiatus or uplift and erosion. Sources of data: 1, Armentrout et al. (1983); 2, Johnson (1984b); 3, Snavely et al. (1958); 4, Gard (1968); 5, Vine (1969); 6, Turner et al. (1983); 7, Lewellen (1983); 8, Tabor et al. (1984); 9, Gresens et al. (1981). Time scale after Palmer (1983).

structural lows in the western North Cascades (Figs. 1, 3). The outcrop pattern suggests that the basin extended eastward to the Straight Creek fault, and Johnson (1984a) has suggested that the inferred Puget fault formed the western margin of the basin. The Boulder Creek and Lummi Island faults (Fig. 5) formed an intermittent northern basin margin (Johnson, 1984b), whereas to the south, the Chuckanut Basin may have merged with the Puget-Naches Basin (discussed below).

The history of the Chuckanut Basin is relatively well understood (Johnson, 1984b, c) and receives the most detailed attention in this paper. Seven stratigraphic members have been recognized in the 60-km-wide main outcrop belt of the Chuckanut Formation near Bellingham. Six distinct

episodes of sedimentation and/or deformation can be inferred from stratigraphic, sedimentologic, lithologic, and structural data (Fig. 6). Age control is provided by fission-track dating on zircons (Johnson et al., 1983; Johnson, 1984b), palynologic studies (Reiswig, 1982), and correlation by mapping (Johnson, 1982). Chuckanut Formation sandstones are predominantly arkosic (Fig. 4).

Eocene Deposition.—The 2,700- to 3,300-m-thick Bellingham Bay Member represents early Eocene deposition of the Chuckanut Formation (Fig. 6A). It consists of repetitive, fining-upward meandering river cycles formed by a lower coarse-grained member consisting of sandstone and minor conglomerate, and an upper fine-grained member consisting of mudstone (Johnson, 1984c). Mean-cycle and

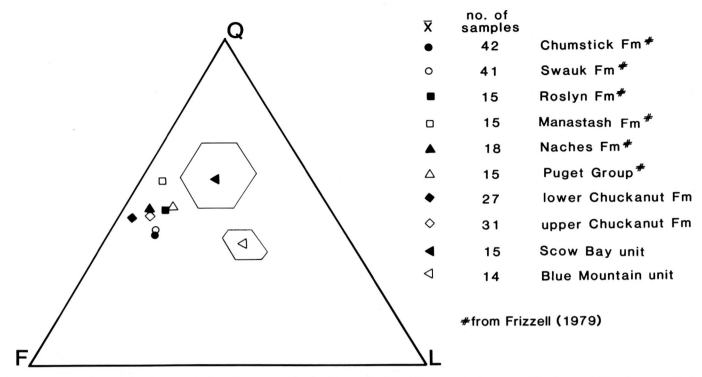

FIG. 4.—QFL diagram showing composition of medium-grained sandstone of selected Eocene units in western Washington. QFL values calculated after Dickinson and Suczek (1979). The lower Chuckanut includes the Bellingham Bay Member only. Polygons show standard deviations for each component for the Blue Mountain and Scow Bay units.

coarse-member thicknesses are 25.6 m and 8.6 m, respectively (Fig. 7). Paleocurrent directions indicate sediment transport toward the southwest (Fig. 8). Fission-track dates indicate that the base of the Bellingham Bay Member can be no older than about 55 Ma, and its upper contact is about 50 Ma (Johnson, 1984b). Sediment accumulation rates calculated from this geochronologic framework are a minimum of 50 cm/1,000 yr (Table 1).

The lower-middle Eocene Governors Point Member (Fig. 6B), consisting of sandstone and conglomerate interpreted as braided-river deposits, overlies the Bellingham Bay Member in the western part of the outcrop belt (Johnson, 1982, 1984b). It has a maximum thickness of 375 m, and thins and disappears to the east. Paleocurrent directions indicate sediment transport to the south-southwest (Fig. 8). Both conglomerates and sandstones contain graywacke, greenstone, and serpentinite detritus derived from an uplifted east-northeast trending fault block on the northern basin margin. The fault is buried for most of its trace, but is exposed on the center of Lummi Island (Fig. 5), and is here termed the Lummi Island fault. In the eastern part of the outcrop belt, the transition from the Bellingham Bay Member to the Slide Member (Figs. 6B, C), both meandering-river deposits, coincides with Governors Point sedimentation to the west.

Like the Governors Point Member, the 3,000-m-thick middle to lower-upper (?) Eocene Padden Member (Johnson, 1982, 1984b) occurs only in the western part of the outcrop belt (Figs. 5, 6C). It conformably overlies the Governors Point Member south of the Lummi Island fault; north

of the fault it rests unconformably on pre-Tertiary bedrock. The Padden Member consists of sandstone, mudstone, and conglomerate interpreted as braided- and coarse-load meandering-river deposits (Johnson, 1982). Paleocurrent data indicate southwest-to-southeast sediment transport. This dispersal pattern and the petrology of Padden Member sandstones are different from that of the east-derived Bellingham Bay and Slide Members, and indicate progradation of a second major fluvial system into the basin.

The 1,960-m-thick Slide Member (Johnson, 1982, 1984b) occurs only in the eastern part of the outcrop belt (Fig. 5, 6B, 6C). Like the underlying Bellingham Bay Member, it consists of repetitive fining-upward cycles interpreted as meandering-river deposits. The Slide Member is much finer grained than the Bellingham Bay Member; conglomerate is almost absent and strata coarser than medium-grained sandstone are rare. This change in grain size reflects a significant decrease in competence of paleo-flow between deposition of the two units. Mean thicknesses of cycles and of coarse members are 14.4 and 3.8 m (Fig. 7), much thinner than those of the underlying Bellingham Bay Member. I assume that mean coarse-member thickness is a crude estimate of paleo-channel depth. If this assumption is correct, and channel depth in meandering rivers is related to channel width in the manner Leeder (1973) has shown, then an abrupt major decrease in fluvial channel size (on the order of 3 to 4 times) occurred between the Bellingham Bay and Slide Members.

Possible explanations for the noted decrease in fluvial channel size and stream competence might include (1) a

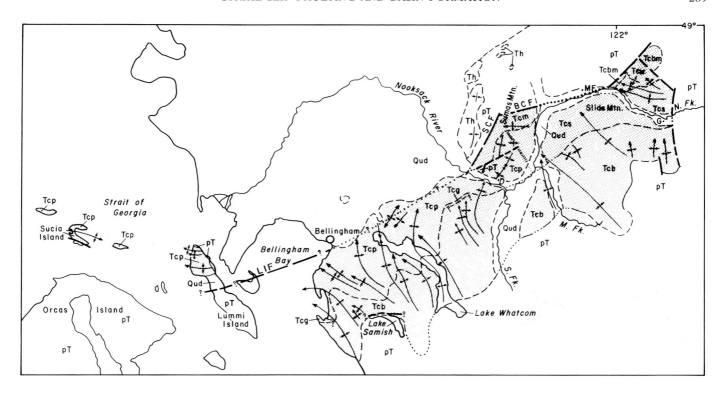

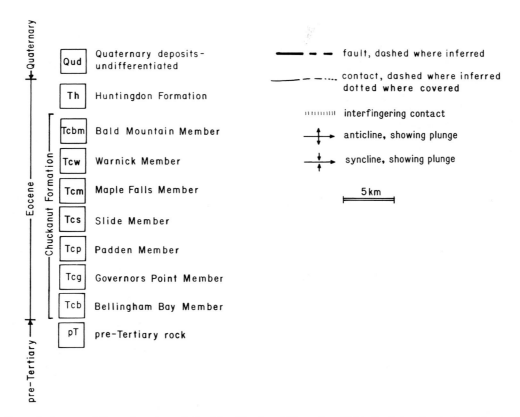

FIG. 5.—Schematic geologic map of the main outcrop belt of the Chuckanut Formation (light stippled pattern). See Johnson (1982) for detailed map. See Figure 2 for correlation of map units. Abbreviations used on map: BCF, Boulder Creek fault; G, Glacier; LIF, Lummi Island fault; SCF, Smith Creek fault.

climate change leading to decreased precipitation and discharge, or (2) a change in the size of the drainage basin. There is no evidence for a major middle Eocene climate change in the Pacific Northwest (Wolfe, 1978); so a decrease in drainage basin size appears likely.

The upper-middle to lower-upper Eocene Maple Falls, Warnick, and Bald Mountain Members (Johnson, 1984b) occur at the top of the section in the northeastern part of

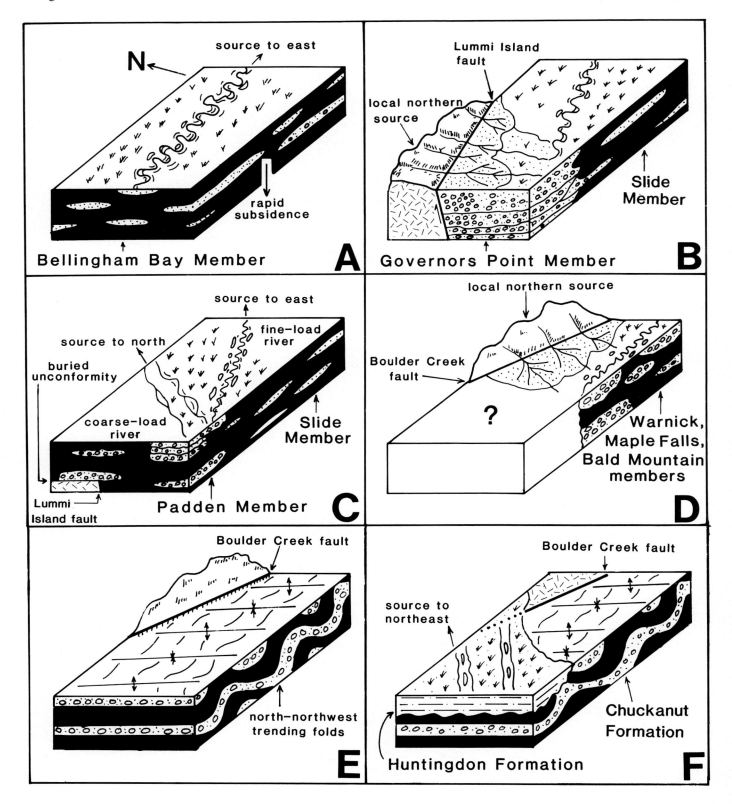

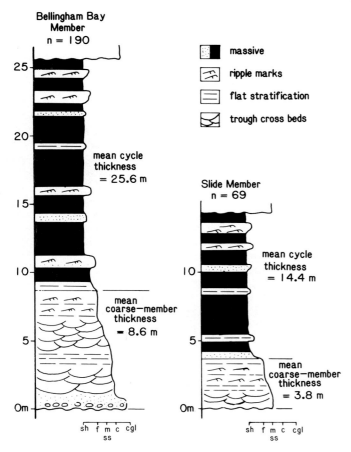

TABLE 1.—Thickness and rates of accumulation of thick, ancient nonmarine sequences

Unit	Thickness (m)	Accumulation Rate (cm/1,000 yr)
Spitsbergen—Old Red Sandstone[1]	2,000	10
Greenland—Old Red Sandstone[1]	2–5,000	30–50
Rocky Mountains—foreland basins[2,3]	2–6,000	10–20
Rocky Mountains—intermontane basins[4,5]	1–3,000	8–20
New Jersey—rift basins[4,6]	6,000	33–40
Columbia—Andes molasse[2]	8,000	15
Spain—Alpine molasse[1,7]	4,000	10–30
Switzerland—Alpine molasse[2,7]	6,000	22–40
Pakistan—Himalayan molasse[2,8,9]	7,000	16–64
Washington—Chickanut Formation	6,000	>50
Washington—Puget Group	4,270	50–1100
Washington—Swauk Formation	7,740	>72
Washington—Chumstick Formation	5,800–9,100	72–114

Sources of data: 1. Friend, 1978; 2. Van Houten, 1969; 3. McLean and Jerzyk-iewicz, 1978; 4. Eardley, 1962; 5. Anderson and Picard, 1974; 6. Fischer, 1975; 7. Van Houten, 1974; 8. Keller et al., 1977; 9. Burbank and Johnson, 1983. See text for sources of Washington data. Rates for Washington basins are calculated based on means of reported isotopic ages. Chuckanut Formation rates calculated for the Bellingham Bay Member only.

FIG. 7.—Average fining-upward cycles of meandering-river origin from the Bellingham Bay and Slide Members. Bellingham Bay data represent 78% of two partly covered measured sections (2,700 and 3,300 m thick). Slide Member data represent 50% of 1,980-m-thick partly covered section. From Johnson (1982).

the main outcrop belt (Fig. 5, 6D). The 800-m-thick Maple Falls Member and the 1,000-m-thick Warnick Member were probably laterally continuous (Fig. 5), but intervening strata have been eroded, or cut out by the Boulder Creek fault. The 500-m-thick Bald Mountain Member unconformably overlies pre-Tertiary basement on the northern basin margin (Fig. 5), and conformably overlies the Warnick Member in the basin. This situation is similar to that previously described for the Padden Member, but occurs at a higher stratigraphic level.

The Maple Falls, Warnick, and Bald Mountain Members consist of conglomerate, sandstone, and mudstone interpreted as interfingering alluvial-plain deposits (Johnson,

1982, 1984b). Paleocurrent directions are southerly. Conglomerate clasts are as large as 70 cm and can be directly matched with local greenstone-chert source terranes north of the east-northeast striking Boulder Creek fault, strongly suggesting that the Boulder Creek fault was active during sedimentation.

Eocene Deformation.—Sedimentologic and stratigraphic evidence discussed above indicate syn-depositional vertical offset on the east-northeast-striking Boulder Creek and Lummi Island faults. Following deposition, but still during the Eocene, the Chuckanut Formation was deformed into north- to northwest-trending folds (Fig. 5, 6E). These folds were subsequently cut by the last movement on the Boulder Creek fault, which is overlain at its western end by the post-Chuckanut Huntingdon Formation of late Eocene age (Fig. 6F; Miller and Misch, 1963; Hopkins, 1966; Reiswig, 1982). Together with Chuckanut stratigraphic data, this overlapping relationship constrains two phases of movement on the east-northeast Boulder Creek fault and an intervening interval of folding about northwest-trending axes to between late-middle and late Eocene time. In Figure 5, the Boulder Creek fault is shown being cut by the Smith Creek fault (Moen, 1961), but the offset on this younger fault is minor and does not significantly affect the trace of the Boulder Creek fault.

Chuckanut Formation Outside the Main Outcrop Belt.—Less is known about the Chuckanut Formation outside the main outcrop belt to the south (Fig. 1). On Higgins Mountain, about 2,500 m of Chuckanut strata comprise a sandstone-rich lower unit and a finer-grained upper unit (Jones, 1959). This transition is similar to that described for the

FIG. 6.—Schematic diagram showing six stages in the evolution of the Chuckanut Basin. A) Early to early-middle Eocene. Bellingham Bay Member deposited in west-flowing meandering-river system. B) Early-middle Eocene. Uplift on Lummi Island fault, deposition in west of Governors Point Member in south-flowing braided-river system, and deposition in east of Slide Member in west-flowing meandering-fluvial system. C) Middle Eocene. Continued Slide Member deposition in east. Padden Member deposited in coarse-load fluvial system to west, unconformably on uplifted fault block on northern basin margin, and conformably on older Chuckanut deposits in the basin. D) Late-middle to early-late Eocene. Uplift on Boulder Creek fault and deposition of Maple Falls, Warnick, and Bald Mountain Members as alluvial fans in east. E) Late-middle to early-late Eocene. Chuckanut Formation deformed into northwest-trending folds, which are subsequently cut by Boulder Creek fault. F) Late Eocene. Huntingdon Formation deposited unconformably over trace of Boulder Creek fault.

PALEOCURRENTS

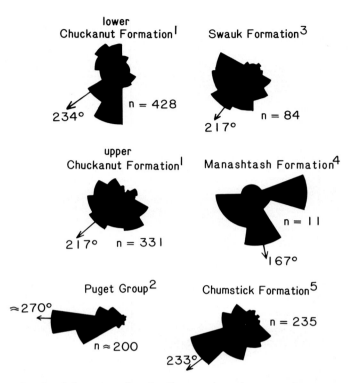

FIG. 8.—Paleocurrent data for Eocene nonmarine strata in western Washington. Sources of data: 1, Johnson (1984b); 2, Buckovic (1979); 3, this paper; 4, Lewellen (1983); 5, Buza (1979). The lower Chuckanut Formation includes the Bellingham Bay Member only (Fig. 3). Swauk data are entirely from the easternmost portion of the main Swauk outcrop belt (Fig. 10).

Bellingham Bay and Slide Members in the main outcrop belt, and may be correlative. On Coal Mountain, 1,700 m of sandstone and mudstone are probably correlative with the Bellingham Bay Member (Robertson, 1981). South of Mount Vernon, the Chuckanut is interbedded with rhyolitic rocks dated at 52.7 ± 2.5 Ma. A bimodal suite of rhyolite flows, tuffs, and intrusive (?) rocks (ages from 38 to 45 Ma) and basaltic dikes (41 to 50 Ma) are also present in the Mount Vernon area and locally intrude the Chuckanut (J. T. Whetten, personal commun., 1981). Frizzell (1979) described the petrology of isolated arkosic bodies still farther south (Fig. 1) that may also be Chuckanut correlatives.

Puget-Naches Basin

Strata of the Raging River Formation, Puget Group, and the Naches Formation (Figs. 1, 3) underlie a large area of the Puget Lowland and western Cascades west of the Straight Creek fault. These strata are discontinuously exposed below an extensive cover of Oligocene and younger volcanic and sedimentary rocks, Quaternary deposits, and dense vegetation. Because the Puget Group and Naches Formation are the same age, have similar sandstone petrology

(Frizzell, 1979), and there is no structural or sedimentological evidence for Eocene uplift in the covered area between outcrops of the Puget Group and Naches Formation, I have assumed that they were deposited in a single basin. The Straight Creek fault probably formed the eastern basin margin (as shown by Tabor et al., 1984, Fig. 9C). To the west, the inferred Puget fault may have formed an intermittent basin margin, or the Puget Group fluvial system may have crossed the inferred fault and connected with Eocene deltaic deposits near Centralia (Snavely et al., 1958; Buckovic, 1979). To the north, the basin may have merged with the Chuckanut Basin (discussed below). Because of cover, the southern limit of the basin is impossible to determine. The thickness of the basin fill varies and is as great as 4,270 m (Vine, 1969). Buckovic constructed a depositional model for the western Puget-Naches Basin that is summarized in Figure 9 and briefly discussed below.

Eocene Deposition.—The Raging River Formation (Vine, 1969) was deposited in the northwestern part of the Puget-Naches Basin during middle Eocene time (Fig. 9A). It consists of more than 1,000 m of shallow(?) marine siltstone, sandstone, and conglomerate of dominantly volcanic derivation. Its base is not exposed.

The Raging River Formation is overlain by the Puget Group (Vine, 1969), which includes several formations of local extent (Fig. 9B, C). These formations comprise delta-plain facies (Tiger Mountain, Carbonado, Renton, and Spiketon Formations) and continental volcanic facies (Tukwila and Northcraft Formations; Snavely et al., 1951; Gard, 1968; Vine, 1969). Delta-plain deposits consist of channel sandstones and interchannel mudstones and coal. Paleocurrent data from delta-plain facies in part of the Puget Group indicate sediment transport to the west (Fig. 8) (Buckovic, 1979). Buckovic (1979) and Frizzell (1979) noted that the sandstones are arkosic (Fig. 4), texturally immature, and variable in composition. They related these characteristics to multiple source terranes and minimal sediment reworking.

The Northcraft and Tukwila Formations (Fig. 9B) consist dominantly of andesitic volcanic flows, agglomerates, and associated volcaniclastic rocks. These two units occupy a similar stratigraphic position, but formed at separate and distinct volcanic centers. Maximum thicknesses are 2,100 m for the Tukwila (Vine, 1969) and 630 m for the Northcraft (Gard, 1968). These units thin abruptly away from their eruptive sources and do not occur in many parts of the western Puget-Naches Basin.

Vine (1969) noted a significant decrease in the thickness of Puget-Naches Basin strata from 4,270 m near Tiger Mountain to 1,890 m along the Green River, about 20 km to the south (Fig. 1). Turner et al. (1983) isotopically dated volcanic ash partings near the upper and lower boundaries of the Green River section at 41.2 ± 1.8 Ma and 45.0 ± 2.1 Ma, respectively. These dates suggest sediment accumulation rates of about 50 cm/1,000 yr (reported by Turner et al. as ≥ 25 cm/1,000 yr). If Vine's correlation (1969, p. 26–27, and plate 2) between the Green River and Tiger Mountain sections is correct, then an even higher sediment accumulation rate (about 110 cm/1,000 yr) is indicated for the Tiger Mountain section. Gard (1968, p. 28) also discussed significant local differences in the subsidence history of the Puget-Naches Basin for the area south of Tiger Mountain and the Green River.

The 1,500- to 3,000-m-thick middle to upper(?) Eocene Naches Formation (Tabor et al., 1984) crops out in the eastern part of the Puget-Naches Basin along the west side of the Straight Creek fault. Tabor et al. (1984) recognized seven lithologic units in the Naches. Its lower part consists largely of fluvial arkosic sandstone (Fig. 4) with common interbeds of volcanic rocks, which are mostly rhyolite. Basalt predominates higher in the section, although arkosic sandstone interbeds are still abundant. The Naches Formation unconformably overlies pre-Tertiary basement. Local unconformities within the Naches are common (Foster, 1960; Tabor et al., 1984).

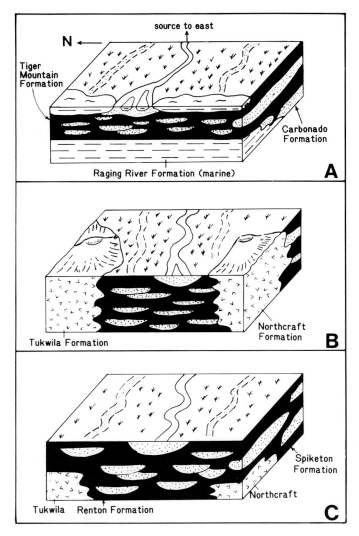

FIG. 9.—Schematic diagram showing three stages in the evolution of the western Puget-Naches Basin. A) Middle Eocene. Fluvial-deltaic deposits of the Carbonado and Tiger Mountain Formations prograde over marine strata of the upper-lower to lower-middle Eocene Raging River Formation. B) Middle Eocene. Tukwila and Northcraft volcanic centers build on delta plain. C) Late-middle to late Eocene. Fluvial-deltaic deposits of Spiketon and Renton Formations bury and prograde over Tukwila and Northcraft volcanic centers. Adopted from Buckovic (1979).

Eocene Deformation.—Evidence for abrupt stratigraphic thinning and thickening in the western Puget-Naches Basin suggests active syn-depositional tectonism (Vine, 1969; Buckovic, 1979). Folds and faults in the Puget Group involve the unconformably overlying Oligocene Ohanapecosh Formation and are thus post-Oligocene in age (Gard, 1968, p. 27).

The Naches Formation in the eastern part of the basin is tightly folded, faulted, and locally overturned adjacent to the Straight Creek fault (Ashleman, 1979; Tabor et al., 1984). Fold axes trend northwest (Fig. 10), and Ashleman (1979) ascribed the fold geometry to right-lateral shear on the Straight Creek. In contrast to the Puget Group in the western part of the basin, the Naches is unconformably overlain by probable correlatives of the Ohanapecosh Formation (Tabor et al., 1984). Local unconformities within the Naches Formation indicate that periods of deformation interrupted Naches deposition. Tabor et al. (1984) have named this structurally complex area, underlain by the Naches Formation and adjacent to the Straight Creek fault, the Cabin Creek structural block.

Swauk Basin

The Swauk Basin (Figs. 1, 10) is located between the Straight Creek fault on the west and the Leavenworth fault on the east, and coincides with the Teanaway River structural block of Tabor et al. (1984). To the north, strata of the Swauk Basin are in depositional and fault contact with pre-Tertiary rock, whereas to the south, these strata are covered by Miocene Columbia River basalt. Three different stages in the evolution of the Swauk Basin are shown in Figure 11.

Eocene Deposition.—The lower to lower-middle Eocene Swauk Formation (Smith and Calkins, 1906; Tabor et al., 1984), consisting of arkosic sandstone (Fig. 4), mudstone, and conglomerate, unconformably overlies pre-Tertiary basement and is the oldest unit in the basin (Figs. 3, 11A). Tabor et al. (1982a) mapped six lithologic units in the Swauk Formation in the eastern part of the basin. The Swauk Formation was deposited in alluvial fans, braided- and meandering-fluvial, and lacustrine environments (Taylor et al., 1984). Tabor et al. (1984) inferred that the Late Cretaceous Mount Stuart batholith (Erikson, 1977) to the north was the major sediment source and that drainage was to the south. The presence of Mount Stuart-derived conglomerate along the northern edge of the Swauk Basin outcrop belt supports this interpretation. Paleocurrent data from the eastern portion of the outcrop belt (Fig. 8) and the presence of east-derived boulder conglomerate (Tsf unit of Tabor et al., 1982a) in the Swauk Formation along the Leavenworth fault, however, suggest that eastern sources were also important. In the eastern portion of the outcrop belt, the top of the Swauk section includes more than 1,180 m of lacustrine strata with east-directed paleocurrents (Taylor et al., 1984), indicating that there was at least local internal drainage.

The Swauk Formation includes andesitic to rhyolitic volcaniclastic rocks and flows of the Silver Pass Volcanic Member (Tabor et al., 1984). The Silver Pass is locally up to 1,800 m thick and has been dated at about 50 to 52 Ma

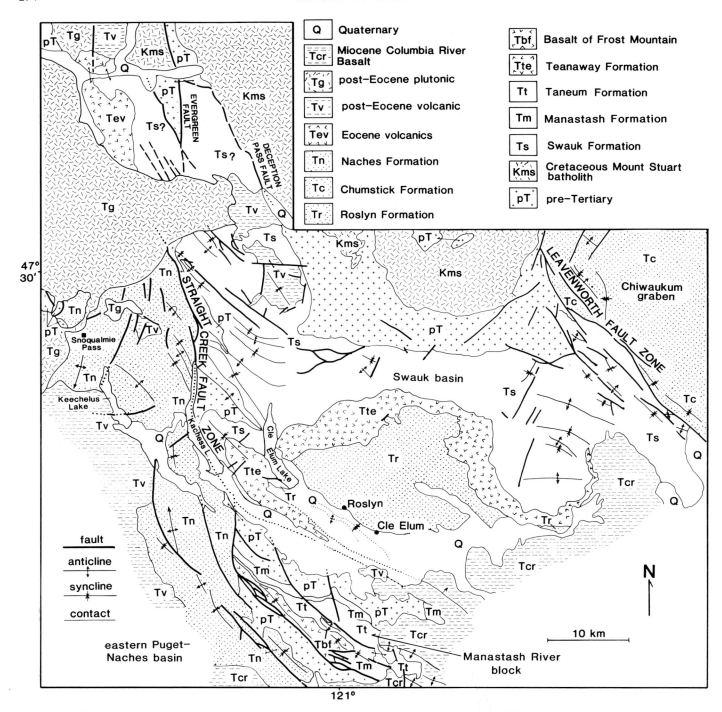

FIG. 10.—Schematic geologic map of a portion of the central Cascades of Washington, showing the Swauk Basin, the Manashtash River block, the eastern portion of the Puget-Naches Basin, and the western portion of the Chiwaukum graben. Map symbols are the same as in Figure 5. From Tabor et al. (1982a, 1984) and McDougall (1980).

(Tabor et al., 1984). Tabor et al. (1984) have estimated a thickness of about 4,800 m for the Swauk Formation below the Silver Pass Member, while G. T. Fraser and J. W. Roberts (personal commun., 1984) have measured more than 2,940 m of Swauk Formation above the Silver Pass. On the basis of palynological data, Newman (1981) suggested that the base of the Swauk does not extend down into the Paleocene. Folds in the Swauk Formation are cut by dikes feeding basalts of the unconformably overlying Teanaway Formation (47 Ma; Smith, 1904; Gresens, 1982b; Tabor et al., 1984), providing an upper maximum age limit for Swauk sedimentation. Given these age brackets and using 57.8 Ma as the Paleocene-Eocene boundary (Palmer, 1983), minimum sediment accumulation rates of about 72 cm/1,000 yr can be calculated (Table 1).

The Swauk Formation was deformed into broad to tight

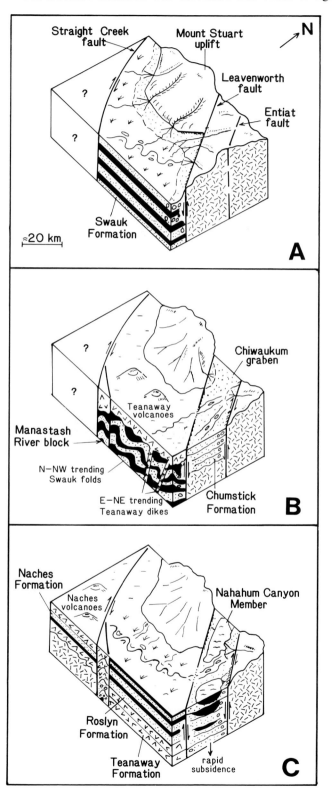

FIG. 11.—Schematic diagram showing three stages in the Eocene history of the central Cascades of Washington. A) Early to early-middle Eocene. Swauk Formation deposited in west- and south-flowing fluvial systems in Swauk Basin. Uplifted source areas include terranes east of the Leavenworth fault and pre-Tertiary rocks of the Mount Stuart uplift. B) Middle Eocene. Swauk Formation has been folded, uplifted, partly eroded, and intruded by dikes feeding the Teanaway Formation. Volcanic rocks and interbedded sediments of the Teanaway are being deposited unconformably on the Swauk. To the east, the sense of offset on the Leavenworth fault has reversed, and the fluvial Chumstick Formation is being deposited in the newly-forming Chiwaukum graben. To the west, small basins (i.e., Manastash River block) are forming and being deformed in the Straight Creek fault zone. C) Late-middle Eocene. Roslyn Formation is deposited in south(?)- and west(?)-flowing fluvial systems. Chiwaukum graben continues to subside, and the Nahahum Canyon Member is deposited in lacustrine environments. Volcanism and sedimentation of the Naches Formation are occurring west of the Straight Creek fault. Tabor et al. (1984, Fig. 9) have also constructed a diagram showing the Eocene history of the central Cascades, from which some of the ideas and features illustrated here have been obtained.

north-northwest-trending folds between deposition of the Silver Pass Member (50 to 52 Ma) and Teanaway Formation (47 Ma; Fig. 3, 11B; Gresens, 1982b, Tabor et al., 1984). Folds in the Swauk Formation are unconformably overlain by the Teanaway, which consists of basaltic, andesitic, and minor rhyolitic flows, breccia, tuff, and minor sandstone. The Teanaway ranges in thickness from about 10 to 2,500 m, presumably as a function of distance from intrabasinal volcanic centers (Tabor et al., 1984).

The Teanaway is conformably overlain by the 2,590-m-thick Roslyn Formation (Fig. 11C; Walker, 1980), which consists largely of arkosic sandstone (Fig. 4), mudstone, coal, and in its lower part, minor rhyolite flows. On the basis of unpublished facies and paleocurrent data, Walker (1980, p. 5) inferred that the Roslyn formed in a west-flowing meandering river system. Alternatively, Tabor et al. (1984) suggested a south-dipping paleoslope. Newman's (1981) palynologic data indicate a middle to late Eocene age.

Eocene Deformation.—Middle Eocene (50 to 47 Ma) folding and faulting of the Swauk Formation was most intense adjacent to the Straight Creek and Leavenworth faults (Fig. 10). Fold axes trend northwest, an orientation consistent with north-south oriented right-lateral shear (Frizzell, 1979). Teanaway feeder dikes have a general northeast trend, perpendicular to the trend of Swauk folds (Gresens, 1982b; G. T. Fraser and S. B. Taylor, personal commun., 1984). Both the Teanaway and Roslyn Formations are tightly folded (northwest-trending fold axes) and faulted near the Straight Creek fault zone (Walker, 1980). Farther from the Straight Creek, the Teanaway and Roslyn have moderate to gentle dips, typically less than 30° (Walker, 1980; Tabor et al., 1982a).

Swauk Formation Outside the Main Swauk Basin Outcrop Belt.—The easternmost protuberance of the Miocene Snoqualmie batholith intrudes the Straight Creek fault and forms the northwesternmost border of the main Swauk outcrop belt (Figs. 1, 10). North of this protuberance, McDougall (1980) mapped a 5-km-wide graben (here referred to as the Foss River-Deception Pass graben) in the broadly defined Straight Creek fault zone (Figs. 1, 10). The graben fill consists of 4,300 m of conglomerate and lithic to arkosic sandstone assigned to the Swauk Formation. The west side of the graben is bounded by the Evergreen fault and the Tonga Ridge horst, which is underlain by the pre-Tertiary Easton Schist. Alluvial fans draining this horst are thought to have been the main source of sediment to the graben. The east side of the graben is bounded by the Deception Pass fault and the Mount Stuart batholith, inferred to be a secondary sediment source. The basin fill was folded along north- to northwest-trending axes prior to eruption of overlying upper Oligocene volcanic rocks (Tabor et al., 1984).

Other small outcrop belts of arkosic sandstone, conglomerate, and mudstone, inferred to be of Eocene age and mapped as the Swauk Formation, occur adjacent to and west of the Straight Creek fault, between the main Swauk Basin outcrop belt and outcrops of the Chuckanut Formation to the northwest (Milnes, 1976; Dotter, 1977; Frizzell, 1979; Tabor et al., 1982b, 1984). The petrology and affinity of these isolated arkosic bodies have been discussed by Frizzell (1979). Limited sedimentologic, stratigraphic, and geochronologic data do not yet permit the history of these strata and their relationships to the Swauk and Chuckanut Basins to be determined.

Manastash River Block

The 10-km-wide Manastash River block (Frizzell et al., 1984; Tabor et al., 1984) is bounded on the west by the Straight Creek fault and the Naches Formation, and on the east by a splay of the Straight Creek fault and the Swauk Basin (Figs. 1, 11). The southern end of the Manastash River block is unconformably overlain by Columbia River Basalt.

Eocene Deposition.—The Manastash River block is underlain by a stratigraphic succession consisting in ascending order of arkosic to quartzose sandstone, mudstone, coal, and minor conglomerate of the 950-m-thick Manastash Formation, andesitic to rhyolitic pyroclastic rocks of the 1,000-m-thick Taneum Formation, and basaltic to andesitic flows of the Basalt of Frost Mountain. Angular unconformities separate the Manastash and the Taneum (Lewellen, 1983, p. 26), and probably the Taneum and Basalt of Frost Mountain (Tabor et al., 1984). Tabor et al. (1984) favor early to early-middle Eocene ages for these strata and correlate them with the Swauk, Silver Pass, and Teanaway Formations in the Swauk Basin (Fig. 3).

Sandstone of the Manastash Formation is more quartz-rich than Swauk Formation sandstone (Fig. 4; Frizzell, 1979) and limited paleocurrent data (Lewellen, 1983) suggest a northern source (Fig. 8). Tabor et al. (1984) suggest that the Manastash may be a down-paleoslope equivalent of the Swauk Formation, and that the petrologic changes might be a function of increased physical and chemical weathering during transport. Tabor et al. (1984) also suggest sources to the west and northwest for the Manastash Formation. Lewellen (1983) suggested that the petrographic contrast between the Swauk and Manastash Formations reflects deposition in separate fluvial systems that drained different sources. In that Johnson (1984b) noted no significant petrologic changes in sandstones over a 60 km down-

paleoslope distance in correlative arkosic Chuckanut Formation sandstone, it seems more likely that differences in Swauk-Manastash petrology reflect derivation from different source terranes. The Manastash may have been deposited in a small graben within the Straight Creek fault zone (Tabor et al., 1984).

Eocene Deformation.—Rocks of the Manastash River block are deformed into en echelon, open, northwest-trending folds (Lewellen, 1983; Frizzell et al., 1984; Tabor et al., 1984). Locally, beds are overturned and deformed into tight, minor folds. The deformation is post-early Eocene (Manastash Formation) and pre-Miocene (Columbia River Basalt) in age. Angular unconformities within and above the Eocene section indicate that some, if not all, of this deformation was Eocene.

Chiwaukum Graben

The Chiwaukum graben (Fig. 1) in the east-central Cascades (Tabor et al., 1980, 1982a) is bounded on the east by the Entiat fault and the pre-Late Cretaceous Swakane Biotite Gneiss (Waters, 1932), and on the west by the Leavenworth fault, the Swauk Formation, the Mount Stuart batholith, and the Jurassic Ingalls ophiolite (Miller and Frost, 1977). A 15-km-long horst block (Eagle Creek horst) of Swakane Biotite Gneiss occurs in the center of the graben. The graben is approximately 20 km wide at its widest point, and more than 80 km long. The southern end of the graben is covered by Miocene Columbia River Basalt.

Eocene Deposition.—The oldest sedimentary rocks in the graben consist of arkosic sandstone, mudstone, and minor pebble to boulder conglomerate. Gresens (1983) assigned these rocks to the lower to middle Eocene Swauk Formation and argued that they were deposited before the main middle Eocene pulse of graben subsidence (Figs. 11B, C). He suggested the presence of a concealed erosional unconformity between these rocks and the middle Eocene Chumstick Formation. If this is the case, the Swauk Formation may have intermittently extended eastward to the Entiat fault or beyond. On the basis of petrographic data, however, Frizzell (1979) and Tabor et al. (1982a) included these lower strata in the overlying and lithologically similar Chumstick Formation. If this is true, then the base of the Chumstick Formation is not exposed (Fig. 3).

The Chumstick Formation, consisting of arkosic sandstone (Fig. 4), mudstone, conglomerate, and minor tuff, forms the main fill of the Chiwaukum graben. Tabor et al. (1982a) mapped five lithologic units in the Chumstick. Locally derived boulder conglomerate occurs adjacent to the Entiat and Leavenworth faults, and along the flanks of the Eagle Creek horst (Cashman and Whetten, 1976; Laravie, 1976; Frizzell and Tabor, 1977; Tabor et al., 1982a). Nine fission-track ages from zircon in Chumstick tuffs range from 41.9 ± 6.8 to 48.8 ± 7.2 Ma (Gresens et al., 1981).

In early Chumstick time (Fig. 11B), alluvial fans on the graben margins fed sediment to fluvial systems in the graben. Limited paleocurrent data from the eastern portion of the graben indicate sediment transport to the southwest (Buza, 1979). In late Chumstick time (Fig. 11C), drainage became more restricted and a thick sequence of lacustrine strata, the Nahahum Canyon Member (Gresens et al., 1981), was deposited over a large portion of the graben.

Gresens et al. (1981) estimated the thickness of the Chumstick Formation as at least 5,800 m. Chappell (1936) noted a 9,100-m-thick homoclinal Chumstick section on the west side of the Eagle Creek horst. He considered this stratigraphic thickness incredible and concluded that it must result from undetected isoclinal folding. However, there are no overturned beds in the section, and the well-documented tuff stratigraphy indicates a continuous section broken only by minor faults (Gresens, 1983). Using 8 m.y. as an estimate for duration of Chumstick deposition, and 5,800 to 9,100 m as Chumstick thickness, sediment accumulation rates of 72 to 114 cm/1,000 yr can be calculated (Table 1).

Igenous activity in the graben synchronous with Chumstick sedimentation is indicated by isotopic dates of 48.3 ± 2.8 and 41.5 ± 2.6 Ma on gabbro dikes, and 43.2 ± 0.4 and 41.4 ± 1.6 Ma on rhyodacite domes (Gresens, 1983). *Eocene Deformation.*—Gresens (1982b) suggested that in the Chiwaukum graben the rocks considered by him to be Swauk Formation were folded in the early-middle Eocene, synchronous with folding of the Swauk Formation outside the graben. Formation of the graben probably began at about 46 to 48 Ma (Gresens, 1982b). The Chumstick Formation was folded and faulted prior to deposition of the unconformably overlying Oligocene Wenatchee Formation (Gresens, 1983).

DISCUSSION

Eocene nonmarine basins in Washington formed within a regional network of faults that have demonstrated or probable Eocene right-lateral offset. These basins variously contain 4,270 to 9,100(?) m of rapidly deposited strata, which form some of the thickest alluvial sequences in North America. Sedimentation was characterized by abrupt facies changes and rapidly reorganized drainages. Intrabasinal unconformities and abrupt stratigraphic thinning and thickening are common. All of the basins contain volcanic rocks, suggesting high heat flow. Pulses of subsidence and uplift alternated in many of the basins. Eocene fold and fault patterns are consistent with simple predictive models for right-lateral shear (e.g., Harding, 1974). Deformation was most pronounced adjacent to faults. Because the sedimentological and deformational characteristics listed above are similar to those reported for other modern and ancient strike-slip basins (e.g., Crowell, 1974a, b; Ballance and Reading, 1980; Hempton et al., 1983; Mann et al., 1983), I conclude that strike-slip faulting was an important control in the evolution of Eocene nonmarine basins in Washington.

Rapid subsidence is typical of strike-slip basins and commonly leads to formation of anomalously thick sedimentary sequences, although in many strike-slip basins the stratigraphic thickness may exceed the depth to the basin floor (e.g., Ridge Basin of southern California; Crowell and Link, 1982). In basins or parts of basins where a geochronologic framework exists, sediment accumulation rates of greater than 50 cm/1,000 yr (Chuckanut Formation), 50–110 cm/1,000 yr (Puget Group), greater than 72 cm/1,000 yr (Swauk Formation), and 72–114 cm/1,000 yr (Chumstick Formation) can be estimated. These rates are higher than those estimated for other ancient thick nonmarine sequences formed in tectonically active regions (Table 1).

Lateral offsets along master faults can produce major sedimentologic changes in strike-slip basins. The prominent decrease in the size of the fluvial system between the Bellingham Bay and Slide Members of the Chuckanut Formation, recorded by cycle parameters (Fig. 7) and grain size, may have resulted from fault-induced drainage reorganization. The transitions from fluvial to lacustrine sedimentation in both the Swauk and Chumstick Formations also require a major change from open to internal or partly internal drainages. Internal drainage into lakes is a characteristic of many basins controlled by strike-slip faults (Hempton et al., 1983).

Master-fault offset can also result in juxtaposition of correlative sequences with contrasting petrology. Chuckanut Formation sandstones are very different from those of coeval marine strata (Scow Bay and Blue Mountain units) on the west side of the inferred Puget fault (Fig. 4; Johnson, 1984a). Sandstones in the apparently coeval and adjacent Swauk and Manastash Formations also differ significantly (Fig. 4; Frizzell, 1979).

Abrupt stratigraphic thinning and thickening (Chuckanut and Puget-Naches Basins), and intrabasinal and basin-margin unconformities (Chuckanut, Puget-Naches, and Swauk Basins) generally indicate local tectonic control and are typical features of strike-slip basins. Local tectonic control commonly results in abrupt lateral and vertical facies changes, similar to those described for Eocene nonmarine basins in Washington.

Eocene deformation in these basins is geometrically consistent with simple models (e.g., Harding, 1974) for right-lateral shear along north-striking strike-slip faults. Eocene folds in the Chuckanut Basin have a north-northwest trend, whereas the extensional northern basin margin has a transverse east-northeast trend. Early-middle Eocene folds in the Swauk Basin have a north-northwest trend, perpendicular to the trend of extensional northeast-oriented Teanaway basaltic feeder dikes. Eocene folds in the Naches Formation and the Manastash River block also have northwesterly trends. In both the Chuckanut and Swauk Basins, alternating intervals of subsidence, folding and uplift, and renewed subsidence can be bracketed geochronologically or stratigraphically to a few million years. Similar short-lived alternating intervals of subsidence and uplift are typical of basins in strike-slip provinces (Reading, 1980; Ballance et al., 1983).

Significantly, Eocene deformation is local and not synchronous throughout western Washington (Fig. 3). The most pronounced deformation occurs near faults. The Manastash-Taneum unconformity is found only in the Manastash River block. The early-middle Eocene folding of the Swauk Formation is not expressed in the Chuckanut Basin west of the Straight Creek fault. The late-middle to early-late Eocene folding of the Chuckanut Formation is not strongly expressed elsewhere. Deformation and deposition clearly overlapped in time throughout western Washington during the Eocene. The control on this deformation was therefore heterogeneous and almost certainly a product of local fault geometry and kinematics.

Extension in strike-slip settings can produce high heat

flow and volcanism (Crowell, 1974a; Mann et al., 1983). Volcanic rocks of possible extensional origin in Eocene nonmarine basins in Washington are voluminous, including unnamed basaltic dikes and rhyolitic flows in the Chuckanut Basin, the andesitic Northcraft-Tukwila volcanics and Naches Formation volcanics in the Puget-Naches Basin, the andesitic-rhyolitic Silver Pass Member and the basaltic to rhyolitic Teanaway Formation in the Swauk Basin, the andesitic Taneum Formation and Basalt of Frost Mountain in the Manastash River block, and rhyolitic domes and gabbro dikes in the Chiwaukum graben. Because these nonmarine basins probably formed within an oblique-convergent margin, the relative importance of strike-slip extension versus subduction-generated magmatism in the production of these volcanic rocks is difficult to evaluate.

PALEOGEOGRAPHY AND BASIN FORMATION

Paleogeographic reconstruction of Eocene nonmarine basins in Washington is hampered by limited outcrop and incomplete knowledge of the timing of faulting. A speculative model for basin formation at approximately 47 Ma is shown in Figure 2. The model assumes that offset on the Straight Creek and Entiat-Leavenworth fault systems after 46 Ma is no more than a few tens of kilometers.

The Chuckanut Basin probably formed as a large pull-apart trough between the Straight Creek fault and the inferred Puget fault. The basin may have been continuous with or coalesced with the Puget-Naches Basin to the south, although outcrops needed to test this assertion are lacking. Basin width (about 90 km) was probably controlled by the magnitude of fault spacing, and basin length ($\geq$200 km) by the amounts of north-south fault overlap and synchronous lateral offsets on the two master faults. Offset on the inferred Puget fault may decrease to the north where it probably bends into a transpressive zone south of the San Juan Islands and on Vancouver Island (MacLeod et al., 1977). Much of the dextral offset on the Puget fault may have been taken up by the Straight Creek fault. The southern end of this composite Chuckanut-Puget-Naches Basin is covered by post-Eocene strata and roughly coincides with the southeasterly bend of the Straight Creek fault. It may be that the Chuckanut and Puget-Naches Basins formed two distinct depocenters within this hypothetical composite basin. On the basis of mathematical models, Rodgers (1980, p. 37) suggested that the development of two depocenters near the ends of master faults is a normal stage in the evolution of pull-apart basins.

Fault-wedge basins form in anastomosed strike-slip systems where the relative displacements on two diverging faults combine to produce an extensional zone between the faults (Crowell, 1974a). If the Straight Creek fault continues to the southeast on the trend of the Olympic-Wallowa lineament until it merges with the inferred southern extension of the Entiat-Leavenworth fault system, then the Swauk Basin could have formed as a fault-wedge basin in a zone of extension between these diverging fault zones. The Swauk Basin outcrop belt occurs in the area of maximum spacing between these two master faults.

Gresens (1982a) suggested that the Chiwaukum graben formed as a pull-apart basin that opened when lateral offset of the Leavenworth fault was transferred to the Entiat fault. The great thickness of the Chumstick Formation might then be related to a migrating depocenter, such as Crowell and Link (1982) have described for the pull-apart Ridge Basin in southern California. The southern end of the basin is now covered with Miocene flood basalt. Using the mean approximate length:width ratio of 3:1 that Aydin and Nur (1982) calculated for modern and ancient pull-apart basins, it seems unlikely that the 80 km-by-20 km Chiwaukum graben would extend far to the south below its cover. However, the presence of other strike-slip basins below this cover, associated with the southern extensions of the Straight Creek-Olympic-Wallowa lineament and the Entiat-Leavenworth fault systems, seems likely.

Origins for the sedimentary accumulations localized within the Straight Creek fault zone (e.g., Manastash River block, Foss River-Deception Pass graben) are difficult to determine. The development of these small basins or basin fragments was undoubtedly significantly affected by strike-slip faulting (Tabor et al., 1984).

If the model for basin formation proposed here is correct, then the Chuckanut-Puget-Naches, and Swauk Basins are notable in that they are quite large (90 km and 50 km wide) with respect to most other ancient pull-apart and fault-wedge basins (e.g., Aydin and Nur, 1982). Their existence would suggest that processes of basin formation operate at a variety of scales, and that strike-slip basins of similar size may occur in other strike-slip provinces. As with Eocene nonmarine basins in Washington, the effects of transpressive deformation and subsequent erosion could make positive identification of their origin and size difficult.

Significantly, the framework sandstone petrography of the Chuckanut, Swauk, Roslyn, Naches, and Chumstick Formations, and the Puget Group, is very similar (Fig. 4; Frizzell, 1979; Johnson, 1984a). Though the faulted margins of many of the basins contributed abundant lithic debris, most of the detritus is arkosic. Johnson (1984b) has inferred that rapidly uplifted crystalline terranes to the east (Mathews, 1981) were the main sediment source. This implies some degree of connection between different basins, and intermittent relief on master faults. Frizzell (1979) postulated that the Chuckanut and Swauk outcrop belts may have been contiguous and were offset by the Straight Creek fault. Stratigraphic and sedimentologic data needed to test this hypothesis rigorously are not yet available.

Finally, it should be noted that although Eocene nonmarine basins in Washington occupied an intra-arc or fore-arc tectonic setting, they are very different in shape, subsidence and sedimentation history, and composition from the range of forearc and intra-arc basin types generally associated with convergence at oceanic-continental plate boundaries (Dickinson, 1974). Dickinson et al. (1979), Howell et al. (1980), and others have previously discussed strike-slip faulting and basin formation in oblique-convergent margins. This paper further emphasizes the importance of oblique convergence in generating strike-slip faults and strike-slip basins.

SUMMARY

Eocene strike-slip basins in Washington include the Chuckanut, Puget-Naches, Chiwaukum graben, and Swauk Basins, and small basins within and adjacent to the Straight Creek fault zone. Strike-slip fault control is suggested by (1) high sediment accumulation rates, implying rapid subsidence; (2) abrupt stratigraphic thickening and thinning; (3) intrabasinal and basin-margin unconformities; (4) abrupt facies changes; (5) evidence for fault-induced (?) drainage reorganization; (6) intermittent internal drainage systems; (7) petrographic mismatches in coeval strata across master faults; (8) interbedded and intrusive relationships with extension-related (?) volcanic rocks; (9) characteristic fold and fault patterns; (10) rapidly alternating pulses of uplift and subsidence; and (11) local and diachronous deformation.

The Chuckanut-Puget-Naches Basin probably formed as a large pull-apart basin between the Straight Creek and Puget faults. The Swauk Basin may be a fault-wedge basin formed in a divergent zone between the Entiat-Leavenworth and Straight Creek faults. The Chiwaukum graben probably opened as a pull-apart basin between the Entiat and Leavenworth faults.

Eocene strike-slip faults in Washington were probably driven by oblique subduction of the Kula plate below North America. Eocene nonmarine basins occupied an intra-arc or forearc tectonic setting in this convergent margin, yet differ significantly from typical intra-arc or forearc basins. The presence of these basins in this tectonic setting reinforces the following concepts: (1) the occurrence of continental strike-slip basins in an ancient margin can be consistent with a history of either ideal transform or oblique convergence along the plate boundary; (2) the absence of a well-developed forearc basin within an ancient continental margin does not indicate that the plate boundary was not convergent.

ACKNOWLEDGMENTS

Work on the Chuckanut Formation was partly funded by the United States Geological Survey, Amoco Oil Company, the Geological Society of America, the Society of Sigma Xi, and the Corporation Fund of the Department of Geological Sciences, University of Washington. The manuscript has been greatly improved by the comments and suggestions of Kevin T. Biddle, Nicholas Christie-Blick, Virgil A. Frizzell, Jr., Parke D. Snavely, Jr., Richard G. Stanley, Joseph A. Vance, and A. John Watkinson. Discussions with Joanne Bourgeois, Mark T. Brandon, Darrel S. Cowan, Gregory T. Fraser, Virgil A. Frizzell, Jr., Randall L. Gresens, Peter Misch, James W. Roberts, Rowland W. Tabor, Stephen B. Taylor, John T. Whetten, and in particular Joseph A. Vance, were helpful in developing the ideas presented here.

REFERENCES

ANDERSON, D. W., AND PICARD, M. D., 1974, Evolution of synorogenic deposits in the intermontane Uinta Basin of Utah, *in* Dickinson, W. R., ed., Tectonics and Sedimentation: Society of Economic Paleon-

tologists and Mineralogists Special Publication No. 22, p. 167–189.

ARMENTROUT, J. M., MALLORY, V. S., AND EASTERBROOK, D. J., 1983, Northeast Olympic Peninsula, *in* Armentrout, J. M., Hull, D. A., Beaulieu, J. D., and Rau, W. W., eds., Correlation of Cenozoic Units of Western Oregon and Washington: Oregon Department of Geology and Mineral Industries, Oil and Gas Investigations 7, p. 71–74 and stratigraphic column.

ARMSTRONG, R. L., 1979, Cenozoic igneous history of the United States Cordillera from Lat. 42° to 49° N, *in* Smith, R. B., and Eaton, G. P., eds., Cenozoic Tectonics and Regional Geophysics of the Western Cordillera: Geological Society of America Memoir 152, p. 263–287.

ASHLEMAN, J. C., 1979, The geology of the western part of the Kachess Lake quadrangle, Washington [unpubl. M.S. thesis]: Seattle, University of Washington, 88 p.

ATWATER, T. M., 1970, Implications of plate tectonics for the Cenozoic tectonic evolution of western North America: Geological Society of America Bulletin, v. 81, p. 3513–3536.

AYDIN, ATILLA, AND NUR, AMOS, 1982, Evolution of pull-apart basins and their scale independence: Tectonics, v. 1, p. 91–106.

BALLANCE, P. F., AND READING, H. G., eds., 1980, Sedimentation in Oblique-Slip Mobile Zones: International Association of Sedimentologists Special Publication No. 4, 265 p.

BALLANCE, P. F., HOWELL, D. G., AND ORT, KATHLEEN, 1983, Late Cenozoic wrench tectonics along the Nacimiento, South Cuyama, and La Panza faults, California, indicated by depositional history of the Simmler Formation, *in* Andersen, D. W., and Rymer, M. J., eds., Tectonics and Sedimentation Along Faults of the San Andreas System: Society of Economic Paleontologists and Mineralogists, Pacific Section, p. 1–10.

BATES, R. G., BECK, M. E., AND BURMESTER, R. F., 1981, Tectonic rotations in the Cascade Range of southern Washington: Geology, v. 9, p. 184–189.

BRANDON, M. T., COWAN, D. S., MULLER, J. A., AND VANCE, J. A., 1983, Pre-Tertiary geology of San Juan Islands, Washington, and southeast Vancouver Island, British Columbia: Geological Association of Canada Field Trip Guidebook 5, 65 p.

BONINI, W. E., HUGHES, D. W., AND DANES, Z. F., 1974, Complete Bouguer gravity anomaly map of Washington: Washington Department of Natural Resources, Geologic Map GM-11.

BUCKOVIC, W. A., 1979, The Eocene deltaic system of west-central Washington, *in* Armentrout, J. M., Cole, M. R., and TerBest, H., Jr., eds., Cenozoic Paleogeography of the Western United States: Society of Economic Paleontologists and Mineralogists, Pacific Section, p. 147–164.

BURBANK, D. W., AND JOHNSON, G. D., 1983, The late Cenozoic chronologic and stratigraphic development of the Kashmir intermontane basin, northwestern Himalaya: Palaeogeography, Palaeoclimatology, Palaeoecology, v. 43, p. 205–235.

BUZA, J. W., 1979, Dispersal patterns and paleogeographic implications of lower and middle Tertiary fluviatile sandstones in the Chiwaukum graben, east-central Cascades, Washington, *in* Armentrout, J. M., Cole, M. R., and TerBest, H. Jr., eds., Cenozoic Paleogeography of the Western United States: Society of Economic Paleontologists and Mineralogists, Pacific Section, p. 63–74.

CADY, W. M., 1975, Tectonic setting of the Tertiary volcanic rocks of the Olympic Peninsula: United States Geological Survey, Journal of Research, v. 3, p. 573–582.

CASHMAN, S. M., AND WHETTEN, J. T., 1976, Low-temperature serpentinization of peridotite fanglomerate on the west margin of the Chiwaukum graben, Washington: Geological Society of America Bulletin, v. 87, p. 1773–1776.

CHAPPELL, W. M., 1936, Geology of the Wenatchee Quadrangle [unpubl. Ph.D. thesis]: Seattle, University of Washington, 249 p.

COWAN, D. S., 1982, Geological evidence for post-40 m.y.B.P. large-scale northwestward displacement of part of southeastern Alaska: Geology, v. 10, p. 309–313.

CROWELL, J. C., 1974a, Sedimentation along the San Andreas fault zone, *in* Dott, R. H., Jr., and Shaver, R. H., eds., Modern and Ancient Geosynclinal Sedimentation: Society of Economic Paleontologists and Mineralogists Special Publication No. 19, p. 292–303.

———, 1974b, Origin of late Cenozoic basins in southern California, *in* Dickinson, W. R., ed., Tectonics and Sedimentation: Society of Economic Paleontologists and Mineralogists Special Publication No. 22, p. 167–189.

CROWELL, J. C., AND LINK, M. H., eds., 1982, Geologic History of Ridge Basin, southern California: Society of Economic Paleontologists and Mineralogists, Pacific Section, 304 p.

DAVIS, G. A., MONGER, J. W. H., AND BURCHFIEL, B. C., 1978, Mesozoic construction of the Cordilleran "collage," central British Columbia to central California, in Howell, D. G., and McDougall, K. A., eds., Mesozoic Paleogeography of the Western United States: Society of Economic Paleontologists and Mineralogists, Pacific Section, p. 1–32.

DICKINSON, W. R., 1974, Plate tectonics and sedimentation, in Dickinson, W. R., ed., Tectonics and Sedimentation: Society of Economic Paleontologists and Mineralogists Special Publication No. 22, p. 1–27.

———, 1979, Cenozoic plate tectonic setting of the Cordilleran region in the United States, in Armentrout, J. M., Cole, M. R., and TerBest, H., Jr., eds., Cenozoic Paleogeography of the Western United States: Society of Economic Paleontologists and Mineralogists, Pacific Section, p. 1–13.

DICKINSON, W. R., INGERSOLL, R. V., AND GRAHAM, S. A., 1979, Paleogene sediment dispersal and paleotectonics in northern California: Geological Society of America Bulletin, Part II, v. 90, p. 1458–1528.

DICKINSON, W. R., AND SUCZEK, C. A., 1979, Plate tectonics and sandstone composition: American Association of Petroleum Geologists Bulletin, v. 63, p. 2164–2182.

DOTTER, J. A., 1977, Prairie Mountain Lakes Area, Skagit County: Structural geology, sedimentary petrography, and magnetics [unpubl. M.S. thesis]: Corvallis, Oregon State University, 105 p.

DUNCAN, R. A., 1982, A captured island chain in the Coast Range of Oregon and Washington: Journal of Geophysical Research, v. 7, p. 10827–10837.

EARDLEY, A. J., 1962, Structural Geology of North America: New York, Harper and Row, 2nd edition, 743 p.

ENGEBRETSON, D. C., 1982, Relative motions between oceanic and continental plates in the Pacific basin [unpubl. Ph.D. thesis]: Stanford, Stanford University, 211 p.

ERIKSON, E. H., JR., 1977, Petrology and petrogenesis of the Mount Stuart Batholith—plutonic equivalent of the high-alumina basalt association?: Contributions to Mineralogy and Petrology, v. 60, p. 183–207.

EWING, T. E., 1980, Paleogene tectonic evolution of the Pacific Northwest: Journal of Geology, v. 8, p. 619–639.

FISCHER, A. G., 1975, Origin and growth of basins, in Fischer, A. G., and Judson, S. J., eds., Petroleum and Global Tectonics: Princeton, Princeton University Press, p. 47–79.

FOSTER, R. J., 1960, Tertiary geology of a portion of the central Cascade Mountains, Washington: Geological Society of America Bulletin, v. 71, p. 99–126.

FRIEND, P. F., 1978, Distinctive features of some ancient river systems, in Miall, A. D., ed., Fluvial Sedimentology: Canadian Society of Petroleum Geologists, Memoir 5, p. 531–542.

FRIZZELL, V. A., JR., 1979, Petrology and stratigraphy of Paleogene nonmarine sandstones, Cascade Range, Washington [unpubl. Ph.D. thesis]: Stanford, Stanford University, 151 p.

FRIZZELL, V. A., JR., AND TABOR, R. W., 1977, Stratigraphy of Tertiary arkoses and their included monolithologic fanglomerates and breccias in the Leavenworth fault zone, central Cascades, Washington: Geological Society of America Abstracts with Programs, v. 9, p. 421.

FRIZZELL, V. A., JR., TABOR, R. W., BOOTH, D. B., ORT, K. M., AND WAITT, R. B., JR., 1984, Preliminary geologic map of the Snoqualmie Pass 1:100,000 quadrangle, Washington: United States Geological Survey Open File Map 84-693.

FRIZZELL, V. A., JR., TABOR, R. W., ZARTMAN, R. E., AND JONES, D. L., 1982, Mesozoic melanges in the western Cascades of Washington: Geological Society of America Abstracts with Programs, v. 14, p. 164.

GABRIELSE, HUBERT, CAMPBELL, R. B., MONGER, J. W. H., RICHARD, T. A., AND TIPPER, H. W., 1977, Major faults and paleogeography in the Canadian Cordillera: Geological Association of Canada, Program with Abstracts, v. 2, p. 20.

GABRIELSE, HUBERT, AND DODDS, C. J., 1977, The structural significance of the Northern Rocky Mountain Trench and related lineaments in north-central British Columbia: Geological Association of Canada, Program with Abstracts, v. 2, p. 19.

GARD, L. M., JR., 1968, Bedrock geology of the Lake Tapps quadrangle, Pierce County, Washington: United States Geological Survey Professional Paper 38-B, 33 p.

GRESENS, R. L., 1982a, Early Cenozoic geology of central Washington state: II. Implications for plate tectonics and alternatives for the origin of the Chiwaukum graben: Northwest Science, v. 56, p. 259–264.

———, 1982b, Early Cenozoic geology of central Washington state. I. Summary of sedimentary, igneous, and tectonic events: Northwest Science, v. 56, p. 218–229.

———, 1983, Geology of the Wenatchee and Monitor quadrangles, Chelan and Douglas Counties, Washington: Washington Division of Geology and Earth Resources, Bulletin 75, 75 p.

GRESENS, R. L., NAESER, C. W., AND WHETTEN, J. T., 1981, Stratigraphy and age of the Chumstick and Wenatchee formations: Tertiary fluvial and lacustrine rocks, Chiwaukum graben, Washington: Summary: Geological Society of America Bulletin, Part I, v. 92, p. 233–236; Part II, p. 841–876.

HAMMOND, P. E., 1979, A tectonic model for evolution of the Cascade Range, in Armentrout, J. M., Cole, M. R., and TerBest, H., Jr., eds., Cenozoic Paleogeography of the Western United States: Society of Economic Paleontologists and Mineralogists, Pacific Section, p. 219–238.

HARDING, T. P., 1974, Petroleum traps associated with wrench faults: American Association of Petroleum Geologists Bulletin, v. 58, p. 1290–1304.

HEMPTON, M. R., DUNNE, L. A., AND DEWEY, J. F., 1983, Sedimentation in an active strike-slip basin, southeastern Turkey: Journal of Geology, v. 91, p. 401–412.

HOPKINS, W. S., JR., 1966, Palynology of Tertiary rocks of the Whatcom Basin, southwestern British Columbia and northwestern Washington [unpubl. Ph.D. thesis]: Vancouver, University of British Columbia, 184 p.

HOPPE, W. J., 1982, Structure and geochronology of the Gabriel Peak orthogneiss and adjacent crystalline rocks, north Cascades, Washington: Geological Society of America Abstracts with Programs, v. 14, p. 173.

HOWELL, D. G., CROUCH, J. K., GREENE, H. G., McCULLOH, D. S., AND VEDDER, J. G., 1980, Basin development along the late Mesozoic and Cainozoic California margin: a plate tectonic margin of subduction, oblique subduction, and transform tectonics, in Ballance, P. F., and Reading, H. G., eds., Sedimentation in Oblique-Slip Mobile Zones: International Association of Sedimentologists, Special Publication No. 4, p. 43–62.

JOHNSON, S. Y., 1982, Stratigraphy, sedimentology, and tectonic setting of the Eocene Chuckanut Formation, northwest Washington [unpubl. Ph.D. thesis]: Seattle, University of Washington, 221 p.

———, 1984a, Evidence for a margin-truncating transcurrent fault (pre-Late Eocene) in western Washington: Geology, v. 12, p. 538–541.

———, 1984b, Stratigraphy, age, and paleogeography of the Eocene Chuckanut Formation, northwest Washington: Canadian Journal of Earth Sciences, v. 21, p. 92–106.

———, 1984c, Cyclic fluvial sedimentation in a rapidly subsiding basin, northwest Washington: Sedimentary Geology, v. 38, p. 361–392.

JOHNSON, S. Y., BRANDON, M. T., AND STEWART, R. J., 1981, Major early Tertiary transcurrent fault in western Washington (abs.): EOS, v. 62, p. 1047.

JOHNSON, S. Y., WHETTEN, J. T., NAESER, C. W., AND ZIMMERMANN, R. A., 1983, Fission track ages from the Chuckanut Formation, northwest Washington: Geological Society of America Abstracts with Programs, v. 15, p. 393.

JOHNSON, S. Y., McLAIN, K. J., BRANDON, M. T., TABER, J. J., LEWIS, B. T. R., AND STEWART, R. J., 1984, Washington geologic cross-sections C-C' and D-D', in Kulm, L. D., et al., eds., Atlas of the Ocean Margin Drilling Program: Region 5: Woods Hole, Massachusetts, Marine Sciences International, sheets 21 and 22.

JONES, R. W., 1959, Geology of the Finney Peak area, northern Cascades of Washington [unpubl. Ph.D. thesis]: Seattle, University of Washington, 186 p.

KELLER, H. M., TAHIRKHELI, R. A. K., MIZRA, M. A., JOHNSON, G. D., JOHNSON, N. M., AND OPDYKE, N. E., 1977, Magnetic polarity stratigraphy of the Upper Siwalik deposits, Pabbi Hills, Pakistan: Earth and Planetary Science Letters, v. 36, p. 187–201.

LARAVIE, J. A., 1976, Geologic field studies along the eastern border of the Chiwaukum graben, central Washington [unpubl. M.S. thesis]: Seattle, University of Washington, 55 p.

LEEDER, M. R., 1973, Fluviatile fining-upward cycles and the magnitude of paleochannels: Geology Magazine, v. 110, p. 265–276.

LEWELLEN, D. G., 1983, The structure and depositional environment of the Manastash Formation, Kittitas County, Washington [unpubl. M.S. thesis]: Cheney, Eastern Washington University, 107 p.

MACLEOD, N. S., TIFFIN, D. L., SNAVELY, P. D., JR., AND CURRIE, R. G., 1977, Geologic interpretations of magnetic and gravity anomalies in the Strait of Juan de Fuca, U.S.—Canada: Canadian Journal of Earth Sciences, v. 14, p. 223–238.

MANN, PAUL, HEMPTON, M. R., BRADLEY, D. C., AND BURKE, KEVIN, 1983, Development of pull-apart basins: Journal of Geology, v. 91, p. 529–554.

MATHEWS, W. H., 1981, Early Cenozoic resetting of potassium-argon dates and geothermal history of north Okanogan area, British Columbia: Canadian Journal of Earth Sciences, v. 8, p. 1310–1319.

MCDOUGALL, J. W., 1980, Geology and structural evolution of the Foss River-Deception Creek area, Cascade Mountains, Washington [unpubl. M.S. thesis]: Corvallis, Oregon State University, 87 p.

MCLEAN, J. R., AND JERZYKIEWICZ, TOMASZ, 1978, Cyclicity, tectonics, and coal: some aspects of fluvial sedimentology in the Brazeau-Paskapoo Formations, Coal Valley area, Alberta, Canada, in Miall, A. D., Fluvial Sedimentology: Canadian Society of Petroleum Geologists, Memoir 5, p. 441–468.

MCTAGGART, K. C., AND THOMPSON, R. M., 1967, Geology of part of the northern Cascades in southern British Columbia: Canadian Journal of Earth Sciences, v. 4, p. 1191–1228.

MILLER, G. M., AND MISCH, PETER, 1963, Early Eocene angular unconformity at western front of northern Cascades, Whatcom County, Washington: American Association of Petroleum Geologists Bulletin, v. 47, p. 163–174.

MILLER, R. B., AND FROST, B. R., 1977, Structure, stratigraphy, plutonism, and volcanism of the central Cascades, Washington: Part 3, Geology of the Ingalls Complex and related pre-Cretaceous rocks of the Mount Stuart uplift, central Cascades, Washington, in Brown, E. H., and Ellis, R. C., eds., Geological Excursions in the Pacific Northwest: Bellingham, Western Washington University, p. 283–291.

MILNES, P. T., 1976, Structural geology and metamorphic petrology of the Ilabot Peaks area, Skagit County, Washington [unpubl. M.S. thesis]: Corvallis, Oregon State University, 118 p.

MISCH, PETER, 1966, Tectonic evolution of the northern Cascades of Washington—a west-Cordilleran case history: Canadian Institute of Mining and Metallurgy, Special Volume 8, p. 101–148.

———, 1977, Dextral displacements at some major strike faults in the North Cascades: Geological Association of Canada, Program with Abstracts, v. 2, p. 37.

MOEN, W. S., 1961, Geology and mineral deposits of the north half of the Van Zandt quadrangle: Washington Division of Mines and Geology Bulletin, No. 50, 129 p.

NEWMAN, K. R., 1981, Palynological biostratigraphy of some early Tertiary nonmarine formations in central and western Washington, in Armentrout, J. M., ed., Pacific Northwest Biostratigraphy: Geological Society of America Special Paper 184, p. 49–65.

OKULITCH, A. V., PRICE, R. A., AND RICHARDS, T. A., 1977, A guide to the geology of the southern Canadian Cordillera: Geological Association of Canada, Field Trip Guidebook 8, 135 p.

PALMER, A. R., 1983, The Decade of North American Geology 1983 geologic time scale: Geology, v. 11, p. 503–504.

RAISZ, ERWIN, 1945, The Olympic-Wallowa lineament: American Journal of Science, v. 243-A, p. 479–485.

READING, H. G., 1980, Characteristics and recognition of strike-slip fault systems, in Ballance, P. F., and Reading, H. G., eds., Sedimentation in Oblique-Slip Mobile Zones: International Association of Sedimentologists Special Publication No. 4, p. 7–26.

REISWIG, K. N., 1982, Palynological differences between the Chuckanut and Huntingdon formations, northwestern Washington [unpubl. M.S. thesis]: Bellingham, Western Washington University, 61 p.

ROBERTSON, C. A., 1981, Petrology, sedimentology, and structure of the Chuckanut Formation, Coal Mountain, Skagit County, Washington [unpubl. M.S. thesis]: Seattle, University of Washington, 41 p.

RODGERS, D. A., 1980, Analysis of pull-apart basin development produced by en echelon strike-slip faults, in Ballance, P. F., and Reading, H. G., eds., Sedimentation in Oblique-Slip Mobile Zones: Interna-

tional Association of Sedimentologists Special Publication No. 4, p. 27–42.

SMITH, G. O., 1904, Description of the Mount Stuart quadrangle, Washington: United States Geological Survey Geologic Atlas, Mount Stuart Folio 106, 10 p.

SMITH, G. O. AND CALKINS, F. C., 1906, Description of the Snoqualmie quadrangle, Washington: United States Geological Survey, Geologic Atlas, Snoqualmie Folio 139, 14 p.

SNAVELY, P. D., JR., BROWN, R. D., JR., ROBERTS, A. E., AND RAU, W. W., 1958, Geology and coal resources of the Centralia-Chehalis district, Washington: United States Geological Survey, Bulletin 1053, 159 p.

TABOR, R. W., AND CADY, W. B., 1978a, Geologic map of the Olympic Peninsula, Washington: United States Geological Survey Miscellaneous Investigations Map I-994.

TABOR, R. W., AND CADY, W. B., 1978b, The structure of the Olympic Mountains, Washington—anatomy of a subduction zone: United States Geological Survey Professional Paper 1033, 38 p.

TABOR, R. W., FRIZZELL, V. A., JR., WHETTEN, J. T., SWANSON, D. A., BYERLY, G. R., BOOTH, D. B., HETHERINGTON, M. J., AND WAITT, R. B., JR., 1980, Preliminary geologic map of the Chelan 1:100,000 quadrangle, Washington: United States Geological Survey Open-File Map 80-841.

TABOR, R. W., WAITT, R. B., JR., FRIZZELL, V. A., JR., SWANSON, D. A., BYERLY, G. R., AND BENTLEY, R. D., 1982a, Geologic map of the Wenatchee 1:100,000 quadrangle, Washington: United States Geological Survey Miscellaneous Investigations Map MI-1311.

TABOR, R. W., FRIZZELL, V. A., JR., BOOTH, D. B., WHETTEN, J. T., WAITT, R. B., JR., AND ZARTMAN, R. E., 1982b, Preliminary geologic map of the Skykomish River 1:100,000 quadrangle, Washington: United States Geological Survey Open-File Map 82-747.

TABOR, R. W., ZARTMAN, R. E., AND FRIZZELL, V. A., JR., 1982c, Possible accreted terranes in the North Cascades crystalline core, Washington: Geological Society of America Abstracts with Programs, v. 14, p. 239.

TABOR, R. W., FRIZZELL, V. A., JR., VANCE, J. A., AND NAESER, C. W., 1984, Ages and stratigraphy of lower and middle Tertiary sedimentary and volcanic rocks of the central Cascades, Washington: Application to the tectonic history of the Straight Creek fault: Geological Society of America Bulletin, v. 95, p. 26–44.

TAYLOR, S. B., FRASER, G. T., ROBERTS, J. W., AND JOHNSON, S. Y., 1984, Reversals of paleo-drainage patterns in Eocene Swauk basin, Washington: Implications for strike-slip basin evolution: Geological Society of America Abstracts with Programs, v. 16, p. 674.

TURNER, D. L., FRIZZELL, V. A., JR., TRIPLEHORN, D. M., AND NAESER, C. W., 1983, Radiometric dating of ash partings in coal of the Eocene Puget Group, Washington: Implications for paleobotanical stages: Geology, v. 11, p. 527–531.

VANCE, J. A., 1957, The geology of the Sauk River area in the northern Cascades of Washington [unpubl. Ph.D. thesis]: Seattle, University of Washington, 312 p.

———, 1979, Early and middle Cenozoic arc magmatism and tectonics in Washington state: Geological Society of America Abstracts with Programs, v. 11, p. 132.

———, 1982, Cenozoic stratigraphy and tectonics of the Washington Cascades: Geological Society of America Abstracts with Programs, v. 14, p. 244.

VANCE, J. A., DUNGAN, M. A., BLANCHARD, D. P., AND RHODES, J. M., 1980, Tectonic setting and trace element geochemistry of Mesozoic ophiolitic rocks in western Washington: American Journal of Science, v. 280-A, p. 359–388.

VANCE, J. A., AND MILLER, R. B., 1981, The movement history of the Straight Creek fault in Washington state, in Monger, J. W. H., ed., The Last 100 Million Years (Mid-Cretaceous to Holocene) of Geology and Mineral Deposits in the Canadian Cordillera: Geological Association of Canada, Cordilleran Section Programme and Abstracts, p. 39–41.

VAN HOUTEN, F. B., 1969, Molasse facies: Records of worldwide crustal stresses: Science, v. 166, p. 1506–1507.

———, 1974, Northern Alpine molasse and similar Cenozoic sequences of southern Europe, in Dott, R. H., Jr., and Shaver, R. H., eds., Modern and Ancient Geosynclinal Sedimentation: Society of Economic Paleontologists and Mineralogists Special Publication No. 19, p. 260–273.

VINE, J. D., 1969, Geology and coal resources of the Cumberland, Hobart, and Maple Valley quadrangles, King County, Washington: United States Geological Survey Professional Paper 624, 67 p.

WALKER, C. W., 1980, Geology and energy resources of the Roslyn-Cle Elum area, Kittitas County, Washington: Washington Department of Natural Resources, Division of Geology and Earth Resources, Open-File Report 80-1, 59 p.

WATERS, A. C., 1932, A petrologic and structural study of the Swakane gneiss, Entiat Mountains, Washington: Journal of Geology, v. 40, p. 604–633.

WELLS, R. E., ENGEBRETSON, D. C., SNAVELY, P. D., JR., AND COE, R. S., 1984, Cenozoic plate motions and the volcano-tectonic evolution of western Oregon and Washington: Tectonics, v. 3, p. 275–294.

WHETTEN, J. T., ZARTMAN, R. E., BLAKELY, R. J., AND JONES, D. L., 1980, Allochthonous Jurassic ophiolite in northwest Washington: Geological Society of America Bulletin, Part 1, v. 91, p. 359–368.

WOLFE, J. A., 1978, A paleobotanical interpretation of Tertiary climate in the northern hemisphere: American Scientist, v. 66, p. 694–703.

PALEOGENE STRIKE-SLIP DEFORMATION AND SEDIMENTATION ALONG THE SOUTHEASTERN MARGIN OF THE EBRO BASIN

PERE ANADÓN

Instituto "Jaime Almera," C.S.I.C., Martí i Franquès s/n, 08028 Barcelona, SPAIN;

AND

LLUÍS CABRERA, JOAN GUIMERÀ, AND PERE SANTANACH

Facultat de Geologia, Universitat de Barcelona, Gran Via 585, 08007 Barcelona, SPAIN

ABSTRACT: The boundary of sites of sedimentation along the southeastern margin of the Ebro Basin during the Paleogene was defined by upthrust structures and by steeply dipping limbs of folds and flexures. These features are related to convergent wrenching along sinistral, right-stepping, en echelon, basement-involved faults located along the Catalan Coastal Range. During Paleogene time, thick alluvial terrigenous deposits accumulated along the fault-controlled basin boundary. The evolution of the alluvial-fan systems and of progressive angular syntectonic unconformities document the timing of slip on the faults. The coarse-grained sediments are diachronous indicating that strike slip was active at different times.

INTRODUCTION

This paper is concerned with the relationships between sedimentation and tectonics observed along the southeastern margin of the Ebro Basin in the northeastern part of Iberian Peninsula (Fig. 1). During the Paleogene this margin was defined by a system of right-stepping, en echelon, left-lateral strike-slip faults. Diachronous slip on these faults is recorded by sedimentary facies that were deposited adjacent to active tectonic structures as well as by the pattern and distribution of unconformities that developed in the sedimentary fill of the basin.

GEOLOGICAL SETTING

The Ebro Basin

The Ebro Basin is located on the northeastern Iberian Peninsula (Fig. 1). It is a Tertiary basin that formed during the Alpine orogeny as the foreland basin of three mountain chains: the Pyrenees, the Iberian Range, and the Catalan Coastal Range. The basin is asymmetric. The main sedimentary troughs are located in the north where the pre-Tertiary substratum ranges from depths of 3,000 to 5,000 m (Riba and Reguant, in press). The overall structure at the surface is very simple. The Tertiary beds are nearly horizontal over most of their extent and significantly deformed only along the margins of the basin. Along the northern margin is the southern Pyrenean thrust belt (Seguret, 1970). Nappes of this belt were displaced southward during the Paleogene, overlapping the main sedimentary troughs of the Ebro Basin (Seguret, 1970; Solé-Sugrañes, 1978; Muñoz et al., in press). The total shortening is several tens of kilometers and locally as much as 50 km. The Iberian Chain stretches along the southwestern margin of the Ebro Basin and consists of an array of northwest-oriented folds and thrusts. The Tertiary beds of this margin are commonly overthrust by the Mesozoic rocks. However, some structures of the chain are unconformably overlapped by upper Tertiary beds. The Catalan Coastal Range along the southeastern margin of the basin, is characterized by an array of en echelon northeast-striking faults. In the area where the Iberian Chain and the Catalan Coastal Range join, there is an array of north-vergent, east-trending folds and thrusts. This area, called the Linking Zone (Guimerà, 1984), differs from the Iberian Range and the Catalan Coastal Range in

having a thicker Mesozoic cover cut by low-angle thrusts of larger displacement.

The Southeastern Margin of the Ebro Basin

The Catalan Coastal Range is located along the southeastern margin of the Ebro Basin (Fig. 2), and is composed of Hercynian basement unconformably overlain by a sedimentary cover of Triassic to Cretaceous terrigenous red beds, carbonates, and evaporitic marine deposits. The upper part of the cover (mainly Jurassic and Cretaceous rocks) is detached from the basement at a *décollement* within Keuper (uppermost Triassic) lutites and evaporitic rocks (Llopis, 1947). The lower part of the cover, however, remains attached to the basement.

The principal structures of the Catalan Coastal Range are nearly vertical, basement-involved, strike-slip faults that constitute a right-stepping, en echelon array (in the sense of Rodgers, 1980). The strike of these faults changes from east-northeast to northeast along the range and is slightly oblique to the overall trend of the range (Figs. 2, 3). There are also some transverse, northwest-striking basement faults with somewhat smaller displacements. All of the basement-involved faults of the Catalan Coastal Range probably developed during the late Paleozoic as strike-slip faults. They were active as normal faults during Mesozoic times as can be inferred from the study of the Mesozoic cover rocks (Esteban and Robles, 1976; Anadón et al., 1979; Marzo, 1980). During the Alpine orogeny the faults were again reactivated as strike-slip faults. Later, during a Neogene extensional phase, they behaved as normal faults associated with the Neogene rifts (Fontboté, 1954).

The most important and intense Alpine deformation in the Catalan Coastal Range took place along these major faults (Julivert, 1978), and led to the development of diverse structures in the basement and the Mesozoic cover. In the basement, motion along the nearly vertical Vallès-Penedès fault, gave rise to the development of fault gouge (up to 500 m wide), as well as to the emplacement of basement slices, which overthrust the Mesozoic cover and even the Paleogene deposits of the Ebro Basin (Fig. 4A). Along other major longitudinal faults (e.g., the Falset fault), fault gouge is less important. The location of basement-involved faults is suggested in the sedimentary cover by flexures and, less commonly, by folds. An example of such folds is the

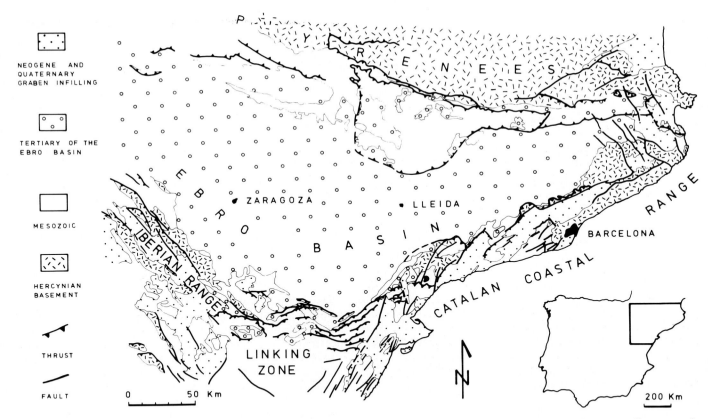

FIG. 1.—Generalized geological map of the northeastern Iberian Peninsula showing the location of the Ebro Basin and the surrounding mountain chains.

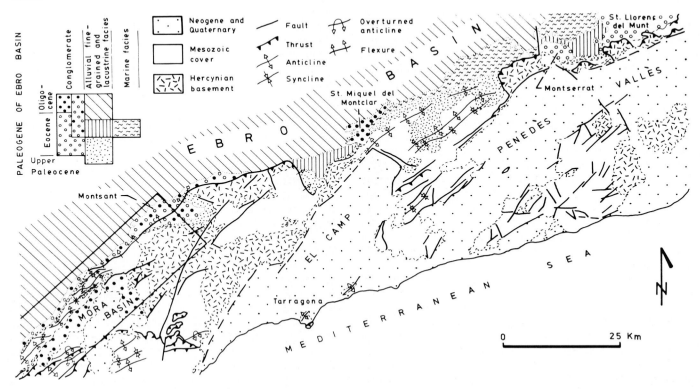

FIG. 2.—Generalized geological map of the Catalan Coastal Range and the adjacent Ebro Basin. Lower Eocene marine facies in the central part have not been indicated for the sake of clarity. See Figure 7 for a more detailed explanation of Paleogene stratigraphy.

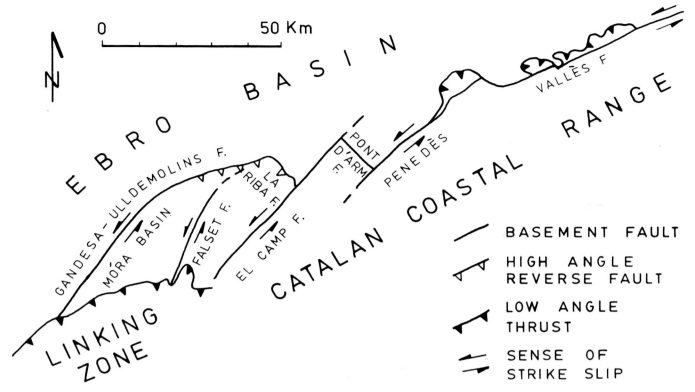

FIG. 3.—Structural pattern of strike-slip faults along the southeastern margin of the Ebro Basin.

Cavalls-Pàndols anticlinorium located between the Ebro Basin and the small Móra Basin (Fig. 2), also of Paleogene age. This complex, elongated (30 km long and 3 km wide), northeast-trending anticlinorium aligns with the Gandesa-Ulldemolins basement fault (Fig. 3), and it contains an array of en echelon, left-stepping folds (Fig. 4B). The folds are confined to the inner part of the anticlinorium, and they die out on the limbs.

The considerable variability in the structural character of the boundary between the Catalan Coastal Range and the Ebro Basin is largely controlled by whether the Keuper décollement horizon and overlying Jurassic and Cretaceous carbonate cover were present or absent during deformation. Where the décollement horizon and the upper part of the Mesozoic cover were absent, basement-involved faults reached the surface and emplaced basement slices over Lower and Middle Triassic rocks. The latter form a monoclinal fold noticeably vergent towards the Ebro Basin (Fig. 4A). However, where the Keuper décollement horizon and the overlying post-Triassic cover were present, basement-involved faults are not obvious at the surface (Fig. 4B). Instead, the cover was deformed into an elongate anticlinorium.

Examples of transverse northwest-striking faults that occur in relay zones between the major northeast-striking faults are the La Riba and El Pont d'Armentera faults (Figs. 2, 3). La Riba fault is a high-angle reverse fault, the El Pont d'Armentera fault is a normal fault.

STRIKE-SLIP DEFORMATION ALONG THE SOUTHEAST BASIN MARGIN

Evidence for Strike Slip

Sinistral slip on the northeast-striking faults is shown by the following features:

1) There are horizontal slickenside striae on kilometers-long fault planes within the Mesozoic cover at the southwestern end of the Vallès-Penedès fault and on the Falset fault (Figs. 2, 3).

2) Abundant rod-shaped, ellipsoidal blocks of vein quartz within fault gouge along the Vallès-Penedès fault are subvertical and encircled by near-horizontal striae (Fig. 5). The gouge contains subvertical banding defined by textural variations corresponding to different degrees of crushing (Julià and Santanach, 1984), which is parallel to secondary east-northeast-striking faults. The fault-slip direction is thought to be subhorizontal, perpendicular to rod axes and to the poles of the planes normal to the surfaces containing the striae (movement-plane poles after Arthaud, 1969; Fig. 5). That the slip along the Vallès-Penedès fault is sinistral is shown by slickenside striae on the fault plane in the Mesozoic cover at its southern end.

3) The en echelon folds in the sedimentary cover of the Cavalls-Pàndols anticlinorium are left-stepping and are consistent with left slip (Fig. 6).

4) The change of strike exhibited by the low-angle thrusts

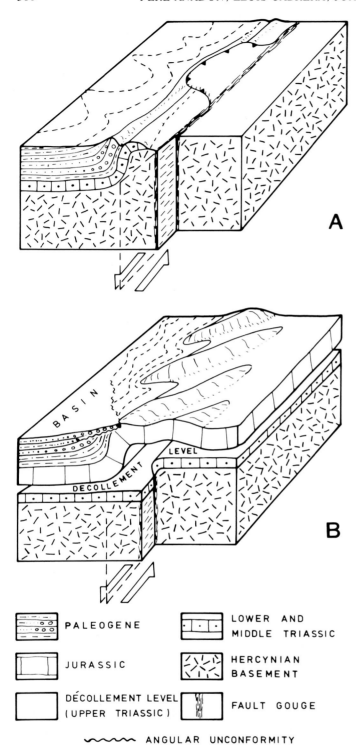

PALEOGENE

JURASSIC

DÉCOLLEMENT LEVEL
(UPPER TRIASSIC)

LOWER AND
MIDDLE TRIASSIC

HERCYNIAN
BASEMENT

FAULT GOUGE

〜〜〜 ANGULAR UNCONFORMITY

FIG. 4.—Two examples of relationships between the Hercynian basement and the Mesozoic cover along the southeastern margin of the Ebro Basin. Deformation in the basement resulted in sinistral slip along preexisting subvertical faults. A) Where décollement levels were missing, basement and cover deformed together. Basement-involved faults reached the surface through the cover and, locally, basement slices were formed. B) Where a décollement level was present, deformation of the cover was different from that in the basement. En echelon folding of the cover took place over the basement-involved strike-slip faults.

and folds of the Linking Zone where they join the major faults of the southernmost Catalan Coastal Range may be explained as a result of sinistral slip along these major basement-involved faults (Figs. 1–3).

Interpretation of Structures

The sinistral slip along the basement faults of the southeastern Ebro Basin margin has been interpreted as a result of regional north-south shortening that affected the northeastern part of the Iberian Peninsula during Paleogene time (Guimerà, 1984). The different orientations and arrangements of basement-involved faults (Fig. 3) suggest different responses to regional shortening.

The east-northeast-striking Vallès-Penedès fault was oriented at an angle of about 60° to the regional shortening direction. As a result, the fault is characterized by oblique slip or convergent wrenching in the manner proposed by Wilcox et al. (1973). To the southwest the El Camp, Falset, and Gandesa-Ulldemolins faults range in strike from N30°E to N45°E and do not show basement slices along their traces. This may be due to their orientations closer to the regional shortening direction. In most cases, slip along the northeast-striking basement faults resulted in uplift of basement blocks southeast of the faults (Fig. 4A, B). Displacement on the Gandesa-Ulldemolins fault was accompanied by folding of the sedimentary cover and development of the Cavalls-Pàndols anticlinorium.

The right-stepping en echelon arrangement of the basement faults and their sinistral slip led to shortening in overlap zones between the faults (see Rodgers, 1980; Xiaohan, 1983). This resulted in reverse slip along the La Riba fault (perpendicular to the northeast-striking faults) and the formation of a push-up zone (Fig. 3). The El Pont d'Armentera fault may be similar, but in this case any possible Paleogene reverse slip has been concealed by Neogene normal slip. The magnitude of displacement on the major northeast-striking basement faults cannot be calculated accurately because there are few markers that cross the faults. The noticeable development of fault gouge in the basement along the Vallès-Penedès fault and the change of strike exhibited by the thrusts and folds of the Linking Zone suggest several kilometers of displacement.

PALEOGENE SEDIMENTATION

Paleogene Sedimentation in the Ebro Basin

In broad outline the fill of the Ebro Basin consists of a thick pile of Tertiary rocks that range in age from Paleocene to Miocene (Riba and Reguant, in press). The rocks that fill the basin were deposited in a variety of depositional settings ranging from deep- and shallow-marine environments (deep-sea fan, carbonate and terrigenous platform) to non-marine settings (alluvial fan, fluvial, and shallow lake). Today, the outcrops of these sedimentary rocks display a peculiar map pattern. The Paleogene rocks crop out mainly in the northern and eastern areas of the basin whereas the Miocene rocks are exposed mainly in the western and central areas. Three main phases of basin evolution may be distinguished on the basis of paleoenvironmental changes in the basin and the structural evolution of its margins.

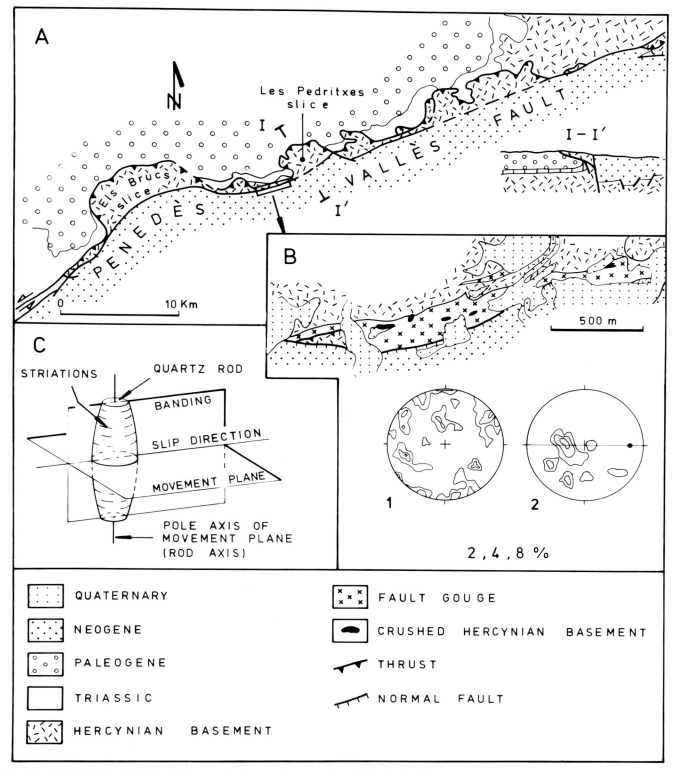

FIG. 5.—The Vallès-Penedès fault. A) Structural scheme and cross-section. B) Attitude of fault-gouge structures within the fault zone and stereo-diagrams (Schmidt net, lower hemisphere projection) of 1—poles of the striated surfaces of quartz blocks showing a dominant subvertical attitude and a great dispersion, and 2—poles of their movement planes showing a nearly horizontal attitude. Solid point in stereo-diagram 2 shows the inferred slip direction. C) Scheme showing the attitude of banding and quartz rods within the fault gouge and their geometrical elements.

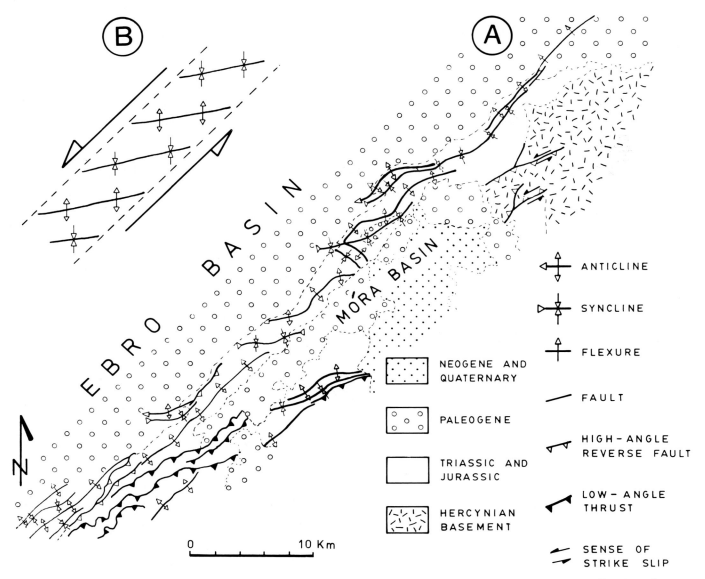

Fig. 6.—Structural scheme of the Cavalls-Pàndols anticlinorium. See Figure 2 for location. A) En echelon left-stepping folds affect the sedimentary cover in a narrow band 30 km long and 3 km wide. B) Scheme showing the interpretation of this fold pattern as a result of sinistral slip along the Gandesa-Ulledmolins basement-involved fault.

Early to Late Eocene.—During this phase the basin was shaped as a sedimentary trough closely associated with uplift of the Pyrenean Chain and the southward emplacement of its nappes. This trough was bounded to the southeast by the Catalan Coastal Range strike-slip system. The Iberian Chain was located along the southern margin of the basin.

Two transgressive-regressive marine cycles occurred within the basin during this phase: the Ilerdian–Cuisian (early Eocene) and the Lutetian–Priabonian (middle to late Eocene) cycles. These transgressions are related to connections with both the Atlantic and Tethys Oceans (Riba and Reguant, in press). During the Ilerdian and Bartonian transgressive maxima, most of the basin was flooded, although the southernmost part remained above sea level.

In Ilerdian (early Eocene) time, the basin was covered by a wide shallow carbonate platform (Plaziat, 1975). During the Cuisian regression there was basinward expansion of alluvial, fluvial, and lacustrine systems from the basin margins, in part related to pronounced tectonic activity along the Pyrenean Chain and the Catalan Coastal Range. This resulted in deposition of a variety of terrigenous deposits, carbonates, and evaporites. During the Bartonian transgressive maximum (Lutetian–Priabonian cycle), alluvial fans and the fluvial systems along the basin margins gave rise to delta and fan-delta systems. Shallow-water siliciclastic sedimentation was widespread, whereas sites of carbonate deposition (including reefs), were relatively restricted (Ferrer, 1971; Puigdefàbregas, 1975). Turbidite deposition occurred throughout the Eocene in the northern part of the basin (Mutti et al., 1972; Rosell and Puigdefàbregas, 1975).

During the Priabonian, however, a pronounced regression created the restricted conditions that led to extensive accumulation of evaporites. Subsequently, non-marine environments prevailed (Riba and Reguant, in press).

Early Oligocene to Middle Miocene.—From earliest Oligocene to middle Miocene time the Ebro Basin was a molasse-like basin, surrounded by the Pyrenees, the Catalan Coastal Range, and the Iberian Chain. Subsidence was greatest in the western part (Puigdefàbregas, 1975). The deposits are predominantly fluvial (including conglomerate and finer-grained sedimentary rocks; Van Houten, 1974), and lacustrine-paludine (including limestones, nonmarine evaporites, and red mudstones).

Late Miocene to Present.—During the late Miocene the drainage of the Ebro Basin was connected again with the sea and rivers debouched into the Mediterranean (Riba and Reguant, in press). Since then, the basin has been subject to erosion.

Paleogene Sedimentation Along the Southeastern Ebro Basin Margin

Recent stratigraphic and sedimentological studies of the Paleogene deposits that crop out in the study area have recognized several lithostratigraphic units (Ferrer et al., 1968; Ferrer, 1971; Rosell et al., 1973; Anadón, 1978a; Colombo, 1980; Plaziat, 1981). Stratigraphic relationships and lateral facies changes between these lithostratigraphic units are shown on Figure 7. Dating and correlation of the non-marine units have been established on the basis of biostratigraphic work on charophyte, gastropoda, and micromammal fossil remains (Rosell et al., 1966; Anadón and Feist, 1981; Anadón et al., 1983). The age of the alluvial coarse-grained facies, despite their scarce content of fossil remains, has been inferred from their relations with units whose ages are well known. The ages and correlation of the marine rocks are based mainly on the study of fossil micro-

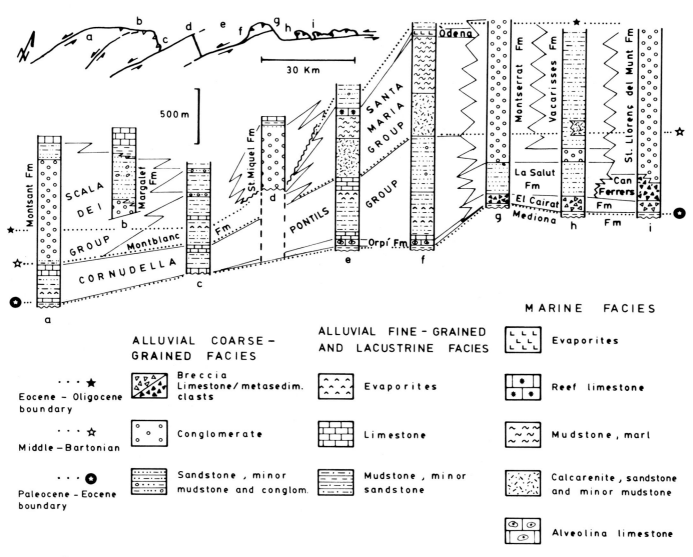

FIG. 7.—Stratigraphic units, correlation, and facies of the Paleogene strata along the southeastern margin of the Ebro Basin.

foraminifers (Ferrer, 1971) and a few data on calcareous nannoplankton (Anadón et al., 1983).

The Paleogene units in the southeastern Ebro Basin can be grouped into three main facies associations: (1) coarse-grained alluvial facies; (2) fine-grained alluvial and lacustrine facies; and (3) marine facies.

Coarse-grained alluvial facies.—In the northern parts of the study area, coarse-grained alluvial facies developed from the early Eocene to early Oligocene. In other areas this facies was deposited from late Eocene to Oligocene, showing a shorter time span of deposition (Fig. 7).

In the northeast the lowest unit is the *Cairat Breccia* (lowermost Eocene). This formation is as much as 200 m thick, and consists of limestone and dolostone breccia with interbedded red sandy mudstones (Anadón, 1978a). The breccia beds range from clast- to matrix-supported, and consist of small pebbles to boulders in a red matrix of sandy mudstone or sandstone. The beds range in thickness from a few centimeters to 4 m, and they tend to be laterally extensive (hundreds of meters). Interbedded red sandy mudstones are from a few centimeters to several meters thick. They are poorly sorted and locally include dispersed or aligned clasts. Paleosols and caliche crusts are common in the breccia, sandstone, and mudstone beds. The Cairat Formation does not display obvious cyclicity and consists mainly of debris-flow, mud-flow and carbonate-slide deposits derived from the Mesozoic and Paleocene cover. These sediments accumulated on alluvial fans and talus cones closely associated with the uplift of the southeastern block of the Vallès-Penedès fault (Anadón, 1980).

The *Can Ferrers Breccia* (lower to middle Eocene) consists mainly of massive, very poorly sorted, clast- to matrix-supported breccias with interbedded lenses of red pebbly sandstones and mudstones. Unlike the Cairat Formation the breccia clasts were derived locally from the basement and consist of crystalline and metasedimentary rocks. The sedimentological features of the Can Ferrers Formation are very similar to those of the Cairat Formation, again suggesting deposition on small alluvial fans or talus cones.

The *Montserrat and Sant Llorenç del Munt Conglomerates* (middle to upper Eocene) consist of massive conglomerates more than 1,000 m thick. Clasts are well to moderately rounded and include carbonate, crystalline, and metasedimentary types, suggesting derivation from both the Mesozoic cover and Hercynian basement. The Montserrat conglomerates are characterized by overall coarsening- and thickening-upward megasequences which consist of thinner coarsening-upward sequences ranging from 50 to 200 m thick. The lower parts of the thinner sequences consist of red mudstones, cross-bedded sandstones, and minor channelized pebble-to-cobble conglomerates. The upper parts of the sequences consist mainly of horizontally stratified, clast-supported, coarse (pebble to cobble in size), imbricated conglomerates with minor channelized and cross-bedded sandstones. Similar coarsening-upward sequences have been observed in the Sant Llorenç del Munt Conglomerate. However, this unit does not display obvious cyclicity. The Montserrat and Sant Llorenç del Munt Conglomerates developed on the inner parts of sizeable, basinward-spreading alluvial fans whose source area was located in internal parts

of the Catalan Coastal Range, where both Mesozoic and Hercynian rocks cropped out over a wide area. These conglomerates thus differ from the locally-derived Can Ferrers Breccia.

The Montserrat and Sant Llorenç del Munt Conglomerates pass laterally into predominantly sandy units, the *Vacarisses* and *La Salut Formations* (Anadón, 1978a). The La Salut Formation consists of medium- to coarse-grained red sandstone, minor mudstone, and pebble conglomerate. Sandstone beds are as thick as 10 m and display a sheet-like or broadly lenticular geometry. The sandstones and conglomerates show trough and planar cross-bedding and represent channel and sheet-flood deposits. Paleosols and some thin lacustrine limestone beds are present locally. The Vacarisses Formation consists of red mudstones and sheet-like, fine-grained sandstone beds, with interbedded channel bodies composed of cross-bedded sandstones and pebble conglomerates. Paleosols are widespread.

The *Sant Miquel del Montclar Conglomerate* (upper Eocene–lowermost Oligocene) extends over the central part of the study area and consists of more than 500 m of massive, conglomerates with interbedded red sandstone, and mudstone lenses (Colombo, 1980). The conglomeratic bodies are sheet-like, and for the most part horizontally stratified with clast-supported pebbles and cobbles that are commonly imbricated. The clasts are rounded and consist largely of Mesozoic and lowermost Eocene marine limestones. Intraformational angular unconformities occur in the upper part of the formation. The map pattern of this unit shows its fan-shaped geometry as well as the basinward transition from massive conglomerates to red sandstone, mudstone, and minor conglomerate (upper Montblanc Formation). Only an overall coarsening- and thickening-upward megasequence overlain by a fining-upward sequence has been described in this unit, which records the development and retreat of a relatively large alluvial-fan system.

In the southwestern part of the area studied, the upper Eocene–Oligocene *Scala Dei Group* (Colombo, 1980) crops out, overlying the fine-grained deposits of the Pontils-Cornudella Group (Fig. 7). The Scala Dei Group consists of alluvial-fan conglomerates of the Montsant Formation and laterally equivalent red sandstones, mudstones, and conglomerates of the Margalef Formation (Allen et al., 1983).

In its type area, the *Montsant Formation* is as much as 900 m thick and displays an overall coarsening- and thickening-upward megasequence (Colombo, 1980). The lower part of this unit consists of cross-bedded sandstone and pebble-to-cobble polymictic channel-filling conglomerate. The channels are cut in red mudstone and sheet-sandstone deposits that include paleosols. Conglomerates are more widespread higher in the section, and the upper part of the megasequence consists of sheet-like conglomeratic bodies as thick as 30 m with interbedded thin sandstone and red mudstone. These thick conglomerates are also polymictic but dominated by Mesozoic carbonate clasts of cobble and boulder size. Crude horizontal bedding, clast imbrication and tabular cross bedding are commonly observed in the conglomeratic beds. Minor channelized conglomerates are also present. In the northwestern areas of outcrop the uppermost conglomerates of the Montsant Formation contain intraformational angular unconformities. Southwest of the type area the formation consists of stacked megasequences for at least 12 km along the basin margin, suggesting progradation and retrogradation of the whole alluvial-fan system or the lateral shifting of distributary channels (Robles, 1982; Colombo and Robles, 1983).

More distal alluvial-fan facies of the Scala Dei Group are represented by the *Margalef Formation*. This largely fluvial unit consists of thick sections of red to mottled mudstones interbedded with cross-bedded conglomerates and sandstone channel bodies up to 8 m thick. The channel-fill conglomerates are polymictic and clast supported, with well rounded and sorted clasts, ranging in size from granules to small cobbles. Channel-fill sandstones range from fine- to medium-grained and show low-angle trough cross-bedding. Variation in the fluvial style within this unit has been described by Allen et al. (1983). The overbank deposits consist mainly of red and highly mottled mudstones with interbedded sheet sandstones and lacustrine limestones. Paleosols are also widespread.

Fine-grained alluvial and lacustrine facies.—In general these facies represent the distal deposits derived from the lateral change of the above-mentioned coarser alluvial facies, but in some cases their proximal equivalents are not known. Sedimentation in those areas located far from the margins of the basin and/or from the areas of coarse-grained deposition, mainly resulted in mudstone-dominated sections interbedded with sandstone, limestone, and evaporites.

The *Mediona Formation* (upper Paleocene) unconformably overlies either the Mesozoic cover or the Hercynian basement. This unit is up to 50 m thick, and consists of red sandy mudstones with interbedded sandstones and calcareous paleosols. Calcareous crusts with *Microcodium* are well developed at the base (Esteban, 1972, 1974) and locally constitute the only record of the unit. The Mediona Formation represents nonmarine sedimentation on very wide alluvial mud flats, before the onset of tectonic activity and of the Ilerdian (early Eocene) marine transgression (Anadón, 1978a).

The Pontils-Cornudella Group directly overlies either the Mediona Formation (in southwestern areas) or the marine limestones of the Orpí Formation (in the central region affected by the Ilerdian transgression). The thickness of this unit ranges from 100 to 800 m, the thickest sections being observed in the central part of the study area. The unit consists mainly of fine-grained terrigenous rocks with minor carbonates and evaporites (Anadón, 1978a; Colombo, 1980).

The terrigenous rocks consist of thick red mudstones interbedded with sheet-like units as thick as a few meters of fine-grained rippled, cross-bedded and horizontally laminated sandstone. In places, the sandstone composes fining-upward sequences with point-bar surfaces that are covered by thin beds of mudstones. Mudstones are also locally interbedded with channelized pebble-conglomerate. The limestones make up either single beds up to 1 m thick or laterally extensive packets several meters thick. They are mainly wackestones with a high content of charophytes, ostracodes, and gastropod fragments. Intraclast packstones and grainstones are also present. These limestones were deposited in a lacustrine environment. Root traces, paleosols and

fine-grained diagenetic dolostones, which characterize shallow lakes and marginal lacustrine areas, are widespread. Gypsum is present as sparse nodules in some red mudstones as well as nodular gypsum bodies up to 40 m thick with stringers of mudstone, dolostone, and limestone ("chicken wire" structure). These nodules resulted from the late diagenetic hydration of anhydrite developed in the vadose zone of either marginal or ephemeral lacustrine areas. Laminated gypsum, probably of lacustrine origin, has also been reported.

The Pontils-Cornudella Group rocks were deposited in a complex of closely related depositional environments including flood plains, dry and ponded mud flats and paludine-lacustrine areas. These environments were located far away from basin margin during the early and middle Eocene (Anadón, 1978a; Colombo, 1980).

The *Montblanc Formation* (upper Eocene to lowermost Oligocene) consists mainly of red mudstones with minor sandstone, pebble conglomerate, and nodular or laminated gypsum. Both the gypsum and the terrigenous beds are interbedded with thick sections of red mudstone containing paleosols (Colombo, 1980). The beds of sandstone and conglomerate are as thick as 3 m and tend to be laterally extensive. They generally display cross-bedding and parallel lamination.

The Montblanc Formation was deposited on wide mudflats marginal to alluvial fans of the Sant Miquel del Montclar Conglomerate and Scala Dei Group. It passes laterally into the marine Santa Maria Group, described below.

Marine Facies.—Marine facies were deposited during the two main marine transgressions that spread across the Ebro Basin from north to south. The first transgression, during the early Eocene (Ilerdian transgression), resulted in the deposition of shallow-water carbonates including *Alveolina* limestone of the *Orpí Formation*. The bulk of this unit consists of foraminiferal grainstones to wackestones locally as much as 100 m thick (Ferrer, 1971). The second transgression was most widespread during the middle Bartonian (middle Eocene; Anadón et al., 1983), and was responsible for deposition of the *Santa Maria Group* far from the basin margin. This unit, 400 to 1,000 m thick, is composed of fine-grained terrigenous rocks and subordinate carbonates (Ferrer, 1971). Transgressive sandstones at the base are overlain by coral and foraminiferal limestones, and in turn, by thick grey mudstones. Near the margins of the basin, marine deposits, including reef limestones, interfinger with deltaic and fan-delta facies characterized by coarsening- and thickening-upward sequences a few meters to more than 50 m thick (Montserrat-Sant Llorenç del Munt area). Such sequences typically grade upward from grey mudstones to coarse-grained sandstones and pebble to boulder conglomerates. The *Òdena Gypsum* accumulated near the basin margin during the late Eocene (Priabonian) regression.

RELATIONSHIP BETWEEN STRIKE-SLIP TECTONICS AND SEDIMENTATION

The Paleogene deposits of the Ebro Basin contain two kinds of features that record strike-slip deformation in the Catalan Coastal Range. These are (1) the distribution pat-

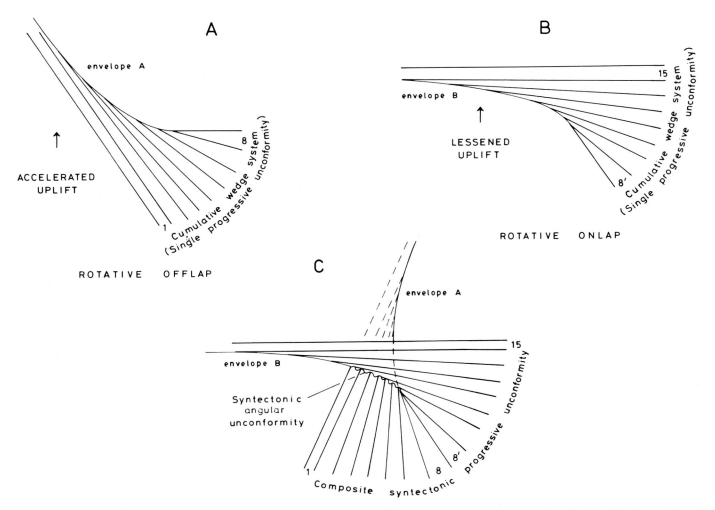

FIG. 8.—Genetic model of progressive and syntectonic angular unconformities (after Riba, 1976). A) Rotative offlap developed during accelerated uplift of a block adjacent to the basin. B) Rotative onlap records a decrease in the rate of uplift. C) Combination of A and B: A composite syntectonic progressive unconformity results in the inner zones of the basin, and a syntectonic angular unconformity develops near the uplifted block.

terns and sequential arrangement of facies, described above, and (2) the presence of what have been termed progressive unconformities (Riba, 1976).

Riba (1976) defined a progressive unconformity as one that forms adjacent to an uplifting structure (anticline, high-angle fault, etc.). Stratal units associated with a progressive unconformity are characterized by rotative offlap or rotative onlap (Figs. 8A, B). Rotative offlap develops during intervals of accelerated uplift of the adjacent block (Fig. 8A); rotative onlap records a decrease in the rate of uplift (Fig. 8B). A sedimentary wedge with rotative offlap is commonly overlain by one with rotative onlap (Fig. 8C), and the two wedges are separated by an angular unconformity (angular syntectonic unconformity of Riba, 1976) that can be laterally persistent parallel to the uplifting structure, but basinward, disappears abruptly into a correlative conformity. Several progressive unconformities have been observed along the convergent strike-slip basement faults of the southeastern margin of the Ebro Basin (Fig. 9). Both the distribution of facies and of progressive unconformities in-

dicate that strike-slip deformation was diachronous, beginning in early Eocene time in the northern part of the study area, but not until middle to late Eocene time in the south. There is also stratigraphic evidence for a basinward (northwestward) migration in fault activity. The youngest deformation appears to be of late Oligocene age.

The earliest tectonic activity along the Catalan Coastal Range is recorded by the development during the early Eocene (Ilerdian–early Cuisian?) of small alluvial fans (Cairat Formation) along the northern part of the Vallès-Penedès fault, and the emplacement of small olistostromes of Triassic carbonate (Anadón, 1980) (Fig. 10B). Correlative deposits in the remainder of the study area include shallow-marine carbonate and alluvial to lacustrine mudstone, sandstone, gypsum, and limestone (Pontils-Cornudella Group). From early to middle Eocene time (Cuisian to early Bartonian), coarse-grained alluvial-fan facies extended southward along the Vallès-Penedès fault, but fine-grained alluvial and lacustrine sediments accumulated over much of the study area (Fig. 10C). In the Montserrat area tectonic

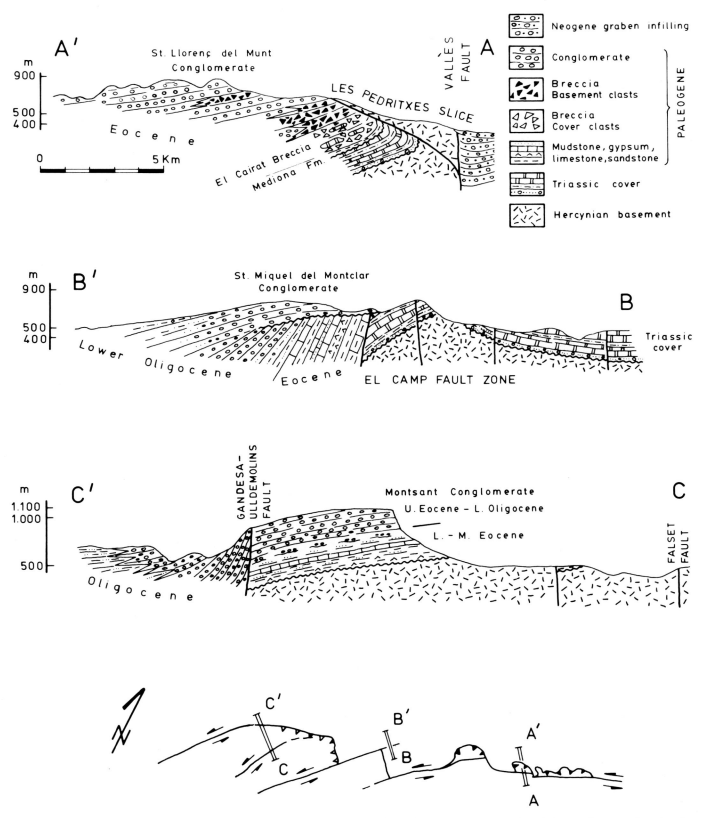

FIG. 9.—Schematic cross-sections through the southeastern margin of the Ebro Basin. Note the diachroneity of the conglomerate deposits, the progressive and angular syntectonic unconformities within the Sant Miquel del Montclar Conglomerate, and the progressive unconformity within the upper levels of the Montsant Conglomerate.

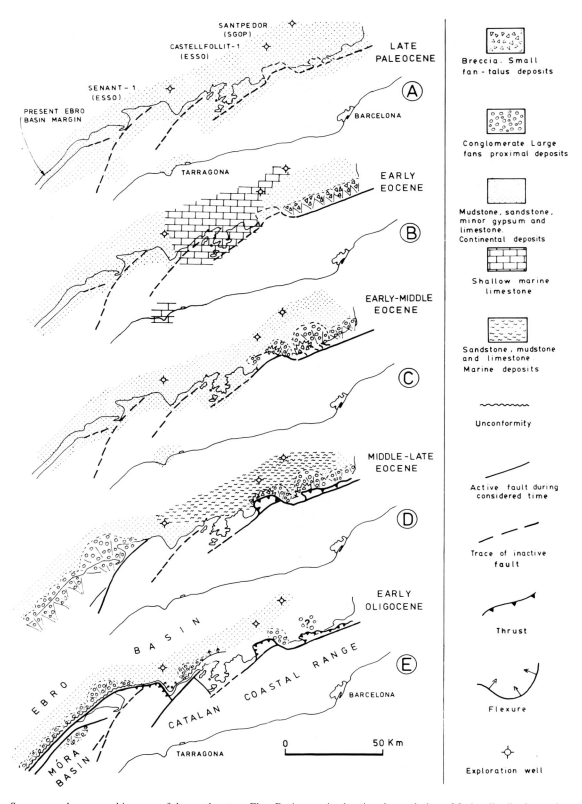

FIG. 10.—Summary paleogeographic maps of the southeastern Ebro Basin margin showing the evolution of facies distribution and tectonic activity during the Paleogene.

activity is recorded by the development of syntectonic unconformities. The alluvial fans appear to be of two types. One type extends into the Ebro Basin for a considerable distance, displays a clear fan geometry and received a great deal of rounded, lithologically varied clasts (e.g., the Sant Llorenç del Munt Conglomerate), originating from the inner parts of the Catalan Coastal Range. The other type is restricted to marginal areas uplifted close to fault-slices of basement rocks. These fans were fed by angular, locally derived clasts (e.g., Can Ferrers Breccia). The first fan type reflects regional uplift of the terrane southeast of the Vallès-Penedès fault, whereas the second records the local emplacement of basement slices.

During middle to late Eocene time (late Bartonian–Priabonian), conglomeratic sedimentation continued in the northern area (Figs. 10D, 11). At the same time, coarse conglomerates were also deposited across the trace of the still inactive Gandesa-Ulldemolins fault (as a result of uplift on the Falset fault (Fig. 3). During the early Oligocene (Fig. 10E), conglomeratic alluvial-fan facies were deposited along

the northwestern and southeastern limbs of the Cavalls-Pàndols anticlinorium (Garcia-Boada, 1974). Sediment was transported both to the northwest (Ebro Basin) and to the southeast (Móra Basin). These observations indicate an early Oligocene age for the earliest slip along the Ulldemolins-Gandesa fault and corresponding uplift of the Cavalls-Pàndols anticlinorium, formation of the Móra Basin, and elevation of the Prades block near the northeastern end of the fault (Fig. 12). The Sant Miquel de Montclar conglomerates were also deposited during the latest Eocene–earliest Oligocene, closely associated with the northern end of the El Camp fault (Fig. 12). These conglomerates display an obvious syntectonic unconformity, and therefore they record the beginning of slip along this fault which had not been previously active. Conglomeratic sedimentation in the northeastern parts of the study area probably ended in the early Oligocene. The cessation of slip along the Vallès-Penedès and Falset faults is difficult to date precisely because the stratigraphic record is inadequate. However, the expansion of carbonate and lutite facies of late Oligocene age from

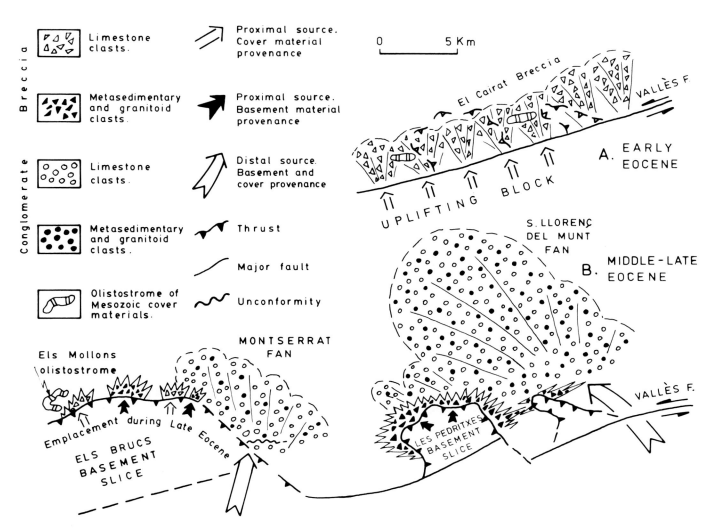

FIG. 11.—Paleogeographic sketch maps of the Montserrat-Sant Llorenç del Munt area, showing distribution of the breccia and conglomerate facies associated with activity of the Vallès fault.

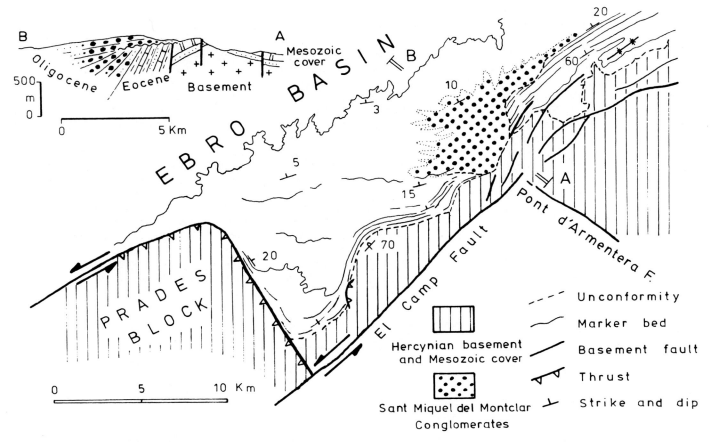

FIG. 12.—Sketch map of the push-up zone of the Prades block. Note the progressive and angular syntectonic unconformities within the Sant Miquel del Montclar Conglomerate at the northern end of the El Camp fault.

basinal areas towards the northwestern limb of the Cavalls-Pàndols anticlinorium suggests that slip along the Ulldemolins-Gandesa fault may have gradually decreased at that time.

In summary, slip along the different basement-involved strike-slip faults started at different times. The northernmost faults were the first to be active, and later on, faults located progressively farther south were successively set in motion. At the same time, the migration of deformation northwestward resulted in the successive initiation of en echelon strike-slip faults in that direction and caused the correlative migration of the basin margin in the same directions (Guimerà, 1984). The development of coarse-grained sedimentation linked to the tectonic activity along the newly created basin boundary is the main record of this process.

CONCLUSIONS

1) The southeastern margin of the Ebro Basin consists of a series of sinistral, right-stepping, en echelon, basement-involved faults. Convergent sinistral slip on these faults resulted in the upthrust of basement rocks (convergent wrenching) and the folding of the Mesozoic cover.

2) Large amounts of terrigenous sediment accumulated in alluvial systems along the basin boundaries. In the northeastern part of the study area these systems developed from the early Eocene to the early Oligocene. Farther from the basin boundaries sediments are predominantly alluvial to lacustrine mudstone, sandstone, limestone, and evaporite. Two marine transgressions spread across the southeastern Ebro Basin from north to south. The first of these resulted in the formation of a shallow-water carbonate platform during the early Eocene (Ilerdian transgression). During the middle-late Eocene the second transgression affected just the northern half of the study area (Bartonian transgression). Under these conditions some of the alluvial fans along the basin margin became fan-deltas in which conglomerate grades basinward into fine-grained terrigenous and carbonate deposits. Late Eocene regression resulted in the deposition of evaporites. Since then the Ebro Basin has been continuously above sea level.

3) The timing of deformation along the southeastern margin of the Ebro Basin is established by the evolution of the alluvial-fan systems which sourced from uplifted blocks and by the occurrence of progressive, angular syntectonic unconformities that developed within the coarsest sediments. Strike-slip deformation was diachronous, beginning in early Eocene time in the northern part of the study area, but not

until middle to late Eocene time in the south. Deformation also migrated basinward (northwestward). The youngest deformation appears to be of late Oligocene age.

ACKNOWLEDGMENTS

We would like to thank K. T. Biddle, N. Christie-Blick, M. Esteban, B. Ward, and an anonymous reviewer for their useful comments on the first draft of this paper. K. T. Biddle and N. Christie-Blick also provided considerable assistance in the preparation of the final version of the manuscript.

REFERENCES

ALLEN, PHILLIP, CABRERA, LLUÍS, COLOMBO, FERRAN, AND MATTER, ALBERT, 1983, Variation in fluvial style of the Eocene-Oligocene alluvial fans of the Scala Dei Group, SE Ebro Basin (Spain): Journal of Geological Society of London, v. 140, p. 113–146.

ANADÓN, PEDRO, 1978a, El Paleógeno continental anterior a la transgresión Biarritziense (Eoceno medio) entre los rios Gaià y Ripoll (prov. de Tarragona y Barcelona): Estudios Geológicos, v. 34, p. 341–440.

——, 1978b, Deslizamientos gravitacionales y depósitos asociados en el Eoceno marino del borde oriental de la Cuenca del Ebro (sector de Igualada): Acta Geologica Hispanica, v. 13, p. 47–53.

——, 1980, Olistostromas asociados a depósitos de cono de deyección en el Eoceno inferior continental del borde oriental de la Cuenca del Ebro (zona de Sant Llorenç del Munt, Prov. de Barcelona): Resumenes y Comunicaciones del IX Congreso Nacional de Sedimentología (Octubre 1980), v. 1, p. 41–42.

ANADÓN, PEDRO, COLOMBO, FERRAN, ESTEBAN, MATEU, MARZO, MARIANO, ROBLES, M. S., SANTANACH, PERE, AND SÓLE-SUGRAÑES, LLUÍS, 1979, Evolución tectonoestratigráfica de los Catalánides: Acta Geologica Hispanica, v. 14, p. 242–270.

ANADÓN, PEDRO, AND FEIST, MONIQUE, 1981, Charophytes et biostratigraphie du Paléogène inférieur du bassin de l'Ebre oriental: Palaeontographica, v. 178 (Ser. B), p. 143–168.

ANADÓN, PEDRO, FEIST, MONIQUE, HARTENBERGER, J. L., MULLER, CARLA, AND VILLALTA, J. F., 1983, Un exemple de correlation biostratigraphique entre échelles marine et continentales dans l'Éocène: la coupe de Pontils (Bassin de l'Ebre, Espagne): Bulletin de la Société géologique de France, v. 25, p. 747–755.

ARTHAUD, FRANÇOIS, 1969, Méthode de détermination graphique des directions de raccourcissement, d'allongement et intermédiaire d'une population de failles: Bulletin de la Société géologique de France, v. 11, p. 729–737.

COLOMBO, FERRAN, 1980, Estratigrafía y sedimentología del Terciario inferior continental de los Catalánides [unpubl. Ph.D. thesis]: Barcelona, Universitat de Barcelona, 609 p.

COLOMBO, FERNANDO, AND ROBLES, SERGIO, 1983, Evolución de los sistemas sedimentarios del borde SE de la Depresión del Ebro entre Gandesa y Horta de St. Joan (Prov. de Tarragona): Comunicaciones del X Congreso Nacional de Sedimentología, Menorca, p. 176–179.

ESTEBAN, MATEU, 1972, Presencia de Caliche fósil en la base del Eoceno de los Catalánides, provincias de Tarragona y Barcelona: Acta Geologica Hispanica, v. 7, p. 164–168.

——, 1974, Caliche textures and "Microcodium": Bolletino della Societa Geologica Italiana, v. 92, p. 105–125.

ESTEBAN, MATEU, AND ROBLES, SERGIO, 1976, Sobre la paleogeografía del Cretácico inferior de los Catalánides entre Barcelona y Tortosa: Acta Geologica Hispanica, v. 11, p. 73–78.

FERRER, JORGE, 1971, El Paleoceno y Eoceno del borde suroriental de la depresión del Ebro (Cataluña): Mémoires suisses de Paléontologie, v. 90, p. 1–70.

FERRER, JORGE, ROSELL, JUAN, AND REGUANT, SALVADOR, 1968, Síntesis litoestratigráfica del Paleógeno del borde oriental de la Depresión del Ebro: Acta Geologica Hispanica, v. 3, p. 2–4.

FONTBOTÉ, J. M., 1954, Las relaciones tectónicas de la depresión del Vallés-Penedés con la cordillera prelitoral catalana y con la depresión

del Ebro: Boletín de la Real Sociedad Española de Historia Natural, v. Homenaje al Profesor E. Hernández-Pacheco, p. 281–310.

GARCIA-BOADA, JUAN, 1974, El Terciario de la Depresión de Mora y su relación con el borde oriental de la Depresión del Ebro (prov. Tarragona): Seminarios de Estratigrafía, v. 9, p. 11–20.

GUIMERÁ, JOAN, 1984, Palaeogene evolution of deformation in the northeastern Iberian Peninsula: Geological Magazine, v. 121, p. 413–420.

JULIÁ, RAMON, AND SANTANACH, PERE, 1984, Estructuras en la salbanda de falla paleógena de la falla del Vallès-Penedès (Cadenas Costeras Catalanas): su relación con el deslizamiento de la falla: Primer Congreso Español de Geología, v. 1, p. 47–59.

JULIVERT, MANUEL, 1978, The areas of alpine cover folding in the Iberian Meseta (Iberian Chain, Catalanides, etc.) in Lemoine, M., ed., Geological Atlas of Alpine Europe and Adjoining Areas: Amsterdam, Elsevier, p. 93–112.

LLOPIS, NOEL, 1947, Contribución al conocimiento de la morfoestructura de los Catalánides: Barcelona, Consejo Superior de Investigaciones Científicas, Instituto Lucas Mallada, 372 p.

MARZO, MARIANO, 1980, El Buntsandstein de los Catalánides: Estratigrafía y procesos de sedimentación [unpubl. Ph.D. thesis]: Barcelona, Universitat de Barcelona, 317 p.

MUÑOZ, JOSEP-ANTON, MARTÍNEZ, ALBERT, AND VERGÉS, JAUME, in press, Thrust sequences in the Spanish eastern Pyrenees: Journal of Structural Geology, v. 8.

MUTTI, EMILIANO, LUTERBACHER, HANS-PETER, FERRER, JORGE, AND ROSELL, JUAN, 1972, Schema stratigrafico e lineamenti di facies del Paleogeno marino nella zona centrale subpirenaica tra Tremp (Catalogna) e Pamplona (Navarra): Memorie della Societa Geologica Italiana, v. 11, p. 391–416.

PLAZIAT, J. C., 1975, L'Ilerdien à l'intérieur du Paleogène languedocien; ses relations avec le Sparnacien, l'Ilerdien sud-pyrénéen, l'Ypresien et le Paléocène: Bulletin de la Société géologique de France, v. 7, p. 168–182.

——, 1981, Late Cretaceous to late Eocene palaeogeographic evolution of southwest Europe: Palaeogeography, Palaeoclimatology, Palaeoecology, v. 36, p. 263–320.

PUIGDEFÀBREGAS, CAI, 1975, La sedimentación molásica en la Cuenca de Jaca: Monografias del Instituto de Estudios Pirenaicos, n. 104, 188 p.

RIBA, ORIOL, 1976, Syntectonic unconformities of the Alto Cardener, Spanish Pyrenees: a genetic interpretation: Sedimentary Geology, v. 15, p. 213–233.

RIBA, ORIOL, AND REGUANT, SALVADOR, in press. Ensayo de síntesis estratigráfica y evolutiva de la Cuenca Terciaria del Ebro, in Libro Jubilar José Ma. Rios, Estudios sobre Geología de España, Chapt. III.3.8.

ROBLES, M. S., 1982, Estudio comparativo del sistema aluvial del borde suroccidental de los Catalánides, en la transversal de Prat de Compte (Tarragona) y los abanicos aluviales de la Pobla de Segur (Prepirineo de Lérida): Acta Geologica Hispanica, v. 17, p. 255–269.

RODGERS, D. A., 1980, Analysis of pull-apart basin development produced by en echelon strike-slip faults, in Ballance, P. F. and Reading, H. G., eds., Sedimentation in Oblique-Slip Mobile Zones: International Association of Sedimentologists Special Publication No. 4, p. 27–41.

ROSELL, JUAN, JULIÀ, RAMON, AND FERRER, JORGE, 1966, Nota sobre la estratigrafía de unos niveles con Carofitas existentes en el tramo rojo de la base del Eoceno al S de los Catalánides (Provincia de Barcelona): Acta Geologica Hispanica, v. 1, p. 17–20.

ROSELL, JUAN, FERRER, JORGE, AND LUTERBACHER, H. P., 1973, El Paleógeno marino del NE de España: XIII Coloquio Europeo de Micropaleontología, Madrid, Enadimsa, p. 29–62.

ROSELL, JUAN, AND PUIGDEFÀBREGAS, CAI, 1975, The sedimentary evolution of the Paleogene South Pyranean Basin: 9th International Congress on Sedimentology, International Association of Sedimentologists, Nice, Guidebook 19, 104 p.

SEGURET, MICHEL, 1970, Étude tectonique des nappes et séries décollées de la partie centrale du versant sud des Pyrénées. Caractère synsédimentaire, rôle de la compression et de la gravité: Publications Université des Sciences et Techniques du Languedoc, Montpellier, série géologie structurale, v. 2, 161 p.

SOLÉ-SUGRAÑES, LUIS, 1978, Gravity and compressive nappes in the Central southern Pyrenees (Spain): American Journal of Science, v. 278, p. 609–637.

VAN HOUTEN, F. B., 1974, Northern Alpine molasse and similar Cenozoic sequences of southern Europe, *in* Dott, R. H., Jr., and Shaver, R. H., eds., Modern and ancient geosynclinal sedimentation, Society of Economic Paleontologists and Mineralogists Special Publication No. 19, p. 260–273.

WILCOX, R. E., HARDING, T. P., AND SEELY, D. R., 1973, Basic wrench tectonics: American Association of Petroleum Geologists Bulletin, v. 57, p. 74–96.

XIAOHAN, LIU, 1983, Partie I: Perturbation de contraintes liées aux structures cassantes dans les calcaires fins du Languedoc; Partie II: Mesure de la déformation finie à l'aide de la méthode Fry. Application aux gneiss de Bornes (Massif des Maures) [unpubl. thesis]: Montpellier, Université des Sciences et Techniques de Languedoc, 152 p.

THE VIENNA BASIN: A THIN-SKINNED PULL-APART BASIN

LEIGH H. ROYDEN

Department of Earth, Atmospheric and Planetary Sciences,
Massachusetts Institute of Technology,
Cambridge, Massachusetts 02139

ABSTRACT: The Vienna Basin was formed by middle Miocene (Karpatian-Badenian, 17.5–13.0 Ma) extension and contains up to 6 km of Miocene to Quaternary sedimentary rocks. This basin is partly superimposed on the north-vergent nappes of the outer West Carpathian flysch belt and partly on the nappes of the inner Carpathian belt. The obvious rhombohedral shape of the Vienna Basin, the left-stepping pattern of en echelon faults within the basin, and the southward migration of basin extension through time strongly suggest that this basin is a pull-apart feature formed during middle Miocene left slip along a northeast-trending fault system. This interpretation is supported by geologic mapping in the Carpathians.

These left-slip faults appear to have functioned mainly as tear faults within the Carpathian nappes and separated the areas of active north-vergent thrusting east of the Vienna Basin from areas west of the basin where thrusting had already been completed. Reflection seismic lines show that the autochthonous European-plate basement continues beneath the allochthonous Carpathian nappes and beneath the Vienna Basin, and that in general the European plate is not significantly disrupted by the normal faults that bound the basin. Thus both the normal faults and the associated strike-slip faults appear to merge into a gently southeast-dipping detachment at depth. In this way, extension of the Vienna Basin appears to have been restricted mainly to shallow crustal levels above that detachment (that is, restricted mainly to the allochthonous nappes of the Carpathians). Detailed analyses of subsidence and heat-flow data indicate that little or no heating of the lithosphere occurred during extension of the Vienna Basin, and support the interpretation that extension was confined to shallow crustal levels. This interpretation explains why hydrocarbons mature at much greater depths in the Vienna Basin (≥5 km) than in the neighboring Pannonian Basin.

INTRODUCTION

The Vienna Basin, located in Austria and Czechoslovakia (Fig. 1), is a classic example of a rhombohedral pull-apart basin formed along an active strike-slip fault system. The basin is about 200 km long and 60 km wide and locally contains up to 6 km of Miocene sedimentary rocks (Fig. 2). In this paper I use isotopic ages for the Paratethyan biostratigraphic stages (Fig. 3) modified after Vass (1978), Vass and Bagdasarian (1978), and Steininger and Rögl (1979). Isotopic ages compiled by other workers differ slightly, but the relative timing of events is dependent on biostratigraphic dating, not on the exact isotopic ages used. Isotopic ages are given here for readers unfamiliar with the Paratethyan time scale. The youth of this basin (early-middle Miocene) and its simple tectonic history have resulted in exceptionally good preservation of many structural and sedimentological features that are typical of such strike-slip regimes.

The Vienna Basin is one of several Miocene basins that together make up the intra-Carpathian or Pannonian basin system (Fig. 2). Like the other basins within this system, the Vienna Basin formed adjacent to the coeval Carpathian thrust belt, and its evolution is intimately related to contemporaneous thrust-belt activity (Royden et al., 1983a). Nowhere is the relation between basin extension and thrusting more clearly illustrated than in the case of the Vienna Basin and adjacent parts of the Alpine-Carpathian thrust belt. This intimate relationship is attributable to the proximity of the basin to the active thrust belt, because part of the basin lies on top of the external nappes of the Eastern Alps and Carpathians (for example, Brix and Schultz, 1980; or Kröll et al., 1981). This paper discusses how the location, magnitude, direction, timing, and style of extension within the Vienna Basin have been largely controlled by contemporaneous thrust-belt activity.

REGIONAL SETTING

The Alpine-Carpathian thrust belt can be subdivided into an inner belt composed largely of carbonate and crystalline rocks deformed mainly in Cretaceous time, and an outer belt deformed mainly in early Tertiary time in the Eastern Alps, and in late Tertiary time in the Carpathians (Fig. 1; for summaries see Andrusov, 1965, 1968; Săndulescu, 1975, 1980; Burchfiel, 1980; and Oberhauser, 1980). A discussion of the inner belt is beyond the scope of this paper. The outer belt consists of thrust sheets composed of Cretaceous to Miocene flysch, and to a lesser extent of molasse with conglomerate beds. Thrusting occurred during subduction of the European lithosphere, on which the flysch had been deposited, southward beneath the inner units of the Alpine-Carpathian chain. Subduction initially involved oceanic lithosphere, but continental collision occurred in Eocene (?) time in the Eastern Alps and early Miocene (?) time in the West Carpathians (Burchfiel, 1980; Fig. 4). Following continental collision, the continental margin of Europe continued to be subducted, and eventually the stable continental platform was over-ridden by the thrust sheets of the Carpathians. Within the outer belt, the age of the thrust faults is progressively younger from the internal to the external parts of the belt and also from west to east (Fig. 5). Thrusting ended diachronously in late Oligocene to early Miocene time in the Eastern Alps and middle to late Miocene time in the Carpathians (for example, Jiřiček, 1979). In the vicinity of the Vienna Basin, thrusting terminated in Karpatian (17.5–16.5 Ma) time. Total shortening across the outer flysch belt is probably several hundred kilometers, (Burchfiel, 1976; Książewicz et al., 1977).

In many respects the outer belt of the Alpine-Carpathian chain is a typical foreland fold and thrust belt. The basement of the Alpine and Carpathian flysch belt has been entirely subducted, so that the flysch nappes are now com-

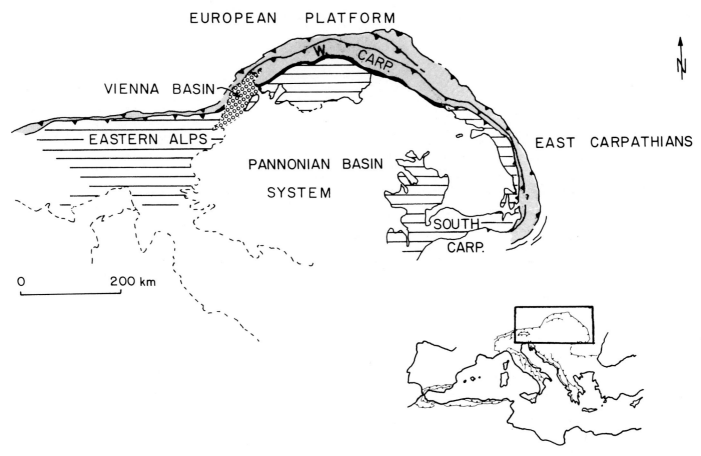

FIG. 1.—Sketch map of the Alpine-Carpathian mountain chain showing the Vienna Basin (open circles). Shading indicates the outer flysch belt of the Alps and Carpathians, deformed in Tertiary time. Horizontal lines show the inner, mainly carbonate and crystalline, parts of the belt, deformed mainly in Cretaceous time. Black indicates the Pieniny Klippen belt (or Klippen belt) that separates the inner and outer West Carpathian belts. The Klippen belt was deformed first in Cretaceous time with the inner West Carpathians, and again in Tertiary time with the outer Carpathian belt.

pletely allochthonous and overlie the autochthonous basement of the European foreland along low-angle thrust faults (Fig. 6, cross sections 1 and 7). The European basement has been followed for more than 30 km beneath the thrust sheets by drilling data (Kröll et al., 1981) and even farther on seismic reflection profiles.

Extension within the Pannonian or intra-Carpathian Basin system occurred in Miocene time. Crustal extension was concentrated in small, well localized zones that are now the deep basin areas (Fig. 2), and the onset of extension and subsidence is of slightly different ages in different basins. These deep basins are separated from one another by less-extended and less-subsided blocks. All are related to strike slip (with the probable exception of the Transylvanian Basin, TS in Fig. 2), and are connected to one another and to the thrust belt by a conjugate system of strike-slip faults (Fig. 7). Royden et al. (1983a) have shown that extension and strike-slip faulting are closely related to diachronous thrusting around the irregularly shaped Carpathian thrust belt. Extension of the Vienna Basin was one of the first events during this gradual, inhomogeneous extension and pulling apart of the continental crust.

In early and middle Miocene time, the Vienna Basin

formed on top of both the external belt, which was deformed in Cenozoic time, and the internal belt, deformed in Cretaceous time (Fig. 8). Cross sections through the basin show that its basement is composed of thrust sheets that overlie the autochthonous sedimentary cover and crystalline basement of Europe along east- to southeast-dipping thrust faults (Fig. 6, cross sections 2–6). Drilling data allow the structural units of the Eastern Alps to be traced eastwards under the Vienna Basin and correlated with those of the Western Carpathians (Brix and Schultz, 1980; Jiříček and Tomek, 1981; Fig. 8).

SUBSIDENCE AND SEDIMENTATION OF THE VIENNA BASIN

Subsidence in the Vienna Basin began in early Miocene time and has continued until the present, but most subsidence occurred in middle and late Miocene time between about 17.5 and 8 Ma (Karpatian through Pannonian time, Fig. 3; for example, see Kröll et al., 1980; Vass, 1982). Basin evolution can be divided into two extensional phases that show different patterns and areal distributions of subsidence and sedimentation. The first phase of subsidence occurred on top of the moving thrust sheets, while the sec-

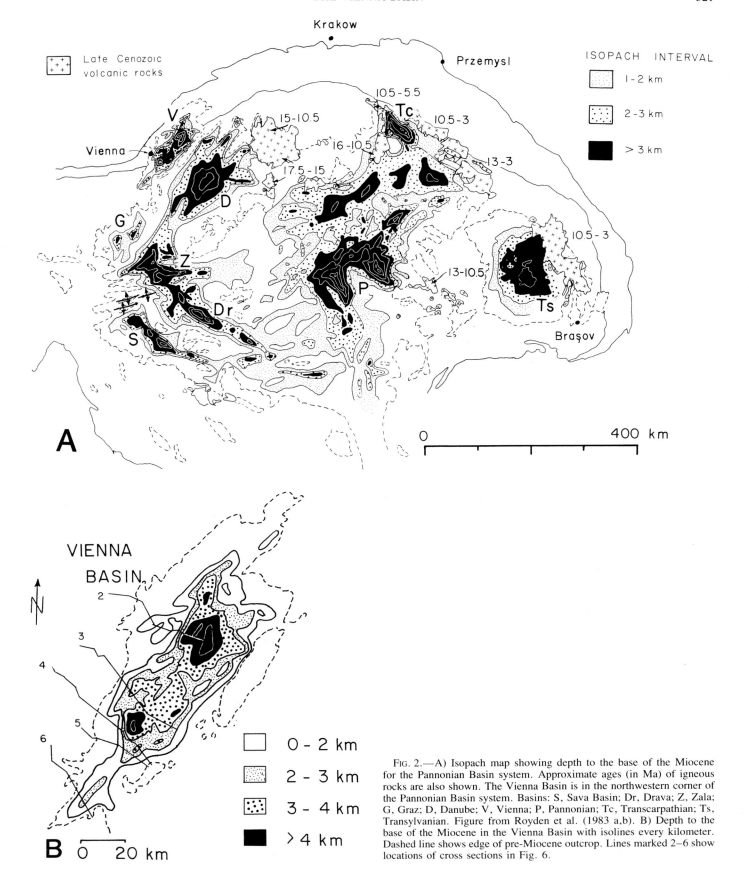

FIG. 2.—A) Isopach map showing depth to the base of the Miocene for the Pannonian Basin system. Approximate ages (in Ma) of igneous rocks are also shown. The Vienna Basin is in the northwestern corner of the Pannonian Basin system. Basins: S, Sava Basin; Dr, Drava; Z, Zala; G, Graz; D, Danube; V, Vienna; P, Pannonian; Tc, Transcarpathian; Ts, Transylvanian. Figure from Royden et al. (1983 a,b). B) Depth to the base of the Miocene in the Vienna Basin with isolines every kilometer. Dashed line shows edge of pre-Miocene outcrop. Lines marked 2–6 show locations of cross sections in Fig. 6.

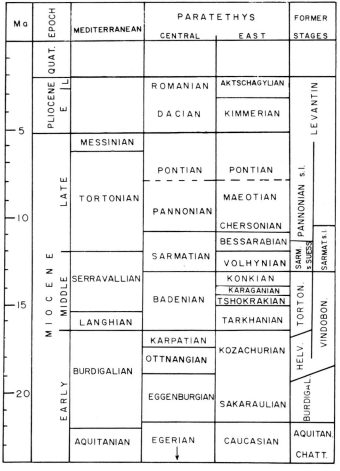

| Ma | EPOCH | | MEDITERRANEAN | PARATETHYS | | FORMER STAGES |
				CENTRAL	EAST	
	QUAT.					
—5	PLIOCENE	L		ROMANIAN	AKTSCHAGYLIAN	
		E		DACIAN	KIMMERIAN	LEVANTIN
			MESSINIAN			
		LATE	TORTONIAN	PONTIAN	PONTIAN	PANNONIAN s.l.
—10				PANNONIAN	MAEOTIAN	
	MIOCENE				CHERSONIAN	
			SERRAVALLIAN	SARMATIAN	BESSARABIAN	SARM. s SUESS / SARMAT s.l.
		MIDDLE			VOLHYNIAN	
—15				BADENIAN	KONKIAN	TORTON. / VINDOBON.
					KARAGANIAN	
					TSHOKRAKIAN	
			LANGHIAN		TARKHANIAN	HELV.
		EARLY	BURDIGALIAN	KARPATIAN	KOZACHURIAN	BURDIGAL.
				OTTNANGIAN		
—20				EGGENBURGIAN	SAKARAULIAN	
			AQUITANIAN	EGERIAN ↓	CAUCASIAN	AQUITAN. / CHATT.

Fig. 3.—Correlation and isotopic age determinations for Mediterranean and Central and East Parathethyan biostratigraphic stages (from Royden et al., 1983b).

ond phase postdates movement of the thrust sheets beneath the basin. Uplift, erosion, and tilting occurred in many parts of the basin in Karpatian or early Badenian time between the two phases of subsidence.

The following summary of Neogene stratigraphy and sedimentation in the Vienna Basin is based largely on recent comprehensive studies by Buday and Cicha (1968), Papp et al. (1973), Steininger et al. (1975), Fuchs (1980), Janoschek and Matura (1980), Kröll (1980), and Jiříček and Tomek (1981).

Lower Miocene sedimentary rocks occur mainly in the northern part of the Vienna Basin (Fig. 9). The original distribution of lower Miocene strata is uncertain because of later erosion, mainly in late Karpatian-early Badenian time. The oldest sediments in the basin are of Egerian age (Fig. 3), but are poorly preserved. They are overlain by a better preserved section of Eggenburgian and Ottnangian sedimentary rocks. The Eggenburgian basal conglomerate (maximum thickness, 300 m) consists of various clastic rocks, including conglomerates that contain cobbles of Mesozoic carbonate rocks, Paleogene sandstone, and dolomite breccia with a calcareous matrix. These beds are overlain

by Eggenburgian-Ottnangian marine claystones, sandstones, and siltstones (Lužice beds). The Eggenburgian part of the Lužice beds is pelitic (maximum thickness, 150 m), but Eggenburgian-age sedimentation ended with an extensive regression, and the lower Ottnangian section consists of sandstones and conglomerates with pebbles of Paleogene sandstone in a clay matrix (maximum thickness, 250 m). The uppermost Lužice beds are composed of consolidated grey, slightly arenaceous, micaceous calcareous marine claystones (Schlier facies). The maximum thickness of the Lužice beds is about 1,200 m. The Eggenburgian- and Ottnangian-age rocks are largely confined to several troughs in the northern Vienna Basin, and their thickness varies greatly. In the southern part of the basin, Ottnangian sediments occur in a brachyhaline facies (the Bockfliess beds, 600 m thick).

Karpatian rocks overlie older lower Miocene rocks unconformably in most places, and onlap pre-Neogene rocks of the Alpine-Carpathian thrust belt. In the northern part of the basin, lower Karpatian rocks are marine, and were deposited mainly in the same troughs as the underlying Lužice beds (Jiříček and Tomek, 1981). During late Karpatian time, however, the environment became abruptly brackish, and sediments began to accumulate over a broader area. These sediments consist of sandstones (the Sastin beds, 100–400 m thick) in the northern part of the Vienna Basin, and in the southern part, of the terrestrial-limnic Gänserndorfer beds, about 300 m of shale, sandstone, and anhydrite overlying a basal conglomerate. The Sastin beds are overlain by marine and brackish claystones and siltstones, and the entire Karpatian sequence in the northern part of the basin is known as the Laa beds (Fig. 9). In the southern part of the basin the Gänserndorfer beds are overlain by the limnic-fluviatile Aderklaa beds (1,000–1,500 m thick) that contain eroded detritus from the alpine napps.

The transition from Karpatian to Badenian time was marked by local uplift, erosion, normal faulting, tilting of fault blocks, and creation of topographic relief. The area of major subsidence and sedimentation moved to the southern part of the Vienna Basin. The Badenian is transgressive over the whole basin, and in the northern part there is a significant angular discordance between Karpatian and Badenian rocks. The lower Badenian is almost totally marine, and the upper Badenian is largely brackish. Badenian rocks consist mainly of sandstones and calcareous claystones and siltstones. In the northeastern part of the basin, coaly claystones and occasional brown coal beds were deposited in latest Badenian time. The thickness of Badenian sedimentary rocks is highly variable, depending upon whether they were deposited on upthrown or downthrown fault blocks, and reaches a maximum of 3,500 m. Sedimentation rates were very high in the Badenian, and even in areas of rapid subsidence the water depth remained shallow.

The final regression of the Parathethyan sea began in Sarmatian time. Sarmatian sedimentary rocks in the Vienna Basin consist mainly of sandstones and calcareous claystones. The lower Sarmatian is brackish in the southern part of the basin, but consists of freshwater claystones in the northern part. By middle Sarmatian time interlayed sandstone and marl was accumulating over large areas. The

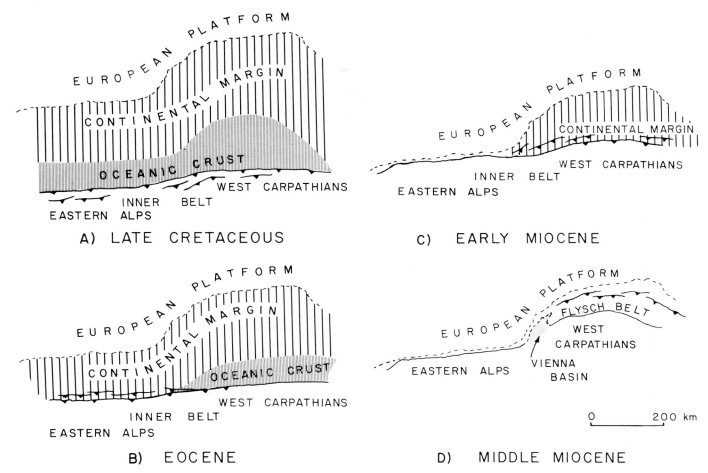

Fig. 4.—Simplified tectonic evolution of the Alpine and Carpathian thrust belts relative to the European platform and its continental margin. Boundary of southern plate in (A) and (B) is schematic. A) In Cretaceous time thrusting was confined to the inner belt and southward directed subduction probably involved only oceanic rocks. B) In Eocene (?) time, continental collision occurred in the Eastern Alps as the oceanic crust was completely subducted (some workers believe collision was a late Cretaceous event). Subduction of the continental margin continued in the Eastern Alps. Oceanic material probably remained in the region of the West Carpathians until Miocene time. C) In early Miocene (or late Oligocene) time, subduction ended in the Eastern Alps after part of the European platform had been overthrust and probably partly subducted. The main phase of deformation began in the outer flysch Carpathians at this time. Continental collision in the Carpathian region probably occurred in early Miocene time, and the leading edge of the European continental margin was subducted. D) In middle Miocene time, the outer flysch belt in the Carpathians over-rode the European platform. Thrusting ended diachronously from west to east and was mostly completed by the end of the middle Miocene. Shading shows location of the Vienna Basin.

thickness of Sarmatian rocks in the Vienna Basin varies from 300 to 600 m.

In Pannonian time, interbedded sandstones and marls continued to be deposited in decreasingly saline conditions. By Pontian time, the depositional environment became freshwater, and mainly sandstones were deposited. The combined thickness of Pannonian plus Pontian sedimentary rocks varies from several tens of meters to several kilometers. The greatest thicknesses of post-Sarmatian rocks occur in the southern part of the basin, and the maximum thickness of the Pannonian to Quaternary section in the northeastern part of the basin is less than about 800 m. Pliocene and Quaternary sedimentary rocks in the Vienna Basin are mainly fluviatile, and as thick as 200 m in some actively subsiding regions.

STRUCTURE OF THE VIENNA BASIN

The fault pattern of the Vienna Basin consists of a series of anastomosing or braided syn-sedimentary faults that define rhombohedral shaped sub-basins separated by less subsided blocks (Fig. 10). Lower Karpatian and older sedimentary rocks are restricted mainly to rhombohedral shaped sub-basins in the northern Vienna Basin, and are bounded by syn-sedimentary faults (Figs. 6, 10). Badenian and younger deposits are restricted mainly to a different set of rhombohedral-shaped sub-basins in the southern part of the basin and are bounded by syn-sedimentary faults that commonly have several kilometers of normal separation (Fig. 6). The same major fault systems within the basin appear to have controlled both the early and late phases of basin

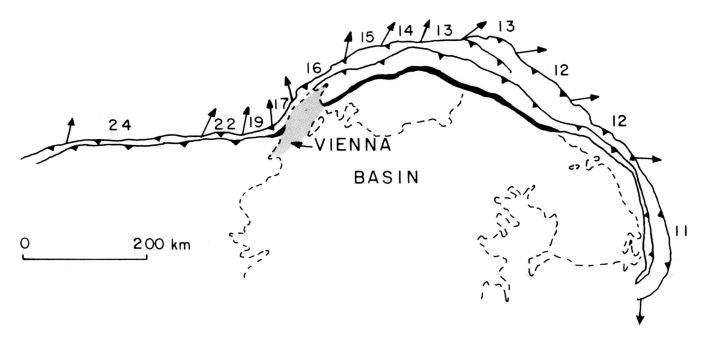

FIG. 5.—Age (in Ma) and direction of the last thrusting events along the outer part of the Alpine-Carpathian chain (after Jiříček, 1979). The regular west to east migration of the end of thrusting is well documented using local biostratigraphic stages. In this figure, the biostratigraphic stages have been replaced by absolute ages using the timescale given in Figure 1. If different absolute age determinations are used for the local biostratigraphic stages, the numbers given in this figure change accordingly, but the regular west to east migration of thrusting is not changed significantly. Shading shows the location of the Vienna Basin.

subsidence, even though the early and late phase depocenters do not coincide.

From examination of the fault patterns within the Vienna Basin and from the development of rhombohedral shaped troughs through time, three major fault zones can be identified within the basin that account for much of its structure (Fig. 10). All are parallel or sub-parallel to the north- to northeast-trending structural units within the underlying nappes (compare Figs. 8 and 10). The first fault zone enters the southern end of the basin and bounds its southeastern side (fault #1, Fig. 10). This fault zone is the northern extension of the Mur-Mürz line, and is seismically active. Fault-plane solutions from the largest shocks indicate mainly strike slip with P axes oriented approximately north-south (Gutdeutsch and Aric, 1976). These solutions indicate that left slip is currently occurring along the Mur-Mürz line south of and in the southern part of the Vienna Basin.

The second fault system (fault #2, Fig. 10) strikes northeast across the central part of the Vienna Basin, and continues beyond the northeastern corner of the basin. In the northern half of the basin this fault zone coincides with the underlying Pieniny "Klippen" belt (compare Figs. 8 and 10). The Pieniny Klippen belt contains no klippe and developed as the erosional outliers of thrust sheets. The word klippe is used for large blocks, as long as several kilometers, of deformed rocks set in a sheared matrix of fine-grained sedimentary rocks. The Klippen belt is a long (600 km), narrow (maximum width, 12 km) belt of imbricated and deformed rocks separating the inner West Carpathians from the outer flysch Carpathians. This belt of rocks has been interpreted as a Cretaceous suture, and as a major left-slip

shear zone (Săndulescu, 1980) active in both early and late Tertiary time (Birkenmajer, 1981; Royden and Báldi, in prep.).

The third fault system strikes northeast beyond the northernmost part of the Vienna Basin and into the flysch belt (fault #3, Fig. 10). Within the basin, this fault zone probably corresponds to the Schrattenburg-Bulhary and Steinburg faults. It is sub-parallel to the structural trend of the underlying flysch nappes. Geologic mapping in the flysch belt north of the Vienna Basin indicates about eighty kilometers of left-slip along this fault in Cenozoic time (Roth, 1980).

Within the Vienna Basin, syn-sedimentary faults with normal separation dip east and west. The largest normal separations occur on east-dipping faults that bound the western sides of the sub-basins or deep troughs (Fig. 6). Near the surface, these faults commonly dip only 40° to 50°. West-dipping faults within the basin generally end against east-dipping faults. In places the Miocene sedimentary rocks within the basin dip gently to the west and the older basin deposits have been rotated more than the younger ones. In cross section the basin fill commonly forms westward thickening wedges bounded to the west by syn-sedimentary faults.

The cross sections through the Vienna Basin shown Figure 6 have been modified after Wessely (1983, and in prep.) and the geometry of the faults has not been changed significantly. In Wessely's interpretation, even where these faults have several kilometers of normal separation at the surface, many of them cause little disruption of the autochthonous cover rocks of the European plate (for example, Fig. 6, cross sections 3, 5, and 6). In many cases, the cumulative near-surface displacement on normal faults is shown to be much greater than the apparent offset of the autochthonous basement. Other faults are shown to intersect the autochthonous basement at low angles (Fig. 6, cross section 4). Only a few faults, such as the large Steinberg

fault with about 4 km of normal separation, have been interpreted as intersecting the autochthonous basement at a steep angle with significant displacement of the crystalline basement (Fig. 6, cross section 2). A seismic reflection profile across the Steinberg fault has been published by Weber (1980).

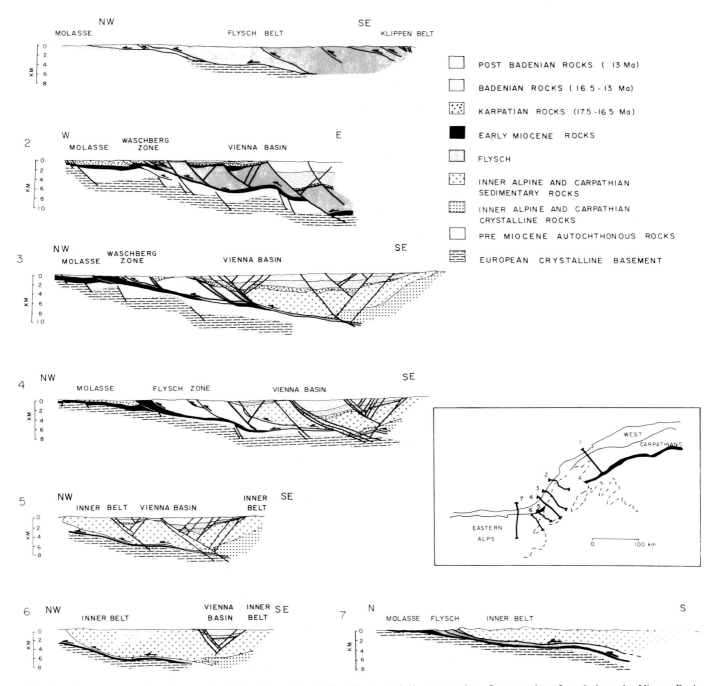

FIG. 6.—Seven cross sections through the Alpine-Carpathian chain with no vertical exaggeration. Cross sections 2 to 6 show the Vienna Basin superimposed on the underlying thrust belt. Tertiary overthrusts are indicated by arrows, and Miocene normal faults are obvious because they displace Miocene rocks at the surface. Normal faults confined to the autochthonous basement are mainly Jurassic syn-sedimentary normal faults associated with Mesozoic rifting. Locations of cross sections 2 to 6 are shown in the inset map and also in Figure 3. Cross sections are simplified after (1) Pesl et al. (1968); (2 to 6), Wessely (1983, and in press); and (7) Kröll et al. (1981).

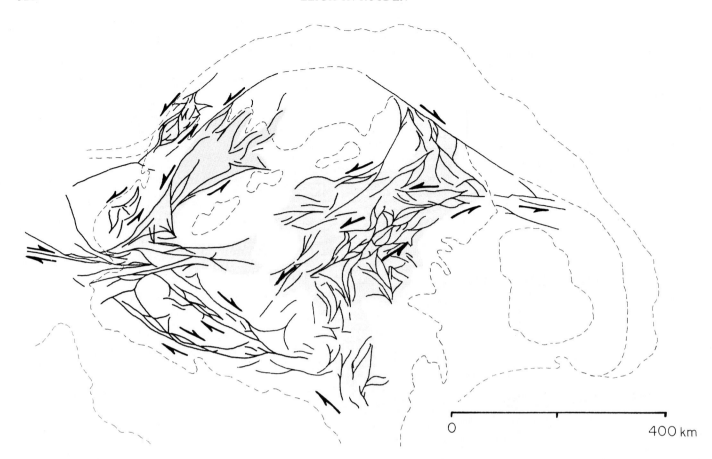

FIG. 7.—Generalized map of Neogene faults in the intra-Carpathian region. Arrows indicate sense of shear along strike-slip zones. Some basins are interpreted as pull-apart basins, whereas others are regions of extension bounded by zones of differential shear. From Royden et al. (1982).

In my interpretation, the apparent decrease in normal displacement at depth on many of the large faults can be explained by transferring some of the displacement to flat or gently eastward-dipping surfaces within the thrust belt. The basal thrust plane beneath the thrust complex and above the autochthonous sedimentary rocks provides a good candidate for such a detachment.

A second system of Jurassic syn-sedimentary normal faults is present in the autochthonous rocks beneath the thrust sheets (Fig. 6, cross-sections 2, 3, and 4). These faults are probably related to Jurassic rifting of Tethys, and they strike parallel or sub-parallel to Neogene faults of the Vienna Basin but they do not cut the Alpine-Carpathian thrust sheets.

THE VIENNA BASIN AS A PULL-APART FEATURE

Royden et al. (1982, 1983a) have interpreted the Vienna Basin as a pull-apart feature (Burchfiel and Stewart, 1966) resulting from left slip along the northeast-striking fault system shown in Figure 10. Left slip along fault #1 in Figure 10 would thus be transferred to faults #2 and #3 across a left-stepping discontinuity at the northern end of fault #1 and the southern ends of faults #2 and #3, and would result in an extension in the area of the Vienna Basin (for example, see Crowell, 1974; Segall and Pollard, 1980). The

obvious rhombohedral shape of the Vienna Basin, the en echelon pattern of faults, and the rapid migration of areas of subsidence (and extension) within the basin are characteristic of pull-apart basins, and are consistent with the northeast-striking fault segments shown in Fig. 10. Thus the major northeast-striking faults within the Vienna Basin would have both normal- and left-slip components of displacement (Fig. 11). This interpretation requires that the strike-slip faulting must be of the same age as basin extension (mainly Karpatian to Pannonian), although the faults may have been active at other times as well.

The thrust sheets beneath the Vienna Basin appear to have been thinned by about 50% during basin extension. For example, on cross section 5 in Figure 6, the thickness of the thrust sheets is about 6–8 km west and east of the basin, but only 3–4 km beneath the basin. The magnitude of strike slip needed to cause 100% extension under the deeper parts of the Vienna Basin by a pull-apart mechanism is probably several tens of kilometers. However, as the major strike-slip faults enter the basin, they are likely to lose their component of strike slip gradually, so that near the end of the fault segments the displacement may be mainly normal (Fig. 12). The north-south striking fault zones within the basin are likely to have mainly normal displacement with little or no strike-slip component.

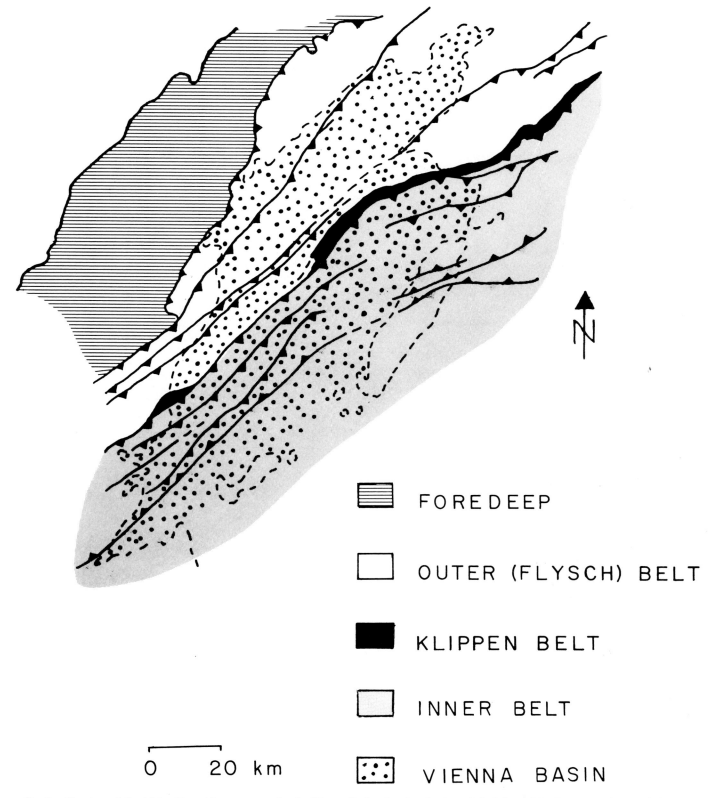

FoREDEEP

OUTER (FLYSCH) BELT

KLIPPEN BELT

INNER BELT

0   20 km

VIENNA BASIN

Fig. 8.—Structure of the Alpine-Carpathian nappes under the Vienna Basin, showing the inner belt deformed in Cretaceous time, and the outer or flysch belt deformed in Tertiary time. Also shown is the Pieniny Klippen belt, a narrow deformed belt that separates the inner and outer Carpathians. Barbs indicate upper plate of thrust faults. Big dots with dashed outline show present extent of Vienna Basin sediments. Modified after Jiříček and Tomek (1981).

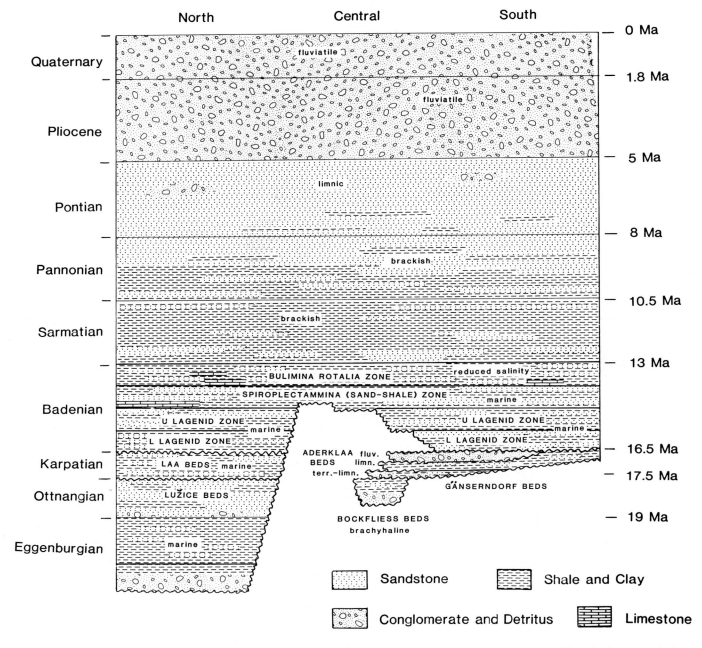

FIG. 9.—Stratigraphic section showing the major rock types in the Vienna Basin (modified after Kröll, 1980; and Wessely, in preparation).

It is difficult to estimate how much left slip has occurred on many of the faults in the Vienna Basin because the major strike-slip faults approximately parallel the northeast structural trend of the underlying nappes of the Alps and Carpathians (compare Figs. 8 and 10). It is clear that little or no strike slip can have occurred on the north-south striking normal faults within the basin, including those that bound the southwestern side of the basin, because they do not offset northeast-trending structural elements below the basin (Fig. 8). This observation is consistent with the fault system presented in Figures 10 and 12.

RELATION OF THE VIENNA BASIN TO THE THRUST BELT

In many respects, the Vienna Basin is a typical pull-apart basin. In some respects, however, it is special because it is superimposed on a thin-skinned thrust belt and because basin extension was roughly contemporaneous with active thrusting in adjacent parts of the belt. In order to understand the evolution of the basin and the associated strike-slip faults better, it is necessary to examine the relationship between thrusting, strike-slip faulting, and basin extension, and to unite all three phenomena within a coherent tectonic frame-

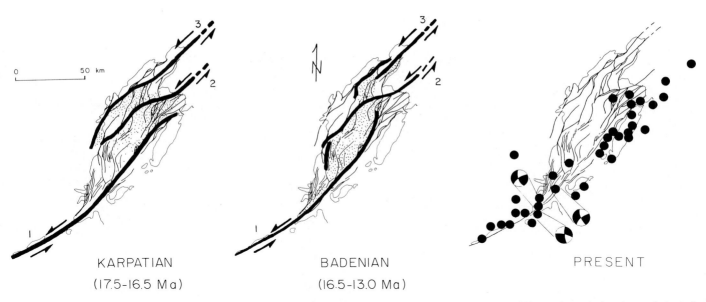

KARPATIAN
(17.5–16.5 Ma)

BADENIAN
(16.5–13.0 Ma)

PRESENT

FIG. 10.—Three major fault systems (bold lines) of the Vienna Basin inferred to have been active at three different times during the evolution of the basin. These three fault zones are meant only as a simplified representation of cumulative movements on many faults, some large and some small. Inferred sense of strike-slip displacement is shown by large arrows. Fine lines indicate other prominent faults of Miocene to Quaternary age within the Vienna Basin (after Wessely, in prep.). Small dots indicate areas of rapid subsidence in Karpatian and Badenian time. For the present day, large dots represent epicenters of earthquakes with seismic intensity greater than 5 (after Gutdeutsch and Aric, in prep.). Fault-plane solutions are after Gutdeutsch and Aric (1976).

work. Key information in determining the interaction between thrust-belt activity, strike-slip faulting and basin extension is given by the relative timing of events and by the geometric relationship between normal, strike-slip, and thrust faults, both at the surface and at depth.

Timing Relationships and Surface Geometry

In the vicinity of the Vienna Basin, as in the rest of the Eastern Alps and Carpathians, thrusting ended diachronously from west to east (Jiříček, 1979; Fig. 5). In the eastern part of the eastern Alps, thrusting ended during Ottnangian time (19.0–17.5 Ma). West of and beneath the Vienna Basin, thrusting ended in Karpatian time (17.5–16.5 Ma). Farther to the northeast (in the north-central part of the West Carpathians), thrusting ended in Badenian time

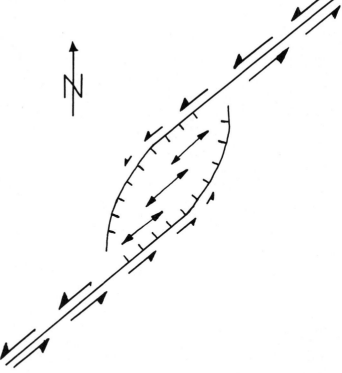

FIG. 12.—Simplified pull-apart interpretation of the Vienna Basin. Displacement on northeast-striking left-slip faults decreases along the faults within the basin, so that near the ends of the faults there is little strike-slip component. Within the basin, the faults have both strike-slip and normal-slip components of displacement. Where the fault segments strike roughly north-south in the basin, displacement is predominantly normal slip. Extension is approximately parallel to the strike-slip faults.

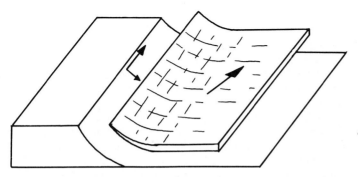

FIG. 11.—Sketch of the geometry of one of the major syn-sedimentary normal faults that bounds the west side of the Vienna Basin. Displacement has both normal- and left-slip components at the surface. As the fault surface flattens at depth, displacement is mainly toward the northeast on a gently dipping detachment surface.

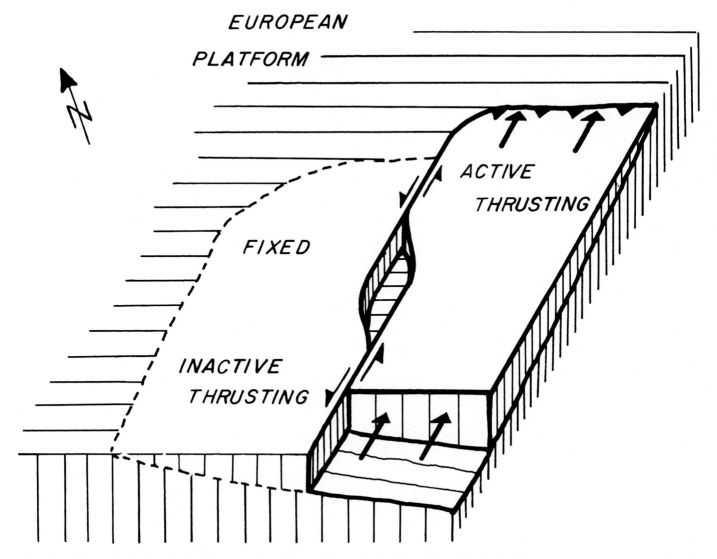

FIG. 13.—Schematic diagram showing how extension within the Vienna Basin is related to strike-slip faulting and contemporaneous thrusting. In Karpatian-Badenian time, nappes to the west of the Vienna Basin were fixed with respect to the European platform; nappes northeast of the basin were still being transported toward the northeast over the European plate. Sinistral strike-slip faulting occurred between the active and inactive nappes and was for the most part confined to the allochthon. The Vienna Basin opened as a pull-apart at a left step in the strike-slip system (see also Figs. 10, 12). In this simplified model, the Vienna Basin is shown as a hole opening up all the way through the thrust sheets. In reality it is an area of up to about 100% extension, so that the nappes are thinned beneath the basin (see for example, Fig. 6, cross section 5).

(16.5–13.0 Ma). This geometry, shown in Figures 4, 5, and 13, indicates that during late Karpatian and Badenian time, the thrust sheets west of the Vienna Basin remained fixed with respect to the European platform while the thrust sheets east of the basin continued to move towards the north or northeast.

The geometry of active and inactive thrust faults requires that relative displacement between thrust sheets of the Eastern Alps and the West Carpathians be accommodated either by sinistral bending of the thrust belt near the Vienna Basin or by left slip along north- or northeast-striking faults within the thrust belt. I suggest that left slip (on faults #1 and #3 in Fig. 10) accommodated part of the displacement between the active and inactive parts of the thrust belt. The age of

displacement along this fault system (inferred from the timing of extension and subsidence within the Vienna Basin) is in good agreement with that required to accommodate continued thrusting towards the northeast in the West Carpathians after termination of thrusting in the Eastern Alps.

The lesser amounts of basin subsidence and extension that occurred prior to Karpatian time are probably related to different rates and directions of early Miocene thrusting around an irregular continental margin in the Eastern Alps and the Carpathians west of the Vienna Basin, and in the Carpathians to the northeast of the basin (Figs. 4, 5). The main phases of basin extension, however, did not occur until thrusting had nearly ended west of the basin in Karpatian time. Subsidence and extension of the Vienna Basin after

Badenian time (when thrusting had ended in the north-central Carpathians) were probably related to thrust-belt activity farther to the east, in the eastern part of the West Carpathians, but it is not clear how the fault systems of the Vienna Basin connect with the eastern part of the Carpathian thrust belt.

The strike-slip fault zones shown in Figures 10 and 13 can be thought of as transform boundaries that separate regions of active thrusting from regions where thrusting had become inactive. These same fault systems connect areas of extension in the Vienna Basin to areas of compression and shortening in the outer Carpathians. Thus extension of the Vienna Basin can be viewed as an immediate result of diachronous thrusting along the Alpine-Carpathian chain.

Geometry at Depth

Cross sections through the Vienna Basin from Wessely (1983, and in prep.) show that many of the normal faults within the Vienna Basin do not significantly displace autochthonous rocks below the basal thrust surface of the Alps and Carpathians. This suggests that extension of the Vienna Basin may be restricted mainly to the allochthonous nappes of the Alpine-Carpathian thrust belt. Such a thin-skinned extension model for the Vienna Basin can be understood by examination of the relation between strike-slip faulting and thrusting within the Carpathian belt. The strike-slip faults that strike northeast beyond the Vienna Basin seem to function as tear faults within the thrust complex. It is unlikely that they extend downwards into the European plate, and they probably end against or merge into thrust faults (Fig. 13). In other words, they are thin-skinned features confined mainly to the thrust complex.

The nature of these strike-slip faults at depth has important implications for the style of extension beneath the Vienna Basin because the basin formed as a pull-apart feature along one such strike-slip system. If the northeast-striking faults that pass through the Vienna Basin merge into a gently dipping decollement (or older thrust fault) at depth (for example, Fig. 6, cross sections 3 and 4) then it is likely that extension of the basin was restricted to shallow crustal levels above the decollement. In simplified terms, the Vienna Basin can be thought of as a small hole that opened up within the thrust complex, without significantly affecting the underlying European basement. In reality, the Vienna Basin is a region where the thrust complex has been thinned by extension to perhaps 50% of its original thickness beneath the deepest parts of the basin (Fig. 6).

According to the cross sections of Wessely (1983, and in prep.), a few normal faults within the basin are interpreted to extend downwards into the autochthonous basement (particularly the large Steinberg fault; Fig. 6, cross section 2). If this interpretation is correct, it need not necessarily be inconsistent with a thin-skinned origin for the basin. One possibility is that not all faults actually flatten into the basal thrust, but may instead flatten eastward within the autochthonous crystalline rocks into gently east-dipping fault surfaces at still greater depth. Wherever such faults root, it appears to be east of the surface expression of the Vienna Basin. A second possibility is that the apparent penetration of the Steinberg fault downwards into autochthonous rocks is the result of interference between Jurassic synsedimentary normal faults that cut authochthonous rocks (Fig. 6, cross sections 2, 3, and 4) and the younger Neogene normal faults within the allochthon. These two fault systems have roughly the same strike. A third possibility is that the Miocene normal faults observed to cut the autochthonous basement may be simple block faults that commonly form in response to down-bending of the subducted slab during thrusting. Thus it is possible that there are two different sets of Miocene normal faults in this region: one set restricted to the allochthonous thrust complex, and the other consisting of simple normal faults within the autochthon, initiated by bending of the foreland during subduction. Interference between these two sets of faults could account for the apparent penetration of some of the extensional features into the autochthonous basement (Fig. 6, cross section 2).

I interpret the extension of the Vienna Basin to be restricted mainly to the over-riding sheet at a convergent boundary. Extensional and strike-slip faulting were thus restricted to shallow crustal levels because extension occurred at the leading edge of the over-riding sheet where it was thin (≤ 10 km). In this way the thin-skinned nature of extension in the Vienna Basin is a result of its proximity to the Alpine-Carpathian structural front. No space problem is inherent in a thin-skinned extensional model for the Vienna Basin because extension was compensated by the over-riding of thrust sheets onto the European platform in the north-central part of the Carpathians (Fig. 13).

INFLUENCES OF OLDER STRUCTURES

The formation of the Vienna Basin can be directly related, in space and time, to contemporaneous thrust belt activity, but its evolution has been influenced by structures inherited from earlier tectonic events. The basin is located at a pronounced bend in the Alpine-Carpathian mountain chain, where east-trending structures of the Eastern Alps change into the northeast-trending structures of the western West Carpathians (Figs. 1, 2). In Mesozoic time, the region east of the Vienna Basin formed an embayment in the European platform (Fig. 4). As the European plate was overridden by the nappes of the Alps and Carpathians during Cenozoic time, collision occurred first in the vicinity of the Eastern Alps and later in the Western Carpathians (Figs. 4, 5). After collision, the continental margin of Europe was subducted southward, and eventually subduction and thrusting ended as part of the continental platform was overridden by thrust sheets. Subduction continued later in the Carpathians than in the Eastern Alps, and the thrust belt advanced much farther to the north in the Carpathian region, an effect attributable to the embayment in the European continent. The resultant sinistral bend in the Alpine-Carpathian thrust belt, and the accompanying diachronous west to east termination of thrusting probably led directly to the geometry shown in Figure 13, and created the tectonic environment for the formation of the Vienna Basin.

On a smaller scale, the strike-slip faults related to basin extension appear to be closely linked to the structure of the underlying thrust belt. The major zones of inferred strike-slip are all parallel or sub-parallel to the structural trends of the underlying nappes. It is difficult to know if the strike of many of the Miocene normal faults within the basin followed that of the older thrust faults, or whether the apparent strike of the thrust planes beneath the basin is the result of later modification by the extensional and strike-slip faults. Both the present structural trend of the thrust belt and the strike-slip faults are parallel or sub-parallel to Jurassic normal faults within the underlying autochthon.

SUBSIDENCE, THERMAL STRUCTURE, AND HEAT FLOW

The nature of Miocene extension beneath the Vienna Basin should be reflected by the regional heat-flow pattern and by the subsidence history of the basin. The Vienna Basin is not a good candidate for analysis of subsidence because in many parts of the basin syn-sedimentary faults reach to the surface (Fig. 6), and there is active seismicity along the southeastern edge of the basin. Together, these observations imply that some extension has been occurring within the Vienna Basin, perhaps episodically, from Miocene time until the present. Thus it is impossible to determine a rate of post-extensional (or thermal) subsidence within the basin. Another factor that complicates subsidence analysis is the small size of the Vienna Basin (60 km by 200 km). For such small basins, lateral conduction of heat, flexure of the lithosphere, and uncertainties about the nature and distribution of extension at depth all make simple subsidence analysis highly speculative.

I propose only to test whether a thin-skinned extensional model is consistent with the observed subsidence and heat-flow data from the Vienna Basin and adjacent areas. The thermal consequences of thin-skinned extension are simple to model because the lower crust and mantle are unaffected by extension near the surface. No thermal anomaly is created beneath the basin at depth, and changes in thermal structure and surface heat flow are all due to near-surface processes. Such a model is easy to test because lateral conduction of heat (except near the basin edges where conductivity contrasts may be important), and lithospheric flexure, etc., can be ignored. Qualitatively, one would expect that if extension of the Vienna Basin involved only shallow crustal levels, as proposed above, the heat flow should be roughly equivalent to that through the adjacent flysch belt. Correspondingly, there should be little or no post-entension (thermal) subsidence.

Some examples of subsidence curves are shown in Figure 14 for three different parts of the Vienna Basin. Curves 2 and 3, from the southern and central areas of the Vienna Basin, correspond to regions where some extension and faulting have probably occurred until recently, or are still

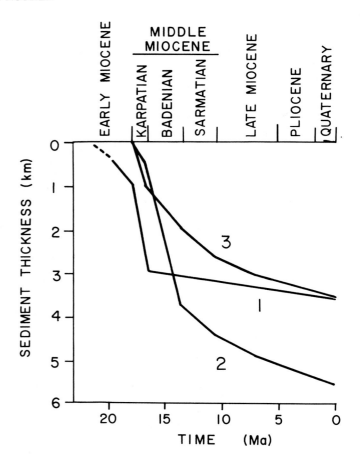

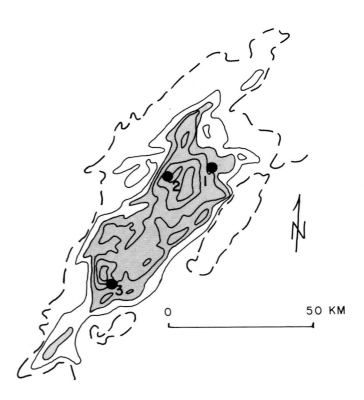

FIG. 14.—Sediment accumulation curves and their locations for three parts of the Vienna Basin. Curve 1 is a generalized curve for the northeastern part of the Vienna Basin (from Jiříček and Tomek, 1981). Curves 2 and 3 represent the central and southern parts of the Vienna Basin, respectively. Location map shows Neogene-Quaternary isopachs (contours interval, 1 km). Shading indicates greater than 2 km of Neogene-Quaternary rocks.

FIG. 15.—Location of heat-flow measurements in the Vienna Basin and adjacent flysch nappes and foredeep (values in mW/m²; data from Čermak, 1979). Shading shows areas where basin sediments are thicker than 2 km.

occurring. Curve 1 from the northeastern corner of the basin corresponds to an area that underwent most of its extension in early-middle Miocene time. Little sedimentation has occurred in the northeastern part of the basin since that time, while over the same interval, the surface of the basin has been uplifted, along with the surrounding area, from sea level to about 200–300 m above sea level.

Heat-flow determinations are available from the northern (Czechoslovakian) part of the Vienna Basin and from the Carpathian flysch belt adjacent to the Vienna Basin (Table 1; Čermak, 1979). These give 50 ± 7 mW/m² for the basin (average of 9 measurements), and 53 ± 7 mW/m² for the adjacent flysch nappes (average of 11 measurements; Fig. 15 and Table 1). Boldizsar (1968) gives 30 thermal gradients for the southern (Austrian) part of the basin based on bottom-hole temperature measurements. The actual temperature measurements and the depth at which they were made are not given by Boldizsar, but he does state that more than half of the measurements were made at depths of between 1,400 and 2,400 m. Because the sediments within the Vienna Basin are lithologically similar to those of the Pannonian Basin, the conductivity-depth relationships used by Royden et al. (1983b) and Dövényi et al. (1983) for sediments of the Pannonian Basin are probably good estimates for the Vienna Basin (Table 2). This relationship gives an average conductivity for sediments between 0 and 2 km depth of about 1.5 to 2.0 W/m °K, yielding an estimated heat flow for the Austrian part of the basin of 45 to 60 mW/m² (Table 1).

EXTENSION AND THERMAL ANALYSIS

Model Used

The model used to compare the calculated thermal effects of thin-skinned extension to observed subsidence and thermal data from the Vienna Basin can be summarized in three steps (Fig. 16): (1) Initial thermal conditions prior to nappe emplacement were calculated from available thermal and other geophysical data from adjacent parts of the European platform; (2) Over-riding of the European basement by flysch and carbonate nappes in Miocene time (about 17.5 Ma) was simulated by instantaneously emplacing onto the European platform 4 km of material to simulate nappes west of the Vienna Basin, and 8 km of material to simulate nappes beneath and along strike with the Vienna Basin; (3) Basin extension was simulated by allowing 100% extension of the nappe pile (assumed originally 8 km thick) and simultaneous deposition of 4 km of basin sediments. The present-day heat flow and thermal subsidence calculated in this way

TABLE 1.—Observed and Calculated (Model) Heat Flow for the Region Around the Vienna Basin

	Observations		Calculated Values	
Location	Heat Flow (mW/m²)	Number of Measurements	Description	Heat Flow[a] (mW/m²)
Bohemian Massif	58 ± 8[b]	12	Initial conditions	58
Flysch Nappes	53 ± 7[b]	11	4-km-thick thrust complex emplaced at 17.5 Ma	52, 57*
			8-km-thick thrust complex emplaced at 17.5 Ma	47, 55*
Northern Vienna Basin* (Czechoslovakia)	50 ± 7[b]	9	8 km thick thrust complex emplaced at 17.5 Ma plus 100% extension at 17.5–16.5 Ma and 4 km of basin sediments	46, 53*
Southern Vienna Basin (Austria)	45–60[c]	30	8 km thick thrust complex emplaced at 17.5 Ma plus 100% extension at 16.15–10.5 Ma and 4 km of basin sediments	46, 53*

[a]Unstarred figures give estimated heat flow without radiogenic component from nappes or basin fill. Starred figures give estimated heat flow assuming a uniform radiogenic contribution from nappes and sediments of 1.0 μW/m³.
[b]Čermak (1979).
[c]Estimated from thermal gradients given by Boldizsar (1968; see text for details).

TABLE 2.—Physical Values Used

Symbol	Value	Definition
K_{sed}	$= K_0 + b \cdot x$	thermal conductivity of sediment, x is depth in meters
K_0	1.3 W/m °K	surface conductivity
b	0.5 mW/m^2 °K	conductivity gradient
K_{flysch}	3.1 W/m °K	conductivity of thrust sheet
Q_R	24 mW/m^2	heat generation from a plate source at depth d below the surface of the European plate
Q_{sed}	1.0 μW/m^3	heat generation from nappes and basin sediment
d	8 km	depth of plane source (Q_R) below upper surface of European plate
ρ_a	3.20 g cm^{-3}	density of asthenosphere
ℓ	125 km	lithospheric thickness
κ	$8 \cdot 10^{-7}$ m^2/sec	thermal diffusivity of the lithosphere
T_0	5° C	surface temperature
T_a	1300° C	fixed temperature at the base of the lithosphere
α		coefficient of thermal expansion
$\dfrac{\alpha(T_a - T_0)\ell\rho_a}{\rho_a - \rho_{water}}$	7,500 m	parameter that describes magnitude of uplift (or subsidence) due to temperature changes in the lithosphere

may then be compared to the observed heat flow in both the flysch belt and in the Vienna Basin and to the observed subsidence in the basin.

Initial Conditions

Heat flow and seismic refraction studies of the European platform in the general vicinity of the Vienna Basin yield estimates of crustal thickness of about 30–40 km and an average surface heat flow of 58 ± 8 mW/m^2 (Čermak, 1979). Thus it is assumed that prior to Miocene nappe emplacement, the European plate near the Vienna Basin, which is a stable platform area, was at or near thermal equilibrium. In this analysis, the European plate was modelled as a lithospheric slab 125 km thick at thermal equilibrium, with a constant basal temperature of 1300° C. The heat generation of radioactive elements was modelled as that produced from a plane source at 8 km depth and contributing 24 mW m^{-2} to the surface heat flux. The total model surface heat flow at equilibrium was 58 mW m^{-2} and is the same as the average measured heat flow for the European platform near the Vienna Basin.

Nappe Emplacement

An average thermal conductivity of 3.1 W/m °K was used for the nappe sequence. Present-day heat flow was calculated, first, neglecting any contributions from radioactive elements within the nappes, and second, by including a contribution from radioactive elements within the nappes assuming a uniform heat production of 1.0 μW/m^3. In the first case, the calculated present-day heat flow was found to be 52 mW/m^2 for the 4 km thick nappe sequence and 47 mW/m^2 for the 8 km thick nappe sequence. The depression in surface heat flow by 10 to 25% relative to the initial heat flow through the Bohemian Massif (58 mW/m^2) is due entirely to the thermal blanketing effect of the nappe pile. When the effects of heat production within the sediments were included, these values increased to 57 mW/m^2 and 55 mW/m^2, respectively. Within the uncertainties

in the calculations and in the observations, the calculated and observed values of heat flow are in good agreement (Table 1). Neither the exact time of nappe emplacement nor the assumed temperature of the nappes had a significant effect on subsequent calculations of basin extension.

Basin Extension

The uncorrected subsidence curves given in Figure 14 were used as guidelines to simulate extension and sedimentation in the Vienna Basin. Two different cases were run: (1) for the northeastern part of the Vienna Basin, where most of the subsidence and sedimentation is of Karpatian age (17.5–16.5 Ma) or older; and (2) for the south-central part of the basin where most of the subsidence is of Badenian through Pannonian age (16.5–8.0 Ma). For simplicity, the total sediment accumulation in the basin was taken to be exactly 4 km. In the northeastern part of the basin all subsidence was assumed to have occurred during Karpatian time (17.5–16.5 Ma). In the south-central part, 3 km of subsidence was assumed for Badenian time (16.5–13.0 Ma) and 1 km for Sarmatian time (13.0–10.5 Ma). Total extension of the flysch nappes (originally 8 km thick) was taken to occur at the same time as the sedimentation. Compaction of sediments is not important for this study, and was neglected.

The model or calculated heat flow obtained for the northeastern and south-central Vienna Basin are identical. Without including the effects of heat production within the sediments or underlying nappes the calculated surface heat flow is 46 mW/m^2. When a uniform heat production of 1.0 μW/m^3 is assumed for both the basin sediments and the underlying nappes, the calculated surface heat flow increases to 53 mW/m^2. As before, the calculated and observed values of heat flow are in good agreement within the uncertainties in the analysis and the observations. The heat flow calculated for the Vienna Basin is nearly the same as that calculated for the adjacent part of the flysch belt (Table 1). This reflects the lack of heating at depth during thin-skinned extension. Observed values of heat flow in the basin and adjacent nappes are also nearly equivalent, and thus in good agreement with calculated results. Variations in the amount of subsidence and extension assumed, and in the original nappe thickness assumed, have little effect on these results; therefore only calculations for 100% extension and 4 km of sedimentation are given here.

Temperatures.—Present-day model temperatures calculated for the Vienna Basin yield temperatures of about 55 to 65° C at 2 km depth and 90 to 100° C at 4 km depth (Fig. 17). These model temperatures are in good agreement with the temperature gradients of 29 ± 3° C/km published by Boldizsar (1968) for the southern part of the Vienna Basin. These temperature gradients are considerably lower than those observed in many other extensional basins of Miocene age. For example, in the nearby Pannonian Basin, temperatures of 100–120° C are found at depths of 2 to 2.5 km (Dövényi et al., 1983). The low heat flow and thermal gradients observed in the Vienna Basin are consistent with a thin-skinned extensional origin for the basin, where no extension of the European lithosphere occurred, so that no

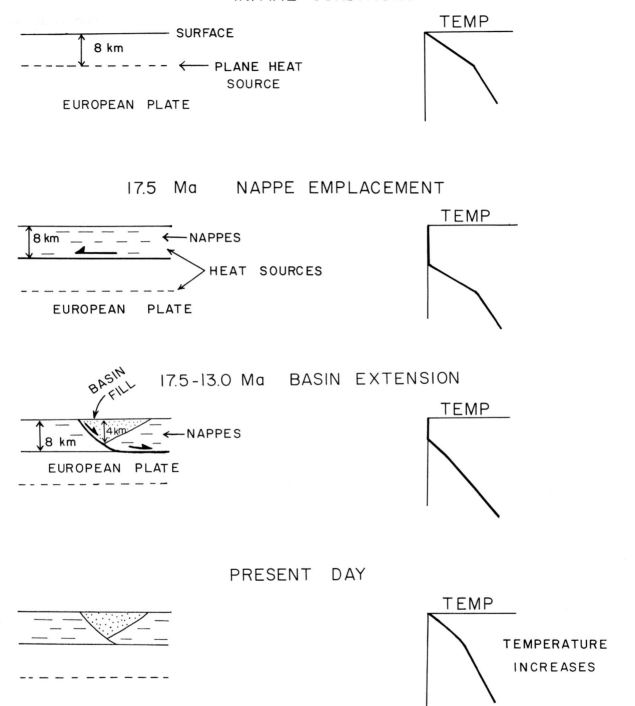

FIG. 16.—Diagram illustrating the thermal model used to calculate the effects of nappe emplacement and subsequent thin-skinned extension of the Vienna Basin. Initial conditions: European plate at thermal equilibrium with a radiogenic plane source at 8 km depth (see text for details). 17.5 Ma: an 8 km thick thrust sheet at 0° C is emplaced over the European platform. 17.5–13.0 Ma: Vienna Basin is created by thin-skinned extension within the thrust complex; structure and temperatures below the nappe sequence are not affected by extension and temperature changes here are due only to thermal conduction. Present day: temperatures increase within the basin by thermal conduction.

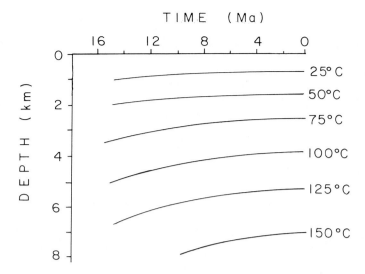

FIG. 17.—Calculated temperatures as a function of depth and age for the Vienna Basin and underlying thrust sheets. This plot corresponds to the case of constant radiogenic heat sources within the basin sediments and the thrust complex of 1.0 µW/m³. (If no heat production is assumed for the basin sediments and underlying nappes, the calculated thermal gradients decrease by about 10%.) Note that the basin temperatures increase slowly with time.

positive thermal anomaly was created under the basin.

Hydrocarbon maturity.—Many authors have published correlation methods relating hydrocarbon maturity to the time and temperature history of source rocks. The low thermal gradients within the Vienna Basin have generally not been sufficient to produce mature source rocks within the basin itself. According to Waples (1980), temperatures of 100–120° C need to be maintained for at least 10 to 20 m.y. in order to produce mature hydrocarbons. Only sedimentary rocks within the very deepest part of the Vienna Basin, at depths greater than 5 km, may have experienced temperature-time conditions compatible with hydrocarbon generation. Indeed, most of the oil and gas found within the Vienna Basin is thought to have originated within the Mesozoic rocks of the thrust belt below the basin or within the autochthonous rocks (Kröll and Wessely, 1973). Two phases of hydrocarbon generation can be determined for the Vienna Basin, one in Karpatian time and one in Badenian-Sarmatian time (Kröll and Wessely, 1973). The former corresponds to the last stage of thrust-sheet emplacement, the latter to the main phase of sedimentation in the south-central Vienna Basin. This timing for hydrocarbon maturation suggests that pre-Miocene rocks passed through the oil window during burial by Karpatian overthrusting and Badenian-Sarmatian sediment deposition. In contrast, in the neighboring Pannonian Basin, where basin extension has clearly involved rocks of the lower crust and upper mantle, upper Miocene source rocks produce mature hydrocarbons at depths of only 2.5 to 3 km (Horváth et al., in prep.). Thus the thin-skinned nature of Miocene extension within the Vienna Basin has probably been responsible for the low thermal gradients observed in the basin and, as a result, for the low level of organic maturity within the basin sediments.

Basin Subsidence Versus Uplift

The calculated temperature history for the nappes and the underlying European lithosphere shows a depression of the surface heat flow a short time after nappe emplacement. For example, without considering the effects of heat production within the nappes, the surface heat flow 5 m.y. after emplacement of an 8 km thick nappe stack with a conductivity of 3.1 W/m °K is only 38 mW/m². This represents a thermal blanketing effect of more than 30% relative to the initial surface heat flow of 58 mW/m². In contrast, the present-day calculated heat flow (or heat flow 17.5 m.y. after thrusting) has increased to 47 mW/m². This reflects the gradual heating of the entire lithosphere after the emplacement of the nappe sequence (Fig. 17). As the lithosphere gradually heats up, it undergoes thermal expansion. For the parameters used in this study, and assuming pointwise isostasy, the amount of uplift calculated to have occurred as a result of lithospheric heating between 17.5 Ma and the present is 70–100 m for a 4 km thick thrust complex and 150–200 m for an 8 km thick thrust complex.

The thermal structure of the lithosphere below the Vienna Basin should be nearly the same as that below the unextended parts of the thrust complex. Thus the lithosphere beneath the Vienna Basin should also have been heating up slightly over the past 15 m.y. This contrasts strongly with the thermal evolution of most extensional sedimentary basins where the underlying lithosphere cools with time, leading to thermal subsidence of the basin. One may infer that the Vienna Basin and the thrust belt are undergoing a small amount of post-extensional uplift because the average temperature of the underlying lithosphere is increasing slowly. In the northeastern corner of the Vienna Basin where Sarmatian (13.0–10.5 Ma) post-extensional sediments were deposited in brackish water, the surface elevation is now about 200–300 m above sea level. Little material has been lost to erosion. This implies that something like 200–300 m of uplift has occurred in this area since the end of extension in middle Miocene time. Such an observation is consistent with the interpretation that the Vienna Basin is the result of thin-skinned extension in the uppermost crust, superimposed on a region that is being slowly uplifted.

DISCUSSION

The Vienna Basin provides an excellent example of how thin-skinned extension can create a sedimentary basin without creating space problems at the edges of the extended region. In this example, strike-slip and normal displacement on faults within the basin were probably for the most part transferred to gently east-dipping detachment faults at depth. Many of these detachment surfaces or decollements probably played an important role in contemporaneous northward transport of thrust sheets east of the basin (Figs. 11 and 13). Basin extension was compensated by continued overthrusting in the northern part of the outer Carpathians.

Although the Vienna Basin is similar to the other intra-Carpathian basins in its strike-slip genesis and general relationship to the Carpathian thrust belt, it differs from the other basins as a result of its proximity to the Carpathian

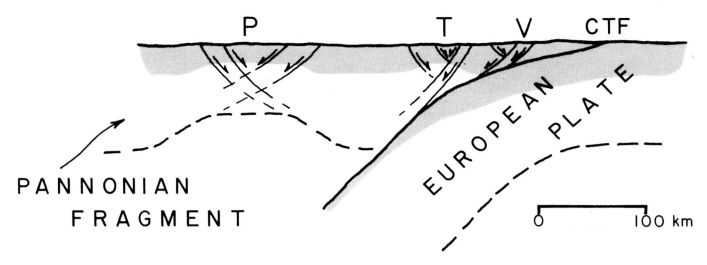

Fig. 18.—Schematic diagram illustrating how the thickness of the overriding plate is related to the distance from the Carpathian thrust front (CTF), and showing the relative positions of the Vienna (V), Transcarpathian (T) and Pannonian (P) Basins. When extension (and strike-slip faulting) are mainly confined to the over-riding plate, the depth to which extension can occur increases with distance from the thrust front. This sketch shows that there are two different zones affected by extension: the outer zone, characterized by the Vienna and Transcarpathian Basins where extension affects the leading (and thin) edge of the Pannonian lithosphere, but does not reach down into the European lithosphere below; and the inner zone, characterized by the Pannonian Basin, where extension involves the Pannonian lithosphere where it is directly underlain by asthenosphere. Shaded interval represents crustal rocks. Sites of extension indicated schematically by normal faults; the fault geometry shown here is not meant to be realistic. Figure from Royden et al. (1983a).

thrust front. Extension of the Vienna Basin and the associated strike-slip faulting were restricted to shallow levels because they were confined to the over-riding sheet at a subduction boundary. In more internal areas of the intra-Carpathian region, where the over-riding (Pannonian) lithosphere was probably thicker, extension and strike-slip faulting should extend to greater depths (Fig. 18). For example, the Transcarpathian Basin, in the northeastern part of the intra-Carpathian region (Fig. 2) seems to have a pull-apart origin similar to that of the Vienna Basin (Horváth and Royden, 1981; Royden et al., 1983a), but because this basin is superimposed on a more internal part of the thrust belt, it is probable that extension in it affected rocks to greater depths than in the Vienna Basin (Fig. 18). In even more internal zones, such as beneath the Pannonian Basin, deformation should extend completely through the lithosphere of the Pannonian fragment. Thus the depth to which deformation occurs may be primarily a function of distance from the active thrust front.

The nature of the Miocene extension beneath the various intra-Carpathian basins is reflected by the regional heat flow, thermal gradients, and the depth at which source rocks mature. Where extension has involved only shallow crustal rocks, such as beneath the Vienna Basin, the heat flow is low and roughly equivalent to that through adjacent parts of the European plate (58 ± 8 mW/m^2; Čermak, 1979); in this case, the hydrocarbon window lies below a depth of about 5 km. Where extension has involved the entire lithosphere, such as beneath the Pannonian Basin, heat-flow values are greatly elevated to more than 100 mW/m^2 (Dövényi et al., 1983), and the generation of hydrocarbons begins at depths as shallow as 2.5 to 3 km.

This is an example of how basins that have formed within one small area (intra-Carpathian area), in the same general tectonic setting (on the over-riding plate at a subduction boundary), at about the same time (middle to late Miocene), and by the same mechanism (strike-slip related extension) can still have important differences. In this example, important differences between the various intra-Carpathian basins can be explained by the different positions that these basins occupy in the Pannonian region, and their distance from the active thrust front. Each basin within a regional tectonic system needs to be examined both for its individual peculiarities, and for its place within the entire tectonic system, in order to understand its formation on a local and on a regional scale.

ACKNOWLEDGMENTS

This research was sponsored by NSF grants INT-7910275 and EAR 8115863, and by Shell Development Company. I would like to thank the Kerr-McGee Foundation and the Kerr Foundation for their generous support of my research, and C. Tomek and G. Wessely for helpful discussions and insights. This manuscript was reviewed by K. T. Bidde, N. Christie-Blick, and C. R. Tapscott.

REFERENCES

ANDRUSOV, D., 1965, Aperçu général sur la géologie des Carpathes occidentales: Bulletin of Geological Society France, v. 7, p. 1029–1062.
———, 1968, Grundriss der Tektonik der Nordlichen Karpeten: Bratislava, Czechoslovakia Verlag der Slowakischen Acadamie der Wissenschaften, 187 p.
BIRKENMAJER, K., 1981, Strike-slip faulting in the Pieniny Klippen belt of Poland: paper presented at the 12th Congress, Carpatho-Balkan Geological Association, Bucharest, Romania, Sept. 8–12, 1981.
BOLDIZSAR, T., 1968, Geothermal data from the Vienna basin: Journal Geophysical Research, v. 73, p. 613–618.
BRIX, F., AND SCHULTZ, O., eds., 1980, Erdoel und erdgas in Oesterreich: Vienna, Verlag Naturhistorisches Museum Wien und F. Berger, Horn, 311 p.

BUDAY, T., AND CICHA, J., 1968, The Vienna basin, *in* Mahel, M. et al., eds., Regional Geology of Czechoslovakia, Part II, the West Carpathians: Stuttgart, E. Schweizerbart'sche Verlagsbuchhandlung, p. 570–583.

BURCHFIEL, B. C., 1976, Geology of Romania: Geological Society of America Special Paper 158, 82 p.

————, 1980, Eastern Alpine system and the Carpathian orocline as an example of collision tectonics: Tectonophysics, v. 63, p. 31–62.

BURCHFIEL, B. C., AND STEWART, J. H., 1966, The "pull-apart" origin of Death Valley, California: Geological Society America Bulletin, v. 77, p. 439–442.

ČERMAK, V., 1979, Review of heat flow measurements in Czechoslovakia, *in* Čermak, V., and Rybach, L., eds., Terrestrial Heat Flow in Europe: New York, Springer-Verlag, p. 152–160.

CROWELL, J. C., 1974, Origin of Late Cenozoic basins in southern California, *in* Dickinson, W. R., ed., Tectonics and Sedimentation: Society of Economic Paleontologists and Mineralogists Special Publication No. 22, p. 190–204.

DÖVÉNYI, P., HORVÁTH, F., LIEBE, P., GÁLFI, J., AND ERKI, I., 1983, Geothermal conditions of Hungary: Geophysical Transactions, Lórand Eötvos Geophysical Institute of Hungary, Budapest, 114 p.

FUCHS, W., 1980, Das Inneralpine Tertiär, *in* Oberhauser, R., ed., Der Geologische aufban Österreichs: New York, Springer-Verlag, p. 452–483.

GUTDEUTSCH, R., AND ARIC, K., 1976, Erdbeben in Ostalpen Raum: Arbeiten aus der Zentralanstalt fur Meteorologie und Geodynamic, v. 19, No. 210, 23 p.

HORVÁTH, F., AND ROYDEN, L., 1981, Mechanism for the formation of the IntraCarpathian Basins: A review: Earth Evolution Sciences, v. 1, p. 307–316.

JANOSCHEK, W. R., AND MATURA, A., 1980, Outline of the geology of Austria: 26th International Geological Congress, v. 34, Wien, p. 7–98.

JIŘIČEK, R., 1979, Tectogenetic development of the Carpathian arc in the Oligocene and Neogene, *in* M. Mahel, ed., Tectonic Profiles through the West Carpathians: Geologicky Ustav Dionyza Stura, Bratislava, Czechoslovakia, p. 205–214.

JIŘIČEK, R., AND TOMEK, Č., 1981, Sedimentary and structural evolution of the Vienna basin: Earth Evolution Sciences, v. 1, p. 195–204.

KRÖLL, A., 1980, Die Österreichischen Erdöl und Erdgasprovinzen: Das Wiener Becken, *in* Brix, F. and Schultz, O., eds., Erdol und Erdgas in Österreich: Vienna, Naturhistorisches Museum Wien und F. Berger, Horn, p. 147–179.

KRÖLL, A., AND WESSELY, G., 1973, Neue Ergebnisse beim Tiefenaufschluss im Wiener Becken: Erdoel, Erdgas Zeitschrift, v. 89, p. 400–413.

KRÖLL, A., SCHIMUNEK, K., AND WESSELY, G., 1981, Ergebnisse und Erfahrungen bei der Exploration in der Kalkalpenzone in Öausterreich: Erdoel, Erdgas Zeitschrift, v. 4, p. 134–148.

KSIAŹKIEWICZ, M., OBERC, J., AND POŻARYSKI, W., 1977, Geology of Poland, v. IV, Tectonics: Warsaw Geological Institute, 718 p.

OBERHAUSER, R., ed., 1980, Der geologische aufbau Oesterreich: New York, Springer-Verlag, 669 p.

PAPP, A., KROBOT, W., AND HLADECEK, K., 1973, Zur Gliederung des Neogens im Zentralen Wiener Becken: Mitteilungen der Gesellschaft der Geologie und Bergbaustudenten in Österreich, v. 22, p. 191–199.

PESL, V., SOLAJ, J., AND VASS, D., 1968, The flysch and Klippen belts, Neogene basins of the West Carpathians: 23rd International Geological Congress, Prague, Guide to Excursion 6AC, 40 p.

ROTH, Z., 1980, Západni Karpati-terciérní struktura Středni Evropy: Ustredny ustav Geologicky, v. 55, 128 p.

ROYDEN, L., 1982, The evolution of the intra-Carpathian basins and their relationship to the Carpathian Mountain system [unpubl. Ph.D. thesis]: Cambridge, Department of Earth and Planetary Science, Massachusetts Institute of Technology, 256 p.

ROYDEN, L. H., HORVÁTH, F., AND BURCHFIEL, B. C., 1982, Transform faulting, extension and subduction in the Carpathian-Pannonian region: Geological Society of America Bulletin, v. 93, p. 717–725.

ROYDEN, L., HORVÁTH, F., AND RUMPLER, J., 1983a, Evolution of the Pannonian basin system, 1. Tectonics: Tectonics, v. 2, p. 63–90.

ROYDEN, L., HORVÁTH, F., NAGYMAROSY, A., AND STEGENA, L., 1983b, Evolution of the Pannonian basin system, 2. Subsidence and thermal history: Tectonics, v. 2, p. 91–137.

SĂNDULESCU, M., 1975, Essai de synthèse structurale des Carpathes: Bulletin of Geological Society France, v. 17, p. 299–358.

————, 1980, Analyse geotectonique des chaînes alpines situées au tour dé la mer noire occidentale: Annuaire de l'Institute de Geologie et de Geophysique, Bucharest, v. 56, p. 5–54.

SEGALL, P., AND POLLARD, D. D., 1980, Mechanics of discontinuous faulting: Journal Geophysical Research, v. 85, p. 4337–4350.

STEININGER, F., PAPP, A., CICHA, J., SENES, J., AND VASS, D., 1975, Excursion A, marine Neogene in Austria and Czechoslovakia: 6th Congress Regional Commission on Mediterranean Neogene Stratigraphy, Bratislava, Czechoslovakia, 96 p.

STEININGER, F. F., AND RÖGL, F., 1979, The Paratethys history—a contribution toward the Neogene geodynamics of the Alpine orogen (abs.): Annales Géologiques des Pays Helleniques, v. 3, p. 1153–1165.

VASS, D., 1978, World Neogene radiometric timescale (estate to the beginning of 1976): Geologicke prace, Geologicky ustav Dionyza Stura, v. 70, p. 197–236.

————, 1982, Explanatory notes to lithotectonic molasse profiles of inner West Carpathian basins in Czechoslovakia (comment to Annex 2–5): Veroffentlichungen des Zentralinstitute für Physik der Erde, v. 66, p. 55–94.

VASS, D., AND BAGDASARIAN, G. P., 1978, A radiometric timescale for the Neogene of the Paratethys region: American Association Petroleum Geologists Bulletin, v. 62, p. 179–204.

WAPLES, D. W., 1980, Time and temperature in petroleum formation: application of Lopatin's method to petroleum exploration: American Association Petroleum Geologists Bulletin, v. 64, p. 916–926.

WEBER, F., 1980, Die rolle angewandten Geophysik beim Erdölaufschluss in Österreich, *in*, Brix, F., and Schultz, O., eds., Erdol und Erdgas in Österreich: Vienna, Naturhistorisches Museum Wien und F. Berger, Horn, p. 139–146.

WESSLEY, G., 1983, Zur Geologie und Hydrodynamik im Südlichen Wiener Becken und seiner Randzone: Mitteilungen geologische Gesellschaft Österreich, v. 76, p. 27–68.

THE TERTIARY STRIKE-SLIP BASINS AND OROGENIC BELT OF SPITSBERGEN

RON STEEL AND JOHN GJELBERG
Norsk Hydro Research Centre, P.O. Box 4313, 5013 Bergen;
AND
WILLIAM HELLAND-HANSEN AND KAREN KLEINSPEHN[1]
University of Bergen, 5014 Bergen, Norway;
AND
ARVID NØTTVEDT
Norsk Hydro Research Centre, P.O. Box 4313, 5013 Bergen;
AND
MORTEN RYE-LARSEN
Esso Norway, P.O. Box 560, 4001 Stavanger, Norway

ABSTRACT: The Svalbard margin evolved to its present rifted configuration through a complex strike-slip history of both transtension and transpression. Because the Paleogene plate boundary, the De Geer Line, lay just west of Spitsbergen, many of the details of this structural evolution are contained in a narrow fold and thrust belt, and within a series of sedimentary basins, on Spitsbergen. The early to mid-Paleocene Central Basin was of extensional (possibly transtensional) origin, and contains more than 800 m of clastic deposits. It evolved from a series of partly connected coal basins to a single, open-marine basin. The late Paleocene to early Eocene Central Basin, of transpressional origin, was infilled by more than 1.5 km of clastic sediments from deltas, which prograded out from the rising orogenic belt. The fold and thrust belt of western Spitsbergen, mainly of late Paleocene to Eocene age, was also a product of transpression. Forlandsundet Graben, infilled by as much as 5 km of alluvial and marine clastics, probably formed from late Eocene collapse of the crest of the orogenic belt, or from extension adjacent to a curved fault zone. Rift basins, up to 7 km deep, developed west of Svalbard as the continental margin changed, beginning in the early Oligocene, from a strike-slip to a rifted regime.

REGIONAL SETTING

The occurrence of sedimentary basins and an orogenic belt of Tertiary age in Spitsbergen (Figs. 1, 2) is well known. Details of the relation between Tertiary tectonics and sedimentation, and of the plate-tectonic causes of these events, have emerged over the years, beginning with papers by Harland (1965, 1969), who first suggested that the 'West Spitsbergen orogeny' was related to large-scale transcurrent movement between Greenland and Eurasia, during the opening of the Norwegian-Greenland Sea. It is therefore important to summarize our present knowledge of the Cenozoic history of the continental margin off northern Norway and Svalbard,[2] before discussing the Tertiary events on Spitsbergen.

A model for the Cenozoic tectonic history of the northeastern Atlantic region and the opening of the Norwegian-Greenland Sea, including a map of the sea-floor magnetic anomalies and the main structural elements, was proposed by Talwani and Eldholm (1977). The main feature of this model is that there were two distinct phases of tectonic development: the first, now generally recognized as beginning about the time of formation of magnetic anomaly 25/24 (58 Ma; Eldholm et al., 1984), involved north-northwesterly motion of Greenland from Eurasia; the second, dating from the time of magnetic anomaly 13 (37 Ma), involved a change in the pole of rotation and caused the relative plate movement to be west-northwesterly. These two tectonic phases thus produced a latest Paleocene to early Oligocene strike-slip regime and an early Oligocene to present rift regime, respectively, off western Svalbard (Fig. 1). The Paleogene transform boundary has been called the De Geer Line (Harland, 1969) or the De Geer-Hornsund Line (Crane et al., 1982), whereas the later plate boundary has been named

Hornsund fault zone (Myhre et al., 1982).

New work has led to modification and refinement of the early tectonic model along the segment of the margin between Norway and Spitsbergen. The Hornsund fault zone was mapped in detail and is now thought to be located at or near the continent-ocean boundary (Myhre et al., 1982; Figs. 1, 2). A deep sedimentary basin (or series of basins) was identified and mapped between the Knipovich Ridge and the Hornsund fault zone (Schluter and Hinz, 1978), and was probably infilled largely during the post-early Oligocene rift phase of the margin's development (Myhre et al., 1982). In the more recent work summarizing the margin off Svalbard, Myhre et al. (1982) and Spencer et al. (1984) tentatively proposed a three-stage series of tectonic events (as shown in Fig. 1) involving: (1) sea-floor spreading south of the Senja Fracture Zone from about 58 Ma; (2) sea-floor spreading between the Senja Fracture Zone and the southern end of the Hornsund fault zone (anomaly 21; 48 Ma; mid-Eocene); and (3) sea-floor spreading opposite the Hornsund fault zone (anomaly 13; 37 Ma; early Oligocene). We emphasize the tentative nature of this scheme as no linear magnetic anomalies have been identified to date in the Greenland Sea.

The data outlined above imply a tectonic regime off Svalbard that evolved from several phases of strike slip to a phase of rifting, as new ocean floor was generated progressively farther north. This is clearly likely to constrain both the timing and the character of the Tertiary tectonic events on Spitsbergen.

The role of Svalbard's Tertiary fold and thrust belt in the development of the continental margin is not yet established in detail, but because the orogenic belt marks the approximate position of the early plate boundary, the De Geer Line (Fig. 1) and is of Paleogene age (Steel and Worsley, 1984), clearly it should be considered part of the framework of the Paleogene events described above (i.e., part of the pre-early Oligocene strike-slip system). This is consistent with the maps and observations of Lowell (1972) and

---

[1]Present address: Department of Geology and Geophysics, University of Minnesota, Minneapolis, Minnesota 55455

[2]Svalbard includes the archipelago of islands in the area 74–81°N, 10–35°E; Spitsbergen is the largest of these islands.

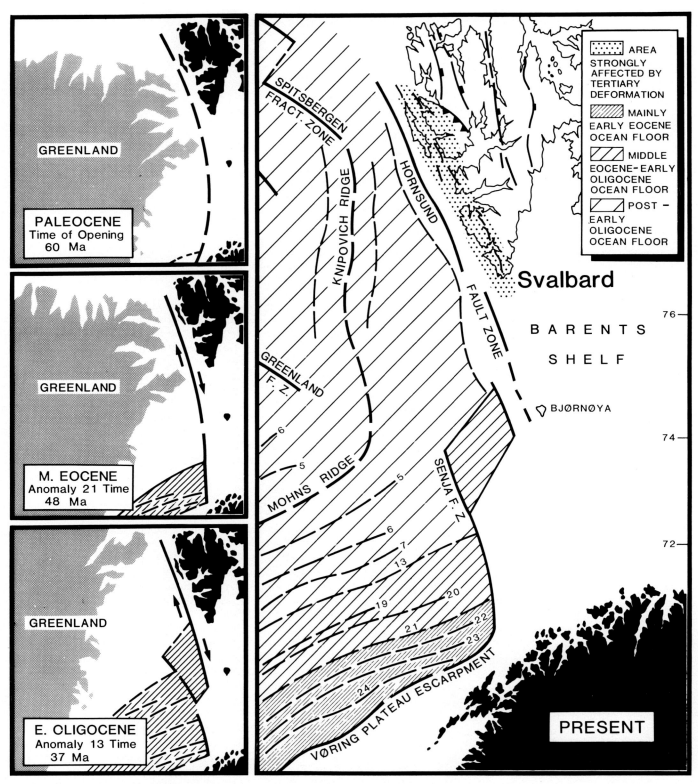

FIG. 1.—Present-day map of Svalbard margin together with an outline of the Tertiary displacement of Svalbard from Greenland during the opening of the Norwegian-Greenland Sea. Data from Grønlie and Talwani (1978), Myhre et al. (1982), Spencer et al. (1984).

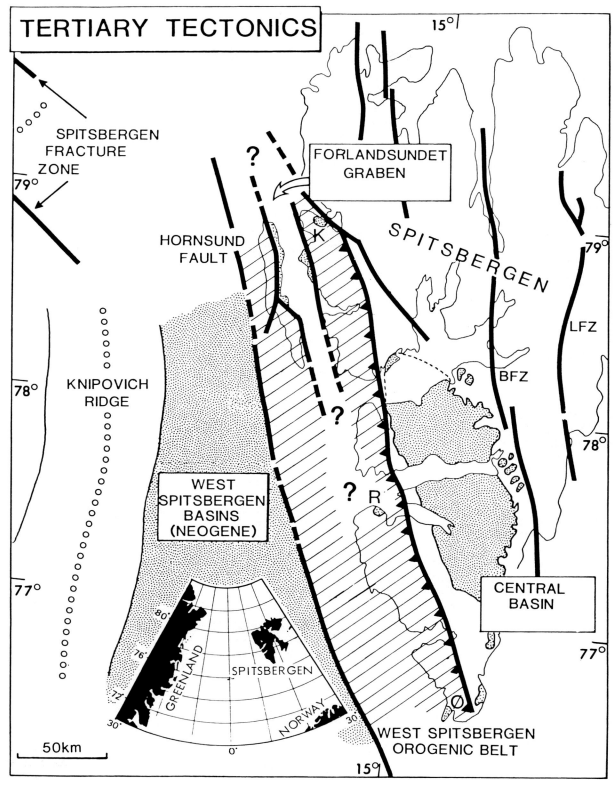

FIG. 2.—The Tertiary basins (stippled) and fold belt (lined) along the segment of the Svalbard margin discussed in text. Tertiary outliers at Kongsfjorden (K), Renardodden (R) and Øyrlandet (Ø) are still of uncertain age. BFZ, Billefjorden fault zone; LFZ, Lomfjorden fault zone.

of Kellogg (1975), which showed that some of the thrust faults steepen with depth and that some wrench faults can be traced laterally into overthrusts. Thus, it is likely that Svalbard's Tertiary orogenic belt is analogous to other major strike-slip fault systems (Lowell, 1972; see below). The timing of the earliest events, or at least of significant uplift along the fold and thrust belt, can be constrained more closely in relation to the infilling of the flanking Central Basin (Fig. 2). Major drainage reversal in the Central Basin, and accompanying influx of metamorphic rock fragments from the emerging orogenic belt, can be dated as late Paleocene (Steel et al., 1981). Thus there is evidence of major tectonic activity along the De Geer Line and adjacent areas somewhat earlier than the creation of the oldest dated ocean floor off northwest Norway. The climax of orogenic activity along western Spitsbergen is likely to have been about mid-Eocene, as judged by stratigraphic evidence from the Central Basin (Steel and Worsley, 1984), and we tentatively suggest that this climax corresponded with the migration of sea-floor spreading from the region south of the Senja Fracture Zone to the region between the Senja Fracture Zone and Bjørnøya (Fig. 1).

Within the framework of these regional constraints we discuss below some of the details of the tectonic and sedimentary events in the Central Basin, in the fold and thrust belt, and in the Forlandsundet Graben (Fig. 2). We contend that it is possible to distinguish times of significant compression or extension superimposed on the regional strike-slip regime (transpression and transtension in the terminology of Harland, 1969). We also consider briefly the post-early Oligocene basins beneath the present continental shelf and slope because they record the important evolution of the plate margin from strike slip to rifting.

THE CENTRAL BASIN

The Tertiary Central Basin of Spitsbergen is 200 km long and 60 km wide (Fig. 2), and contains as much as 2.3 km of clastic deposits, referred to as the Van Mijenfjorden Group by Harland et al. (1976). According to the vitrinite reflection studies of Manum and Throndsen (1978a), an additional overburden of 1.7 km has been eroded since the Eocene. Despite a considerable amount of paleontological research, the precise ages of the various formations in the Tertiary succession are not well established. However, there is a consensus of opinion that the lower part of the succession (Firkanten, Basilika, and Grumantbyen Formations; Fig. 3) is of early to mid-Paleocene age, on the basis of mollusc and foraminifera evidence (Ravn, 1922; Vonderbank, 1970; Birkenmajer, 1972). The upper part of the succession (Hollendardalen, Gilsonryggen, Battfjellet, and Aspelintoppen Formations; Fig. 3) is likely to be late Paleocene and early Eocene in age, on the basis of a latest Paleocene palynological dating at the base of Gilsonryggen Formation (Manum and Throndsen, 1978b). This two-fold division of the succession is adopted below.

Early to Mid-Paleocene Development

Stratigraphy and Sedimentation.—The lower part of the Central Basin succession, which thickens from less than 300 m in the northeast to more than 800 m in the west, consists of four main sequences (Fig. 4): (1) a sequence of alternating coals, shales, and sandstones (Todalen Member of the Firkanten Formation), which are of mixed fluvial and marine origin, and are interpreted as shallow-water deltaic deposits (Steel et al., 1981; Nøttvedt, 1985); (2) a sequence of alternating marine siltstone and sandstone sheets (Endalen and Kolthoffberget Members of the Firkanten Formation), with abundant low-angle and hummocky cross-stratification, and interpreted as shoreline and inner-shelf deposits, which are in part transgressive (Nemec and Steel, 1985); (3) a sequence of marine siltstone and shale units (Basilika Formation), showing repeated upward-coarsening motifs in some areas, and interpreted as outer shelf deposits; and (4) gradationally developing upwards from the Basilika Formation, a sequence of well-bioturbated fine-grained marine sandstones with some siltstones (Grumantbyen Formation), interpreted as possible inner-shelf deposits.

Figure 5 documents grain-size data from 10 laterally equivalent profiles through the Todalen and Endalen/Kolthoffberget Members of the Firkanten Formation, and shows how the coal-bearing sequence appears to occur in sub-basins separated by a structurally high block. In contrast, the upper members are sheet-like, but become finer grained southward. Paleocurrent indicators in both of these sequences, together with the lateral facies changes evident in Figure 5, suggest that the early Paleocene basin was infilled mainly from the north, northeast, and east. The lower part of the succession also thickens toward the southwest and the Hornsund fault zone (Fig. 4). The upward change from coal-bearing deposits to marine sheet sandstones implies a relative rise of sea level. The Basilika and Grumantbyen Formations are less well understood than the underlying succession, but they appear to be in part correlative (Fig. 3), and to thicken significantly to the west. The basin continued to be filled from the northeast during the deposition of these two formations.

Tectonic Setting.—We suggest that the tectonic setting for the early to mid-Paleocene Central Basin was extensional for the following reasons: (1) the sedimentary succession is of considerable thickness and thickens towards the De Geer Line; (2) igneous activity is indicated by a number of volcanic ash layers in the Firkanten Formation (Major and Nagy, 1972); and (3) there is no evidence for uplift (compression) in the west at this time. Assuming a constant rate of sediment input and increasingly asymmetric basin subsidence (Fig. 4), the overall transgressive to regressive character of the Firkanten to Grumantbyen Formation succession can be explained by a eustatic rise in sea level followed by a eustatic fall. We tentatively propose that the sea-level fall corresponds to the mid-late Paleocene eustatic sea-level drop proposed by Vail et al. (1977) on the basis of global coastal onlap curves.

Two lines of evidence suggest that extension may have been accompanied by a component of strike slip. The early Paleocene Central Basin appears to have been partitioned into several, partly connected, coal sub-basins that become slightly younger toward the north (Figs. 5, 6). Although by no means diagnostic of strike-slip deformation, age variation of this sort is consistent with such an interpretation.

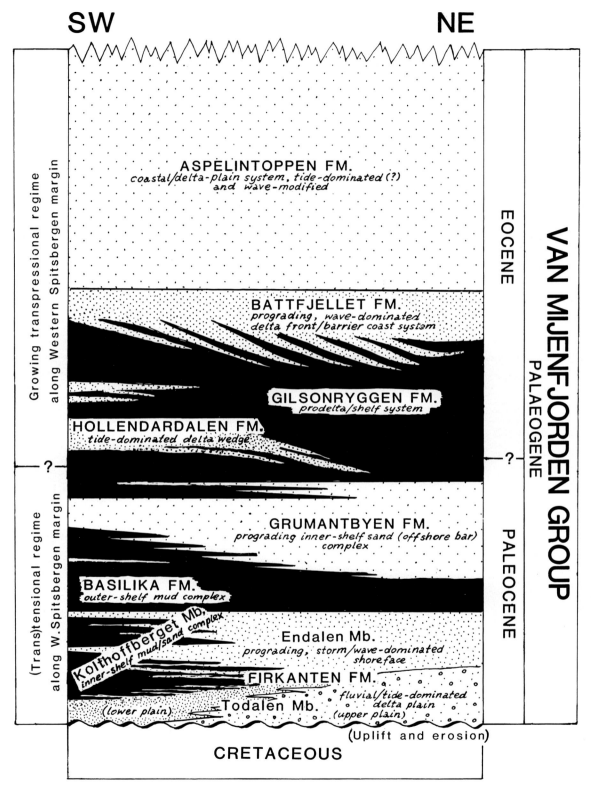

FIG. 3.—Summary of the stratigraphy of the Central Basin of Spitsbergen (Wojtek Nemec, personal commun., 1985). The geometry of the sandstone units is shown schematically.

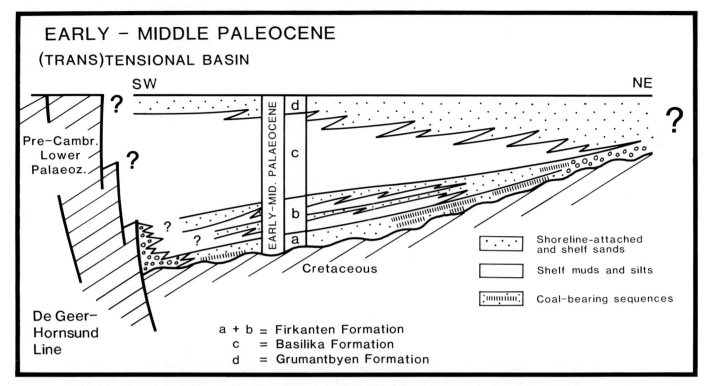

FIG. 4.—Schematic diagram of the main stratigraphic elements in the early to mid-Paleocene Central Basin, between Isfjorden and Van Mijenfjorden. Note that the asymmetry of the basin developed mainly after mid-Paleocene time, and that infilling was mainly from the east and northeast.

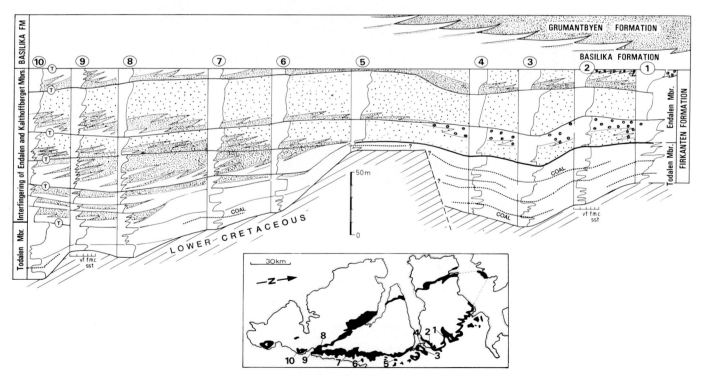

FIG. 5.—Ten correlated, measured sections across part of the Central Basin. The early coaly strata formed in several sub-basins whereas the later wave-generated sandstones (stippled) are sheet-like across the basin (the denser shading delineates more distal facies). The correlations suggest possible northwards younging of the coaly strata.

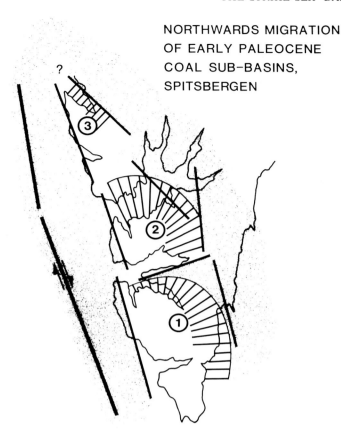

NORTHWARDS MIGRATION OF EARLY PALEOCENE COAL SUB-BASINS, SPITSBERGEN

FIG. 6.—Speculative model of coal-basin development on Spitsbergen implying northwards migration and onlap of strata with time (1–3). The sub-basins are unshaded; the likely coal-forming areas are lined. The hinterland is shaded. Thick black lines show zones of faulting or flexure.

The other line of evidence consists of early Paleocene strike-slip structural features in the Wandel Sea Basin of northern Greenland (Hakansson and Schack Pederdsen, 1982). At that time, northern Greenland was adjacent to Svalbard, west of the De Geer Line.

Coal.—The importance of coal deposits (currently exploited by both Norwegian and Soviet communities) in the oldest part of the Paleocene succession is well known (e.g., Major and Nagy, 1972; Harland et al., 1976). Coal has a particular significance in the model outlined above. The coal-bearing strata formed during the earliest subsidence and extensional break-up along the eastern flank of the De Geer Line, and coal seams are best developed along the margins of the various sub-basins. Potential coal-forming conditions then disappeared in each sub-basin with relative sea-level rise, and the gradual 'drowning' of the early structural highs, and the onset of more open-marine conditions.

Late Paleocene–Early Eocene Development

Stratigraphy and Sedimentation.—The upper part of the succession in the axial region of the Central Basin is more than 1.5 km thick and consists of the Hollenderdalen, Gilsonryggen, Battfjellet, and Aspelintoggen Formations (Fig. 3). Palynological dating of the lowermost part of this succession (the lower part of the Gilsonryggen Formation) suggests a latest Paleocene age (Manum and Throndsen, 1978b).

The Hollenderdalen Formation consists of as much as 150 m of sandstone in the western part of the basin but thins toward the basin center. The sequence has been interpreted in terms of an eastward-prograding shallow-water deltaic system on account of paleocurrent indicators, wave- and tide-generated stratification, and occasional thin coals (Dalland, 1979).

Gilsonryggen Formation shales and the overlying Battfjellet Formation sandstones form a large-scale, coarsening-upward sequence which thins from more than 900 m in the west to less than 300 m in the eastern part of the Central Basin (Kellogg, 1975). Thin sandstone interbeds within the Gilsonryggen Formation appear to be turbidites (T_{abc} type) and tend to pinch out eastwards (Steel et al., 1981).

The Battfjellet Formation is 60 to 200 m thick and consists mainly of siltstones and sandstones, with abundant hummocky cross-strata and other wave-generated sedimentary structures. On account of such facies, the upward-coarsening nature of the sequence, and the presence of coalbearing deposits above, the Battfjellet Formation is interpreted as the product of a prograding deltaic and barrier coastline (Steel, 1977). An additional striking aspect of the formation is the presence, in the western part of the basin, of large-scale (200 m amplitude) clinoforms, which can be seen on mountainsides to cut obliquely through the formation and to continue into the underlying shales (Kellogg, 1975). This demonstrates both eastward-offlapping and the broad lateral time equivalence of upper Gilsonryggen, Battfjellet, and lowermost Aspelintoppen Formations. These features as well as a general eastward migration of the depocenter during late Paleocene and Eocene time, are illustrated in Figure 7.

The Aspelintoppen Formation is more than 1 km thick in places and consists of alternations of sandstones, siltstones, shales, and coals. Plant debris is abundant, as well as petrified tree fragments. Soft sediment deformation is common throughout the succession. The coal seams, lack of marine fauna, and presence of channelized sandstones and fining-upward sequences suggest a deltaic or coastal plain origin for the formation (Steel et al., 1981).

Tectonic Setting.—There was a significant change in the tectonic setting of the Central Basin during the late Paleocene. Sandstones of the Hollenderdalen and Battfjellet Formations were clearly derived from the western side of the basin, in contrast to the pre-late Paleocene deposits, which were transported into the Central Basin from the east and northeast. This drainage reversal, coupled with the evidence of increasingly abundant metamorphic rock fragments in the upper part of the succession, suggests pronounced uplift of the western margin of the Central Basin, associated with initiation of the fold and thrust belt of western Spitsbergen. Regional arguments suggest transpressional deformation (Myhre et al., 1982).

The Central Basin, from late Paleocene time, was thus analogous to a foreland basin, depressed by flexural loading of the thrust sheets. Differential compaction of the thick shale succession, which is thickest in the west, may have

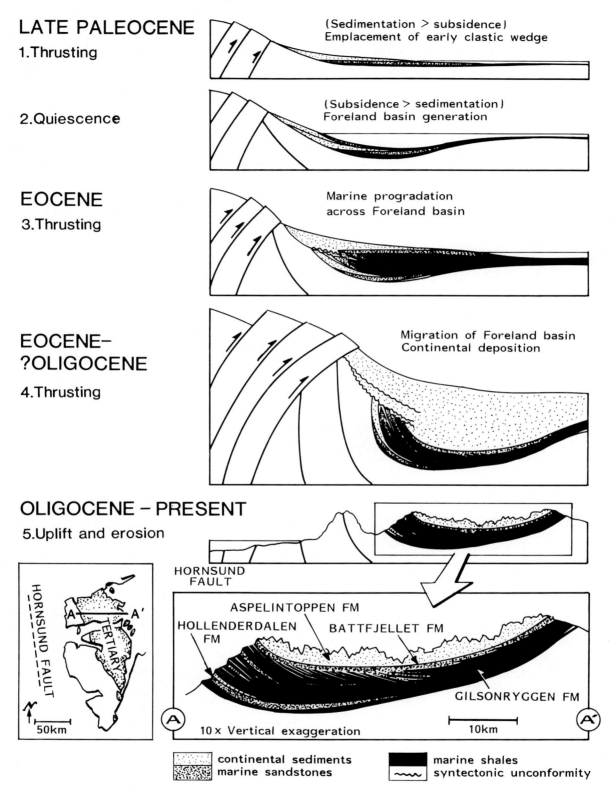

FIG. 7.—Series of diagrams showing the suggested relationship of the late Paleocene–?Oligocene development of the Central Basin and tectonic movements in the fold and thrust belt of western Spitsbergen. Note the large-scale clinoforms in the Battfjellet Formation, one of the more striking responses to hinterland uplift and thrusting. There is an eastward migration of the depocenter with time of at least 20 km.

permitted additional accumulation of sand. There is also evidence of eastward migration of the basin's depocenter during continued eastward movement of the thrust sheets. The eastward shift is on the order of 20 km, as inferred from a comparison of the position of the eastern margin of the latest Paleocene depocenter and the position of the present basin margin (Fig. 7). Manum and Throndsen (1978a) reached a similar conclusion from vitrinite studies.

THE FOLD AND THRUST BELT OF SPITSBERGEN

The fold and thrust belt of western Spitsbergen is some 300 km long and less than 50 km wide (Fig. 2), and is characterized by wrench faults, thrust faults, and asymmetric folds. The fold belt terminates somewhere north of Bjørnøya (Fig. 1) where only extensional, post-Paleozoic tectonism has been recorded (Horn and Orvin, 1928; Gjelberg, 1981). Figure 8 illustrates the present-day fault pattern on Spitsbergen. Most of the faults of western Spitsbergen are of Tertiary age or were reactivated in Tertiary time. The cross sections of Figure 9 illustrate the internal character of the fold belt in the vicinity of Hornsund. The severe deformation along the fold belt has been a subject of study for a long time (Nordenskiold, 1866; Orvin, 1940), but the first clear suggestions that it was related to the opening of the Norwegian-Greenland Sea were made by Harland (1965, 1969), who interpreted the deformation in terms of compression generated by the northward movement of Greenland against Spitsbergen. Lowell (1972) emphasized the strike-slip regime of this orogenic belt and concluded that such belts differ from those related to subduction "in having a discrete pattern of *en echelon* folds, in having a narrow zone of deformed sedimentary cover with a much greater degree of basement involvement, in having presumably different cross-sectional profiles of thrusts, upthrust versus downward flattening, in being shorter in length, and probably in lacking alpine ophiolites and lacking metamorphism." Kellogg (1975) noted that the fold belt consists of three zones that have reacted somewhat differently to stress (Fig. 8): a southern zone of thrusting (Sørkapp to Bellsund), a middle zone of folding (Bellsund to Isfjorden), and a northern zone of thrusting (Isfjorden to Kongsfjorden).

The southern zone (south of Bellsund) is dominated by gently to moderately dipping thrust faults (Fig. 9; CC' in Fig. 10) and normal faults (generally orientated parallel with the coast). Thrusting was towards the northeast or east-northeast and involved rocks of Precambrian to Paleogene age (Birkenmajer, 1981). Some of the thrust surfaces can be mapped as becoming steeper at depth (Fig. 10). Normal faults (probably younger than the thrust faults) with down-to-the-west displacements are common in western Sørkapp Land (Fig. 8). Minor northeast-striking normal faults are also present in the same area, as well as farther east.

In the central zone (Nordenskiold Land), thrusts are generally absent (BB' in Fig. 10), but steeply dipping, basement-involved reverse faults are probably responsible for a series of en echelon asymmetric folds developed along the steep, eastward-dipping western flank of the Central Basin. The only obvious thrust fault is located near Grumantbyen.

It is, however, possible that the normal faults present in the Lower Paleozoic–Precambrian basement of western Nordenskiold Land were thrust faults during the main orogenic phase (most associated structures indicate so), but have later been reactivated as normal faults during post-orogenic extension. Northeast-striking normal faults are also present in the area.

In the northern zone (between Isfjorden and Brøggerhalvøya; AA' in Fig. 10), the rocks are tightly folded about northwest-trending axes, with amplitudes and wavelengths of 0.5 to 1 km (Harland and Horsfield, 1974). Prominent faults, also striking to the northwest, have been observed, some of which are thrust faults (Fig. 8). North- and northeast or east-striking faults are present, mainly in the areas where pre-Caledonian rocks are exposed (Kellogg, 1975). On Brøggerhalvøya, west-northwest-striking thrust faults are associated with tight folds suggesting significant northerly movements (Challinor, 1967). The most prominent tectonic element in the area is the Forlandsundet Graben, located along north-northwest-striking faults. The sedimentary succession in the graben is slightly deformed, suggesting that it developed during or prior to the last deformation phase.

The thrusting in the fold belt was directed mainly toward the east or east-northeast (Birkenmajer, 1981), but in the Brøggerhalvøya area it was towards the northeast. West-directed thrusting has been recorded only from the St. Jonsfjorden and Bellsund area (Fig. 8), but magnitudes are relatively small (<1 km). The crustal shortening due to folding and thrusting has been estimated as about 10 km in the north and 15 km in the south (Birkenmajer, 1981). Birkenmajer also suggested that the entire west coast of Spitsbergen, between Kongsfjorden and Sørkapp has been translated to the north-northwest from a southerly location, possibly by some 30 km, and that at Kongsfjorden this zone collided with the rigid mass of northwest Spitsbergen, causing the anomalous orientation of the structural trends on Brøggerhalvøya.

The notion that thrusting along western Spitsbergen was a high-level expression of wrench-faulting along the De Geer Line was first implied by Lowell (1972). Later mapping has shown more clearly that some of the thrust faults steepen with depth and that some wrench faults can be traced laterally into thrusts (Kellogg, 1975). This hypothesis, that the thrusts are connected with wrench faults, for example, as part of a large flower structure along western Spitsbergen, is consistent with the results of the recent marine geophysical studies (Myhre et al., 1982) west and south of Spitsbergen. The presence of latest Paleocene and Eocene oceanic crust south of Spitsbergen implies contemporaneous, large-scale strike slip along western Spitsbergen. Furthermore, the climax of transpression in the orogenic belt, dated as Eocene from evidence discussed from the Central Basin, may be related to the mid-Eocene sea-floor events reported by Myhre et al. (1982), namely the northwestward migration of the new rift margin to the segment north of the Senja Fracture Zone (Fig. 1). However, we emphasize that the amount and large scale of the thrusting seen in western Spitsbergen primarily indicates that there was significant crustal shortening across parts of western

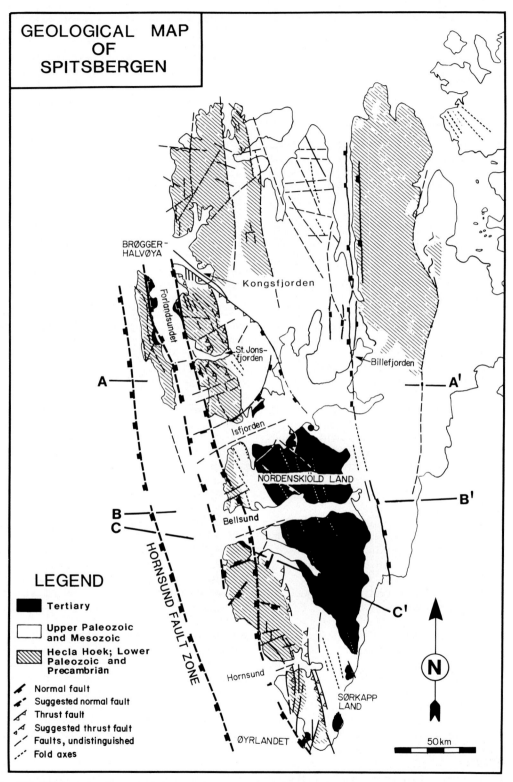

FIG. 8.-—Simplified structural map of Spitsbergen. Lines AA', BB' and CC' are the lines of section shown in Figure 10.

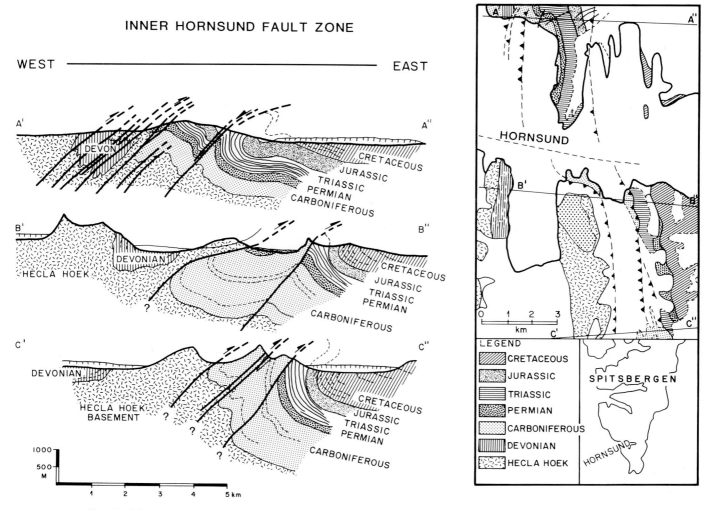

FIG. 9.—Map and representative cross sections of the fold belt in the inner Hornsund area of southern Spitsbergen.

Spitsbergen. This may have been caused by significant curvature along parts of the plate boundary between Greenland and Spitsbergen or by a non-orthogonal relationship between the De Geer Line and the bounding spreading ridges (Crane et al., 1982). There has been insufficient detailed mapping of the fold-and-thrust belt, as yet, to comment further on the likelihood of *most* of the thrusts being attached to wrench faults, or whether there is significant detachment at depth.

A STRIKE-SLIP BASIN WITHIN THE FOLD-AND-THRUST BELT

Tertiary rocks are exposed in three areas within the zone of pronounced Tertiary deformation in western Spitsbergen: Forlandsundet, Renardodden (R in Fig. 2), and Øyrlandet (Ø in Fig. 2). The largest and best exposed area, the fault-bounded Forlandsundet Graben (Atkinson, 1962), contrasts with the Central Basin both in the character of the contained sedimentary succession and in its structural style.

Stratigraphy and Sedimentation

Forlandsundet Graben is some 80 km in length and 25 km wide, but shows only scattered exposures along both margins (Fig. 11). Despite the fragmentary nature of the stratigraphic evidence, a composite succession with an apparent thickness of as much as 5 km has been established (Rye-Larsen, 1982). Along the eastern and southwestern margins of the basin, alluvial-fan deposits, characterized by poorly sorted, unstratified conglomerate beds of debris-flow origin, and finer-grained, flat-stratified and cross-stratified conglomerates, originating from streamflow, have been documented. In the southwestern area, the fan deposits appear to grade laterally (basinwards) into sequences of black shales, siltstones, and cross-stratified/ripple-laminated sandstones, which have been interpreted as fan-delta and nearshore deposits on account of their structures and facies motifs (Rye-Larsen, 1982). In the more than 3-km-thick succession in the northwestern region, there are black shales with associated turbidite and conglomerate beds, inter-

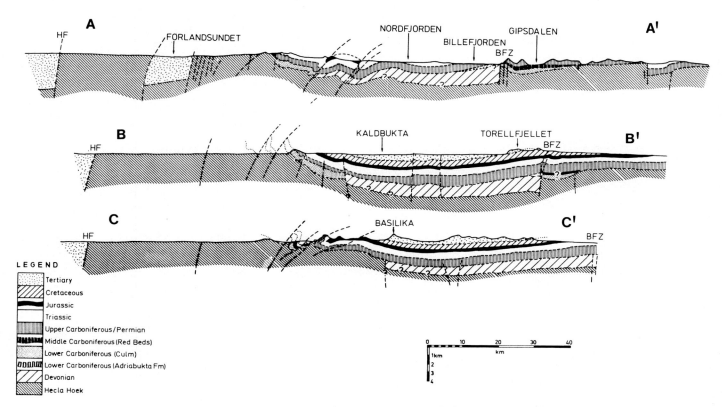

FIG. 10.—Large-scale cross sections across Spitsbergen along the lines shown in Figure 8.

preted as a submarine-fan association. The marginal allu-vial-fan sequences (Selvagen, Sarsbukta, and Sarstangen Formations) are laterally equivalent to the nearshore and shallow-marine deposits (Sesshøgda, Reinhardpyten, Kro-kodillen, and Marchaise Lagune Formations) and to the submarine-fan succession (Aberdeenflya Formation; Fig. 17 below).

Dating of the Forlandsundet Group by Manum (1962) and Livshits (1965; 1974) has suggested Eocene to early Oli-gocene ages. Recent palynological dating on samples from the eastern margin of the graben, from the Sarstangen For-mation, indicates an age not younger than Eocene. This is based mostly on the occurrence of the dinoflagellate *Sval-bardella Cooksonia* (W. Morgan, personal commun., 1984). Additional recent dating using foraminifera (Feyling-Han-sen and Ulleberg, 1984) suggests an Oligocene age for Sarsbukta Formation.

Origin of Forlandsundet Graben

Available dating suggests that the Forlandsundet Graben developed partly during and partly after the main interval of deformation. However, as in most other ancient strike-slip basins, it is relatively easy to demonstrate syn-depo-sitional, dip-slip faulting, but difficult to find evidence of the strike-slip component of movement. Active subsidence during sedimentation is reflected both in the great thickness and coarseness of this small basin's succession. Lateral fa-cies changes are abrupt adjacent to the bounding faults. The most persuasive evidence for strike slip is the regional tec-tonic setting, less than 20 km from the strike-slip De Geer Line. Although the local evidence is inconclusive, the fol-lowing features also suggest strike slip along the margins of the graben:

(1) Metaconglomerate clasts in Selvagen Formation are offset by at least 3 km from their source area (Rye-Larsen, 1982).
(2) The measured stratigraphic thickness of the succes-sion within the graben is at least three times the max-imum vertical stratal thickness inferred from marine geophysical surveys along the basin axis (Guterch et al., 1978). This is a feature typical of other strike-slip basins. Because of depocenter migration in strike-slip basins, and progressive basement onlap of strata, stratigraphic thicknesses are commonly many times greater than true vertical thicknesses (Crowell, 1974; Steel and Gloppen, 1980).
(3) Horizontal slickenside striae are abundant along faults cutting the Forlandsundet Group. Paleostress azi-muths have been calculated from micro-structural data in the southwestern part of the basin. These suggest a principal direction of extension of N60°W, which is consistent with right-lateral transpression on north-northwest-striking faults.

Regional and local evidence for strike slip combined with

FORLANDSUNDET BASIN

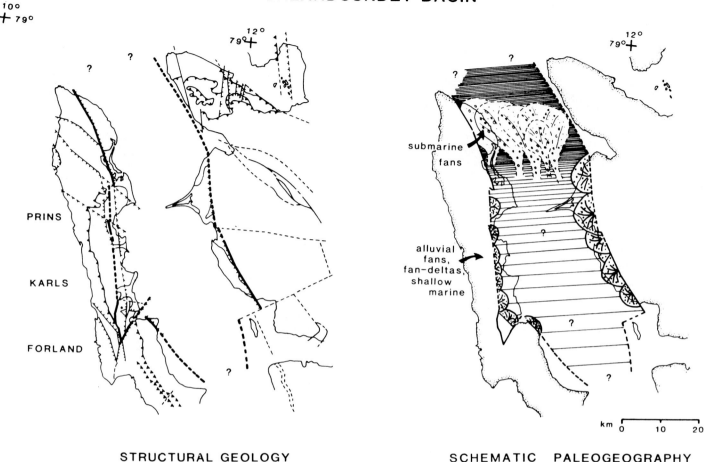

STRUCTURAL GEOLOGY   SCHEMATIC PALEOGEOGRAPHY

FIG. 11.—Maps of the Forlandsundet Graben showing structure and schematic paleogeography. Bold lines are the major faults: fine lines are minor faults. Thrust faults indicated by sawteeth in upper plate. The more densely spaced lines in the graben's axial region schematically indicate deeper water.

available age control thus suggests that Forlandsundet Graben originated and developed in a transpressional strike-slip regime (Steel et al., 1981). This conclusion is at odds with earlier opinion (e.g., Harland and Horsfield, 1974), which assumed that Forlandsundet Graben was entirely post-orogenic, because it appears to cut the orogenic belt.

The co-existence of compressional and extensional features along strike-slip zones is well known (e.g., Crowell, 1974). However, the development of the Forlandsundet Graben in a transpressional regime is still difficult to explain, especially considering the size of the graben (>25 km wide and several kilometers deep).

There are two possible hypotheses for the early development of the Forlandsundet Graben within the setting of the orogenic belt: (1) The graben is related to extension adjacent to a curved strike-slip fault zone (Fig. 12A). This hypothesis is based on plaster cast modelling carried out in cooperation with John Sales at the Dallas Research Division of Mobil Oil Company. It shows that deep and narrow elongated grabens may develop behind bends in the strike-slip system owing to stress release, whereas overthrusting and folding occur around the bends. (2) The graben originated as a collapse graben in the central part of the uplifted and arched orogenic belt (Fig. 12B).

Later extension and deepening of the graben probably took place during the extensional phase from earliest Oligocene time as suggested by Harland and Horsfield (1974). This is, however, difficult to document as only the marginal parts of the graben are exposed today. The proposed sequence of events in the fold belt is summarized in Figure 13.

The other small areas on western Spitsbergen with Tertiary strata, Renardodden and Øyrlandet (Fig. 2), may also be fragments of strike-slip basins or they may be related to post-early Oligocene rifting on the Svalbard margin. The former setting is likely for Renardodden strata, as suggested by recent dating of the strata as Paleocene (Theidig et al., 1980) and late Eocene to early Oligocene (Head, 1984).

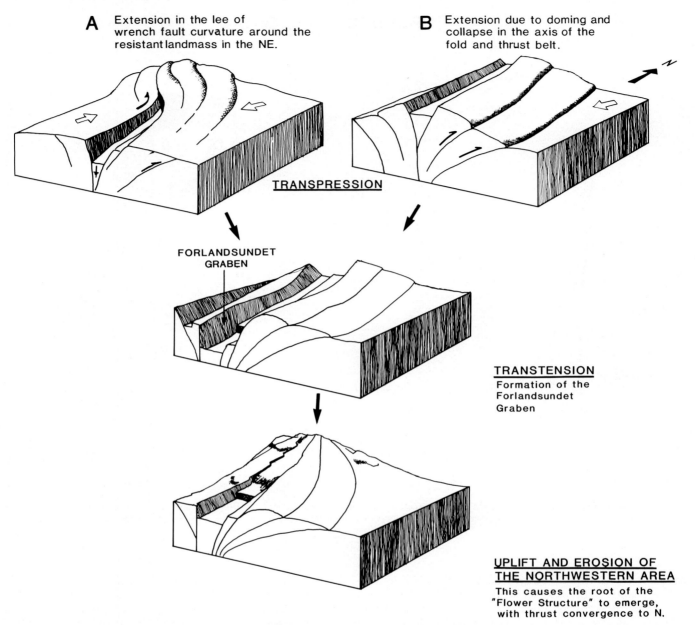

FIG. 12.—Schematic block diagrams showing the possible structural development of the Forlandsundet Graben in relation to the fold and thrust belt of western Spitsbergen. Hypothesis A suggests the graben developed from stress release in relation to fault-zone curvature and to a rigid northwest Spitsbergen landmass. Hypothesis B suggests that the graben may be a collapse feature in the fold belt.

THE BASINS OF THE RIFT MARGIN

The evolution of the continental margin of Svalbard from an oblique-slip to a dip-slip regime, took place in a diachronous manner. The southernmost segment of the margin was subject to rifting by latest Paleocene time, but this did not happen in the north until the early to mid-Oligocene (see Myhre et al., 1982; Eldholm et al., 1984). The post-early Oligocene rifting off Svalbard produced a characteristic suite of sedimentary basins, which partly cut and partly overlap the western edge of the older oblique-slip basins and the orogenic belt of western Spitsbergen. These sedi-

mentary basins, locally up to 7 km deep, have long been recognized to cover an extensive area of the present continental shelf and slope (Eldholm and Ewing, 1971; Sundvor, 1974; Eldholm and Talwani, 1977; Sundvor et al., 1977; Schülter and Hinz, 1978). This region of young Tertiary deposits is bounded by the Knipovich Ridge to the west and the Hornsund fault zone to the east (Figs. 2, 14), and is floored predominantly by oceanic crust. The continuity of the basement reflection suggests that the continent-ocean boundary is located at or near the Hornsund fault zone (Myhre et al., 1982).

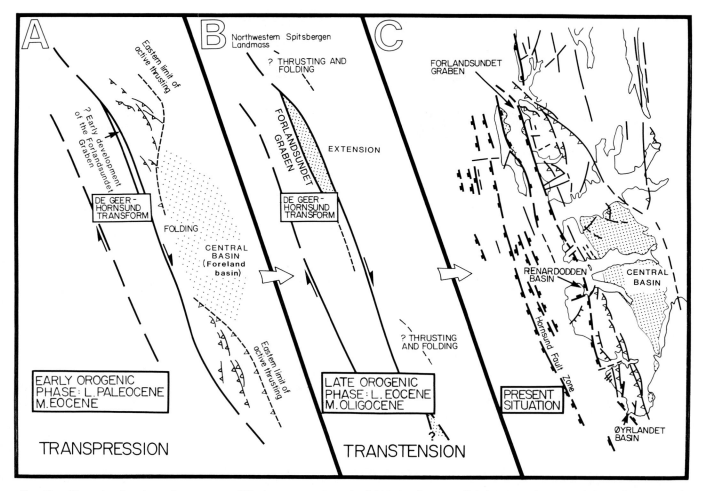

FIG. 13.—Maps showing the main sequence of Tertiary events along the fold belt of western Spitsbergen and a map of the major present-day structural features. A) Late Paleocene to mid-Eocene, B) late Eocene to mid-Oligocene, C) present.

The Stratigraphic Sequences

The sedimentary succession in the rifted region has been divided into three main sequences (SPI 1–3 in Figs. 14, 15), separated by two regional unconformities (U1, U2; Schlüter and Hinz, 1978). An upper low-velocity sequence, SPI 1 (1.7–2.8 km/sec), forms a prograding wedge beneath the outer shelf and slope. It is thickest in the south (2–2.5 km), where it is characterized by sloping, sub-parallel high-continuity reflections (Figs. 14, 15). Northward it becomes thinner (1–1.3 km) and increasingly complex, showing evidence of faulting and erosion (Fig. 15, line A). In general, sequence SPI 1 terminates westward against the Knipovich Ridge (Fig. 14), but in the narrow, northernmost part of the basin, the sediments have overflowed into the present rift valley (Eldholm et al., 1984).

An unconformity, U1, separates sequence SPI 1 from the underlying SPI 2, which is characterized by medium-range velocites (2.4–3.1 km/sec) and an irregular, discontinuous to chaotic reflection pattern. The sequence thickens seaward from the shelf edge, with the maximum development in the southern part of the basin (0.6–0.8 km).

The lowermost sequence SPI 3 is separated from sequence SPI 2 by an unconformity, U2. Sequence SPI-3 is a high-velocity unit (2.9–4.8 km/sec), and although it is masked by multiples to a large extent, it shows a sub-parallel reflection pattern and evidence of onlap against both the Knipovich Ridge and the Hornsund fault zone (Fig. 15, line A).

Discussion

From correlation by reflection tracing from the DSDP site 344 (Fig. 14; Talwani and Udintsev, 1976) the upper sequence SPI 1 has been dated as Pleistocene-Pliocene. The DSDP site 344 cores show sequence SPI 1 to consist of terrigenous muds, silty and sandy muds, and some muddy sandstones of inferred gravity-transported (lower part) and glacial marine (upper part) origin.

The seismic signature of SPI 1 is that of a typical progradational shelf-slope sequence, deposited in response to a changing glacial regime, and with little evidence of major tectonic influence. Some normal faulting can be seen on line A, however, which clearly extends up into sequence

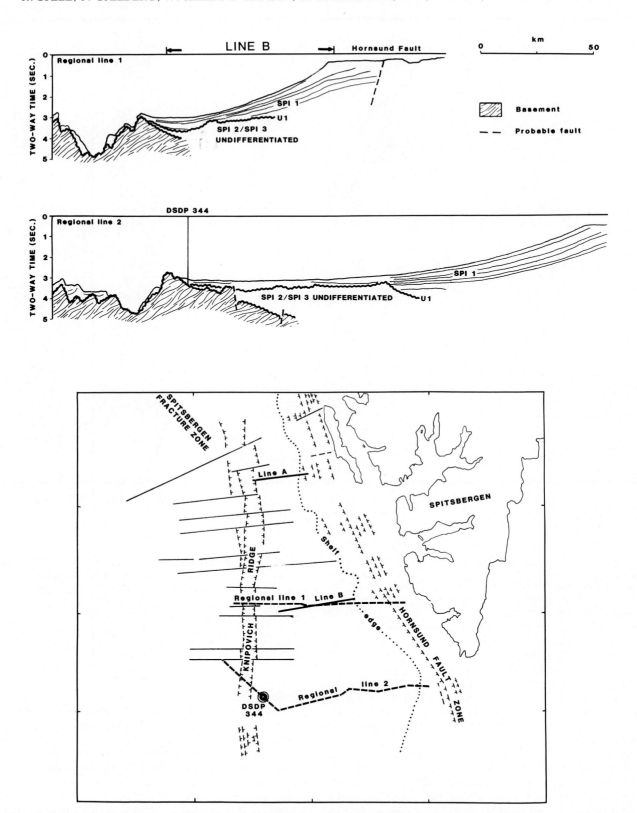

FIG. 14.—Two regional cross sections of the Svalbard margin off Spitsbergen. Data from Schlüter and Hinz (1978). The map shows the location of the regional lines and lines A and B shown in Fig. 15.

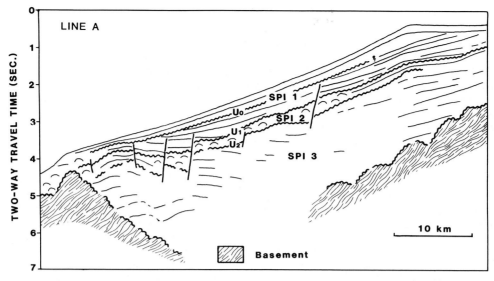

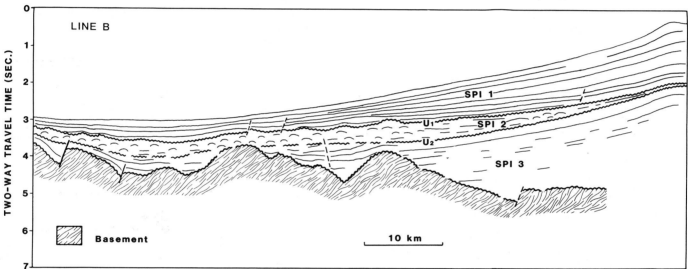

FIG. 15.—Details of seismic sequences SPI-1, 2, and 3 on the Svalbard margin, along lines A and B shown on Figure 14. Data from Schluter and Hinz (1978). See text for discussion.

SPI 1, but terminates against a local unconformity within the sequence.

The unconformity U1 clearly cuts the sequence SPI-2, but shows no evidence of major erosion. Schlüter and Hinz (1978) interpreted sequence SPI-2 as an "allochthonous wedge of gravity-driven masses" on the basis of the chaotic reflection patterns (Fig. 15). We agree with this interpretation but emphasize that the sequence most likely consists of an alternation of all types of gravity-transported deposits, and not a continuous package of slump deposits alone. Furthermore, the distinct character of this sequence, together with the evidence of syn-sedimentary normal fault activity within the sequence in some areas (there is stratal thickening on the downthrown sides of the faults in Line A, Fig. 15), suggests that sequence SPI-2 may reflect a period of significant, tectonic activity on the margin, compared to the intervals represented by the underlying and overlying sequences. We suggest that this tectonic activity may have been related to the eastward shift of the Hornsund transform to the Spitsbergen transform and the incipient development of the Knipovich Ridge in this area, at about 16-10 Ma (Crane et al., 1982).

Sequence SPI-3 has been interpreted from velocity analysis as highly consolidated interbedded sandstones and shales (Schlüter and Hinz, 1978). The lack of large-scale deformation suggests that it post-dates the main transpressional tectonic phase along western Spitsbergen. This is consistent with the suggestion of Myhre et al. (1982) that sequence SPI 3, in front of the Hornsund fault, was deposited beginning in mid-Oligocene time (from 37 Ma) in response to a prominent change in plate motion, the onset of rifting and the early opening of the northern Greenland Sea.

The presence of only a single age determination from the seismic sequences remains an outstanding problem in our understanding of the rift basins, and has been the cause of some debate (compare Schlüter and Hinz, 1978, with Myhre et al., 1982). However, the argument that the rifting was initiated by the major plate-tectonic change at 37 Ma, is persuasive. Based on this hypothesis, the two lower seismic sequences span mid-Oligocene to late Miocene time, and the rift basins were sourced from the post-mid-Oligocene, rising Spitsbergen landmass. Figure 16 summarizes the possible stratigraphic relationships between the rift basin succession and the earlier Tertiary sequences on Svalbard.

CONCLUSIONS

The Paleogene plate boundary between northeastern Greenland and Svalbard, the De Geer Line, lay just west of western Spitsbergen. Sea-floor data indicate strike slip along this fault zone from latest Paleocene time, whereas data from the Central Basin on Spitsbergen suggest major fault activity with a possible strike-slip component, from early Paleocene time. Data from the Central Basin and from the fold and thrust belt on Spitsbergen suggest a transpressive regime beginning in late Paleocene time. Marine geophysical data indicate an early to mid-Oligocene change in

the relative plate motion and a shift from a strike-slip to a rift regime along the margin.

The early to mid-Paleocene Central Basin on Spitsbergen evolved from a series of partially connected coal basins to a single, open marine basin in an extensional, possibly transtensional, regime to the east of the De Geer Line. Subsidence was increasingly asymmetric, and greatest (>800 m) towards the De Geer Line during this phase of basin development (Fig. 17I).

The late Paleocene–Eocene development of the Central Basin was characterized by a reversal of drainage direction and an influx of metamorphic rock fragments. The region along and east of the De Geer Line was uplifted and became the main sediment source for the Central Basin (Fig. 17II). The sedimentary sequence which accumulated in this phase was more than 1.5 km thick and of overall regressive character. Shelf sediments pass upwards through deltaic and shoreline deposits to continental deposits.

The fold and thrust belt of western Spitsbergen developed through late Paleocene–Eocene time and was partly contemporaneous with the late phase of infillng of the Central Basin. Extensive east-directed thrusting and folding suggests crustal shortening of 10 to 15 km in places. Al-

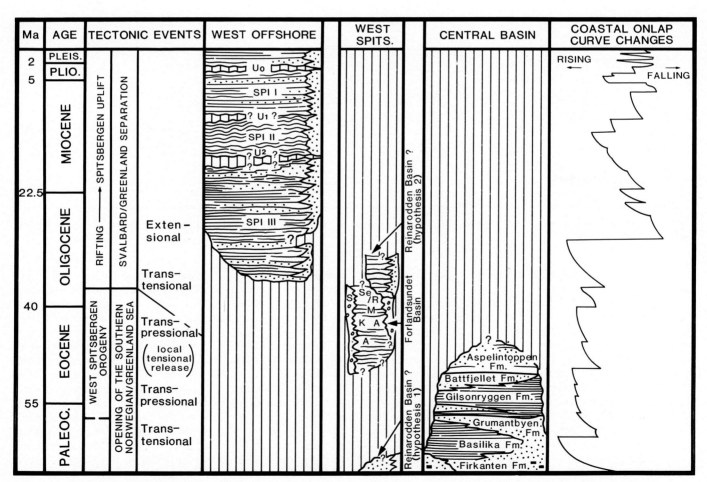

Fig. 16.—Summary of the stratigraphy of the Tertiary basins of Svalbard in relation to the proposed tectonic setting and a curve of global coastal onlap (from Vail et al., 1977).

SPITSBERGEN TERTIARY BASIN EVOLUTION

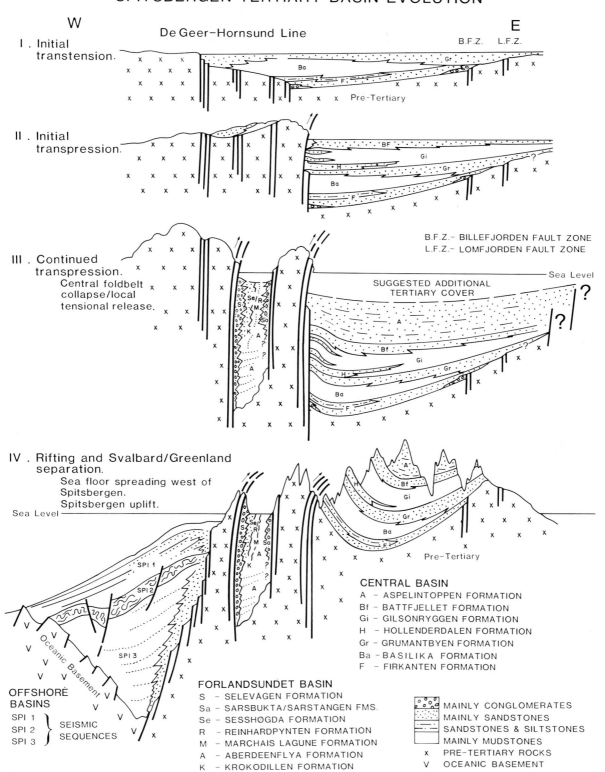

FIG. 17.—Schematic summary of basin development in relation to tectonic evolution of the Svalbard margin. Large vertical exaggeration has been necessary to illustrate stratigraphical details in the basins. See text for discussion.

though some of the wrench faults and thrust are clearly related, there has not been enough detailed structural mapping to confirm the suggestion that most of the thrust faults steepen with depth, and are therefore part of a giant flower structure. The extent of the thrusting suggests major compression, and implies significant curvature along parts of the De Geer Line or a temporal change in the relative plate motion.

The Forlandsundet Graben, located within the fold and thrust belt, developed mainly in Eocene to early Oligocene time, and contains a thick succession of alluvial, shallow-marine and deeper-marine strata (Fig. 17III). Regional evidence suggests that this graben developed in transpressive and later transtensional regimes. Local evidence within the graben is not conclusive as regards strike slip along the boundary faults, but pebble mismatches, some small-scale structural data, and the anomalous stratigraphic thickness of the succession (suggesting depocenter migration) support the notion.

From early Oligocene time, the Svalbard margin was rifted and basins up to 7 km deep developed (Fig. 17IV).

The combination of marine geophysical data with data from the sedimentary basins and fold and thrust belt of Spitsbergen thus suggest that the Svalbard margin off Spitsbergen evolved through regimes of transtension, transpression, and transtension during its strike-slip history, before being rifted beginning at 37 Ma.

ACKNOWLEDGMENTS

We are grateful to our colleagues at Norsk Hydro, the University of Bergen, and Statoil for much discussion of the themes presented here. This work evolved from a Norwegian Applied Science Research Council and Statoil-supported project with the University of Bergen, aimed at a better understanding of the coal basins and of Tertiary tectonics and sedimentation on Svalbard. We thank Paul Gnanalingam for drafting the figures, and Wojtek Nemec for the use of the original version of Figure 3. The manuscript was reviewed by K. T. Biddle, N. Christie-Blick, and John Mutter.

REFERENCES

ATKINSON, D. J., 1962, Tectonic control of sedimentation and the interpretation of sediment alternations in the Tertiary of Prince Charles Foreland, Spitsbergen: Geological Society of America Bulletin, v. 73, p. 343–363.
BIRKENMAJER, K., 1972, Tertiary history of Spitsbergen and continental drift: Acta Geologica Polonica, v. 11, p. 47–123.
———, 1981, The geology of Svalbard, the western part of the Barents Sea and the continental margin of Scandinavia, in Nairn A. E. M., Churkin, M. and Stehli, F. G., eds., The Ocean Basins and Margins, v. 5, The Arctic Ocean: Plenum Press, New York, p. 265–329.
CHALLINOR, A., 1967, The structure of Brøggerhalvøya, Vestspitsbergen: Geological Magazine, v. 104, p. 322–336.
CRANE, K., ELDHOLM, O., MYHRE, A., AND SUNDVOR, E., 1982, Thermal implications for the evolution of the Spitsbergen Transform Fault: Tectonophysics, v. 89, p. 1–32.
CROWELL, J. C., 1974, Sedimentation along the San Andreas Fault, California, in Dott, R. H., Jr., and Shaver, R. H., eds., Modern and Ancient Geosynclinal Sedimentation: Society of Economic Paleontologists and Mineralogists Special Publication No. 19, p. 292–203.
DALLAND, A., 1979, Structural geology and petroleum potential of Nor-

denskiold Land, Svalbard, in Norwegian Sea Symposium, Norwegian Petroleum Society, paper 30, p. NSS/30-1-NSS/30-20.
ELDHOLM, O., AND EWING, J., 1971, Marine geophysical survey in the southwestern Barents Sea: Journal of Geophysical Research, v. 76, p. 3832–3841.
ELDHOLM, O., AND TALWANI, M., 1977, Sediment distribution and structural framework of the Barents Sea: Geological Society of America Bulletin, v. 88, p. 1015–1029.
ELDHOLM, O., SUNDVOR, E. MYHRE, A., AND FALEIDE, J. I., 1984, Cenozoic evolution of the continental margin off Norway and western Svalbard, in Spencer, A. M., ed., Petroleum Geology of the North European Margin: Norwegian Petroleum Society, London, Graham and Trotman, p. 3–19.
FEYLING-HANSEN, R. W., AND ULLEBERG, K., 1984, A Tertiary-Quaternary section at Sarsbukta, Spitsbergen, Svalbard, and its foraminifera: Polar Research, v. 2, p. 77–106.
GJELBERG, J., 1981, Upper Devonian (Famennian) to Middle Carboniferous succession of Bjørnøya: Norsk Polarinstitutt Skrifter 174, 67 p.
GRØNLIE, G., AND TALWANI, M., 1978, Geophysical atlas of the Norwegian-Greenland Sea: Vema Research Series IV, Lamont-Doherty Geological Observatory of Columbia University, 26 p.
GUTERCH, A., PAJCHEL, J., PERCHUC, E., KOWALSKI, J., DUDA, S., KOMBER, J., BOJDYS, G., AND SELLEVOLD, M. A., 1978, Seismic reconnaissance measurements of the crustal structure in the Spitsbergen region: Report of the Seismological Observatory for 1978, University of Bergen, 61 p.
HARLAND, W. B., 1965, The tectonic evolution of the Arctic-North Atlantic Region: Philosophical Transactions of the Royal Society of London, Series B., v. 258, p. 59–75.
HARLAND, W. B., 1969, Contribution of Spitsbergen to understanding of tectonic evolution of the North Atlantic Region, in Kay, M., ed., North Atlantic Geology and Continental Drift: American Association of Petroleum Geologists Memoir, 12, p. 817–851.
HARLAND, W. B., AND HORSFIELD, W. T., 1974, West Spitsbergen Orogen, in Spencer, A. M., ed., Mesozoic-Cenozoic Orogenic Belts, Data for Orogenic Studies: Geological Society of London Special Publication No. 4, p. 747–755.
HARLAND, W. B., PICKTON, C. A. G., WRIGHT, N. J. R., CROXTON, C. A., SMITH D. G., CUTBILL, J. L., AND HENDERSON, W. S., 1976, Some coal-bearing strata of Svalbard: Norsk Polarinstitutt Skrifter 164, 89 p.
HEAD, M., 1984, A palynological investigation of Tertiary strata at Renardodden, W. Spitsbergen (abs): 6th International Palynological Conference.
HORN, G., AND ORVIN, A. K., 1928, Geology of Bear Island with special reference to the coal deposits and an account of the history of the island: Skrifter om Svalbard og Ishavet, 15, 152 p.
HÅKANSSON, E., AND SCHACK PEDERSEN, S. A., 1982, Late Paleozoic to Tertiary tectonic evolution of the continental margin in North Greenland, in Embry, A. F., and Balkwill, H. R., eds., Arctic Geology and Geophysics: Canadian Society of Petroleum Geology, Memoir 8, p. 331–348.
KELLOGG, H. E., 1975, Tertiary stratigraphy and tectonism in Svalbard and continental drift: American Association of Petroleum Geologists Bulletin, v. 59, p. 465–485.
LIVSHITS, Ju. JA., 1965, Paleogene deposits of Nordenskiold Land, Vestspitsbergen, in Sokolov, V. N., ed. Materialy po geologii Shpitsberena: Leningrad (Engl. translation: Boston Spa, Yorkshire, England, National Lending Library of Science and Technology, 1970, p. 193–215).
———, 1974, Paleogene deposits and the platform structure of Svalbard: Norsk Polarinstitutt Skrifter 164, 50 p.
LOWELL, J. D., 1972, Spitsbergen Tertiary orogenic belt and the Spitsbergen fracture zone: Geological Society of America Bulletin, v. 83, p. 3091–3102.
MAJOR, H., AND NAGY, J., 1972, Geology of the Adventdalen map area: Norsk Polarinstitutt Skrifter 138, 58 p.
MANUM, S. B., 1962, Studies in the Tertiary flora of Spitsbergen, with notes on Tertiary floras of Ellesmere Island, Greenland and Iceland: Norsk Polarinstitutt Skrifter 125, 127 p.
MANUM, S. B., AND THRONDSEN, T., 1978a, Rank of coal and dispersed organic matter and its geological bearing on the Spitsbergen Tertiary: Norsk Polarinstitutt Arbok for 1977, p. 159–177.
———, 1978b, Dispersed organic matter in the Spitsbergen Tertiary: Norsk

Polarinstitutt Årbok for 1977, p. 179–187.

MYHRE, A. M., ELDHOLM, O., AND SUNDVOR, E., 1982, The margin between Senja and Spitsbergen fracture zones: implications from plate tectonics: Tectonophysics, v. 89, p. 33–50.

NEMEC, W., AND STEEL, J. R., 1985, Stacked, shore-attached, sheet sandstones in a Paleocene inland seaway succession (Endalen Member, Spitsbergen) (abs.): International Association of Sedimentologists, 6th European Regional Meeting, Lleida, Spain, p. 321–324.

NORDENDKIOLD, A. E., 1866, Utkast till Spetsbergens geologi: Proceedings of the Royal Swedish Academy of Science, v. 6, No. 7, Stockholm.

NØTTVEDT, A., 1985, Askeladden delta sequence (Paleocene) on Spitsbergen—sedimentation and controls on delta formation: Polar Research, v. 3, p. 21–48.

ORVIN, A. K., 1940, Outline of the geological history of Spitsbergen: Skrifter om Svalbard og Ishavet, No. 78, Oslo.

RAVN, J. P. J., 1922, On the Mollusca of the Tertiary of Spitsbergen: Result Norske Spitsbergen-ekspedisjon, v. 1, no. 2, Kristiania.

RYE-LARSEN, M., 1982, Forlandsundet Graben (Paleogene), an oblique-slip basin on Svalbard's western margin (abs.): International Association of Sedimentologist, 3rd European Regional Meeting, Copenhagen, p. 31–34.

SCHLÜTER, H. U., AND HINZ, K., 1978, The geological structure of the western Barents Sea: Marine Geology, v. 26, p. 199–230.

SPENCER, A. M., HOME, P. C., AND BERGLUND, L. T., 1984, Tertiary structural development of the western Barents Shelf, Troms to Svalbard, *in* Spencer, A. M., ed., Petroleum Geology of the North European Margin: Norwegian Petroleum Society, London, Graham and Trotman, p. 199–209.

STEEL, R. J., 1977, Observations on some Cretacous and Tertiary sandstone bodies on Nordenskiold Land, Svalbard: Norsk Polarinstitutt Arbok for 1976, p. 43–68.

STEEL, R. J., AND GLOPPEN, T. G., 1980, Late Caledonian (Devonian) basin formation, western Norway: signs of strike-slip tectonics during infilling, *in* Ballance, P. F., and Reading, H. G., eds., Sedimentation in Oblique-Slip Mobile Zones: International Association of Sedimentologists, Special Publication No. 4, p. 79–103.

STEEL, R. J., DALLAND, A., KALGRAFF, K., AND LARSEN, V., 1981, The Central Tertiary Basin of Spitsbergen, sedimentary development of a sheared margin basin, *in*, Kerr, J. W., and Ferguson, A. J., eds., Geology of the North Atlantic Borderland: Canadian Society of Petroleum Geologists Memoir 7, p. 647–664.

STEEL, R. J., AND WORSLEY, D., 1984, Svalbard's post-Caledonian strata—an atlas of sedimentational patterns and palaeogeographic evolution, *in* Spencer, A. M., ed., Petroleum Geology of the North European Margin: Norwegian Petroleum Society, London, Graham and Trotman, p. 109–135.

SUNDVOR, E., 1974, Seismic refraction and reflection measurements in the southern Barents Sea: Marine Geology, v. 16, p. 255–273.

SUNDVOR, E., ELDHOLM, O., GISKEHAUG, A., AND MYHRE, A., 1977, Marine geophysical survey of the western and northern continental margin of Svalbard: University of Bergen, Seismological Observatory, Scientific Report 4, 35 p.

TALWANI, M., AND ELDHOLM, O., 1977, Evolution of the Norwegian-Greenland Sea: Geological Society of America Bulletin, v. 88, p. 969–999.

TALWANI, M., UDINTSEV, G., et al., 1976, Initial reports of the Deep Sea Drilling Project, v. 38: Washington D.C., United States Government Printing Office, 1256 p.

THEIDIG, F., PICKTON, C. A. G., LEHMANN, U., HARLAND, W. B., AND ANDERSON, H. J., 1980, Das Tertiär von Renardodden (Ostlich Kapp Lyell, Westspitzbergen, Svalbard): Mitteilung Geologisch-Paläontologisch Institutt, University of Hamburg, v. 49, p. 135–146.

VAIL, P. R., MITCHUM, R. M., AND THOMPSON, S., 1977, Global cycles of relative changes of sea level, *in* Payton, C. E., ed., Seismic Stratigraphy—Applications to Hydrocarbon Exploration: American Association of Petroleum Geologists Memoir 26, p. 83–97.

VONDERBANK, K., 1970, Geologie und Fauna der Tertiaren Ablagerungen Zentral-Spitsbergens: Norsk Polarinstitutt Skrifter nr. 153, 156 p.

STRATIGRAPHIC AND STRUCTURAL PREDICTIONS FROM A PLATE-TECTONIC MODEL OF AN OBLIQUE-SLIP OROGEN: THE EUREKA SOUND FORMATION (CAMPANIAN-OLIGOCENE), NORTHEAST CANADIAN ARCTIC ISLANDS

ANDREW D. MIALL

Department of Geology, University of Toronto, Toronto, Ontario M5S 1A1, Canada

ABSTRACT: A recent synthesis of Cretaceous-Cenozoic plate kinematics in the North Atlantic region by Srivastava and Tapscott (in press) provides a framework for a detailed interpretation of the geology of the northeast Canadian Arctic Islands.

The plate model indicates a pattern of successive transpression, transtension, and near-orthogonal convergence between Greenland and Ellesmere Island (northeasternmost of the Arctic Islands) during Campanian to early Oligocene time. Earlier interpretations suggested that this displacement was accommodated by a transcurrent fault along Nares Strait, the waterway dividing these land masses. The alternative model discussed here is that displacement was diffused across the Eurekan Orogen, a zone of open folds, low- to high-angle reverse faults and normal faults that marks the plate boundary. Structural analyses demonstrate 25% to 50% crustal shortening, but the location and style of lateral offset (discrete strike-slip faults, block rotation, pervasive shear, or a combination of all three) remains uncertain.

The plate model can account for the geometry, facies, and age of syn-tectonic stratigraphic units deposited across the Eurekan Orogen. Initial transpressive movements (Campanian to late Paleocene) caused the fragmentation of Sverdrup Basin into smaller depocenters and intervening, fault-bounded uplifts, from which fluvial-deltaic clastic wedges prograded. This phase ended with thrust faulting and the growth of flanking fanglomerate wedges in eastern Ellesmere Island.

Transgression and retrogression accompanied a brief episode of transtensional plate movement (late Paleocene to early Eocene). The pole of rotation of Greenland then gradually underwent a major change, and movement across the plate boundary changed to transpression and then to near orthogonal convergence. Renewed clastic progradation culminated in a major episode of fan progradation during the main phase of Eurekan folding and thrust faulting (latest Eocene to earliest Oligocene).

Eurekan tectonism and sedimentation ended when sea-floor spreading in the Labrador Sea and Baffin Bay ceased in the early Oligocene, and Greenland became coupled to North America.

INTRODUCTION

The Eurekan Orogen is a belt of open folds, low- to high-angle reverse faults and normal faults of Late Cretaceous to Oligocene age that trends north to northeastward across the northeast Arctic Islands (Balkwill, 1978). The structures continue across Nares Strait into Greenland, as the North Greenland fold belt (Fig. 1). The Canadian Arctic is divided from Greenland by Nares Strait, a waterway, which, it has long been assumed, is the site of a transcurrent fault between the Greenland and North American plates (Bullard et al., 1965).

A recent synthesis of sea-floor spreading kinematics for the North Atlantic region indicates that approximately 170 km of left-lateral displacement and up to 200 km of crustal shortening accumulated between the two plates between magnetic anomalies 34 and 13 (84 to 36 Ma; Srivastava and Tapscott, in press; anomalies dated using the time scale of Harland et al., 1982). This plate movement provides a framework for interpreting the Eurekan Orogen, although several problems remain. The most important of these concerns Nares Strait. Geological mapping along the borders of the strait has revealed numerous cross-strait structural and stratigraphic markers that constrain strike slip to less than 50 km (Dawes and Kerr, 1982). Miall (1983) proposed a resolution of this "Nares Strait problem" suggesting that the offset was distributed across a diffuse, oblique-slip plate boundary, of which the Eurekan Orogen is the record.

Several detailed local structural studies of the Eurekan Orogen have been undertaken (Okulitch, 1982; Osadetz, 1982; Higgins and Soper, 1983; Van Berkel et al., 1983), but no regional synthesis has been attempted since that of Balkwill (1978). Existing data confirm tens of kilometers of crustal shortening, locally up to 50% according to Okulitch (1982), but most workers to date have focussed on compressional tectonics, and much research remains to be done to document lateral offset. Whether this includes discrete strike-slip faults, block rotation, pervasive shear, or a combination of all three remains to be determined. Hugon (1983) proposed that the orogen functioned as a ductile megashear zone.

Several basins within the Eurekan Orogen contain successions of nonmarine to shallow-marine clastic sediments that were deposited and deformed during the orogeny (Fig. 2). The deposits are locally as much as 3.2 km thick, and are all assigned to the Eureka Sound Formation, although recent research has revealed many locally mappable units of formation rank, and the Eureka Sound is shortly to be raised to group status (Miall, in press).

Stratigraphic and sedimentological analyses of the syn-tectonic Eureka Sound Formation have yielded many useful data bearing on the tectonic history of the Canadian Arctic Islands. Episodes of transpressional and orthogonal convergence predicted from plate kinematics are documented by coarsening-upward sequences, culminating in fanglomerate wedges that prograded from rising fault blocks. Conversely, an interval of transtensional movement coincided with transgression and the development of relatively fine-grained, low-energy deposits in a marine embayment. An outline of the tectonic and stratigraphic history of the Arctic Islands since 80 Ma has been published elsewhere (Miall, 1984a). The details are summarized in Table 1. The purpose of the present paper is to focus on the Eurekan Orogen. New paleocurrent and petrographic data from the Eureka Sound Formation are presented, and a series of paleogeographic maps document the tectonic and stratigraphic development of the area. These are not drawn on a palinspastic base because of a lack of information on the distribution of displacement within the orogen.

The sedimentary basins described here do not have much

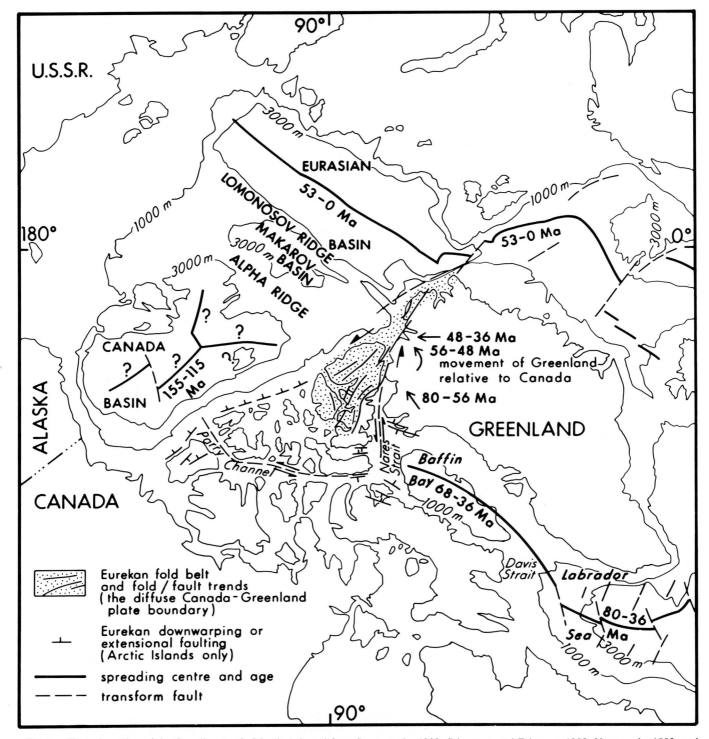

FIG. 1.—Tectonic setting of the Canadian Arctic Islands (adapted from Soper et al., 1982; Srivastava and Falconer, 1982; Vogt et al., 1982; and Miall, 1984a).

in common with the classic wrench-fault style of the Ridge Basin (Crowell and Link, 1982) or Hornelen Basin (Steel and Gloppen, 1980). They do not show the expected facies patterns, and the evidence of the oblique-slip structural style is, as yet, sparse, and by no means convincingly documented. The main evidence for oblique-slip tectonics is the regional plate tectonic pattern deduced from sea-floor spreading kinematics, and the fact that the stratigraphic and structural evidence that is available is entirely consistent with this pattern.

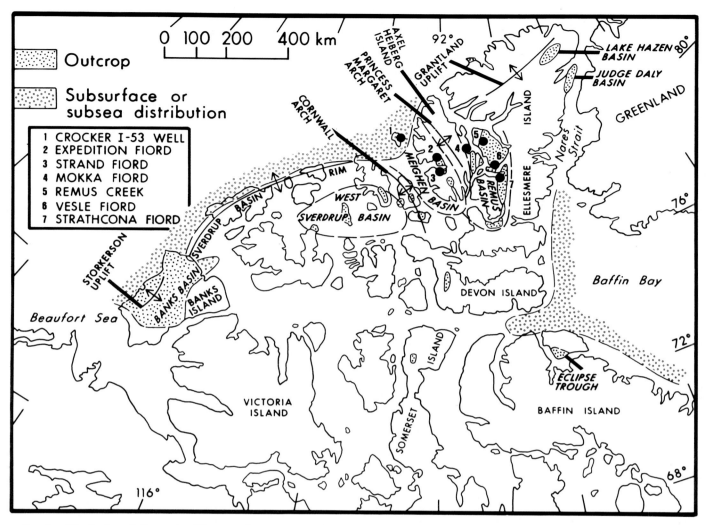

FIG. 2.—Distribution of Campanian-Oligocene depocenters and intrabasin uplifts in the Arctic Islands, showing location of important sections mentioned in the text.

This is perhaps an important general point. In the present case, oblique slip was diffused over a belt at least 100 km wide (possibly double this width) and there probably was no single master fault. Our classic models of wrench-fault basins are based on rather small basins that had high subsidence and sedimentation rates and showed a very sensitive response to tectonic events. Within broader, more slowly evolving oblique-slip zones, sedimentary patterns may be less distinctive. Reading (1980) has pointed out that there is an imperceptible gradation between transtensional and extensional tectonics, and between transpressional and compressional tectonics. In this case, also, the structural and stratigraphic products of oblique slip during the late Paleocene to mid-Eocene were overprinted by the major effects of near-orthogonal compression during the culminating phase of the Eurekan Orogeny.

THE PLATE MODEL

The history of sea-floor spreading in the North Atlantic and Arctic Oceans includes a relatively brief period of spreading in the small subsidiary oceans now constituting Baffin Bay and the Labrador Sea. The history of the Eurekan Orogeny can be explained almost entirely by reference to the plate motions accompanying this spreading activity. The chronology is summarized in Table 1, and the principal tectonic features are shown in Figure 1. Plate kinematics are based on the work of Srivastava and Tapscott (in press). Note that the development of the Canada Basin was probably complete by about 115 Ma (Vogt et al., 1982). The Arctic borderland adjacent to the Canada Basin has therefore functioned as a divergent margin since the Cretaceous and played a passive role during the Eurekan Orogeny. Spreading events in the Makarov Basin may have complicated this picture, but very little is known about this area (some speculations are discussed by Miall, 1984a).

Movement of Greenland relative to Canada began when spreading within the proto-Atlantic Ocean extended into the Labrador Sea at about anomaly 34 time (84 Ma). Greenland rotated about a pole in the central Arctic, accumulating approximately 10° of angular displacement by anomaly 25 time

TABLE 1.—CORRELATION OF TECTONIC AND STRATIGRAPHIC EVENTS IN THE CANADIAN ARCTIC

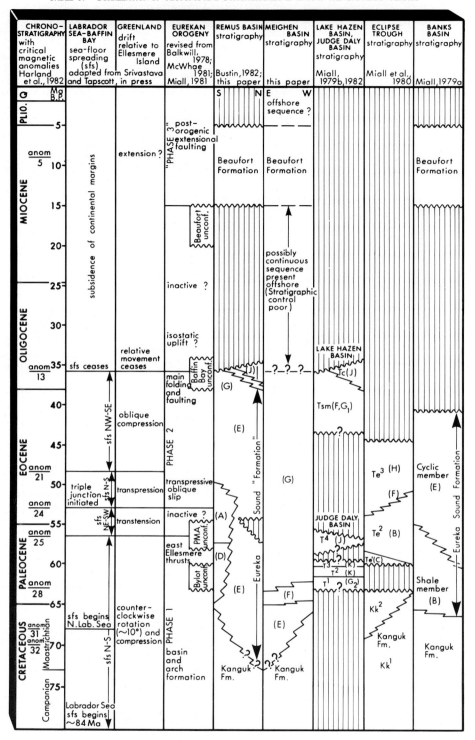

(56 Ma). The sense of movement of Greenland at the Greenland-Canada plate boundary is shown by the appropriate arrow in Figure 1.

At about 53 Ma a ridge-ridge-ridge triple junction was initiated off what is now the southern tip of Greenland, and spreading commenced in the North Atlantic Ocean and Eurasian Basin. The pole of rotation of Greenland changed, probably continuously, so that motion at the plate boundary with Canada changed successively from transtension, through transpression, to near-orthogonal convergence. This is suggested by the two upper arrows in Figure 1. At anomaly 13 time (36 Ma) sea-floor spreading in Baffin Bay-Labrador Sea ceased, and from that time Greenland and Canada have acted as a single plate.

The corresponding tectonic activity of the Eurekan Orogeny, which can be related to this kinematic history, is shown in Table 1. Note the three numbered "phases" that were defined by Balkwill (1978) before the detailed plate model had become available. The timing of these phases was refined by Miall (1984a) using the plate model, and much confirming evidence is now available from the stratigraphic record, as discussed below.

FACIES ANALYSIS

A wide variety of nonmarine to shallow-marine facies occur in the Eureka Sound Formation (Fig. 3). They can be grouped into ten facies assemblages using mainly lithofacies criteria (Table 2), and each assemblage forms a locally mappable stratigraphic unit. Detailed descriptions and illustrations of these assemblages have been given elsewhere (Miall, 1981, 1984b), and need not be repeated in this paper. Assemblages B and C do not occur in the project area and are not discussed here. The stratigraphic relationships between the assemblages are discussed below.

Estuarine to Shallow-Marine Lithofacies Assemblage (A)

This assemblage occurs only in Remus Basin. It constitutes member III of West et al. (1981), and reaches a thickness of 1,400 m in the Strathcona Fiord area (section 6 in Fig. 4). It also forms several major outliers in the northeast part of Remus Basin. The sediments consist mainly of monotonous, very fine grained sandstone with rare current ripple marks and low-angle planar-tabular crossbedding. West et al. (1981) recorded a fauna of marine scaphopods, sharks and bony fish, and marine to brackish-water pelecypods. The assemblage is interpreted as the deposit of a low-energy marine embayment, which occupied much of Remus Basin for part of the Paleocene and Eocene.

Deltaic Lithofacies Assemblages (D, E)

Lithofacies assemblage D comprises a monotonous succession of mudstone and siltstone with thin sandstone and coal beds. Current ripples, groove and load casts are the most common sedimentary structures. Coarsening-upward sequences are common (Fig. 3A). The assemblage was probably deposited in a quiet, fresh- to brackish-water environment fed by sluggish river-dominated deltas. The assemblage composes member II of the section at Strathcona Fiord (section 6 in Fig. 4; West et al., 1981) and underlies the northeast part of Remus Basin.

Lithofacies assemblage E is one of the most widespread in the Eureka Sound Formation. It is the proximal equivalent of assemblage D, representing distributary-mouth, crevasse-splay and lower delta-plain deposits of river-dominated deltas (Fig. 3D). Coarsening-upward sequences up to 80 m thick are common, and record the progradation of individual distributaries or delta lobes. The thicker sequences are interpreted as stacked delta lobes. High-energy

crossbed structures, including planar-tabular sets up to 2 m thick, are common, and large exposures reveal broad, shallow channels as much as 100 m wide and 3 m deep. Delta plain conditions were established locally, as recorded by the rare presence of fluvial point bars and coal seams.

Most of the Eureka Sound Formation of Meighen Basin is composed of this assemblage, which reaches 3.2 km in thickness at Strand Fiord (section 2 in Fig. 4). In Remus Basin the assemblage constitutes member IV of the section at Strathcona Fiord (section 6 in Fig. 4; West et al., 1981). Here coal seams are much more abundant than elsewhere in the Eureka Sound Formation (23 seams in 430 m of section), and the rocks contain an unusual, rich fauna of vertebrates indicating warm, probably subtropical conditions (West et al., 1981). The climatic significance of this fauna has been discussed elsewhere (Miall, 1984b). This assemblage may also underlie parts of the northwestern Remus Basin, in the Remus Creek area, although exposures there are poor, and it is difficult to distinguish fluvial and deltaic successions.

Fluvial Lithofacies Assemblages (F, G, H, J)

A variety of fluvial styles are exhibited by the Eureka Sound Formation (Miall, 1984b).

High-sinuosity (meandering) rivers are distinguished by the point bar (epsilon) crossbedding, fining-upward sequences, and thin coals of assemblage F. The best exposures of this lithofacies assemblage occur in Lake Hazen Basin (Miall, 1979b), and a well exposed section 215 m thick near the base of the Remus Creek succession is also assigned to this assemblage (Fig. 3B). Braided fluvial deposits (lithofacies assemblage G) occur in two main locations in the northeast Arctic, Mokka Fiord and Lake Hazen. They are characterized by pebbly sandstones and thin conglomerates containing abundant crossbedding, and by numerous thin fining-upward sequences (Fig. 3C). At both locations the assemblage overlies finer-grained fluvial or deltaic sediments and, in turn, is overlain by fanglomerates of assemblage J. An interval of assemblage H occurs in the Lake Hazen section. The overall coarsening-upward succession is about 700 m thick in both locations (Fig. 5), and is interpreted as syn-tectonic in origin. The sections provide important documentation of the climactic phase of Eurekan tectonism (Miall, 1984a, b) and are discussed below.

Lacustrine Lithofacies Assemblage (K)

This assemblage occurs only in Judge Daly Basin, where it is characterized by thin- to medium-bedded sandstone, siltstone, and mudstone with current ripple marks and well preserved plant leaves (Miall, 1981, 1982).

STRATIGRAPHY

A reconstructed stratigraphic cross section through Meighen and Remus Basins (Fig. 4) shows that the Eureka Sound Formation may locally have exceeded 4 km in thickness, prior to late Eocene to Oligocene tectonism, uplift, and erosion. The chronostratigraphy of the formation in these

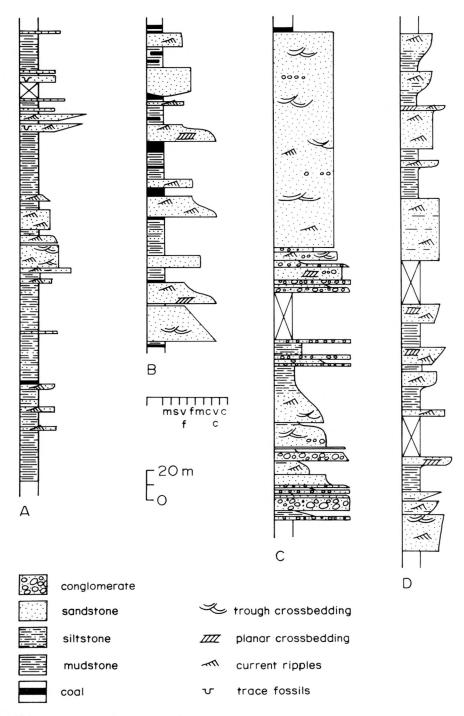

conglomerate

sandstone

siltstone

mudstone

coal

⌢ trough crossbedding

▱ planar crossbedding

⌒ current ripples

∪ trace fossils

FIG. 3.—Examples of partial stratigraphic sections through the Eureka Sound Formation, illustrating characteristic lithofacies and cyclic relationships. A) Assemblage D, north of Vesle Fiord. B) Assemblage F, Remus Creek. C) Assemblage G, Mokka Fiord. D) Assemblage E, Strand Fiord. Grain size is shown by column width (m, mudstone; s, siltstone; vf, f, m, c, vc are very fine to very coarse sand; c, conglomerate).

two basins, and in other major basins in the Arctic Islands, is shown in Table 1. Age relations are based on palynostratigraphy (see references in Table 1, and Miall, 1984a).

At Strathcona Fiord (section 6 in Fig. 4) the Eureka Sound Formation has been subdivided into four members (West et al., 1981; these members are shortly to be redefined as for-

mations within a new Eureka Sound Group). A tongue of deltaic sediments constitutes member I, and is followed by the retrogradational distal deltaic (assemblage D) and marine (assemblage A) sediments of members II and III, respectively. Renewed deltaic progradation formed member IV. This stratigraphic subdivision can be traced northeast-

TABLE 2.—SUMMARY OF LITHOFACIES ASSEMBLAGES

Assemblage	Description	Interpretation
A	Mainly very fine-grained sandstone, marine fish and invertebrates	estuarine to shallow marine
B	Mudstone, siltstone, marine microplankton	prodeltaic
C	Fine-grained glauconitic sandstone	marine shoreline
D	Mudstone, siltstone, thin sandstones, thin coals, coarsening-upward sequences	distal delta front
E	Mudstone, siltstone, fine to very fine sandstone, coal rare to abundant, coarsening-upward sequences	proximal delta front
F	Very fine to very coarse sandstone, mudstone, siltstone, rare conglomerate, crossbedding common, including lateral accretion sets, fining-upward sequences	high sinuosity fluvial
G	Fine to coarse crossbedded sandstone, minor conglomerate, siltstone, mudstone, coal, fining-upward sequences locally present	low sinuosity fluvial (Platte- and Donjek-type)
H	Fine to medium sandstone predominant, plane lamination common	flood deposits of ephemeral streams (Bijou Creek-type)
J	Conglomerate, breccia, minor sandstone, siltstone	proximal alluvial fan (Scott-type braided)
K	Thin bedded fine clastics, well-preserved plants	distal alluvial plain and lacustrine

Summarized from Miall (1981, 1984b)

ward for 100 km to Cañon Fiord. To the northwest there is a major facies change into fluvial deposits, as characterized by the section at Remus Creek (section 4 in Fig. 4). The detailed stratigraphy of this facies change is not known because of poor exposure, and the presence of a major re-

verse fault, along which the amount of shortening and/or lateral displacement has yet to be determined (Vesle Fiord thrust; see discussion below).

The youngest sediments in Remus Basin are fluvial deposits constituting a major coarsening-upward sequence at

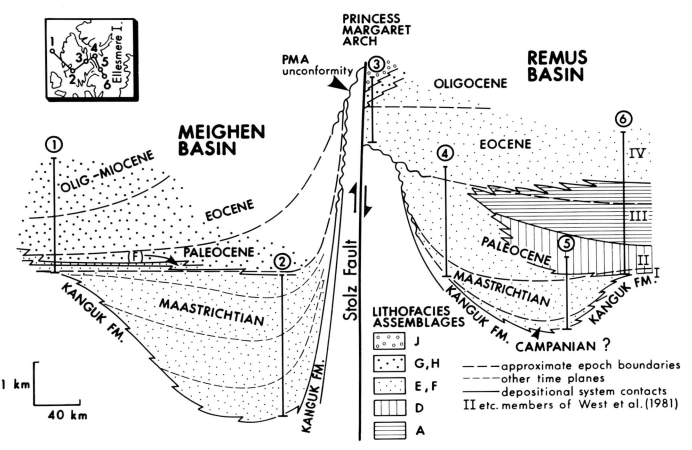

FIG. 4.—Restored stratigraphic cross section, Meighen and Remus Basins. Localities: 1, Crocker I-53 well, Meighen Island; 2, Strand Fiord; 3, Mokka Fiord; 4, Remus Creek; 5, Vesle Fiord; 6, Strathcona Fiord. Lithofacies assemblages are given in Table 2. See Miall (1984a) for data sources.

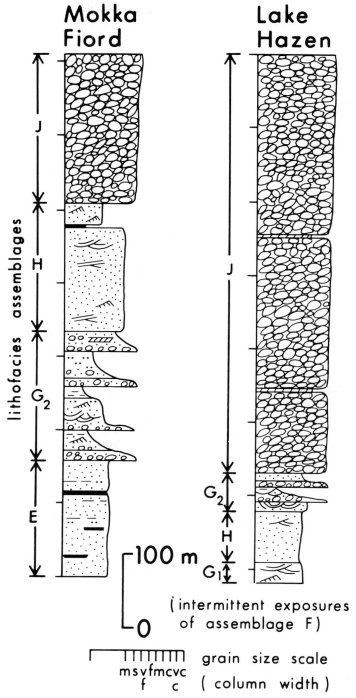

FIG. 5.—Coarsening-upward megasequences at Mokka Fiord and on the northwest flank of Lake Hazen Basin. Legend as in Fig. 3 (from Miall, 1984b).

Mokka Fiord, flanking the Stolz fault (section 3 in Fig. 4; Fig. 5). The sequence is interpreted as syn-tectonic, formed during the uplift of Princess Margaret Arch and a major episode of displacement on the Stolz fault.

The Eureka Sound Formation of Meighen Basin has not been stratigraphically subdivided. At Strand Fiord (section 2 in Fig. 4), 3.2 km of sediment is preserved, all assigned to facies assemblage E. Paleocurrent data (discussed below) indicate a westerly to northerly progradation of this clastic wedge.

The two other major Cenozoic basins within the Eurekan Orogen are at Lake Hazen and Judge Daly Promontory (Fig. 2). The succession at Lake Hazen is similar in lithology, thickness, and age to that at Mokka Fiord (Fig. 5, Table 1). That at Judge Daly Promontory is also capped by fanglomerates of assemblage J. There, however, the conglomerates rest unconformably on a fluvial-lacustrine sequence, and the entire basin-fill is Paleocene in age (Table 1). At least 2,400 m of Eureka Sound sediments are preserved in Judge Daly Basin, with neither the top nor the base of the succession exposed.

PALEOCURRENT ANALYSIS

The deltaic and fluvial lithofacies contain numerous hydrodynamic sedimentary structures, including large-scale trough and planar crossbedding, ripple marks and primary current lineation. Clast imbrication is well developed in many of the conglomerates. A total of 278 orientation measurements were made on these structures in Meighen and Remus Basins, most of them on planar crossbed sets ranging up to 1 m in thickness. The direction of indicated paleoflow was recorded for each structure and, where necessary, corrected for structural dip. Owing to a lack of detailed structural data no corrections for fold plunge were attempted, and it is not known whether orientations have been affected by horizontal rotation.

The data have been grouped for each locality in order to investigate regional, long-term paleoflow patterns. Summary statistics are given in Table 3, calculated using the methods of Curray (1956), and indicated orientations are shown in Figure 6.

Additional paleocurrent data from Lake Hazen and Judge Daly Basins have been reported elsewhere (Miall, 1979b, 1982). All this information is incorporated into the paleogeographic reconstructions set out below.

PETROGRAPHIC ANALYSIS

Point count analyses of selected thin sections are shown in Figure 7. Sandstones from Meighen and Remus Basins are rather mature, most falling in the quartzose arenite field. As noted by Miall (1981, p. 254), most of the detritus in these basins is probably recycled from Carboniferous to Cretaceous sediments of the Sverdrup Basin, and older Paleozoic rocks exposed along the east flank by Remus Basin. A study of palynomorph assemblages in these rocks by G. Norris (personal commun., 1984) led to a similar conclusion—indigenous Cenozoic species commonly are swamped by derived Cretaceous forms. There are few indications of primary igneous or metamorphic material in the Eureka Sound sediments of Meighen and Remus Basins.

These interpretations are supported by the paleocurrent data (Fig. 6), which indicate source directions mainly from within the Eurekan Orogen. Princess Margaret Arch (Fig. 2), occupying the center of what is now Axel Heiberg Island, exposes thick sequences of Triassic and Jurassic sed-

TABLE 3.—PALEOCURRENT DATA

Location Name	No.	Lithofacies Assemblage	n	Structure Type	$\bar{\theta}$	L	P
Strand Fiord	1	E	60	T,P	289	77	3.6×10^{-16}
Expedition Fiord	2	E	32	P	183	87	3.0×10^{-11}
Expedition Fiord	3	E	9	P	350	90	6.8×10^{-4}
Wolf Valley	4	G	47	T,P	224	39	7.8×10^{-4}
Mokka Fiord	5	G	40	P,I	081	77	5.0×10^{-11}
Remus Creek	6	F	5	T,P	060	56	2.1×10^{-1}
Eureka Sound	7	G	26	P	182	88	1.8×10^{-9}
Vesle Fiord	8	E	12	T,P	243	70	2.8×10^{-3}
South Bay	9	G,E	15	T,P	285	46	4.2×10^{-1}
Slidre Anticline	10	G	24	P	218	71	5.6×10^{-6}
Strathcona Fiord	11	E	9	P	033	70	1.0×10^{-2}

n = number of readings, $\bar{\theta}$ = vector mean azimuth, L = vector strength, p = probability of randomness, Structure type: T = trough crossbedding, P = planar crossbedding, I = conglomerate clast imbrication. Location numbers are those shown in Fig. 6

iments. It probably was a major sediment source for Meighen Basin and for the western part of Remus Basin. This is undoubtedly the case at Mokka Fiord where fanglomerates capping the Eureka Sound Formation are clearly locally derived. Other sediments within Remus Basin were probably derived from areas to the north and northeast, including Grantland Uplift (Fig. 2).

Sandstones in Lake Hazen Basin fall into two distinct groups (Fig. 7). Quartzose arenites were derived from Lower Paleozoic rocks to the south of the basin, as indicated by paleocurrent data (Miall, 1979b), whereas the lithic arenites are a younger suite of sandstones collected from the middle part of the coarsening-upward sequence shown in Figure 5, and were derived from Sverdrup Basin sediments and basic

intrusions exposed in the Grantland Uplift to the northwest of the basin. Paleocurrent directions underwent a near reversal in the upper part of the Eureka Sound succession, at the same time as the detrital petrography shows a sharp decrease in maturity.

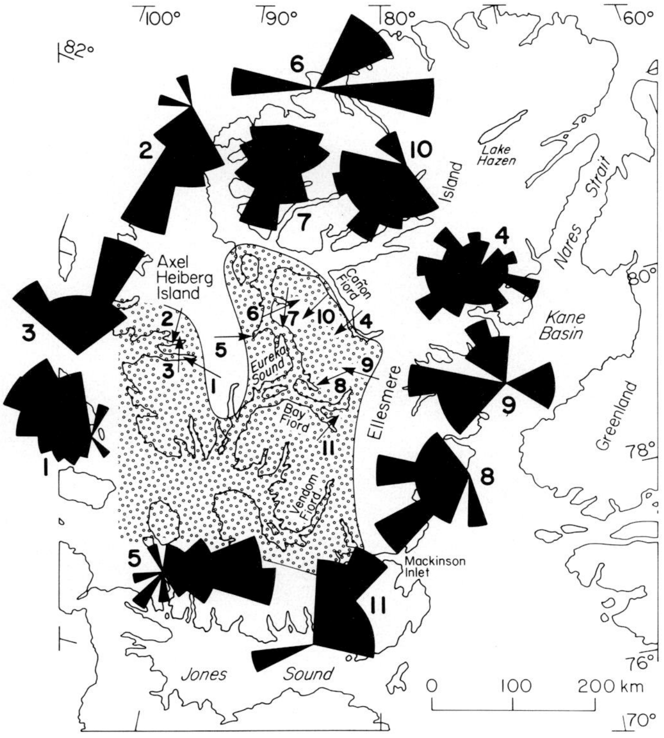

TECTONIC AND PALEOGEOGRAPHIC EVOLUTION

Campanian-Maastrichtian

Sea-floor spreading commenced in Labrador Sea at about anomaly 34 time (Campanian). Initially, spreading occurred about a pole located in the central Canadian Arctic, and produced about 10° of angular displacement (Fig. 1, Table 1). Regional crustal shortening in the northeast Arctic Islands is predicted from this kinematic pattern, which agrees well with the events of phase 1 of the Eurekan Orogeny, as defined earlier by Balkwill (1978). The Sverdrup Basin was broken up into a series of smaller depocenters and intervening uplifts from which prograded the first coarse clastic sediments of the Eureka Sound Formation.

In the northeast Arctic, Princess Margaret Arch, Grantland Uplift, and the craton to the west of Remus Basin were uplifted. The structural style of these uplifts is unknown,

FIG. 6.—Paleocurrent data, showing vector mean azimuth arrows and current rose diagrams. Station numbers correspond to location numbers in Table 3.

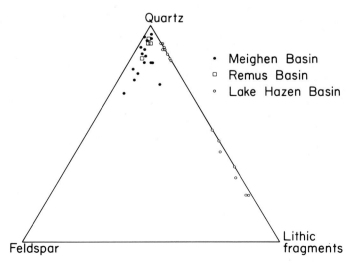

FIG. 7.—Detrital composition of sandstones from three basins in the northeast Arctic.

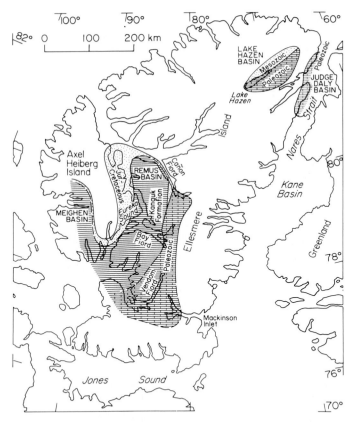

FIG. 8.—Subcrop map of rocks occurring beneath the Eureka Sound Formation.

but may involve deep-seated, reverse faults (Balkwill and Bustin, 1980; Miall, 1981; Balkwill, 1983). The Eureka Sound Formation onlaps Jurassic strata on the east flank of Princess Margaret Arch, and rests on lower Paleozoic rocks on the east side of Remus Basin (Fig. 8). Prior to Eocene time, however, Eureka Sound sedimentation was confined to the center of Meighen and Remus Basins (Fig. 9), where the formation rests conformably on mudstones and siltstones of the Kanguk Formation (Upper Cretaceous).

A broad fluvial plain filled the north end of Remus Basin, fed by sources to the west, north, and northeast. Downstream, this graded into a delta complex which, at first, may not have extended much farther south than Strathcona Fiord. On the west side of Princess Margaret Arch, a similar fluvial-deltaic complex prograded west and southwest into the Strand Fiord-Expedition Fiord area (Fig. 9). Cornwall Arch was probably undergoing uplift and erosion at this time (Balkwill, 1983), and defined the western margin of Meighen Basin.

Paleocene

In Meighen Basin, fluvial-deltaic progradation probably continued as before, with Princess Margaret Arch continuing to supply abundant sand, silt, and mud detritus (Fig. 10).

In Remus Basin the main depocenter shifted northward to the area of Remus Creek, where a fluvial-deltaic coastal

plain complex continued to accumulate. This resulted in a transgressive (retrograding), distal deltaic sequence (assemblage D) occupying most of the southern and eastern parts of Remus Basin (Fig. 10). At Strathcona Fiord this corresponds to member II of the Eureka Sound Formation. Toward the end of the Paleocene, the transgression had reached its maximum extent, with the basin occupied by a low-energy marine gulf or estuary, within which the fine-grained deposits of assemblage A (member III) were deposited. Presumably sediment sources to the north, and possibly to the west and east of Remus Basin, were continuing to supply sediment, but the transgressive nature of members II and III would seem to suggest a lull in source area rejuvenation. A possible reason for this is noted below.

Judge Daly Basin was initiated during the Paleocene, presumably as part of the response to Eurekan phase 1 transpressive movements. Coarse fluvial sediments of assemblage G were deposited by rivers flowing along the axis of the basin toward the northeast. These gradually became

LEGEND (for figures 9-12)

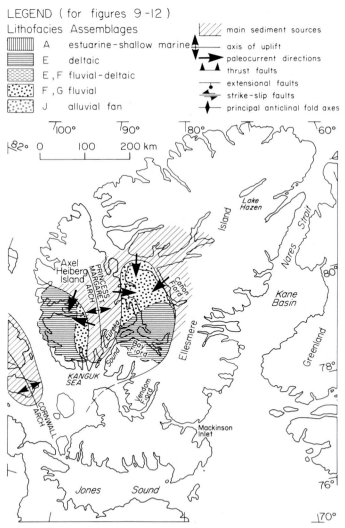

FIG. 9.—Paleogeography of the northeastern Arctic during Maastrichtian time.

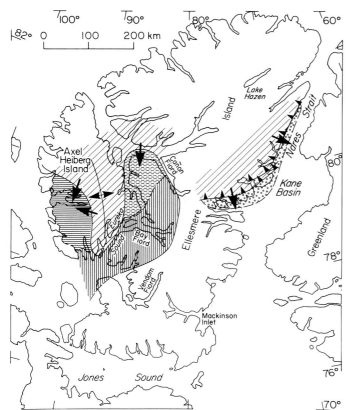

FIG. 10.—Paleogeography of the northeastern Arctic during Paleocene time. Legend given in Fig. 9.

more sluggish, and the beds become finer-grained upward, passing into a lacustrine facies (assemblage K).

Phase 1 of the Eurekan Orogeny culminated in uplift and localized thrust faulting in eastern Ellesmere Island (Parrish Glacier, Rawlings Bay, and Cape Back thrusts of Mayr and de Vries, 1982). Thick fanglomerate wedges (assemblage J) flank these faults and are cut by them, suggesting that the alluvial fans were overridden by their own source rocks at the close of this tectonic episode. Localization of Eurekan strain along the thrust faults may have been accompanied by cessation of compression to the west, permitting subsidence and transgression in Remus Basin. However, the precise age relationships of these events are not yet known.

Eocene

At anomaly 25 time (56 Ma) plate trajectories underwent a major change, and until anomaly 21 time (48 Ma) the sense of motion between Greenland and Canada was one of left-lateral oblique slip. There was a gradual shift from transtension to transpression.

The sea-floor spreading model of Srivastava and Tapscott (in press) indicates that 170 km of left-lateral offset accumulated between the Greenland and North American plates during the anomaly 25 to 21 interval. This is an average displacement rate of only 2.1 cm/yr, which compares with

an average post-Oligocene displacement rate of 4 cm/yr deduced for the San Andreas transform system (Atwater and Molnar, 1973; Crowell, 1979). Shear may have been distributed across a belt at least 100 km wide, and possibly double this width (Miall, 1983, 1984a). Localized strike-slip fault movement develops a distinctive style of sedimentary basin (Ballance and Reading, 1980; other papers in this volume), as exemplified by Ridge Basin, California (Crowell and Link, 1982) and Hornelen Basin, Norway (Steel and Gloppen, 1980). The typical stratigraphic patterns do not appear to be present in the northeast Arctic, possibly because of the absence of a single master fault, and because of the slow rate of movement.

The presence of a cumulative 170 km of left-lateral offset within the Eurekan fold belt should be detectable from a detailed structural analysis. The required work has not yet been done, although there are some preliminary data that bear on this point, as discussed below.

Most of the major structures in the fold belt (e.g., see description by Okulitch, 1982; Osadetz, 1982; Higgins and Soper, 1983) were probably developed toward the end of phase 2 of the Eurekan Orogeny, when the sense of motion between Greenland and Canada was more nearly orthogonal compression (Fig. 1, Table 1). Such tectonism would have overprinted any older structures formed during the episode of oblique slip, rendering them more difficult to recognize.

Okulitch (1982) made several transects across the fold belt, but he focussed on a documentation of crustal shortening and structural style perpendicular to the structural grain and did not discuss the possibility of strike-slip faulting. Wynne et al. (1983) determined that Permian volcanic rocks in northern Ellesmere Island had been rotated 36 ± 10° anticlockwise, and Miall (1983) suggested that this may have been caused partly by local fault-block rotation or shear during the Eocene interval of oblique slip.

An examination of 1:250,000 geological maps of the central Ellesmere Island area (particularly Geological Survey of Canada maps 1300A, 1302A, 1307A and 1308A by R. Thorsteinsson) suggests several structures that may have originated by oblique-slip tectonism during the Eocene. These are indicated in Figure 11. They include steep, straight, or braided faults, oriented north-northeast and which may be synthetic strike-slip faults; and normal faults, oriented north-northwest. According to the analysis of Harding (1974), this is a fault pattern that would be expected to result from the imposition of a northeast-southwest shear couple. A proposed fault along Cañon Fiord, that bounds Remus Basin to the northeast, is suggested by the abrupt truncation of the basin on the south side of the fiord. The Vesle Fiord thrust appears to mark a boundary between outcrops of fluvial-deltaic facies to the northwest and distal-deltaic to marine facies to the south and east (see particularly the Paleocene map, Fig. 10). Whether this facies boundary has been affected by strike slip along the fault remains to be determined. Unfortunately, outcrops of the Eureka Sound Formation are rather sparse near the fault.

Mayr and de Vries (1982) reported evidence of strike-slip faulting in Judge Daly Promontory, indicating a shear couple oriented approximately parallel to Nares Strait. It is

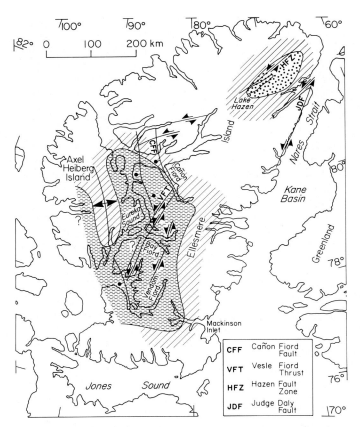

FIG. 11.—Paleogeography and hypothesized active faults during early to middle Eocene time. Legend given in Fig. 9.

Oligocene

Phase 2 of the Eurekan Orogeny culminated in the latest Eocene and earliest Oligocene. During this time (anomalies 21 to 13; 48–36 Ma), the rotation pole changed so that Greenland began to move northwestward toward Canada (Srivastava and Tapscott, in press). This resulted in near-orthogonal shortening, and most of the folds, reverse faults, and thrust faults that make up the Eurekan fold belt (Balkwill, 1978; Okulitch, 1982; Osadetz, 1982) are thought to have developed at this time. Over much of Ellesmere Island the stress field predicted from plate kinematics was not precisely perpendicular to what are now Eurekan structural trends (Fig. 1), trends that may have been initiated during the phase of oblique slip. This slight obliquity would have resolved into a minor dextral shear along Eurekan trends and could have led to small dextral offsets accompanying compression. This predictable pattern has in fact been observed on the Hazen fault zone (Higgins and Soper, 1983). Miall (1982, p. 4) stated that fold patterns in Eureka Sound sediments of Judge Daly Basin indicate minor dextral wrench movement, and this observation can also be related to the stress pattern that developed at the close of phase 2 movement.

A structural-stratigraphic model of useful predictive value is that in many fold belts major episodes of crustal shortening are documented by coarse clastic wedges that prograded from fault-bounded uplifts. Of all the major structures that are thought to have developed during this final

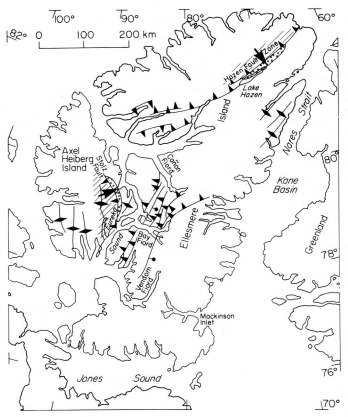

FIG. 12.—Paleogeography and major structures generated during latest Eocene to earliest Oligocene time. Legend given in Fig. 9.

suggested that this deformation was probably part of the same Eocene episode of oblique slip.

The paleogeography of the project area during the Eocene is shown in Figure 11. Remus Basin became much enlarged, with sediments onlapping the flanks of Princess Margaret Arch to the west, the cratonic Paleozoic rocks to the east, and extending south to the Vendom Fiord-Mackinson Inlet area. The marine embayment was filled by a widespread deltaic regression, forming member IV in the Strathcona Fiord area (West et al., 1981). This is a succession of sandstone and mudstone with numerous coal seams and a rich vertebrate fauna. Very little paleocurrent evidence is available from these rocks, and it is unclear where the principal sediment sources were located. It is possible that multiple sources were rejuvenated by the oblique-slip tectonics, although these cannot be identified because of the absence of clearly related facies changes (such as alluvial-fan conglomerates banked against active faults). Rocks of Eocene age are not preserved in the Strand Fiord-Expedition Fiord area of Meighen Basin, and so it is not known whether sedimentation continued on the west side of Princess Margaret Arch. Lake Hazen Basin became active during the Eocene, and was filled by high-sinuosity streams depositing point bars and floodplain deposits of lithofacies assemblage F. Paleocurrent directions were toward the north, oblique to the basin trend.

phase of movement, two have preserved some stratigraphic evidence of tectonism. The Hazen fault zone and the Stolz fault are flanked by thick wedges of boulder conglomerate deposited on alluvial fans (Fig. 12). These conglomerates cap coarsening-upward megasequences (Fig. 5), the sedimentological interpretation of which has been discussed in detail elsewhere (Miall, 1984b). Both sequences have been dated tentatively as late Eocene or Oligocene (W. S. Hopkins, Jr., in Miall, 1979b; Sepulveda and Norris, 1982), which provides an important confirmation of the age relationship between regional plate tectonic patterns and local sedimentary-tectonic events.

Following the culmination of phase 2, the Arctic Islands were uplifted and the main locus of sedimentation shifted to the offshore continental margin wedges (Miall, 1984a).

DISCUSSION

It is now a standard procedure to use a documented history of sea-floor spreading kinematics as a basis for interpreting the geology of adjacent continental areas. Structural styles may be predicted from plate trajectories, and stratigraphic styles depend largely on the processes occurring at nearby plate boundaries. These data provide the basis for a suite of basin models that can be used to infer plate tectonic history even in the absence of a sea-floor spreading record (Miall, 1984c).

In the present case, sea-floor spreading kinematics predict an oblique-slip model for the Eurekan Orogen. At present, the regional tectonic framework is better understood than the local structural geology, and it is to be hoped that the kinematic model will provide a basis for a detailed structural field investigation of this deformed belt. The kinematic model also yields predictions about the timing and style of stratigraphic processes, many of which have now been documented in the stratigraphic record. Episodes of transpression and orthogonal shortening were marked by deltaic progradation, culminating in the development of fanglomerate wedges flanking rising fault-bounded uplifts. An episode of transtension was marked by local deltaic retrogression and a marine transgression.

The Eurekan Orogeny may be considered as an example of an intraplate orogeny, because the Greenland and North American continents probably were in contact at all times. The structure and stratigraphy of the project area do not conform to existing models for oblique-slip mobile zones (Ballance and Reading, 1980; Miall, 1984c). That the plate trajectories changed several times during the orogeny is certainly not unusual plate boundary behavior, and the almost infinite variety of possible drift patterns that can occur across an oblique-slip mobile zone means that it may never be possible to describe the geology of such zones in terms of a few simple models.

ACKNOWLEDGMENTS

Research in Arctic Cenozoic sedimentation and tectonics has been supported by operating grants and a Strategic Grant in Energy awarded by the Natural Sciences and Engineering Research Council. Logistic support for fieldwork during many field seasons was provided by the Geological Survey of Canada and the Polar Continental Shelf Project.

Grateful acknowledgments are extended to S. P. Srivastava and C. R. Tapscott for permission to quote from their unpublished study of plate kinematics. R. Jackson, G. Norris, and J. R. H. McWhae are thanked for helpful discussions.

This paper has benefitted considerably from critical reading by K. Biddle, N. Christie-Blick, K. O. Stanley, and F. L. Wehr.

REFERENCES

ATWATER, T., AND MOLNAR, P., 1973, Relative motion of the Pacific and North American plates deduced from sea-floor spreading in the Atlantic, Indian and South Pacific Oceans, *in* Kovach, R. L. and Nur, A., eds., Proceedings of Conference on Tectonic Problems of the San Andreas Fault System: Stanford University Publications in Geological Science, v. 13, p. 126–148.

BALKWILL, H. R., 1978, Evolution of Sverdrup Basin, Arctic Canada: American Association of Petroleum Geologists Bulletin, v. 62, p. 1004–1028.

———, 1983, Geology of Amund Ringnes, Cornwall and Haig-Thomas Islands, District of Franklin: Geological Survey of Canada Memoir 390, 76 p.

BALKWILL, H. R., AND BUSTIN, R. M., 1980, Late Phanerozoic structures, Canadian Arctic Archipelago: Palaeogeography, Palaeoclimatology, Palaeoecology, v. 30, p. 219–227.

BALLANCE, P. F., AND READING, H. G., eds., 1980, Sedimentation in Oblique-Slip Mobile Zones: International Association of Sedimentologists Special Publication No. 4, 265 p.

BULLARD, E., EVERETT, J. E., AND SMITH, A. G., 1965, The fit of continents around the Atlantic: Philosophical Transactions, Royal Society of London, Series A, v. 258, p. 41–51.

BUSTIN, R. M., 1982, Beaufort Formation, eastern Axel Heiberg Island, Canadian Arctic Archipelago: Bulletin of Canadian Petroleum Geology, v. 30, p. 140–149.

CROWELL, J. C., 1979, The San Andreas fault system through time: Journal of Geological Society of London, v. 136, p. 293–302.

CROWELL, J. C., AND LINK, M. H., eds., 1982, Geologic History of Ridge Basin, Southern California: Society of Economic Paleontologists and Mineralogists, Pacific Section, 304 p.

CURRAY, J. R., 1956, The analysis of two-dimensional orientation data: Journal of Geology, v. 64, p. 117–131.

DAWES, P. R., AND KERR, J. W., eds., 1982, Nares Strait and the drift of Greenland: a conflict in plate tectonics: Meddelelser om Grønland 8, 392 p.

HARDING, T. P., 1974, Petroleum traps associated with wrench faults: American Association of Petroleum Geologists Bulletin, v. 58, p. 1290–1304.

HARLAND, W. B., COX, A. V., LLEWELLYN, P. G., PICKTON, C. A. G., SMITH, A. G., AND WALTERS, R., 1982, A Geologic Time Scale: Cambridge, Cambridge University Press, 131 p.

HIGGINS, A. K., AND SOPER, N. J., 1983, The Lake Hazen fault zone, Ellesmere Island: a transpressional upthrust?: Geological Survey of Canada Paper 83-1B, p. 215–221.

HUGON, H., 1983, Ellesmere-Greenland fold belt: structural evidence for left-lateral shearing: Tectonophysics, v.100, p. 215–225.

MAYR, U., AND DE VRIES, C. D. S., 1982, Reconnaissance of Tertiary structures along Nares Strait, Ellesmere Island, Canadian Arctic Archipelago, *in* Dawes, P. R., and Kerr, J. W., eds., Nares Strait and the Drift of Greenland: A Conflict in Plate Tectonics: Meddelelser om Grønland 8, p. 167–175.

McWHAE, J. R. H., 1981, Structure and spreading history of the northwestern Atlantic region from the Scotian Shelf to Baffin Bay, *in* Kerr, J. W., and Fergusson, A. J., eds., Geology of the North Atlantic Borderlands: Canadian Society of Petroleum Geologists, Memoir 7, p. 299–332.

MIALL, A. D., 1979a, Mesozoic and Tertiary geology of Banks Island, Arctic Canada: the history of an unstable craton margin: Geological Survey of Canada Memoir 387, 235 p.

———, 1979b, Tertiary fluvial sediments in the Lake Hazen intermontane basin, Arctic Canada: Geological Survey of Canada Paper 79-9, 25 p.

———, 1981, Late Cretaceous and Paleogene sedimentation and tectonics in the Canadian Arctic Islands, *in* Miall, A. D., ed., Sedimentation and Tectonics in Alluvial Basins: Geological Association of Canada Special Paper 23, p. 221–272.

———, 1982, Tertiary sedimentation and tectonics in the Judge Daly Basin, northeast Ellesmere Island, Arctic Canada: Geological Survey of Canada Paper 80-30, 17 p.

———, 1983, The Nares Strait problem: a re-evaluation of the geological evidence in terms of a diffuse, oblique-slip plate boundary between Greenland and the Canadian Arctic Islands: Tectonophysics, v. 100, p. 227–239.

———, 1984a, Sedimentation and tectonics of a diffuse plate boundary: the Canadian Arctic Islands from 80 Ma B.P. to the present: Tectonophysics, v. 107, p. 261–277.

———, 1984b, Variations in fluvial style in the early Cenozoic synorogenic sediments of the Canadian Arctic Islands: Sedimentary Geology, v. 38, p. 499–523.

———, 1984c, Principles of Sedimentary Basin Analysis: New York, Springer-Verlag, 490 p.

———, in press, The Eureka Sound Group (Upper Cretaceous-Oligocene), Canadian Arctic Islands: Bulletin of Canadian Petroleum Geology, v. 34.

OKULITCH, A. V., 1982, Preliminary structure sections, southern Ellesmere Island, District of Franklin: Geological Survey of Canada Paper 82-1A, p. 55-62.

OSADETZ, K. G., 1982, Eurekan structure of the Ekblaw Lake area, Ellesmere Island, Canada, *in* Embry, A. F., and Balkwill, H. R., eds., Arctic Geology and Geophysics: Canadian Society of Petroleum Geologists, Memoir 8, p. 219–232.

READING, H. G., 1980, Characteristics and recognition of strike-slip fault systems, *in* Ballance, P. F., and Reading, H. G., eds., Sedimentation in Oblique-Slip Mobile Zones: International Association of Sedimen-

tologists Special Publication No. 4, p. 7–26.

SEPULVEDA, E. G., AND NORRIS, G., 1982, A comparison of Paleogene fungal spores from northern Canada and Patagonia, Argentina: Armeghiniana, v. 19, p. 319–334.

SRIVASTAVA, S. P., AND FALCONER, R. K. H., 1982, Nares Strait: a conflict between plate tectonic predictions and geological interpretation, *in* Dawes, P. R., and Kerr, J. W., eds., Nares Strait and the Drift of Greenland: A Conflict in Plate Tectonics: Meddelelser om Grønland 8, p. 339–352.

SRIVASTAVA, S. P., AND TAPSCOTT, C. R., in press, Plate kinematics in the North Atlantic, *in* Geology of the western North Atlantic: Geological Society of America, Decade of North American Geology, v. M.

STEEL, R. J., AND GLOPPEN, T. G., 1980, Late Caledonian (Devonian) basin formation, western Norway: signs of strike-slip tectonics during infilling, *in* Ballance, P. F., and Reading, H. G., eds., Sedimentation in Oblique-Slip Mobile Zones: International Association of Sedimentologists Special Publication No. 4, p. 79–104.

VOGT P. R., TAYLOR, P. T., KOVACS, L. C., AND JOHNSON, G. L., 1982, The Canada Basin: aeromagnetic constraints on structure and evolution: Tectonophysics, v. 89, p. 295–336.

WEST, R. M., DAWSON, M. R., HICKEY, L. J., AND MIALL, A. D., 1981, Upper Cretaceous and Paleogene sedimentary rocks, eastern Canadian Arctic and related North Atlantic areas, *in* Kerr, J. W., and Ferguson, A. J., eds., Geology of the North Atlantic borderlands: Canadian Society of Petroleum Geologists, Memoir 7, p. 279–298.

WYNNE, P. J., IRVING, E., AND OSADETZ, K., 1983, Paleomagnetism of the Esayoo Formation (Permian) of northern Ellesmere Island: possible clue to the solution of the Nares Strait Dilemma: Tectonophysics, v. 100, p. 241–256.

VAN BERKEL, J. T., HUGON, H., SCHWERDTNER, W. M., AND BOUCHEZ, J. L., 1983, Study of anticlines, faults and diapirs in the central Eureka Sound Fold Belt: Bulletin of Canadian Petroleum Geology, v. 31, p. 109–116.

GLOSSARY—STRIKE-SLIP DEFORMATION, BASIN FORMATION, AND SEDIMENTATION[1]

KEVIN T. BIDDLE

Exxon Production Research Company, P. O. Box 2189, Houston, Texas 77252-2189;

AND

NICHOLAS CHRISTIE-BLICK

Department of Geological Sciences and Lamont-Doherty Geological Observatory of Columbia University, Palisades, New York 10964

INTRODUCTION

Many of the geological terms having to do with strike-slip deformation, basin formation, and sedimentation are used in a variety of ways by different authors (e.g., pull-apart basin), or they are synonymous with other words (e.g., left-lateral, sinistral). Rather than enforcing a rigorously uniform terminology in this book, we decided to set down our preferred definitions in a glossary, and where appropriate to indicate alternative usage. In selecting terms for definition, we have tried to steer a course between being overly encyclopedic and providing a list useful to those having little familiarity with the geology of strike-slip basins, especially those described in this volume. Some words (e.g., cycle) have additional meanings in the geological sciences not included here, and this glossary should therefore be used in the context of strike-slip basins. The references cited are those from which we obtained definitions, or which illustrate the concept embodied by a particular term. We have not attempted to provide original references for every term, especially for those long used in the geological literature.

THE GLOSSARY

Anastomosing—Pertaining to a network of branching and rejoining surfaces or surface traces. Commonly used to describe braided fault systems.

Antithetic fault—Originally defined by H. Cloos (1928, 1936) to describe faults that dip in a direction opposite to the dip of the rocks displaced, and that rotate fault-bounded blocks so that the net slip on each fault is greater than it would be without rotation (Dennis, 1967, p. 3). Many authors now use the term to describe faults that (1) are subsidiary to a major fault and have less displacement than that fault, (2) formed in the same stress regime as the major fault with which they are associated, (3) are oriented at a high angle to the major fault (in map view for strike-slip faults, in cross-sectional view for normal faults), and (4) for strike-slip faults, have a sense of displacement opposite that of the major fault, or for normal faults, dip in the opposite direction. Antithetic strike-up faults compose the R′ set of Reidel shears formed in simple shear (Fig. 1).

Aulacogen—A term introduced by Shatski (1946a, b) to describe narrow, elongate sedimentary basins that extend into cratons from either a geosyncline or a mountain belt that formed from a geosyncline (for a discussion of genesis in terms of plate tectonics, see Hoffman et al., 1974).

---

[1]Lamont-Doherty Geological Observatory Contribution No. 3913.

Basin—(1) A site of pronounced sediment accumulation; (2) a relatively thick accumulation of sedimentary rock (for a discussion of the history and usage of the word, see Dennis, 1967, p. 9; Bates and Jackson, 1980, p. 55).

Bubnoff curve—A plot of subsidence versus time (Fischer, 1974).

Bubnoff unit—A standard measure of geologic rates, such as subsidence rates, defined as 1 m/m.y. (Fischer, 1969; Bates and Jackson, 1980, p. 84).

Burial history curve—A plot, for a given location, of the cumulative thickness of sediments overlying a surface versus time (Philipp, 1961; van Hinte, 1978).

Closing bend—See restraining bend.

Compaction—(1) The reduction in bulk volume or thickness of, or the pore space within, a body of sediment in response to the increasing weight of a superimposed load; (2) the physical process by which fine-grained sediment is converted to consolidated rock (modified from Bates and Jackson, 1980, p. 127).

Compression—(1) A system of stresses that tends to shorten or decrease the volume of a substance (preferred definition, modified from Bates and Jackson, 1980, p. 130). Uniaxial compression involves one nonzero principal stress, which is compressive; in triaxial compression, all three principal stresses are nonzero (Means, 1976, p. 80). It is also possible for a compressive principal stress to occur with one or more tensile principal stresses. (2) A state of strain in which material lines become shorter under compressive stress (J. T. Engelder, personal commun., 1985; Aydin and Nur, 1985 this volume; see contraction).

Compressional bend—See restraining bend.

Compressional overstep—See restraining overstep.

Conjugate Riedel shear—Synonymous with R′ Riedel or antithetic shear (Fig. 1). See Riedel shear, synthetic and antithetic faults.

Contraction—A strain involving (1) a reduction in volume (e.g., thermal contraction), or (2) a reduction of length (e.g., contraction fault of Norris, 1958; McClay, 1981). Contraction has been gaining popularity as the general strain term associated with compressive stress, much like the relationship between the stress term tension and the strain

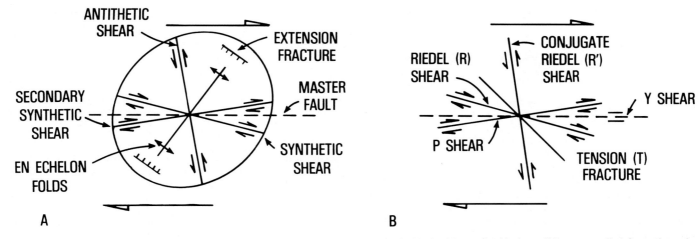

FIG. 1.—The angular relations between structures that tend to form in right-lateral simple shear under ideal conditions, compiled from clay-cake models and from geological examples. Arrangements for structures along left-slip faults may be determined by viewing the figure in reverse image. A) Terminology largely from Wilcox et al. (1973), superimposed on a strain ellipse for the overall deformation. B) Riedel shear terminology, modified from Tchalenko and Ambraseys (1970) and Bartlett et al. (1981). See glossary for definitions of terms.

term extension. The word shortening, as used by Hobbs et al. (1976, p. 27), may be a better choice for general use because it does not imply a volume change.

Convergent bend—A bend in a strike-slip fault that results in overall crustal shortening in the vicinity of the bend (synonymous with restraining bend of Crowell, 1974a).

Convergent overstep—See restraining overstep.

Convergent (transpressional) strike-slip or wrench fault—A strike-slip or wrench fault along which strike-slip deformation is accompanied by a component of shortening transverse to the fault (Wilcox et al., 1973).

Cycle—(1) An interval of time during which one series of recurrent events is completed (preferred definition); (2) a sequence of sediment or rock units repeated in a succession (for additional discussion, see Bates and Jackson, 1980, p. 156).

Depositional sequence—A stratigraphic unit composed of a relatively conformable succession of genetically related strata and bounded at its top and base by unconformities or their correlative conformities (Mitchum, 1977, p. 206).

Dextral—Pertaining to the right (e.g., dextral slip is right slip).

Dip—The acute angle between an inclined surface and the horizontal, measured in a vertical plane perpendicular to strike.

Dip separation—Separation measured parallel to the dip of a fault (modified from Crowell, 1959, p. 2662; Bates and Jackson, 1980, p. 177). See separation.

Dip slip—The component of slip measured parallel to the dip of a fault (Crowell, 1959, p. 2655; Bates and Jackson, 1980, p. 177). See slip.

Dip-slip fault—A fault along which most of the displacement is accomplished by dip slip (modified from Bates and Jackson, 1980, p. 177).

Divergent bend—A bend in a strike-slip fault that results in overall crustal extension in the vicinity of the bend (synonymous with releasing bend of Crowell, 1974a).

Divergent overstep—See releasing overstep.

Divergent (transtensional) strike-slip or wrench fault—A strike-slip or wrench fault along which strike-slip deformation is accompanied by a component of extension transverse to the fault (Wilcox et al., 1973; Harding et al., 1985 this volume).

Downlap—A base-discordant relation in which initially inclined strata terminate downdip against an initially horizontal or inclined surface (Mitchum, 1977, p. 206).

Drag fold—(1) A fold produced by movement along a fault (see normal drag and reverse drag). In this context, the term is somewhat misleading because folding commonly initiates before faulting (Hobbs et al., 1976, p. 306). (2) A minor fold formed in a less competent bed between more competent beds by movement of the competent beds in opposite directions relative to one another (Bates and Jackson, 1980, p. 186).

Drape fold—A fold in a sedimentary layer that conforms passively to the configuration of underlying structures (Friedman et al., 1976, p. 1049). A fold formed by differential compaction is an example of a drape fold.

Dynamic analysis—The study of kinematics and kinetics that relates strains to the evolution of stress fields.

Echelon—Step (e.g., echelon faults of Clayton, 1966, and of Segall and Pollard, 1980, meaning overstepping faults).

En echelon—(1) A stepped arrangement of relatively short, consistently overlapping or underlapping structural elements such as faults or folds that are approximately parallel to each other but oblique to the linear or relatively narrow zone in which they occur (preferred definition, modified from Campbell, 1958; Harding and Lowell, 1979; see Fig. 2). En echelon arrangements can occur in both map view and cross section (Shelton, 1984; Aydin and Nur, 1985 this volume). (2) Any stepped arrangement of two or more overlapping or underlapping structural elements such as faults or folds that are approximately parallel to each other and to the zone in which they occur, without reference to whether the sense of overstep is consistent or inconsistent (e.g., for strike-slip deformation, D. A. Rodgers, 1980; Aydin and Nur, 1985 this volume; for thrust and fold belts, J. Rodgers, 1963; Armstrong, 1968; Dahlstrom, 1970). See oblique, relay pattern.

Extension—A strain involving an increase in length.

Extension fault—A fault that results in lengthening of an arbitrary datum, commonly but not necessarily bedding (synonymous with one usage of normal fault; Suppe 1985, p. 269). The term may be applied to faults of any dip (Christie-Blick, 1983).

Extension fracture—A mode I crack, or one that shows no motion in the plane of the crack (Lawn and Wilshaw, 1975, p. 52; J. T. Engelder, personal commun., 1985). Extension fractures form when effective stresses are tensile (i.e., when pore-fluid pressure exceeds lithostatic pressure). Partly synonymous with T fracture of Tchalenko and Ambraseys (1970). See tension fracture. In strike-slip systems, extension fractures and tension fractures form in response to simple shear at about 45° to the master fault (Fig. 1).

Extensional bend—See releasing bend.

Extensional overstep—See releasing overstep.

External rotation—A change in the orientation of structural features during deformation with reference to coor-

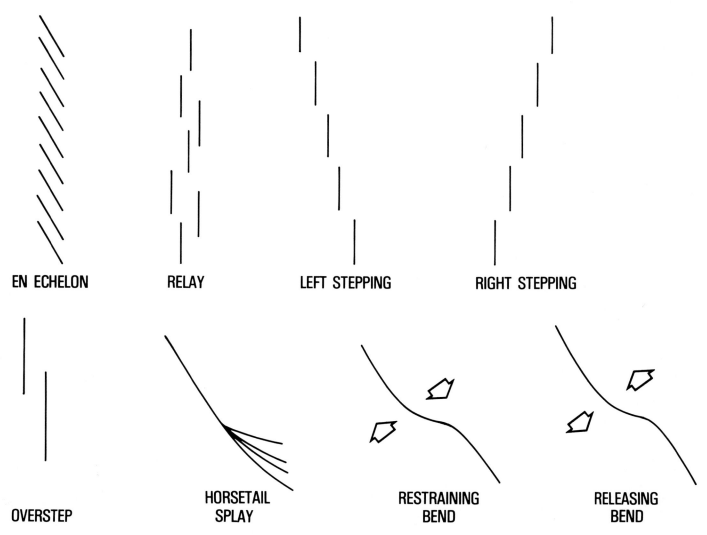

EN ECHELON RELAY LEFT STEPPING RIGHT STEPPING

OVERSTEP HORSETAIL SPLAY RESTRAINING BEND RELEASING BEND

Fig. 2.—Some simple structural patterns.

dinate axes external to the deformed body (Bates and Jackson, 1980, p. 218).

Facies—(1) Laterally or sequentially associated bodies of sediment or sedimentary rock distinguished on the basis of objective lithologic and paleontologic characteristics that reflect the processes and environments of deposition and/or diagenesis; (2) distinctive, adjacent, coeval rock units (White, 1980); (3) the lithologic and paleontologic characteristics of a body of sediment or sedimentary rock that reflect the processes and environments of deposition (original sense of Gressly, 1838). See Walker (1984) for further discussion and references to numerous reviews.

Fault-angle depression—A subsiding depression parallel to the trace of an oblique-slip fault (Ballance, 1980, p. 232).

Fault-flank depression—A depression between subsidiary folds of a strike-slip fault system (Crowell, 1976).

Fault-slice ridge—A linear topographic high associated with a fault-bounded uplifted block within a fault zone (Crowell, 1974b; synonymous with pressure ridge).

Fault splay—A subsidiary fault that merges with and is genetically related to a more prominent fault. Fault splays are common near the termination of a major strike-slip fault (Fig. 2), unless this is at an intersection with another strike-slip fault.

Fault strand—An individual fault of a set of closely spaced, parallel or subparallel faults of a fault system.

Fault-wedge basin—A basin formed by extension at a releasing junction between two predominantly strike-slip faults having the same sense of offset (Crowell, 1974a; synonymous with wedge graben of Freund, 1982). See releasing fault junction.

Flexure—(1) A fold produced by a force couple applied parallel to the direction of deflection (Suppe, 1985, p. 360); (2) a mechanism of regional isostatic compensation in which loads are supported by broad deflection of the lithosphere as a result of lithospheric rigidity (Watts and Ryan, 1976; Watts, 1983; M. S. Steckler, personal commun., 1985).

Flower structure—An array of upward-diverging fault splays within a strike-slip zone (attributed to R. F. Gregory by Harding and Lowell, 1979; see positive flower structure and negative flower structure; synonymous with palm-tree structure of Sylvester, 1984, but preferred for reasons of precedence).

Forced fold—A fold whose overall shape and trend are dominated by the shape and trend of an underlying forcing member (Stearns, 1978).

Foreland—A more-or-less stable area underlain by continental crust, and adjacent to an orogenic belt, toward which rocks of the belt were tectonically transported (Bates and Jackson, 1980, p.241).

Fracture zone—An extension of a transform fault beyond its intersection with an oceanic ridge. Fracture zones are characterized by dip slip, especially where juxtaposed oceanic crust is of markedly different age, and usually they do not experience strike slip (see Freund, 1974; Fox and Gallo, 1984).

Graben—An elongate, relatively depressed block bounded by normal faults (Bates and Jackson, 1980, p. 268).

Heat flow—The product of a thermal gradient and the thermal conductivity of the material across which the thermal gradient is measured.

Horsetail splay—One of a set of curved fault splays near the end of a strike-slip fault that merge with that fault. The set forms an array that crudely resembles a horse's tail (Figs. 2, 3).

Internal rotation—A change in the orientation of structural features during deformation with reference to coordinate axes internal to the deformed body (Bates and Jackson, 1980, p. 322).

Kinematic analysis—The analysis of a movement pattern based on displacement without reference to force or stress (modified from Spencer, 1977, p. 39).

Leaky transform—A transform plate boundary characterized by significant volcanism and/or intrusion along its length. See Garfunkel (1981) for a continental example. See transform fault, transform margin.

Left-hand overstep or stepover—An overstep (stepover) in which one fault or fold segment occurs to the left of the adjacent segment from which it is being viewed (Campbell, 1958; Wilcox et al., 1973). See left-stepping, overstep, stepover. For oversteps in cross section, it is necessary to specify the direction from which the overstep is being viewed.

Left-lateral—Refers to an offset along a fault in map view, in which the far side is apparently displaced to the left with respect to the near side.

Left separation—Strike separation in which the far side of a fault is apparently displaced to the left with respect to the near side. See separation, strike separation.

Left slip—The component of slip measured parallel to the strike of a fault in which the far side of the fault is displaced to the left with respect to the near side. See slip.

Left-stepping—Refers to an overstep in which one fault or fold segment occurs to the left of the adjacent segment from which it is being viewed (Fig. 2). See left-hand overstep.

Lineament—A linear topographic feature of regional extent that is thought to reflect crustal structure (Hobbs et al., 1976, p. 267).

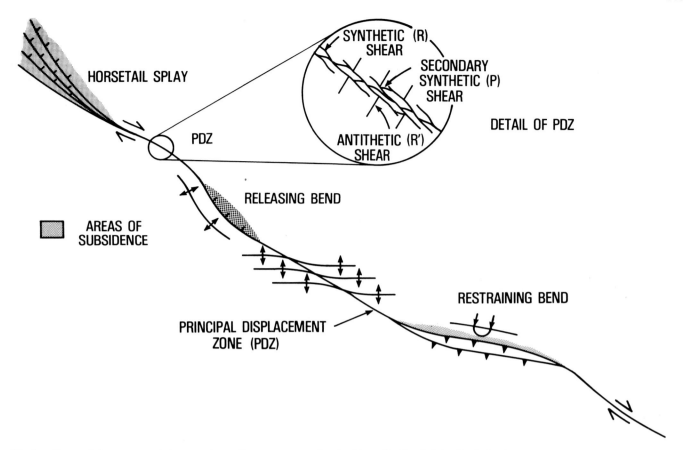

Fig. 3.—The spatial arrangement, in map view, of structures associated with an idealized right-slip fault. See glossary for definitions of terms.

Listric fault—A curved, generally concave-upward, downward-flattening fault (see Bally et al., 1981). Listric faults may be characterized by normal or reverse separation.

Lithosphere—The outer shell of the Earth consisting of crust and upper mantle, and characterized by strength relative to the underlying asthenosphere for deformation at geologic rates (Bates and Jackson, 1980, p. 364; Watts, 1983). The base of the lithosphere can be defined by a number of different properties that reflect rheology, such as temperature, seismic velocity and degree of seismic attenuation.

Marginal basin—A semi-isolated basin lying behind the volcanic chain of an island arc system (Karig, 1971, p. 2542).

Master fault—A major fault in a fault system (Wilcox et al., 1973; Rodgers, 1980; nearly synonymous with principal displacement zone of Tchalenko and Ambraseys, 1970).

Megashear—A strike-slip fault with horizontal displacement that significantly exceeds the thickness of the crust (Carey, 1958, 1976, p. 85).

Multiple overstep—A series of discontinuities between approximately parallel overlapping or underlapping strike-slip faults (new term). See overstep, overlap, and underlapping faults.

Negative flower structure—A flower structure in which the upward-diverging fault splays are predominantly of normal separation and commonly associated with a prominent synformal structure or structures in strata above, or cut by, the faults (Harding, 1983, 1985; Harding et al., 1985 this volume). See flower structure.

Net slip—The displacement vector connecting formerly adjacent points on opposite sides of a fault (modified from Hobbs et al., 1976, p. 300).

Normal drag—Folding near a fault resulting from resistance to slip along the fault. Folded strata are convex toward the slip direction on both sides of the fault. See drag fold.

Normal fault—(1) A fault with normal separation across which the hanging wall is apparently lowered with respect to the footwall (preferred definition; Hill, 1959); (2) a fault generated by normal slip (see Gill, 1941, 1971). The term may be applied to faults of any dip (Christie-Blick, 1983).

Normal separation—Separation measured parallel to the dip of a fault across which the hanging wall is apparently lowered with respect to the footwall. See separation.

Normal slip—The component of slip measured parallel to the dip of a fault across which the hanging wall is lowered with respect to the footwall. See slip.

Oblique—Not parallel; intersecting at an acute angle. En echelon elements are oblique to the zone in which they occur, but not all oblique elements are en echelon. See en echelon.

Oblique slip—The relative displacement of formerly adjacent points on opposite sides of a fault, involving components of both dip slip and strike slip. See slip, dip slip, strike slip.

Oblique-slip fault—A fault along which displacement is accomplished by a combination of strike slip and dip slip. See slip-oblique fault.

Oblique subduction—The relative displacement of one lithospheric plate beneath another plate such that in map view the displacement vector is oblique to the plate boundary.

Onlap—A base-discordant stratigraphic relation in which initially horizontal or inclined strata terminate updip against an initially inclined surface (modified from Mitchum, 1977, p. 208).

Opening bend—See releasing bend.

Orogeny—Profound deformation of rock bodies along restricted zones and within a limited time interval (Dennis, 1967, p. 112).

Overlap—(1) The distance between the ends of overlapping parallel faults, measured parallel to the faults (Rodgers, 1980; Mann et al., 1983; Aydin and Nur, 1985 this volume; generally applied to strike-slip faults in map view; nearly synonymous with separation of Segall and Pollard, 1980); (2) a relation between two superimposed stratigraphic units onlapping a given surface, in which the upper unit extends beyond the line of pinch-out in the lower unit.

Overstep—(1) a discontinuity between two approximately parallel overlapping or underlapping faults (Fig. 2; synonymous with stepover of Aydin and Nur, 1982a, b; 1985 this volume). Oversteps can occur in both map view and cross section, and on both strike-slip and dip-slip faults (Aydin and Nur, 1985 this volume), but the term is commonly applied to strike-slip faults in map view. See solitary overstep, multiple overstep, releasing overstep, and restraining overstep. (2) A stratigraphic relation in which one or more stratigraphic units unconformably overlie the eroded edge of older, generally tilted or folded sedimentary rocks.

Palm-tree structure—An array of upward-diverging fault splays within a strike-slip zone (nomenclature of A. G. Sylvester and R. R. Smith; Sylvester, 1984; synonymous with flower structure of Harding and Lowell, 1979, a term that has precedence).

P shear—One of a set of faults that develop in simple shear generally after the formation of Riedel shears. P shears have the same sense of displacement as Riedel R shears, and form at an angle to the principal displacement zone that is of about the same magnitude but of opposite sign (Skempton, 1966; Tchalenko and Ambraseys, 1970; Figs. 1, 3). Synonymous with secondary synthetic strike-slip fault.

Piercing points—The points of intersection of formerly contiguous linear features (real or constructed) on opposite sides of a fault, by means of which net slip on the fault may be determined (Crowell, 1959, p. 2656). Examples of such linear features are the pinchout line of a sedimentary wedge, offset streams, and facies boundaries used in conjunction with structure contours.

Plunge—The acute angle between an inclined line and the horizontal, measured in a vertical plane containing the line.

Pop-up—A relatively uplifted block between thrusts verging in opposite directions. Originally applied to structures in thrust and fold belts (Butler, 1982).

Positive flower structure—A flower structure in which the upward-diverging fault splays are predominantly of reverse separation and commonly associated with a prominent antiformal structure or structures in strata above, or cut by, the faults (Harding et al., 1983; Harding, 1985). See flower structure.

Pressure ridge—A linear topographic high associated with a fault-bounded uplifted block within a fault zone (Tchalenko and Ambraseys, 1970; synonymous with fault-slice ridge).

Principal displacement zone—A relatively narrow zone that accounts for most of the slip on a given fault (Tchalenko and Ambraseys, 1970, p. 43; Fig. 3). See master fault.

Progradation—The outward building of sediment in the direction of transport, generally, but not exclusively, from a shoreline toward a body of water.

Pull-apart basin—(1) A basin formed by crustal extension at a releasing bend or releasing overstep along a strike-slip fault zone (preferred definition; Burchfiel and Stewart, 1966; Crowell, 1974b; Mann et al., 1983; nearly synonymous with rhomb graben); (2) any basin resulting from crustal extension (Klemme, 1980; Bois et al., 1982).

Pure shear—A homogeneous deformation involving either a plane strain or a general strain in which lines of particles that are parallel to the principal axes of the strain ellipsoid have the same orientation before and after deformation (Hobbs et al., 1976, p. 28). Also referred to as an irrotational deformation or strain. See shear, simple shear.

Push-up—A block elevated by crustal shortening at a restraining bend or restraining overstep along a strike-slip fault zone (Aydin and Nur, 1982a; Mann et al., 1983). See rhomb horst.

Ramp valley—A topographic basin bounded by reverse faults (Willis, 1928, p. 493; Burke et al., 1982). Not all ramp valleys are related to strike-slip deformation.

Regression—A seaward retreat of a shoreline, generally expressed as a seaward migration of shallow-marine facies (modified from Mitchum, 1977, p. 209).

Relay pattern—A shingled arrangement of inconsistently overlapping or underlapping structural elements such as faults or folds that are approximately parallel to each other and to the elongate zone in which they occur (modified from Harding and Lowell, 1979; Fig. 2). Many authors do not distinguish between en echelon and relay arrangements. See Christie-Blick and Biddle (1985 this volume).

Releasing bend—A bend in a strike-slip fault associated with overall crustal extension in the vicinity of the bend (Figs. 2, 3; Crowell, 1974a; synonymous with divergent bend).

Releasing fault junction—A junction between two strike-slip faults associated with overall crustal extension and basin formation between the faults (Christie-Blick and Biddle, 1985 this volume). See fault-wedge basin, wedge graben.

Releasing overstep—A right overstep between right-slip faults or a left overstep between left-slip faults associated with overall crustal extension and basin formation between the faults (Christie-Blick and Biddle, 1985 this volume).

Restraining bend—A bend in a strike-slip fault associated with overall crustal shortening and uplift in the vicinity of the bend (Figs. 2, 3; Crowell, 1974a; synonymous with convergent bend).

Restraining fault junction—A junction between two strike-slip faults associated with overall crustal shortening and uplift between the faults (Christie-Blick and Biddle, 1985 this volume).

Restraining overstep—A right overstep between left-slip faults or a left overstep between right-slip faults associated with overall crustal shortening and uplift between the faults (Christie-Blick and Biddle, 1985 this volume).

Retrogradation—The landward backstepping of sedimentary units usually but not exclusively from a shoreline, as expressed by a landward migration of facies belts.

Reverse drag—Deformation along a fault that creates a fold or set of folds whose curvature is opposite that which would be formed by normal drag folding. Reverse drag is a common feature of listric normal faults where hanging-wall folds are concave toward the slip direction.

Reverse fault—(1) A fault with reverse separation across which the hanging wall is apparently elevated with respect to the footwall (preferred definition; Hill, 1959); (2) a fault generated by reverse slip (see Gill, 1941, 1971). The term may be applied to faults of any dip (Christie-Blick, 1983).

Reverse separation—Separation measured parallel to the dip of a fault across which the hanging wall is apparently elevated with respect to the footwall. See separation.

Reverse slip—The component of slip measured parallel to the dip of a fault across which the hanging wall is elevated with respect to the footwall. See slip.

Rhombochasm—A parallel-sided gap in sialic (continental) crust occupied by simatic (oceanic) crust (modified from Carey, 1976, p. 81). One of S. W. Carey's type examples is the Gulf of California. For original definition and discussion, see Carey (1958).

Rhomb graben—A basin formed by crustal extension at a releasing bend or releasing overstep in a strike-slip fault zone (Freund, 1971; Aydin and Nur, 1982b; synonymous with pull-apart basin of Burchfiel and Stewart, 1966, and Crowell, 1974a, b, particularly sharp pull-aparts, or ones that are angular in map view).

Rhomb horst—A block elevated by crustal shortening at a restraining bend or restraining overstep in a strike-slip fault zone (Aydin and Nur, 1982b; nearly synonymous with push-up of Aydin and Nur, 1982a, and Mann et al., 1983, particularly those that are angular in map view).

Riedel shear—In simple shear, two sets of shear fractures tend to form, oriented at $\phi/2$ and $90°-\phi/2$ to the principal displacement zone (where ϕ is the internal coefficient of friction commonly taken to be about 30°). Shear fractures oriented at $\phi/2$ are called R shears whereas those formed at $90°-\phi/2$ are termed R′ shears (modified from Tchalenko and Ambraseys, 1970). See synthetic and antithetic faults (Figs. 1, 3).

Right-hand overstep or stepover—An overstep (stepover) in which one fault or fold segment occurs to the right of the adjacent segment from which it is being viewed (Campbell, 1958; Wilcox et al., 1973). See right-stepping, overstep, stepover. For oversteps in cross section, it is necessary to specify the direction from which the overstep is being viewed.

Right-lateral—Refers to an offset along a fault in map view, in which the far side is apparently displaced to the right with respect to the near side.

Right separation—Strike separation in which the far side of a fault is apparently displaced to the right with respect to the near side. See separation, strike separation.

Right slip—The component of slip measured parallel to the strike of a fault in which the far side of the fault is displaced to the right with respect to the near side. See slip.

Right-stepping—Refers to an overstep in which one fault or fold segment occurs to the right of the adjacent segment from which it is being viewed (Fig. 2). See right-hand overstep.

Rotation—Motion in which the path of a point in the moving object defines an arc around a specified axis.

Rotational strain—Strain in which the orientation of the strain axes is different before and after deformation (Bates and Jackson, 1980, p. 546). See simple shear.

Secondary synthetic fault—One of a set of faults that develop in simple shear, generally after the formation of synthetic faults (Riedel shears). Secondary synthetic faults have the same sense of displacement as the synthetic faults and form an angle to the principal displacement zone that is of about the same magnitude as the synthetic faults but of opposite sign (Figs. 1, 3). Synonymous with P shear. See synthetic and antithetic fault, Riedel shear.

Separation—(1) The apparent displacement of formerly contiguous surfaces on opposite sides of a fault, measured in any given direction (modified from Reid et al., 1913, p. 169; Crowell, 1959, p. 2661); (2) the perpendicular distance between overlapping parallel strike-slip faults (Rodgers, 1980; Mann et al., 1983; Aydin and Nur, 1985 this volume); (3) the distance between overstepping parallel strike-slip faults (either overlapping or underlapping), measured parallel to the faults (Segall and Pollard, 1980; nearly synonymous with overlap of Rodgers, 1980).

Shear—A strain resulting from stresses that cause, or tend to cause, parts of a body to move relatively to each other in a direction parallel to their plane of contact (modified from Bates and Jackson, 1980, p. 575). See pure shear, simple shear.

Simple shear—A constant volume, homogeneous deformation involving plane strain, in which a single family of parallel material planes is undistorted in the deformed state and parallel to the same family of planes in the undeformed state (Hobbs et al., 1976, p. 29). Also referred to as a rotational deformation or strain. See shear, pure shear.

Simple strike-slip or wrench fault—A strike-slip or wrench fault along which adjacent blocks move laterally with no component of shortening or extension transverse to the fault (Christie-Blick and Biddle, 1985 this volume; synonymous with simple parallel strike-slip or wrench fault of Wilcox et al., 1973; and with slip-parallel fault of Mann et al., 1983).

Sinistral—Pertaining to the left (e.g., sinistral slip is left slip).

Slickenside—A polished or smoothly striated surface on either side of a fault that results from motion along the fault (Bates and Jackson, 1980, p. 587).

Slip—The relative displacement of formerly adjacent points on opposite sides of a fault measured along the fault surface (modified from Reid et al., 1913, p. 168; Crowell, 1959, p. 2655).

Slip-oblique fault—A strike-slip fault along which strike-slip deformation is accompanied by a component of either shortening or extension transverse to the fault (Mann et al., 1983; includes convergent and divergent strike-slip or wrench faults of Wilcox et al., 1973; and transpressional and transtensional faults of Harland, 1971; nearly synonymous with oblique-slip fault).

Slip-parallel fault—A fault that strikes parallel to the azimuth of the slip direction (Mann et al., 1983; synonymous with simple parallel strike-slip or wrench fault of Wilcox et al., 1973; and with simple strike-slip fault of Christie-Blick and Biddle, 1985 this volume).

Solitary overstep—An isolated discontinuity, between two approximately parallel overlapping or underlapping faults (Guiraud and Seguret, 1985 this volume; Fig. 2).

Sphenochasm—A triangular gap of oceanic crust separating two continental blocks with fault margins converging to a point, and interpreted as having originated by the rotation of one block with respect to the other (modified from Carey, 1976, p. 81). The Bay of Biscay is one of S. W. Carey's examples of a sphenochasm. See Carey (1958) for original definition.

Splay—Generally synonymous with fault splay, a subsidiary fault that merges with, and is genetically related to, a more prominent fault.

Stepover—A discontinuity between two approximately parallel overlapping or underlapping faults (Aydin and Nur, 1982a, b; 1985 this volume; Aydin and Page, 1984; synonymous with overstep). Stepovers can occur in both map view and cross section, and on both strike-slip and dip-slip faults (Aydin and Nur, 1985 this volume), but the term is commonly applied to discontinuities on strike-slip faults in map view.

Stratigraphic separation—The stratigraphic thickness either cut out or repeated by a fault (modified from Crowell, 1959, p. 2663).

Strand—See fault strand.

Strike—The azimuth of the line of intersection of an inclined surface with a horizontal plane.

Strike separation—Separation measured parallel to the strike of a fault (modified from Crowell, 1959, p. 2662; Bates and Jackson, 1980, p. 618). See separation.

Strike slip—The component of slip measured parallel to the strike of a fault (Crowell, 1959, p. 2655; Bates and Jackson, 1980, p. 618). See slip.

Strike-slip fault—A fault along which most of the displacement is accomplished by strike slip (modified from Bates and Jackson, 1980, p. 618).

Strike-slip basin—Any basin in which sedimentation is accompanied by significant strike slip (modified from Mann et al., 1983).

Subsidence—The depression of an area of the Earth's crust with respect to surrounding areas.

Synthetic fault—Originally defined by H. Cloos (1928, 1936) to describe faults that dip in the same direction as the rocks displaced and that rotate fault-bounded blocks so that the net slip on each fault is less than it would be without rotation (Dennis, 1967, p. 148–149). Many authors now use the term to describe faults that (1) are subsidiary to a major fault and have less displacement than that fault; (2) formed in the same stress regime as the major fault with which they are associated; (3) are oriented at a low angle to the major fault (in map view for strike-slip faults, in cross-sectional view for normal faults); and (4) for strike-slip faults, have the same sense of displacement as the major fault with which they are associated, or for normal faults, dip in the same direction. The R set of Riedel shears and the P shears of Tchalenko and Ambraseys (1970) are synthetic faults (Figs. 1, 3).

Tear fault—A strike-slip or oblique-slip fault within or bounding an allochthon produced by either regional extension or regional shortening. Tear faults accommodate differential displacement within a given allochthon, or between the allochthon and adjacent structural units.

Tectonic depression—(1) Any structurally produced topographic low; (2) a topographic low produced by strike-slip deformation (Clayton, 1966).

Tectonic subsidence—That part of the subsidence at a given point in a sedimentary basin caused by a tectonic driving mechanism. Tectonic subsidence is calculated by removing the component of subsidence produced by non-tectonic processes such as sediment loading, sediment compaction, and water-depth changes (Watts and Ryan, 1976; Steckler and Watts, 1978; Keen, 1979; Bond and Kominz, 1984).

Tension—A system of stresses that tends to lengthen or increase the volume of a substance. Uniaxial tension involves one nonzero principal stress, which is tensile; in general tension, two principal stresses are tensile (Means, 1976, p. 79). It is possible for a tensile principal stress to occur with one or more compressive principal stresses.

Tension (T) fracture—A mode I crack that forms when lithostatic loads become negative (Lawn and Wilshaw, 1975, p. 52; J. T. Engelder, personal commun., 1985). See extension fracture. In strike-slip systems, extension fractures and tension fractures form in response to simple shear at about 45° to the master fault (Fig. 1; Tchalenko and Ambraseys, 1970).

Thermal subsidence—That part of the tectonic subsidence at a given point in a sedimentary basin caused by thermal contraction (Sleep, 1971; Parsons and Sclater, 1977). See tectonic subsidence.

Thrust fault—A map-scale contraction fault that shortens an arbitrary datum, commonly but not necessarily bedding (McClay, 1981). The term may be applied to faults of any dip, although thrust faults tend to dip less than 30° during active slip.

Trace slip—The component of slip measured parallel with the trace of a bed, vein, or other surface on the fault plane (Reid et al., 1913, p. 170; Beckwith, 1941, p. 2182).

Transcurrent fault—(1) A strike-slip fault, typically subvertical at depth and commonly involving igneous and metamorphic basement as well as supracrustal sediments and sedimentary rocks (see Moody and Hill, 1956; Freund, 1974; nearly synonymous with wrench fault); (2) a long, subvertical strike-slip fault that cuts strata approximately perpendicular to strike (original sense of Geikie, 1905, p. 169; modified from Dennis, 1967; p. 57).

Transform fault—A strike-slip fault that acts as a lithospheric plate boundary and terminates at both ends against major tectonic features (such as oceanic ridges, subduction zones, or rarely, other transform faults) that are also plate boundaries (Wilson, 1965; Freund, 1974).

Transform margin—A plate margin formed by a transform fault or system of transform faults and dominated by strike-slip deformation.

Transgression—A landward movement of a shoreline, generally expressed as a landward migration of shallow-marine facies (modified from Mitchum, 1977, p. 211).

Transtension—A system of stresses that operates in zones of oblique extension (modified from Harland, 1971). See divergent strike-slip or wrench fault, transpression.

Transpression—A system of stresses that operates in zones of oblique shortening (modified from Harland, 1971; Sylvester and Smith, 1976). See convergent strike-slip or wrench fault, transtension.

Trend—The azimuth of an inclined or horizontal line.

Unconformity—A buried surface of erosion or non-deposition (modified from J. C. Crowell, personal commun., 1975).

Underlapping faults—Approximately parallel faults that overstep without overlapping (applied to oceanic ridge segments by Pollard and Aydin, 1984).

Wedge graben—A basin formed by extension at a releasing junction between two predominantly strike-slip faults having the same sense of offset (Freund, 1982; synonymous with fault-wedge basin of Crowell, 1974a). See releasing fault junction.

Wrench fault—A strike-slip fault, typically sub-vertical at depth, involving igneous and metamorphic basement rocks as well as supracrustal sediments and sedimentary rocks (modified from Moody and Hill, 1956, p. 1208; Wilcox et al., 1973; nearly synonymous with transcurrent fault).

Y-shear—A fault that forms in response to simple shear, and as deformation continues gradually accommodates most of the movement along the principal displacement zone (Bartlett et al., 1981).

ACKNOWLEDGMENTS

In compiling this glossary we have been inspired by J. C. Crowell's insistence on the value of precise terminology for the effective communication of both observations and ideas. The manuscript was reviewed at various stages by A. Aydin, J. T. Engelder, A. M. Grunow, D. R. Seely, and C. C. Wielchowsky. We also thank I. W. D. Dalziel, K. A. Kastens, C. Nicholson, and M. S. Steckler for helpful suggestions. Remaining errors or omissions are, however, the responsibility of the authors. Logistical support was provided by Exxon Production Research Company, and by an ARCO Foundation Fellowship to Christie-Blick.

REFERENCES

ARMSTRONG, R. L., 1968, Sevier orogenic belt in Nevada and Utah: Geological Society of America Bulletin, v. 79, p. 429–458.

AYDIN, A., AND NUR, A., 1982a, Evolution of pull-apart basins and push-up ranges: Pacific Petroleum Geologists Newsletter, American Association of Petroleum Geologists, Pacific Section, Nov. 1982, p. 2–4.

———, 1982b, Evolution of pull-apart basins and their scale independence: Tectonics, v. 1, p. 91–105.

———, 1985, The types and role of stepovers in strike-slip tectonics, in Biddle, K. T., and Christie-Blick, N., Strike-Slip Deformation, Basin Formation, and Sedimentation: Society of Economic Paleontologists and Mineralogists Special Publication No. 37, p. 35–44.

AYDIN, A., AND PAGE, B. M., 1984, Diverse Pliocene-Quaternary tectonics in a transform environment, San Francisco Bay region, California: Geological Society of America Bulletin, v. 95, p. 1303–1317.

BALLANCE, P. F., 1980, Models of sediment distribution in non-marine and shallow marine environments in oblique-slip fault zones, in Ballance, P. F., and Reading, H. G., eds., Sedimentation in Oblique-Slip Mobile Zones: International Association of Sedimentologists Special Publication No. 4, p. 229–236.

BALLY, A. W., BERNOULLI, D., DAVIS, G. A., AND MONTADERT, L., 1981, Listric normal faults: Oceanologica Acta, Proceedings, 26th International Geological Congress, Geology of Continental Margins Symposium, Paris, p. 87–101.

BARTLETT, W. L., FRIEDMAN, M., AND LOGAN, J. M., 1981, Experimental folding and faulting of rocks under confining pressure. Part IX. Wrench faults in limestone layers: Tectonophysics, v. 79, p. 255–277.

BATES, R. L., AND JACKSON, J. A., 1980, eds., Glossary of Geology: American Geological Institute, 749 p.

BECKWITH, R. H., 1941, Trace-slip faults: American Association of Petroleum Geologists Bulletin, v. 25, p. 2181–2193.

BOIS, C., BOUCHE, P., AND PELET, R., 1982, Global geologic history and distribution of hydrocarbon reserves: American Association of Petroleum Geologists Bulletin, v. 66, p. 1248–1270.

BOND, G. C., AND KOMINZ, M. A., 1984, Construction of tectonic subsidence curves for the early Paleozoic miogeocline, southern Canadian Rocky Mountains: Implications for subsidence mechanisms, age of breakup, and crustal thinning: Geological Society of America Bulletin, v. 95, p. 155–173.

BURCHFIEL, B. C., AND STEWART, J. H., 1966, "Pull-apart" origin of the central segment of Death Valley, California: Geological Society of America Bulletin, v. 77, p. 439–442.

BURKE, K., MANN, P., AND KIDD, W., 1982, What is a ramp valley?: 11th International Congress on Sedimentology, Hamilton, Ontario, International Association of Sedimentologists, Abstracts of Papers, p. 40.

BUTLER, R. W. H., 1982, The terminology of structures in thrust belts: Journal of Structural Geology, v. 4, p. 239–245.

CAMPBELL, J. D., 1958, En echelon folding: Economic Geology, v. 53, p. 448–472.

CAREY, S. W., 1958, A tectonic approach to continental drift, in Carey, S. W., convenor, Continental Drift: a Symposium: Hobart, University of Tasmania, 177–355.

———, 1976, The Expanding Earth: Amsterdam, Elsevier Scientific Publishing Company, Developments in Geotectonics 10, 488 p.

CHRISTIE-BLICK, N., 1983, Structural geology of the southern Sheeprock Mountains, Utah: Regional significance, in Miller, D. M., Todd, V. R., and Howard, K. A., eds., Tectonic and Stratigraphic Studies in the Eastern Great Basin: Geological Society of American Memoir 157, p. 101–124.

CHRISTIE-BLICK N., AND BIDDLE, K. T., 1985, Deformation and basin formation along strike-slip faults, in Biddle, K. T., and Christie-Blick, N., eds., Strike-Slip Deformation, Basin Formation, and Sedimentation: Society of Economic Paleontologists and Mineralogists Special Publication 37, p. 1–34.

CLAYTON, L., 1966, Tectonic depressions along the Hope Fault, a transcurrent fault in North Canterbury, New Zealand: New Zealand Journal of Geology and Geophysics, v. 9, p. 95–104.

CLOOS, H., 1928, Über antithetische Bewegungen: Geologische Rundschau, v. 19, p. 246–251.

———, 1936, Einführung in die Geologie: Berlin, Borntraeger, 503 p.

CROWELL, J. C., 1959, Problems of fault nomenclature: American Association of Petroleum Geologists Bulletin, v. 43, p. 2653–2674.

———, 1974a, Origin of late Cenozoic basins in southern California, in Dickinson, W. R., ed., Tectonics and Sedimentation: Society of Economic Paleontologists and Mineralogists Special Publication No. 22, p. 190–204.

———, 1974b, Sedimentation along the San Andreas fault, California, in Dott, R. H., Jr., and Shaver, R. H., eds., Modern and Ancient Geosynclinal Sedimentation: Society of Economic Paleontologists and Mineralogists Special Publication No. 19, p. 292–303.

———, 1976, Implications of crustal stretching and shortening of coastal Ventura Basin, California, in Howell, D. G., ed., Aspects of the Geologic History of the California Continental Borderland: American Association of Petroleum Geologists, Pacific Section, Miscellaneous Publication 24, p. 365–382.

DAHLSTROM, C. D. A., 1970, Structural geology in the eastern margin of the Canadian Rocky Mountains: Bulletin of Canadian Petroleum Geology, v. 18, p. 332–406.

DENNIS, J. G., ed., 1967, International Tectonic Dictionary: American Association of Petroleum Geologists Memoir 7, 196 p.

FISCHER, A. G., 1969, Geologic time-distance rates: the Bubnoff unit: Geological Society of America Bulletin, v. 80, p. 549–551.

———, 1974, Origin and growth of basins, in Fischer, A. G., and Judson, S., ed., Petroleum and Global Tectonics: Princeton, New Jersey, Princeton University Press, p. 47–82.

FOX, P. J., AND GALLO, D. G., 1984, A tectonic model for ridge-transform-ridge plate boundaries: Implications for the structure of oceanic lithosphere: Tectonophysics, v. 104, p. 205–242.

FREUND, R., 1971, The Hope Fault, a strike-slip fault in New Zealand: New Zealand Geological Survey Bulletin, v. 86, p. 1–49.

———, 1974, Kinematics of transform and transcurrent faults: Tectonophysics, v. 21, p. 93–134.

———, 1982, The role of shear in rifting, in Pálmason, G., ed., Continental and Oceanic Rifts: American Geophysical Union Geodynamics Series, v. 8, p. 33–39.

FRIEDMAN, M., HANDIN, J., LOGAN, J. M., MIN, K. D., AND STEARNS, D. W., 1976, Experimental folding of rocks under confining pressure: Part III. Faulted drape folds in multilithologic layered specimens: Geological Society of America Bulletin, v. 87, p. 1049–1066.

GARFUNKEL, Z., 1981, Internal structure of the Dead Sea leaky transform (rift) in relation to plate kinematics: Tectonophysics, v. 80, p. 81–108.

GEIKIE, J., 1905, Structural and Field Geology: Edinburgh, Oliver and Boyd, 435 p.

GILL, J. E., 1941, Fault nomenclature: Royal Society of Canada Transactions, v. 35, p. 71–85.

———, 1971, Continued confusion in the classification of faults: Geological Society of America Bulletin, v. 82, p. 1389–1392.

GRESSLY, A., 1838, Observations géologique sur le Jura Soleurois: Nouveaux Mémoires de la Société Helvétique des Sciences Naturelles, v. 2, p. 1–112.

GUIRAUD, M., AND SEGURET, M., 1985, A releasing solitary overstep model for the late Jurassic—early Cretaceous (Wealdian) Soria strike-slip basin (northern Spain), in Biddle, K. T., and Christie-Blick, N., eds., Strike-Slip Deformation, Basin Formation, and Sedimentation: Society of Economic Paleontologists and Mineralogists Special Publication No. 37, p. 159–177.

HARDING, T. P., 1983, Divergent wrench fault and negative flower structure, Andaman Sea, in Bally, A. W., ed., Seismic Expression of Structural Styles, v. 3: American Association of Petroleum Geologists Studies in Geology Series 15, v. 3, p. 4.2-1 to 4.2-8.

———, 1985, Seismic characteristics and identification of negative flower structures, positive flower structures, and positive structural inversion: American Association Petroleum Geologists Bulletin, v. 69, p. 582–600.

HARDING, T. P., AND LOWELL, J. D., 1979, Structural styles, their plate-tectonic habitats, and hydrocarbon traps in petroleum provinces: American Association of Petroleum Geologists Bulletin, v. 63, p. 1016–1058.

HARDING, T. P., GREGORY, R. F., AND STEPHENS, L. H., 1983, Convergent wrench fault and positive flower structure, Ardmore Basin, Oklahoma, in Bally, A. W., ed., Seismic Expression of Structural Styles, v. 3, American Association of Petroleum Geologists Studies in Geology, Series 15, p. 4.2-13 to 4.2-17.

HARDING, T. P., VIERBUCHEN, R. C., AND CHRISTIE-BLICK, N., 1985, Structural styles, plate-tectonic settings, and hydrocarbon traps of divergent (transtensional) wrench faults, in Biddle, K. T., and Christie-Blick, N., eds., Strike-Slip Deformation, Basin Formation, and Sedimentation: Society of Economic Paleontologists and Mineralogists Special Publication No. 37, p. 51–78.

HARLAND, W. B., 1971, Tectonic transpression in Caledonian Spitsbergen: Geological Magazine, v. 108, p. 27–42.

HILL, M. L., 1959, Dual classification of faults: American Association of Petroleum Geologists Bulletin, v. 43, p. 217–237.

HOBBS, B. E., MEANS, W. D., AND WILLIAMS, P. F., 1976, An Outline of Structural Geology: New York, John Wiley and Sons, 571 p.

HOFFMAN, P., DEWEY, J. F., AND BURKE, K., 1974, Aulacogens and their genetic relation to geosynclines, with a Proterozoic example from Great Slave Lake, Canada, in Dott, R. H., Jr., and Shaver, R. H., eds., Modern and Ancient Geosynclinal Sedimentation: Society of Economic Paleontologists and Mineralogists Special Publication No. 19, p. 38–55.

KARIG, D. E., 1971, Origin and development of marginal basins in the western Pacific: Journal of Geophysical Research, v. 76, p. 2542–2561.

KEEN, C. E., 1979, Thermal history and subsidence of rifted continental margins—Evidence from wells on the Nova Scotian and Labrador Shelves: Canadian Journal of Earth Sciences, v. 16, p. 505–522.

KLEMME, H. D., 1980, Petroleum basins—classifications and characteristics: Journal of Petroleum Geology, v. 3, p. 187–207.

LAWN, B. R., AND WILSHAW, T. R., 1975, Fracture of Brittle Solids: New York, Cambridge University Press, 204 p.

MANN, P., HEMPTON, M. R., BRADLEY, D. C., AND BURKE, K., 1983, Development of pull-apart basins: Journal of Geology, v. 91, p. 529–554.

MCCLAY, K. R., 1981, What is a thrust? What is a nappe?, in McClay, K. R., and Price, N. J., eds., Thrust and Nappe Tectonics: Geological Society of London Special Publication No. 9, p. 7–9.

MEANS, W. D., 1976, Stress and Strain: New York, Springer-Verlag, 339 p.

MITCHUM, R. M., JR., 1977, Seismic stratigraphy and global changes of sea level, Part 11: Glossary of terms used in seismic stratigraphy, in

Payton, C. E., ed., Seismic Stratigraphy—Applications to Hydrocarbon Exploration: American Association of Petroleum Geologists Memoir 26, p. 205–212.

MOODY, J. D., AND HILL, M. L., 1956, Wrench-fault tectonics: Geological Society of America Bulletin, v. 67, p. 1207–1246.

NORRIS, D. K., 1958, Structural conditions in Canadian coal mines: Geological Survey of Canada Bulletin 44, 54 p.

PARSONS, B., AND SCLATER, J. G., 1977, An analysis of the variation of ocean floor bathymetry and heat flow with age: Journal of Geophysical Research, v. 82, p. 803–827.

PHILIPP, W., 1961, Struktur- und Lagerstattengeschichte des Erdolfeldes Eldingen: Deutsche Geologische Gesellschaft Zeitschrift, v. 112, p. 414–482.

POLLARD, D. D., AND AYDIN, A., 1984, Propagation and linkage of oceanic ridge segments: Journal of Geophysical Research, v. 89, p. 10,017–10,028.

REID, H. F., et al., 1913, Report of the Committee on the Nomenclature of Faults: Geological Society of America Bulletin, v. 24, p. 163–186.

RODGERS, D. A., 1980, Analysis of pull-apart basin development produced by en echelon strike-slip faults, in Ballance, P. F., and Reading, H. G., eds., Sedimentation in Oblique-Slip Mobile Zones: International Association of Sedimentologists Special Publication No. 4, p. 27–41.

RODGERS, J., 1963, Mechanics of Appalachian foreland folding in Pennsylvania and West Virginia: American Association of Petroleum Geologists Bulletin, v. 47, p. 1527–1536.

SEGALL, P., AND POLLARD, D. D., 1980, Mechanics of discontinuous faults: Journal of Geophysical Research, v. 85, p. 4337–4350

SHATSKI, N. S., 1946a, Basic features of the structures and development of the East European platform. Comparative tectonics of ancient platforms: Izvestiya Akademii Nauk SSSR, Ser. Geol. No. 1, p. 5–62.

———, 1946b, The Great Donets Basin and the Wichita System. Comparative tectonics of ancient platforms: Izvestiya Akademii Nauk SSSR, Ser. Geol. No. 6, p. 57–90.

SHELTON, J. W., 1984, Listric normal faults: An illustrated summary: American Association of Petroleum Geologists Bulletin, v. 68, p. 801–815.

SLEEP, N. H., 1971, Thermal effects of the formation of Atlantic continental margins by continental breakup: Royal Astronomical Society Geophysical Journal, v. 24, p. 325–350.

SKEMPTON, A. W., 1966, Some observations on tectonic shear zones: 1st Congress of International Society of Rock Mechanics, Lisbon, Proceedings, v. 1, p. 329–335.

SPENCER, E. W., 1977, Introduction to the Structure of the Earth: New York, McGraw-Hill Book Company, 2nd edition, 640 p.

STEARNS, D. W., 1978, Faulting and forced folding in the Rocky Mountains foreland, in Matthews, V., III, ed., Laramide Folding Associated With Basement Block Faulting in the Western United States: Geological Society of America Memoir 151, p. 1–37.

STECKLER, M. S., AND WATTS, A. B., 1978, Subsidence of the Atlantic-type continental margin off New York: Earth and Planetary Science Letters, v. 41, p. 1–13.

SUPPE, J., 1985, Principles of Structural Geology: Englewood Cliffs, New Jersey, Prentice-Hall, 537 p.

SYLVESTER, A. G., compiler, 1984, Wrench Fault Tectonics: American Association of Petroleum Geologists Reprint Series, No. 28, 374 p.

SYLVESTER, A. G., AND SMITH, R. R., 1976, Tectonic transpression and basement-controlled deformation in the San Andreas fault zone, Salton Trough, California: American Association of Petroleum Geologists Bulletin, v. 60, p. 2081–2102.

TCHALENKO, J. S., AND AMBRASEYS, N. N., 1970, Structural analysis of the Dasht-e Baȳaz (Iran) earthquake fractures: Geological Society of America Bulletin, v. 81, p. 41–60.

VAN HINTE, J. E., 1978, Geohistory analysis—application of micropaleontology in exploration geology: American Association of Petroleum Geologists Bulletin, v. 62, p. 201–222.

WALKER, R. G., 1984, General introduction: Facies, facies sequences and facies models, in Walker, R. G., ed., Facies Models: Geoscience Canada Reprint Series 1, 2nd edition, p. 1–9.

WATTS, A. B., 1983, The strength of the Earth's crust: Marine Technology Society Journal, v. 17, p. 5–17.

WATTS, A. B., AND RYAN, W. B. F., 1976, Flexure of the lithosphere and continental margin basins: Tectonophysics, v. 36, p. 25–44.

WHITE, D. A., 1980, Assessing oil and gas plays in facies-cycle wedges: American Association of Petroleum Geologists Bulletin, v. 64, p. 1158–1178.

WILCOX, R. E., HARDING, T. P., AND SEELY, D. R., 1973, Basic wrench tectonics: American Association of Petroleum Geologists Bulletin, v. 57, p. 74–96.

WILLIS, Bailey, 1928, Dead Sea problem: rift valley or ramp valley?: Geological Society of America Bulletin, v. 39, p. 490–542.

WILSON, J. T., 1965, A new class of faults and their bearing on continental drift: Nature, v. 207, p. 343–347.